W9-BVF-851

IMPORTANT SYMBOLS

LINEAR ALGEBRA

WITH APPLICATIONS

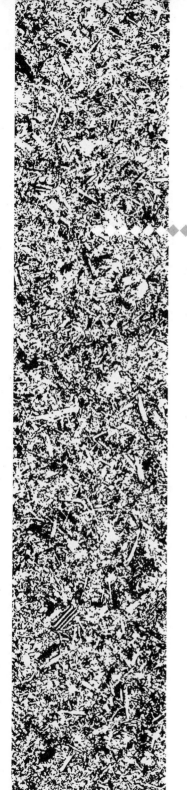

LINEAR
ALGEBRA
WITH APPLICATIONS

◆◆

Third Edition

W. KEITH NICHOLSON
University of Calgary

PWS PUBLISHING COMPANY
Boston

PWS PUBLISHING COMPANY
20 Park Plaza, Boston, MA 02116-4324

PWS Publishing Company is a division of Wadsworth, Inc.

I(T)P ™
International Thomson Publishing
The trademark ITP is used under license

 This book is printed on recycled, acid-free paper.

Library of Congress Cataloging-in-Publication Data

Nicholson, W. Keith.
    Linear algebra with applications / W. Keith Nicholson.—3rd ed.
        p. cm.
    Updated ed. of: Elementary linear algebra, with applications.
© 1990.
    Includes index.
    ISBN 0-534-93666-0
    1. Algebras, Linear. I. Nicholson, W. Keith. Elementary linear algebra, with applications. II. Title.
QA184.N53 1993   93-28879
512′ .5—dc20   CIP

Sponsoring Editor:	Steve Quigley
Production Coordinator:	Susan M. C. Caffey
Manufacturing Coordinator:	Marcia A. Locke
Marketing Manager:	Marianne C. P. Rutter
Assistant Editor:	Marnie Pommett
Editorial Assistant:	John V. Ward
Production:	Bookman Productions/Hal Lockwood
Interior Illustrator:	Deborah Doherty
Interior/Cover Designer:	Susan M. C. Caffey
Compositor:	Weimer Graphics, Inc.
Cover Art:	*Odyssey* by Vivian Angel
Cover Printer:	John P. Pow Company, Inc.
Text Printer and Binder:	RR Donnelley/Harrisonburg

Printed and bound in the United States of America.
    95 96 97 98 — 10 9 8 7 6 5 4

Contents

◆◆

v

Preface

◆◆

This textbook is a basic introduction to the ideas and techniques of linear algebra for first- or second-year students who have a working knowledge of high school algebra. Its aim is to achieve a balance among the computational skills, theory, and applications of linear algebra, while keeping the level suitable for beginning students. The contents are arranged to permit enough flexibility to allow the presentation of a traditional introduction to the subject, or to allow a more applied course. Calculus is not a prerequisite; places where it is mentioned are clearly marked and may be omitted.

Linear algebra has wide application to the mathematical and natural sciences, to engineering, to computer science, and (increasingly) to management and the social sciences. As a rule, students of linear algebra learn the subject by studying examples and solving problems. More than 330 solved examples are included here, many of a computational nature, together with a wide variety of exercises. In addition, a number of sections are devoted to applications and to the computational side of the subject. These are optional, but they are included at the end of the relevant chapters (rather than at the end of the book) to encourage students to browse.

The examples also play a role in motivating theorems, although most proofs are included at a level appropriate to the student. This means that the book can be used to give a course emphasizing computation and examples (and omitting many proofs) or to give a more rigorous treatment. Some longer proofs are omitted altogether or are deferred to the end of the chapter.

The third edition continues the trend toward spending more time on matrix computations as well as applications, a view supported by the Linear Algebra Curriculum Study Group.[1] For example, the chapter on abstract inner product spaces has been moved to the end and replaced by a discussion of the dot product in $\mathbb{R}^n$. This allows diagonalization, with its wealth of applications, to be introduced earlier and also shifts linear transformations toward the beginning of the book. The net effect is an overall reduction in the level of abstraction. A good example of this effect is

---

[1]*College Mathematics Journal 24*, Jan. 1993, p. 41.

evidenced in Chapter 6, where students will find the more "hands-on" treatment a refreshing change from the abstract material in Chapter 5.

New Features in the Third Edition

◆ Greater emphasis on matrix computations achieves an overall reduction in the level of abstraction in the text. For example, there are new sections in this edition on positive definite matrices and the block upper triangular form of a matrix.

◆ To provide a seamless, more interesting coverage of topics, subsections have been eliminated and biographical footnotes have been added.

◆ New exercises have been inserted, and exercises with answers that appear in the text or in the student manual are marked with a ◆.

◆ Revisions of some parts of the text now result in improved exposition. Examples include the discussions of Gaussian elimination and vector geometry, and the section on quadratic forms, now with a more thorough discussion of congruence.

◆ Changes in organization and location of some topics now allow for more flexible coverage: Eigenvalues and diagonalization and their applications now appear earlier in the text (in Chapter 6); and the discussion of operators (formerly Chapter 9) is now divided between Chapters 7 and 8. Abstract inner products now appear in Chapter 8, where a new section on isometries has been added.

The following characteristics, popular with users of the first two editions of this text, have been retained in the third edition:

◆ Presentation of techniques in examples, with emphasis on concrete computations and on the algorithmic nature of some techniques, allows students to master new skills readily. The text has more than 330 solved examples, all keyed to the exercises in the book.

◆ A wide variety of exercises, which start with routine computational problems and progress to more theoretical exercises, helps students develop skills in an appropriate, logically paced fashion.

◆ Optional applications at the end of each chapter offer students specific examples of where linear algebra yields new insight into problems rather than merely playing a descriptive role.

◆ Material is organized for maximum flexibility of coverage, giving instructors alternative paths through the topics. (See the "Chapter Dependencies" chart on page xii.) Optional sections include LU-factorization, LP-factorization, computing eigenvalues, and complex matrices, plus appendices on linear programming, complex numbers, and mathematical induction. For example, the material on LU-factorization requires only Chapter 2; the optional appendix on linear programming depends only upon Chapter 1.

◆ Answers to even-numbered computational exercises and to selected others enable students to check the accuracy of their computation immediately.

Ancillary Materials

◆ The *Partial Solutions Manual* now contains solutions to selected even-numbered exercises.

◆ The *Complete Answers and Solutions Manual* contains answers or solutions to all the exercises found in the book.

◆ A computer package (Pascal program MAX: MAtriX Algebra Calculator) is available upon request.

◆ *EXPtest* is a testing program for IBM PCs.

◆ The *Testbank* is a printed version of *EXPtest.*

Chapter Summaries

Chapter 1: A standard treatment of Gaussian elimination is given. The rank of a matrix is introduced via the row-echelon form.

Chapter 2: The operations of matrix algebra (including transposition) are introduced, and matrix inverses are defined and studied through the use of elementary matrices. The relationship of matrix algebra to linear equations is emphasized, and block multiplication is introduced to simplify matrix computations. An optional section on LU-factorization is included for more applied courses.

Chapter 3: Determinants are defined inductively. The Laplace expansion is stated first (motivated by examples and the 2×2 case), so the students begin by computing determinants using familiar row and column operations (the proof is given later). The usual rules are deduced from the Laplace expansion, and the adjoint formula is given.

Chapter 4: Vector operations are defined (motivated by examples) and used to solve (primarily geometric) problems. Then coordinates are introduced in $\mathbb{R}^2$ and $\mathbb{R}^3$, and straight lines and planes are described via the dot and cross products.

Chapter 5: The basic theory of finite dimensional vector spaces is given. The prototype example throughout is $\mathbb{R}^n$. Many examples are given to motivate and describe such concepts as subspaces, spanning sets, linear independence, and dimension. Examples involving matrices and polynomials are also given, as are examples of spaces of functions (examples requiring calculus are clearly marked). The pace is slow because this is the first acquaintance many students have had with an abstract system.

Chapter 6: Eigenvalues and the characteristic polynomial are introduced, similarity is defined, and the diagonalization algorithm is given. The Gram–Schmidt algorithm is presented in $\mathbb{R}^n$ and orthogonal projections are discussed. Then the principal axes theorem is proved and the Cholesky decomposition for positive definite matrices is given. Three applications are covered. Linear transformations are not required. An optional section on diagonalization of complex matrices is included.

Chapter 7: Linear transformations are introduced, motivated by many examples from geometry, matrix theory, and calculus (clearly marked). The kernel and image are defined, the dimension theorem is proved, and isomorphisms are discussed. Invariant subspaces are introduced and used to derive the block triangular form.

Chapter 8: General inner products are introduced (the prototype example being $\mathbb{R}^n$), and distance, norm, and the Schwarz inequality are discussed. The Gram–Schmidt algorithm is given, projections are introduced, and the approximation theorem is proved. Another proof of the principal axes theorem is given and isometries are characterized.

Chapter Dependencies

The following chart suggests how the material introduced in each chapter draws on concepts covered in certain earlier chapters. A solid arrow means that ready assimilation of ideas and techniques presented in the later chapter depends on familiarity with the earlier chapter. A broken arrow indicates that some reference to the earlier chapter is made but the chapter need not be covered.

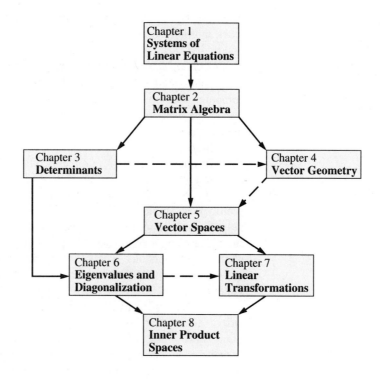

Suggested Course Outlines

1. *Two-Semester Course.* Much of the book can be covered in two 35-lecture semesters, with time left for some applications. The following outline is based on class experience and includes three applications. The pace in the first semester is more leisurely.

Chapter 1	Sections 1.1–1.3	4 lectures
Applications	Sections 1.4–1.5	1 lecture
Chapter 2	Sections 2.1–2.4	9 lectures
Application	Section 2.7	3 lectures
Chapter 3	Sections 3.1–3.2	6 lectures
Chapter 4	Sections 4.1–4.3	10 lectures
Application	Section 4.4	2 lectures
Chapter 5	Sections 5.1–5.5	10 lectures
Chapter 6	Sections 6.1–6.4	8 lectures
Application	Section 6.9	3 lectures
Chapter 7	Sections 7.1–7.6	7 lectures
Chapter 8	Sections 8.1–8.3	7 lectures

2. *One-Semester Applied Course.* This 35-lecture outline goes directly to diagonalization and its applications. The sections marked with an asterisk are intended as alternatives.

Chapter 1	Sections 1.1–1.3	3 lectures
Chapter 2	Sections 2.1–2.4	6 lectures
LU-factorization*	Section 2.5	2 lectures*
Chapter 3	Sections 3.1–3.2	4 lectures
Chapter 5	Sections 5.1–5.5	9 lectures
Chapter 6	Sections 6.1–6.7	9 lectures
Complex Matrices*	Section 6.8	2 lectures*
Application	Section 6.9, 6.10, or 6.11	2 lectures

Acknowledgments

I would like to thank the following people for their useful comments and suggestions:

G. D. Allen
Texas A & M University

R. F. V. Anderson
University of British Columbia

Robert Beezer
University of Puget Sound

E. R. Bishop
Acadia University

S. Bulman-Fleming
Wilfred Laurier University

D. Burbulla
University of Toronto

M. Chacron
Carleton University

G. A. Chambers
University of Alberta

Peter Colwell
Iowa State University

Francis J. Conlan
Santa Clara University

V. F. Connolly
Worcester Polytechnic Institute

P. Cook II
Furman University

H. Cunsolo
University of Guelph

H. P. Decell, Jr.
University of Houston

James B. Derr
University of Southern Colorado

G. J. Etgen
University of Houston

Marjorie A. Fitting
San Jose State University

J. B. Florence
University of Western Ontario

Chris Gardiner
Eastern Michigan University

I. Gombos
Dawson College

Barry Jessup
University of Ottawa

E. W. Johnson
University of Iowa

Hans Joshi
University of Toronto

Toni Kasper
Borough of Manhattan Community College—CUNY

B. J. Kirby
Queens University

S. O. Kochman
York University

John Labute
McGill University

P. Lancaster
University of Calgary

A. H. Low
University of New South Wales

Roger Marty
Cleveland State University

J. Petro
Western Michigan University

Irwin Pressman
Carleton University

J. Repka
University of Toronto

L. G. Roberts
University of British Columbia

Maziar Shirvani
University of Alberta

P. N. Stewart
Dalhousie University

B. F. Wyman
Ohio State University

I would like to thank several colleagues at the University of Calgary for their suggestions, especially P. A. Binding, N. H. Choksy, P. F. Elhers, D. Gunderson, A. Schaer, J. Schaer, and M. G. Stone.

I would also like to thank Steve Quigley for his work on the third edition. Thanks are also due to the production staff at Bookman Productions and at PWS, to Joanne Longworth and Gisele Vezina who typed the first edition, to Jason Brown who helped with the exercises, to Nasli Choksy for proofreading, and to Jason and Mark Nicholson for typing and proofreading. Finally I want to thank my wife, Kathleen, for her unfailing support.

W. Keith Nicholson

A Note to the Student

◆◆

This text was written with you, the student, in mind, and it is my sincere hope that you will benefit from it. Many comments from previous students have been incorporated into the text, which I feel will reflect your own needs.

Linear algebra is an old subject and had its beginnings in the last century as a tool for studying geometry. In recent years it has begun to rival calculus as the most commonly used subject in mathematics. Linear algebra is used in engineering, medicine, computer science, economics, statistics, and biology to name a few areas, as well as in other parts of mathematics. Though I cannot demonstrate examples of all these applications in this text, enough are included to give you a good idea of the power of linear algebra. What I can promise is that this book will give you the information and skills you need to start to learn these various applications.

It is said that there are three vital things to remember when buying a house: First there is location, second you must consider location, and we cannot forget the third, location. Similarly, there are three things you must do to pass a mathematics course: First do exercises, second do exercises, and third (you guessed it) do exercises. Mathematics is a bit like swimming—you cannot learn the butterfly stroke if you stand on the edge of the pool and watch your instructor in the water. You have to get in there and *do it*. Doing it in a mathematics course means doing exercises.

Of course you may falter now and then. That is why so many examples are included in the text. Many of these examples are prototypes for the exercises and will reveal the techniques you need. However, the examples are not the entire text! You should think of each theorem as a sort of super exercise that distills facts and information into a compact form applicable to a wide variety of special cases. Hence it should not be surprising that the proofs of the theorems involve techniques and ways of thinking that are useful in all these cases.

Actually, there is another, more subtle, aspect of the theorems which is also important. The proofs of theorems (and the solutions of examples) are partly designed to train you to think logically or analytically. To some extent this will come automatically as you study this subject (or any other mathematics course for that matter), but you can foster its development by writing the solutions to the exercises as clearly as possible in a step-by-step manner, with enough explanation to make it understandable. It has been said that you do not really understand mathematics until you can explain it to the first person you meet on the street. The reader of your solution should be considered as such a person.

Finally, mathematics is a tough taskmaster and demands work from anyone seeking to master it. You may become discouraged. But it is not *that difficult!* Thousands of students have passed this course, and you can too. So let us begin.

W. Keith Nicholson

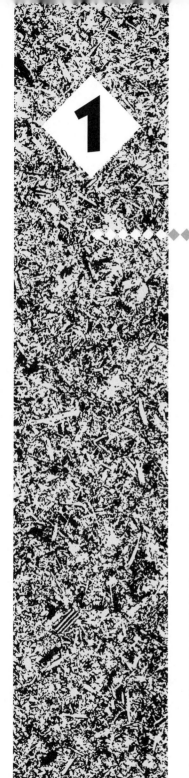

Systems of Linear Equations

Section 1.1
Solutions and Elementary Operations

Practical problems in many fields of study—such as biology, business, chemistry, economics, electronics, engineering, and the social sciences—can often be reduced to solving a system of linear equations. Linear algebra arose from attempts to find systematic methods for solving these systems, so it is natural to begin this book by studying linear equations.

If a, b, and c are real numbers, the graph of an equation of the form

$$ax + by = c$$

is a straight line (provided that a and b are not both zero). Accordingly, such an equation is called a linear equation in the variables x and y. When only two or three variables are present, they are usually denoted by x, y, and z. However, it is often convenient to write the variables as $x_1, x_2, \ldots, x_n$, particularly when more than three variables are involved.

> **DEFINITION**
>
> An equation of the form
>
> $$a_1x_1 + a_2x_2 + \cdots + a_nx_n = b$$
>
> is called a **linear equation** in the n variables $x_1, x_2, \ldots, x_n$. Here $a_1, a_2, \ldots, a_n$ denote real numbers (called the **coefficients** of $x_1, x_2, \ldots, x_n$, respectively) and b is also a number (called the **constant term** of the equation). A finite collection of linear equations in the variables $x_1, x_2, \ldots, x_n$ is called a **system of linear equations** in these variables.

Hence,

$$2x_1 - 3x_2 + 5x_3 = 7$$

and

$$x_1 + x_2 + x_3 + x_4 = 0$$

are both linear equations; the coefficients of the first equation are 2, –3, and 5, and the constant term is 7. Note that each variable in a linear equation occurs to the first power only. The following are *not* linear equations.

$$x_1^2 + 3x_2 - 2x_3 = 5$$
$$x_1 + x_1x_2 + 2x_3 = 1$$
$$\sqrt{x_1} + x_2 - x_3 = 0$$

DEFINITION

Given a linear equation $a_1x_1 + a_2x_2 + \cdots + a_nx_n = b$, a sequence $s_1, s_2, \ldots,$ s_n of n numbers is called a **solution** to the equation if

$$a_1s_1 + a_2s_2 + \cdots + a_ns_n = b$$

that is, if the equation is satisfied when the substitutions $x_1 = s_1$, $x_2 = s_2, \ldots,$ $x_n = s_n$ are made. A sequence of numbers is called a **solution to a system** of equations if it is a solution to every equation in the system.

For example, $x = -2$, $y = 5$, $z = 0$ and $x = 0$, $y = 4$, $z = -1$ are both solutions to the system

$$x + y + z = 3$$
$$2x + y + 3z = 1$$

A system may have no solution at all, or it may have an infinite family of solutions. For instance, the system $x + y = 2$, $x + y = 3$ has no solution because the sum of two numbers cannot be 2 and 3. On the other hand, the system in Example 1 has infinitely many solutions.

EXAMPLE 1

Show that, for arbitrary values of s and t,

$$x_1 = t - s + 1$$
$$x_2 = t + s + 2$$
$$x_3 = s$$
$$x_4 = t$$

is a solution to the system

$$x_1 - 2x_2 + 3x_3 + x_4 = -3$$

$$2x_1 - x_2 + 3x_3 - x_4 = 0$$

Solution   Simply substitute these values for x_1, x_2, x_3, and x_4 in each equation.

$$x_1 - 2x_2 + 3x_3 + x_4 = (t - s + 1) - 2(t + s + 2) + 3s + t = -3$$
$$2x_1 - x_2 + 3x_3 - x_4 = 2(t - s + 1) - (t + s + 2) + 3s - t = \;\; 0$$

Because both equations are satisfied, it is a solution for all s and t.

                  ◆◆◆

The quantities s and t in Example 1 are called **parameters**, and the set of solutions, described in this way, is said to be given in **parametric form** and is called the **general solution** to the system. It turns out that the solutions to *every* system of equations (if there *are* solutions) can be given in parametric form. The following examples show how this happens in the simplest systems where only one equation is present.

EXAMPLE 2

Describe all solutions to $3x - y = 4$ in parametric form.

Solution   The equation can be written in the form

$$y = 3x - 4$$

Thus, if t denotes *any* number at all, we can set $x = t$ and then obtain $y = 3t - 4$. This is clearly a solution for any value of t. On the other hand, *every* solution to $3x - y = 4$ arises in this way (t is just the value of x). Hence the set of *all* solutions can be described parametrically as

$$\begin{aligned} x &= t \\ y &= 3t - 4 \end{aligned} \qquad t \text{ arbitrary}$$

Note that there are *infinitely* many distinct solutions, one for each choice of the parameter t.

It is important to realize that the solutions to $3x - y = 4$ can be given in parametric form in several ways. Instead of writing $y = 3x - 4$ as before, we could have found x in terms of y

$$x = \tfrac{1}{3}(y + 4)$$

and then chosen $y = s$ (s a parameter). Hence the solutions are

$$\begin{aligned} x &= \tfrac{1}{3}(s + 4) \\ y &= s \end{aligned} \qquad s \text{ arbitrary}$$

This is also a correct parametric representation of the solutions to $3x - y = 4$. In fact, the parameters are related by $s = 3t - 4$ (or $t = \tfrac{1}{3}(s + 4)$).

                  ◆◆◆

EXAMPLE 3

Describe all solutions to $3x - y + 2z = 6$ in parametric form.

Solution

Solving the equation for y in terms of x and z, we get $y = 3x + 2z - 6$. If s and t are arbitrary, then, setting $x = s, z = t$, we get solutions

$$x = s$$
$$y = 3s + 2t - 6 \qquad s \text{ and } t \text{ arbitrary}$$
$$z = t$$

Of course we could have solved for x: $x = \frac{1}{3}(y - 2z + 6)$. Then, if we take $y = p$, $z = q$, the solutions are represented as follows:

$$x = \tfrac{1}{3}(p - 2q + 6)$$
$$y = p \qquad\qquad p \text{ and } q \text{ arbitrary}$$
$$z = q$$

The same family of solutions can "look" quite different!

When only two variables are involved, the solutions to systems of linear equations can be described geometrically because the graph of a linear equation $ax + by = c$ is a straight line. Moreover, a point $P(s, t)$ with coordinates s and t lies on the line if and only if $as + bt = c$ — that is, when $x = s, y = t$ is a solution to the equation. Hence the solutions to a *system* of linear equations correspond to the points $P(s, t)$ that lie on *all* the lines in question.

In particular, if the system consists of just one equation (as in Example 2), there must be infinitely many solutions because there are infinitely many points on a line. If the system has two equations, there are three possibilities for the corresponding straight lines:

1. The lines intersect in a single point. Then the system has a *unique solution* corresponding to that point.

2. The lines are parallel (and distinct) and so do not intersect. Then the system has *no solution*.

3. The lines are identical. Then the system has *infinitely many solutions* — one for each point on the (common) line.

These three situations are illustrated in Figure 1.1. In each case the graphs of two specific lines are plotted and the corresponding equations indicated. In the last case, the equations are $3x - y = 4$ (treated in Example 2) and $-6x + 2y = -8$, which have identical graphs.

When three variables are present, the graph of an equation $ax + by + cz = d$ can be shown to be a plane and so again provides a "picture" of the set of solutions. However, this graphical method has its limitations: When more than three variables

FIGURE 1.1 (a) Unique solution ($x = 2$, $y = 1$), (b) No solution, (c) Infinitely many solutions ($x = t$, $y = 3t - 4$)

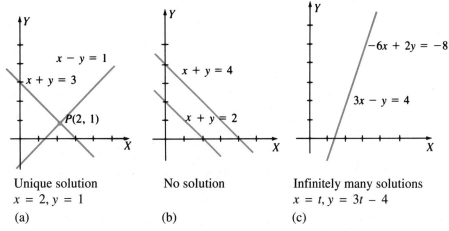

Unique solution
$x = 2$, $y = 1$
(a)

No solution
(b)

Infinitely many solutions
$x = t$, $y = 3t - 4$
(c)

are involved, no physical image of the graphs (called hyperplanes) is possible. It is necessary to turn to a more "algebraic" method of solution.

Before describing the method, we introduce a concept that simplifies the computations involved. Consider the following system

$$
\begin{aligned}
3x_1 + 2x_2 - x_3 + x_4 &= -1 \\
2x_1 \quad\;\; - x_3 + 2x_4 &= 0 \\
3x_1 + x_2 + 2x_3 + 5x_4 &= 2
\end{aligned}
$$

of three equations in four variables. The array of numbers[1]

$$
\begin{bmatrix}
3 & 2 & -1 & 1 & -1 \\
2 & 0 & -1 & 2 & 0 \\
3 & 1 & 2 & 5 & 2
\end{bmatrix}
$$

occurring in the system is called the **augmented matrix** of the system. Each row of the matrix consists of the coefficients of the variables (in order) from the corresponding equation, together with the constant term. For clarity, the constants are separated by a vertical line. The augmented matrix is just a different way of describing the system of equations. The array of coefficients of the variables

$$
\begin{bmatrix}
3 & 2 & -1 & 1 \\
2 & 0 & -1 & 2 \\
3 & 1 & 2 & 5
\end{bmatrix}
$$

is called the **coefficient matrix** of the system and $\begin{bmatrix} -1 \\ 0 \\ 2 \end{bmatrix}$ is called the **constant matrix** of the system.

---

[1]A rectangular array of numbers is called a **matrix**. Matrices will be discussed in more detail in Chapter 2.

Elementary Operations

The algebraic method for solving systems of linear equations is described as follows. Two such systems are said to be **equivalent** if each has the same set of solutions. A system is solved by writing a series of systems, one by one, each equivalent to the previous system. Each of these systems has the same set of solutions as the original one; the aim is to end up with a system that is easy to solve. Each system in the series is obtained from the preceding system by a simple manipulation chosen so that it does not change the set of solutions.

As an illustration, we solve the system $x + 2y = -2$, $2x + y = 7$ in this manner. At each stage, the corresponding augmented matrix is displayed. The original system is

$$\begin{aligned} x + 2y &= -2 \\ 2x + y &= 7 \end{aligned} \qquad \begin{bmatrix} 1 & 2 & | & -2 \\ 2 & 1 & | & 7 \end{bmatrix}$$

First, subtract twice the first equation from the second. The resulting system is

$$\begin{aligned} x + 2y &= -2 \\ -3y &= 11 \end{aligned} \qquad \begin{bmatrix} 1 & 2 & | & -2 \\ 0 & -3 & | & 11 \end{bmatrix}$$

which is equivalent to the original (see Theorem 1). At this stage we obtain $y = -\frac{11}{3}$ by multiplying the second equation by $-\frac{1}{3}$. The result is the equivalent system

$$\begin{aligned} x + 2y &= -2 \\ y &= -\tfrac{11}{3} \end{aligned} \qquad \begin{bmatrix} 1 & 2 & | & -2 \\ 0 & 1 & | & -\tfrac{11}{3} \end{bmatrix}$$

Finally, we subtract twice the second equation from the first to get another equivalent system.

$$\begin{aligned} x &= \tfrac{16}{3} \\ y &= -\tfrac{11}{3} \end{aligned} \qquad \begin{bmatrix} 1 & 0 & | & \tfrac{16}{3} \\ 0 & 1 & | & -\tfrac{11}{3} \end{bmatrix}$$

Now *this* system is easy to solve! And because it is equivalent to the original system, it provides the solution to that system.

Observe that, at each stage, a certain operation is performed on the system (and thus on the augmented matrix) to produce an equivalent system. The following operations, called **elementary operations**, can routinely be performed on systems of linear equations to produce equivalent systems.

    I.   Interchange two equations.

   II.   Multiply one equation by a nonzero number.

  III.   Add a multiple of one equation to a different equation.

> **THEOREM 1**
>
> Suppose that an elementary operation is performed on a system of linear equations. Then the resulting system has the same set of solutions as the original, so the two systems are equivalent.

Proof   We prove this theorem only for operations of type III and leave the proofs for the other operations as exercises. Let

$$c_1 x_1 + c_2 x_2 + \cdots + c_n x_n = d \tag{*}$$

and

$$a_1 x_1 + a_2 x_2 + \cdots + a_n x_n = b \tag{**}$$

denote two different equations in the system in question. Suppose a new system is formed by replacing (**) by equation (***)

$$(a_1 + kc_1)x_1 + \cdots + (a_n + kc_n)x_n = b + kd \tag{***}$$

obtained by adding k times equation (*) to equation (**). If $s_1, s_2, \ldots, s_n$ is a solution of the original system, then

$$c_1 s_1 + c_2 s_2 + \cdots + c_n s_n = d$$

and

$$a_1 s_1 + a_2 s_2 + \cdots + a_n s_n = b$$

By multiplication and addition, these give

$$(a_1 + kc_1)s_1 + (a_2 + kc_2)s_2 + \cdots + (a_n + kc_n)s_n = b + kd$$

Hence $s_1, s_2, \ldots, s_n$ is a solution of the new system. Moreover, no additional solutions have been introduced because the process is reversible. In fact, because (*) and (**) are *different* equations in the original system, we can subtract k times equation (*) from equation (***) in the new system, and so retrieve the original system. Thus every solution of the new system is a solution of the original system.   ◆

Elementary operations performed on a system of equations produce corresponding manipulations of the *rows* of the augmented matrix. Thus multiplying a row of a matrix by a number k means multiplying *every entry* of the row by k. Adding one row to another row means adding *each entry* of that row to the corresponding entry of the other row. Subtracting two rows is done similarly.

In hand calculations (and in computer programs) we usually manipulate the rows of the augmented matrix rather than the equations. For this reason we restate these elementary operations for matrices.

> **DEFINITION**
>
> The following are called **elementary row operations** on a matrix.
>
> I. Interchange two rows.
>
> II. Multiply one row by a nonzero number.
>
> III. Add a multiple of one row to a different row.

In the illustration preceding Theorem 1 these operations led to a matrix of the form

$$\begin{bmatrix} 1 & 0 & | & * \\ 0 & 1 & | & * \end{bmatrix}$$

where the asterisks represent numbers. In the case of three equations in three variables, the goal is to produce a matrix of the form

$$\begin{bmatrix} 1 & 0 & 0 & | & * \\ 0 & 1 & 0 & | & * \\ 0 & 0 & 1 & | & * \end{bmatrix}$$

This does not always happen, as we will see in the next section. Here is an example in which it does happen.

EXAMPLE 4

Find all solutions to the following system of equations.

$$\begin{aligned} 3x + 4y + z &= 1 \\ 2x + 3y &= 0 \\ 4x + 3y - z &= -2 \end{aligned}$$

Solution   The augmented matrix of the original system is

$$\begin{bmatrix} 3 & 4 & 1 & | & 1 \\ 2 & 3 & 0 & | & 0 \\ 4 & 3 & -1 & | & -2 \end{bmatrix}$$

To create a 1 in the upper left corner we could multiply row 1 through by $\frac{1}{3}$. However, the 1 can be obtained without fractions by subtracting row 2 from row 1. The result is

$$\begin{bmatrix} 1 & 1 & 1 & | & 1 \\ 2 & 3 & 0 & | & 0 \\ 4 & 3 & -1 & | & -2 \end{bmatrix}$$

The upper left 1 is now used to "clean up" the first column. First subtract 2 times row

1 from row 2 to obtain

$$\begin{bmatrix} 1 & 1 & 1 & | & 1 \\ 0 & 1 & -2 & | & -2 \\ 4 & 3 & -1 & | & -2 \end{bmatrix}$$

Next subtract 4 times row 1 from row 3. The result is

$$\begin{bmatrix} 1 & 1 & 1 & | & 1 \\ 0 & 1 & -2 & | & -2 \\ 0 & -1 & -5 & | & -6 \end{bmatrix}$$

This completes the work on column 1. We now use the 1 in the second position of the second row to clean up the second column by subtracting row 2 from row 1 and adding row 2 to row 3. For convenience, both row operations are done in one step. The result is

$$\begin{bmatrix} 1 & 0 & 3 & | & 3 \\ 0 & 1 & -2 & | & -2 \\ 0 & 0 & -7 & | & -8 \end{bmatrix}$$

Note that these manipulations *did not affect* the first column (the second row has a zero there), so our previous effort there has not been undermined. Finally we clean up the third column. Begin by multiplying row 3 by $-\frac{1}{7}$ to obtain

$$\begin{bmatrix} 1 & 0 & 3 & | & 3 \\ 0 & 1 & -2 & | & -2 \\ 0 & 0 & 1 & | & \frac{8}{7} \end{bmatrix}$$

Now subtract 3 times row 3 from row 1, and add 2 times row 3 to row 2 to get

$$\begin{bmatrix} 1 & 0 & 0 & | & -\frac{3}{7} \\ 0 & 1 & 0 & | & \frac{2}{7} \\ 0 & 0 & 1 & | & \frac{8}{7} \end{bmatrix}$$

The corresponding equations are $x = -\frac{3}{7}$, $y = \frac{2}{7}$, and $z = \frac{8}{7}$, which give the solution.

$\blacklozenge\blacklozenge\blacklozenge$

EXERCISES 1.1

1. In each case verify that the following are solutions for all values of s and t.

(a) $x = 19t - 35$
$y = 25 - 13t$ is a solution of
$z = t$

$$2x + 3y + z = 5$$
$$5x + 7y - 4z = 0$$

(b) $x_1 = 2s + 12t + 13$
$x_2 = s$ is a
$x_3 = -s - 3t - 3$ solution of
$x_4 = t$

$$2x_1 + 5x_2 + 9x_3 + 3x_4 = -1$$
$$x_1 + 2x_2 + 4x_3 = 1$$

2. Find all solutions to the following in parametric form in two ways.[2]
(**a**) $3x + y = 2$ ◆(**b**) $2x + 3y = 1$
(**c**) $3x - y + 2z = 5$ ◆(**d**) $x - 2y + 5z = 1$

3. Regarding $2x = 5$ as the equation $2x + 0y = 5$ in two variables, find all solutions in parametric form.

◆**4.** Regarding $4x - 2y = 3$ as the equation $4x - 2y + 0z = 3$ in three variables, find all solutions in parametric form.

◆**5.** Find all solutions to the general system $ax = b$ of one equation in one variable (**a**) when $a = 0$ and (**b**) when $a \neq 0$.

6. Show that a system consisting of exactly one linear equation can have no solution, one solution, or infinitely many solutions. Give examples.

7. Write the augmented matrix for each of the following systems of linear equations.
(**a**) $x - 3y = 5$ ◆(**b**) $x + 2y = 0$
    $2x + y = 1$ $y = 1$
(**c**) $x - y + z = 2$ ◆(**d**) $x + y = 1$
    $x - z = 1$ $y + z = 0$
    $y + 2x = 0$ $z - x = 2$

8. Write a system of linear equations that has each of the following augmented matrices.
(**a**) $\begin{bmatrix} 1 & -1 & 6 & | & 0 \\ 0 & 1 & 0 & | & 3 \\ 2 & -1 & 0 & | & 1 \end{bmatrix}$ ◆(**b**) $\begin{bmatrix} 2 & -1 & 0 & | & -1 \\ -3 & 2 & 1 & | & 0 \\ 0 & 1 & 1 & | & 3 \end{bmatrix}$

9. Find the solution of each of the following systems of linear equations using augmented matrices.
(**a**) $x - 3y = 1$ ◆(**b**) $x + 2y = 1$
    $2x - 7y = 3$ $3x + 4y = -1$
(**c**) $2x + 3y = -1$ ◆(**d**) $3x + 4y = 1$
    $3x + 4y = 2$ $4x + 5y = -3$

10. Find the solution of each of the following systems of linear equations using augmented matrices.
(**a**) $x + y + 2z = -1$ ◆(**b**) $2x + y + z = -1$
    $2x + y + 3z = 0$ $x + 2y + z = 0$
    $-2y + z = 2$ $3x - 2z = 5$

11. Find all solutions (if any) of the following systems of linear equations.
(**a**) $3x - 2y = 5$ ◆(**b**) $3x - 2y = 5$
    $-12x + 8y = -20$ $-12x + 8y = 16$

12. Show that $2x + y + 3z = b$
    $x - 4y + 9z = c$
has no solution unless $c = 2b - 3a$.

13. By examining the possible positions of lines in the plane, show that three equations in two variables can have zero, one, or infinitely many solutions.

◆**14.** Solve the system $\begin{matrix} 3x + 2y = 5 \\ 7x + 5y = 1 \end{matrix}$ by changing variables $\begin{matrix} x = 5x' - 2y' \\ y = -7x' - 3y' \end{matrix}$, and solving the resulting equations for x' and y'.

◆**15.** Find a, b, and c such that
$$\frac{x^2 - x + 3}{(x^2 + 2)(2x - 1)} = \frac{ax + b}{x^2 + 2} + \frac{c}{2x - 1}$$
[*Hint*: Multiply through by $(x^2 + 2)(2x - 1)$ and equate coefficients of powers of x.]

Section 1.2   Gaussian Elimination

The algebraic method introduced in the preceding section can be summarized as follows: Given a system of linear equations, use a series of elementary row operations to carry the augmented matrix to a "nice" matrix (meaning that the corresponding equations are easy to solve). In Example 4§1.1,[3] this nice matrix took the form

$$\begin{bmatrix} 1 & 0 & 0 & | & * \\ 0 & 1 & 0 & | & * \\ 0 & 0 & 1 & | & * \end{bmatrix}$$

---

[2] ◆Indicates that the exercise has an answer at the end of the book or in the student solution manual.
[3] This means Example 4 in Section 1.1.

The following definitions identify the nice matrices that arise in this process.

DEFINITION

A matrix is said to be in **row-echelon form** (and will be called a **row-echelon matrix**) if it satisfies the following three conditions:

1. All **zero rows** (consisting entirely of zeros) are at the bottom.
2. The first nonzero entry from the left in each nonzero row is a 1, called the **leading** 1 for that row.
3. Each leading 1 is to the right of all leading 1's in the rows above it.

The row-echelon matrices have a "staircase" form, as indicated by the following example (the asterisks indicate arbitrary numbers).

$$
\begin{bmatrix}
0 & 1 & * & * & * & * & * \\
0 & 0 & 0 & 1 & * & * & * \\
0 & 0 & 0 & 0 & 1 & * & * \\
0 & 0 & 0 & 0 & 0 & 0 & 1 \\
0 & 0 & 0 & 0 & 0 & 0 & 0
\end{bmatrix}
$$

The leading 1's proceed "down and to the right" through the matrix. Entries above and to the right of the leading 1's are arbitrary, but all entries below and to the left of them are zero.

DEFINITION

A row-echelon matrix is said to be in **reduced row-echelon form** (and will be called a **reduced row-echelon matrix**) if it also satisfies the following condition:

4. Each leading 1 is the only nonzero entry in its column.

Hence a matrix in row-echelon form is in reduced form if, in addition, the entries directly above each leading 1 are zero. Note that a matrix in row-echelon form can, with a few more row operations, be carried to reduced form.

EXAMPLE I

The following matrices are in row-echelon form (for any choice of numbers in ∗-positions).

$$\begin{bmatrix} 1 & * & * \\ 0 & 0 & 1 \end{bmatrix} \quad \begin{bmatrix} 0 & 1 & * & * \\ 0 & 0 & 1 & * \\ 0 & 0 & 0 & 0 \end{bmatrix} \quad \begin{bmatrix} 1 & * & * & * \\ 0 & 1 & * & * \\ 0 & 0 & 0 & 1 \end{bmatrix} \quad \begin{bmatrix} 1 & * & * \\ 0 & 1 & * \\ 0 & 0 & 1 \end{bmatrix}$$

The following, on the other hand, are in reduced row-echelon form.

$$\begin{bmatrix} 1 & * & 0 \\ 0 & 0 & 1 \end{bmatrix} \quad \begin{bmatrix} 0 & 1 & 0 & * \\ 0 & 0 & 1 & * \\ 0 & 0 & 0 & 0 \end{bmatrix} \quad \begin{bmatrix} 1 & 0 & * & 0 \\ 0 & 1 & * & 0 \\ 0 & 0 & 0 & 1 \end{bmatrix} \quad \begin{bmatrix} 1 & 0 & 0 \\ 0 & 1 & 0 \\ 0 & 0 & 1 \end{bmatrix}$$

Clearly the choice of the positions for the leading 1's determines the (reduced) row-echelon form (apart from the numbers in ∗-positions).

◆◆◆

The importance of row-echelon matrices stems from the following theorem.

THEOREM 1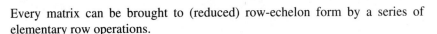

Every matrix can be brought to (reduced) row-echelon form by a series of elementary row operations.

The proof is given later. In fact a step-by-step procedure (called the **Gaussian algorithm**) is given for actually finding the row-echelon matrix.

Theorem 1 reduces the problem of finding solutions of systems of linear equations to the case in which the augmented matrix is in (reduced) row-echelon form. This case is easy as the next example shows.

EXAMPLE 2

In each case assume that the augmented matrix of a system of linear equations has been carried to the given reduced row-echelon matrix by row operations. Then solve the system.

(a) $\begin{bmatrix} 1 & 0 & 0 & | & -1 \\ 0 & 1 & 0 & | & 0 \\ 0 & 0 & 1 & | & 2 \end{bmatrix}$ **(b)** $\begin{bmatrix} 1 & 3 & 0 & 1 & -3 & | & 0 \\ 0 & 0 & 1 & 2 & 5 & | & 0 \\ 0 & 0 & 0 & 0 & 0 & | & 1 \end{bmatrix}$ **(c)** $\begin{bmatrix} 1 & 5 & 0 & 0 & -2 & | & 3 \\ 0 & 0 & 1 & 0 & 4 & | & -5 \\ 0 & 0 & 0 & 1 & 2 & | & 6 \\ 0 & 0 & 0 & 0 & 0 & | & 0 \end{bmatrix}$

Solution   In case **(a)** the corresponding system of equations is

$$\begin{aligned} x_1 \qquad\qquad &= -1 \\ x_2 \qquad &= 0 \\ x_3 &= 2 \end{aligned}$$

and the solution is apparent.

In case **(b)** the last equation in the corresponding system is

$$0x_1 + 0x_2 + 0x_3 + 0x_4 + 0x_5 = 1$$

There are certainly no values of x_1, x_2, x_3, x_4, and x_5 that satisfy this equation, so the system has *no* solution in this case.

In case **(c)** the corresponding system of equations is

$$\begin{aligned} x_1 + 5x_2 \qquad\qquad - 2x_5 &= 3 \\ x_3 \qquad + 4x_5 &= -5 \\ x_4 + 2x_5 &= 6 \\ 0 &= 0 \end{aligned}$$

Use these equations to solve for the variables x_1, x_3, and x_4 (corresponding to the leading 1's) in terms of the other variables.

$$\begin{aligned} x_1 &= 3 - 5x_2 + 2x_5 \\ x_3 &= -5 - 4x_5 \\ x_4 &= 6 - 2x_5 \end{aligned}$$

Now x_2 and x_5 can be assigned arbitrary values, say $x_2 = s$ and $x_5 = t$, where s and t are parameters. This gives the solution

$$x_1 = 3 - 5s + 2t \qquad x_2 = s \qquad x_3 = -5 - 4t \qquad x_4 = 6 - 2t \qquad x_5 = t$$

in parametric form.

◆◆◆

In general, when the augmented matrix of a system of linear equations has been carried to (reduced) row-echelon form, variables corresponding to columns containing a leading 1 are called **leading variables**. The nonleading variables (if any) end up as parameters in the final solution, and the leading variables are given (by the equations) in terms of these parameters.

Gaussian Elimination

There remains the task of giving a systematic procedure by which the augmented matrix can be carried by row operations to reduced row-echelon form. One such reduction was carried out in Example 4§1.1 where the solution turned out to be unique. Other situations are described in the following examples.

EXAMPLE 3

Solve the following system of equations.

$$\begin{aligned} x + 2y - z &= 2 \\ 2x + 5y + 2z &= -1 \\ 7x + 17y + 5z &= -1 \end{aligned}$$

Solution   The augmented matrix

$$\begin{bmatrix} 1 & 2 & -1 & | & 2 \\ 2 & 5 & 2 & | & -1 \\ 7 & 17 & 5 & | & -1 \end{bmatrix}$$

already has the first leading 1 in place. Subtract 2 times row 1 from row 2, and subtract 7 times row 1 from row 3, to obtain

$$\begin{bmatrix} 1 & 2 & -1 & | & 2 \\ 0 & 1 & 4 & | & -5 \\ 0 & 3 & 12 & | & -15 \end{bmatrix}$$

Now subtract 2 times row 2 from row 1, and subtract 3 times row 2 from row 3, to get the reduced row-echelon form.

$$\begin{bmatrix} 1 & 0 & -9 & | & 12 \\ 0 & 1 & 4 & | & -5 \\ 0 & 0 & 0 & | & 0 \end{bmatrix}$$

The corresponding system of equations is

$$\begin{aligned} x \quad - 9z &= 12 \\ y + 4z &= -5 \\ 0 &= 0 \end{aligned}$$

If $z = t$ where t is an arbitrary parameter, we obtain the solution

$$x = 9t + 12 \qquad y = -4t - 5 \qquad z = t$$

in parametric form.

◆◆◆

The row of 0's in the row-echelon form in Example 3 means that one of the equations in the original system is *redundant* in the sense that it provides no new information about the solutions to the system. In fact, the third equation is just the sum of the first equation plus three times the second, so any solution to the first two equations is *automatically* a solution to the third. It is instructive to see how one can *discover* this fact. We do this by "keeping track" of the row operations performed in the solution to Example 3. Let R_1, R_2, R_3 denote the three rows of the original augmented matrix. Then the first manipulation in the solution can be described as follows:

$$\begin{bmatrix} 1 & 2 & -1 & | & 2 \\ 2 & 5 & 2 & | & -1 \\ 7 & 17 & 5 & | & -1 \end{bmatrix} \quad \begin{matrix} R_1 \\ R_2 \\ R_3 \end{matrix}$$

$$\begin{bmatrix} 1 & 2 & -1 & | & 2 \\ 0 & 1 & 4 & | & -5 \\ 0 & 3 & 12 & | & -15 \end{bmatrix} \quad \begin{matrix} R_1 \\ R_2 - 2R_1 \\ R_3 - 7R_1 \end{matrix}$$

Now observe that, in this last matrix, the last row is three times the second row. Hence

$$R_3 - 7R_1 = 3(R_2 - 2R_1)$$

This can be manipulated algebraically to give

$$R_3 = 3R_2 + R_1$$

In other words, the original third row, R_3, equals the first row plus three times the second row, as asserted. Note that we have been manipulating the symbols R_1, R_2, and R_3 as though they were numbers or variables. This is a typical matrix calculation. Matrix calculations will be treated in detail later in this book.

EXAMPLE 4

Solve the following system of equations.

$$\begin{aligned} x \quad\quad + 10z &= 5 \\ 3x + y - 4z &= -1 \\ 4x + y + 6z &= 1 \end{aligned}$$

Solution

We manipulate the augmented matrix, keeping track of the manipulations for reference. As before, we use R_1, R_2, R_3 to denote the three rows. The augmented matrix is

$$\begin{bmatrix} 1 & 0 & 10 & | & 5 \\ 3 & 1 & -4 & | & -1 \\ 4 & 1 & 6 & | & 1 \end{bmatrix} \quad \begin{matrix} R_1 \\ R_2 \\ R_3 \end{matrix}$$

The indicated row operations give

$$\begin{bmatrix} 1 & 0 & 10 & | & 5 \\ 0 & 1 & -34 & | & -16 \\ 0 & 1 & -34 & | & -19 \end{bmatrix} \quad \begin{matrix} R_1 \\ R_2 - 3R_1 \\ R_3 - 4R_1 \end{matrix}$$

$$\begin{bmatrix} 1 & 0 & 10 & | & 5 \\ 0 & 1 & -34 & | & -16 \\ 0 & 0 & 0 & | & -3 \end{bmatrix} \quad \begin{matrix} R_1 \\ R_2 - 3R_1 \\ (R_3 - 4R_1) - (R_2 - 3R_1) \end{matrix}$$

But this means that the following system of equations

$$\begin{aligned} x \qquad + 10z &= \quad 5 \\ y - 34z &= -16 \\ 0 &= -3 \end{aligned}$$

is equivalent to the original system. In other words, the two have the *same solutions*. This last system clearly has *no* solution (the last equation requires x, y, and z to have the property that $0x + 0y + 0z = -3$ and no such x, y, and z exist). Hence the original system has *no* solution.

◆◆◆

Because we have kept track of the operations performed in Example 4, it is possible to give a clear explanation of *why* there is no solution. The offending equation $0 = -3$ corresponds to the row $[0\ 0\ 0\ -3]$ in the last matrix. In terms of the rows R_1, R_2, and R_3 of the original matrix, the last row is

$$[0 \quad 0 \quad 0 \quad -3] = (R_3 - 4R_1) - (R_2 - 3R_1) = R_3 - (R_1 + R_2)$$

The fact that this is "almost zero" suggests that we compare R_3 with $R_1 + R_2$ or, what is the same thing, that we compare the third equation with the first plus the second.

Third equation:	$4x + y + 6z = 1$
First equation plus the second:	$4x + y + 6z = 4$

Since no numbers x, y, and z can satisfy both these equations, the system has no solution.

The key to all these examples is carrying the augmented matrix to reduced row-echelon form. Suppose an arbitrary matrix is presented to you, even one that is not the augmented matrix of a system of linear equations. Here is a step-by-step procedure by which it can be brought to row-echelon form.

> ### GAUSSIAN ALGORITHM[4]
>
> Step 1. If the matrix consists entirely of zeros, stop — it is *already* in row-echelon form.
>
> Step 2. Otherwise, find the first column from the left containing a nonzero entry (call it a), and move the row containing that entry to the top position.
>
> Step 3. Now multiply that row by $\frac{1}{a}$ to create a leading 1.
>
> Step 4. By subtracting multiples of that row from rows below it, make each entry below the leading 1 zero.
>
> This completes the first row, and all further row operations are carried out on the other rows.
>
> Step 5. Repeat steps 1–4 on the matrix consisting of the remaining rows.
>
> The process stops when either no rows remain at step 5 or the remaining rows consist of zeros.

Observe that the Gaussian algorithm is recursive: When the first leading 1 has been obtained, the procedure is repeated on the remaining rows of the matrix. This makes the algorithm easy to use on a computer.

EXAMPLE 5

Bring the following matrix to row-echelon form, using the Gaussian algorithm.

$$\begin{bmatrix} 0 & 0 & 0 & 2 & 1 & 9 \\ 0 & -2 & -6 & 2 & 0 & 2 \\ 0 & 2 & 6 & -2 & 2 & 0 \\ 0 & 3 & 9 & 2 & 2 & 19 \end{bmatrix}$$

Solution   We follow the steps in the algorithm.

Step 1. The matrix has nonzero entries, so proceed to step 2.

Step 2. The first nonzero column is column 2. Choose $a = -2$ in row 2 ($a = 2$ in row 3 or $a = 3$ in row 4 would do as well). Interchange rows 1 and 2 to obtain

---

[4]Carl Friedrich Gauss (1777–1855) ranks with Archimedes and Newton as one of the three greatest mathematicians of all time. He was a child prodigy; and at the age of 21 he gave the first proof that every polynomial has a complex root. In 1801 he published a timeless masterpiece, *Disquisitiones Arithmeticae,* in which he founded modern number theory. He went on to make ground-breaking contributions to nearly every branch of mathematics, often well before others rediscovered and published the results. In addition, he did fundamental work in both physics and astronomy. Gauss is said to have been the last mathematician to know everything in his subject; it is no wonder he is called "the prince of mathematicians."

$$\begin{bmatrix} 0 & -2 & -6 & 2 & 0 & 2 \\ 0 & 0 & 0 & 2 & 1 & 9 \\ 0 & 2 & 6 & -2 & 2 & 0 \\ 0 & 3 & 9 & 2 & 2 & 19 \end{bmatrix}$$

Step 3. Multiply row 1 through by $-\frac{1}{2}$ to obtain

$$\begin{bmatrix} 0 & 1 & 3 & -1 & 0 & -1 \\ 0 & 0 & 0 & 2 & 1 & 9 \\ 0 & 2 & 6 & -2 & 2 & 0 \\ 0 & 3 & 9 & 2 & 2 & 19 \end{bmatrix}$$

Step 4. Subtract 2 times row 1 from row 3, and subtract 3 times row 1 from row 4.
The result is

$$\begin{bmatrix} 0 & 1 & 3 & -1 & 0 & -1 \\ 0 & 0 & 0 & 2 & 1 & 9 \\ 0 & 0 & 0 & 0 & 2 & 2 \\ 0 & 0 & 0 & 5 & 2 & 22 \end{bmatrix}$$

Step 5. This completes the first row (now shaded), and we ignore it from now
on and repeat steps 1–4 on the new matrix (unshaded) consisting of the
remaining three rows. Steps 1–3 amount to multiplying row 2 by $\frac{1}{2}$
(creating the second leading 1), and step 4 then subtracts 5 times this
new row 2 from row 4. This produces the matrix

$$\begin{bmatrix} 0 & 1 & 3 & -1 & 0 & -1 \\ 0 & 0 & 0 & 1 & \frac{1}{2} & \frac{9}{2} \\ 0 & 0 & 0 & 0 & 2 & 2 \\ 0 & 0 & 0 & 0 & -\frac{1}{2} & -\frac{1}{2} \end{bmatrix}$$

where the second row (now completed) has been shaded.

Now repeat steps 1–4 on the last two rows to get

$$\begin{bmatrix} 0 & 1 & 3 & -1 & 0 & -1 \\ 0 & 0 & 0 & 1 & \frac{1}{2} & \frac{9}{2} \\ 0 & 0 & 0 & 0 & 1 & 1 \\ 0 & 0 & 0 & 0 & 0 & 0 \end{bmatrix}$$

where the completed third row has been shaded. This completes the procedure
since the remaining (1–row) matrix consists of zeros and the algorithm stops at
step 1. Of course the matrix is now in row-echelon form.

◆◆◆

In all our earlier examples the augmented matrix was carried to *reduced* row-echelon form by modifying step 4 in the Gaussian algorithm to introduce zeros *above* (as well as below) each leading 1 as soon as the leading 1 is created. However, when solving a large system it is numerically more efficient to go to row-echelon form first and *then* create the zeros above the leading 1's, starting with the first leading 1 from the right and working from right to left.[5]

This is the matrix version of a technique called **back-substitution** for solving a system of linear equations with a row-echelon augmented matrix. The idea is to use the last equation to find the last leading variable (in terms of the nonleading variables) and then substitute it back into all the earlier equations. Then the second last equation yields the second last leading variable (which is also substituted back), and so on. Here is an example.

EXAMPLE 6

Solve the following system of equations.

$$
\begin{aligned}
x_1 - 3x_2 + x_3 - x_4 &= -1 \\
-x_1 + 3x_2 + 3x_4 + x_5 &= 3 \\
2x_1 - 6x_2 + 3x_3 - x_5 &= 2 \\
-x_1 + 3x_2 + x_3 + 5x_4 + x_5 &= 6
\end{aligned}
$$

using the Gaussian algorithm and back-substitution.

Solution

The reduction of the augmented matrix to row-echelon form is as follows (the details are omitted).

$$
\left[\begin{array}{ccccc|c}
1 & -3 & 1 & -1 & 0 & -1 \\
-1 & 3 & 0 & 3 & 1 & 3 \\
2 & -6 & 3 & 0 & -1 & 2 \\
-1 & 3 & 1 & 5 & 1 & 6
\end{array}\right]
\rightarrow
\left[\begin{array}{ccccc|c}
1 & -3 & 1 & -1 & 0 & -1 \\
0 & 0 & 1 & 2 & 1 & 2 \\
0 & 0 & 0 & 0 & 1 & -1 \\
0 & 0 & 0 & 0 & 0 & 0
\end{array}\right]
$$

The corresponding equations are

$$
\begin{aligned}
x_1 - 3x_2 + x_3 - x_4 &= -1 \\
x_3 + 2x_4 + x_5 &= 2 \\
x_5 &= -1
\end{aligned}
$$

The leading variables are x_1, x_3, and x_5. Solving the last equation for x_5 is easy:

$$
x_5 = -1
$$

---

[5]With n equations where n is large, this procedure requires roughly $\frac{n^3}{3}$ multiplications and divisions, whereas this number is roughly $\frac{n^3}{2}$ if the zeros above each leading 1 are introduced when the leading 1 is created. This is an important consideration when solving a large system.

Now substitute this into the second to last equation. The result is

$$x_3 = 3 - 2x_4$$

Finally, substituting both x_5 and x_3 into the first equation gives

$$x_1 = -4 + 3x_2 + 3x_4$$

If the nonleading variables are assigned as parameters $x_2 = s$, $x_4 = t$, the solution is

$$x_1 = -4 + 3s + 3t \qquad x_2 = s \qquad x_3 = 3 - 2t \qquad x_4 = t \qquad x_5 = -1$$

◆◆◆

Systems of linear equations that have no solution are called **inconsistent systems;** systems that have at least one solution are said to be **consistent.**

EXAMPLE 7

Find a condition on the numbers a, b, and c such that the following system of equations is consistent. When that condition is satisfied, find all solutions (in terms of a, b, and c).

$$x + 3y + z = a$$
$$-x - 2y + z = b$$
$$3x + 7y - z = c$$

Solution   The Gaussian algorithm applies except that now the augmented matrix has entries a, b, and c as well as known numbers.

$$\begin{bmatrix} 1 & 3 & 1 & a \\ -1 & -2 & 1 & b \\ 3 & 7 & -1 & c \end{bmatrix}$$

$$\begin{bmatrix} 1 & 3 & 1 & a \\ 0 & 1 & 2 & b + a \\ 0 & -2 & -4 & c - 3a \end{bmatrix}$$

$$\begin{bmatrix} 1 & 0 & -5 & -2a - 3b \\ 0 & 1 & 2 & b + a \\ 0 & 0 & 0 & c - a + 2b \end{bmatrix}$$

Now the whole thing depends on the quantity $c - a + 2b$. The last row corresponds to an equation $0 = c - a + 2b$. Thus if $c - a + 2b$ is *not* zero, there is no solution (just as in Example 4). Hence the condition for consistency is $c = a - 2b$. Then the last matrix becomes

$$\begin{bmatrix} 1 & 0 & -5 & -2a - 3b \\ 0 & 1 & 2 & a + b \\ 0 & 0 & 0 & 0 \end{bmatrix}$$

Thus if $c = a - 2b$, taking $z = t$, t a parameter, gives the solutions

$$x = 5t - (2a + 3b) \qquad y = (a + b) - 2t \qquad z = t$$

Rank

It can be proven that the *reduced* row-echelon form of a matrix A is uniquely determined by A. That is, no matter which series of row operations is used to carry A to a reduced row-echelon matrix, the result will always be the same matrix. (See Supplementary Exercise 7 in Chapter 5.) By contrast, this is not true for row-echelon matrices: Different series of row operations can carry the same matrix A to *different* row-echelon matrices. However, it *is* true that the number of leading 1's must be the same in each of these row-echelon matrices (this will be proved in Chapter 5). Hence the number of leading 1's depends only on A and not on the way in which A is carried to row-echelon form.

DEFINITION

If a matrix A is carried to a row-echelon matrix R by elementary row operations, the number of leading 1's in R is called the **rank** of A and is denoted rank A.

EXAMPLE 8

Compute the rank of $\qquad A = \begin{bmatrix} 1 & 1 & -1 & 4 \\ 2 & 1 & 3 & 0 \\ 0 & 1 & -5 & 8 \end{bmatrix}$.

Solution   The reduction of A to row-echelon form is

$$A = \begin{bmatrix} 1 & 1 & -1 & 4 \\ 2 & 1 & 3 & 0 \\ 0 & 1 & -5 & 8 \end{bmatrix} \rightarrow \begin{bmatrix} 1 & 1 & -1 & 4 \\ 0 & -1 & 5 & -8 \\ 0 & 1 & -5 & 8 \end{bmatrix} \rightarrow \begin{bmatrix} 1 & 1 & -1 & 4 \\ 0 & 1 & -5 & 8 \\ 0 & 0 & 0 & 0 \end{bmatrix}$$

Because this row-echelon matrix has two leading 1's, rank $A = 2$.

The notion of rank of a matrix has a useful application to equations.

THEOREM 2

Suppose a system of m equations in n variables has a solution. If the rank of the augmented matrix is r, the set of solutions involves exactly $n - r$ parameters.

Proof The fact that the rank of the augmented matrix is r means there are exactly r leading variables, and hence exactly $n - r$ nonleading variables. These nonleading variables are all assigned as parameters, so the set of solutions involves exactly $n - r$ parameters. ◆

In particular, this shows that, for any system of linear equations, exactly three possibilities exist:

1. *No solution.*
2. *A unique solution.* This occurs when *every* variable is a leading variable.
3. *Infinitely many solutions.* This occurs when there is at least one nonleading variable, so a parameter is involved.

EXAMPLE 9

Suppose the matrix A in Example 8 is the augmented matrix of a system of $m = 3$ linear equations in $n = 3$ variables. As rank $A = r = 2$, the set of solutions will have $n - r = 1$ parameter. The reader can verify this fact directly.

◆◆◆

EXERCISES 1.2

1. Which of the following matrices are in reduced row-echelon form? Which are in row-echelon form?

(a) $\begin{bmatrix} 1 & -1 & 2 \\ 0 & 0 & 0 \\ 0 & 0 & 1 \end{bmatrix}$
◆**(b)** $\begin{bmatrix} 2 & 1 & -1 & 3 \\ 0 & 0 & 0 & 0 \end{bmatrix}$

(c) $\begin{bmatrix} 1 & -2 & 3 & 5 \\ 0 & 0 & 0 & 1 \end{bmatrix}$
◆**(d)** $\begin{bmatrix} 1 & 0 & 0 & 3 & 1 \\ 0 & 0 & 0 & 1 & 1 \\ 0 & 0 & 0 & 0 & 1 \end{bmatrix}$

(e) $\begin{bmatrix} 1 & 1 \\ 0 & 1 \end{bmatrix}$
◆**(f)** $\begin{bmatrix} 0 & 0 & 1 \\ 0 & 0 & 1 \\ 0 & 0 & 1 \end{bmatrix}$

2. Carry each of the following matrices to reduced row-echelon form.

(a) $\begin{bmatrix} 0 & -1 & 2 & 1 & 2 & 1 & -1 \\ 0 & 1 & -2 & 2 & 7 & 2 & 4 \\ 0 & -2 & 4 & 3 & 7 & 1 & 0 \\ 0 & 3 & -6 & 1 & 6 & 4 & 1 \end{bmatrix}$

◆**(b)** $\begin{bmatrix} 0 & -1 & 3 & 1 & 3 & 2 & 1 \\ 0 & -2 & 6 & 1 & -5 & 0 & -1 \\ 0 & 3 & -9 & 2 & 4 & 1 & -1 \\ 0 & 1 & -3 & -1 & 3 & 0 & 1 \end{bmatrix}$

3. The augmented matrix of a system of linear equations has been carried to the following by row operations. In each case solve the system.

(a) $\left[\begin{array}{cccccc|c} 1 & 2 & 0 & 3 & 1 & 0 & -1 \\ 0 & 0 & 1 & -1 & 1 & 0 & 2 \\ 0 & 0 & 0 & 0 & 0 & 1 & 3 \\ 0 & 0 & 0 & 0 & 0 & 0 & 0 \end{array}\right]$

◆**(b)** $\left[\begin{array}{cccccc|c} 1 & -2 & 0 & 2 & 0 & 1 & 1 \\ 0 & 0 & 1 & 5 & 0 & -3 & -1 \\ 0 & 0 & 0 & 0 & 1 & 6 & 1 \\ 0 & 0 & 0 & 0 & 0 & 0 & 0 \end{array}\right]$

(c) $\left[\begin{array}{ccccc|c} 1 & 2 & 1 & 3 & 1 & 1 \\ 0 & 1 & -1 & 0 & 1 & 1 \\ 0 & 0 & 0 & 1 & -1 & 0 \\ 0 & 0 & 0 & 0 & 0 & 0 \end{array}\right]$

◆**(d)** $\left[\begin{array}{ccccc|c} 1 & -1 & 2 & 4 & 6 & 2 \\ 0 & 1 & 2 & 1 & -1 & -1 \\ 0 & 0 & 0 & 1 & 0 & 1 \\ 0 & 0 & 0 & 0 & 0 & 0 \end{array}\right]$

4. Find all solutions (if any) to each of the following systems of linear equations.

(a) $\begin{aligned} x - 2y &= 1 \\ 4y - x &= -2 \end{aligned}$
◆**(b)** $\begin{aligned} 3x - y &= 0 \\ 2x - 3y &= 1 \end{aligned}$

(c) $\begin{aligned} 2x + y &= 5 \\ 3x + 2y &= 6 \end{aligned}$
◆**(d)** $\begin{aligned} 3x - y &= 2 \\ 2y - 6x &= -4 \end{aligned}$

(e) $\begin{aligned} 3x - y &= 4 \\ 2y - 6x &= 1 \end{aligned}$
◆**(f)** $\begin{aligned} 2x - 3y &= 5 \\ 3y - 2x &= 2 \end{aligned}$

5. Find all solutions (if any) to each of the following systems of linear equations.

(a) $\begin{aligned} x + y + 2z &= 1 \\ 3x - y + z &= -1 \\ -x + 3y + 4z &= 1 \end{aligned}$
◆(b) $\begin{aligned} -2x + 3y + 3z &= -9 \\ 3x - 4y + z &= 5 \\ -5x + 7y + 2z &= -14 \end{aligned}$

(c) $\begin{aligned} x + y - z &= 3 \\ -x + 4y + 5z &= -2 \\ x + 6y + 3z &= 4 \end{aligned}$
◆(d) $\begin{aligned} x + 2y - z &= 2 \\ 2x + 5y - 3z &= 1 \\ x + 4y - 3z &= 3 \end{aligned}$

(e) $\begin{aligned} 5x + y &= 2 \\ 3x - y + 2z &= 1 \\ x + y - z &= 5 \end{aligned}$
(f) $\begin{aligned} 3x - 2y + z &= -2 \\ x - y + 3z &= 5 \\ -x + y + z &= -1 \end{aligned}$

(g) $\begin{aligned} x + y + z &= 4 \\ x + z &= 5 \\ 2x + 5y + 2z &= 5 \end{aligned}$
◆(h) $\begin{aligned} x + 2y - 4z &= 10 \\ 2x - y + 2z &= 5 \\ x + y - 2z &= 7 \end{aligned}$

6. Find all solutions to each of the following systems and express the last equation of each system as a sum of multiples of the first two.

(a) $\begin{aligned} x_1 + x_2 + x_3 &= 1 \\ 2x_1 - x_2 + 3x_3 &= 3 \\ 3x_1 - 3x_2 + 5x_3 &= 5 \end{aligned}$
◆(b) $\begin{aligned} 2x_1 + x_2 - 3x_3 &= -3 \\ 3x_1 + x_2 - 5x_3 &= 5 \\ -2x_1 + x_2 + 5x_3 &= -35 \end{aligned}$

7. Find all solutions to the following systems.

(a) $\begin{aligned} 3x_1 + 8x_2 - 3x_3 - 14x_4 &= 1 \\ 2x_1 + 3x_2 - x_3 - 2x_4 &= 2 \\ x_1 - 2x_2 + x_3 + 10x_4 &= 3 \\ x_1 + 5x_2 - 2x_3 - 12x_4 &= -1 \end{aligned}$

◆(b) $\begin{aligned} x_1 - x_2 + x_3 - x_4 &= 0 \\ -x_1 + x_2 + x_3 + x_4 &= 0 \\ x_1 + x_2 - x_3 + x_4 &= 0 \\ x_1 + x_2 + x_3 + x_4 &= 0 \end{aligned}$

(c) $\begin{aligned} x_1 - x_2 + x_3 - 2x_4 &= 3 \\ -x_1 + x_2 + x_3 + x_4 &= 2 \\ -x_1 + 2x_2 + 3x_3 - x_4 &= 9 \\ x_1 - x_2 + 2x_3 + x_4 &= 2 \end{aligned}$

◆(d) $\begin{aligned} x_1 + x_2 + 2x_3 - x_4 &= 4 \\ 3x_2 - x_3 + 4x_4 &= 2 \\ x_1 + 2x_2 - 3x_3 + 5x_4 &= 0 \\ x_1 + x_2 - 5x_3 + 6x_4 &= -3 \end{aligned}$

8. In each of the following, find conditions on a and b such that the system has no solution, one solution, and infinitely many solutions.

(a) $\begin{aligned} x - 2y &= 1 \\ ax + by &= 5 \end{aligned}$
◆(b) $\begin{aligned} x + by &= -1 \\ ax + 2y &= 5 \end{aligned}$

(c) $\begin{aligned} x - by &= -1 \\ x + ay &= 3 \end{aligned}$
◆(d) $\begin{aligned} ax + y &= 1 \\ 2x + y &= b \end{aligned}$

9. In each of the following, find (if possible) conditions on a, b, and c such that the system has no solution, one solution, or infinitely many solutions.

(a) $\begin{aligned} 3x + y - z &= a \\ x - y + 2z &= b \\ 5x + 3y - 4z &= c \end{aligned}$
◆(b) $\begin{aligned} 2x + y - z &= a \\ 2y + 3z &= b \\ x - z &= c \end{aligned}$

(c) $\begin{aligned} -x + 3y + 2z &= -8 \\ x + z &= 2 \\ 3x + 3y + az &= b \end{aligned}$
◆(d) $\begin{aligned} x + ay &= 0 \\ y + bz &= 0 \\ z + cx &= 0 \end{aligned}$

(e) $\begin{aligned} 3x - y + 2z &= 3 \\ x + y - z &= 2 \\ 2x - 2y + 3z &= b \end{aligned}$

◆(f) $\begin{aligned} x + ay - z &= 1 \\ -x + (a - 2)y + z &= -1 \\ 2x + 2y + (a - 2)z &= 1 \end{aligned}$

10. Find the rank of each of the matrices in Exercise 1.

11. Find the rank of each of the following matrices.

(a) $\begin{bmatrix} 1 & 1 & 2 \\ 3 & -1 & 1 \\ -1 & 3 & 4 \end{bmatrix}$
◆(b) $\begin{bmatrix} -2 & 3 & 3 \\ 3 & -4 & 1 \\ -5 & 7 & 2 \end{bmatrix}$

(c) $\begin{bmatrix} 1 & 1 & -1 & 3 \\ -1 & 4 & 5 & -2 \\ 1 & 6 & 3 & 4 \end{bmatrix}$
◆(d) $\begin{bmatrix} 3 & -2 & 1 & -2 \\ 1 & -1 & 3 & 5 \\ -1 & 1 & 1 & -1 \end{bmatrix}$

(e) $\begin{bmatrix} 1 & 2 & -1 & 0 \\ 0 & a & 1-a & a^2+1 \\ 1 & 2-a & -1 & -2a^2 \end{bmatrix}$

◆(f) $\begin{bmatrix} 1 & 1 & 2 & a^2 \\ 1 & 1-a & 2 & 0 \\ 2 & 2-a & 6-a & 4 \end{bmatrix}$

12. Consider a system of linear equations with augmented matrix A and coefficient matrix C. In each case either prove the statement or give an example showing that it is false.

(a) If there is more than one solution, A has a row of zeros.

◆(b) If A has a row of zeros, there is more than one solution.

(c) If there is no solution, the row-echelon form of C has a row of zeros.

◆(d) If the row-echelon form of C has a row of zeros, there is no solution.

Now assume that A has 3 rows and 5 columns.

(e) If the system is consistent, there is more than one solution.

◆ **(f)** The rank of A is at most 3.

(g) If rank $A = 3$, the system is consistent.

(h) If rank $C = 3$, the system is consistent.

13. Find a sequence of row operations carrying

$$\begin{bmatrix} b_1 + c_1 & b_2 + c_2 & b_3 + c_3 \\ c_1 + a_1 & c_2 + a_2 & c_3 + a_3 \\ a_1 + b_1 & a_2 + b_2 & a_3 + b_3 \end{bmatrix} \text{ to } \begin{bmatrix} a_1 & a_2 & a_3 \\ b_1 & b_2 & b_3 \\ c_1 & c_2 & c_3 \end{bmatrix}$$

14. In each case, show that the reduced row-echelon form is as given.

(a) $\begin{bmatrix} p & 0 & a \\ b & 0 & 0 \\ q & c & r \end{bmatrix}$ with $abc \neq 0$ $\begin{bmatrix} 1 & 0 & 0 \\ 0 & 1 & 0 \\ 0 & 0 & 1 \end{bmatrix}$

◆ **(b)** $\begin{bmatrix} 1 & a & b+c \\ 1 & b & c+a \\ 1 & c & a+b \end{bmatrix}$ where $c \neq a$ or $b \neq a$ $\begin{bmatrix} 1 & 0 & * \\ 0 & 1 & * \\ 0 & 0 & 0 \end{bmatrix}$

15. Show that $\begin{cases} ax + by + cz = 0 \\ a_1x + b_1y + c_1z = 0 \end{cases}$ always has a solution other than $x = 0, y = 0, z = 0$.

16. Find the circle $x^2 + y^2 + ax + by + c = 0$ passing through the following points.

(a) $(-2, 1), (5, 0),$ and $(4, 1)$

◆ **(b)** $(1, 1), (5, -3),$ and $(-3, -3)$

17. Three Nissans, two Fords, and four Chevrolets can be rented for $106 per day. At the same rates two Nissans, four Fords, and three Chevrolets cost $107 per day, whereas four Nissans, three Fords, and two Chevrolets cost $102 per day. Find the rental rates for all three kinds of cars.

◆ **18.** A school has three clubs, and each student is required to belong to exactly one club. One year the students switched club membership as follows:

Club A. $\frac{4}{10}$ remain in A, $\frac{1}{10}$ switch to B, $\frac{5}{10}$ switch to C.

Club B. $\frac{7}{10}$ remain in B, $\frac{2}{10}$ switch to A, $\frac{1}{10}$ switch to C.

Club C. $\frac{6}{10}$ remain in C, $\frac{2}{10}$ switch to A, $\frac{2}{10}$ switch to B.

If the fraction of the student population in each club is unchanged, find each of these fractions.

19. Given points $(p_1, q_1), (p_2, q_2),$ and (p_3, q_3) in the plane with $p_1, p_2,$ and p_3 distinct, show that they lie on some curve with equation $y = a + bx + cx^2$. [*Hint*: Solve for a, b, and c.]

20. The scores of three players in a tournament have been lost. The only information available is the total of the scores for players 1 and 2, the total for players 2 and 3, and the total for players 3 and 1.

(a) Show that the individual scores can be rediscovered.

◆ **(b)** Is this true with four players (knowing the totals for players 1 and 2, 2 and 3, 3 and 4, and 4 and 1)?

21. A boy finds $1.05 in dimes, nickels, and pennies. If there are 17 coins in all, how many coins of each type can he have?

Section 1.3   Homogeneous Equations

A system of equations in the variables $x_1, x_2, \ldots, x_n$ is called **homogeneous** if all the constant terms are zero—that is, if each equation of the system has the form

$$a_1x_1 + a_2x_2 + \cdots + a_nx_n = 0$$

Clearly $x_1 = 0, x_2 = 0, \ldots, x_n = 0$ is a solution to such a system; it is called the **trivial** solution. Any solution in which at least one variable has a nonzero value is called a **nontrivial** solution. Our chief goal in this section is to give a useful condition for a homogeneous system to have nontrivial solutions. The following example is instructive.

EXAMPLE 1

Show that the following homogeneous system has nontrivial solutions.

$$x_1 - x_2 + 2x_3 + x_4 = 0$$
$$2x_1 + 2x_2 \qquad - x_4 = 0$$
$$3x_1 + x_2 + 2x_3 + x_4 = 0$$

Solution   The reduction of the augmented matrix to reduced row-echelon form is outlined below.

$$\begin{bmatrix} 1 & -1 & 2 & 1 & | & 0 \\ 2 & 2 & 0 & -1 & | & 0 \\ 3 & 1 & 2 & 1 & | & 0 \end{bmatrix} \rightarrow \begin{bmatrix} 1 & -1 & 2 & 1 & | & 0 \\ 0 & 4 & -4 & -3 & | & 0 \\ 0 & 4 & -4 & -2 & | & 0 \end{bmatrix} \rightarrow \begin{bmatrix} 1 & 0 & 1 & 0 & | & 0 \\ 0 & 1 & -1 & 0 & | & 0 \\ 0 & 0 & 0 & 1 & | & 0 \end{bmatrix}$$

The leading variables are x_1, x_2, and x_4, so x_3 is assigned as a parameter—say $x_3 = t$. Then the general solution is $x_1 = -t$, $x_2 = t$, $x_3 = t$, $x_4 = 0$; so, taking $t = 1$ (say), we get a nontrivial solution.

◆◆◆

The existence of a nontrivial solution in Example 1 is ensured by the presence of a parameter in the solution. This is due to the fact that there is a *nonleading* variable (x_3 in this case). But there *must* be a nonleading variable here because there are four variables and only three equations (and hence at *most* three leading variables). This discussion generalizes to a proof of the following useful theorem.

THEOREM 1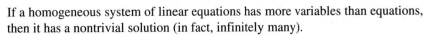

If a homogeneous system of linear equations has more variables than equations, then it has a nontrivial solution (in fact, infinitely many).

Proof   Suppose there are m equations in n variables where $n > m$, and let R denote the reduced row-echelon form of the augmented matrix. If there are r leading variables, there are $n - r$ nonleading variables, and so $n - r$ parameters. Hence it suffices to show that $r < n$. But $r \leq m$ because R has r leading 1's and m rows, and $m < n$ by hypothesis.   ◆

Note that the converse of Theorem 1 is not true (Exercise 3(b)). The next example provides an illustration of how Theorem 1 is used.

EXAMPLE 2

We call the graph of an equation $ax^2 + bxy + cy^2 + dx + ey + f = 0$ a **conic** if the numbers a, b, and c are not all zero. Show that there is at least one conic through any five points in the plane that are not all on a line.

Solution   Let the coordinates of the five points be (p_1, q_1), (p_2, q_2), (p_3, q_3), (p_4, q_4), and (p_5, q_5).

The graph of $ax^2 + bxy + cy^2 + dx + ey + f = 0$ passes through (p_i, q_i) if

$$ap_i^2 + bp_iq_i + cq_i^2 + dp_i + eq_i + f = 0$$

This gives five equations, linear in the six variables a, b, c, d, e, and f. Hence there is a nontrivial solution by Theorem 1. If $a = b = c = 0$, the five points all lie on the line $dx + ey + f = 0$, contrary to assumption. Hence one of a, b, c is nonzero.

◆◆◆

EXERCISES 1.3

1. Consider the following statements about a system of linear equations with augmented matrix A. In each case either prove the statement or give an example for which it is false.

(a) If the system is homogeneous, every solution is trivial.

◆ (b) If the system has a nontrivial solution, it cannot be homogeneous.

(c) If there exists a trivial solution, the system is homogeneous.

◆ (d) If the system is consistent, it must be homogeneous. Now assume that the system is homogeneous.

(e) If there exists a nontrivial solution, there is no trivial solution.

◆ (f) If there exists a solution, there are infinitely many solutions.

(g) If there exist nontrivial solutions, the row-echelon form of A has a row of zeros.

◆ (h) If the row-echelon form of A has a row of zeros, there exist nontrivial solutions.

2. In each of the following, find all values of a for which the system has nontrivial solutions, and determine all solutions in each case.

(a)
$\begin{aligned} x - 2y + z &= 0 \\ x + ay - 3z &= 0 \\ -x + 6y - 5z &= 0 \end{aligned}$

◆ (b)
$\begin{aligned} x + 2y + z &= 0 \\ x + 3y + 6z &= 0 \\ 2x + 3y + az &= 0 \end{aligned}$

(c)
$\begin{aligned} x + y - z &= 0 \\ ay - z &= 0 \\ x + y + az &= 0 \end{aligned}$

◆ (d)
$\begin{aligned} ax + y + z &= 0 \\ x + y - z &= 0 \\ x + y + az &= 0 \end{aligned}$

3. (a) Does Theorem 1 imply that the system $\begin{aligned} -x + 3y &= 0 \\ 2x - 6y &= 0 \end{aligned}$ has nontrivial solutions? Explain.

◆ (b) Show that the converse to Theorem 1 is not true. That is, show that the existence of nontrivial solutions does *not* imply that there are more variables than equations.

4. In each case determine how many solutions (and how many parameters) are possible for a homogeneous system of four linear equations in six variables with augmented matrix A. Assume that A has nonzero entries. Give all possibilities.

(a) Rank $A = 2$.

◆ (b) Rank $A = 1$.

(c) A has a row of zeros.

◆ (d) The row-echelon form of A has a row of zeros.

5. The graph of an equation $ax + by + cz = 0$ is a plane through the origin (provided that not all of a, b, and c are zero). Use Theorem 1 to show that two planes through the origin have a point in common other than the origin $(0, 0, 0)$.

6. (a) Show that there is a line through any pair of points in the plane. [*Hint:* Every line has equation $ax + by + c = 0$, where a, b, and c are not all zero.]

◆ (b) Generalize and show that there is a plane $ax + by + cz + d = 0$ through any three points in space.

7. The graph of an equation $a(x^2 + y^2) + bx + cy + d = 0$ is a circle if $a \neq 0$. Show that there is a circle through any three points in the plane that are not all on a line.

8. Consider a homogeneous system of linear equations in n variables, and suppose that the augmented matrix has rank r. Show that the system has nontrivial solutions if and only if $n > r$.[6]

---

[6] If p and q are statements, "p if and only if q" means that both p implies q and q implies p.

An Application to Network Flow (Optional)

There are many types of problems that concern a network of conductors along which some sort of flow is observed. Examples of these include an irrigation network and a network of streets or freeways. There are often points in the system at which a net flow either enters or leaves the system. The basic principle behind the analysis of such systems is that the total flow into the system must equal the total flow out. In fact we apply this principle at every junction in the system.

> **JUNCTION RULE**
>
> At each of the junctions in the network, the total flow into that junction must equal the total flow out.

This requirement gives a linear equation relating the flows in conductors emanating from the junction.

EXAMPLE 1

A network of one-way streets is shown in the accompanying diagram. The rate of flow of cars into intersection A is 500 cars per hour, and 400 and 100 cars per hour emerge from B and C, respectively. Find the possible flows along each street.

Solution

Suppose the flows along the streets are $f_1, f_2, f_3, f_4, f_5,$ and f_6 cars per hour in the directions shown. Then, equating the flow in with the flow out at each intersection, we get

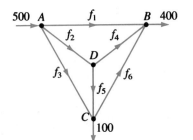

Intersection A	$500 = f_1 + f_2 + f_3$
Intersection B	$f_1 + f_4 + f_6 = 400$
Intersection C	$f_3 + f_5 = f_6 + 100$
Intersection D	$f_2 = f_4 + f_5$

These give four equations in the six variables $f_1, f_2, \ldots, f_6$.

$$
\begin{aligned}
f_1 + f_2 + f_3 \qquad\qquad\qquad &= 500 \\
f_1 \qquad\quad + f_4 \qquad + f_6 &= 400 \\
f_3 \qquad + f_5 - f_6 &= 100 \\
f_2 \qquad - f_4 - f_5 \qquad &= 0
\end{aligned}
$$

The reduction of the augmented matrix is

$$
\begin{bmatrix}
1 & 1 & 1 & 0 & 0 & 0 & 500 \\
1 & 0 & 0 & 1 & 0 & 1 & 400 \\
0 & 0 & 1 & 0 & 1 & -1 & 100 \\
0 & 1 & 0 & -1 & -1 & 0 & 0
\end{bmatrix}
\rightarrow
\begin{bmatrix}
1 & 0 & 0 & 1 & 0 & 1 & 400 \\
0 & 1 & 0 & -1 & -1 & 0 & 0 \\
0 & 0 & 1 & 0 & 1 & -1 & 100 \\
0 & 0 & 0 & 0 & 0 & 0 & 0
\end{bmatrix}
$$

Hence, when we use $f_4, f_5,$ and f_6 as parameters, the general solution is

$$f_1 = 400 - f_4 - f_6$$
$$f_2 = f_4 + f_5$$
$$f_3 = 100 - f_5 + f_6$$

This gives all solutions to the system of equations and hence all the possible flows.

Of course, not all these solutions may be acceptable in the real situation. For example, the flows $f_1, f_2, \ldots, f_6$ are all *positive* in the present context (if one came out negative, it would mean traffic flowed in the opposite direction). This imposes constraints on the flows: $f_1 \geq 0$ and $f_3 \geq 0$ become

$$f_4 + f_6 \leq 400$$
$$f_5 - f_6 \leq 100$$

Further constraints might be imposed by insisting on maximum values on the flow in each street.

EXERCISES 1.4

1. Find the possible flows in each of the following networks of pipes.

(a)

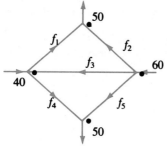

(b)

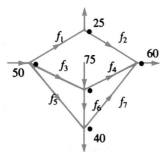

2. A proposed network of irrigation canals is described in the accompanying diagram. At peak demand, the flows at interchanges A, B, C, and D are as shown.

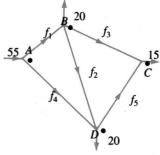

(a) Find the possible flows.
♦ **(b)** If canal BC is closed, what range of flow on AD must be maintained so that no canal carries a flow of more than 30?

3. A traffic circle has five one-way streets, and vehicles enter and leave as shown in the accompanying diagram.

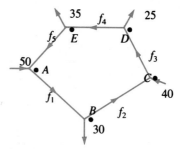

(a) Compute the possible flows.
♦ **(b)** What road in the circle has the heaviest flow?

Section 1.5   An Application to Electrical Networks (Optional)[7]

In an electrical network it is often necessary to find the current in amperes (A) flowing in various parts of the network. These networks usually contain resistors that retard the current. The resistors are indicated by a symbol $-\!\!\bigvee\!\!\bigvee\!\!-$, and the resistance is measured in ohms (Ω). Also, the current is increased at various points by voltage sources (for example, a battery). The voltage of these sources is measured in volts (V), and they are represented by the symbol $-\!\!\mid\!\!\mid-$. We assume these voltage sources have no resistance. The flow of current is governed by the following principles.

OHM'S LAW

The current I and the voltage drop V across a resistance R are related by the equation $V = RI$.

KIRCHHOFF'S LAWS

1. (Junction Rule) The current flow into a junction equals the current flow out of that junction.
2. (Circuit Rule) The algebraic sum of the voltage drops (due to resistances) around any closed circuit of the network must equal the sum of the voltage increases around the circuit.

When applying rule 2, select a direction (clockwise or counterclockwise) around the closed circuit and then consider all voltages and currents positive when in this direction and negative when in the opposite direction. This is why the term *algebraic sum* is used in rule 2. Here is an example.

EXAMPLE 1

Find the various currents in the circuit shown.

Solution   First apply the junction rule at junctions A, B, C, and D to obtain

Junction A	$I_1 = I_2 + I_3$
Junction B	$I_6 = I_1 + I_5$
Junction C	$I_2 + I_4 = I_6$
Junction D	$I_3 + I_5 = I_4$

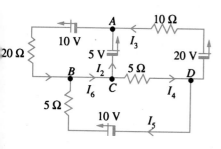

Note that these equations are not independent (in fact, the third is an easy consequence of the other three).

Next, the circuit rule insists that the sum of the voltage increases (due to the sources) around a closed circuit must equal the sum of the voltage drops (due to resistances). By Ohm's law, the voltage loss across a resistance R (in the direction of

---

[7]This section is independent of Section 1.4.

the current I) is RI. Going counterclockwise around three closed circuits yields

Upper left	$10 + 5 = 20I_1$
Upper right	$-5 + 20 = 10I_3 + 5I_4$
Lower	$-10 = -5I_5 - 5I_4$

Hence, disregarding the redundant equation obtained at junction C, we have six equations in the six unknowns $I_1, \ldots, I_6$. The solution is

$$I_1 = \tfrac{15}{20} \qquad\qquad I_4 = \tfrac{28}{20}$$
$$I_2 = \tfrac{-1}{20} \qquad\qquad I_5 = \tfrac{12}{20}$$
$$I_3 = \tfrac{16}{20} \qquad\qquad I_6 = \tfrac{27}{20}$$

The fact that I_2 is negative means, of course, that this current is in the opposite direction, with a magnitude of $\tfrac{1}{20}$ amperes.

EXERCISES 1.5

In Exercises 1–4, find the currents in the circuits.

1.

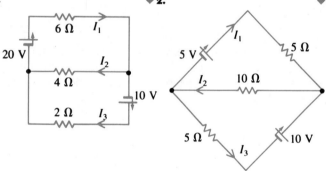

◆2.

◆4. All resistances are 10Ω

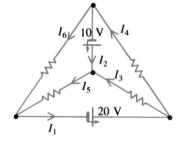

3.

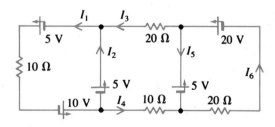

5. Find the voltage x such that the current $I_1 = 0$.

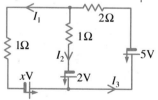

SUPPLEMENTARY EXERCISES FOR CHAPTER 1

1. We show in Chapter 4 that the graph of an equation $ax + by + cz = d$ is a plane in space when not all of a, b, and c are zero.

(a) By examining the possible positions of planes in space, show that three equations in three variables can have zero, one, or infinitely many solutions.

(b) Can two equations in three variables have a unique solution? Give reasons for your answer.

2. Find all solutions to the following systems of linear equations.

(a)
$$\begin{aligned}
x_1 + x_2 + x_3 - x_4 &= 3 \\
3x_1 + 5x_2 - 2x_3 + x_4 &= 1 \\
-3x_1 - 7x_2 + 7x_3 - 5x_4 &= 7 \\
x_1 + 3x_2 - 4x_3 + 3x_4 &= -5
\end{aligned}$$

◆ **(b)**
$$\begin{aligned}
x_1 + 4x_2 - x_3 + x_4 &= 2 \\
3x_1 + 2x_2 + x_3 + 2x_4 &= 5 \\
x_1 - 6x_2 + 3x_3 &= 1 \\
x_1 + 14x_2 - 5x_3 + 2x_4 &= 3
\end{aligned}$$

3. In each case find (if possible) conditions on a, b, and c such that the system has zero, one, or infinitely many solutions.

(a)
$$\begin{aligned}
x + 2y - 4z &= 4 \\
3x - y + 13z &= 2 \\
4x + y + a^2z &= a + 3
\end{aligned}$$

◆ **(b)**
$$\begin{aligned}
x + y + 3z &= a \\
ax + y + 5z &= 4 \\
x + ay + 4z &= a
\end{aligned}$$

◆ **4.** Show that any two rows of a matrix can be interchanged by elementary row transformations of the other two types.

5. If $ad \neq bc$, show that $\begin{bmatrix} a & b \\ c & d \end{bmatrix}$ has reduced row-echelon form $\begin{bmatrix} 1 & 0 \\ 0 & 1 \end{bmatrix}$.

◆ **6.** Find a, b, and c so that the system
$$\begin{aligned}
x + ay + cz &= 0 \\
bx + cy - 3z &= 1 \\
ax + 2y + bz &= 5
\end{aligned}$$
has the solution $x = 3$, $y = -1$, $z = 2$.

7. Solve the system $\begin{cases} x + 2y + 2z = -3 \\ 2x + y + z = 0 \\ x - y - iz = i \end{cases}$ where $i^2 = -1$.

[See Appendix A.]

◆ **8.** Show that the *real* system $\begin{cases} x + y + z = 5 \\ 2x - y - z = 1 \\ -3x + 2y + 2z = 0 \end{cases}$ has a *complex* solution: $x = 2$, $y = i$, $z = 3 - i$ where $i^2 = -1$. Explain. What happens when such a real system has a unique solution?

9. A man is ordered by his doctor to take 5 units of vitamin A, 13 units of vitamin B, and 23 units of vitamin C each day. Three brands of vitamin pills are available, and the numbers of units of each vitamin per pill are shown in the accompanying table.

Brand	Vitamin		
	A	B	C
1	1	2	4
2	1	1	3
3	0	1	1

(a) Find all combinations of pills that provide exactly the required amount of vitamins (no partial pills allowed).

◆ **(b)** If brands 1, 2, and 3 cost 3¢, 2¢, and 5¢ per pill, respectively, find the least expensive treatment.

10. A restaurant owner plans to use x tables seating 4, y tables seating 6, and z tables seating 8, for a total of 20 tables. When fully occupied, the tables seat 108 customers. If only half of the x tables, half of the y tables, and one-fourth of the z tables are used, each fully occupied, then 46 customers will be seated. Find x, y, and z.

11. (a) Show that a matrix with two rows and two columns that is in reduced row-echelon form must have one of the following forms:

$$\begin{bmatrix} 1 & 0 \\ 0 & 1 \end{bmatrix} \quad \begin{bmatrix} 0 & 1 \\ 0 & 0 \end{bmatrix} \quad \begin{bmatrix} 0 & 0 \\ 0 & 0 \end{bmatrix} \quad \begin{bmatrix} 1 & * \\ 0 & 0 \end{bmatrix}$$

[*Hint:* The leading 1 in the first row must be in column 1 or 2 or not exist.]

(b) List the seven reduced row-echelon forms for matrices with two rows and three columns.

(c) List the four reduced row-echelon forms for matrices with three rows and two columns.

12. An amusement park charges $7 for adults, $2 for youths, and $.50 for children. If 150 people enter and pay a total of $100, find the numbers of adults, youths, and children. [*Hint:* These numbers are nonnegative *integers.*]

13. Solve the following system of equations for x and y.
$$\begin{aligned}
x^2 + xy - y^2 &= 1 \\
2x^2 - xy + 3y^2 &= 13 \\
x^2 + 3xy + 2y^2 &= 0
\end{aligned}$$

[*Hint:* These equations are linear in the new variables $x_1 = x^2$, $x_2 = xy$, and $x_3 = y^2$.]

14. If a consistent (possibly nonhomogeneous) system of linear equations has more variables than equations, prove that it has more than one solution.

2

Matrix Algebra

◆◆

Section 2.1
Matrix Addition, Scalar Multiplication, and Transposition

In the study of systems of linear equations in Chapter 1, we found it convenient to manipulate the augmented matrix of the system. Our aim was to reduce it to row-echelon form (using elementary row operations) and hence to write down all solutions to the system. In the present chapter, we will consider matrices for their own sake, although some of the motivation comes from linear equations. This subject is quite old and was first studied systematically in 1858 by Arthur Cayley.[1]

> **DEFINITION**
>
> A rectangular array of numbers is called a **matrix** (the plural is **matrices**), and the numbers are called the **entries** of the matrix.

Matrices are usually denoted by uppercase letters: *A, B, C,* and so on. Hence

$$A = \begin{bmatrix} 1 & 2 & -1 \\ 0 & 5 & 6 \end{bmatrix} \qquad B = \begin{bmatrix} 1 & -1 \\ 0 & 2 \end{bmatrix} \qquad C = \begin{bmatrix} 1 \\ 3 \\ 2 \end{bmatrix}$$

are matrices.

---

[1]Arthur Cayley (1821–1895) showed his mathematical talent early and graduated from Cambridge in 1842 as senior wrangler. With no employment in mathematics in view, he took legal training and worked as a lawyer while continuing to do mathematics, publishing nearly 300 papers in fourteen years. Finally, in 1863, he accepted the Sadlerian professorship at Cambridge and remained there for the rest of his life, valued for his administrative and teaching skills as well as for his scholarship. His mathematical achievements were of the first rank. In addition to originating matrix theory and the theory of determinants, he did fundamental work in group theory, in higher-dimensional geometry, and in the theory of invariants. He was one of the most prolific mathematicians of all time and produced 966 papers filling thirteen volumes of 600 pages each.

Clearly matrices come in various shapes depending on the number of **rows** and **columns**. For example, the matrix A shown has 2 rows and 3 columns. In general, a matrix with m rows and n columns is referred to as an $m \times n$ **matrix** or as having **size $m \times n$**. Thus matrices A, B, and C above have sizes 2×3, 2×2, and 3×1, respectively. A matrix of size $1 \times n$ is called a **row matrix**, whereas one of size $n \times 1$ is called a **column matrix**.

Each entry of a matrix is identified by the row and column in which it lies. The rows are numbered from the top down, and the columns are numbered from left to right. Then the **(i, j)-entry** of a matrix is the number lying simultaneously in row i and column j. For example,

$$\text{The } (1, \ 2)\text{-entry of } \begin{bmatrix} 1 & -1 \\ 0 & 1 \end{bmatrix} \text{ is } -1$$

$$\text{The } (2, \ 3)\text{-entry of } \begin{bmatrix} 1 & 2 & -1 \\ 0 & 5 & 6 \end{bmatrix} \text{ is } 6$$

A special notation has been devised for the entries of a matrix. If A is an $m \times n$ matrix, and if the (i, j)-entry of A is denoted as a_{ij}, then A is displayed as follows:

$$A = \begin{bmatrix} a_{11} & a_{12} & a_{13} & \cdots & a_{1n} \\ a_{21} & a_{22} & a_{23} & \cdots & a_{2n} \\ \vdots & \vdots & \vdots & & \vdots \\ a_{m1} & a_{m2} & a_{m3} & \cdots & a_{mn} \end{bmatrix}$$

This is usually denoted simply as $A = [a_{ij}]$. Thus a_{ij} is the entry in row i and column j of A. For example, a 3×4 matrix in this notation is written

$$A = \begin{bmatrix} a_{11} & a_{12} & a_{13} & a_{14} \\ a_{21} & a_{22} & a_{23} & a_{24} \\ a_{31} & a_{32} & a_{33} & a_{34} \end{bmatrix}$$

An $n \times n$ matrix A is called a **square matrix**. For a square matrix $A = [a_{ij}]$, the entries $a_{11}, a_{22}, a_{33}, \ldots, a_{nn}$ are said to lie on the **main diagonal** of the matrix A. Hence the main diagonal extends from the upper left corner of A to the lower right corner (shaded in the following 3×3 matrix):

$$\begin{bmatrix} a_{11} & a_{12} & a_{13} \\ a_{21} & a_{22} & a_{23} \\ a_{31} & a_{32} & a_{33} \end{bmatrix}$$

It is worth pointing out a convention regarding rows and columns:

> **CONVENTION**
>
> Rows are mentioned before columns.

For example: If a matrix has size $m \times n$, it has m rows and n columns.

If we speak of the (i, j)-entry of a matrix, it lies in row i and column j.

If an entry is denoted a_{ij}, the first subscript i refers to the row and the second subscript j to the column in which a_{ij} lies.

Two matrices A and B are called **equal** (written $A = B$) if and only if:

1. They have the same size.

2. Corresponding entries are equal.

If the entries of A and B are written in the form $A = [a_{ij}]$, $B = [b_{ij}]$, described earlier, then the second condition takes the following form:

$$[a_{ij}] = [b_{ij}] \qquad \text{means} \qquad a_{ij} = b_{ij} \text{ for all } i \text{ and } j.$$

EXAMPLE 1

Given $A = \begin{bmatrix} a & b \\ c & d \end{bmatrix}$, $B = \begin{bmatrix} 1 & 2 & -1 \\ 3 & 0 & 1 \end{bmatrix}$, and $C = \begin{bmatrix} 1 & 0 \\ -1 & 2 \end{bmatrix}$, discuss the possibility that $A = B$, $B = C$, $A = C$.

Solution $A = B$ is impossible because A and B are of different sizes: A is 2×2 whereas B is 2×3. Similarly, $B = C$ is impossible. $A = C$ is possible provided that corresponding entries are equal: $\begin{bmatrix} a & b \\ c & d \end{bmatrix} = \begin{bmatrix} 1 & 0 \\ -1 & 2 \end{bmatrix}$ means $a = 1$, $b = 0$, $c = -1$, and $d = 2$.

♦♦♦

Matrix Addition

> **DEFINITION**
>
> If A and B are matrices of the same size, their **sum** $A + B$ is the matrix formed by adding corresponding entries. If $A = [a_{ij}]$ and $B = [b_{ij}]$, this takes the form
>
> $$A + B = [a_{ij} + b_{ij}]$$

Note that addition is *not* defined for matrices of different sizes.

EXAMPLE 2

If $A = \begin{bmatrix} 2 & 1 & 3 \\ -1 & 2 & 0 \end{bmatrix}$ and $B = \begin{bmatrix} 1 & 1 & -1 \\ 2 & 0 & 6 \end{bmatrix}$, compute $A + B$.

Solution $A + B = \begin{bmatrix} 2+1 & 1+1 & 3-1 \\ -1+2 & 2+0 & 0+6 \end{bmatrix} = \begin{bmatrix} 3 & 2 & 2 \\ 1 & 2 & 6 \end{bmatrix}$ ◆◆◆

EXAMPLE 3

Find a, b, and c if $[a \quad b \quad c] + [c \quad a \quad b] = [3 \quad 2 \quad -1]$.

Solution   Add the matrices on the left side to obtain

$$[a + c \quad b + a \quad c + b] = [3 \quad 2 \quad -1]$$

Because corresponding entries must be equal, this gives three equations: $a + c = 3$, $b + a = 2$, and $c + b = -1$. Solving these yields $a = 3$, $b = -1$, $c = 0$.   ◆◆◆

If A, B, and C are any matrices *of the same size*, then

$$A + B = B + A \qquad \text{(commutative law)}$$
$$A + (B + C) = (A + B) + C \qquad \text{(associative law)}$$

In fact, if $A = [a_{ij}]$ and $B = [b_{ij}]$, then the (i, j)-entries of $A + B$ and $B + A$ are, respectively, $a_{ij} + b_{ij}$ and $b_{ij} + a_{ij}$. Since these are equal, we get

$$A + B = [a_{ij} + b_{ij}] = [b_{ij} + a_{ij}] = B + A$$

The associative law is verified similarly.

The $m \times n$ matrix in which every entry is zero is called the **zero matrix** and is denoted as 0 (or 0_{mn} if it is important to emphasize the size). Hence

$$0 + X = X$$

holds for all $m \times n$ matrices X. The **negative** of an $m \times n$ matrix A (written $-A$) is defined to be the $m \times n$ matrix obtained by multiplying each entry of A by -1. If $A = [a_{ij}]$, this becomes $-A = [-a_{ij}]$. Hence

$$A + (-A) = 0$$

holds for all matrices A where, of course, 0 is the zero matrix of the same size as A.

A closely related notion is that of subtracting matrices. If A and B are two $m \times n$ matrices, their **difference** $A - B$ is defined by

$$A - B = A + (-B)$$

Note that if $A = [a_{ij}]$ and $B = [b_{ij}]$, then

$$A - B = [a_{ij}] + [-b_{ij}] = [a_{ij} - b_{ij}]$$

is the $m \times n$ matrix formed by subtracting corresponding entries.

EXAMPLE 4

$A = \begin{bmatrix} 3 & -1 & 0 \\ 1 & 2 & -4 \end{bmatrix}$, $B = \begin{bmatrix} 1 & -1 & 1 \\ -2 & 0 & 6 \end{bmatrix}$, and $C = \begin{bmatrix} 1 & 0 & -2 \\ 3 & 1 & 1 \end{bmatrix}$. Compute $-A$, $A - B$, and $A + B - C$.

Solution

$$-A = \begin{bmatrix} -3 & 1 & 0 \\ -1 & -2 & 4 \end{bmatrix}$$

$$A - B = \begin{bmatrix} 3-1 & -1-(-1) & 0-1 \\ 1-(-2) & 2-0 & -4-6 \end{bmatrix} = \begin{bmatrix} 2 & 0 & -1 \\ 3 & 2 & -10 \end{bmatrix}$$

$$A + B - C = \begin{bmatrix} 3+1-1 & -1-1-0 & 0+1+2 \\ 1-2-3 & 2+0-1 & -4+6-1 \end{bmatrix} = \begin{bmatrix} 3 & -2 & 3 \\ -4 & 1 & 1 \end{bmatrix}$$ ◆◆◆

EXAMPLE 5

Solve $\begin{bmatrix} 3 & 2 \\ -1 & 1 \end{bmatrix} + X = \begin{bmatrix} 1 & 0 \\ -1 & 2 \end{bmatrix}$, where X is a matrix.

Solution 1

X must be a 2×2 matrix. If $X = \begin{bmatrix} x & y \\ z & w \end{bmatrix}$, the equation reads

$$\begin{bmatrix} 1 & 0 \\ -1 & 2 \end{bmatrix} = \begin{bmatrix} 3 & 2 \\ -1 & 1 \end{bmatrix} + \begin{bmatrix} x & y \\ z & w \end{bmatrix} = \begin{bmatrix} 3+x & 2+y \\ -1+z & 1+w \end{bmatrix}$$

The rule of matrix equality gives $1 = 3 + x$, $0 = 2 + y$, $-1 = -1 + z$, and $2 = 1 + w$. Thus $X = \begin{bmatrix} -2 & -2 \\ 0 & 1 \end{bmatrix}$.

Solution 2

We solve a numerical equation $a + x = b$ by subtracting the number a from both sides to obtain $x = b - a$. This also works for matrices. To solve $\begin{bmatrix} 3 & 2 \\ -1 & 1 \end{bmatrix} + X = \begin{bmatrix} 1 & 0 \\ -1 & 2 \end{bmatrix}$, simply subtract the matrix $\begin{bmatrix} 3 & 2 \\ -1 & 1 \end{bmatrix}$ from both sides to get

$$X = \begin{bmatrix} 1 & 0 \\ -1 & 2 \end{bmatrix} - \begin{bmatrix} 3 & 2 \\ -1 & 1 \end{bmatrix} = \begin{bmatrix} 1-3 & 0-2 \\ -1+1 & 2-1 \end{bmatrix} = \begin{bmatrix} -2 & -2 \\ 0 & 1 \end{bmatrix}$$

This is the same solution as obtained in Solution 1.

◆◆◆

The two solutions in Example 5 are really different ways of doing the same thing. However, the first obtains four numerical equations, one for each entry, and solves them to get the four entries of X. The second solution solves the single matrix equation directly via matrix subtraction, and manipulation of entries comes in only at the end. The matrices themselves are manipulated. This ability to work with matrices as entities lies at the heart of matrix algebra.

It is important to note that the size of X in Example 5 was inferred *from the context*: X had to be a 2×2 matrix because otherwise the equation would not make sense. This type of situation occurs frequently; the sizes of matrices involved in some calculations are often determined by the context. For example, if

$$A + C = \begin{bmatrix} 1 & 3 & -1 \\ 2 & 0 & 1 \end{bmatrix}$$

then A and C must be the same size (so that $A + C$ makes sense), and that size must be 2×3 (so that the sum is 2×3). For simplicity we shall often omit reference to such facts when they are clear from the context.

Scalar Multiplication

DEFINITION

If A is any matrix and k is any number, the **scalar multiple** kA is the matrix obtained from A by multiplying each entry of A by k. If $A = [a_{ij}]$, this is

$$kA = [ka_{ij}]$$

The term *scalar* arises here because the set of numbers from which the entries are drawn is usually referred to as the set of scalars. We have been using real numbers as scalars, but we could equally well have been using complex numbers.

EXAMPLE 6

If $A = \begin{bmatrix} 3 & -1 & 4 \\ 2 & 0 & 6 \end{bmatrix}$ and $B = \begin{bmatrix} 1 & 2 & -1 \\ 0 & 3 & 2 \end{bmatrix}$, compute $5A$, $\frac{1}{2}B$, and $3A - 2B$.

Solution

$$5A = \begin{bmatrix} 15 & -5 & 20 \\ 10 & 0 & 30 \end{bmatrix}, \qquad \frac{1}{2}B = \begin{bmatrix} \frac{1}{2} & 1 & -\frac{1}{2} \\ 0 & \frac{3}{2} & 1 \end{bmatrix}.$$

$$3A - 2B = \begin{bmatrix} 9 & -3 & 12 \\ 6 & 0 & 18 \end{bmatrix} - \begin{bmatrix} 2 & 4 & -2 \\ 0 & 6 & 4 \end{bmatrix} = \begin{bmatrix} 7 & -7 & 14 \\ 6 & -6 & 14 \end{bmatrix}$$

◆◆◆

If A is any matrix, note that kA is the same size as A for all scalars k. We also have

$$0A = 0 \qquad \text{and} \qquad k0 = 0$$

because the zero matrix has every entry zero. In other words, $kA = 0$ if either $k = 0$ or $A = 0$. The converse of these properties is also true, as Example 7 shows.

EXAMPLE 7

If $kA = 0$, show that either $k = 0$ or $A = 0$.

Solution

Write $A = [a_{ij}]$ so that $kA = 0$ means $ka_{ij} = 0$ for all i and j. If $k = 0$, there is nothing to do. If $k \neq 0$, then $ka_{ij} = 0$ implies that $a_{ij} = 0$ for all i and j; that is, $A = 0$.

◆◆◆

For future reference, the basic properties of matrix addition and scalar multiplication are listed in Theorem 1.

THEOREM 1

Let A, B, and C denote arbitrary $m \times n$ matrices where m and n are fixed. Let k and p denote arbitrary real numbers. Then

1. $A + B = B + A$.

2. $A + (B + C) = (A + B) + C$

3. There is an $m \times n$ matrix 0 such that $0 + A = A$ for each A.

4. For each A there is an $m \times n$ matrix $-A$ such that $A + (-A) = 0$.

5. $k(A + B) = kA + kB$

6. $(k + p)A = kA + pA$

7. $(kp)A = k(pA)$

8. $1A = A$

Proof   Properties 1–4 were given previously. To check property 5, let $A = [a_{ij}]$ and $B = [b_{ij}]$ denote matrices of the same size. Then $A + B = [a_{ij} + b_{ij}]$, as before, so the (i, j)-entry of $k(A + B)$ is

$$k(a_{ij} + b_{ij}) = ka_{ij} + kb_{ij}$$

But this is just the (i, j)-entry of $kA + kB$, and it follows that $k(A + B) = kA + kB$. The other properties can be similarly verified; the details are left to the reader.   ◆

These properties enable us to do calculations with matrices in much the same way that numerical calculations are carried out. To begin with, property 2 implies that the sum $(A + B) + C = A + (B + C)$ is the same no matter how it is formed and so is written as $A + B + C$. Similarly, the sum $A + B + C + D$ is independent of how it is formed; for example, it equals both $(A + B) + (C + D)$

and $A + [B + (C + D)]$. Furthermore, property 1 ensures that, for example, $B + D + A + C = A + B + C + D$. In other words, the *order* in which the matrices are added does not matter. A similar remark applies to sums of five (or more) matrices.

Properties 5 and 6 in Theorem 1 extend to sums of more than two terms. For example,

$$k(A + B + C) = kA + kB + kC$$

$$(k + p + m)A = kA + pA + mA$$

Similar observations hold for more than three summands. These facts, together with properties 7 and 8, enable us to simplify expressions by collecting like terms, expanding, and taking common factors in exactly the same way that algebraic expressions involving variables are manipulated. The following examples illustrate these techniques.

EXAMPLE 8

Simplify $2(A + 3C) - 3(2C - B) - 3[2(2A + B - 4C) - 4(A - 2C)]$ where A, B and C are all matrices of the same size.

Solution The reduction proceeds as though A, B, and C were variables.

$$2(A + 3C) - 3(2C - B) - 3[2(2A + B - 4C) - 4(A - 2C)]$$
$$= 2A + 6C - 6C + 3B - 3[4A + 2B - 8C - 4A + 8C]$$
$$= 2A + 3B - 3[2B]$$
$$= 2A - 3B$$

◆◆◆

EXAMPLE 9

Find 1×3 matrices X and Y such that

$$X + 2Y = \begin{bmatrix} 1 & 3 & -2 \end{bmatrix}$$
$$X + Y = \begin{bmatrix} 2 & 0 & 1 \end{bmatrix}$$

Solution If we write $A = [1\ 3\ -2]$ and $B = [2\ 0\ 1]$, the equations become $X + 2Y = A$ and $X + Y = B$. The manipulations used to solve these equations when X, Y, A, and B are numbers all apply in the present context. Hence subtracting the second equation from the first gives $Y = A - B = [-1\ 3\ -3]$. Similarly, subtracting the first equation from twice the second gives $X = 2B - A = [3\ -3\ 4]$.

◆◆◆

Transpose

Many results about a matrix A involve the *rows* of A, and the corresponding result for columns is derived in an analogous way, essentially by replacing the word *row* by the word *column* throughout. The following definition is made with such applications in mind.

DEFINITION

If A is an $m \times n$ matrix, the **transpose** of A, written A^T, is the $n \times m$ matrix whose rows are just the columns of A in the same order.

In other words, the first row of A^T is the first column of A, the second row of A^T is the second column of A, and so on.

EXAMPLE 10

Write down the transpose of each of the following matrices.

$$A = \begin{bmatrix} 1 \\ 3 \\ 2 \end{bmatrix} \qquad B = \begin{bmatrix} 5 & 2 & 6 \end{bmatrix} \qquad C = \begin{bmatrix} 1 & 2 \\ 3 & 4 \\ 5 & 6 \end{bmatrix} \qquad D = \begin{bmatrix} 3 & 1 & -1 \\ 1 & 3 & 2 \\ -1 & 2 & 1 \end{bmatrix}$$

Solution $A^T = \begin{bmatrix} 1 & 3 & 2 \end{bmatrix}$, $B^T = \begin{bmatrix} 5 \\ 2 \\ 6 \end{bmatrix}$, $C^T = \begin{bmatrix} 1 & 3 & 5 \\ 2 & 4 & 6 \end{bmatrix}$, and $D^T = D$.

If $A = [a_{ij}]$ is a matrix, write $A^T = [b_{ij}]$. Then b_{ij} is the jth element of the ith row of A^T and so is the jth element of the ith *column* of A. This means $b_{ij} = a_{ji}$ so the definition of A^T can be stated as follows:

$$\text{If } A = [a_{ij}], \text{ then } A^T = [a_{ji}]$$

This is useful in verifying the following properties of transposition.

THEOREM 2

Let A and B denote matrices of the same size, and let k denote a scalar.

1. If A is an $m \times n$ matrix, then A^T is an $n \times m$ matrix.
2. $(A^T)^T = A$
3. $(kA)^T = kA^T$
4. $(A + B)^T = A^T + B^T$

Proof   We prove only property 3. If $A = [a_{ij}]$, then $kA = [ka_{ij}]$, so

$$(kA)^T = [ka_{ji}] = k[a_{ji}] = kA^T$$

which proves property 3.   ◆

The matrix D in Example 10 has the property that $D = D^T$. Such matrices are important.

DEFINITION

A matrix A is called **symmetric** if $A = A^T$.

A symmetric matrix A is necessarily square (if A is $m \times n$, then A^T is $n \times m$, so $A = A^T$ forces $n = m$). The name comes from the fact that these matrices exhibit a symmetry about the main diagonal. That is, entries that are directly across the main diagonal from each other are equal. For example, $\begin{bmatrix} a & b & c \\ b' & d & e \\ c' & e' & f \end{bmatrix}$ is symmetric when

$b = b'$ $c = c'$ and $e = e'$.

EXAMPLE 11

If A and B are symmetric $n \times n$ matrices, show that $A + B$ is symmetric.

Solution   We have $A^T = A$ and $B^T = B$, so, by Theorem 2, $(A + B)^T = A^T + B^T = A + B$. Hence $A + B$ is symmetric.

◆◆◆

EXERCISES 2.1

1. Find a, b, c, and d if:

(a) $\begin{bmatrix} a & b \\ c & d \end{bmatrix} = \begin{bmatrix} c - 3d & -d \\ 2a + d & a + b \end{bmatrix}$

◆(b) $\begin{bmatrix} a - b & b - c \\ c - d & d - a \end{bmatrix} = 2\begin{bmatrix} 1 & 1 \\ -3 & 1 \end{bmatrix}$

(c) $3\begin{bmatrix} a \\ b \end{bmatrix} + 2\begin{bmatrix} b \\ a \end{bmatrix} = \begin{bmatrix} 1 \\ 2 \end{bmatrix}$ ◆(d) $\begin{bmatrix} a & b \\ c & d \end{bmatrix} = \begin{bmatrix} b & c \\ d & a \end{bmatrix}$

2. Compute the following:

(a) $\begin{bmatrix} 3 & 2 & 1 \\ 5 & 1 & 0 \end{bmatrix} - 5\begin{bmatrix} 3 & 0 & -2 \\ 1 & -1 & 2 \end{bmatrix}$

◆(b) $3\begin{bmatrix} 3 \\ -1 \end{bmatrix} - 5\begin{bmatrix} 6 \\ 2 \end{bmatrix} + 7\begin{bmatrix} 1 \\ -1 \end{bmatrix}$

(c) $\begin{bmatrix} -2 & 1 \\ 3 & 2 \end{bmatrix} - 4\begin{bmatrix} 1 & -2 \\ 0 & -1 \end{bmatrix} + 3\begin{bmatrix} 2 & -3 \\ -1 & -2 \end{bmatrix}$

◆(d) $[3 \ -1 \ 2] - 2[9 \ 3 \ 4] + [3 \ 11 \ -6]$

(e) $\begin{bmatrix} 1 & -5 & 4 & 0 \\ 2 & 1 & 0 & 6 \end{bmatrix}^T$ ◆(f) $\begin{bmatrix} 0 & -1 & 2 \\ 1 & 0 & -4 \\ -2 & 4 & 0 \end{bmatrix}^T$

(g) $\begin{bmatrix} 3 & -1 \\ 2 & 1 \end{bmatrix} - 2\begin{bmatrix} 1 & -2 \\ 1 & 1 \end{bmatrix}^T$ ◆(h) $3\begin{bmatrix} 2 & 1 \\ -1 & 0 \end{bmatrix}^T - 2\begin{bmatrix} 1 & -1 \\ 2 & 3 \end{bmatrix}$

3. Let $A = \begin{bmatrix} 2 & 1 \\ 0 & -1 \end{bmatrix}$, $B = \begin{bmatrix} 3 & -1 & 2 \\ 0 & 1 & 4 \end{bmatrix}$, $C = \begin{bmatrix} 3 & -1 \\ 2 & 0 \end{bmatrix}$,

$D = \begin{bmatrix} 1 & 3 \\ -1 & 0 \\ 1 & 4 \end{bmatrix}$, and $E = \begin{bmatrix} 1 & 0 & 1 \\ 0 & 1 & 0 \end{bmatrix}$. Compute the following (where possible).

(a) $3A - 2B$ ◆(b) $5C$ (c) $3E^T$

◆(d) $B + D$ (e) $4A^T - 3C$ ◆(f) $(A + C)^T$

(g) $2B - 3E$ ◆(h) $A - D$ (i) $(B - 2E)^T$

4. Find A if:

(a) $5A - \begin{bmatrix} 1 & 0 \\ 2 & 3 \end{bmatrix} = 3A - \begin{bmatrix} 5 & 2 \\ 6 & 1 \end{bmatrix}$

◆(b) $3A + \begin{bmatrix} 2 \\ 1 \end{bmatrix} = 5A - 2\begin{bmatrix} 3 \\ 0 \end{bmatrix}$

5. Find A in terms of B if:

(a) $A + B = 3A + 2B$ ◆(b) $2A - B = 5(A + 2B)$

6. If X, Y, A, and B are matrices of the same size, solve the following equations to obtain X and Y in terms of A and B.

(a) $5X + 3Y = A$
$\quad\, 2X + \ Y = B$

◆(b) $4X + 3Y = A$
$\quad\,\ 5X + 4Y = B$

7. Find all matrices X and Y such that:

(a) $3X - 2Y = [3 \ \ -1]$ ◆(b) $2X - 5Y = [1 \ \ 2]$

8. Simplify the following expressions where A, B, and C are matrices.

(a) $2[9(A - B) + 7(2B - A)] - 2[3(2B + A) - 2(A + 3B) - 5(A + B)]$

◆(b) $5[3(A - B + 2C) - 2(3C - B) - A] + 2[3(3A - B + C) + 2(B - 2A) - 2C]$

9. If A is any 2×2 matrix, show that:

(a) $A = a\begin{bmatrix} 1 & 0 \\ 0 & 0 \end{bmatrix} + b\begin{bmatrix} 0 & 1 \\ 0 & 0 \end{bmatrix} + c\begin{bmatrix} 0 & 0 \\ 1 & 0 \end{bmatrix} + d\begin{bmatrix} 0 & 0 \\ 0 & 1 \end{bmatrix}$
for some numbers a, b, c, and d.

◆(b) $A = p\begin{bmatrix} 1 & 0 \\ 0 & 1 \end{bmatrix} + q\begin{bmatrix} 1 & 1 \\ 0 & 0 \end{bmatrix} + r\begin{bmatrix} 1 & 0 \\ 1 & 0 \end{bmatrix} + s\begin{bmatrix} 0 & 1 \\ 1 & 0 \end{bmatrix}$
for some numbers p, q, r, and s.

10. Let $A = [1 \ 1 \ -1]$, $B = [0 \ 1 \ 2]$, and $C = [3 \ 0 \ 1]$. If $rA + sB + tC = 0$ for some scalars r, s, and t, show that necessarily $r = s = t = 0$.

11. (a) If $Q + A = A$ holds for every $m \times n$ matrix A, show that $Q = 0_{mn}$.

◆(b) If A is an $m \times n$ matrix and $A + A' = 0_{mn}$, show that $A' = -A$.

12. If A denotes an $m \times n$ matrix, show that $A = -A$ if and only if $A = 0$.

13. A square matrix is called a **diagonal** matrix if all the entries off the main diagonal are zero. If A and B are diagonal matrices, show that the following matrices are also diagonal.

(a) $A + B$ ◆(b) $A - B$ (c) kA for any number k

14. In each case determine all s and t such that the given matrix is symmetric:

(a) $\begin{bmatrix} 1 & s \\ -2 & t \end{bmatrix}$ ◆(b) $\begin{bmatrix} s & t \\ st & 1 \end{bmatrix}$

(c) $\begin{bmatrix} s & 2s & st \\ t & -1 & s \\ t & s^2 & s \end{bmatrix}$ ◆(d) $\begin{bmatrix} 2 & s & t \\ 2s & 0 & s+t \\ 3 & 3 & t \end{bmatrix}$

15. In each case find the matrix A.

(a) $\left(A + 3\begin{bmatrix} 1 & -1 & 0 \\ 1 & 2 & 4 \end{bmatrix} \right)^T = \begin{bmatrix} 2 & 1 \\ 0 & 5 \\ 3 & 8 \end{bmatrix}$

◆(b) $\left(3A^T + 2\begin{bmatrix} 1 & 0 \\ 0 & 2 \end{bmatrix} \right)^T = \begin{bmatrix} 8 & 0 \\ 3 & 1 \end{bmatrix}$

(c) $\left(2A - 3[1 \ 2 \ 0] \right)^T = 3A^T + [2 \ 1 \ -1]^T$

◆(d) $\left(2A^T - 5\begin{bmatrix} 1 & 0 \\ -1 & 2 \end{bmatrix} \right)^T = 4A - 9\begin{bmatrix} 1 & 1 \\ -1 & 0 \end{bmatrix}$

16. Let A and B be symmetric (of the same size). Show that each of the following is symmetric.

(a) $A - B$ ◆(b) kA for any scalar k

17. Show that $A + A^T$ is symmetric for *any* square matrix A.

18. A square matrix W is called **skew-symmetric** if $W^T = -W$. Let A be any square matrix.

(a) Show that $A - A^T$ is skew-symmetric.

(b) Find a symmetric matrix S and a skew-symmetric matrix W such that $A = S + W$.

◆(c) Show that S and W in part (b) are uniquely determined by A.

19. If W is skew-symmetric (Exercise 18), show that the entries on the main diagonal are zero.

20. Prove the following parts of Theorem 1.

(a) $(k + p)A = kA + pA$ ◆(b) $(kp)A = k(pA)$

21. Let A, A_1, A_2, . . . , A_n denote matrices of the same size. Use induction on n to verify the following extensions of properties 5 and 6 of Theorem 1.

(a) $k(A_1 + A_2 + \cdots + A_n) = kA_1 + kA_2 + \cdots + kA_n$ for any number k

(b) $(k_1 + k_2 + \cdots + k_n)A = k_1A + k_2A + \cdots + k_nA$ for any numbers $k_1, k_2, \ldots, k_n$

Section 2.2   **Matrix Multiplication**

Matrix multiplication is a little more complicated than matrix addition or scalar multiplication, but it is well worth the extra effort. It provides a new way to look at systems of linear equations and has a wide variety of other applications as well (see, for example, Sections 2.6 and 2.7).

DEFINITION

If A is an $m \times n$ matrix and B is an $n \times k$ matrix, the **product** AB of A and B is the $m \times k$ matrix whose (i, j)-entry is computed as follows:

Multiply each entry of *row i* of A by the corresponding entry of *column j* of B, and add the results.

This is called the **dot product** of row i of A and column j of B.

EXAMPLE 1

Compute the $(1, 3)$- and $(2, 4)$-entries of AB where

$$A = \begin{bmatrix} 3 & -1 & 2 \\ 0 & 1 & 4 \end{bmatrix} \quad \text{and} \quad B = \begin{bmatrix} 2 & 1 & 6 & 0 \\ 0 & 2 & 3 & 4 \\ -1 & 0 & 5 & 8 \end{bmatrix}$$

Then compute AB.

Solution   The $(1, 3)$-entry of AB is the dot product of row 1 of A and column 3 of B (highlighted in the following display), computed by multiplying corresponding entries and adding the results.

$$\begin{bmatrix} 3 & -1 & 2 \\ 0 & 1 & 4 \end{bmatrix} \begin{bmatrix} 2 & 1 & 6 & 0 \\ 0 & 2 & 3 & 4 \\ -1 & 0 & 5 & 8 \end{bmatrix} \qquad (1, 3)\text{-entry} = 3 \cdot 6 + (-1) \cdot 3 + 2 \cdot 5 = 25$$

Similarly the $(2, 4)$ entry of AB involves row 2 of A and column 4 of B.

$$\begin{bmatrix} 3 & -1 & 2 \\ 0 & 1 & 4 \end{bmatrix} \begin{bmatrix} 2 & 1 & 6 & 0 \\ 0 & 2 & 3 & 4 \\ -1 & 0 & 5 & 8 \end{bmatrix} \qquad (2, 4)\text{-entry} = 0 \cdot 0 + 1 \cdot 4 + 4 \cdot 8 = 36$$

Since A is 2×3 and B is 3×4, the product is 2×4.

$$\begin{bmatrix} 3 & -1 & 2 \\ 0 & 1 & 4 \end{bmatrix} \begin{bmatrix} 2 & 1 & 6 & 0 \\ 0 & 2 & 3 & 4 \\ -1 & 0 & 5 & 8 \end{bmatrix} = \begin{bmatrix} 4 & 1 & 25 & 12 \\ -4 & 2 & 23 & 36 \end{bmatrix}$$

◆◆◆

Computing the (i, j)-entry of AB involves going *across* row i of A and *down* column j of B, multiplying corresponding entries, and adding the results. This requires that the rows of A and the columns of B be the same length. The following rule is a useful way to remember when the product of A and B can be formed and what size of the product matrix is.

RULE

Suppose A and B have sizes:

$$\begin{array}{cc} A & B \\ m \times (n \qquad n') \times p \end{array}$$

The product AB is defined only when $n = n'$; in this case, the product matrix AB is of size $m \times p$.

EXAMPLE 2

If $A = [1 \ 3 \ 2]$ and $B = \begin{bmatrix} 5 \\ 6 \\ 4 \end{bmatrix}$, compute A^2, AB, BA, and B^2 when they are defined.

Solution   Here A is a 1×3 matrix and B is a 3×1 matrix so A^2 and B^2 are not defined. However, the rule reads

$$\begin{array}{cc} A & B \\ 1 \times 3 & 3 \times 1 \end{array} \quad \text{and} \quad \begin{array}{cc} B & A \\ 3 \times 1 & 1 \times 3 \end{array}$$

so both AB and BA can be formed and these are 1×1 and 3×3 matrices, respectively.

$$AB = \begin{bmatrix} 1 & 3 & 2 \end{bmatrix} \begin{bmatrix} 5 \\ 6 \\ 4 \end{bmatrix} = \begin{bmatrix} 1 \cdot 5 + 3 \cdot 6 + 2 \cdot 4 \end{bmatrix} = \begin{bmatrix} 31 \end{bmatrix}$$

$$BA = \begin{bmatrix} 5 \\ 6 \\ 4 \end{bmatrix} \begin{bmatrix} 1 & 3 & 2 \end{bmatrix} = \begin{bmatrix} 5 \cdot 1 & 5 \cdot 3 & 5 \cdot 2 \\ 6 \cdot 1 & 6 \cdot 3 & 6 \cdot 2 \\ 4 \cdot 1 & 4 \cdot 3 & 4 \cdot 2 \end{bmatrix} = \begin{bmatrix} 5 & 15 & 10 \\ 6 & 18 & 12 \\ 4 & 12 & 8 \end{bmatrix}$$

Unlike numerical multiplication, matrix products AB and BA *need not be equal*. In fact they need not even be the same size, as Example 2 shows. It turns out to be rare that $AB = BA$ (although it is by no means impossible), and A and B are said to **commute** when this happens.

EXAMPLE 3

$$\text{Let } A = \begin{bmatrix} 6 & 9 \\ -4 & -6 \end{bmatrix} \text{ and } B = \begin{bmatrix} 1 & 2 \\ -1 & 0 \end{bmatrix}. \text{ Compute } A^2, AB, \text{ and } BA.$$

Solution $A^2 = \begin{bmatrix} 6 & 9 \\ -4 & -6 \end{bmatrix}\begin{bmatrix} 6 & 9 \\ -4 & -6 \end{bmatrix} = \begin{bmatrix} 0 & 0 \\ 0 & 0 \end{bmatrix}$, so $A^2 = 0$ can occur even if $A \neq 0$. Next,

$$AB = \begin{bmatrix} 6 & 9 \\ -4 & -6 \end{bmatrix}\begin{bmatrix} 1 & 2 \\ -1 & 0 \end{bmatrix} = \begin{bmatrix} -3 & 12 \\ 2 & -8 \end{bmatrix}$$

$$BA = \begin{bmatrix} 1 & 2 \\ -1 & 0 \end{bmatrix}\begin{bmatrix} 6 & 9 \\ -4 & -6 \end{bmatrix} = \begin{bmatrix} -2 & -3 \\ -6 & -9 \end{bmatrix}$$

Hence $AB \neq BA$, even though AB and BA are the same size.

◆◆◆

The number 1 plays a neutral role in numerical multiplication in the sense that $1 \cdot a = a$ and $a \cdot 1 = a$ for all numbers a. An analogous role for matrix multiplication is played by square matrices of the following types:

$$\begin{bmatrix} 1 & 0 \\ 0 & 1 \end{bmatrix} \quad \begin{bmatrix} 1 & 0 & 0 \\ 0 & 1 & 0 \\ 0 & 0 & 1 \end{bmatrix} \quad \begin{bmatrix} 1 & 0 & 0 & 0 \\ 0 & 1 & 0 & 0 \\ 0 & 0 & 1 & 0 \\ 0 & 0 & 0 & 1 \end{bmatrix} \quad \text{and so on}$$

In general, an **identity matrix** I is a *square matrix with 1's on the main diagonal and zeros elsewhere.* If it is important to stress the size of an $n \times n$ identity matrix, we shall denote it by I_n; however, these matrices are usually written simply as I. Identity matrices play a neutral role with respect to matrix multiplication in the sense that

$$AI = A \quad \text{and} \quad IB = B$$

whenever the products are defined.

Before proceeding, let us state the definition of matrix multiplication more formally. If $A = [a_{ij}]$ is $m \times n$ and $B = [b_{ij}]$ is $n \times p$, the ith row of A and the jth column of B are, respectively,

$$\begin{bmatrix} a_{i1} & a_{i2} & \cdots & a_{in} \end{bmatrix} \quad \text{and} \quad \begin{bmatrix} b_{1j} \\ b_{2j} \\ \vdots \\ b_{nj} \end{bmatrix}$$

Hence the (i, j)-entry of the product matrix AB is the dot product

$$a_{i1}b_{1j} + a_{i2}b_{2j} + \cdots + a_{in}b_{nj} = \sum_{k=1}^{n} a_{ik}b_{kj}.$$

where summation notation has been introduced for convenience.[2] This is useful in verifying facts about matrix multiplication.

THEOREM 1

Assume that k is an arbitrary scalar and that A, B, and C are matrices of sizes such that the indicated operations can be performed.

1. $IA = A, \qquad BI = B$

2. $A(BC) = (AB)C$

3. $A(B + C) = AB + AC, \qquad A(B - C) = AB - AC$

4. $(B + C)A = BA + CA, \qquad (B - C)A = BA - CA$

5. $k(AB) = (kA)B = A(kB)$

6. $(AB)^T = B^T A^T$

Proof   We prove properties 3 and 6, leaving the rest as exercises.

Property 3. Write $A = [a_{ij}]$, $B = [b_{ij}]$, and $C = [c_{ij}]$ and assume that A is $m \times n$ and that B and C are $n \times p$. Then $B + C = [b_{ij} + c_{ij}]$, so the (i, j)-entry of $A(B + C)$ is

$$\sum_{k=1}^{n} a_{ik}\left(b_{kj} + c_{kj}\right) = \sum_{k=1}^{n} \left(a_{ik}b_{kj} + a_{ik}c_{kj}\right) = \sum_{k=1}^{n} a_{ik}b_{kj} + \sum_{k=1}^{n} a_{ik}c_{kj}$$

This is the (i, j)-entry of $AB + AC$ because the sums on the right are the (i, j)-entries of AB and AC, respectively. Hence $A(B + C) = AB + AC$.

Property 6. Write $A^T = [a'_{ij}]$ and $B^T = [b'_{ij}]$, where $a'_{ij} = a_{ji}$ and $b'_{ij} = b_{ji}$. If B^T and A^T are $p \times n$ and $n \times m$, respectively, the (i, j)-entry of $B^T A^T$ is

$$\sum_{k=1}^{n} b'_{ik}a'_{kj} = \sum_{k=1}^{n} b_{ki}a_{jk} = \sum_{k=1}^{n} a_{jk}b_{ki}$$

This is the (j, i)-entry of AB—that is, the (i, j)-entry of $(AB)^T$. Hence $B^T A^T = (AB)^T$. ◆

Property 2 in Theorem 1 asserts that the **associative law** $A(BC) = (AB)C$ holds for all matrices (if the products are defined). Hence the product is the same no matter

---

[2]Summation notation is a convenient shorthand way to write sums. For example, $a_1 + a_2 + a_3 + a_4 = \sum_{k=1}^{4} a_k$,

$a_5 x_5 + a_6 x_6 + a_7 x_7 = \sum_{i=5}^{7} a_i x_i$, and $1^2 + 2^2 + 3^2 + 4^2 + 5^2 = \sum_{j=1}^{5} j^2$.

how it is formed and so is simply written *ABC*. This extends: The product *ABCD* of four matrices can be formed several ways — for example, (*AB*)(*CD*), [*A*(*BC*)]*D*, and *A*[*B*(*CD*)] — but property 2 implies that they are all equal and so are written simply as *ABCD*. A similar remark applies in general: Matrix products can be written unambiguously with no parentheses.

However, a note of caution about matrix multiplication *is* in order. The fact that *AB* and *BA* need *not* be equal means that the *order* of the factors is important in a product of matrices. For example, *ABCD* and *ADCB* may *not* be equal.

◆◆◆

Warning If the order of the factors in a product of matrices is changed, the product matrix may change (or may not exist).

◆◆◆

Ignoring this warning is a source of many errors by students of linear algebra!

Properties 3 and 4 in Theorem 1 are called the **distributive laws**, and they extend to more than two terms. For example,

$$A(B - C + D - E) = AB - AC + AD - AE$$

$$(A + C - D)B = AB + CB - DB$$

Note again that the warning is in effect: For example, $A(B - C)$ need *not* equal $AB - CA$. Together with property 5 of Theorem 1, the distributive laws make possible a lot of simplification of matrix expressions.

EXAMPLE 4

Simplify the expression $A(BC - CD) + A(C - B)D - AB(C - D)$.

Solution
$$A(BC - CD) + A(C - B)D - AB(C - D)$$
$$= ABC - ACD + (AC - AB)D - ABC + ABD$$
$$= ABC - ACD + ACD - ABD - ABC + ABD$$
$$= 0$$
◆◆◆

Examples 5 and 6 show how we can use the properties in Theorem 1 to deduce facts about matrix multiplication.

EXAMPLE 5

Suppose that *A*, *B*, and *C* are $n \times n$ matrices and that both *A* and *B* commute with *C*; that is, $AC = CA$ and $BC = CB$. Show that *AB* commutes with *C*.

Solution Showing that *AB* commutes with *C* means verifying that $(AB)C = C(AB)$. The computation uses property 2 of Theorem 1 and the given facts that $AC = CA$ and $BC = CB$.

$$(AB)C = A(BC) = A(CB) = (AC)B = (CA)B = C(AB)$$
◆◆◆

EXAMPLE 6

Show that $AB = BA$ if and only if $(A - B)(A + B) = A^2 - B^2$.

Solution   Theorem 1 shows that the following always holds:

$$(A - B)(A + B) = A(A + B) - B(A + B) = A^2 + AB - BA - B^2 \qquad (*)$$

Hence if $AB = BA$, then $(A - B)(A + B) = A^2 - B^2$ follows. Conversely, if this last equation holds, then equation $(*)$ becomes

$$A^2 - B^2 = A^2 + AB - BA - B^2$$

This gives $0 = AB - BA$, and $AB = BA$ follows.

◆◆◆

Matrices and Linear Equations

One of the most important motivations for matrix multiplication results from its close connection with linear equations.

EXAMPLE 7

Write the following system of linear equations as a single matrix equation.

$$3x_1 - 2x_2 + x_3 = b_1$$
$$2x_1 + x_2 - x_3 = b_2$$

Solution   The two linear equations can be written as a single matrix equation as follows:

$$\begin{bmatrix} 3x_1 - 2x_2 + x_3 \\ 2x_1 + x_2 - x_3 \end{bmatrix} = \begin{bmatrix} b_1 \\ b_2 \end{bmatrix} \qquad (*)$$

The matrix on the left can be factored as a product of matrices:

$$\begin{bmatrix} 3 & -2 & 1 \\ 2 & 1 & -1 \end{bmatrix} \begin{bmatrix} x_1 \\ x_2 \\ x_3 \end{bmatrix} = \begin{bmatrix} b_1 \\ b_2 \end{bmatrix}$$

If these matrices are denoted by A, X, and B, respectively, the system of equations becomes the matrix equation $AX = B$.

◆◆◆

In the same way, consider *any* system of linear equations:

$$a_{11}x_1 + a_{12}x_2 + \cdots + a_{1n}x_n = b_1$$
$$a_{21}x_1 + a_{22}x_2 + \cdots + a_{2n}x_n = b_2$$
$$\vdots \qquad \vdots \qquad \qquad \vdots \qquad \vdots$$
$$a_{m1}x_1 + a_{m2}x_2 + \cdots + a_{mn}x_n = b_m$$

If $A = \begin{bmatrix} a_{11} & a_{12} & \cdots & a_{1n} \\ a_{21} & a_{22} & \cdots & a_{2n} \\ \vdots & \vdots & & \vdots \\ a_{m1} & a_{m2} & \cdots & a_{mn} \end{bmatrix}$, $X = \begin{bmatrix} x_1 \\ x_2 \\ \vdots \\ x_n \end{bmatrix}$, and $B = \begin{bmatrix} b_1 \\ b_2 \\ \vdots \\ b_m \end{bmatrix}$, these equations become

the single matrix equation

$$AX = B$$

This is called the **matrix form** of the system of equations, and B is called the **constant matrix**. As in Section 1.1, A is called the **coefficient matrix** of the system, and a column matrix X_1 is called a **solution** to the system if $AX_1 = B$.

The matrix form is useful for formulating results about solutions of systems of linear equations. For example, if X_1 is a solution to $AX = B$ and if X_0 is a solution to the **associated homogeneous system** $AX = 0$, then $X_1 + X_0$ is a solution to $AX = B$. Indeed, $AX_1 = B$ and $AX_0 = 0$, so

$$A(X_1 + X_0) = AX_1 + AX_0 = B + 0 = B$$

This observation has a useful converse.

THEOREM 2

Suppose X_1 is a particular solution to the system $AX = B$ of linear equations. Then every solution X_2 to $AX = B$ has the form

$$X_2 = X_0 + X_1$$

for some solution X_0 of the associated homogeneous system $AX = 0$.

Proof   Suppose that X_2 is *any* solution to $AX = B$ so that $AX_2 = B$. Write $X_0 = X_2 - X_1$. Then $X_2 = X_0 + X_1$, and we compute:

$$AX_0 = A(X_2 - X_1) = AX_2 - AX_1 = B - B = 0$$

Thus X_0 is a solution to the associated homogeneous system $AX = 0$.   ◆

EXAMPLE 8

Express every solution to the following system as the sum of a specific solution plus a solution to the associated homogeneous system.

$$\begin{aligned} x - y - \ z &= 2 \\ 2x - y - 3z &= 6 \\ x \quad\ - 2z &= 4 \end{aligned}$$

Solution   Gaussian elimination gives $x = 4 + 2t$, $y = 2 + t$, $z = t$, t arbitrary. Hence the general solution is

$$X = \begin{bmatrix} x \\ y \\ z \end{bmatrix} = \begin{bmatrix} 4 + 2t \\ 2 + t \\ t \end{bmatrix} = \begin{bmatrix} 4 \\ 2 \\ 0 \end{bmatrix} + t \begin{bmatrix} 2 \\ 1 \\ 1 \end{bmatrix}$$

Thus $X_0 = \begin{bmatrix} 4 \\ 2 \\ 0 \end{bmatrix}$ is a specific solution, and it is easily verified that $X_1 = t \begin{bmatrix} 2 \\ 1 \\ 1 \end{bmatrix}$ gives *all* solutions to the associated homogeneous system.

◆◆◆

Block Multiplication

It is sometimes useful to be able to compute a particular row or column of a matrix product AB without having to find the whole product. The next example shows how.

EXAMPLE 9

Find row 2 and column 3 of AB where $A = \begin{bmatrix} 3 & 1 & 2 \\ 4 & 8 & 0 \\ 0 & 1 & 2 \end{bmatrix}$ and $B = \begin{bmatrix} 0 & 5 & 2 & 1 \\ 1 & 8 & 0 & -6 \\ 1 & 4 & 3 & 7 \end{bmatrix}$.

Solution   Write $R_2 = [4\ 8\ 0]$ for row 2 of A. By the definition of matrix multiplication, row 2 of AB is obtained by multiplying R_2 by the various columns of B:

$$\text{row 2 of } AB = R_2 B = \begin{bmatrix} 4 & 8 & 0 \end{bmatrix} \begin{bmatrix} 0 & 5 & 2 & 1 \\ 1 & 8 & 0 & -6 \\ 1 & 4 & 3 & 7 \end{bmatrix} = \begin{bmatrix} 8 & 84 & 8 & -44 \end{bmatrix}$$

Similarly, if we write C_3 for column 3 of B, then column 3 of AB is

$$\text{column 3 of } AB = AC_3 = \begin{bmatrix} 3 & 1 & 2 \\ 4 & 8 & 0 \\ 0 & 1 & 2 \end{bmatrix} \begin{bmatrix} 2 \\ 0 \\ 3 \end{bmatrix} = \begin{bmatrix} 12 \\ 8 \\ 6 \end{bmatrix}$$

◆◆◆

In general, when forming matrix products YA and AX, it is often convenient to view the matrix A as a column of rows or as a row of columns. If A is $m \times n$, and if $R_1, R_2 \ldots, R_m$ are the rows of A and $C_1, C_2, \ldots, C_n$ are the columns, we write

$$A = \begin{bmatrix} R_1 \\ R_2 \\ \vdots \\ R_m \end{bmatrix} \qquad \text{and} \qquad A = \begin{bmatrix} C_1 & C_2 & \cdots & C_n \end{bmatrix}$$

Then the definition of matrix multiplication shows that

$$AX = \begin{bmatrix} R_1 \\ R_2 \\ \vdots \\ R_m \end{bmatrix} X = \begin{bmatrix} R_1 X \\ R_2 X \\ \vdots \\ R_m X \end{bmatrix} \quad \text{and} \quad YA = Y\begin{bmatrix} C_1 & C_2 & \cdots & C_n \end{bmatrix} = \begin{bmatrix} YC_1 & YC_2 & \cdots & YC_n \end{bmatrix}(*)$$

This gives AX in terms of its rows and YA in terms of its columns. In other words, the rows of AX are $R_1 X, R_2 X, \ldots, R_m X$ and the columns of YA are $YC_1, YC_2, \ldots, YC_n$.

These are special cases of a more general way of looking at matrices that, among its other uses, can greatly simplify matrix multiplications. The idea is to partition a matrix A into smaller matrices (called **blocks**) by inserting vertical lines between the columns and horizontal lines between the rows.[3] As an example, consider the matrices

$$A = \begin{bmatrix} 1 & 0 & 0 & 0 & 0 \\ 0 & 1 & 0 & 0 & 0 \\ 2 & -1 & 4 & 2 & 1 \\ 3 & 1 & -1 & 7 & 5 \end{bmatrix} = \begin{bmatrix} I_2 & 0_{23} \\ P & Q \end{bmatrix} \quad \text{and} \quad B = \begin{bmatrix} 4 & -2 \\ 5 & 6 \\ 7 & 3 \\ -1 & 0 \\ 1 & 6 \end{bmatrix} = \begin{bmatrix} X \\ Y \end{bmatrix}$$

where the blocks have been labeled as indicated. This is a natural way to think of A in view of the blocks I_2 and 0_{23} that occur.

This notation is particularly useful when we are multiplying the matrices A and B because the product AB can be computed in block form as follows:

$$AB = \begin{bmatrix} I & 0 \\ P & Q \end{bmatrix} \begin{bmatrix} X \\ Y \end{bmatrix} = \begin{bmatrix} IX + 0Y \\ PX + QY \end{bmatrix} = \begin{bmatrix} X \\ PX + QY \end{bmatrix} = \begin{bmatrix} 4 & -2 \\ 5 & 6 \\ 30 & 8 \\ 8 & 27 \end{bmatrix}$$

This is easily checked to be the product AB, computed in the conventional manner. In other words, *we can compute the product by ordinary matrix multiplication, using blocks as entries*. The only requirement is that the blocks be **compatible**. That is, *the sizes of the blocks must be such that all (matrix) products of blocks that occur make sense*. This means that the number of columns in each block of A must equal the number of rows in the corresponding block of B.

Block Multiplication

If matrices A and B are partitioned compatibly into blocks, the product AB can be computed by matrix multiplication using blocks as entries.

We omit the proof.

---
[3]We have already been doing this with the augmented matrices arising from systems of linear equations.

One of the most important examples of block multiplication is given in equation (∗) earlier where one factor is partitioned into its rows or columns. Here is another illustration.

EXAMPLE 10

If $C_1, C_2, \ldots, C_n$ are the columns of an $m \times n$ matrix A, and if $X = \begin{bmatrix} x_1 \\ x_2 \\ \vdots \\ x_n \end{bmatrix}$ is a column matrix, then

$$AX = \begin{bmatrix} C_1 & C_2 & \cdots & C_n \end{bmatrix} \begin{bmatrix} x_1 \\ x_2 \\ \vdots \\ x_n \end{bmatrix} = x_1 C_1 + x_2 C_2 + \cdots + x_n C_n.$$

◆◆◆

Any compatible partitioning can be used in block multiplication. We conclude with an example in which block multiplication is used to compute higher powers of a matrix.

EXAMPLE 11

Compute A^8 using block multiplication, where $A = \begin{bmatrix} 1 & -1 & 0 \\ -1 & 1 & 0 \\ 1 & -1 & -2 \end{bmatrix}$.

Solution   Write $X = \begin{bmatrix} 1 & -1 \\ -1 & 1 \end{bmatrix}$, $Y = \begin{bmatrix} 1 & -1 \end{bmatrix}$, and $Z = \begin{bmatrix} -2 \end{bmatrix}$. Then

$$A^2 = \begin{bmatrix} X & 0 \\ Y & Z \end{bmatrix} \begin{bmatrix} X & 0 \\ Y & Z \end{bmatrix} = \begin{bmatrix} X^2 & 0 \\ YX + ZY & Z^2 \end{bmatrix}$$

Now $YX + ZY = \begin{bmatrix} 2 & -2 \end{bmatrix} + \begin{bmatrix} -2 & 2 \end{bmatrix} = 0$, so

$$A^2 = \begin{bmatrix} X^2 & 0 \\ 0 & Z^2 \end{bmatrix}$$

Powers of this are easy to compute:

$$A^4 = (A^2)^2 = \begin{bmatrix} X^2 & 0 \\ 0 & Z^2 \end{bmatrix} \begin{bmatrix} X^2 & 0 \\ 0 & Z^2 \end{bmatrix} = \begin{bmatrix} X^4 & 0 \\ 0 & Z^4 \end{bmatrix}$$

$$A^8 = (A^4)^2 = \begin{bmatrix} X^4 & 0 \\ 0 & Z^4 \end{bmatrix} \begin{bmatrix} X^4 & 0 \\ 0 & Z^4 \end{bmatrix} = \begin{bmatrix} X^8 & 0 \\ 0 & Z^8 \end{bmatrix}$$

We have $Z^8 = [-2]^8 = [256]$. Also $X^2 = 2X$ is easily verified, so $X^4 = (2X)^2 = 4X^2 = 8X$, and $X^8 = (8X)^2 = 64X^2 = 128X$. Finally:

$$A^8 = \begin{bmatrix} 128X & 0 \\ 0 & 256 \end{bmatrix} = 128 \begin{bmatrix} X & 0 \\ 0 & 2 \end{bmatrix} = 128 \begin{bmatrix} 1 & -1 & 0 \\ -1 & 1 & 0 \\ 0 & 0 & 2 \end{bmatrix}$$

◆◆◆

Block multiplication is useful in computing products of matrices in a computer with limited memory capacity. The matrices are partitioned into blocks in such a way that each product of blocks can be handled. Then the blocks are stored in auxiliary memory (on tape, for example), and the products are computed one by one.

EXERCISES 2.2

1. Compute the following matrix products.

(a) $\begin{bmatrix} 1 & 3 \\ 0 & -2 \end{bmatrix} \begin{bmatrix} 2 & -1 \\ 0 & 1 \end{bmatrix}$
◆**(b)** $\begin{bmatrix} 1 & -1 & 2 \\ 2 & 0 & 4 \end{bmatrix} \begin{bmatrix} 2 & 3 & 1 \\ 1 & 9 & 7 \\ -1 & 0 & 2 \end{bmatrix}$

(c) $\begin{bmatrix} 5 & 0 & -7 \\ 1 & 5 & 9 \end{bmatrix} \begin{bmatrix} 3 \\ 1 \\ -1 \end{bmatrix}$
◆**(d)** $\begin{bmatrix} 1 & 3 & -3 \end{bmatrix} \begin{bmatrix} 3 & 0 \\ -2 & 1 \\ 0 & 6 \end{bmatrix}$

(e) $\begin{bmatrix} 1 & 0 & 0 \\ 0 & 1 & 0 \\ 0 & 0 & 1 \end{bmatrix} \begin{bmatrix} 3 & -2 \\ 5 & -7 \\ 9 & 7 \end{bmatrix}$
◆**(f)** $\begin{bmatrix} 1 & -1 & 3 \end{bmatrix} \begin{bmatrix} 2 \\ 1 \\ -8 \end{bmatrix}$

(g) $\begin{bmatrix} 2 \\ 1 \\ -7 \end{bmatrix} \begin{bmatrix} 1 & -1 & 3 \end{bmatrix}$
◆**(h)** $\begin{bmatrix} 3 & 1 \\ 5 & 2 \end{bmatrix} \begin{bmatrix} 2 & -1 \\ -5 & 3 \end{bmatrix}$

(i) $\begin{bmatrix} 2 & 3 & 1 \\ 5 & 7 & 4 \end{bmatrix} \begin{bmatrix} a & 0 & 0 \\ 0 & b & 0 \\ 0 & 0 & c \end{bmatrix}$
◆**(j)** $\begin{bmatrix} a & 0 & 0 \\ 0 & b & 0 \\ 0 & 0 & c \end{bmatrix} \begin{bmatrix} a' & 0 & 0 \\ 0 & b' & 0 \\ 0 & 0 & c' \end{bmatrix}$

2. In each of the following cases, find all possible products A^2, AB, AC, and so on.

(a) $A = \begin{bmatrix} 1 & 2 & 3 \\ -1 & 0 & 0 \end{bmatrix}$ $B = \begin{bmatrix} 1 & -2 \\ \frac{1}{2} & 3 \end{bmatrix}$ $C = \begin{bmatrix} -1 & 0 \\ 2 & 5 \\ 0 & 3 \end{bmatrix}$

◆**(b)** $A = \begin{bmatrix} 1 & 2 & 4 \\ 0 & 1 & -1 \end{bmatrix}$ $B = \begin{bmatrix} -1 & 6 \\ 1 & 0 \end{bmatrix}$ $C = \begin{bmatrix} 2 & 0 \\ -1 & 1 \\ 1 & 2 \end{bmatrix}$

3. Find a, b, a_1, and b_1 if:

(a) $\begin{bmatrix} a & b \\ a_1 & b_1 \end{bmatrix} \begin{bmatrix} 3 & -5 \\ -1 & 2 \end{bmatrix} = \begin{bmatrix} 1 & -1 \\ 2 & 0 \end{bmatrix}$

◆**(b)** $\begin{bmatrix} 2 & 1 \\ -1 & 2 \end{bmatrix} \begin{bmatrix} a & b \\ a_1 & b_1 \end{bmatrix} = \begin{bmatrix} 7 & 2 \\ -1 & 4 \end{bmatrix}$

4. Verify that $A^2 - A - 6I = 0$ if:

(a) $A = \begin{bmatrix} 3 & -1 \\ 0 & -2 \end{bmatrix}$
◆**(b)** $A = \begin{bmatrix} 2 & 2 \\ 2 & -1 \end{bmatrix}$

5. Given $A = \begin{bmatrix} 1 & -1 \\ 0 & 1 \end{bmatrix}$, $B = \begin{bmatrix} 1 & 0 & -2 \\ 3 & 1 & 0 \end{bmatrix}$, $C = \begin{bmatrix} 1 & 0 \\ 2 & 1 \\ 5 & 8 \end{bmatrix}$,

and $D = \begin{bmatrix} 3 & -1 & 2 \\ 1 & 0 & 5 \end{bmatrix}$, verify the following facts from Theorem 1.

(a) $A(B - D) = AB - AD$
◆**(b)** $A(BC) = (AB)C$
(c) $(CD)^T = D^T C^T$

6. Let A be a 2×2 matrix.

(a) If A commutes with $\begin{bmatrix} 0 & 1 \\ 0 & 0 \end{bmatrix}$, show that $A = \begin{bmatrix} a & b \\ 0 & a \end{bmatrix}$ for some a and b.

◆**(b)** If A commutes with $\begin{bmatrix} 0 & 0 \\ 1 & 0 \end{bmatrix}$, show that $A = \begin{bmatrix} a & 0 \\ c & a \end{bmatrix}$ for some a and c.

(c) Show that A commutes with *every* 2×2 matrix if and only if $A = \begin{bmatrix} a & 0 \\ 0 & a \end{bmatrix}$ for some a.

7. Write each of the following systems of linear equations in matrix form.

(a) $\begin{aligned} 3x_1 + 2x_2 - x_3 + x_4 &= 1 \\ x_1 - x_2 \qquad\quad + 3x_4 &= 0 \\ 2x_1 - x_2 \quad\; - x_3 \qquad &= 5 \end{aligned}$

◆**(b)** $\begin{aligned} -x_1 + 2x_2 - x_3 + x_4 &= 6 \\ 2x_1 + x_2 - x_3 + 2x_4 &= 1 \\ 3x_1 - 2x_2 \qquad\quad + x_4 &= 0 \end{aligned}$

8. In each case, express every solution of the system as a sum of a specific solution plus a solution of the associated homogeneous system.

(a) $\begin{aligned} x + y + z &= 2 \\ 2x + y \qquad &= 3 \\ x - y - 3z &= 0 \end{aligned}$
◆**(b)** $\begin{aligned} x - y - 4z &= -4 \\ x + 2y + 5z &= 2 \\ x + y + 2z &= 0 \end{aligned}$

(c) $\begin{aligned} x_1 + x_2 - x_3 \qquad\quad -5x_5 &= 2 \\ x_2 + x_3 \qquad\quad -4x_5 &= -1 \\ x_2 + x_3 + x_4 - x_5 &= -1 \\ 2x_1 \qquad -4x_3 + x_4 + x_5 &= 6 \end{aligned}$

(d) $\begin{aligned} 2x_1 + x_2 - x_3 - x_4 &= -1 \\ 3x_1 + x_2 + x_3 - 2x_4 &= -2 \\ -x_1 - x_2 + 2x_3 + x_4 &= 2 \\ -2x_1 - x_2 \qquad\quad + 2x_4 &= 3 \end{aligned}$

9. If X_0 and X_1 are solutions to the homogeneous system of equations $AX = 0$, show that $sX_0 + tX_1$ is also a solution for any scalars s and t (called a **linear combination** of X_0 and X_1).

10. In each of the following, write the general solution as a linear combination (Exercise 9) of specific solutions.

(a) $\begin{aligned} x_1 + 2x_2 + x_3 - x_4 + 3x_5 &= 0 \\ x_1 + 2x_2 + 2x_3 + x_4 + 2x_5 &= 0 \\ 2x_1 + 4x_2 + 2x_3 - x_4 + 7x_5 &= 0 \end{aligned}$

(b) $\begin{aligned} x_1 + x_2 - 2x_3 + 3x_4 + 2x_5 &= 0 \\ 2x_1 - x_2 + 3x_3 + 4x_4 + x_5 &= 0 \\ -x_1 - 2x_2 + 3x_3 + \quad\ x_4 &= 0 \\ 3x_1 \qquad\quad + x_3 + 7x_4 + 3x_5 &= 0 \end{aligned}$

11. Assume that $A \begin{bmatrix} 1 \\ -1 \\ 2 \end{bmatrix} = 0 = A \begin{bmatrix} 2 \\ 0 \\ 3 \end{bmatrix}$, and that $AX = B$ has a solution $X_0 = \begin{bmatrix} 2 \\ -1 \\ 3 \end{bmatrix}$. Find a two-parameter family of solutions to $AX = B$.

12. (a) If A^2 can be formed, what can be said about the size of A?

(b) If AB and BA can both be formed, describe the sizes of A and B.

(c) If ABC can be formed, A is 3×3, and C is 5×5, what size is B?

13. (a) Find two 2×2 matrices A such that $A^2 = 0$.

(b) Find three 2×2 matrices A such that **(i)** $A^2 = I$ and **(ii)** $A^2 = A$.

(c) Find 2×2 matrices A and B such that $AB = 0$ but $BA \neq 0$.

14. Write $P = \begin{bmatrix} 1 & 0 & 0 \\ 0 & 0 & 1 \\ 0 & 1 & 0 \end{bmatrix}$, and let A be $3 \times n$ and B be $m \times 3$.

(a) Describe PA in terms of the rows of A.

(b) Describe BP in terms of the columns of B.

15. Let A, B, and C be as in Exercise 5. Find the $(3, 1)$-entry of CAB using exactly six numerical multiplications.

16. (a) Compute AB, using the indicated block partitioning.

$$A = \begin{bmatrix} 2 & -1 & 3 & 1 \\ 1 & 0 & 1 & 2 \\ 0 & 0 & 1 & 0 \\ 0 & 0 & 0 & 1 \end{bmatrix} \qquad B = \begin{bmatrix} 1 & 2 & 0 \\ -1 & 0 & 0 \\ 0 & 5 & 1 \\ 1 & -1 & 0 \end{bmatrix}$$

(b) Partition A and B in part **(a)** differently and compute AB again.

(c) Find A^2 using the partitioning in part **(a)** and then again using a different partitioning.

17. In each case compute all powers of A using the block decomposition indicated.

(a) $A = \begin{bmatrix} 1 & 0 & 0 \\ 1 & 1 & -1 \\ 1 & -1 & 1 \end{bmatrix}$

(b) $A = \begin{bmatrix} 1 & -1 & 2 & -1 \\ 0 & 1 & 0 & 0 \\ 0 & 0 & -1 & 1 \\ 0 & 0 & 0 & 1 \end{bmatrix}$

18. Compute the following using block multiplication (all blocks $k \times k$).

(a) $\begin{bmatrix} I & X \\ -Y & I \end{bmatrix} \begin{bmatrix} I & 0 \\ Y & I \end{bmatrix}$

(b) $\begin{bmatrix} I & X \\ 0 & I \end{bmatrix} \begin{bmatrix} I & -X \\ 0 & I \end{bmatrix}$

(c) $\begin{bmatrix} I & X \end{bmatrix} \begin{bmatrix} I & X \end{bmatrix}^T$

(d) $\begin{bmatrix} I & X^T \end{bmatrix} \begin{bmatrix} -X & I \end{bmatrix}^T$

(e) $\begin{bmatrix} I & X \\ 0 & -I \end{bmatrix}^n$, any $n \geq 1$

(f) $\begin{bmatrix} 0 & X \\ I & 0 \end{bmatrix}^n$, any $n \geq 1$

19. (a) If A has a row of zeros, show that the same is true of AB for any B.

(b) If B has a column of zeros, show that the same is true of AB for any A.

20. Let A denote an $m \times n$ matrix.

(a) If $AX = 0$ for every $n \times 1$ matrix X, show that $A = 0$.

(b) If $YA = 0$ for every $1 \times m$ matrix Y, show that $A = 0$.

21. (a) If $U = \begin{bmatrix} 1 & 2 \\ 0 & -1 \end{bmatrix}$, and $AU = 0$, show that $A = 0$.

(b) Let U be such that $AU = 0$ implies that $A = 0$. If $AU = BU$, show that $A = B$.

22. Simplify the following expressions where A, B, and C represent matrices.

(a) $A(3B - C) + (A - 2B)C + 2B(C + 2A)$

(b) $A(B + C - D) + B(C - A + D) - (A + B)C + (A - B)D$

(c) $AB(BC - CB) + (CA - AB)BC + CA(A - B)C$

(d) $(A - B)(C - A) + (C - B)(A - C) + (C - A)^2$

23. If $A = \begin{bmatrix} a & b \\ c & d \end{bmatrix}$ where $a \neq 0$, show that A factors in the form $A = \begin{bmatrix} 1 & 0 \\ x & 1 \end{bmatrix} \begin{bmatrix} y & z \\ 0 & w \end{bmatrix}$

24. If A and B commute with C, show that the same is true of:
   (a) $A + B$ ◆**(b)** kA, k any scalar

25. If A is any matrix, show that AA^T and A^TA are symmetric.

◆**26.** If A and B are symmetric, show that AB is symmetric if and only if $AB = BA$.

27. If A is a 2×2 matrix, show that $A^TA = AA^T$ if and only if A is symmetric or $A = \begin{bmatrix} a & b \\ -b & a \end{bmatrix}$ for some a and b.

28. **(a)** Find all symmetric 2×2 matrices A such that $A^2 = 0$.
   ◆**(b)** Repeat **(a)** if A is 3×3.
   (c) Repeat **(a)** if A is $n \times n$.

29. Show that there exist no 2×2 matrices A and B such that $AB - BA = I$. [*Hint:* Examine the $(1, 1)$- and $(2, 2)$-entries.]

◆**30.** Let B be an $n \times n$ matrix. Suppose $AB = 0$ for some nonzero $m \times n$ matrix A. Show that no $n \times n$ matrix C exists such that $BC = I$.

31. **(a)** If A and B are 2×2 matrices whose rows sum to 1, show that the rows of AB also sum to 1.
   ◆**(b)** Repeat part **(a)** for the case where A and B are $n \times n$.

32. Let A and B be $n \times n$ matrices for which the systems of equations $AX = 0$ and $BX = 0$ each have only the trivial solution $X = 0$. Show that the system $(AB)X = 0$ has only the trivial solution.

33. The **trace** of a square matrix A, denoted tr A, is the sum of the elements on the main diagonal of A. Show that, if A and B are $n \times n$ matrices:
   (a) $\text{tr}(A + B) = \text{tr } A + \text{tr } B$
   ◆**(b)** $\text{tr }(kA) = k \text{ tr }(A)$ for any number k
   (c) $\text{tr}(A^T) = \text{tr}(A)$

   (d) $\text{tr}(AB) = \text{tr}(BA)$
   ◆**(e)** $\text{tr}(AA^T)$ is the sum of the squares of all entries of A.

34. Show $AB - BA = I$ is impossible. [*Hint:* See the preceding exercise.]

35. A square matrix P is called an **idempotent** if $P^2 = P$. Show that:
   (a) 0 and I are idempotents.
   (b) $\begin{bmatrix} 1 & 1 \\ 0 & 0 \end{bmatrix}$, $\begin{bmatrix} 1 & 0 \\ 1 & 0 \end{bmatrix}$, and $\dfrac{1}{\sqrt{2}}\begin{bmatrix} 1 & 1 \\ 1 & 1 \end{bmatrix}$ are idempotents.
   (c) If P is an idempotent, so is $I - P$, and $P(I - P) = 0$.
   (d) If P is an idempotent, so is P^T.
   ◆**(e)** If P is an idempotent, so is $Q = P + AP - PAP$ for any square matrix A (of the same size as P).
   (f) If A is $n \times m$ and B is $m \times n$, and if $AB = I_n$, then BA is an idempotent.

36. Let A and B be $n \times n$ **diagonal matrices** (all entries off the main diagonal are zero).
   (a) Show that AB is diagonal and $AB = BA$.
   (b) Formulate a rule for calculating XA if X is $m \times n$.
   (c) Formulate a rule for calculating AY if Y is $n \times k$.

37. If A and B are $n \times n$ matrices, show that:
   (a) $AB = BA$ if and only if $(A + B)^2 = A^2 + 2AB + B^2$.
   ◆**(b)** $AB = BA$ if and only if $(A + B)(A - B) = (A - B)(A + B)$.
   (c) $AB = BA$ if and only if $A^TB^T = B^TA^T$.

38. Prove the following parts of Theorem 1.
   (a) Part 1
   ◆**(b)** Part 2
   (c) Part 4
   (d) Part 5

Section 2.3   Matrix Inverses

Three basic operations on matrices, addition, multiplication, and subtraction, are analogues for matrices of the same operations for numbers. In this section we introduce the matrix analogue of numerical division.

To begin, consider how a numerical equation

$$ax = b$$

is solved when a and b are known numbers. If $a = 0$, there is no solution (unless $b = 0$). But if $a \neq 0$, we can multiply both sides by the inverse a^{-1} to obtain the solution $x = a^{-1}b$. This multiplication by a^{-1} is commonly called dividing by a, and the property of a^{-1} that makes this work is that $a^{-1}a = 1$. Moreover, we saw in Section 2.2 that the role that 1 plays in arithmetic is played in matrix algebra by the identity matrix I. This suggests the following definition.

DEFINITION

If A is a square matrix, a matrix B is called an **inverse** of A if and only if

$$AB = I \qquad \text{and} \qquad BA = I$$

A matrix A that has an inverse is called an **invertible matrix**.[4]

EXAMPLE 1

Show that $B = \begin{bmatrix} -1 & 1 \\ 1 & 0 \end{bmatrix}$ is an inverse of $A = \begin{bmatrix} 0 & 1 \\ 1 & 1 \end{bmatrix}$.

Solution   Compute AB and BA.

$$AB = \begin{bmatrix} 0 & 1 \\ 1 & 1 \end{bmatrix} \begin{bmatrix} -1 & 1 \\ 1 & 0 \end{bmatrix} = \begin{bmatrix} 1 & 0 \\ 0 & 1 \end{bmatrix} \qquad BA = \begin{bmatrix} -1 & 1 \\ 1 & 0 \end{bmatrix} \begin{bmatrix} 0 & 1 \\ 1 & 1 \end{bmatrix} = \begin{bmatrix} 1 & 0 \\ 0 & 1 \end{bmatrix}$$

Hence $AB = I = BA$, so B is indeed an inverse of A.   ◆◆◆

EXAMPLE 2

Show that $A = \begin{bmatrix} 0 & 0 \\ 1 & 3 \end{bmatrix}$ has no inverse.

Solution   Let $B = \begin{bmatrix} a & b \\ c & d \end{bmatrix}$ denote an arbitrary 2×2 matrix. Then

$$AB = \begin{bmatrix} 0 & 0 \\ 1 & 3 \end{bmatrix} \begin{bmatrix} a & b \\ c & d \end{bmatrix} = \begin{bmatrix} 0 & 0 \\ a + 3c & b + 3d \end{bmatrix}$$

so AB has a row of zeros. Hence AB cannot equal I for any B.

◆◆◆

Example 2 shows that *it is possible for a nonzero matrix to have no inverse*. But if a matrix *does* have an inverse, it has only one.

---

[4]No nonsquare matrix is invertible (see Exercise 24§2.4).

THEOREM 1

If B and C are both inverses of A, then $B = C$.

Proof Since B and C are both inverses of A, $CA = I = AB$. Hence $B = IB = (CA)B = C(AB) = CI = C$. ◆

If A is an invertible matrix, the (unique) inverse of A is denoted as A^{-1}. Hence A^{-1} (when it exists) is a square matrix of the same size as A with the property that

$$AA^{-1} = I \qquad \text{and} \qquad A^{-1}A = I$$

These equations characterize A^{-1} in the following sense: If somehow a matrix B can be found such that $AB = I = BA$, then A is invertible and B is the inverse of A; in symbols, $B = A^{-1}$. This gives us a way of verifying that the inverse of a matrix exists. Examples 3 and 4 offer illustrations.

EXAMPLE 3

If $A = \begin{bmatrix} 0 & -1 \\ 1 & -1 \end{bmatrix}$, show that $A^3 = I$ and so find A^{-1}.

Solution We have $A^2 = \begin{bmatrix} 0 & -1 \\ 1 & -1 \end{bmatrix}\begin{bmatrix} 0 & -1 \\ 1 & -1 \end{bmatrix} = \begin{bmatrix} -1 & 1 \\ -1 & 0 \end{bmatrix}$, and so

$$A^3 = A^2A = \begin{bmatrix} -1 & 1 \\ -1 & 0 \end{bmatrix}\begin{bmatrix} 0 & -1 \\ 1 & -1 \end{bmatrix} = \begin{bmatrix} 1 & 0 \\ 0 & 1 \end{bmatrix} = I$$

Hence $A^3 = I$, as asserted. This can be written as $A^2A = I = AA^2$, so it shows that A^2 is the inverse of A. That is, $A^{-1} = A^2 = \begin{bmatrix} -1 & 1 \\ -1 & 0 \end{bmatrix}$.

◆◆◆

The following example gives a useful formula for the inverse of a 2×2 matrix.

EXAMPLE 4

If $A = \begin{bmatrix} a & b \\ c & d \end{bmatrix}$ and $ad - bc \neq 0$, show that $A^{-1} = \dfrac{1}{ad - bc}\begin{bmatrix} d & -b \\ -c & a \end{bmatrix}$

Solution   We verify that $AA^{-1} = I$ and leave $A^{-1}A = I$ to the reader.

$$\begin{bmatrix} a & b \\ c & d \end{bmatrix} \left(\frac{1}{ad-bc} \begin{bmatrix} d & -b \\ -c & a \end{bmatrix} \right) = \frac{1}{ad-bc} \begin{bmatrix} a & b \\ c & d \end{bmatrix} \begin{bmatrix} d & -b \\ -c & a \end{bmatrix}$$

$$= \frac{1}{ad-bc} \begin{bmatrix} ad-bc & 0 \\ 0 & ad-bc \end{bmatrix} = I$$

$\blacklozenge\blacklozenge\blacklozenge$

Matrix inverses can be used to solve certain systems of linear equations. Recall (Example 7§2.2) that a *system* of linear equations can be written as a *single* matrix equation

$$AX = B$$

where A and B are known matrices and X is to be determined. If A is invertible, we multiply each side of the equation on the left by A^{-1} to get

$$A^{-1}AX = A^{-1}B$$
$$IX = A^{-1}B$$
$$X = A^{-1}B$$

This gives the solution to the system of equations (the reader should verify that $X = A^{-1}B$ really does satisfy $AX = B$). Furthermore, the argument shows that if X is *any* solution, then necessarily $X = A^{-1}B$, so the solution is unique. Of course the technique works only when the coefficient matrix A has an inverse. This proves Theorem 2.

THEOREM 2

Suppose a system of n equations in n variables is written in matrix form as

$$AX = B$$

If the $n \times n$ coefficient matrix A is invertible, the system has the unique solution

$$X = A^{-1}B$$

EXAMPLE 5

If $A = \begin{bmatrix} 1 & -2 & 2 \\ 2 & 1 & 1 \\ 1 & 0 & 1 \end{bmatrix}$, show that $A^{-1} = \begin{bmatrix} 1 & 2 & -4 \\ -1 & -1 & 3 \\ -1 & -2 & 5 \end{bmatrix}$, and use it to solve the following system of linear equations.

$$x_1 - 2x_2 + 2x_3 = 3$$
$$2x_1 + x_2 + x_3 = 0$$
$$x_1 \qquad + x_3 = -2$$

Solution   Verification that $AA^{-1} = I$ and $A^{-1}A = I$ is left to the reader. The matrix form of the system of equations is $AX = B$, where A is as before and

$$X = \begin{bmatrix} x_1 \\ x_2 \\ x_3 \end{bmatrix} \qquad B = \begin{bmatrix} 3 \\ 0 \\ -2 \end{bmatrix}$$

Theorem 2 gives the solution

$$X = A^{-1}B = \begin{bmatrix} 1 & 2 & -4 \\ -1 & -1 & 3 \\ -1 & -2 & 5 \end{bmatrix} \begin{bmatrix} 3 \\ 0 \\ -2 \end{bmatrix} = \begin{bmatrix} 11 \\ -9 \\ -13 \end{bmatrix}$$

Thus, $x_1 = 11$, $x_2 = -9$, and $x_3 = -13$.

◆◆◆

An Inversion Method

Given a particular $n \times n$ matrix A, it is desirable to have an efficient technique to determine whether A has an inverse and, if so, to find that inverse. For simplicity, we shall derive the technique for 2×2 matrices; the $n \times n$ case is entirely analogous.

Given the invertible 2×2 matrix A, we determine A^{-1} from the equation $AA^{-1} = I$. Write

$$A^{-1} = \begin{bmatrix} x_1 & x_2 \\ y_1 & y_2 \end{bmatrix}$$

where x_1, y_1, x_2, and y_2 are to be determined. Equating columns in the equation $AA^{-1} = I$ gives

$$A\begin{bmatrix} x_1 \\ y_1 \end{bmatrix} = \begin{bmatrix} 1 \\ 0 \end{bmatrix} \qquad \text{and} \qquad A\begin{bmatrix} x_2 \\ y_2 \end{bmatrix} = \begin{bmatrix} 0 \\ 1 \end{bmatrix}$$

These are systems of linear equations, each with A as coefficient matrix. Since A is invertible, each system has a unique solution by Theorem 2. But this means there is a sequence of elementary row operations carrying A to the 2×2 identity matrix I. This sequence carries the augmented matrices of both systems to reduced row-echelon form and so solves the systems:

$$\begin{bmatrix} A & \Big| & 1 \\ & & 0 \end{bmatrix} \to \begin{bmatrix} I & \Big| & x_1 \\ & & y_1 \end{bmatrix} \qquad \begin{bmatrix} A & \Big| & 0 \\ & & 1 \end{bmatrix} \to \begin{bmatrix} I & \Big| & x_2 \\ & & y_2 \end{bmatrix}$$

Hence we can do *both* calculations simultaneously.

$$\left[A \;\middle|\; \begin{matrix} 1 & 0 \\ 0 & 1 \end{matrix} \right] \rightarrow \left[I \;\middle|\; \begin{matrix} x_1 & x_2 \\ y_1 & y_2 \end{matrix} \right]$$

This can be written more compactly as follows:

$$\begin{bmatrix} A & I \end{bmatrix} \rightarrow \begin{bmatrix} I & A^{-1} \end{bmatrix}$$

In other words, the sequence of row operations that carries A to I also carries I to A^{-1}. This is the desired algorithm.

MATRIX INVERSION ALGORITHM

If A is a (square) invertible matrix, there exists a sequence of elementary row operations that carry A to the identity matrix I of the same size, written $A \rightarrow I$. This same series of row operations carries I to A^{-1}; that is, $I \rightarrow A^{-1}$. The algorithm can be summarized as follows:

$$\begin{bmatrix} A & I \end{bmatrix} \rightarrow \begin{bmatrix} I & A^{-1} \end{bmatrix}$$

where the row operations on A and I are carried out simultaneously.

EXAMPLE 6

Use the inversion algorithm to find the inverse of the matrix

$$A = \begin{bmatrix} 2 & 7 & 1 \\ 1 & 4 & -1 \\ 1 & 3 & 0 \end{bmatrix}$$

Solution   Apply elementary row operations to the double matrix

$$\begin{bmatrix} A & I \end{bmatrix} = \left[\begin{array}{ccc|ccc} 2 & 7 & 1 & 1 & 0 & 0 \\ 1 & 4 & -1 & 0 & 1 & 0 \\ 1 & 3 & 0 & 0 & 0 & 1 \end{array} \right]$$

so as to carry A to I. First interchange rows 1 and 2.

$$\left[\begin{array}{ccc|ccc} 1 & 4 & -1 & 0 & 1 & 0 \\ 2 & 7 & 1 & 1 & 0 & 0 \\ 1 & 3 & 0 & 0 & 0 & 1 \end{array} \right]$$

Next subtract 2 times row 1 from row 2, and subtract row 1 from row 3.

$$\begin{bmatrix} 1 & 4 & -1 & | & 0 & 1 & 0 \\ 0 & -1 & 3 & | & 1 & -2 & 0 \\ 0 & -1 & 1 & | & 0 & -1 & 1 \end{bmatrix}$$

Continue to reduced row-echelon form.

$$\begin{bmatrix} 1 & 0 & 11 & | & 4 & -7 & 0 \\ 0 & 1 & -3 & | & -1 & 2 & 0 \\ 0 & 0 & -2 & | & -1 & 1 & 1 \end{bmatrix}$$

$$\begin{bmatrix} 1 & 0 & 0 & | & \frac{-3}{2} & \frac{-3}{2} & \frac{11}{2} \\ 0 & 1 & 0 & | & \frac{1}{2} & \frac{1}{2} & \frac{-3}{2} \\ 0 & 0 & 1 & | & \frac{1}{2} & \frac{-1}{2} & \frac{-1}{2} \end{bmatrix}$$

Hence $A^{-1} = \frac{1}{2}\begin{bmatrix} -3 & -3 & 11 \\ 1 & 1 & -3 \\ 1 & -1 & -1 \end{bmatrix}$, as is readily verified.

◆◆◆

Given any $n \times n$ matrix A, Theorem 1§1.2[5] shows that A can be carried by elementary row operations to a matrix R in reduced row-echelon form. If $R = I$, the matrix A is invertible (this will be proved in the next section), so the algorithm produces A^{-1}. If $R \neq I$, then R has a row of zeros (it is square), so no system of linear equations $AX = B$ can have a unique solution. But then A is not invertible by Theorem 2. Hence the algorithm is effective in the sense conveyed in Theorem 3.

THEOREM 3

If A is an $n \times n$ matrix, either A can be reduced to I by elementary row operations or it cannot. In the first case, the algorithm produces A^{-1}; in the second case, A^{-1} does not exist.

Properties of Inverses

Sometimes the inverse of a matrix is given by a formula. Example 4 is one illustration, Examples 7 and 8 provide two more.

---

[5]This means Theorem 1 in Section 1.2.

EXAMPLE 7

If A is an invertible matrix, show that the transpose A^T is also invertible. Show further that the inverse of A^T is just the transpose of A^{-1}; in symbols, $(A^T)^{-1} = (A^{-1})^T$.

Solution A^{-1} exists (by assumption). Its transpose $(A^{-1})^T$ is the candidate proposed for the inverse of A^T. We test it as follows:

$$A^T(A^{-1})^T = (A^{-1}A)^T = I^T = I$$
$$(A^{-1})^TA^T = (AA^{-1})^T = I^T = I$$

Hence $(A^{-1})^T$ is indeed the inverse of A^T; that is, $(A^T)^{-1} = (A^{-1})^T$.   ◆◆◆

EXAMPLE 8

If A and B are invertible $n \times n$ matrices, show that their product AB is also invertible and $(AB)^{-1} = B^{-1}A^{-1}$.

Solution   We are given a candidate for the inverse of AB, namely $B^{-1}A^{-1}$. We test it as follows:

$$(B^{-1}A^{-1})(AB) = B^{-1}(A^{-1}A)B = B^{-1}IB = B^{-1}B = I$$
$$(AB)(B^{-1}A^{-1}) = A(BB^{-1})A^{-1} = AIA^{-1} = AA^{-1} = I$$

Hence $B^{-1}A^{-1}$ is the inverse of AB; in symbols, $(AB)^{-1} = B^{-1}A^{-1}$.   ◆◆◆

We now collect several basic properties of matrix inverses for reference.

THEOREM 4

All the following matrices are square matrices of the same size.

1. I is invertible and $I^{-1} = I$.
2. If A is invertible, so is A^{-1}, and $(A^{-1})^{-1} = A$.
3. If A and B are invertible, so is AB, and $(AB)^{-1} = B^{-1}A^{-1}$.
4. If $A_1, A_2, \ldots, A_k$ are all invertible, so is their product $A_1A_2 \cdots A_k$, and $(A_1A_2 \cdots A_k)^{-1} = A_k^{-1} \cdots A_2^{-1}A_1^{-1}$.
5. If A is invertible, so is A^k for $k \geq 1$, and $(A^k)^{-1} = (A^{-1})^k$.
6. If A is invertible and $a \neq 0$ is a number, then aA is invertible and $(aA)^{-1} = \frac{1}{a}A^{-1}$.
7. If A is invertible, so is its transpose A^T, and $(A^T)^{-1} = (A^{-1})^T$.

Proof

1. This is an immediate consequence of the formula $I^2 = I$.
2. The equations $AA^{-1} = I = A^{-1}A$ show that A is the inverse of A^{-1}; in symbols, $(A^{-1})^{-1} = A$.
3. This is Example 8.

4. Use induction on k. If $k = 1$, there is nothing to prove because the conclusion reads $(A_1)^{-1} = A_1^{-1}$. If $k = 2$, the result is just property 3. If $k > 2$, assume inductively that $(A_1 A_2 \cdots A_{k-1})^{-1} = A_{k-1}^{-1} \cdots A_2^{-1} A_1^{-1}$. We apply this fact together with property 3 as follows:

$$\begin{aligned}
\left[A_1 A_2 \cdots A_{k-1} A_k \right]^{-1} &= \left[(A_1 A_2 \cdots A_{k-1}) A_k \right]^{-1} \\
&= A_k^{-1} (A_1 A_2 \cdots A_{k-1})^{-1} \\
&= A_k^{-1} (A_{k-1}^{-1} \cdots A_2^{-1} A_1^{-1})
\end{aligned}$$

Here property 3 is applied to get the second equality. This is the conclusion for k matrices, so the proof by induction is complete.

5. This is property 4 with $A_1 = A_2 = \cdots = A_k = A$.

6. This is left as Exercise 28.

7. This is Example 7.   ◆

EXAMPLE 9

Find A if $(A^T - 2I)^{-1} = \begin{bmatrix} 2 & 1 \\ -1 & 0 \end{bmatrix}$

Solution   By Theorem 4(2) and Example 4

$$(A^T - 2I) = \left[(A^T - 2I)^{-1} \right]^{-1} = \begin{bmatrix} 2 & 1 \\ -1 & 0 \end{bmatrix}^{-1} = \begin{bmatrix} 0 & -1 \\ 1 & 2 \end{bmatrix}$$

Hence $A^T = 2I + \begin{bmatrix} 0 & -1 \\ 1 & 2 \end{bmatrix} = \begin{bmatrix} 2 & 0 \\ 0 & 2 \end{bmatrix} + \begin{bmatrix} 0 & -1 \\ 1 & 2 \end{bmatrix}$, so $A = \begin{bmatrix} 2 & 1 \\ -1 & 4 \end{bmatrix}$.   ◆◆◆

The reversal of the order of the inverses in properties 3 and 4 of Theorem 4 is a consequence of the fact that matrix multiplication is not commutative. Another manifestation of this comes when matrix equations are dealt with. If a matrix equation $B = C$ is given, it can be *left-multiplied* by a matrix A to yield $AB = AC$. Similarly, *right-multiplication* gives $BA = CA$. However, we cannot mix the two: If $B = C$, it need *not* be the case that $AB = CA$. Examples 10 and 11 illustrate how such manipulations are used.

EXAMPLE 10
Cancellation Laws

Let A be an invertible matrix. Show that:

(a) If $AB = AC$, then $B = C$.

(b) If $BA = CA$, then $B = C$.

Solution   Given the equation $AB = AC$, left-multiply both sides by A^{-1} to obtain $A^{-1}AB = A^{-1}AC$. This gives $IB = IC$ — that is, $B = C$. This proves part (a), and the proof of part (b) is similar.

♦♦♦

One application of cancellation is as follows: If A is invertible, then the only matrix X such that $AX = 0$ is $X = 0$. This follows directly from Example 10 because $AX = 0$ can be written $AX = A0$. (Alternatively, left-multiply $AX = 0$ by A^{-1} to get $X = A^{-1}(AX) = A^{-1}0 = 0$.) Of course, $YA = 0$ implies $Y = 0$ in the same way, and these facts give a useful method of showing that a matrix is *not* invertible.

EXAMPLE 11

Show that $A = \begin{bmatrix} 6 & 8 \\ 3 & 4 \end{bmatrix}$ has no inverse.

Solution   Observe that $AX = 0$ if $X = \begin{bmatrix} 4 \\ -3 \end{bmatrix}$. Hence if A had an inverse, it would mean $X = 0$, which is not the case. So A has no inverse.

♦♦♦

EXERCISES 2.3

1. In each case, show that the matrices are inverses of each other.

(a) $\begin{bmatrix} 3 & 5 \\ 1 & 2 \end{bmatrix}, \begin{bmatrix} 2 & -5 \\ -1 & 3 \end{bmatrix}$

(b) $\begin{bmatrix} 3 & 0 \\ 1 & -4 \end{bmatrix}, \frac{1}{12}\begin{bmatrix} 4 & 0 \\ 1 & -3 \end{bmatrix}$

(c) $\begin{bmatrix} 1 & 2 & 0 \\ 0 & 2 & 3 \\ 1 & 3 & 1 \end{bmatrix}, \begin{bmatrix} 7 & 2 & -6 \\ -3 & -1 & 3 \\ 2 & 1 & -2 \end{bmatrix}$

(d) $\begin{bmatrix} 3 & 0 \\ 0 & 5 \end{bmatrix}, \begin{bmatrix} \frac{1}{3} & 0 \\ 0 & \frac{1}{5} \end{bmatrix}$

♦**(j)** $\begin{bmatrix} -1 & 4 & 5 & 2 \\ 0 & 0 & 0 & -1 \\ 1 & -2 & -2 & 0 \\ 0 & -1 & -1 & 0 \end{bmatrix}$

(k) $\begin{bmatrix} 1 & -1 & 5 & 2 \\ 1 & 0 & 7 & 5 \\ 0 & 1 & 3 & 6 \\ 1 & -1 & 5 & 1 \end{bmatrix}$

♦**(l)** $\begin{bmatrix} 1 & 2 & 0 & 0 & 0 \\ 0 & 1 & 3 & 0 & 0 \\ 0 & 0 & 1 & 5 & 0 \\ 0 & 0 & 0 & 1 & 7 \\ 0 & 0 & 0 & 0 & 1 \end{bmatrix}$

2. Find the inverse of each of the following matrices.

(a) $\begin{bmatrix} 3 & -1 \\ -3 & 2 \end{bmatrix}$

♦**(b)** $\begin{bmatrix} 4 & 1 \\ 3 & 2 \end{bmatrix}$

(c) $\begin{bmatrix} 1 & -1 & 0 \\ 3 & 0 & 2 \\ -1 & 0 & -1 \end{bmatrix}$

♦**(d)** $\begin{bmatrix} 1 & -1 & 2 \\ -5 & 7 & -11 \\ -2 & 3 & -5 \end{bmatrix}$

(e) $\begin{bmatrix} 3 & 5 & 0 \\ 1 & 2 & 1 \\ 3 & 7 & 1 \end{bmatrix}$

♦**(f)** $\begin{bmatrix} 3 & 1 & -1 \\ 2 & 1 & 0 \\ 1 & 5 & -1 \end{bmatrix}$

(g) $\begin{bmatrix} 2 & 3 & 4 \\ 4 & 3 & 1 \\ 1 & 2 & 4 \end{bmatrix}$

♦**(h)** $\begin{bmatrix} 3 & 1 & -1 \\ 5 & 2 & 0 \\ 1 & 1 & -1 \end{bmatrix}$

(i) $\begin{bmatrix} 2 & 1 & 3 \\ 3 & -1 & 1 \\ 4 & 2 & 1 \end{bmatrix}$

3. In each case, solve the equations by finding the inverse of the coefficient matrix.

(a) $\begin{aligned} 3x - y &= 5 \\ 2x + 3y &= 1 \end{aligned}$

♦**(b)** $\begin{aligned} 2x - 3y &= 0 \\ x - 4y &= 1 \end{aligned}$

(c) $\begin{aligned} x + y + 2z &= 5 \\ x + y + z &= 0 \\ x + 2y + 4z &= -2 \end{aligned}$

♦**(d)** $\begin{aligned} x + 4y + 2z &= 1 \\ 2x + 3y + 3z &= -1 \\ 4x + y + 4z &= 0 \end{aligned}$

(e) $\begin{aligned} x + y \quad - w &= 1 \\ -x + y - z \quad &= -1 \\ y + z + w &= 0 \\ x \quad - z + w &= 1 \end{aligned}$

♦**(f)** $\begin{aligned} x + y + z + w &= 1 \\ x + y \quad &= 0 \\ y \quad + w &= -1 \\ x \quad + w &= 2 \end{aligned}$

4. Given $A^{-1} = \begin{bmatrix} 1 & -1 & 3 \\ 2 & 0 & 5 \\ -1 & 1 & 0 \end{bmatrix}$:

  (a) Solve the system of equations $AX = \begin{bmatrix} 1 \\ -1 \\ 3 \end{bmatrix}$.

  ◆**(b)** Find a matrix B such that $AB = \begin{bmatrix} 1 & -1 & 2 \\ 0 & 1 & 1 \\ 1 & 0 & 0 \end{bmatrix}$.

  (c) Find a matrix C such that $CA = \begin{bmatrix} 1 & 2 & -1 \\ 3 & 1 & 1 \end{bmatrix}$.

5. Find A when:

  (a) $(3A)^{-1} = \begin{bmatrix} 1 & -1 \\ 0 & 1 \end{bmatrix}$ ◆**(b)** $(2A)^T = \begin{bmatrix} 1 & -1 \\ 2 & 3 \end{bmatrix}^{-1}$

  (c) $(I + 2A)^{-1} = \begin{bmatrix} 2 & 0 \\ 1 & 1 \end{bmatrix}$

  ◆**(d)** $(I - 2A^T)^{-1} = \begin{bmatrix} 2 & 1 \\ 1 & 1 \end{bmatrix}$

  (e) $\left(A\begin{bmatrix} 1 & -1 \\ 0 & 1 \end{bmatrix} \right)^{-1} = \begin{bmatrix} 2 & 3 \\ 1 & 2 \end{bmatrix}$

  ◆**(f)** $\left(\begin{bmatrix} 1 & 0 \\ 2 & 1 \end{bmatrix} A \right)^{-1} = \begin{bmatrix} 1 & 0 \\ 2 & 2 \end{bmatrix}$

  (g) $(A^T - 5I)^{-1} = 2\begin{bmatrix} 1 & 1 \\ 2 & 3 \end{bmatrix}$

  ◆**(h)** $(A^{-1} - 2I)^T = -2\begin{bmatrix} 1 & 1 \\ 1 & 0 \end{bmatrix}$

6. Find A when:

  (a) $A^{-1} = \begin{bmatrix} 1 & -1 & 3 \\ 2 & 1 & 1 \\ 0 & 2 & -2 \end{bmatrix}$ ◆**(b)** $A^{-1} = \begin{bmatrix} 0 & 1 & -1 \\ 1 & 2 & 1 \\ 1 & 0 & 1 \end{bmatrix}$

7. Given $\begin{bmatrix} x_1 \\ x_2 \\ x_3 \end{bmatrix} = \begin{bmatrix} 3 & -1 & 2 \\ 1 & 0 & 4 \\ 2 & 1 & 0 \end{bmatrix}\begin{bmatrix} y_1 \\ y_2 \\ y_3 \end{bmatrix}$ and $\begin{bmatrix} z_1 \\ z_2 \\ z_3 \end{bmatrix} =$

$\begin{bmatrix} 1 & -1 & 1 \\ 2 & -3 & 0 \\ -1 & 1 & -2 \end{bmatrix}\begin{bmatrix} y_1 \\ y_2 \\ y_3 \end{bmatrix}$, express the variables x_1, x_2, and

x_3 in terms of z_1, z_2, and z_3.

8. (a) In the system $\begin{array}{r} 3x + 4y = 7 \\ 4x + 5y = 1 \end{array}$, substitute the new

    variables x' and y' given by $\begin{array}{r} x = -5x' + 4y' \\ y = 4x' - 3y' \end{array}$. Then

    find x and y.

  ◆**(b)** Explain part **(a)** by writing the equations as $A\begin{bmatrix} x \\ y \end{bmatrix} = \begin{bmatrix} 7 \\ 1 \end{bmatrix}$ and $\begin{bmatrix} x \\ y \end{bmatrix} = B\begin{bmatrix} x' \\ y' \end{bmatrix}$. What is the relationship between A and B? Generalize.

9. In each case either prove the assertion or give an example showing that it is false.

  (a) If $A \neq 0$ is a square matrix, then A is invertible.

  ◆**(b)** If A and B are both invertible, then $A + B$ is invertible.

  (c) If A and B are both invertible, then $(A^{-1}B)^T$ is invertible.

  ◆**(d)** If $A^4 = 3I$, then A is invertible.

  (e) If $A^2 = A$ and $A \neq 0$, then A is invertible.

  ◆**(f)** If $AB = B$ for some $B \neq 0$, then A is invertible.

  (g) If A is invertible and skew symmetric ($A^T = -A$), the same is true of A^{-1}.

◆**10.** If A, B, and C are square matrices and $AB = I = CA$, show that A is invertible and $B = C = A^{-1}$.

11. Suppose $CA = I_m$, where C is $m \times n$ and A is $n \times m$. Consider the system $AX = B$ of n equations in m variables.

  (a) Show that this system has a unique solution CB if it is consistent.

  ◆**(b)** If $C = \begin{bmatrix} 0 & -5 & 1 \\ 3 & 0 & -1 \end{bmatrix}$ and $A = \begin{bmatrix} 2 & -3 \\ 1 & -2 \\ 6 & -10 \end{bmatrix}$, find X

    (if it exists) when **(i)** $B = \begin{bmatrix} 1 \\ 0 \\ 3 \end{bmatrix}$; and **(ii)** $B = \begin{bmatrix} 7 \\ 4 \\ 22 \end{bmatrix}$

12. Verify that $A = \begin{bmatrix} 1 & -1 \\ 0 & 2 \end{bmatrix}$ satisfies $A^2 - 3A + 2I = 0$, and use this fact to show that $A^{-1} = \frac{1}{2}(3I - A)$.

13. Let $Q = \begin{bmatrix} a & -b & -c & -d \\ b & a & -d & c \\ c & d & a & -b \\ d & -c & b & a \end{bmatrix}$. Compute QQ^T and so find Q^{-1}.

14. Let $U = \begin{bmatrix} 0 & 1 \\ 1 & 0 \end{bmatrix}$. Show that each of U, $-U$, and $-I_2$ is its own inverse and that the product of any two of these is the third.

15. Consider $A = \begin{bmatrix} 1 & 1 \\ -1 & 0 \end{bmatrix}$, $B = \begin{bmatrix} 0 & -1 \\ 1 & 0 \end{bmatrix}$, $C = \begin{bmatrix} 0 & 1 & 0 \\ 0 & 0 & 1 \\ 5 & 0 & 0 \end{bmatrix}$.

Find the inverses by computing **(a)** A^6; ◆ **(b)** B^4; and **(c)** C^3.

16. In each case, find A^{-1} in terms of c.

  (a) $\begin{bmatrix} c & 1 \\ -1 & c \end{bmatrix}$ ◆**(b)** $\begin{bmatrix} 2 & -c \\ c & 3 \end{bmatrix}$

  (c) $\begin{bmatrix} 1 & c & 0 \\ c & -1 & c \\ 2 & c & 1 \end{bmatrix}$ ◆**(d)** $\begin{bmatrix} 1 & 0 & 1 \\ c & 1 & c \\ 3 & c & 2 \end{bmatrix}$

17. If $c \neq 0$, find the inverse of $\begin{bmatrix} 1 & -1 & 1 \\ 2 & -1 & 2 \\ 0 & 2 & c \end{bmatrix}$ in terms of c.

♦ **18.** Find the inverse of $\begin{bmatrix} \sin\theta & \cos\theta \\ -\cos\theta & \sin\theta \end{bmatrix}$ for any real number θ.

19. Show that A has no inverse when **(a)** A has a row of zeros; ♦ **(b)** A has a column of zeros; **(c)** each row of A sums to 0; ♦ **(d)** each column of A sums to 0.

20. Let A denote a square matrix.

　(a) Let $YA = 0$ for some matrix $Y \neq 0$. Show that A has no inverse.

　(b) Use part **(a)** to show that **(i)** $\begin{bmatrix} 1 & -1 & 1 \\ 0 & 1 & 1 \\ 1 & 0 & 2 \end{bmatrix}$; and

　♦ **(ii)** $\begin{bmatrix} 2 & 1 & -1 \\ 1 & 1 & 0 \\ 1 & 0 & -1 \end{bmatrix}$ have no inverse. [*Hint:* For part **(ii)** compare row 3 with the difference between row 1 and row 2.]

21. If A is invertible, show that **(a)** $A^2 \neq 0$; **(b)** $A^k \neq 0$ for all $k = 1, 2, \ldots$.

22. Suppose $AB = 0$, where A and B are square matrices. Show that:

　(a) If one of A and B has an inverse, the other is zero.

　♦ **(b)** It is impossible for both A and B to have inverses.

　(c) $(BA)^2 = 0$

23. **(a)** Show that $\begin{bmatrix} a & 0 \\ 0 & b \end{bmatrix}$ is invertible if and only if $a \neq 0$ and $b \neq 0$. Describe the inverse.

　(b) Show that a diagonal matrix is invertible if and only if all the main diagonal entries are nonzero. Describe the inverse.

　(c) If A and B are square matrices, show that **(i)** the block matrix $\begin{bmatrix} A & 0 \\ 0 & B \end{bmatrix}$ is invertible if and only if A and B are both invertible; and **(ii)** $\begin{bmatrix} A & 0 \\ 0 & B \end{bmatrix}^{-1} = \begin{bmatrix} A^{-1} & 0 \\ 0 & B^{-1} \end{bmatrix}$

　(d) Use part **(c)** to find the inverses of:

　　(i) $\begin{bmatrix} 1 & 0 & 0 \\ 0 & 2 & -1 \\ 0 & 1 & -1 \end{bmatrix}$　♦ **(ii)** $\begin{bmatrix} 3 & 1 & 0 \\ 5 & 2 & 0 \\ 0 & 0 & -1 \end{bmatrix}$

　　(iii) $\begin{bmatrix} 2 & 1 & 0 & 0 \\ 1 & 1 & 0 & 0 \\ 0 & 0 & 1 & -1 \\ 0 & 0 & 1 & -2 \end{bmatrix}$　♦ **(iv)** $\begin{bmatrix} 3 & 4 & 0 & 0 \\ 2 & 3 & 0 & 0 \\ 0 & 0 & 1 & 3 \\ 0 & 0 & 0 & -1 \end{bmatrix}$

　(e) Extend part **(c)** to **block diagonal matrices** — that is, matrices with square blocks down the main diagonal and zero blocks elsewhere.

24. **(a)** Show that $\begin{bmatrix} a & x \\ 0 & b \end{bmatrix}$ is invertible if and only if $a \neq 0$ and $b \neq 0$.

　♦ **(b)** If A and B are square and invertible, show that **(i)** the block matrix $\begin{bmatrix} A & X \\ 0 & B \end{bmatrix}$ is invertible for any X; and

　(ii) $\begin{bmatrix} A & X \\ 0 & B \end{bmatrix}^{-1} = \begin{bmatrix} A^{-1} & -A^{-1}XB^{-1} \\ 0 & B^{-1} \end{bmatrix}$.

　(c) Use part **(b)** to invert **(i)** $\begin{bmatrix} 1 & 2 & 1 & 1 \\ 0 & -1 & -1 & 0 \\ 0 & 0 & 2 & 1 \\ 0 & 0 & 1 & 1 \end{bmatrix}$; and

　♦ **(ii)** $\begin{bmatrix} 3 & 1 & 3 & 0 \\ 2 & 1 & -1 & 1 \\ 0 & 0 & 5 & 2 \\ 0 & 0 & 3 & 1 \end{bmatrix}$.

25. If A and B are invertible symmetric matrices such that $AB = BA$, show that A^{-1}, AB, AB^{-1}, and $A^{-1}B^{-1}$ are also invertible and symmetric.

26. **(a)** Let $A = \begin{bmatrix} 1 & 1 \\ 0 & 1 \end{bmatrix}$, $B = \begin{bmatrix} 0 & 0 \\ 1 & 2 \end{bmatrix}$, and $C = \begin{bmatrix} 1 & 1 \\ 1 & 1 \end{bmatrix}$. Verify that $AB = CA$ and that A is invertible but $B \neq C$. (Compare with Example 10.)

　♦ **(b)** Find 2×2 matrices P, Q, and R such that $PQ = PR$, P is not invertible, and $Q \neq R$. (Compare with Example 10.)

27. Let A be an $n \times n$ matrix and let I be the $n \times n$ identity matrix.

　(a) If $A^2 = 0$, verify that $(I - A)^{-1} = I + A$.

　(b) If $A^3 = 0$, verify that $(I - A)^{-1} = I + A + A^2$.

　(c) Using part **(b)**, find the inverse of $\begin{bmatrix} 1 & 2 & -1 \\ 0 & 1 & 3 \\ 0 & 0 & 1 \end{bmatrix}$.

　♦ **(d)** If $A^n = 0$, find the formula for $(I - A)^{-1}$.

28. Prove property 6 of Theorem 4: If A is invertible and $a \neq 0$, then aA is invertible and $(aA)^{-1} = \frac{1}{a}A^{-1}$.

29. Let A, B, and C denote $n \times n$ matrices. Show that:

　(a) If A and AB are both invertible, B is invertible.

　♦ **(b)** If AB and BA are both invertible, A and B are both invertible. [*Hint:* See Exercise 10.]

　(c) If A, C, and ABC are all invertible, B is invertible.

30. Let A and B denote invertible $n \times n$ matrices.

　(a) If $A^{-1} = B^{-1}$, does it mean that $A = B$? Explain.

　♦ **(b)** Show that $A = B$ if and only if $A^{-1}B = I$.

31. Let A, B, and C be $n \times n$ matrices, with A and B invertible. Show that:

    ◆**(a)** If A commutes with C, then A^{-1} commutes with C.

    (b) If A commutes with B, then A^{-1} commutes with B^{-1}.

32. Let A and B be square matrices of the same size.

    (a) Show that $(AB)^2 = A^2B^2$ if $AB = BA$.

    ◆**(b)** If A and B are invertible and $(AB)^2 = A^2B^2$, show that $AB = BA$.

    (c) If $A = \begin{bmatrix} 1 & 0 \\ 0 & 0 \end{bmatrix}$ and $B = \begin{bmatrix} 1 & 1 \\ 0 & 0 \end{bmatrix}$, show that $(AB)^2 = A^2B^2$ but $AB \neq BA$.

33. If $U^2 = I$, show that $I + U$ is not invertible unless $U = I$.

34. **(a)** If J is the 4×4 matrix with every entry 1, show that $I - \frac{1}{2}J$ is self-inverse and symmetric.

    (b) If X is $n \times m$ and satisfies $X^TX = I_m$, show that $I_n - 2XX^T$ is self-inverse and symmetric.

35. An $n \times n$ matrix P is called an idempotent if $P^2 = P$. Show that:

    (a) I is the only invertible idempotent.

    ◆**(b)** P is an idempotent if and only if $I - 2P$ is self-inverse.

    (c) U is self-inverse if and only if $U = I - 2P$ for some idempotent P.

    (d) $I - aP$ is invertible for any $a \neq 1$, and $(I - aP)^{-1} = I + \left(\dfrac{a}{1-a} \right) P$.

36. If $A^2 = kA$, where $k \neq 0$, show that A is invertible if and only if $A = kI$.

37. Let A and B denote $n \times n$ invertible matrices.

    (a) Show that $A^{-1} + B^{-1} = A^{-1}(A + B)B^{-1}$.

    ◆**(b)** If $A + B$ is also invertible, show that $A^{-1} + B^{-1}$ is invertible and find a formula for $(A^{-1} + B^{-1})^{-1}$.

38. Let A and B be $n \times n$ matrices, and let I be the $n \times n$ identity matrix.

    (a) Verify that $A(I + BA) = (I + AB)A$ and that $(I + BA)B = B(I + AB)$.

    (b) If $I + AB$ is invertible, verify that $I + BA$ is also invertible and that $(I + BA)^{-1} = I - B(I + AB)^{-1}A$.

Section 2.4   Elementary Matrices

It is now evident that elementary row operations play a fundamental role in linear algebra by providing a general method for solving systems of linear equations. This leads to the matrix inversion algorithm. It turns out that these elementary row operations can be performed by left-multiplication by certain invertible matrices (called elementary matrices). This section is devoted to a discussion of this useful fact and some of its consequences.

Recall that the elementary row operations are of three types:

    Type I: Interchange two rows.

    Type II: Multiply a row by a nonzero number.

    Type III: Add a multiple of a row to a different row.

> **DEFINITION**
>
> An $n \times n$ matrix is called an **elementary matrix** if it is obtained from the $n \times n$ identity matrix by an elementary row operation.

The elementary matrix so constructed is said to be of type I, II, or III when the corresponding row operation is of type I, II, or III.

EXAMPLE 1

Verify that $E_1 = \begin{bmatrix} 0 & 1 & 0 \\ 1 & 0 & 0 \\ 0 & 0 & 1 \end{bmatrix}$, $E_2 = \begin{bmatrix} 1 & 0 & 0 \\ 0 & 1 & 0 \\ 0 & 0 & 9 \end{bmatrix}$, and $E_3 = \begin{bmatrix} 1 & 0 & 5 \\ 0 & 1 & 0 \\ 0 & 0 & 1 \end{bmatrix}$, are elementary matrices.

Solution E_1 is obtained from the 3×3 identity I_3 by interchanging the first two rows, so it is an elementary matrix of type I. Similarly, E_2 comes from multiplying the third row of I_3 by 9 and so is an elementary matrix of type II. Finally, E_3 is an elementary matrix of type III; it is obtained by adding 5 times the third row of I_3 to the first row.

◆◆◆

Now consider the following three 2×2 elementary matrices E_1, E_2, and E_3 obtained by doing the indicated elementary row operations to I_2.

$$E_1 = \begin{bmatrix} 0 & 1 \\ 1 & 0 \end{bmatrix} \qquad \text{Interchange rows 1 and 2 of } I_2$$

$$E_2 = \begin{bmatrix} 1 & 0 \\ 0 & k \end{bmatrix} \qquad \text{Multiply row 2 of } I_2 \text{ by } k \neq 0.$$

$$E_3 = \begin{bmatrix} 1 & k \\ 0 & 1 \end{bmatrix} \qquad \text{Add } k \text{ times row 2 of } I_2 \text{ to row 1.}$$

If $A = \begin{bmatrix} a & b & c \\ p & q & r \end{bmatrix}$ is any 2×3 matrix, we compute $E_1 A$, $E_2 A$, and $E_3 A$:

$$E_1 A = \begin{bmatrix} 0 & 1 \\ 1 & 0 \end{bmatrix} \begin{bmatrix} a & b & c \\ p & q & r \end{bmatrix} = \begin{bmatrix} p & q & r \\ a & b & c \end{bmatrix}$$

$$E_2 A = \begin{bmatrix} 1 & 0 \\ 0 & k \end{bmatrix} \begin{bmatrix} a & b & c \\ p & q & r \end{bmatrix} = \begin{bmatrix} a & b & c \\ kp & kq & kr \end{bmatrix}$$

$$E_3 A = \begin{bmatrix} 1 & k \\ 0 & 1 \end{bmatrix} \begin{bmatrix} a & b & c \\ p & q & r \end{bmatrix} = \begin{bmatrix} a + kp & b + kq & c + kr \\ p & q & r \end{bmatrix}$$

Observe that $E_1 A$ is the matrix resulting from interchanging rows 1 and 2 of A and that this row operation is the one that was used to produce E_1 from I_2. Similarly, $E_2 A$ is obtained from A by the same row operation that produced E_2 from I_2 (multiplying row 2 by k). Finally, the same is true of $E_3 A$: It is obtained from A by the same operation that produced E_3 from I_2 (adding k times row 2 to row 1). This phenomenon holds for arbitrary $m \times n$ matrices A.

THEOREM 1

Let A denote any $m \times n$ matrix and let E be the $m \times m$ elementary matrix obtained by performing some elementary row operation on the $m \times m$ identity matrix I. If the same elementary row operation is performed on A, the resulting matrix is EA.

Proof We prove it only for E of type III (types I and II are left as Exercise 16). If E is obtained by adding k times row p of I_m to row q, we must show that EA is obtained from A in the same way. Let $R_1, R_2, \ldots, R_m$ and $K_1, K_2, \ldots, K_m$ denote the rows of E and I, respectively. Then, by the definition of E,

$$R_i = K_i \qquad \text{if } i \neq q$$
$$R_q = K_q + kK_p$$

Hence

$$\text{row } i \text{ of } EA = R_iA = K_iA = \text{row } i \text{ of } A \qquad \text{if } i \neq q$$

whereas

$$\text{row } q \text{ of } EA = R_qA = (K_q + kK_p)A = K_qA + kK_pA$$

This is row q of A plus k times row p of A, as required. $\blacklozenge$

EXAMPLE 2

Given $A = \begin{bmatrix} 4 & 1 & 2 & 1 \\ 3 & 0 & 1 & 6 \\ 5 & 7 & 9 & 8 \end{bmatrix}$, find an elementary matrix E such that EA is the result of subtracting 7 times row 1 from row 3.

Solution The elementary matrix is $E = \begin{bmatrix} 1 & 0 & 0 \\ 0 & 1 & 0 \\ -7 & 0 & 1 \end{bmatrix}$, obtained by doing the given row operation to I_3. The product

$$EA = \begin{bmatrix} 1 & 0 & 0 \\ 0 & 1 & 0 \\ -7 & 0 & 1 \end{bmatrix}\begin{bmatrix} 4 & 1 & 2 & 1 \\ 3 & 0 & 1 & 6 \\ 5 & 7 & 9 & 8 \end{bmatrix} = \begin{bmatrix} 4 & 1 & 2 & 1 \\ 3 & 0 & 1 & 6 \\ -23 & 0 & -5 & 1 \end{bmatrix}$$

is indeed the result of applying the operation to A.

Given any elementary row operation, there is another row operation (called its **inverse**) that reverses the effect of the first operation. The inverses are described in the accompanying table.

Type	Operation	Inverse operation
I	Interchange rows p and q	Interchange rows p and q
II	Multiply row p by $c \neq 0$	Multiply row p by $\frac{1}{c}$
III	Add k times row p to row q $(p \neq q)$	Subtract k times row p from row q

Note that type I operations are self-inverse.

THEOREM 2

Every elementary matrix E is invertible, and the inverse is an elementary matrix of the same type. More precisely:

E^{-1} is the elementary matrix obtained from I by the inverse of the row operation that produced E from I.

Proof E is the result of applying a row operation ρ to I. Let E' denote the matrix obtained by applying the inverse operation ρ' to I. By Theorem 1, applying ρ to a matrix A produces EA; then applying ρ' to EA gives $E'(EA)$:

$$A \xrightarrow{\ \rho\ } EA \xrightarrow{\ \rho'\ } E'EA$$

But ρ' reverses the effect of ρ, so applying ρ followed by ρ' does not change A. Hence $E'EA = A$. In particular, taking $A = I$ gives $E'E = I$. A similar argument shows $EE' = I$, so $E^{-1} = E'$ as required. ◆

EXAMPLE 3

Write down the inverses of the elementary matrices E_1, E_2, and E_3 in Example 1.

Solution The matrices are $E_1 = \begin{bmatrix} 0 & 1 & 0 \\ 1 & 0 & 0 \\ 0 & 0 & 1 \end{bmatrix}$, $E_2 = \begin{bmatrix} 1 & 0 & 0 \\ 0 & 1 & 0 \\ 0 & 0 & 9 \end{bmatrix}$ and $E_3 = \begin{bmatrix} 1 & 0 & 5 \\ 0 & 1 & 0 \\ 0 & 0 & 1 \end{bmatrix}$, so they are of types I, II, and III, respectively. Hence Theorem 2 gives

$$E_1^{-1} = E_1 = \begin{bmatrix} 0 & 1 & 0 \\ 1 & 0 & 0 \\ 0 & 0 & 1 \end{bmatrix} \qquad E_2^{-1} = \begin{bmatrix} 1 & 0 & 0 \\ 0 & 1 & 0 \\ 0 & 0 & \frac{1}{9} \end{bmatrix} \qquad E_3^{-1} = \begin{bmatrix} 1 & 0 & -5 \\ 0 & 1 & 0 \\ 0 & 0 & 1 \end{bmatrix}$$

◆◆◆

Now suppose a sequence of elementary row operations are performed on an $m \times n$ matrix A, and let E_1, E_2, . . . , E_k denote the corresponding elementary matrices. Theorem 1 asserts that A is carried to E_1A under the first operation; in symbols, $A \rightarrow E_1A$. Then the second row operation is applied to E_1A (not to A) and the result is $E_2(E_1A)$, again by Theorem 1. Hence the reduction can be described as follows:

$$A \rightarrow E_1A \rightarrow E_2E_1A \rightarrow E_3E_2E_1A \rightarrow \cdots \rightarrow E_k \cdots E_2E_1A$$

In other words, the net effect of the *sequence* of elementary row operations is to left-multiply by the *product* $U = E_k \cdots E_2E_1$ of the corresponding elementary matrices (note the order). The result is

$$A \rightarrow UA \qquad \text{where} \qquad U = E_k \cdots E_2E_1$$

Moreover, the matrix U can be easily constructed. Apply the same sequence of elementary operations to the $n \times n$ identity matrix I in place of A:

$$I \rightarrow UI = U$$

In other words, the sequence of elementary row operations that carries $A \rightarrow UA$ also carries $I \rightarrow U$. Hence it carries the double matrix $[A \quad I]$ to $[UA \quad U]$:

$$[A \quad I] \rightarrow [UA \quad U]$$

just as in the matrix inversion algorithm. This simple observation is surprisingly useful, and we record it as Theorem 3.

THEOREM 3

Let A be an $m \times n$ matrix and assume that A can be carried to a matrix B by elementary row operations. Then:

1. $B = UA$ where U is an invertible $m \times m$ matrix.
2. $U = E_kE_{k-1} \cdots E_2E_1$ where $E_1, E_2, \ldots, E_{k-1}, E_k$ are the elementary matrices corresponding (in order) to the elementary row operations that carry $A \rightarrow B$.
3. U can be constructed without finding the E_i by

$$[A \quad I] \rightarrow [UA \quad U]$$

In other words, the operations that carry $A \rightarrow UA$ also carry $I \rightarrow U$.

Proof All that remains is to verify that U is invertible. Since U is a product of elementary matrices, this follows by Theorem 2.   ◆

EXAMPLE 4

Find the reduced row-echelon form R of $A = \begin{bmatrix} 2 & 3 & 1 \\ 1 & 2 & 1 \end{bmatrix}$ and express it as $R = UA$, where U is invertible.

Solution   Use the usual row reduction $A \rightarrow R$ but carry out $I \rightarrow U$ simultaneously in the format $[A \quad I] \rightarrow [R \quad U]$.

$$\begin{bmatrix} 2 & 3 & 1 & | & 1 & 0 \\ 1 & 2 & 1 & | & 0 & 1 \end{bmatrix} \rightarrow \begin{bmatrix} 1 & 2 & 1 & | & 0 & 1 \\ 2 & 3 & 1 & | & 1 & 0 \end{bmatrix}$$

$$\rightarrow \begin{bmatrix} 1 & 2 & 1 & | & 0 & 1 \\ 0 & -1 & -1 & | & 1 & -2 \end{bmatrix}$$

$$\rightarrow \begin{bmatrix} 1 & 0 & -1 & | & 2 & -3 \\ 0 & 1 & 1 & | & -1 & 2 \end{bmatrix}$$

Hence $R = \begin{bmatrix} 1 & 0 & -1 \\ 0 & 1 & 1 \end{bmatrix}$ and $U = \begin{bmatrix} 2 & -3 \\ -1 & 2 \end{bmatrix}$.

◆◆◆

The next example shows how the factorization of U into elementary matrices (as in property 2 of Theorem 3) can be obtained.

EXAMPLE 5

Bring $A = \begin{bmatrix} 1 & 2 & -1 \\ 1 & 1 & 5 \end{bmatrix}$ to a reduced row-echelon matrix R by elementary row operations and find elementary matrices E_1, E_2, and E_3 such that $R = E_3 E_2 E_1 A$.

Solution   The reduction is as follows:

$$A = \begin{bmatrix} 1 & 2 & -1 \\ 1 & 1 & 5 \end{bmatrix} \rightarrow \begin{bmatrix} 1 & 2 & -1 \\ 0 & -1 & 6 \end{bmatrix} \rightarrow \begin{bmatrix} 1 & 2 & -1 \\ 0 & 1 & -6 \end{bmatrix} \rightarrow \begin{bmatrix} 1 & 0 & 11 \\ 0 & 1 & -6 \end{bmatrix} = R$$

Each row operation can be carried out by left-multiplication by the elementary matrix obtained by performing that row operation on the 2×2 identity matrix. The three reductions and the corresponding elementary matrices are

$$\begin{bmatrix} 1 & 2 & -1 \\ 1 & 1 & 5 \end{bmatrix} = A$$

$$\downarrow$$

$$\begin{bmatrix} 1 & 2 & -1 \\ 0 & -1 & 6 \end{bmatrix} = E_1 A \qquad \text{where } E_1 = \begin{bmatrix} 1 & 0 \\ -1 & 1 \end{bmatrix}$$

$$\downarrow$$

$$\begin{bmatrix} 1 & 2 & -1 \\ 0 & 1 & -6 \end{bmatrix} = E_2(E_1 A) \quad \text{where } E_2 = \begin{bmatrix} 1 & 0 \\ 0 & -1 \end{bmatrix}$$

$$\downarrow$$

$$\begin{bmatrix} 1 & 0 & 11 \\ 0 & 1 & -6 \end{bmatrix} = E_3(E_2 E_1 A) \text{ where } E_3 = \begin{bmatrix} 1 & -2 \\ 0 & 1 \end{bmatrix}$$

This gives $R = E_3 E_2 E_1 A$, as required.

◆◆◆

These techniques are very useful when applied to square matrices. In particular, Theorem 3 provides a way to deduce several conditions for invertibility that will be referred to later. Recall that the rank of a matrix A is the number of nonzero rows in any row-echelon matrix to which it can be carried (Section 1.2).[6]

THEOREM 4

The following conditions are equivalent for an $n \times n$ matrix A:

1. A is invertible.
2. If $YA = 0$ where Y is $1 \times n$, then necessarily $Y = 0$.
3. A has rank n.
4. A can be carried to the $n \times n$ identity matrix by elementary row operations.
5. A is a product of elementary matrices.

Proof   We show that if any of the statements is true, then the next one is necessarily true, and also that the truth of property 5 implies the truth of property 1. Hence, if any statement is true, they all are true.

(1) *implies* (2). Assume A is invertible so that A^{-1} exists. If $YA = 0$, right-multiplication by A^{-1} gives $Y = YI = YAA^{-1} = 0A^{-1} = 0$.

(2) *implies* (3). Assume property 2 holds. Now A can be carried to a matrix R in reduced row-echelon form that, by Theorem 3, can be written as $R = UA$ for some invertible matrix U. We must show that R has n nonzero rows. If not, the last row of R consists of zeros (R is $n \times n$), so $YR = 0$ where $Y = [0 \ \ 0 \ \cdots \ 0 \ \ 1]$. But then $YUA = 0$, so $YU = 0$ by property 2. Because U is invertible, this implies $Y = 0$, a contradiction. Hence R has n nonzero rows, and property 3 follows.

(3) *implies* (4). A can be carried to a matrix R in reduced row-echelon form, and R has n nonzero rows by property 3. Since R is $n \times n$, this means $R = I$, and property 4 follows.

(4) *implies* (5). Given property 4, Theorem 3 implies that $I = UA$, where U is an invertible matrix that can be factored as a product $U = E_k \cdots E_2 E_1$ of elementary matrices. Hence

$$A = U^{-1} = (E_k \cdots E_2 E_1)^{-1} = E_1^{-1} E_2^{-1} \cdots E_k^{-1}$$

and property 5 follows from Theorem 2.

(5) *implies* (1). This follows from Theorem 2 and the fact that the product of invertible matrices is invertible.   ◆

---

[6]The proof that the number of nonzero rows is the same in each row-echelon matrix to which A can be carried will be given in Section 5.5.

EXAMPLE 6

Express $A = \begin{bmatrix} -2 & 3 \\ 1 & 0 \end{bmatrix}$ as a product of elementary matrices.

Solution   We reduce A to I and write the elementary matrix at each stage.

$$\begin{bmatrix} -2 & 3 \\ 1 & 0 \end{bmatrix} = A$$

$$\downarrow$$

$$\begin{bmatrix} 1 & 0 \\ -2 & 3 \end{bmatrix} = E_1A \qquad \text{where } E_1 = \begin{bmatrix} 0 & 1 \\ 1 & 0 \end{bmatrix}$$

$$\downarrow$$

$$\begin{bmatrix} 1 & 0 \\ 0 & 3 \end{bmatrix} = E_2(E_1A) \qquad \text{where } E_2 = \begin{bmatrix} 1 & 0 \\ 2 & 1 \end{bmatrix}$$

$$\downarrow$$

$$\begin{bmatrix} 1 & 0 \\ 0 & 1 \end{bmatrix} = E_3(E_2E_1A) \qquad \text{where } E_3 = \begin{bmatrix} 1 & 0 \\ 0 & \frac{1}{3} \end{bmatrix}$$

Hence $E_3E_2E_1A = I$ and so $A = (E_3E_2E_1)^{-1}$. This means that

$$A = E_1^{-1}E_2^{-1}E_3^{-1} = \begin{bmatrix} 0 & 1 \\ 1 & 0 \end{bmatrix}\begin{bmatrix} 1 & 0 \\ -2 & 1 \end{bmatrix}\begin{bmatrix} 1 & 0 \\ 0 & 3 \end{bmatrix}$$

by Theorem 2. This is the desired factorization.

◆◆◆

We have seen that the matrix products AB and BA need *not* be equal. However, if $AB = I$ where A and B are both square matrices, then necessarily $BA = I$ as well. (This fails if A and B are not both square; see Exercise 12.) The proof requires Theorem 4.

THEOREM 5

Let A and B be $n \times n$ matrices. If $AB = I$, then also $BA = I$, so A and B are invertible, $A = B^{-1}$ and $B = A^{-1}$.

Proof   It suffices to show that A is invertible (then left-multiplying $AB = I$ by A^{-1} gives $B = A^{-1}$). We use property 2 of Theorem 4. If $YA = 0$, then right-multiplication by B gives

$$0 = (YA)B = Y(AB) = YI = Y$$

Hence $YA = 0$ implies $Y = 0$, so A is invertible by Theorem 4.   ◆

Finally, we give two fundamental conditions for invertibility of a square matrix A in terms of systems of equations with A as coefficient matrix.

THEOREM 6

Let A be an $n \times n$ matrix. The following conditions are equivalent.

1. A is invertible.
2. The homogeneous system $AX = 0$ has only the trivial solution $X = 0$.
3. The system $AX = B$ has a solution for every $n \times 1$ matrix B.

Proof   Theorem 2§2.3 shows that (1) implies both (2) and (3), so we show that either (2) or (3) implies that A is invertible. If (2) holds, it suffices to show that A^T is invertible. We use Theorem 4 by verifying that $YA^T = 0$, Y a $1 \times n$ matrix, implies that $Y = 0$. Taking transposes gives $AY^T = (YA^T)^T = 0^T = 0$. Hence $Y^T = 0$ by (2), so $Y = 0$ as required.

Now assume that (3) holds. If B_j denotes column j of I, (3) gives a column X_j such that $AX_j = B_j$. If $X = [X_1 \quad X_2 \quad \cdots \quad X_n]$ is the $n \times n$ matrix with the X_j as columns, then

$$AX = A[X_1 \quad X_2 \quad \cdots \quad X_n] = [AX_1 \quad AX_2 \quad \cdots \quad AX_n] = [B_1 \quad B_2 \quad \cdots \quad B_n] = I$$

Hence A is invertible by Theorem 5.   ◆

EXERCISES 2.4

1. For each of the following elementary matrices, describe the corresponding elementary row operation and write the inverse.

   (a) $E = \begin{bmatrix} 1 & 0 & 3 \\ 0 & 1 & 0 \\ 0 & 0 & 1 \end{bmatrix}$ ◆(b) $E = \begin{bmatrix} 0 & 0 & 1 \\ 0 & 1 & 0 \\ 1 & 0 & 0 \end{bmatrix}$

   (c) $E = \begin{bmatrix} 1 & 0 & 0 \\ 0 & \frac{1}{2} & 0 \\ 0 & 0 & 1 \end{bmatrix}$ ◆(d) $E = \begin{bmatrix} 1 & 0 & 0 \\ -2 & 1 & 0 \\ 0 & 0 & 1 \end{bmatrix}$

   (e) $E = \begin{bmatrix} 0 & 1 & 0 \\ 1 & 0 & 0 \\ 0 & 0 & 1 \end{bmatrix}$ ◆(f) $E = \begin{bmatrix} 1 & 0 & 0 \\ 0 & 1 & 0 \\ 0 & 0 & 5 \end{bmatrix}$

2. In each case find an elementary matrix E such that $B = EA$.

   (a) $A = \begin{bmatrix} 2 & 1 \\ 3 & -1 \end{bmatrix}$, $B = \begin{bmatrix} 2 & 1 \\ 1 & -2 \end{bmatrix}$

   ◆(b) $A = \begin{bmatrix} -1 & 2 \\ 0 & 1 \end{bmatrix}$, $B = \begin{bmatrix} 1 & -2 \\ 0 & 1 \end{bmatrix}$

   (c) $A = \begin{bmatrix} 1 & 1 \\ -1 & 2 \end{bmatrix}$, $B = \begin{bmatrix} -1 & 2 \\ 1 & 1 \end{bmatrix}$

   ◆(d) $A = \begin{bmatrix} 4 & 1 \\ 3 & 2 \end{bmatrix}$, $B = \begin{bmatrix} 1 & -1 \\ 3 & 2 \end{bmatrix}$

   (e) $A = \begin{bmatrix} -1 & 1 \\ 1 & -1 \end{bmatrix}$, $B = \begin{bmatrix} -1 & 1 \\ -1 & 1 \end{bmatrix}$

   ◆(f) $A = \begin{bmatrix} 2 & 1 \\ -1 & 3 \end{bmatrix}$, $B = \begin{bmatrix} -1 & 3 \\ 2 & 1 \end{bmatrix}$

3. Let $A = \begin{bmatrix} 1 & 2 \\ -1 & 1 \end{bmatrix}$ and $C = \begin{bmatrix} -1 & 1 \\ 2 & 1 \end{bmatrix}$.

   (a) Find elementary matrices E_1 and E_2 such that $C = E_2 E_1 A$.

   ◆(b) Show that there is *no* elementary matrix E such that $C = EA$.

◆4. If E is elementary, show that A and EA differ in at most two rows.

5. **(a)** Is I an elementary matrix? Explain.
   ◆**(b)** Is 0 an elementary matrix? Explain.

6. In each case find an invertible matrix U such that $UA = R$ is in reduced row-echelon form, and express U as a product of elementary matrices.

(a) $A = \begin{bmatrix} 1 & -1 & 2 \\ -2 & 1 & 0 \end{bmatrix}$

◆**(b)** $A = \begin{bmatrix} 1 & 2 & 1 \\ 5 & 12 & -1 \end{bmatrix}$

(c) $A = \begin{bmatrix} 1 & 2 & -1 & 0 \\ 3 & 1 & 1 & 2 \\ 1 & -3 & 3 & 2 \end{bmatrix}$

◆**(d)** $A = \begin{bmatrix} 2 & 1 & -1 & 0 \\ 3 & -1 & 2 & 1 \\ 1 & -2 & 3 & 1 \end{bmatrix}$

7. In each case find an invertible matrix U such that $UA = B$, and express U as a product of elementary matrices.

(a) $A = \begin{bmatrix} 2 & 1 & 3 \\ -1 & 1 & 2 \end{bmatrix}$, $B = \begin{bmatrix} 1 & -1 & -2 \\ 3 & 0 & 1 \end{bmatrix}$

◆**(b)** $A = \begin{bmatrix} 2 & -1 & 0 \\ 1 & 1 & 1 \end{bmatrix}$, $B = \begin{bmatrix} 3 & 0 & 1 \\ 2 & -1 & 0 \end{bmatrix}$

8. In each case factor A as a product of elementary matrices.

(a) $A = \begin{bmatrix} 1 & 1 \\ 2 & 1 \end{bmatrix}$ ◆**(b)** $A = \begin{bmatrix} 2 & 3 \\ 1 & 2 \end{bmatrix}$

(c) $A = \begin{bmatrix} 1 & 0 & 2 \\ 0 & 1 & 1 \\ 2 & 1 & 6 \end{bmatrix}$ ◆**(d)** $A = \begin{bmatrix} 1 & 0 & -3 \\ 0 & 1 & 4 \\ -2 & 2 & 15 \end{bmatrix}$

9. Let E be an elementary matrix.
   (a) Show that E^T is also elementary of the same type.
   (b) Show that $E^T = E$ if E is of type I or II.

◆**10.** Show that every matrix A can be factored as $A = UR$ where U is invertible and R is in reduced row-echelon form.

11. If $A = \begin{bmatrix} 1 & 2 \\ 1 & -3 \end{bmatrix}$ and $B = \begin{bmatrix} 5 & 2 \\ -5 & -3 \end{bmatrix}$, find an elementary matrix F such that $AF = B$. [*Hint:* See Exercise 9.]

12. Let $A = \frac{1}{7}\begin{bmatrix} 2 & 6 & -3 \\ 3 & 2 & 6 \end{bmatrix}$. Show that $AA^T = I_2$ but $A^TA \neq I_3$.

13. If $A = \begin{bmatrix} 1 & 3 & 2 \\ 1 & 2 & 2 \end{bmatrix}$ and $B = \begin{bmatrix} 0 & 3 \\ 1 & -1 \\ -1 & 0 \end{bmatrix}$, verify that $AB = I_2$ but $BA \neq I_3$.

14. Show that the following are equivalent for $n \times n$ matrices A and B:
   (i) A and B are both invertible.
   ◆**(ii)** AB is invertible.

15. Consider $A = \begin{bmatrix} 1 & 3 & -1 \\ 2 & 1 & 5 \\ 1 & -7 & 13 \end{bmatrix}$, $B = \begin{bmatrix} 1 & 1 & 2 \\ 3 & 0 & -3 \\ -2 & 5 & 17 \end{bmatrix}$.

   (a) Show that A is not invertible by finding a nonzero 1×3 matrix Y such that $YA = 0$. [*Hint:* Row 3 of A equals 2(row 2) − 3(row 1).]
   ◆**(b)** Show that B is not invertible. [*Hint:* Column 3 = 3(column 2) − column 1.]

16. Prove Theorem 1 for elementary matrices of: **(a)** type I; **(b)** type II.

17. While trying to invert A, $[A \ \ I]$ is carried to $[P \ \ Q]$ by row operations. Show that $P = QA$.

◆**18.** Show that a square matrix A is invertible if and only if it can be left-cancelled: $AB = AC$ implies $B = C$.

19. If A and B are $n \times n$ matrices and AB is a product of elementary matrices, show that the same is true of A.

◆**20.** If U is invertible, show that the reduced row-echelon form of a matrix $[U \ \ A]$ is $[I \ \ U^{-1}A]$.

21. Two matrices A and B are called **row-equivalent** (written $A \underset{r}{\sim} B$) if there is a sequence of elementary row operations carrying A to B.
   (a) Show that $A \underset{r}{\sim} B$ if and only if $A = UB$ for some invertible matrix U.
   ◆**(b)** Show that: **(i)** $A \underset{r}{\sim} A$ for all matrices A.
      (ii) If $A \underset{r}{\sim} B$, then $B \underset{r}{\sim} A$.
      (iii) If $A \underset{r}{\sim} B$ and $B \underset{r}{\sim} C$, then $A \underset{r}{\sim} C$.
   (c) Show that, if A and B are both row-equivalent to some third matrix, then $A \underset{r}{\sim} B$.
   (d) Show that $\begin{bmatrix} 1 & -1 & 3 & 2 \\ 0 & 1 & 4 & 1 \\ 1 & 0 & 8 & 6 \end{bmatrix}$ and $\begin{bmatrix} 1 & -1 & 4 & 5 \\ -2 & 1 & -11 & -8 \\ -1 & 2 & 2 & 2 \end{bmatrix}$ are row-equivalent. [*Hint:* Consider part **(c)** and Theorem 1§1.2.]

22. If U and V are invertible $n \times n$ matrices, show that $U \underset{r}{\sim} V$. (See Exercise 21.)

23. (See Exercise 21.) Find all matrices that are row-equivalent to:

(a) $\begin{bmatrix} 0 & 0 & 0 \\ 0 & 0 & 0 \end{bmatrix}$ ◆**(b)** $\begin{bmatrix} 0 & 0 & 0 \\ 0 & 0 & 1 \end{bmatrix}$

(c) $\begin{bmatrix} 1 & 0 & 0 \\ 0 & 1 & 0 \end{bmatrix}$ **(d)** $\begin{bmatrix} 1 & 2 & 0 \\ 0 & 0 & 1 \end{bmatrix}$

24. Let A and B be $m \times n$ and $n \times m$ matrices, respectively.

   (a) If $m > n$, show that AB is not invertible. [*Hint:* Use Theorem 1§1.3 to find $Y \neq 0$ with $YA = 0$.]

   (b) Show that the only invertible matrices are square; that is, $AB = I_m$ and $BA = I_n$ imply $m = n$. (But see Exercises 12 and 13.)

25. Define an *elementary column operation* on a matrix to be one of the following: (I) Interchange two columns. (II) Multiply a column by a nonzero scalar. (III) Add a multiple of a column to another column. Show that:

   (a) If an elementary column operation is done to an $m \times n$ matrix A, the result is AF, where F is an $n \times n$ elementary matrix.

   (b) Given any $m \times n$ matrix A, there exist $m \times m$ elementary matrices $E_1, \ldots, E_k$ and $n \times n$ elementary matrices $F_1, \ldots, F_p$ such that, in block form,

$$E_k \cdots E_1 A F_1 \cdots F_p = \begin{bmatrix} I_r & 0 \\ 0 & 0 \end{bmatrix}$$

26. Suppose B is obtained from A by: **(a)** interchanging rows i and j; ◆**(b)** multiplying row i by $k \neq 0$; **(c)** adding k times row i to row j ($i \neq j$). In each case describe how to obtain B^{-1} from A^{-1}. [*Hint:* See part **(a)** of the preceding exercise.]

▧▧▧▧▧ Section 2.5   LU-Factorization (Optional)[7]

In this section the Gaussian algorithm is used to show that any matrix A can be written as a product of matrices of a particularly nice type. This is used in computer programs to solve systems of linear equations.

An $m \times n$ matrix A is called **upper triangular** if each entry of A below and to the left of the main diagonal is zero. Here, as for square matrices, the elements a_{11}, $a_{22}, \ldots$ are called the **main diagonal** of A. Hence the matrices

$$\begin{bmatrix} 1 & -1 & 0 & 3 \\ 0 & 2 & 1 & 1 \\ 0 & 0 & -3 & 0 \end{bmatrix} \quad \begin{bmatrix} 0 & 2 & 1 & 0 & 5 \\ 0 & 0 & 0 & 3 & 1 \\ 0 & 0 & 1 & 0 & 1 \end{bmatrix} \quad \begin{bmatrix} 1 & 1 & 1 \\ 0 & -1 & 1 \\ 0 & 0 & 0 \\ 0 & 0 & 0 \end{bmatrix}$$

are all upper triangular. Note that each row-echelon matrix is upper triangular.

By analogy a matrix is called **lower triangular** if its transpose is upper triangular—that is, each entry above and to the right of the main diagonal is zero. A matrix is called **triangular** if it is either upper or lower triangular.

One reason for the importance of triangular matrices is the ease with which systems of linear equations can be solved when the coefficient matrix is triangular.

EXAMPLE 1

Solve the system

$$\begin{aligned} x_1 + 2x_2 - 3x_3 - x_4 + 5x_5 &= 3 \\ 5x_3 + x_4 + x_5 &= 8 \\ 2x_5 &= 6 \end{aligned}$$

where the coefficient matrix is upper triangular.

---

[7]This section is not used later, so it may be omitted with no loss of continuity.

Solution As for a row-echelon matrix, let $x_2 = s$ and $x_4 = t$. Then solve for x_5, x_3, and x_1 in that order as follows:

$$x_5 = 6/2 = 3$$

Substitution into the second equation gives

$$x_3 = 1 - \tfrac{1}{5}t$$

Finally substitution of both x_5 and x_3 into the first equation gives

$$x_1 = -9 - 2s + \tfrac{2}{5}t$$

The method used in Example 1 is called **back substitution** for obvious reasons. It works because the matrix is upper triangular, and it provides an efficient method for finding the solutions (when they exist). In particular, it can be used in Gaussian elimination because row-echelon matrices are upper triangular. Similarly, if the matrix of a system of equations is lower triangular, the system can be solved (if a solution exists) by **forward substitution**. Here each equation is used to solve for one variable by substituting values already found for earlier variables.

Suppose now that an arbitrary matrix A is given and consider the system

$$AX = B$$

of linear equations with A as coefficient matrix. If A can be factored as $A = LU$, where L is lower triangular and U is upper triangular, the system can be solved in two stages as follows:

1. Solve $LY = B$ for Y by forward substitution.
2. Solve $UX = Y$ for X by back substitution.

Then X is a solution to $AX = B$ because $AX = LUX = LY = B$. Moreover, every solution arises in this way (take $Y = UX$). This focuses attention on obtaining such factorizations $A = LU$ of matrices.

The Gaussian algorithm provides a method of obtaining these factorizations. The method exploits the following facts about triangular matrices.

LEMMA 1

The product of two lower triangular matrices (or two upper triangular matrices) is again lower triangular (upper triangular).

LEMMA 2

Let A be an $n \times n$ lower triangular (or upper triangular) matrix. Then A is invertible if and only if no main diagonal entry is zero. In this case, A^{-1} is also lower (upper) triangular.

The proofs are straightforward and are left as Exercises 8 and 9.

Now let A by any $m \times n$ matrix. The Gaussian algorithm produces a sequence of row operations that carry A to a row-echelon matrix U. However, no multiple of a row is ever added to a row *above* it (because we are not insisting on *reduced* row-echelon form). The point is that, apart from row interchanges,[8] the only row operations needed are those that make the corresponding elementary matrix *lower triangular*. This observation gives the following theorem.

THEOREM 1

Suppose that, via the Gaussian algorithm, a matrix A can be carried to a row-echelon matrix U using no row interchanges. Then

$$A = LU$$

where L is lower triangular and invertible and U is row-echelon (and upper triangular).

Proof   The hypotheses imply that there exist lower triangular, elementary matrices $E_1, E_2, \ldots, E_k$ such that $U = (E_k \cdots E_2 E_1) A$. Hence $A = LU$, where $L = E_1^{-1} E_2^{-1} \cdots E_k^{-1}$ is lower triangular and invertible by Lemmas 1 and 2.   ◆

DEFINITION

A factorization $A = LU$ as in Theorem 1 is called an **LU-factorization** of the matrix A.

Such a factorization may not exist (Exercise 4) because at least one row interchange is required in the Gaussian algorithm. A procedure for dealing with this situation will be outlined later. However, if an LU-factorization does exist, the row-echelon matrix U in Theorem 1 is obtained by Gaussian elimination and the algorithm also yields a simple procedure for writing down the matrix L. The following example illustrates the technique.

EXAMPLE 2

Find an LU-factorization of the matrix $A = \begin{bmatrix} 0 & 2 & -6 & -2 & 4 \\ 0 & -1 & 3 & 3 & 2 \\ 0 & -1 & 3 & 7 & 10 \end{bmatrix}$.

---

[8]Any row interchange can actually be accomplished by row operations of other types (Exercise 5), but one of these must involve adding a multiple of some row to a row *above* it.

Solution   We are assuming that we can carry A to a row-echelon matrix U as before, using no row interchanges. The steps in the Gaussian algorithm are shown, and at each stage the corresponding elementary matrix is computed. The reason for the circled entries will be apparent shortly.

$$\begin{bmatrix} 0 & 2 & -6 & -2 & 4 \\ 0 & -1 & 3 & 3 & 2 \\ 0 & -1 & 3 & 7 & 10 \end{bmatrix} = A$$

$$\begin{bmatrix} 0 & 1 & -3 & -1 & 2 \\ 0 & -1 & 3 & 3 & 2 \\ 0 & -1 & 3 & 7 & 10 \end{bmatrix} = E_1 A \qquad E_1 = \begin{bmatrix} \frac{1}{2} & 0 & 0 \\ 0 & 1 & 0 \\ 0 & 0 & 1 \end{bmatrix}$$

$$\begin{bmatrix} 0 & 1 & -3 & -1 & 2 \\ 0 & 0 & 0 & 2 & 4 \\ 0 & -1 & 3 & 7 & 10 \end{bmatrix} = E_2 E_1 A \qquad E_2 = \begin{bmatrix} 1 & 0 & 0 \\ 1 & 1 & 0 \\ 0 & 0 & 1 \end{bmatrix}$$

$$\begin{bmatrix} 0 & 1 & -3 & -1 & 2 \\ 0 & 0 & 0 & 2 & 4 \\ 0 & 0 & 0 & 6 & 12 \end{bmatrix} = E_3 E_2 E_1 A \qquad E_3 = \begin{bmatrix} 1 & 0 & 0 \\ 0 & 1 & 0 \\ 1 & 0 & 1 \end{bmatrix}$$

$$\begin{bmatrix} 0 & 1 & -3 & -1 & 2 \\ 0 & 0 & 0 & 1 & 2 \\ 0 & 0 & 0 & 6 & 12 \end{bmatrix} = E_4 E_3 E_2 E_1 A \qquad E_4 = \begin{bmatrix} 1 & 0 & 0 \\ 0 & \frac{1}{2} & 0 \\ 0 & 0 & 1 \end{bmatrix}$$

$$U = \begin{bmatrix} 0 & 1 & -3 & -1 & 2 \\ 0 & 0 & 0 & 1 & 2 \\ 0 & 0 & 0 & 0 & 0 \end{bmatrix} = E_5 E_4 E_3 E_2 E_1 A \qquad E_5 = \begin{bmatrix} 1 & 0 & 0 \\ 0 & 1 & 0 \\ 0 & -6 & 1 \end{bmatrix}$$

Thus (as in the proof of Theorem 1), the LU-factorization of A is $A = LU$, where

$$L = E_1^{-1} E_2^{-1} E_3^{-1} E_4^{-1} E_5^{-1} = \begin{bmatrix} 2 & 0 & 0 \\ -1 & 2 & 0 \\ -1 & 6 & 1 \end{bmatrix}$$

Now observe that the first two columns of L can be obtained from the columns circled during execution of the algorithm.

◆◆◆

The procedure in Example 2 works in general. Moreover, we can construct the matrix L as we go along, one (circled) column at a time, starting from the left. There is no need to compute the elementary matrices E_k, and the method is suitable for use in a computer program because the circled columns can be stored in memory as they are created.

To describe the process in general, the following notation is useful. Given positive integers $m \geq r$, let $C_1, C_2, \ldots, C_r$ be columns of decreasing lengths m, $m - 1, \ldots, m - r + 1$. Then let

$$L_m[C_1, \ldots, C_r] \tag{*}$$

denote the $m \times m$ lower triangular matrix obtained from the identity matrix by replacing the bottom $m - j + 1$ entries of column j by C_j for each $j = 1, 2, \ldots, r$. Thus the matrix L in Example 2 has this form:

$$L = L_3[C_1, C_2] = \begin{bmatrix} 2 & 0 & 0 \\ -1 & 2 & 0 \\ -1 & 6 & 1 \end{bmatrix} \quad \text{where } C_1 = \begin{bmatrix} 2 \\ -1 \\ -1 \end{bmatrix} \text{ and } C_2 = \begin{bmatrix} 2 \\ 6 \end{bmatrix}.$$

Here is another example.

EXAMPLE 3

If If $C_1 = \begin{bmatrix} 3 \\ -1 \\ 0 \\ 2 \end{bmatrix}$ and $C_2 = \begin{bmatrix} 5 \\ -1 \\ 7 \end{bmatrix}$, then $L_4[C_1, C_2] = \begin{bmatrix} 3 & 0 & 0 & 0 \\ -1 & 5 & 0 & 0 \\ 0 & -1 & 1 & 0 \\ 2 & 7 & 0 & 1 \end{bmatrix}$.

◆◆◆

Note that if $r < m$, the last $m - r$ columns of $L_m[C_1, \ldots, C_r]$ are the corresponding columns of the identity matrix I_m.

Now the general version of the procedure in Example 2 can be stated. Given a nonzero matrix A, call the first nonzero column of A (from the left) the **leading column** of A.

LU-ALGORITHM

Let A be an $m \times n$ matrix that can be carried to a row-echelon matrix U using no row interchanges. An LU-factorization $A = LU$ can be obtained as follows:

Step 1. If $A = 0$, take $L = I_m$ and $U = 0$.

Step 2. If $A \neq 0$, let C_1 be the leading column of A and do row operations (with no row interchanges) to create the first leading 1 and bring A to the following block form:

$$A \rightarrow \left[\begin{array}{c|c|c} 0 & 1 & X_2 \\ \hline 0 & 0 & A_2 \end{array} \right]$$

Step 3. If $A_2 \neq 0$, let C_2 be the leading column of A_2 and apply step 2 to bring A_2 to block form:

$$A_2 \rightarrow \left[\begin{array}{c|c|c} 0 & 1 & X_3 \\ \hline 0 & 0 & A_3 \end{array}\right]$$

Step 4. Continue in this way until all the rows below the last leading 1 created consist of zeros. Take U to be the (row-echelon) matrix just created, and take [see preceding equation $(*)$]

$$L = L_m[C_1, C_2, \ldots, C_r]$$

where C_1, C_2, C_3, ... are the leading columns of the matrices A, A_2, A_3, ...

The proof is given at the end of this section.

Of course the integer r in the LU-algorithm is the number of leading 1's in the row-echelon matrix U, so it is the rank of A.

EXAMPLE 4

Find an LU-factorization for $A = \begin{bmatrix} 5 & -5 & 10 & 0 & 5 \\ -3 & 3 & 2 & 2 & 1 \\ -2 & 2 & 0 & -1 & 0 \\ 1 & -1 & 10 & 2 & 5 \end{bmatrix}$.

Solution   The reduction to row-echelon form is

$$\begin{bmatrix} 5 & -5 & 10 & 0 & 5 \\ -3 & 3 & 2 & 2 & 1 \\ -2 & 2 & 0 & -1 & 0 \\ 1 & -1 & 10 & 2 & 5 \end{bmatrix} \rightarrow \begin{bmatrix} 1 & -1 & 2 & 0 & 1 \\ 0 & 0 & 8 & 2 & 4 \\ 0 & 0 & 4 & -1 & 2 \\ 0 & 0 & 8 & 2 & 4 \end{bmatrix}$$

$$\rightarrow \begin{bmatrix} 1 & -1 & 2 & 0 & 1 \\ 0 & 0 & 1 & \frac{1}{4} & \frac{1}{2} \\ 0 & 0 & 0 & -2 & 0 \\ 0 & 0 & 0 & 0 & 0 \end{bmatrix}$$

$$\rightarrow \begin{bmatrix} 1 & -1 & 2 & 0 & 1 \\ 0 & 0 & 1 & \frac{1}{4} & \frac{1}{2} \\ 0 & 0 & 0 & 1 & 0 \\ 0 & 0 & 0 & 0 & 0 \end{bmatrix}.$$

If U denotes this row-echelon matrix, then $A = LU$, where

$$L = \begin{bmatrix} 5 & 0 & 0 & 0 \\ -3 & 8 & 0 & 0 \\ -2 & 4 & -2 & 0 \\ 1 & 8 & 0 & 1 \end{bmatrix}$$

◆◆◆

The next example deals with a case where no row of zeros is present in U (in fact, A is invertible).

EXAMPLE 5

Find an LU-factorization for $A = \begin{bmatrix} 2 & 4 & 2 \\ 1 & 1 & 2 \\ -1 & 0 & 2 \end{bmatrix}$.

Solution   The reduction to row-echelon form is

$$\begin{bmatrix} 2 & 4 & 2 \\ 1 & 1 & 2 \\ -1 & 0 & 2 \end{bmatrix} \rightarrow \begin{bmatrix} 1 & 2 & 1 \\ 0 & -1 & 1 \\ 0 & 2 & 3 \end{bmatrix} \rightarrow \begin{bmatrix} 1 & 2 & 1 \\ 0 & 1 & -1 \\ 0 & 0 & 5 \end{bmatrix} \rightarrow \begin{bmatrix} 1 & 2 & 1 \\ 0 & 1 & -1 \\ 0 & 0 & 1 \end{bmatrix} = U$$

so $L = \begin{bmatrix} 2 & 0 & 0 \\ 1 & -1 & 0 \\ -1 & 2 & 5 \end{bmatrix}$.

◆◆◆

There are matrices (for example $\begin{bmatrix} 0 & 1 \\ 1 & 0 \end{bmatrix}$) that have no LU-factorization and so require at least one row interchange when being carried to row-echelon form via the Gaussian algorithm. However, it turns out that if all the row interchanges encountered in the algorithm are carried out first, the resulting matrix requires no interchanges and so has an LU-factorization. Here is the precise result.

THEOREM 2

Suppose an $m \times n$ matrix A is carried to a row-echelon matrix U via the Gaussian algorithm. Let $P_1, P_2, \ldots, P_s$ be the elementary matrices corresponding (in order) to the row interchanges used and write $P = P_s \cdots P_2 P_1$. (If no interchanges are used take $P = I_m$.)

Then:

1. PA is the matrix obtained from A by doing these interchanges (in order) to A.
2. PA has an LU-factorization.

The proof is given at the end of this section.

A matrix P that is the product of elementary matrices corresponding to row interchanges is called a **permutation matrix**. Such a matrix is obtained from the identity matrix by arranging the rows in a different order, so it has exactly one 1 in each row and each column, and has zeros elsewhere. We regard the identity matrix as a permutation matrix. The elementary permutation matrices are those obtained from I by a single row interchange, and every permutation matrix is a product of elementary ones.

EXAMPLE 6

If $A = \begin{bmatrix} 0 & 0 & -1 & 2 \\ -1 & -1 & 1 & 2 \\ 2 & 1 & -3 & 6 \\ 0 & 1 & -1 & 4 \end{bmatrix}$, find a permutation matrix P such that PA has an LU-

factorization, and then find the factorization.

Solution   Apply the Gaussian algorithm to A:

$$A \to \begin{bmatrix} -1 & -1 & 1 & 2 \\ 0 & 0 & -1 & 2 \\ 2 & 1 & -3 & 6 \\ 0 & 1 & -1 & 4 \end{bmatrix} \to \begin{bmatrix} 1 & 1 & -1 & -2 \\ 0 & 0 & -1 & 2 \\ 0 & -1 & -1 & 10 \\ 0 & 1 & -1 & 4 \end{bmatrix} \to \begin{bmatrix} 1 & 1 & -1 & -2 \\ 0 & -1 & -1 & 10 \\ 0 & 0 & -1 & 2 \\ 0 & 1 & -1 & 4 \end{bmatrix}$$

$$\to \begin{bmatrix} 1 & 1 & -1 & -2 \\ 0 & 1 & 1 & -10 \\ 0 & 0 & -1 & 2 \\ 0 & 0 & -2 & 14 \end{bmatrix} \to \begin{bmatrix} 1 & 1 & -1 & -2 \\ 0 & 1 & 1 & -10 \\ 0 & 0 & 1 & -2 \\ 0 & 0 & 0 & 10 \end{bmatrix}$$

Two row interchanges were needed, first rows 1 and 2 and then rows 2 and 3. Hence, as in Theorem 2,

$$P = \begin{bmatrix} 1 & 0 & 0 & 0 \\ 0 & 0 & 1 & 0 \\ 0 & 1 & 0 & 0 \\ 0 & 0 & 0 & 1 \end{bmatrix} \begin{bmatrix} 0 & 1 & 0 & 0 \\ 1 & 0 & 0 & 0 \\ 0 & 0 & 1 & 0 \\ 0 & 0 & 0 & 1 \end{bmatrix} = \begin{bmatrix} 0 & 1 & 0 & 0 \\ 0 & 0 & 1 & 0 \\ 1 & 0 & 0 & 0 \\ 0 & 0 & 0 & 1 \end{bmatrix}$$

If we do these interchanges (in order) to A, the result is PA. Now apply the LU-algorithm to PA:

$$PA = \begin{bmatrix} -1 & -1 & 1 & 2 \\ 2 & 1 & -3 & 6 \\ 0 & 0 & -1 & 2 \\ 0 & 1 & -1 & 4 \end{bmatrix} \to \begin{bmatrix} 1 & 1 & -1 & -2 \\ 0 & -1 & -1 & 10 \\ 0 & 0 & -1 & 2 \\ 0 & 1 & -1 & 4 \end{bmatrix} \to \begin{bmatrix} 1 & 1 & -1 & -2 \\ 0 & 1 & 1 & -10 \\ 0 & 0 & -1 & 2 \\ 0 & 0 & -2 & 14 \end{bmatrix}$$

$$\rightarrow \begin{bmatrix} 1 & 1 & -1 & -2 \\ 0 & 1 & 1 & -10 \\ 0 & 0 & 1 & -2 \\ 0 & 0 & 0 & \boxed{10} \end{bmatrix} \rightarrow \begin{bmatrix} 1 & 1 & -1 & -2 \\ 0 & 1 & 1 & -10 \\ 0 & 0 & 1 & -2 \\ 0 & 0 & 0 & 1 \end{bmatrix} = U$$

Hence $PA = LU$, where $L = \begin{bmatrix} -1 & 0 & 0 & 0 \\ 2 & -1 & 0 & 0 \\ 0 & 0 & -1 & 0 \\ 0 & 1 & -2 & 10 \end{bmatrix}$ and $U = \begin{bmatrix} 1 & 1 & -1 & -2 \\ 0 & 1 & 1 & -10 \\ 0 & 0 & 1 & -2 \\ 0 & 0 & 0 & 1 \end{bmatrix}$

◆◆◆

Theorem 2 provides an important general factorization theorem for matrices. If A is any $m \times n$ matrix, it asserts that there exists a permutation matrix P and an LU-factorization $PA = LU$. Moreover, it shows that either $P = I$ or $P = P_s \cdots P_2 P_1$, where $P_1, P_2, \cdots, P_s$ are the elementary permutation matrices arising in the reduction of A to row-echelon form. Now observe that $P_i^{-1} = P_i$ for each i. Thus $P^{-1} = P_1 P_2 \cdots P_s$, so the matrix A can be factored as

$$A = P^{-1}LU$$

where P^{-1} is a permutation matrix, L is lower triangular and invertible, and U is a row-echelon matrix. This is called a **PLU-factorization** of A.

The LU-factorization in Theorem 1 is not unique. For example,

$$\begin{bmatrix} 1 & 0 \\ 3 & 2 \end{bmatrix} \begin{bmatrix} 1 & -2 & 3 \\ 0 & 0 & 0 \end{bmatrix} = \begin{bmatrix} 1 & 0 \\ 3 & 1 \end{bmatrix} \begin{bmatrix} 1 & -2 & 3 \\ 0 & 0 & 0 \end{bmatrix}$$

However, the fact that the row-echelon matrix here has a row of zeros is necessary. Recall that the rank of a matrix A is the number of nonzero rows in any row-echelon matrix U to which A can be carried by row operations. Thus if A is $m \times n$, the matrix U has no row of zeros if and only if A has rank m.

THEOREM 3

Let A be an $m \times n$ matrix that has an LU-factorization

$$A = LU$$

If A has rank m (that is, U has no row of zeros), then L and U are uniquely determined by A.

Proof   Suppose $A = MV$ is another LU-factorization of A, so M is lower triangular and invertible and V is row-echelon. Hence $LU = MV$, and we must show that $L = M$ and $U = V$. We write $N = M^{-1}L$. Then N is lower triangular and invertible (Lemmas 1 and 2) and $NU = V$, so it suffices to prove that $N = I$. If N is $m \times m$, we use induction on m. The case $m = 1$ is left to the reader. If $m > 1$, observe first that column 1 of V is N times column 1 of U. Thus if either column is zero, so is the other (N is invertible). Hence we can assume (by deleting zero columns) that the $(1, 1)$-entry is 1 in both U and V. Now we write $N = \begin{bmatrix} a & 0 \\ X & N_1 \end{bmatrix}$, $U = \begin{bmatrix} 1 & Y \\ 0 & U_1 \end{bmatrix}$, and $V = \begin{bmatrix} 1 & Z \\ 0 & V_1 \end{bmatrix}$ in block form. Then $NU = V$ becomes $\begin{bmatrix} a & aY \\ X & XY + N_1 U_1 \end{bmatrix} = \begin{bmatrix} 1 & Z \\ 0 & V_1 \end{bmatrix}$. Hence $a = 1$, $Y = Z$, $X = 0$, and $N_1 U_1 = V_1$. But $N_1 U_1 = V_1$ implies $N_1 = I$ by induction, whence $N = I$.   ◆

If A is an $m \times m$ invertible matrix, then A has rank m by Theorem 4§2.4. Hence we get the following important special case of Theorem 3.

COROLLARY

If an invertible matrix A has an LU-factorization $A = LU$, then L and U are uniquely determined by A.

Of course, in this case U is an upper triangular matrix with 1s along the main diagonal.

Proofs of Theorems

Proof of the LU-algorithm   Proceed by induction on n. If $n = 1$, it is left to the reader. If $n > 1$, let C_1 denote the leading column of A and let K_1 denote the first column of the $m \times m$ identity matrix. There exist elementary matrices $E_1, \ldots, E_k$ such that, in block form,

$$(E_k \cdots E_2 E_1)A = \left[\, 0 \mid K_1 \mid \begin{matrix} X_1 \\ \hline A_1 \end{matrix} \,\right] \qquad \text{where } (E_k \cdots E_2 E_1)C_1 = K_1.$$

Moreover, each E_j can be taken to be lower triangular (by assumption). Write

$$L_0 = (E_k \cdots E_2 E_1)^{-1} = E_1^{-1} E_2^{-1} \cdots E_k^{-1}$$

Then L_0 is lower triangular, and $L_0 K_1 = C_1$. Also, each E_j (and so each E_j^{-1}) is the result of either multiplying row 1 of I_m by a constant or adding a multiple of row 1 to another row. Hence

$$L_0 = (E_1^{-1}E_2^{-1} \cdots E_k^{-1})I_m = \left[C_1 \left| \begin{array}{c} 0 \\ \hline I_{m-1} \end{array} \right. \right]$$

in block form. Now, by induction, let $A_1 = L_1U_1$ be an LU-factorization of A_1, where $L_1 = L_{m-1}[C_2, \ldots, C_r]$ and U_1 is row-echelon. Then block multiplication gives

$$L_0^{-1}A = \left[\begin{array}{c|c} 0 & K_1 \end{array} \left| \begin{array}{c} X_1 \\ \hline L_1U_1 \end{array} \right. \right] = \left[\begin{array}{c|c} 1 & 0 \\ \hline 0 & L_1 \end{array} \right] \left[\begin{array}{cc|c} 0 & 1 & X_1 \\ \hline 0 & 0 & U_1 \end{array} \right]$$

Hence $A = LU$, where $U = \left[\begin{array}{cc|c} 0 & 1 & X_1 \\ \hline 0 & 0 & U_1 \end{array} \right]$ is row-echelon and

$$L = \left[C_1 \left| \begin{array}{c} 0 \\ \hline I_{m-1} \end{array} \right. \right] \left[\begin{array}{c|c} 1 & 0 \\ \hline 0 & L_1 \end{array} \right] = \left[C_1 \left| \begin{array}{c} 0 \\ \hline L_1 \end{array} \right. \right] = L_m[C_1, C_2, \ldots, C_r].$$

This completes the proof.   ◆

Proof of Theorem 2   Let A be a nonzero $m \times n$ matrix and let K_j denote column j of I_m. There is a permutation matrix P_1 (where either P_1 is elementary or $P_1 = I_m$) such that the first nonzero column C_1 of P_1A has a nonzero entry on top. Hence, as in the LU-algorithm,

$$L_m[C_1]^{-1} \cdot P_1 \cdot A = \left[\begin{array}{cc|c} 0 & 1 & X_1 \\ \hline 0 & 0 & A_1 \end{array} \right]$$

in block form. Then let P_2 be a permutation matrix (either elementary or I_m) such that

$$P_2 \cdot L_m[C_1]^{-1} \cdot P_1 \cdot A = \left[\begin{array}{cc|c} 0 & 1 & X_1 \\ \hline 0 & 0 & A_1' \end{array} \right]$$

and the first nonzero column C_2 of A_1' has a nonzero entry on top. Thus

$$L_m[K_1, \ C_2]^{-1} \cdot P_2 \cdot L_m[C_1]^{-1} \cdot P_1 \cdot A = \left[\begin{array}{cc|cc} 0 & 1 & & X_1 \\ \hline 0 & 0 & \begin{array}{cc} 0 & 1 \\ 0 & 0 \end{array} & \begin{array}{c} X_2 \\ A_2 \end{array} \end{array} \right]$$

in block form. Continue to obtain elementary permutation matrices $P_1, P_2, \ldots, P_r$ and columns $C_1, C_2, \ldots, C_r$ of lengths $m, m-1, \ldots$, such that

$$(L_rP_rL_{r-1}P_{r-1} \cdots L_2P_2L_1P_1)A = U$$

where U is a row-echelon matrix and $L_j = L_m[K_1, \ldots, K_{j-1}, C_j]^{-1}$ for each j, where the notation means the first $j-1$ columns are those of I_m. It is not hard to verify that each L_j has the form $L_j = L_m[K_1, \ldots, K_{j-1}, C_j']$ where C_j' is a column of length $m-j + 1$. We now claim that each permutation matrix P_k can be "moved past" each matrix L_j to the right of it, in the sense that

$$P_kL_j = L_j'P_k$$

where $L'_j = L_m[K_1, \ldots, K_{j-1}, C''_j]$ for some column C''_j of length $m - j + 1$. Given that this is true, we obtain a factorization of the form

$$(L_r L'_{r-1} \cdots L'_2 L'_1)(P_r P_{r-1} \cdots P_2 P_1)A = U$$

If we write $P = P_r P_{r-1} \cdots P_2 P_1$, this shows that PA has an LU-factorization because $L_r L'_{r-1} \cdots L'_2 L'_1$ is lower triangular and invertible. All that remains is to prove the following rather technical result.   ◆

LEMMA 3

Let P_k result from interchanging row k of I_m with a row below it. If $j < k$, let C_j be a column of length $m - j + 1$. Then there is another column C'_j of length $m - j + 1$ such that

$$P_k \cdot L_m[K_1 \cdots K_{j-1}C_j] = L_m[K_1 \cdots K_{j-1}C'_j] \cdot P_k$$

The proof is left as Exercise 12.

EXERCISES 2.5

1. Find an LU-factorization of the following matrices.

(a) $\begin{bmatrix} 2 & 6 & -2 & 0 & 2 \\ 3 & 9 & -3 & 3 & 1 \\ -1 & -3 & 1 & -3 & 1 \end{bmatrix}$ ◆**(b)** $\begin{bmatrix} 2 & 4 & 2 \\ 1 & -1 & 3 \\ -1 & 7 & -7 \end{bmatrix}$

(c) $\begin{bmatrix} 2 & 6 & -2 & 0 & 2 \\ 1 & 5 & -1 & 2 & 5 \\ 3 & 7 & -3 & -2 & 5 \\ -1 & -1 & 1 & 2 & 3 \end{bmatrix}$ ◆**(d)** $\begin{bmatrix} -1 & -3 & 1 & 0 & -1 \\ 1 & 4 & 1 & 1 & 1 \\ 1 & 2 & -3 & -1 & 1 \\ 0 & -2 & -4 & -2 & 0 \end{bmatrix}$

(e) $\begin{bmatrix} 2 & 2 & 4 & 6 & 0 & 2 \\ 1 & -1 & 2 & 1 & 3 & 1 \\ -2 & 2 & -4 & -1 & 1 & 6 \\ 0 & 2 & 0 & 3 & 4 & 8 \\ -2 & 4 & -4 & 1 & -2 & 6 \end{bmatrix}$

◆**(f)** $\begin{bmatrix} 2 & 2 & -2 & 4 & 2 \\ 1 & -1 & 0 & 2 & 1 \\ 3 & 1 & -2 & 6 & 3 \\ 1 & 3 & -2 & 2 & 1 \end{bmatrix}$

2. Find a permutation matrix P and an LU-factorization of PA if A is:

(a) $\begin{bmatrix} 0 & 0 & 2 \\ 0 & -1 & 4 \\ 3 & 5 & 1 \end{bmatrix}$ ◆**(b)** $\begin{bmatrix} 0 & -1 & 2 \\ 0 & 0 & 4 \\ -1 & 2 & 1 \end{bmatrix}$

(c) $\begin{bmatrix} 0 & -1 & 2 & 1 & 3 \\ -1 & 1 & 3 & 1 & 4 \\ 1 & -1 & -3 & 6 & 2 \\ 2 & -2 & -4 & 1 & 0 \end{bmatrix}$ ◆**(d)** $\begin{bmatrix} -1 & -2 & 3 & 0 \\ 2 & 4 & -6 & 5 \\ 1 & 1 & -1 & 3 \\ 2 & 5 & -10 & 1 \end{bmatrix}$

3. In each case use the given LU-decomposition of A to solve the system $AX = B$ by finding Y such that $LY = B$, and then X such that $UX = Y$:

(a) $A = \begin{bmatrix} 2 & 0 & 0 \\ 0 & -1 & 0 \\ 1 & 1 & 3 \end{bmatrix} \begin{bmatrix} 1 & 0 & 0 & 1 \\ 0 & 0 & 1 & 2 \\ 0 & 0 & 0 & 1 \end{bmatrix}$; $B = \begin{bmatrix} 1 \\ -1 \\ 2 \end{bmatrix}$

◆**(b)** $A = \begin{bmatrix} 2 & 0 & 0 \\ 1 & 3 & 0 \\ -1 & 2 & 1 \end{bmatrix} \begin{bmatrix} 1 & 1 & 0 & -1 \\ 0 & 1 & 0 & 1 \\ 0 & 0 & 0 & 0 \end{bmatrix}$; $B = \begin{bmatrix} -2 \\ -1 \\ 1 \end{bmatrix}$

(c) $A = \begin{bmatrix} -2 & 0 & 0 & 0 \\ 1 & -1 & 0 & 0 \\ -1 & 0 & 2 & 0 \\ 0 & 1 & 0 & 2 \end{bmatrix} \begin{bmatrix} 1 & -1 & 2 & -1 \\ 0 & 1 & 1 & -4 \\ 0 & 0 & 1 & -\frac{1}{2} \\ 0 & 0 & 0 & 1 \end{bmatrix}$; $B = \begin{bmatrix} 1 \\ -1 \\ 2 \\ 0 \end{bmatrix}$

◆**(d)** $A = \begin{bmatrix} 2 & 0 & 0 & 0 \\ 1 & -1 & 0 & 0 \\ -1 & 1 & 2 & 0 \\ 3 & 0 & 1 & -1 \end{bmatrix} \begin{bmatrix} 1 & -1 & 0 & 1 \\ 0 & 1 & -2 & -1 \\ 0 & 0 & 1 & 1 \\ 0 & 0 & 0 & 0 \end{bmatrix}$; $B = \begin{bmatrix} 4 \\ -6 \\ 4 \\ 5 \end{bmatrix}$

4. Show that $\begin{bmatrix} 0 & 1 \\ 1 & 0 \end{bmatrix} = LU$ is impossible where L is lower triangular and U is upper triangular.

5. Let E and F be the elementary matrices obtained from the identity matrix by adding multiples of row k to rows p and q. If $k \neq p$ and $k \neq q$, show that $EF = FE$.

◆**6.** Show that we can accomplish any row interchange by using only row operations of other types.

7. **(a)** Let L and L_1 be invertible lower triangular matrices, and let U and U_1 be invertible upper triangular matrices. Show that $LU = L_1U_1$ if and only if there exists an invertible diagonal matrix D such that $L_1 = LD$ and $U_1 = D^{-1}U$. [*Hint:* Scrutinize $L^{-1}L_1 = U U_1^{-1}$.]

   ◆**(b)** Use part **(a)** to prove Theorem 3 in the case that A is invertible.

◆**8.** Prove Lemma 1. [*Hint:* Use block multiplication and induction.]

9. Prove Lemma 2. [*Hint:* Use block multiplication and induction.]

10. A triangular matrix is called **unit triangular** if it is square and every main diagonal element is a 1.

   (a) If A can be carried by the Gaussian algorithm to row-echelon form using no row interchanges, show that $A = LU$ where L is unit lower triangular and U is upper triangular.

   (b) Show that the factorization in **(a)** is unique.

11. Let $C_1, C_2, \ldots, C_r$ be columns of lengths $m, m - 1, \ldots, m - r + 1$. If K_j denotes column j of I_m, show that $L_m[C_1, C_2, \ldots, C_r] = L_m[C_1] \, L_m[K_1, C_2] \, L_m[K_1, K_2, C_3] \cdots L_m[K_1, K_2, \ldots, K_{r-1}, C_r]$. The notation is as in the proof of Theorem 2. [*Hint:* Use induction on m and block multiplication.]

12. Prove Lemma 3. [*Hint:* $P_k^{-1} = P_k$. Write $P_k = \begin{bmatrix} I_k & 0 \\ 0 & P_0 \end{bmatrix}$ in block form where P_0 is an $(m - k) \times (m - k)$ permutation matrix.]

Section 2.6   An Application to Input-Output Economic Models (Optional)[9]

In 1973 Wassily Leontief was awarded the Nobel prize in economics for his work on mathematical models.[10] Roughly speaking, an economic system in this model consists of several industries, each of which produces a product and each of which uses some of the production of the other industries. The following example is typical.

EXAMPLE 1

A primitive society has three basic needs: food, shelter, and clothing. There are thus three industries in the society—the farming, housing, and garment industries—that produce these commodities. Each of these industries consumes a certain proportion of the total output of each commodity according to the following table.

		OUTPUT		
		Farming	Housing	Garment
	Farming	.4	.2	.3
Consumption	Housing	.2	.6	.4
	Garment	.4	.2	.3

Find the annual prices that each industry must charge for its income to equal its expenditures.

Solution   Let p_1, p_2, and p_3 be the prices charged per year by the farming, housing, and garment industries, respectively, for their total output. To see how these prices are determined, consider the farming industry. It receives p_1 for its production in any year. But it *con-*

---

sumes products from all these industries in the following amounts (from row 1 of the table): 40% of the food, 20% of the housing, and 30% of the clothing. Hence the expenditures of the farming industry are $.4p_1 + .2p_2 + .3p_3$, so

$$.4p_1 + .2p_2 + .3p_3 = p_1$$

A similar analysis of the other two industries leads to the following system of equations.

$$.4p_1 + .2p_2 + .3p_3 = p_1$$
$$.2p_1 + .6p_2 + .4p_3 = p_2$$
$$.4p_1 + .2p_2 + .3p_3 = p_3$$

This has the matrix form $EP = P$, where

$$E = \begin{bmatrix} .4 & .2 & .3 \\ .2 & .6 & .4 \\ .4 & .2 & .3 \end{bmatrix} \quad \text{and} \quad P = \begin{bmatrix} p_1 \\ p_2 \\ p_3 \end{bmatrix}$$

The equations can be written as the homogeneous system

$$(I - E)P = 0$$

where I is the 3×3 identity matrix, and the solutions are

$$P = \begin{bmatrix} 2t \\ 3t \\ 2t \end{bmatrix}$$

where t is a parameter. Thus the pricing must be such that the total output of the farming industry has the same value as the total output of the garment industry, whereas the total value of the housing industry must be $\frac{3}{2}$ as much. ◆◆◆

In general, suppose an economy has n industries, each of which uses some (possibly none) of the production of every industry. We assume first that the economy is **closed** (that is, no product is exported or imported) and that all product is used. Given two industries i and j, let e_{ij} denote the proportion of the total annual output of industry j that is consumed by industry i. Then $E = [e_{ij}]$ is called the **input-output** matrix for the economy. Clearly,

$$0 \le e_{ij} \le 1 \qquad \text{for all } i \text{ and } j \tag{1}$$

Moreover, all the output from industry j is used by *some* industry (the model is closed), so

$$e_{1j} + e_{2j} + \cdots + e_{nj} = 1 \qquad \text{for each } j \tag{2}$$

Condition 2 asserts that each column of E sums to 1. Matrices satisfying conditions 1 and 2 are called **stochastic matrices**.

As in Example 1, let p_i denote the price of the total annual production of industry i. Then p_i is the annual revenue of industry i. On the other hand, industry i spends $e_{i1}p_1 + e_{i2}p_2 + \cdots + e_{in}p_n$ annually for the product it uses ($e_{ij}p_j$ is the cost for product from industry j). The closed economic system is said to be in **equilibrium** if the annual expenditure equals the annual revenue for each industry—that is, if

$$e_{i1}p_1 + e_{i2}p_2 + \cdots + e_{in}p_n = p_i \qquad \text{for each } i = 1, 2, \ldots, n$$

If we write $P = \begin{bmatrix} p_1 \\ p_2 \\ \vdots \\ p_n \end{bmatrix}$, these equations can be written as the matrix equation

$$EP = P$$

This is called the **equilibrium condition**, and the solutions P are called **equilibrium price structures**. The equilibrium condition can be written as

$$(I - E)P = 0$$

which is a system of homogeneous equations for P. Moreover, there is always a nontrivial solution P. Indeed, the column sums of $I - E$ are all 0 (because E is stochastic), so the row-echelon form of $I - E$ has a row of zeros. In fact, more is true:

THEOREM 1

Let E be any $n \times n$ stochastic matrix. Then there is a nonzero $n \times 1$ matrix P with nonnegative entries such that $EP = P$. If all the entries of E are positive, the matrix P can be chosen with all entries positive.

Theorem 1 guarantees the existence of an equilibrium price structure for any closed input-output system of the type discussed here. The proof is beyond the scope of this book.[11]

EXAMPLE 2

Find the equilibrium price structures for four industries if the input-output matrix is

$$E = \begin{bmatrix} .6 & .2 & .1 & 1 \\ .3 & .4 & .2 & 0 \\ .1 & .3 & .5 & .2 \\ 0 & .1 & .2 & .7 \end{bmatrix}$$

---

[11]The interested reader is referred to P. Lancaster's *Theory of Matrices* (New York: Academic Press, 1969) or to E. Seneta's *Non-negative Matrices* (New York: Wiley, 1973).

Find the prices if the total value of business is $1000.

Solution   If $P = \begin{bmatrix} p_1 \\ p_2 \\ p_3 \\ p_4 \end{bmatrix}$ is the equilibrium price structure, then the equilibrium condition is

$EP = P$. When we write this as $(I - E)P = 0$, the methods of Chapter 1 yield the following family of solutions:

$$P = \begin{bmatrix} 44t \\ 39t \\ 51t \\ 47t \end{bmatrix}$$

where t is a parameter. If we insist that $p_1 + p_2 + p_3 + p_4 = \$1000$, then $t = 5.525$ (to four figures). Hence

$$P = \begin{bmatrix} 243.09 \\ 215.47 \\ 281.77 \\ 259.67 \end{bmatrix}$$

to five figures.

◆◆◆

The Open Model

We now assume that there is a demand for products in the **open sector** of the economy, which is the part of the economy other than the producing industries (for example, consumers). Let d_i denote the total value of the demand for product i in the open sector. If p_i and e_{ij} are as before, the value of the annual demand for product i by the producing industries themselves is $e_{i1}p_1 + e_{i2}p_2 + \cdots + e_{in}p_n$, so the total annual revenue p_i of industry i breaks down as follows:

$$p_i = (e_{i1}p_1 + e_{i2}p_2 + \cdots + e_{in}p_n) + d_i \qquad \text{for each } i = 1, 2, \ldots, n$$

The column $D = \begin{bmatrix} d_1 \\ \vdots \\ d_n \end{bmatrix}$ is called the **demand matrix**, and this gives a matrix equation

$$P = EP + D$$

or

$$(I - E)P = D \qquad\qquad (*)$$

This is a system of linear equations for P, and we ask for a solution P with every entry nonnegative. Note that every entry of E is between 0 and 1, but the column sums of E need not equal 1 as in the closed model.

Before proceeding, it is convenient to introduce a useful notation. If $A = [a_{ij}]$ and $B = [b_{ij}]$ are matrices of the same size, we write $A > B$ if $a_{ij} > b_{ij}$ for all i and j, and we write $A \geq B$ if $a_{ij} \geq b_{ij}$ for all i and j. Thus $P \geq 0$ means that every entry of P is nonnegative. Note that $A \geq 0$ and $B \geq 0$ implies that $AB \geq 0$.

Now, given a demand matrix $D \geq 0$, we look for a production matrix $P \geq 0$ satisfying equation (∗). This certainly exists if $I - E$ is invertible and $(I - E)^{-1} \geq 0$. On the other hand, the fact that $D \geq 0$ means any solution P to equation (∗) satisfies $P \geq EP$. Hence the following theorem is not too surprising.

THEOREM 2

Let $E \geq 0$ be a square matrix. Then $I - E$ is invertible and $(I - E)^{-1} \geq 0$ if and only if there exists a column $P > 0$ such that $P > EP$.

Heuristic Proof   If $(I - E)^{-1} \geq 0$, the existence of $P > 0$ with $EP > P$ is left as Exercise 11. Conversely, suppose such a column P exists. Observe that

$$(I - E)(I + E + E^2 + \cdots + E^{k-1}) = I - E^k$$

holds for all $k \geq 2$. If we can show that every entry of E^k approaches 0 as k becomes large, then, intuitively, the infinite matrix sum

$$U = I + E + E^2 + \cdots$$

exists and $(I - E)U = I$. Since $U \geq 0$, this does it. To show that E^k approaches 0, it suffices to show that $EP < \mu P$ for some number μ with $0 < \mu < 1$ (then $E^k P < \mu^k P$ for all $k \geq 1$ by induction). The existence of μ is left as Exercise 12.   ◆

The condition $P > EP$ in Theorem 2 has a simple economic interpretation. If P is a production matrix, entry i of EP is the total value of all product used by industry i in a year. Hence the condition $P > EP$ means that, for each i, the value of product produced by industry i exceeds the value of the product it uses. In other words, each industry runs at a profit.

EXAMPLE 3

If $E = \begin{bmatrix} .6 & .2 & .3 \\ .1 & .4 & .2 \\ .2 & .5 & .1 \end{bmatrix}$, show that $I - E$ is invertible and $(I - E)^{-1} \geq 0$.

Solution   Use $P = [3 \quad 2 \quad 2]^T$ in Theorem 2.

◆◆◆

If $P_0 = [1 \quad 1 \quad \cdots \quad 1]^T$, the entries of EP_0 are the row sums of E. Hence $P_0 > EP_0$ holds if the row sums of E are all less than 1. This proves the first of the following useful facts (the second is Exercise 10).

COROLLARY

Let $E \geq 0$ be a square matrix. In each of the following cases, $I - E$ is invertible and $(I - E)^{-1} \geq 0$:

1. All row sums of E are less than 1.

2. All column sums of E are less than 1.

EXERCISES 2.6

1. Find the possible equilibrium price structures when the input-output matrices are:

   (a) $\begin{bmatrix} .1 & .2 & .3 \\ .6 & .2 & .3 \\ .3 & .6 & .4 \end{bmatrix}$ ◆(b) $\begin{bmatrix} .5 & 0 & .5 \\ .1 & .9 & .2 \\ .4 & .1 & .3 \end{bmatrix}$

   (c) $\begin{bmatrix} .3 & .1 & .1 & .2 \\ .2 & .3 & .1 & 0 \\ .3 & .3 & .2 & .3 \\ .2 & .3 & .6 & .5 \end{bmatrix}$ ◆(d) $\begin{bmatrix} .5 & 0 & .1 & .1 \\ .2 & .7 & 0 & .1 \\ .1 & .2 & .8 & .2 \\ .2 & .1 & .1 & .6 \end{bmatrix}$

◆2. Three industries A, B, and C are such that all the output of A is used by B, all the output of B is used by C, and all the output of C is used by A. Find the possible equilibrium price structures.

3. Find the possible equilibrium price structures for three industries where the input-output matrix is $\begin{bmatrix} 1 & 0 & 0 \\ 0 & 0 & 1 \\ 0 & 1 & 0 \end{bmatrix}$.

   Discuss why there are two parameters here.

◆4. Prove Theorem 1 for a 2×2 stochastic matrix E by first writing it in the form $E = \begin{bmatrix} a & b \\ 1-a & 1-b \end{bmatrix}$, where $0 \leq a \leq 1$ and $0 \leq b \leq 1$.

5. If E is an $n \times n$ stochastic matrix and C is an $n \times 1$ matrix, show that the sum of the entries of C equals the sum of the entries of the $n \times 1$ matrix EC.

6. Let $W = [1 \quad 1 \quad 1 \quad \cdots \quad 1]$. Let E and F denote $n \times n$ matrices with nonnegative entries.

   (a) Show that E is a stochastic matrix if and only if $WE = W$.

   (b) Use part (a) to deduce that, if E and F are both stochastic matrices, then EF is also stochastic.

7. Find a 2×2 matrix E with entries between 0 and 1 such that:

   (a) $I - E$ has no inverse.

   ◆(b) $I - E$ has an inverse but not all entries of $(I - E)^{-1}$ are nonnegative.

8. If E is a 2×2 matrix with entries between 0 and 1, show that $I - E$ is invertible and $(I - E)^{-1} \geq 0$ if and only if tr $E < 1 + \det E$. Here, if $E = \begin{bmatrix} a & b \\ c & d \end{bmatrix}$ then tr $E = a + d$ and $\det E = ad - bc$.

9. In each case show that $I - E$ is invertible and $(I - E)^{-1} \geq 0$.

   (a) $\begin{bmatrix} .6 & .5 & .1 \\ .1 & .3 & .3 \\ .2 & .1 & .4 \end{bmatrix}$ ◆(b) $\begin{bmatrix} .7 & .1 & .3 \\ .2 & .5 & .2 \\ .1 & .1 & .4 \end{bmatrix}$

   (c) $\begin{bmatrix} .6 & .2 & .1 \\ .3 & .4 & .2 \\ .2 & .5 & .1 \end{bmatrix}$ (d) $\begin{bmatrix} .8 & .1 & .1 \\ .3 & .1 & .2 \\ .3 & .3 & .2 \end{bmatrix}$

10. Prove that (1) implies (2) in the Corollary to Theorem 2.

11. If $(I - E)^{-1} \geq 0$, find $P > 0$ such that $EP < P$.

12. If $EP < P$ where $E \geq 0$ and $P > 0$, find a number μ such that $EP < \mu P$ and $0 < \mu < 1$.

Section 2.7   An Application to Markov Chains (Optional)

Many natural phenomena progress through various stages and can be in a variety of states at each stage. For example, the weather in a given city progresses day by day and, on any given day, may be sunny or rainy. Here the states are "sun" and "rain," and the weather progresses from one state to another in daily stages. Another example might be a football team: The stages of its evolution are the games it plays, and the possible states are "win," "draw," and "loss."

The general setup is as follows: A "system" evolves through a series of "stages," and at any stage it can be in any one of a finite number of "states." At any given stage, the state to which it will go at the next stage depends on the past and present history of the system—that is, on the sequence of states it has occupied to date. A **Markov chain** is such an evolving system wherein the state to which it will go next depends *only* on its *present* state and does not depend on the earlier history of the system.[12]

Even in the case of a Markov chain, the state the system will occupy at any stage is determined only in terms of probabilities. In other words, chance plays a role. For example, if a football team wins a particular game, we do not know whether it will win, draw, or lose the next game. On the other hand, we may know that the team tends to persist in winning streaks; for example, if it wins one game it may win the next game $\frac{1}{2}$ of the time, lose $\frac{4}{10}$ of the time and draw $\frac{1}{10}$ of the time. These fractions are called the **probabilities** of these various possibilities. Similarly, if the team loses, it may lose the next game with probability $\frac{1}{2}$ (that is, half the time), win with probability $\frac{1}{4}$, and draw with probability $\frac{1}{4}$. The probabilities of the various outcomes after a drawn game will also be known.

We shall treat probabilities informally here: *The probability that a given event will occur is the long-run proportion of the time that the event does indeed occur.* Hence all probabilities are numbers between 0 and 1. A probability of 0 means the event is impossible and never occurs; events with probability 1 are certain to occur.

If a Markov chain is in a particular state, the probabilities that it goes to the various states at the next stage of its evolution are called the **transition probabilities** for the chain, and they are assumed to be known quantities. To motivate the general conditions that follow, consider the following simple example. Here the system is a man, the stages are his successive lunches, and the states are the two restaurants he chooses.

EXAMPLE 1

A man always eats lunch at one of two restaurants, *A* and *B*. He never eats at *A* twice in a row. However, if he eats at *B*, he is three times as likely to eat at *B* next time as at *A*. Initially he is equally likely to eat at either restaurant.

---

[12]The name honors Andrei Andreyevich Markov (1856–1922) who was a professor at the university in St. Petersburg, Russia.

(a) What is the probability that he eats at A on the third day after the initial one?

(b) What proportion of his lunches does he eat at A?

Solution   The table of transition probabilities follows. The A column indicates that if he eats at A on one day, he never eats there again on the next day and so is certain to go to B.

		Present lunch	
		A	**B**
Next lunch	A	0	.25
	B	1	.75

The B column shows that, if he eats at B on one day, he will eat there on the next day $\frac{3}{4}$ of the time and switches to A only $\frac{1}{4}$ of the time.

The restaurant he visits on a given day is not determined. The most that we can expect is to know the probability that he will visit A or B on that day. Let $S_m = \begin{bmatrix} s_1^{(m)} \\ s_2^{(m)} \end{bmatrix}$ denote the *state vector* for day m. Here $s_1^{(m)}$ denotes the probability that he eats at A on day m, and $s_2^{(m)}$ is the probability that he eats at B. It is convenient to let S_0 correspond to the initial day. Because he is equally likely to eat at A or B on that initial day, $s_1^{(0)} = .5$ and $s_2^{(0)} = .5$, so $S_0 = \begin{bmatrix} .5 \\ .5 \end{bmatrix}$.

Now let

$$P = \begin{bmatrix} 0 & .25 \\ 1 & .75 \end{bmatrix}$$

denote the *transition matrix*. We claim that the relationship

$$S_{m+1} = PS_m$$

holds for all m. This will be derived later; for now, we use it as follows to successively compute $S_1, S_2, S_3, \ldots$.

$$S_1 = PS_0 = \begin{bmatrix} 0 & .25 \\ 1 & .75 \end{bmatrix} \begin{bmatrix} .5 \\ .5 \end{bmatrix} = \begin{bmatrix} .125 \\ .875 \end{bmatrix}$$

$$S_2 = PS_1 = \begin{bmatrix} 0 & .25 \\ 1 & .75 \end{bmatrix} \begin{bmatrix} .125 \\ .875 \end{bmatrix} = \begin{bmatrix} .21875 \\ .78125 \end{bmatrix}$$

$$S_3 = PS_2 = \begin{bmatrix} 0 & .25 \\ 1 & .75 \end{bmatrix} \begin{bmatrix} .21875 \\ .78125 \end{bmatrix} = \begin{bmatrix} .1953125 \\ .8046875 \end{bmatrix}$$

Hence the probability that his third lunch (after the initial one) is at A is approximately .195, whereas the probability that it is at B is .805.

If we carry these calculations on, the next state vectors are (to five figures)

$$S_4 = \begin{bmatrix} .20117 \\ .79883 \end{bmatrix} \qquad S_5 = \begin{bmatrix} .19971 \\ .80029 \end{bmatrix}$$

$$S_6 = \begin{bmatrix} .20007 \\ .79993 \end{bmatrix} \qquad S_7 = \begin{bmatrix} .19998 \\ .80002 \end{bmatrix}$$

Moreover, the higher values of S_m get closer and closer to $\begin{bmatrix} .2 \\ .8 \end{bmatrix}$. Hence, in the long run, he eats 20% of his lunches at A and 80% at B.

◆◆◆

Example 1 incorporates most of the essential features of all Markov chains. The general model is as follows: The system evolves through various stages and at each stage can be in exactly one of n distinct states. It progresses through a sequence of states as time goes on.

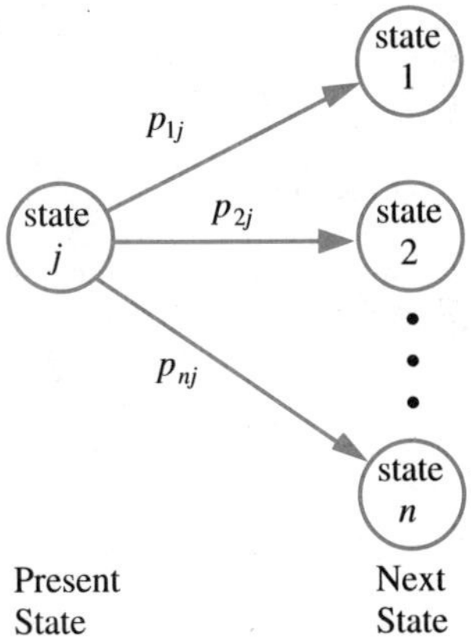

DEFINITION

If a Markov chain is in state j at a particular stage of its development, the probability p_{ij} that it goes to state i at the next stage is called the **transition probability**. The $n \times n$ matrix $P = [p_{ij}]$ is called the **transition matrix** for the Markov chain.

FIGURE 2.1

The situation is depicted graphically in Figure 2.1.

We make one important assumption about the transition matrix $P = [p_{ij}]$: It does *not* depend on which stage the process is in. This assumption means that the transition probabilities are *independent of time* — that is, they do not change as time goes on. It is this assumption that distinguishes Markov chains in the literature of this subject.

EXAMPLE 2

Suppose the transition matrix of a three-state Markov chain is

$$
\begin{array}{cc}
 & \text{Present state} \\
 & \begin{array}{ccc} 1 & 2 & 3 \end{array}
\end{array}
$$

$$
P = \begin{bmatrix} p_{11} & p_{12} & p_{13} \\ p_{21} & p_{22} & p_{23} \\ p_{31} & p_{32} & p_{33} \end{bmatrix} = \begin{bmatrix} .3 & .1 & .6 \\ .5 & .9 & .2 \\ .2 & 0 & .2 \end{bmatrix} \begin{array}{c} 1 \\ 2 \\ 3 \end{array} \quad \text{Next state}
$$

If, for example, the system is in state 2, column 2 lists the probabilities of where it goes next. Thus the probability is $p_{12} = .1$ that it goes from state 2 to state 1, and the probability is $p_{22} = .9$ that it goes from state 2 to state 2. The fact that $p_{32} = 0$ means that it is impossible for it to go from state 2 to state 3 at the next stage.

◆◆◆

Consider the jth column of the transition matrix P.

$$\begin{bmatrix} p_{1j} \\ p_{2j} \\ \vdots \\ p_{nj} \end{bmatrix}$$

If the system is in state j at some stage of its evolution, the transition probabilities p_{1j}, $p_{2j}, \ldots, p_{nj}$ represent the fraction of the time that the system will move to state 1, state 2, $\ldots$, state n, respectively, at the next stage. We assume that it has to go to *some* state at each transition, so the sum of these probabilities equals 1:

$$p_{1j} + p_{2j} + \cdots + p_{nj} = 1 \qquad \text{for each } j$$

Thus the columns of P all sum to 1 and the entries of P lie between 0 and 1. A matrix with these properties is called a **stochastic matrix**.

As in Example 1, we introduce the following notation: Let $s_i^{(m)}$ denote the probability that the system is in state i after m transitions. The $n \times 1$ matrices

$$S_m = \begin{bmatrix} s_1^{(m)} \\ s_2^{(m)} \\ \vdots \\ s_n^{(m)} \end{bmatrix} \qquad m = 0, 1, 2, \ldots$$

are called the **state vectors** for the Markov chain. Note that the sum of the entries of S_m must equal 1, because the system must be in *some* state after m transitions. The matrix S_0 is called the **initial state vector** for the Markov chain and is given as part of the data of the particular chain. For example, if the chain has only two states, then an initial vector $S_0 = \begin{bmatrix} 1 \\ 0 \end{bmatrix}$ means that it started in state 1. If it started in state 2, the initial vector would be $S_0 = \begin{bmatrix} 0 \\ 1 \end{bmatrix}$. If $S_0 = \begin{bmatrix} .5 \\ .5 \end{bmatrix}$, it is equally likely that the system started in state 1 or in state 2.

THEOREM 1

Let P be the transition matrix for an n-state Markov chain. If S_m is the state vector at stage m, then

$$S_{m+1} = PS_m$$

for each $m = 0, 1, 2, \ldots$.

Heuristic Proof   Suppose that the Markov chain has been run N times, each time starting with the same initial state vector. Recall that p_{ij} is the proportion of the time the system goes from state j at some stage to state i at the next stage, whereas $s_i^{(m)}$ is the proportion of the time it is in state i at stage m. Hence

$$s_i^{(m+1)}N$$

is (approximately) the number of times the system is in state i at stage $m + 1$. We are going to calculate this number another way. The system got to state i at stage $m + 1$ through *some* other state (say state j) at stage m. The number of times it was *in* state j at that stage is (approximately) $s_j^{(m)}N$, so the number of times it got to state i via state j is $p_{ij}(s_j^{(m)}N)$. Summing over j gives the number of times the system is in state i (at stage $m + 1$). This is the number we calculated before, so

$$s_i^{(m+1)}N = p_{i1}s_1^{(m)}N + p_{i2}s_2^{(m)}N + \cdots + p_{in}s_n^{(m)}N$$

Cancelling N gives $s_i^{(m+1)} = p_{i1}s_1^{(m)} + p_{i2}s_2^{(m)} + \cdots + p_{in}s_n^{(m)}$ for each i, and this can be expressed as the matrix equation $S_{m+1} = PS_m$. ◆

If the initial probability vector S_0 and the transition matrix P are given, Theorem 1 gives $S_1, S_2, S_3, \ldots$, one after the other, as follows:

$$S_1 = PS_0$$
$$S_2 = PS_1$$
$$S_3 = PS_2$$
$$\vdots$$

Hence the state vector S_m is completely determined for each $m = 0, 1, 2, \ldots$ by P and S_0.

EXAMPLE 3

A wolf pack always hunts in one of three regions R_1, R_2, and R_3. Its hunting habits are as follows:

1. If it hunts in one region one day, it is as likely as not to hunt there again the next day.

2. If it hunts in R_1, it never hunts in R_2 the next day.

3. If it hunts in R_2 or R_3, it is equally likely to hunt in each of the other regions the next day.

If the pack hunts in R_1 on Monday, find the probability that it hunts there on Thursday.

	R_1	R_2	R_3
R_1	$\frac{1}{2}$	$\frac{1}{4}$	$\frac{1}{4}$
R_2	0	$\frac{1}{2}$	$\frac{1}{4}$
R_3	$\frac{1}{2}$	$\frac{1}{4}$	$\frac{1}{2}$

Solution   The stages of this process are the successive days; the states are the three regions. The transition matrix P is determined as follows (see the table): The first habit asserts that $p_{11} = p_{22} = p_{33} = \frac{1}{2}$. Now column 1 displays what happens when the pack starts in R_1: It never goes to state 2, so $p_{21} = 0$ and, because the column must sum to 1, $p_{31} = \frac{1}{2}$. Column 2 describes what happens if it starts in R_2: $p_{22} = \frac{1}{2}$ and p_{12} and p_{32} are equal (by habit 3), so $p_{12} = p_{32} = \frac{1}{4}$ because the column sum must equal 1. Column 3 is filled in a similar way.

Now let Monday be the initial stage. Then $S_0 = \begin{bmatrix} 1 \\ 0 \\ 0 \end{bmatrix}$ because the pack hunts in R_1 on that day. Then S_1, S_2, and S_3 describe Tuesday, Wednesday, and Thursday, respectively, and we compute them using Theorem 1.

$$S_1 = PS_0 = \begin{bmatrix} \frac{1}{2} \\ 0 \\ \frac{1}{2} \end{bmatrix} \qquad S_2 = PS_1 = \begin{bmatrix} \frac{3}{8} \\ \frac{1}{8} \\ \frac{4}{8} \end{bmatrix} \qquad S_2 = PS_2 = \begin{bmatrix} \frac{11}{32} \\ \frac{6}{32} \\ \frac{15}{32} \end{bmatrix}$$

Hence the probability that the pack hunts in Region R_1 on Thursday is $\frac{11}{32}$.

◆◆◆

Another phenomenon that was observed in Example 1 can be expressed in general terms. The state vectors $S_0, S_1, S_2, \ldots$ were calculated in that example and were found to "approach" $S = \begin{bmatrix} .2 \\ .8 \end{bmatrix}$. That means that the first component of S_m becomes and remains very close to .2 as m becomes large, whereas the second component approaches .8 as m increases. When this is the case, we say that S_m **converges** to S. For large m, then, there is very little error in taking $S_m = S$, so the long-term probability that the system is in state 1 is .2, whereas the probability that it is in state 2 is .8. In Example 1, enough state vectors were computed for the limiting vector S to be apparent. However, there is a better way to do this that works in most cases.

Suppose P is the transition matrix of a Markov chain, and assume that the state vectors S_m converge to a limiting vector S. Then S_m is very close to S for sufficiently large m, so S_{m+1} is also very close to S. Thus the equation $S_{m+1} = PS_m$ from Theorem 1 is closely approximated by

$$S = PS$$

so it is not surprising that S should be a solution to this matrix equation. Moreover, it is easily solved because it can be written as a system of linear equations

$$(I - P)S = 0$$

with the entries of S as variables.

In Example 1, where $P = \begin{bmatrix} 0 & .25 \\ 1 & .75 \end{bmatrix}$, the general solution to $(I - P)S = 0$ is

$S = \begin{bmatrix} t \\ 4t \end{bmatrix}$, where t is a parameter. But if we insist that the entries of S sum to 1

(as must be true of all state vectors), we find $t = .2$ and so $S = \begin{bmatrix} .2 \\ .8 \end{bmatrix}$ as before.

All this is predicated on the existence of a limiting vector for the sequence of state vectors of the Markov chain, and such a vector may not always exist. However, it does exist in one commonly occurring situation. A stochastic matrix P is called **regular** if some power P^m of P has every entry positive. The matrix $P = \begin{bmatrix} 0 & .25 \\ 1 & .75 \end{bmatrix}$ of Example 1 is regular (in this case, each entry of P^2 is positive), and the general theorem is as follows:

THEOREM 2

Let P be the transition matrix of a Markov chain and assume that P is regular. Then there is a unique column matrix S satisfying the following conditions.

1. $PS = S$.
2. The entries of S are positive and sum to 1.

Moreover, condition 1 can be written as

$$(I - P)S = 0$$

and so gives a homogeneous system of linear equations for S. Finally, the sequence of state vectors $S_0, S_1, S_2, \ldots$ converges to S in the sense that if m is large enough, each entry of S_m is closely approximated by the corresponding entry of S.

This theorem will not be proved here.[13]

If P is the regular transition matrix of a Markov chain, the column S satisfying conditions 1 and 2 of Theorem 2 is called the **steady-state vector** for the Markov chain. The entries of S are the long-term probabilities that the chain will be in each of the various states.

---

[13]The interested reader can find an elementary proof in J. Kemeny, H. Mirkil, J. Snell, and G. Thompson, *Finite Mathematical Structures* (Englewood Cliffs, N.J.: Prentice-Hall, 1958).

EXAMPLE 4

A man eats one of three soups — beef, chicken, and vegetable — each day. He never eats the same soup two days in a row. If he eats beef soup on a certain day, he is equally likely to eat each of the others the next day; if he does not eat beef soup, he is twice as likely to eat it the next day as the alternative.

(a)   If he has beef soup one day, what is the probability that he has it again two days later?

(b)   What are the long-run probabilities that he eats each of the three soups?

	B	C	V
B	0	$\frac{2}{3}$	$\frac{2}{3}$
C	$\frac{1}{2}$	0	$\frac{1}{3}$
V	$\frac{1}{2}$	$\frac{1}{3}$	0

Solution   The states here are B, C, and V, the three soups. The transition matrix P is given in the table. (Recall that for each state, the corresponding column lists the probabilities for the next state.) If he has beef soup initially, then the initial state vector is

$$S_0 = \begin{bmatrix} 1 \\ 0 \\ 0 \end{bmatrix}$$

Then two days later the state vector is S_2. If P is the transition matrix, then

$$S_1 = PS_0 = \frac{1}{2}\begin{bmatrix} 0 \\ 1 \\ 1 \end{bmatrix}, \qquad S_2 = PS_1 = \frac{1}{6}\begin{bmatrix} 4 \\ 1 \\ 1 \end{bmatrix}$$

so he eats beef soup two days later with probability $\frac{2}{3}$. This answers (a) and also shows that he eats chicken and vegetable soup each with probability $\frac{1}{6}$.

To find the long-run probabilities, we must find the steady-state vector S. Theorem 2 applies because P is regular (P^2 has positive entries), so S satisfies $PS = S$. That is, $(I - P)S = 0$ where

$$I - P = \frac{1}{6}\begin{bmatrix} 6 & -4 & -4 \\ -3 & 6 & -2 \\ -3 & -2 & 6 \end{bmatrix}$$

The solution is $S = \begin{bmatrix} 4t \\ 3t \\ 3t \end{bmatrix}$, where t is a parameter, and we use $S = \begin{bmatrix} .4 \\ .3 \\ .3 \end{bmatrix}$ because the entries of S must sum to 1. Hence, in the long run, he eats beef soup 40% of the time and eats chicken soup and vegetable soup each 30% of the time.

$\blacklozenge\blacklozenge\blacklozenge$

EXERCISES 2.7

1. Which of the following stochastic matrices is regular?

(a) $\begin{bmatrix} 0 & 0 & \frac{1}{2} \\ 1 & 0 & \frac{1}{2} \\ 0 & 1 & 0 \end{bmatrix}$ ◆**(b)** $\begin{bmatrix} \frac{1}{2} & 0 & \frac{1}{3} \\ \frac{1}{4} & 1 & \frac{1}{3} \\ \frac{1}{4} & 0 & \frac{1}{3} \end{bmatrix}$

2. In each case find the steady-state vector and, assuming that it starts in state 1, find the probability that it is in state 2 after 3 transitions.

(a) $\begin{bmatrix} .5 & .3 \\ .5 & .7 \end{bmatrix}$ ◆**(b)** $\begin{bmatrix} \frac{1}{2} & 1 \\ \frac{1}{2} & 0 \end{bmatrix}$ **(c)** $\begin{bmatrix} 0 & \frac{1}{2} & \frac{1}{4} \\ 1 & 0 & \frac{1}{4} \\ 0 & \frac{1}{2} & \frac{1}{2} \end{bmatrix}$

◆**(d)** $\begin{bmatrix} .4 & .1 & .5 \\ .2 & .6 & .2 \\ .4 & .3 & .3 \end{bmatrix}$ **(e)** $\begin{bmatrix} .8 & 0 & .2 \\ .1 & .6 & .1 \\ .1 & .4 & .7 \end{bmatrix}$ ◆**(f)** $\begin{bmatrix} .1 & .3 & .3 \\ .3 & .1 & .6 \\ .6 & .6 & .1 \end{bmatrix}$

3. A fox hunts in three territories A, B, and C. He never hunts in the same territory on two successive days. If he hunts in A, then he hunts in C the next day. If he hunts in B or C, he is twice as likely to hunt in A the next day as in the other territory.

(a) What proportion of his time does he spend in A, in B, and in C?

(b) If he hunts in A on Monday (C on Monday), what is the probability that he will hunt in B on Thursday?

4. Assume that there are three classes — upper, middle, and lower — and that social mobility behaves as follows:
1. Of the children of upper-class parents, 70% remain upper-class, whereas 10% become middle-class and 20% become lower-class.
2. Of the children of middle-class parents, 80% remain middle-class, whereas the others are evenly split between the upper class and the lower class.
3. For the children of lower-class parents, 60% remain lower-class, whereas 30% become middle-class and 10% upper-class.

(a) Find the probability that the grandchild of lower-class parents becomes upper-class.

◆**(b)** Find the long-term breakdown of society into classes.

5. The Prime Minister says she will call an election. This gossip is passed from person to person with a probability $p \neq 0$ that the information is passed incorrectly at any stage. Assume that when a person hears the gossip he or she passes it to one person who does not know. Find the long-term probability that a person will hear that there is going to be an election.

◆**6.** John makes it to work on time one Monday out of four. On other work days his behavior is as follows: If he is late one day, he is twice as likely to come to work on time the next day as to be late. If he is on time one day, he is as likely to be late as not the next day. Find the probability of his being late and that of his being on time Wednesdays.

7. Suppose you have 1¢ and match coins with a friend. At each match you either win or lose 1¢ with equal probability. If you go broke or ever get 4¢, you quit. Assume your friend never quits. If the states are 0, 1, 2, 3, and 4 representing your wealth, show that the corresponding transition matrix P is not regular. Find the probability that you will go broke after 3 matches.

8. A mouse is put into a maze of compartments, as in the diagram. Assume that he always leaves any compartment he enters and that he is equally likely to take any tunnel entry.

◆**(a)** If he starts in compartment 1, find the probability that he is in compartment 4 after 3 moves.

◆**(b)** Find the compartment in which he spends most of his time if he is left for a long time.

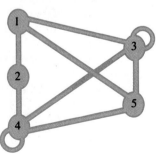

9. If a stochastic matrix has a 1 on its main diagonal, show that it cannot be regular. Assume it is not 1×1.

10. If S_m is the stage-m state vector for a Markov chain, show that $S_{m+k} = P^k S_m$ holds for all $m \geq 1$ and $k \geq 1$ (where P is the transition matrix).

11. A stochastic matrix is **doubly stochastic** if all the row sums also equal 1. Find the steady-state vector for a doubly stochastic matrix.

◆12. Consider the 2×2 stochastic matrix $P = \begin{bmatrix} 1-p & q \\ p & 1-q \end{bmatrix}$, where $0 < p < 1$ and $0 < q < 1$.

(a) Show that $\dfrac{1}{p+q} \begin{bmatrix} q \\ p \end{bmatrix}$ is the steady-state vector for P.

(b) Show that P^m converges to the matrix $\dfrac{1}{p+q} \begin{bmatrix} q & q \\ p & p \end{bmatrix}$ by first verifying inductively that

$$P^m = \frac{1}{p+q} \begin{bmatrix} q & q \\ p & p \end{bmatrix} + \frac{(1-p-q)^m}{p+q} \begin{bmatrix} p & -q \\ -p & q \end{bmatrix}$$

for $m = 1, 2, \ldots$. (It can be shown that the sequence of powers $P, P^2, P^3, \ldots$ of any regular transition matrix converges to the matrix each of whose columns equals the steady-state vector for P.)

SUPPLEMENTARY EXERCISES FOR CHAPTER 2

1. Solve for the matrix X if: **(a)** $PXQ = R$; **(b)** $XP = S$; where

$$P = \begin{bmatrix} 1 & 0 \\ 2 & -1 \\ 0 & 3 \end{bmatrix}, \quad Q = \begin{bmatrix} 1 & 1 & -1 \\ 2 & 0 & 3 \end{bmatrix},$$

$$R = \begin{bmatrix} -1 & 1 & -4 \\ -4 & 0 & -6 \\ 6 & 6 & -6 \end{bmatrix}, \quad S = \begin{bmatrix} 1 & 6 \\ 3 & 1 \end{bmatrix}$$

2. Consider $p(X) = X^3 - 5X^2 + 11X - 4I$.

(a) If $p(A) = \begin{bmatrix} 1 & 3 \\ -1 & 0 \end{bmatrix}$, compute $p(A^T)$.

◆(b) If $p(U) = 0$ where U is $n \times n$, find U^{-1} in terms of U.

3. Show that, if a (possibly nonhomogeneous) system of equations is consistent and has more variables than equations, then it must have infinitely many solutions. [*Hint:* Use Theorem 2§2.2 and Theorem 1§1.3.]

4. Assume that a system $AX = B$ of linear equations has at least two distinct solutions Y and Z.

(a) Show that $X_k = Y + k(Y - Z)$ is a solution for every k.

◆(b) Show that $X_k = X_m$ implies $k = m$. [*Hint:* See Example 7§2.1.]

(c) Deduce that $AX = B$ has infinitely many solutions.

5. (a) Let A be a 3×3 matrix with all entries on and below the main diagonal zero. Show that $A^3 = 0$.

(b) Generalize to the $n \times n$ case and prove your answer.

6. Let I_{pq} denote the $n \times n$ matrix with (p, q)-entry equal to 1 and all other entries 0. Show that:

(a) $I_n = I_{11} + I_{22} + \cdots + I_{nn}$

(b) $I_{pq} I_{rs} = \begin{cases} I_{ps} \\ 0 \end{cases}$

(c) If $A = [a_{ij}]$ is $n \times n$, then $A = \displaystyle\sum_{i=1}^{n} \sum_{j=1}^{n} a_{ij} I_{ij}$.

◆(d) If $A = [a_{ij}]$, then $I_{pq} A I_{rs} = a_{qr} I_{ps}$ for all p, q, r, and s.

7. A matrix of the form aI_n, where a is a number, is called an $n \times n$ **scalar matrix.**

(a) Show that each $n \times n$ scalar matrix commutes with every $n \times n$ matrix.

◆(b) Show that A is a scalar matrix if it commutes with every $n \times n$ matrix. [*Hint:* See part **(d)** of Exercise 6.]

8. Let $M = \begin{bmatrix} A & B \\ C & D \end{bmatrix}$, where A, B, C, and D are all $n \times n$ and each commutes with all the others. If $M^2 = 0$, show that $(A + D)^3 = 0$. [*Hint:* First show that $A^2 = -BC = D^2$ and that $B(A + D) = 0 = C(A + D)$.]

9. If A is 2×2, show that $A^{-1} = A^T$ if and only if $A = \begin{bmatrix} \cos\theta & \sin\theta \\ -\sin\theta & \cos\theta \end{bmatrix}$ for some θ or $A = \begin{bmatrix} \cos\theta & \sin\theta \\ \sin\theta & -\cos\theta \end{bmatrix}$ for some θ. [*Hint:* If $a^2 + b^2 = 1$, then $a = \cos\theta$, $b = \sin\theta$ for some θ. Use $\cos(\theta - \varphi) = \cos\theta \cos\varphi + \sin\theta \sin\varphi$.]

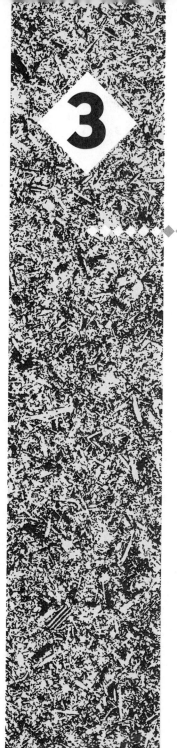

3

Determinants

With each square matrix we can associate a number called the determinant of the matrix. The determinant of a matrix tells us whether or not the matrix is invertible; in fact, determinants can be used to give a formula for the inverse of a matrix. Determinants also give us a method (called Cramer's rule) for solving linear equations. Another important application of determinants to matrix theory will be found in Chapter 6.

The theory of determinants is older than the theory of matrices (Cramer's rule was published in 1750), and determinants were extensively studied in the eighteenth and nineteenth centuries. Although they are somewhat less important today, they still play a role in the theory and applications of matrices.

Section 3.1
The Laplace Expansion

It is a well-known fact that division by a number a is allowed only if $a \neq 0$. In other words a^{-1} exists if and only if $a \neq 0$. Let A denote a square matrix. The **determinant** of A is a number, denoted det A or $|A|$, that can be computed from the entries of A and which enables us to tell whether A is invertible. In fact, we show later that A is invertible if and only if det $A \neq 0$. In this section det A will be defined, and some methods for computing it will be given.

There is no difficulty if A is 1×1.

DEFINITION

Given a 1×1 matrix $[a]$, define $\det[a] = a$.

Then $[a]$ has an inverse if and only if $\det[a] \neq 0$. In fact, the inverse is $[1/a]$.

DEFINITION

If $A = \begin{bmatrix} a & b \\ c & d \end{bmatrix}$, define $\det A = \begin{vmatrix} a & b \\ c & d \end{vmatrix} = ad - bc$.

EXAMPLE 1

$\begin{vmatrix} 3 & 5 \\ 2 & 4 \end{vmatrix} = 12 - 10 = 2 \qquad$ and $\qquad \begin{vmatrix} -1 & -1 \\ 0 & 3 \end{vmatrix} = -3 - 0 = -3.$

◆◆◆

We now show that a 2×2 matrix A has an inverse if and only if $\det A \neq 0$. Write $A = \begin{bmatrix} a & b \\ c & d \end{bmatrix}$ and consider the matrix $B = \begin{bmatrix} d & -b \\ -c & a \end{bmatrix}$, called the **adjoint** of A. Direct computation gives

$$AB = (\det A)I = BA$$

If $\det A \neq 0$, we can multiply through by $\dfrac{1}{\det A}$ to obtain the formula

$$A^{-1} = \frac{1}{\det A} B$$

On the other hand, if A^{-1} exists, we claim that $\det A \neq 0$. For if $\det A = 0$, then $AB = (\det A)I = 0$. This implies that $B = A^{-1}AB = A^{-1}0 = 0$, and hence that $a = b = c = d = 0$. But then $A = 0$, contrary to the assumption that A is invertible.

EXAMPLE 2

In each case, calculate the determinant of A and find the inverse (if it exists) from the formula.

(a) $A = \begin{bmatrix} 5 & 4 \\ 2 & 3 \end{bmatrix}$ (b) $A = \begin{bmatrix} 2 & 3 \\ 6 & 9 \end{bmatrix}$

Solution   (a) $\det A = \begin{vmatrix} 5 & 4 \\ 2 & 3 \end{vmatrix} = 5 \cdot 3 - 4 \cdot 2 = 7$, so A^{-1} exists. The adjoint of A is $B = \begin{bmatrix} 3 & -4 \\ -2 & 5 \end{bmatrix}$, so $A^{-1} = \dfrac{1}{\det A} B = \dfrac{1}{7} \begin{bmatrix} 3 & -4 \\ -2 & 5 \end{bmatrix}$.

(b) In this case, $\det A = \begin{vmatrix} 2 & 3 \\ 6 & 9 \end{vmatrix} = 2 \cdot 9 - 3 \cdot 6 = 0$. Hence A has no inverse.

◆◆◆

Here is a procedure for defining the determinant of any $n \times n$ matrix. Once we know how to define determinants of 2×2 matrices, we give a rule by which the determinant of any 3×3 matrix can be defined (in terms of certain determinants of 2×2 matrices). Next we do 4×4 matrices in terms of 3×3 matrices, and so on. At each stage we give a rule for defining the determinant of a square matrix in terms of determinants of square matrices one size smaller.

Before stating this rule, we must introduce minors and cofactors of a square matrix. Assume that it has been specified how to compute determinants of $(n - 1) \times (n - 1)$ matrices.

DEFINITION

The **(i, j)-minor** of an $n \times n$ matrix A, denoted $M_{ij}(A)$, is defined to be the determinant of the $(n - 1) \times (n - 1)$ matrix formed from A by deleting row i and column j. Next the number $C_{ij}(A) = (-1)^{i+j}M_{ij}(A)$ is called the **(i, j)-cofactor** of A and $(-1)^{i+j}$ is called the **sign** of the (i, j)-position.

Clearly $C_{ij}(A)$ equals either $M_{ij}(A)$ or $-M_{ij}(A)$, depending on the choice of i and j. The following sign diagram is a useful device for remembering the sign of a position.

$$\begin{bmatrix} +1 & -1 & +1 & -1 & +1 & \cdots \\ -1 & +1 & -1 & +1 & -1 & \cdots \\ +1 & -1 & +1 & -1 & +1 & \cdots \\ -1 & +1 & -1 & +1 & -1 & \cdots \\ \vdots & \vdots & \vdots & \vdots & \vdots & \end{bmatrix}$$

Note that the signs alternate along each row and column and that the sign of position $(1, 1)$ is $+1$. We have already decided how to compute the determinant of a 2×2 matrix, so we can find the minors and cofactors for any 3×3 matrix.

EXAMPLE 3

Find the minors and cofactors of positions $(1, 2)$, $(3, 1)$, and $(2, 3)$ in the following matrix.

$$A = \begin{bmatrix} 3 & -1 & 6 \\ 5 & 2 & 7 \\ 8 & 9 & 4 \end{bmatrix}$$

Solution   The $(1, 2)$-minor is the determinant of the matrix $\begin{bmatrix} 5 & 7 \\ 8 & 4 \end{bmatrix}$ that remains when row 1 and column 2 are deleted. The sign of position $(1, 2)$ is $(-1)^{1+2} = -1$ (this is also the $(1, 2)$-entry of the sign diagram), so the $(1, 2)$-minor and the $(1, 2)$-cofactor are

$$M_{12}(A) = \begin{vmatrix} 5 & 7 \\ 8 & 4 \end{vmatrix} = 5 \cdot 4 - 7 \cdot 8 = -36$$

$$C_{12}(A) = (-1)^{1+2} M_{12}(A) = (-1)(-36) = 36$$

Turning to position $(3, 1)$, we find

$$M_{31}(A) = \begin{vmatrix} -1 & 6 \\ 2 & 7 \end{vmatrix} = (-1) \cdot 7 - 6 \cdot 2 = -19$$

$$C_{31}(A) = (-1)^{3+1} M_{31}(A) = (+1)(-19) = -19$$

Finally, the $(2, 3)$-minor and the $(2, 3)$-cofactor are

$$M_{23}(A) = \begin{vmatrix} 3 & -1 \\ 8 & 9 \end{vmatrix} = 3 \cdot 9 - (-1) \cdot 8 = 35$$

$$C_{23}(A) = (-1)^{2+3} M_{23}(A) = (-1) \cdot 53 = -35$$

Clearly other minors and cofactors can be found—there are nine in all, one for each position in the matrix.

◆◆◆

With the notion of minor and cofactor in hand, we can formulate the rule for finding the determinant of an $n \times n$ matrix A. Recall that the idea is to do this inductively—that is, to find a way to compute $\det A$ in terms of determinants of certain $(n - 1) \times (n - 1)$ matrices.

DEFINITION

Given $n \geq 2$, assume that $\det M$ has been defined for any $(n - 1) \times (n - 1)$ matrix M. If A is $n \times n$, define

$$\det A = a_{11}C_{11}(A) + a_{21}C_{21}(A) + \cdots + a_{n1}C_{n1}(A)$$

In other words, the definition says that $\det A$ can be found by multiplying each entry a_{i1} in the first column by the corresponding cofactor $C_{i1}(A)$ and adding the results. This is called the **Laplace expansion**[1] of $\det A$ along the first column. The astonishing thing is that $\det A$ can be computed by taking the Laplace expansion along *any* column: Simply multiply the entries of the column by the corresponding cofactors and add the results. Even more remarkably, $\det A$ can also be found from the Laplace expansion along any row.

---

[1]This expansion was first used by Pierre Simon de Laplace (1749 –1827). He is most remembered for his work on celestial mechanics and probability theory.

THEOREM 1
Laplace Expansion

The determinant of an $n \times n$ matrix A can be computed by using the Laplace expansion along any row or column of A. More precisely, if $A = [a_{ij}]$ so that a_{ij} is the (i, j)-entry of A, then the expansion along row i is

$$\det A = a_{i1}C_{i1}(A) + a_{i2}C_{i2}(A) + a_{i3}C_{i3}(A) + \cdots + a_{in}C_{in}(A)$$

The expansion along column j is given by

$$\det A = a_{1j}C_{1j}(A) + a_{2j}C_{2j}(A) + a_{3j}C_{3j}(A) + \cdots + a_{nj}C_{nj}(A)$$

The proof will be given in Section 3.4.

EXAMPLE 4

Compute the determinant of $A = \begin{bmatrix} 3 & 4 & 5 \\ 1 & 7 & 2 \\ 9 & 8 & -6 \end{bmatrix}$.

Solution   The Laplace expansion along the first row is as follows:

$$\begin{aligned}
\det A &= 3C_{11}(A) + 4C_{12}(A) + 5C_{13}(A) \\
&= 3M_{11}(A) - 4M_{12}(A) + 5M_{13}(A) \\
&= 3\begin{vmatrix} 7 & 2 \\ 8 & -6 \end{vmatrix} - 4\begin{vmatrix} 1 & 2 \\ 9 & -6 \end{vmatrix} + 5\begin{vmatrix} 1 & 7 \\ 9 & 8 \end{vmatrix} \\
&= 3(-58) - 4(-24) + 5(-55) \\
&= -353
\end{aligned}$$

Now we compute $\det A$ by expanding along the first column.

$$\begin{aligned}
\det A &= 3C_{11}(A) + 1C_{21}(A) + 9C_{31}(A) \\
&= 3M_{11}(A) - M_{21}(A) + 9M_{31}(A) \\
&= 3\begin{vmatrix} 7 & 2 \\ 8 & -6 \end{vmatrix} - \begin{vmatrix} 4 & 5 \\ 8 & -6 \end{vmatrix} + 9\begin{vmatrix} 4 & 5 \\ 7 & 2 \end{vmatrix} \\
&= 3(-58) - (-64) + 9(-27) \\
&= -353
\end{aligned}$$

The reader is invited to verify that $\det A$ can be computed by expanding along the second or third row or along the second or third column.

◆◆◆

The fact that the Laplace expansion along *any row or column* of a matrix A always gives the same result (the determinant of A) is remarkable, to say the least. The choice of a particular row or column can simplify the calculation.

EXAMPLE 5

Compute det A where $A = \begin{bmatrix} 3 & 0 & 0 & 0 \\ 5 & 1 & 2 & 0 \\ 2 & 6 & 0 & -1 \\ -6 & 3 & 1 & 0 \end{bmatrix}$.

Solution
The first choice we must make is which row or column to use in the Laplace expansion. The expansion involves multiplying entries by cofactors, so the work is minimized when the row or column contains as many zeros as possible. Row 1 is a best choice in this matrix (column 4 would do as well), and the expansion is

$$\det A = 3C_{11}(A) + 0C_{12}(A) + 0C_{13}(A) + 0C_{14}(A)$$
$$= 3M_{11}(A) + 0 + 0 + 0$$
$$= 3 \begin{vmatrix} 1 & 2 & 0 \\ 6 & 0 & -1 \\ 3 & 1 & 0 \end{vmatrix}$$

This is the first stage of the calculation, and we have succeeded in expressing the determinant of (the 4×4 matrix) A in terms of the determinant of a 3×3 matrix. The next stage involves this 3×3 matrix. Again, we can use any row or column for the Laplace expansion. The third column is preferred (with two zeros), so

$$\det A = 3 \left(0 \begin{vmatrix} 6 & 0 \\ 3 & 1 \end{vmatrix} - (-1) \begin{vmatrix} 1 & 2 \\ 3 & 1 \end{vmatrix} + 0 \begin{vmatrix} 1 & 2 \\ 6 & 0 \end{vmatrix} \right)$$
$$= 3[0 + 1(-5) + 0]$$
$$= -15$$

This completes the calculation.

◆◆◆

Computing the determinant of matrix A can be tedious, even using the Laplace expansion. For example, if A is a 4×4 matrix, the Laplace expansion along any row or column involves calculating four minors, each of which is itself the determinant of a 3×3 matrix. And if A is 5×5, the expansion involves five determinants of 4×4 matrices! There is a clear need for some techniques to cut down the work.

The motivation for the method (see Example 5) is the observation that calculating a determinant is simplified a great deal when a row or column consists mostly of zeros. (In fact, when a row or column consists *entirely* of zeros, the determinant is zero — simply expand along that row or column.)

Recall next that one method of *creating* zeros in a matrix is to apply elementary row operations to it. Hence a natural question to ask is what effect such a row operation has on the determinant of the matrix. It turns out that the effect is easy to determine and that elementary *column* operations can be used in the same way. These observations lead to a technique for evaluating determinants that greatly reduces the labor involved. The necessary information is given in Theorem 2.

THEOREM 2

Let A denote an $n \times n$ matrix.

1. If A has a row or column of zeros, det $A = 0$.

2. If two distinct rows (or columns) of A are interchanged, the determinant of the resulting matrix is $-\det A$.

3. If a row (or column) of A is multiplied by a constant u, the determinant of the resulting matrix is $u(\det A)$.

4. If two distinct rows (or columns) of A are identical, det $A = 0$.

5. If a multiple of one row of A is added to a different row (or if a multiple of a column is added to a different column), the determinant of the resulting matrix is det A.

Proof   We prove properties 2, 4, and 5 and leave the rest as exercises.

Property 2. If A is $n \times n$, this follows by induction on n. If $n = 2$, the verification is left to the reader. If $n > 2$ and two rows are interchanged, let B denote the resulting matrix. Expand det A and det B along a row *other than* the two that were interchanged. The entries in this row are the same for both A and B, but the cofactors in B are the negatives of those in A (by induction) because the corresponding $(n - 1) \times (n - 1)$ matrices have two rows interchanged. Hence det $B = -\det A$, as required. A similar argument works if two columns are interchanged.

Property 4. If two rows of A are equal, let B be the matrix obtained by interchanging them. Then $B = A$, so det $B = \det A$. But det $B = -\det A$ by property 2, so det $A = \det B = 0$. Again, the same argument works for columns.

Property 5. Let B be obtained from $A = [a_{ij}]$ by adding u times row p to row q. Then row q of B is $(a_{q1} + ua_{p1}, a_{q2} + ua_{p2}, \ldots, a_{qn} + ua_{pn})$. The cofactors of these elements in B are the same as in A (they do not involve row q); in symbols, $C_{qj}(B) = C_{qj}(A)$ for each j. Hence, expanding B along row q gives

$$\det B = \sum_{j=1}^{n} (a_{qj} + u a_{pj}) C_{qj}(B)$$

$$= \sum_{j=1}^{n} a_{qj} C_{qj}(A) + u \sum_{j=1}^{n} a_{pj} C_{qj}(A)$$

$$= \det A + u \det C$$

where C is the matrix obtained from A by replacing row q by row p (and both expansions are along row q). Because rows p and q of C are equal, $\det C = 0$ by property 4. Hence $\det B = \det A$, as required. As before, a similar proof holds for columns. ◆

To illustrate Theorem 2, consider the following matrices.

$$\begin{vmatrix} 3 & -1 & 2 \\ 2 & 5 & 1 \\ 0 & 0 & 0 \end{vmatrix} = 0$$

(because the last row consists of zeros)

$$\begin{vmatrix} 3 & -1 & 5 \\ 2 & 8 & 7 \\ 1 & 2 & -1 \end{vmatrix} = - \begin{vmatrix} 5 & -1 & 3 \\ 7 & 8 & 2 \\ -1 & 2 & 1 \end{vmatrix}$$

(because two columns are interchanged)

$$\begin{vmatrix} 8 & 1 & 2 \\ 3 & 0 & 9 \\ 1 & 2 & -1 \end{vmatrix} = 3 \begin{vmatrix} 8 & 1 & 2 \\ 1 & 0 & 3 \\ 1 & 2 & -1 \end{vmatrix}$$

(because the second row of the matrix on the left is 3 times the second row of the matrix on the right)

$$\begin{vmatrix} 2 & 1 & 2 \\ 4 & 0 & 4 \\ 1 & 3 & 1 \end{vmatrix} = 0$$

(because two columns are identical)

$$\begin{vmatrix} 2 & 5 & 2 \\ -1 & 2 & 9 \\ 3 & 1 & 1 \end{vmatrix} = \begin{vmatrix} 0 & 9 & 20 \\ -1 & 2 & 9 \\ 3 & 1 & 1 \end{vmatrix}$$

(because twice the second row of the matrix on the left was added to the first row)

The following four examples illustrate how Theorem 2 is used to evaluate determinants.

EXAMPLE 6

Evaluate $\det A$ when $A = \begin{bmatrix} 1 & -1 & 3 \\ 1 & 0 & -1 \\ 2 & 1 & 6 \end{bmatrix}$.

Solution   The matrix does have zero entries, so expansion along (say) the second row would involve somewhat less work. However, a column operation can be used to get a zero in position $(2, 3)$—namely, add column 1 to column 3. Because this does not change the value of the determinant, we obtain

$$\det A = \begin{vmatrix} 1 & -1 & 3 \\ 1 & 0 & -1 \\ 2 & 1 & 6 \end{vmatrix} = \begin{vmatrix} 1 & -1 & 4 \\ 1 & 0 & 0 \\ 2 & 1 & 8 \end{vmatrix} = -\begin{vmatrix} -1 & 4 \\ 1 & 8 \end{vmatrix} = 12$$

where we expanded the second 3×3 matrix along row 2. ◆◆◆

EXAMPLE 7

If $\det \begin{bmatrix} a & b & c \\ p & q & r \\ x & y & z \end{bmatrix} = 6$, evaluate det A where $A = \begin{bmatrix} a + x & b + y & c + z \\ 3x & 3y & 3z \\ -p & -q & -r \end{bmatrix}$.

Solution   First take common factors out of rows 2 and 3.

$$\det A = 3(-1) \det \begin{bmatrix} a + x & b + y & c + z \\ x & y & z \\ p & q & r \end{bmatrix}$$

Now subtract the second row from the first and interchange the last two rows.

$$\det A = -3 \det \begin{bmatrix} a & b & c \\ x & y & z \\ p & q & r \end{bmatrix} = 3 \det \begin{bmatrix} a & b & c \\ p & q & r \\ x & y & z \end{bmatrix} = 3 \cdot 6 = 18$$

◆◆◆

The determinant of a matrix is a sum of products of its entries. In particular, if these entries are polynomials in x, then the determinant itself is a polynomial in x. It is often of interest to determine which values of x make the determinant zero, so it is very useful if the determinant is given in factored form. Theorem 2 can help.

EXAMPLE 8

Find the values of x for which det $A = 0$, where $A = \begin{bmatrix} 1 & x & x \\ x & 1 & x \\ x & x & 1 \end{bmatrix}$.

Solution   To evaluate det A, subtract x times row 1 from rows 2 and 3.

$$\det A = \begin{vmatrix} 1 & x & x \\ x & 1 & x \\ x & x & 1 \end{vmatrix} = \begin{vmatrix} 1 & x & x \\ 0 & 1 - x^2 & x - x^2 \\ 0 & x - x^2 & 1 - x^2 \end{vmatrix} = \begin{vmatrix} 1 - x^2 & x - x^2 \\ x - x^2 & 1 - x^2 \end{vmatrix}$$

At this stage we could simply evaluate the determinant (the result is $2x^3 - 3x^2 + 1$). Then we would have to factor this polynomial to find the values of x that make it zero. However, this factorization can be obtained directly by first factoring each entry in the determinant and taking a common factor of $(1 - x)$ from each row.

$$\det A = \begin{vmatrix} (1 - x)(1 + x) & x(1 - x) \\ x(1 - x) & (1 - x)(1 + x) \end{vmatrix} = (1 - x)^2 \begin{vmatrix} 1 + x & x \\ x & 1 + x \end{vmatrix}$$

$$= (1 - x)^2 (2x + 1)$$

Hence $\det A = 0$ means $(1 - x)^2(2x + 1) = 0$, so $x = 1$ or $x = -\frac{1}{2}$.   ◆◆◆

EXAMPLE 9

If a_1, a_2, and a_3 are given, show that

$$\det \begin{bmatrix} 1 & 1 & 1 \\ a_1 & a_2 & a_3 \\ a_2^2 & a_2^2 & a_3^2 \end{bmatrix} = (a_3 - a_2)(a_3 - a_1)(a_2 - a_1)$$

Solution   Begin by subtracting the second column from the third, and then subtract the first from the second.

$$\det \begin{bmatrix} 1 & 1 & 1 \\ a_1 & a_2 & a_3 \\ a_2^2 & a_2^2 & a_3^2 \end{bmatrix} = \det \begin{bmatrix} 1 & 0 & 0 \\ a_1 & a_2 - a_1 & a_3 - a_2 \\ a_1^2 & a_2^2 - a_1^2 & a_3^2 - a_2^2 \end{bmatrix}$$

$$= \det \begin{bmatrix} a_2 - a_1 & a_3 - a_2 \\ a_2^2 - a_1^2 & a_3^2 - a_2^2 \end{bmatrix}$$

Now $(a_2 - a_1)$ and $(a_3 - a_2)$ are common factors in the first and second columns, so

$$\det \begin{bmatrix} 1 & 1 & 1 \\ a_1 & a_2 & a_3 \\ a_2^2 & a_2^2 & a_3^2 \end{bmatrix} = (a_3 - a_2)(a_2 - a_1) \det \begin{bmatrix} 1 & 1 \\ a_2 + a_1 & a_3 + a_2 \end{bmatrix}$$

$$= (a_3 - a_2)(a_2 - a_1)(a_3 - a_1)$$

◆◆◆

The matrix in Example 9 is called a Vandermonde matrix, and the formula for its determinant can be generalized to the $n \times n$ case (see Theorem 2§3.3).

If A is an $n \times n$ matrix, forming uA means multiplying *every* row of A by u. Applying property 3 of Theorem 2 to each row gives the following useful result.

THEOREM 3

If A is an $n \times n$ matrix, then $\det(uA) = u^n \det A$ for any number u.

The next example displays a type of matrix whose determinant is easy to compute.

EXAMPLE 10

Evaluate det A if $A = \begin{bmatrix} a & 0 & 0 & 0 \\ u & b & 0 & 0 \\ v & w & c & 0 \\ x & y & z & d \end{bmatrix}$.

Solution   Expand along row 1 to get det $A = a\begin{vmatrix} b & 0 & 0 \\ w & c & 0 \\ y & z & d \end{vmatrix}$. Now expand this along the top row

to get det $A = ab\begin{vmatrix} c & 0 \\ z & d \end{vmatrix} = abcd$, the product of the main diagonal entries.

◆◆◆

A square matrix is called a **lower triangular matrix** if all entries above the main diagonal are zero (as in Example 10). Similarly, an **upper triangular matrix** is one for which all entries below the main diagonal are zero. A **triangular matrix** is one that is either upper or lower triangular. Theorem 4 gives an easy rule for calculating the determinant of any triangular matrix. The proof is like the solution to Example 10.

THEOREM 4

If A is a triangular matrix, then det A is the product of the entries on the main diagonal.

This theorem is useful in computer calculations because it is a routine matter to carry a matrix to triangular form using row operations.

Block matrices such as those in the next theorem arise frequently in practice, and the theorem gives an easy method of computing their determinants.

THEOREM 5

Consider matrices $\begin{bmatrix} A & X \\ 0 & B \end{bmatrix}$ and $\begin{bmatrix} A & 0 \\ Y & B \end{bmatrix}$ in block form, where A and B are square matrices. Then

$$\det\begin{bmatrix} A & X \\ 0 & B \end{bmatrix} = \det A \det B \qquad \text{and} \qquad \det\begin{bmatrix} A & 0 \\ Y & B \end{bmatrix} = \det A \det B$$

Proof Write $T = \begin{bmatrix} A & X \\ 0 & B \end{bmatrix}$ and proceed by induction on k where A is $k \times k$. If $k = 1$ it is easily verified. In general, compute det T using the Laplace expansion along the first column.

$$\det T = a_{11}M_{11}(T) - a_{21}M_{21}(T) + \cdots \pm a_{k1}M_{k1}(T) \qquad (*)$$

where $a_{11}, a_{21}, \ldots, a_{k1}$ denote the entries in the first column of A. The minor $M_{i1}(T)$ is the determinant of the submatrix $S_i(T)$ of T obtained by deleting row i and column 1 of T. Hence $S_i(T) = \begin{bmatrix} S_i(A) & X_i \\ 0 & B \end{bmatrix}$ in block form, so

$$M_{i1}(T) = \det\begin{bmatrix} S_i(A) & X_i \\ 0 & B \end{bmatrix} = \det[S_i(A)] \cdot \det B = M_{i1}(A) \cdot \det B$$

by induction. Hence det B is a common factor in equation $(*)$, and so

$$\det T = (a_{11}M_{11}(A) - a_{21}M_{21}(A) + \cdots \pm a_{k1}M_{k1}(A)) \det B$$
$$= (\det A)\ \det B$$

The proof of the other case is similar.   ◆

EXAMPLE 11

$$\det\begin{bmatrix} 3 & 1 & 0 & 0 & 0 \\ 2 & 1 & 0 & 0 & 0 \\ 0 & 1 & 3 & -1 & 2 \\ 2 & -1 & 5 & 0 & 0 \\ 3 & 1 & 1 & 0 & 1 \end{bmatrix} = \det\begin{bmatrix} 3 & 1 \\ 2 & 1 \end{bmatrix} \det\begin{bmatrix} 3 & -1 & 2 \\ 5 & 0 & 0 \\ 1 & 0 & 1 \end{bmatrix} = 1 \cdot 5 = 5.$$

◆◆◆

Theorem 5 extends to **block upper (lower) triangular matrices,** where each "diagonal block" is a square matrix and all blocks below (above) the main diagonal are zero. Then the determinant is the product of the determinants of the diagonal blocks.

EXERCISES 3.1

1. Compute the determinants of the following matrices.

(a) $\begin{bmatrix} 2 & -1 \\ 3 & 2 \end{bmatrix}$

◆(b) $\begin{bmatrix} 6 & 9 \\ 8 & 12 \end{bmatrix}$

(c) $\begin{bmatrix} a^2 & ab \\ ab & b^2 \end{bmatrix}$

◆(d) $\begin{bmatrix} a+1 & a \\ a & a-1 \end{bmatrix}$

(e) $\begin{bmatrix} \cos\theta & -\sin\theta \\ \sin\theta & \cos\theta \end{bmatrix}$

◆(f) $\begin{bmatrix} 2 & 0 & -3 \\ 1 & 2 & 5 \\ 0 & 3 & 0 \end{bmatrix}$

(g) $\begin{bmatrix} 1 & 2 & 3 \\ 4 & 5 & 6 \\ 7 & 8 & 9 \end{bmatrix}$

(i) $\begin{bmatrix} 1 & b & c \\ b & c & 1 \\ c & 1 & b \end{bmatrix}$

◆(h) $\begin{bmatrix} 0 & a & 0 \\ b & c & d \\ 0 & e & 0 \end{bmatrix}$

◆(j) $\begin{bmatrix} 0 & a & b \\ a & 0 & c \\ b & c & 0 \end{bmatrix}$

(k) $\begin{bmatrix} 1 & 0 & -1 & 0 \\ 0 & 3 & 0 & 2 \\ 1 & 0 & 2 & 1 \\ 0 & 5 & 0 & 7 \end{bmatrix}$ ◆ **(l)** $\begin{bmatrix} 1 & 0 & 3 & 1 \\ 2 & 2 & 6 & 0 \\ -1 & 0 & -3 & 1 \\ 4 & 1 & 12 & 0 \end{bmatrix}$

(m) $\begin{bmatrix} 3 & 1 & -5 & 2 \\ 1 & 0 & 5 & 2 \\ 1 & 3 & 0 & 1 \\ 1 & 1 & 2 & -1 \end{bmatrix}$ ◆ **(n)** $\begin{bmatrix} 4 & -1 & 3 & -1 \\ 3 & 1 & 0 & 2 \\ 0 & 1 & 2 & 2 \\ 1 & 2 & -1 & 1 \end{bmatrix}$

(o) $\begin{bmatrix} 1 & -1 & 5 & 2 \\ 3 & 1 & 2 & 4 \\ -1 & -3 & 8 & 0 \\ 1 & 1 & 2 & -1 \end{bmatrix}$ ◆ **(p)** $\begin{bmatrix} 0 & 0 & 0 & a \\ 0 & 0 & b & p \\ 0 & c & q & k \\ d & s & t & u \end{bmatrix}$

2. Show that det $A = 0$ if A has a row or column consisting of zeros.

3. Show that the sign of the position in the last row and the last column is always $+1$.

◆ **4.** Show that det $I = 1$ for any identity matrix I.

5. Evaluate each determinant by reducing it to upper triangular form.

(a) $\begin{bmatrix} 1 & -1 & 2 \\ 3 & 1 & 1 \\ 2 & -1 & 3 \end{bmatrix}$ ◆ **(b)** $\begin{bmatrix} -1 & 3 & 1 \\ 2 & 5 & 3 \\ 1 & -2 & 1 \end{bmatrix}$

(c) $\begin{bmatrix} -1 & -1 & 1 & 0 \\ 2 & 1 & 1 & 3 \\ 0 & 1 & 1 & 2 \\ 1 & 3 & -1 & 2 \end{bmatrix}$ ◆ **(d)** $\begin{bmatrix} 2 & 3 & 1 & 1 \\ 0 & 2 & -1 & 3 \\ 0 & 5 & 1 & 1 \\ 1 & 1 & 2 & 5 \end{bmatrix}$

6. Evaluate by inspection: **(a)** $\det \begin{bmatrix} a & b & c \\ a+1 & b+1 & c+1 \\ a-1 & b-1 & c-1 \end{bmatrix}$

◆ **(b)** $\det \begin{bmatrix} a & b & c \\ a+b & 2b & c+b \\ 2 & 2 & 2 \end{bmatrix}$

7. If $\det \begin{bmatrix} a & b & c \\ p & q & r \\ x & y & z \end{bmatrix} = -1$, compute:

(a) $\det \begin{bmatrix} -x & -y & -z \\ 3p+a & 3q+b & 3r+c \\ 2p & 2q & 2r \end{bmatrix}$

◆ **(b)** $\det \begin{bmatrix} -2a & -2b & -2c \\ 2p+x & 2q+y & 2r+z \\ 3x & 3y & 3z \end{bmatrix}$

8. Show that:

(a) $\det \begin{bmatrix} p+x & q+y & r+z \\ a+x & b+y & c+z \\ a+p & b+q & c+r \end{bmatrix} = 2 \det \begin{bmatrix} a & b & c \\ p & q & r \\ x & y & z \end{bmatrix}$

◆ **(b)** $\det \begin{bmatrix} 2a+p & 2b+q & 2c+r \\ 2p+x & 2q+y & 2r+z \\ 2x+a & 2y+b & 2z+c \end{bmatrix} = 9 \det \begin{bmatrix} a & b & c \\ p & q & r \\ x & y & z \end{bmatrix}$

9. Compute the determinants of each matrix, using Theorem 5.

(a) $\begin{bmatrix} 1 & -1 & 2 & 0 & -2 \\ 0 & 1 & 0 & 4 & 1 \\ 1 & 1 & 3 & 0 & 0 \\ 0 & 0 & 0 & 3 & -1 \\ 0 & 0 & 1 & 1 & 1 \end{bmatrix}$ ◆ **(b)** $\begin{bmatrix} 1 & 2 & 0 & 3 & 0 \\ -1 & 3 & 1 & 4 & 0 \\ 0 & 0 & 2 & 1 & 1 \\ 0 & 0 & -1 & 0 & 2 \\ 0 & 0 & 3 & 0 & 1 \end{bmatrix}$

10. If det $A = 2$, det $B = -1$, and det $C = 3$, find:

(a) $\det \begin{bmatrix} A & X & Y \\ 0 & B & Z \\ 0 & 0 & C \end{bmatrix}$ ◆ **(b)** $\det \begin{bmatrix} A & 0 & 0 \\ X & B & 0 \\ Y & Z & C \end{bmatrix}$

(c) $\det \begin{bmatrix} A & X & Y \\ 0 & B & 0 \\ 0 & Z & C \end{bmatrix}$ ◆ **(d)** $\det \begin{bmatrix} A & X & 0 \\ 0 & B & 0 \\ Y & Z & C \end{bmatrix}$

11. **(a)** Find det A if A is 3×3 and det $(2A) = 6$.
(b) Under what conditions is det $(-A) = $ det A?

12. Evaluate by first adding all other rows to the first row.

(a) $\det \begin{bmatrix} x-1 & 2 & 3 \\ 2 & -3 & x-2 \\ -2 & x & -2 \end{bmatrix}$

◆ **(b)** $\det \begin{bmatrix} x-1 & -3 & 1 \\ 2 & -1 & x-1 \\ -3 & x+2 & -2 \end{bmatrix}$

13. **(a)** Find b if $\det \begin{bmatrix} 3 & -1 & x \\ 2 & 6 & y \\ -5 & 4 & z \end{bmatrix} = ax + by + cz$.

◆ **(b)** Find c if $\det \begin{bmatrix} 2 & x & -1 \\ 1 & y & 3 \\ -3 & z & 4 \end{bmatrix} = ax + by + cz$.

14. Find the real numbers x^2 and y such that det $A = 0$ if:

(a) $A = \begin{bmatrix} 0 & x & y \\ y & 0 & x \\ x & y & 0 \end{bmatrix}$ ◆ **(b)** $A = \begin{bmatrix} 1 & x & x \\ -x & -2 & x \\ -x & -x & -3 \end{bmatrix}$

(c) $A = \begin{bmatrix} 1 & x & x^2 & x^3 \\ x & x^2 & x^3 & 1 \\ x^2 & x^3 & 1 & x \\ x^3 & 1 & x & x^2 \end{bmatrix}$ ◆ **(d)** $A = \begin{bmatrix} x & y & 0 & 0 \\ 0 & x & y & 0 \\ 0 & 0 & x & y \\ y & 0 & 0 & x \end{bmatrix}$

15. Show that $\det \begin{bmatrix} 0 & 1 & 1 & 1 \\ 1 & 0 & x & x \\ 1 & x & 0 & x \\ 1 & x & x & 0 \end{bmatrix} = -3x^2$.

16. Show that $\det \begin{bmatrix} 1 & x & x^2 & x^3 \\ a & 1 & x & x^2 \\ p & b & 1 & x \\ q & r & c & 1 \end{bmatrix} = (1 - ax)(1 - bx)(1 - cx)$.

17. Show that $\det \begin{bmatrix} x & -1 & 0 & 0 \\ 0 & x & -1 & 0 \\ 0 & 0 & x & -1 \\ a & b & c & x + d \end{bmatrix} = a + bx + cx^2 +$

$dx^3 + x^4$.

(This matrix is called the **companion matrix** of the polynomial $a + bx + cx^2 + dx^3 + x^4$.)

◆ **18.** Show that $\det \begin{bmatrix} a + x & b + x & c + x \\ b + x & c + x & a + x \\ c + x & a + x & b + x \end{bmatrix} =$

$(a + b + c + 3x)[(ab + ac + bc) - (a^2 + b^2 + c^2)]$.

19. If $C_1, C_2, \ldots, C_n$ denote the columns of a matrix A, write $A = [C_1 \, C_2 \cdots C_n]$ in block form. Show that:

$\det[C_1 + C_1' \, C_2 \cdots C_n] = \det[C_1 \, C_2 \cdots C_n]$
$+ \det[C_1' \, C_2 \cdots C_n]$

20. Show that

$\det \begin{bmatrix} 0 & 0 & \cdots & 0 & a_1 \\ 0 & 0 & \cdots & a_2 & * \\ \vdots & \vdots & & \vdots & \vdots \\ 0 & a_{n-1} & \cdots & * & * \\ a_n & * & \cdots & * & * \end{bmatrix} = (-1)^k a_1 a_2 \cdots a_n$

where either $n = 2k$ or $n = 2k + 1$, and $*$-entries are arbitrary.

21. By expanding along the first column, show that:

$\det \begin{bmatrix} 1 & 1 & 0 & 0 & \cdots & 0 & 0 \\ 0 & 1 & 1 & 0 & \cdots & 0 & 0 \\ 0 & 0 & 1 & 1 & \cdots & 0 & 0 \\ \vdots & \vdots & \vdots & \vdots & & \vdots & \vdots \\ 0 & 0 & 0 & 0 & \cdots & 1 & 1 \\ 1 & 0 & 0 & 0 & \cdots & 0 & 1 \end{bmatrix} = 1 + (-1)^{n+1}$

if the matrix is $n \times n$, $n \geq 2$.

◆ **22.** Form matrix B from a matrix A by writing the columns of A in reverse order. Express $\det B$ in terms of $\det A$.

23. Prove property 3 of Theorem 2 by expanding along the row (or column) in question.

24. Show that the line through two distinct points (x_1, y_1) and (x_2, y_2) in the plane has equation $\det \begin{bmatrix} x & y & 1 \\ x_1 & y_1 & 1 \\ x_2 & y_2 & 1 \end{bmatrix} = 0$.

25. Let A be an $n \times n$ matrix. Given a polynomial $p(x) = a_0 + a_1 x + \cdots + a_m x^m$, we write $p(A) = a_0 I + a_1 A + \cdots + a_m A^m$. For example, if $p(x) = 2 - 3x + 5x^2$, then $p(A) = 2I - 3A + 5A^2$. The *characteristic polynomial* of A is defined to be $c_A(x) = \det [xI - A]$ and the Cayley-Hamilton theorem asserts that $c_A(A) = 0$ for any matrix A.

(a) Verify the theorem for (i) $A = \begin{bmatrix} 3 & 2 \\ 1 & -1 \end{bmatrix}$ and

(ii) $A = \begin{bmatrix} 1 & -1 & 1 \\ 0 & 1 & 0 \\ 8 & 2 & 2 \end{bmatrix}$.

(b) Prove the theorem for $A = \begin{bmatrix} a & b \\ c & d \end{bmatrix}$.

Section 3.2   Determinants and Matrix Inverses

In this section, several theorems about determinants are derived. One consequence of these theorems is that a square matrix A is invertible if and only if $\det A \neq 0$. Moreover, determinants are used to give a formula for A^{-1} that, in turn, yields a formula (called Cramer's rule) for the solution of any system of linear equations with an invertible coefficient matrix.

We begin with a remarkable theorem about the determinant of a product of matrices. The proof is given at the end of this section.

THEOREM 1
Product Theorem

If A and B are $n \times n$ matrices, then $\det(AB) = \det A \det B$.

The complexity of matrix multiplication makes the product theorem quite unexpected. (The reader should verify it for arbitrary 2×2 matrices A and B.)

EXAMPLE 1

If $A = \begin{bmatrix} a & b \\ -b & a \end{bmatrix}$ and $B = \begin{bmatrix} c & d \\ -d & c \end{bmatrix}$, then $AB = \begin{bmatrix} ac - bd & ad + bc \\ -(ad + bc) & ac - bd \end{bmatrix}$.

Hence det A det B = det(AB) gives the identity

$$(a^2 + b^2)(c^2 + d^2) = (ac - bd)^2 + (ad + bc)^2$$

This identity is important for complex numbers.

◆◆◆

Theorem 1 extends easily to det(ABC) = det A det B det C. In fact, induction gives

$$\det(A_1 A_2 \cdots A_{k-1} A_k) = \det A_1 \det A_2 \cdots \det A_{k-1} \det A_k$$

for any square matrices $A_1, \ldots, A_k$ of the same size. In particular, if each $A_i = A$, we obtain

$$\det(A^k) = (\det A)^k \quad \text{for any } k \geq 1$$

We can now give the invertibility condition.

THEOREM 2

An $n \times n$ matrix A is invertible if and only if det $A \neq 0$. When this is the case,

$$\det (A^{-1}) = \frac{1}{\det A} .$$

Proof   If A is invertible, then $AA^{-1} = I$; so, using Theorem 1,

$$1 = \det I = \det(AA^{-1}) = \det A \det A^{-1}$$

Hence det $A \neq 0$ and det $A^{-1} = \dfrac{1}{\det A}$.

Conversely, if det $A \neq 0$, we show that A can be carried to I by elementary row operations (and invoke Theorem 4§2.4). Certainly, A can be carried to its reduced row-echelon form R, so $R = E_k \cdots E_2 E_1 A$ where the E_i are elementary matrices (Theorem 3§2.4). Hence

$$\det R = \det E_k \cdots \det E_2 \det E_1 \det A$$

Since det $E \neq 0$ for all elementary matrices E, this shows det $R \neq 0$. In particular R has no row of zeros, so $R = I$ (R is square). This is what we wanted.   ◆

EXAMPLE 2

For which values of c does $A = \begin{bmatrix} 1 & 0 & -c \\ -1 & 3 & 1 \\ 0 & 2c & -4 \end{bmatrix}$ have an inverse?

Solution   Compute $\det A$ by first adding c times column 1 to column 3 and then expanding along row 1.

$$\det A = \det \begin{bmatrix} 1 & 0 & -c \\ -1 & 3 & 1 \\ 0 & 2c & -4 \end{bmatrix} = \det \begin{bmatrix} 1 & 0 & 0 \\ -1 & 3 & 1-c \\ 0 & 2c & -4 \end{bmatrix} = 2(c+2)(c-3).$$

Hence $\det A = 0$ if $c = -2$ or $c = 3$, and A has an inverse if $c \neq -2$ and $c \neq 3$. ◆◆◆

EXAMPLE 3

Reprove Theorem 5§2.4: If $AB = I$, then A and B are invertible, $A = B^{-1}$, $B = A^{-1}$, and $BA = I$.

Solution   Take determinants to obtain $1 = \det I = \det(AB) = \det A \det B$. This implies $\det A \neq 0$ and $\det B \neq 0$, so A^{-1} and B^{-1} exist. Then left-multiplying $AB = I$ by A^{-1} yields $B = A^{-1}$, and $A = B^{-1}$ is derived similarly. Finally, $BA = BB^{-1} = I$. ◆◆◆

The next theorem is not too surprising in view of the similar way that row and column operations affect the determinant.

THEOREM 3

If A is any square matrix, $\det A^T = \det A$.

Proof   Consider first the case of an elementary matrix E. If E is of type I or II, then $E^T = E$; so certainly $\det E^T = \det E$. If E is of type III, then E^T is also of type III; so $\det E^T = 1 = \det E$ by Theorem 2§3.1. Hence $\det E^T = \det E$ for every elementary matrix E.

Now let A be any square matrix. If A is not invertible, then neither is A^T; so $\det A^T = 0 = \det A$ by Theorem 2. If A is invertible, then $A = E_k \cdots E_2 E_1$, where the E_i are elementary matrices (Theorem 4§2.4).

Hence $A^T = E_1^T E_2^T \cdots E_k^T$, so the product theorem gives

$$\det A^T = \det E_1^T \det E_2^T \cdots \det E_k^T = \det E_1 \det E_2 \cdots \det E_k$$
$$= \det E_k \cdots \det E_2 \det E_1$$
$$= \det A$$

This completes the proof. ◆

EXAMPLE 4

If $\det A = 2$ and $\det B = 5$, calculate $\det(A^3 B^{-1} A^T B^2)$.

Solution

We use several of the facts just derived.

$$\det(A^3 B^{-1} A^T B^2) = \det(A^3) \det B^{-1} \det A^T \det(B^2)$$

$$= (\det A)^3 \frac{1}{\det B} \det A \, (\det B)^2$$

$$= 2^3 \cdot \frac{1}{5} \cdot 2 \cdot 5^2$$

$$= 80$$

◆◆◆

EXAMPLE 5

A square matrix is called **orthogonal** if $A^{-1} = A^T$. What are the possible values of $\det A$ if A is orthogonal?

Solution

If A is orthogonal, we have $I = AA^T$. Take determinants to obtain $1 = \det I = \det(AA^T) = \det A \det A^T = (\det A)^2$. Hence $\det A = \pm 1$.

◆◆◆

Adjoints

At the beginning of Section 3.1 we defined the adjoint of a 2×2 matrix $A = \begin{bmatrix} a & b \\ c & d \end{bmatrix}$ to be $\text{adj}(A) = \begin{bmatrix} d & -b \\ -c & a \end{bmatrix}$. Then we verified that $A(\text{adj } A) = (\det A)I = (\text{adj } A)A$ and hence that, if $\det A \neq 0$, $A^{-1} = \frac{1}{\det A} \text{adj } A$. It is now possible to define the adjoint of an arbitrary square matrix and to show that this formula for the inverse is valid (when the inverse exists).

Recall that the (i, j)-cofactor $C_{ij}(A)$ of a square matrix A is a number defined for each position (i, j) in the matrix.

> **DEFINITION**
>
> If A is a square matrix, the **cofactor matrix of A** is defined to be the matrix $[C_{ij}(A)]$ whose (i, j)-entry is the (i, j)-cofactor of A. The **adjoint** of A, denoted $\text{adj}(A)$, is the transpose of this cofactor matrix; in symbols,
>
> $$\text{adj}(A) = [C_{ij}(A)]^T$$

This agrees with the earlier definition for a 2×2 matrix A.

EXAMPLE 6

Compute the adjoint of $A = \begin{bmatrix} 1 & 3 & -2 \\ 0 & 1 & 5 \\ -2 & -6 & 7 \end{bmatrix}$ and calculate $A(\text{adj } A)$ and $(\text{adj } A)A$.

Solution We first find the cofactor matrix.

$$\begin{bmatrix} C_{11}(A) & C_{12}(A) & C_{13}(A) \\ C_{21}(A) & C_{22}(A) & C_{23}(A) \\ C_{31}(A) & C_{32}(A) & C_{33}(A) \end{bmatrix} = \begin{bmatrix} \begin{vmatrix} 1 & 5 \\ -6 & 7 \end{vmatrix} & -\begin{vmatrix} 0 & 5 \\ -2 & 7 \end{vmatrix} & \begin{vmatrix} 0 & 1 \\ -2 & -6 \end{vmatrix} \\ -\begin{vmatrix} 3 & -2 \\ -6 & 7 \end{vmatrix} & \begin{vmatrix} 1 & -2 \\ -2 & 7 \end{vmatrix} & -\begin{vmatrix} 1 & 3 \\ -2 & -6 \end{vmatrix} \\ \begin{vmatrix} 3 & -2 \\ 1 & 5 \end{vmatrix} & -\begin{vmatrix} 1 & -2 \\ 0 & 5 \end{vmatrix} & \begin{vmatrix} 1 & 3 \\ 0 & 1 \end{vmatrix} \end{bmatrix}$$

$$= \begin{bmatrix} 37 & -10 & 2 \\ -9 & 3 & 0 \\ 17 & -5 & 1 \end{bmatrix}$$

Then the adjoint of A is the transpose of this cofactor matrix.

$$\text{adj } A = \begin{bmatrix} 37 & -10 & 2 \\ -9 & 3 & 0 \\ 17 & -5 & 1 \end{bmatrix}^T = \begin{bmatrix} 37 & -9 & 17 \\ -10 & 3 & -5 \\ 2 & 0 & 1 \end{bmatrix}$$

The computation of $A(\text{adj } A)$ gives

$$A(\text{adj } A) = \begin{bmatrix} 1 & 3 & -2 \\ 0 & 1 & 5 \\ -2 & -6 & 7 \end{bmatrix} \begin{bmatrix} 37 & -9 & 17 \\ -10 & 3 & -5 \\ 2 & 0 & 1 \end{bmatrix} = \begin{bmatrix} 3 & 0 & 0 \\ 0 & 3 & 0 \\ 0 & 0 & 3 \end{bmatrix} = 3I$$

and the reader can verify that also $(\text{adj } A)A = 3I$. Hence analogy with the 2×2 case would indicate that $\det A = 3$; this is in fact the case.

◆◆◆

The relationship $A(\text{adj } A) = (\det A)I$ holds for any square matrix A. To see why this is so, consider the general 3×3 case. Writing $C_{ij}(A) = C_{ij}$ for short, we have

$$\text{adj } A = \begin{bmatrix} C_{11} & C_{12} & C_{13} \\ C_{21} & C_{22} & C_{23} \\ C_{31} & C_{32} & C_{33} \end{bmatrix}^T = \begin{bmatrix} C_{11} & C_{21} & C_{31} \\ C_{12} & C_{22} & C_{32} \\ C_{13} & C_{23} & C_{33} \end{bmatrix}$$

If $A = [a_{ij}]$ in the usual notation, we are to verify that $A(\text{adj } A) = (\det A)I$. That is,

$$A(\text{adj } A) = \begin{bmatrix} a_{11} & a_{12} & a_{13} \\ a_{21} & a_{22} & a_{23} \\ a_{31} & a_{32} & a_{33} \end{bmatrix} \begin{bmatrix} C_{11} & C_{21} & C_{31} \\ C_{12} & C_{22} & C_{32} \\ C_{13} & C_{23} & C_{33} \end{bmatrix} = \begin{bmatrix} \det A & 0 & 0 \\ 0 & \det A & 0 \\ 0 & 0 & \det A \end{bmatrix}$$

Consider the (1, 1)-entry in the product. It is given by $a_{11}C_{11} + a_{12}C_{12} + a_{13}C_{13}$, and this is just the Laplace expansion of $\det A$ along the first row of A. Similarly, the (2, 2)-entry and the (3, 3)-entry are the Laplace expansions of $\det A$ along rows 2 and 3, respectively.

So it remains to be seen why the off-diagonal elements in the matrix product $A(\text{adj } A)$ are all zero. Consider the (1, 2)-entry of the product. It is given by $a_{11}C_{21} + a_{12}C_{22} + a_{13}C_{23}$. This *looks* like the Laplace expansion of the determinant of *some* matrix. To see which, observe that C_{21}, C_{22}, and C_{23} are all computed by *deleting* row 2 of A (and one of the columns), so they remain the same if row 2 of A is changed. In particular, if row 2 of A is replaced by row 1, we obtain

$$a_{11}C_{21} + a_{12}C_{22} + a_{13}C_{23} = \det \begin{bmatrix} a_{11} & a_{12} & a_{13} \\ a_{11} & a_{12} & a_{13} \\ a_{31} & a_{32} & a_{33} \end{bmatrix} = 0$$

where the expansion is along row 2 and where the determinant is zero because two rows are identical. A similar argument shows that the other off-diagonal entries are zero.

This argument works in general and yields the first part of Theorem 4. The second assertion follows from the first by multiplying through by the scalar $\dfrac{1}{\det A}$.

THEOREM 4
Adjoint Formula

If A is any square matrix, then

$$A(\text{adj } A) = (\det A)I = (\text{adj } A)A$$

In particular, if $\det A \neq 0$, the inverse of A is given by

$$A^{-1} = \frac{1}{\det A} \text{ adj } A$$

It is important to note that this theorem is *not* an efficient way to find the inverse of matrix A. For example, if A were 10×10, the calculation of adj A would require computing $10^2 = 100$ determinants of 9×9 matrices! On the other hand, the matrix inversion algorithm would find A^{-1} with about the same effort as finding $\det A$. Clearly Theorem 4 is not a *practical* result; its main virtue is that it gives a formula for A^{-1} that is useful for *theoretical* purposes.

EXAMPLE 7

Use Theorem 4 to find the inverse of $A = \begin{bmatrix} 1 & 1 & a \\ -a & 1 & -a \\ a & -1 & 1 \end{bmatrix}$ for the values of a for which it exists.

Solution   The adjoint is computed as follows:

$$\text{adj } A = \begin{bmatrix} C_{11} & C_{12} & C_{13} \\ C_{21} & C_{22} & C_{23} \\ C_{31} & C_{32} & C_{33} \end{bmatrix}^T = \begin{bmatrix} 1-a & a-a^2 & 0 \\ -1-a & 1-a^2 & 1+a \\ -2a & a-a^2 & 1+a \end{bmatrix}^T$$

$$= \begin{bmatrix} 1-a & -1-a & -2a \\ a-a^2 & 1-a^2 & a-a^2 \\ 0 & 1+a & 1+a \end{bmatrix}$$

The reader can verify that $A(\text{adj } A) = (1 - a^2)I = (\text{adj } A)A$, and this shows that det $A = 1 - a^2$ (as can be separately verified). Hence A^{-1} exists if $1 - a^2 \neq 0$ (that is, $a \neq \pm 1$), and in this case

$$A^{-1} = \frac{1}{1-a^2} \begin{bmatrix} 1-a & -1-a & -2a \\ a-a^2 & 1-a^2 & a-a^2 \\ 0 & 1+a & 1+a \end{bmatrix}$$

◆◆◆

EXAMPLE 8

If A is $n \times n$, $n \geq 2$, show that $\det(\text{adj } A) = (\det A)^{n-1}$.

Solution   Write $d = \det A$ so that $A(\text{adj } A) = dI$ by Theorem 4. Taking determinants gives $d \det(\text{adj } A) = d^n$, so we are done if $d \neq 0$. So assume $d = 0$; we must show that $\det(\text{adj } A) = 0$, that is adj A is not invertible. If $A \neq 0$, this follows from $A(\text{adj } A) = dI = 0$; if $A = 0$, it follows because adj $A = 0$.

◆◆◆

Theorem 4 has a nice application to linear equations. Suppose

$$AX = B$$

is a system of n equations in n variables $x_1, x_2, \ldots, x_n$. Here A is the $n \times n$ coefficient matrix, and X and B are the columns

$$X = \begin{bmatrix} x_1 \\ x_2 \\ \vdots \\ x_n \end{bmatrix} \qquad B = \begin{bmatrix} b_1 \\ b_2 \\ \vdots \\ b_n \end{bmatrix}$$

of variables and constants, respectively. If $\det A \neq 0$, we left-multiply by A^{-1} to obtain the solution $X = A^{-1}B$. When we use the adjoint formula, this becomes

$$\begin{bmatrix} x_1 \\ x_2 \\ \vdots \\ x_n \end{bmatrix} = \frac{1}{\det A}(\text{adj } A)B$$

$$= \frac{1}{\det A}\begin{bmatrix} C_{11}(A) & C_{21}(A) & \cdots & C_{n1}(A) \\ C_{12}(A) & C_{22}(A) & \cdots & C_{n2}(A) \\ \vdots & \vdots & & \vdots \\ C_{1n}(A) & C_{2n}(A) & \cdots & C_{nn}(A) \end{bmatrix}\begin{bmatrix} b_1 \\ b_2 \\ \vdots \\ b_n \end{bmatrix}$$

Hence the variables $x_1, x_2, \ldots, x_n$ are given by

$$x_1 = \frac{1}{\det A}[b_1 C_{11}(A) + b_2 C_{21}(A) + \cdots + b_n C_{n1}(A)]$$

$$x_2 = \frac{1}{\det A}[b_1 C_{12}(A) + b_2 C_{22}(A) + \cdots + b_n C_{n2}(A)]$$

$$\vdots \qquad\qquad\qquad\qquad\qquad\qquad \vdots$$

$$x_n = \frac{1}{\det A}[b_1 C_{1n}(A) + b_2 C_{2n}(A) + \cdots + b_n C_{nn}(A)]$$

Now the quantity $b_1 C_{11}(A) + b_2 C_{21}(A) + \cdots + b_n C_{n1}(A)$ occurring in the formula for x_1 looks like the Laplace expansion of the determinant of a matrix. The cofactors involved are $C_{11}(A), C_{21}(A), \ldots, C_{n1}(A)$, corresponding to the first column of A. If A_1 is obtained from A by replacing the first column of A by B, then $C_{i1}(A_1) = C_{i1}(A)$ for each i. Hence expanding $\det(A_1)$ by the first column gives

$$\det A_1 = b_1 C_{11}(A_1) + b_2 C_{21}(A_1) + \cdots + b_n C_{n1}(A_1)$$
$$= b_1 C_{11}(A) + b_2 C_{21}(A) + \cdots + b_n C_{n1}(A)$$
$$= (\det A)x_1$$

Hence $x_1 = \dfrac{\det A_1}{\det A}$, and similar results hold for the other variables.

THEOREM 5
Cramer's Rule[2]

If A is an invertible $n \times n$ matrix, the solution to the system

$$AX = B$$

of n equations in the variables $x_1, x_2, \ldots, x_n$ is given by

---

[2]Gabriel Cramer (1704 –1752) was a Swiss mathematician who wrote an introductory work on algebraic curves. He popularized the rule that bears his name, but the idea was known earlier.

$$x_1 = \frac{\det A_1}{\det A}, \qquad x_2 = \frac{\det A_2}{\det A}, \qquad \cdots, \qquad x_n = \frac{\det A_n}{\det A}$$

where, for each k, A_k is the matrix obtained from A by replacing column k by B.

EXAMPLE 9

Find x_1, given the following system of equations.

$$5x_1 + x_2 - x_3 = 4$$
$$9x_1 + x_2 - x_3 = 1$$
$$x_1 - x_2 + 5x_3 = 2$$

Solution Compute the determinant of the coefficient matrix A and the matrix A_1 obtained from it by replacing the first column by the column of constants.

$$\det A = \det \begin{bmatrix} 5 & 1 & -1 \\ 9 & 1 & -1 \\ 1 & -1 & 5 \end{bmatrix} = -16$$

$$\det A_1 = \det \begin{bmatrix} 4 & 1 & -1 \\ 1 & 1 & -1 \\ 2 & -1 & 5 \end{bmatrix} = 12$$

Hence $x_1 = (\det A_1)/\det A = -\frac{3}{4}$.

◆◆◆

Note that Cramer's rule enabled us to calculate x_1 here without computing x_2 or x_3. Although this might seem an advantage, the truth of the matter is that, for large systems of equations, the number of computations needed to find *all* the variables by the Gaussian algorithm is comparable to the number required to find *one* of the determinants involved in Cramer's rule. Furthermore, the algorithm works when the matrix of the system is not invertible and even when the coefficient matrix is not square. Like the adjoint formula, then, Cramer's rule is *not* a practical numerical technique; its virtue is theoretical.

Proof of Theorem 1

Theorem 1 asserts that $\det AB = \det A \det B$ holds for any two $n \times n$ matrices A and B. The proof we give involves elementary matrices.

Recall that each elementary matrix E is obtained by performing an elementary row operation on the identity matrix I. Because $\det I = 1$, Theorem 2§3.1 provides the following information:

$$\det E = \begin{cases} -1 \text{ if } E \text{ results from interchanging two rows of } I \text{ (type I)} \\ u \text{ if } E \text{ results from multiplying a row of } I \text{ by } u \neq 0 \text{ (type II)} \\ 1 \text{ if } E \text{ results from adding a multiple of one row of } I \text{ to} \\ \qquad \text{another row (type III)} \end{cases}$$

Next recall that when the elementary row operation that produced E from I is performed on any $n \times n$ matrix A, the resulting matrix is EA (Theorem 1§2.4). But then, $\det(EA)$ equals $-\det A$, $u \det A$, and $\det A$, respectively (again by Theorem 2§3.1). Combining this with the formulas for $\det E$, we have

$$\det(EA) = \det E \det A$$

This can be extended as follows: If E_1 and E_2 are both elementary, then

$$\det(E_2 E_1 A) = \det E_2 \det(E_1 A) = \det E_2 \det E_1 \det A$$

This process continues to produce Lemma 1.

LEMMA 1

If $E_1, E_2, \ldots, E_k$ are $n \times n$ elementary matrices, then

$$\det(E_k \cdots E_2 E_1 A) = \det E_k \cdots \det E_2 \det E_1 \det A$$

for any $n \times n$ matrix A.

This formula is the key to proving Theorem 1. The following preliminary result is needed.

LEMMA 2

If A is a noninvertible square matrix, then $\det A = 0$.

Proof Because A is *not* invertible, it *cannot* be carried to the identity matrix by row operations (Theorem 4§2.4). Therefore, if R is the reduced row-echelon form of A, R must have a row of zeros. Hence $\det R = 0$. On the other hand, R can be obtained from A by row operations, so (by Theorem 3§2.4) there exist elementary matrices $E_1, E_2, \ldots, E_k$ such that $R = E_k \cdots E_2 E_1 A$. But then Lemma 1 gives

$$0 = \det R = \det E_k \cdots \det E_2 \det E_1 \det A$$

so $\det A = 0$ (each $\det E_i \neq 0$ by the remark above).   ◆

Proof of Theorem 1 We must show that $\det AB = \det A \det B$. We split the argument into two cases.

Case 1. A is not invertible. In this case, det $A = 0$ by Lemma 2. Now observe that AB is not invertible either (if $(AB)^{-1}$ did exist, then $A[B(AB)^{-1}] = I$, so A would be invertible by Theorem 5§2.4). But then $\det(AB) = 0$ by Lemma 2 (applied to AB in place of A), so $\det AB = 0 = \det A \det B$, as required.

Case 2. A is invertible. Then A is a product of elementary matrices by Theorem 4§2.4, say $A = E_k \cdots E_2 E_1$. Hence Lemma 1 with $A = I$ gives

$$\det A = \det(E_k \cdots E_2 E_1) = \det E_k \cdots \det E_2 \det E_1.$$

Now use Lemma 1 once more to obtain

$$\det(AB) = \det(E_k \cdots E_2 E_1 B)$$
$$= \det E_k \cdots \det E_2 \det E_1 \det B$$
$$= \det A \det B$$

Hence the result holds in this case too. ◆

EXERCISES 3.2

1. Find the adjoint of each of the following matrices.
(a) $\begin{bmatrix} 3 & 1 & 2 \\ -1 & 1 & 3 \\ 1 & 3 & 8 \end{bmatrix}$ ◆(b) $\begin{bmatrix} 1 & -1 & 2 \\ 3 & 1 & 0 \\ 0 & -1 & 1 \end{bmatrix}$

(c) $\begin{bmatrix} 1 & 0 & 1 \\ 1 & 1 & 0 \\ 0 & 1 & 1 \end{bmatrix}$ ◆(d) $\frac{1}{3}\begin{bmatrix} -1 & 2 & 2 \\ 2 & -1 & 2 \\ 2 & 2 & -1 \end{bmatrix}$

2. Use determinants to find which real values of c make each of the following matrices invertible.
(a) $\begin{bmatrix} 1 & 0 & 2 \\ 3 & -4 & c \\ 2 & 5 & 8 \end{bmatrix}$ ◆(b) $\begin{bmatrix} 0 & c & -c \\ -1 & 2 & -1 \\ c & -c & c \end{bmatrix}$

(c) $\begin{bmatrix} c & 1 & 0 \\ 0 & 2 & c \\ -1 & c & 1 \end{bmatrix}$ ◆(d) $\begin{bmatrix} 4 & c & 3 \\ c & 2 & c \\ 5 & c & 4 \end{bmatrix}$

(e) $\begin{bmatrix} 1 & 2 & 4 \\ -1 & -1 & c \\ 2 & c & 1 \end{bmatrix}$ ◆(f) $\begin{bmatrix} 1 & c & -1 \\ c & 1 & 1 \\ 0 & 1 & c \end{bmatrix}$

3. Let A, B, and C denote $n \times n$ matrices and assume that $\det A = -1$, $\det B = 2$, and $\det C = 3$. Evaluate:
(a) $\det(A^2 B C^T B^{-1})$ ◆(b) $\det(B^2 C^{-1} A B^{-1} C^T)$

4. Let A and B be invertible $n \times n$ matrices. Evaluate:
(a) $\det(B^{-1}AB)$ ◆(b) $\det(A^{-1}B^{-1}AB)$

5. If A is 3×3 and $\det(2A^{-1}) = -4 = \det(A^3(B^{-1})^T)$, find $\det A$ and $\det B$.

6. Let $A = \begin{bmatrix} a & b & c \\ p & q & r \\ u & v & w \end{bmatrix}$ and assume that $\det A = 3$.
Compute:
(a) $\det(3B^{-1})$ where $B = \begin{bmatrix} 4u & 2a & -p \\ 4v & 2b & -q \\ 4w & 2c & -r \end{bmatrix}$

◆(b) $\det(2C^{-1})$ where $C = \begin{bmatrix} 2p & -a+u & 3u \\ 2q & -b+v & 3v \\ 2r & -c+w & 3w \end{bmatrix}$

7. If $\det\begin{bmatrix} a & b \\ c & d \end{bmatrix} = -2$, calculate:
(a) $\det\begin{bmatrix} 2 & -2 & 0 \\ c+1 & -1 & 2a \\ d-2 & 2 & 2b \end{bmatrix}$

◆(b) $\det\begin{bmatrix} 2b & 0 & 4d \\ 1 & 2 & -2 \\ a+1 & 2 & 2(c-1) \end{bmatrix}$

8. Solve each of the following by Cramer's rule:
(a) $2x + y = 1$
     $3x + 7y = -2$
◆(b) $3x + 4y = 9$
     $2x - y = -1$
(c) $5x + y - z = -7$
     $2x - y - 2z = 6$
     $3x + 2z = -7$
◆(d) $4x - y + 3z = 1$
     $6x + 2y - z = 0$
     $3x + 3y + 2z = -1$

9. Use Theorem 4 to find the $(2, 3)$-entry of A^{-1} if:
(a) $A = \begin{bmatrix} 3 & 2 & 1 \\ 1 & 1 & 2 \\ -1 & 2 & 1 \end{bmatrix}$ ◆(b) $A = \begin{bmatrix} 1 & 2 & -1 \\ 3 & 1 & 1 \\ 0 & 4 & 7 \end{bmatrix}$

10. Explain what can be said about det A if:

   (a) $A^2 = A$ ◆ **(b)** $A^2 = I$

   (c) $A^3 = A$ ◆ **(d)** $PA = P$, P invertible

   (e) $A^2 = uA$, A is $n \times n$ ◆ **(f)** $A = -A^T$, A is $n \times n$

   (g) $A^2 + I = 0$, A is $n \times n$

11. Let A be $n \times n$. Show that $uA = (uI)A$, and use this with Theorem 1 to deduce the result in Theorem 3§3.1: $\det(uA) = u^n \det A$.

12. If A and B are $n \times n$ matrices, $AB = -BA$, and n is odd, show that either A or B has no inverse.

13. Show that det $AB =$ det BA holds for any two $n \times n$ matrices A and B.

14. If $A^k = 0$ for some $k \geq 1$, show that A is not invertible.

15. If ABC is invertible (A, B, and C all square), show that B is invertible.

◆ **16.** If $A^{-1} = A^T$, describe the cofactor matrix of A in terms of A.

17. Show that no 3×3 matrix A exists such that $A^2 + I = 0$. Find a 2×2 matrix A with this property.

18. Show that $\det(A + B^T) = \det(A^T + B)$ for any $n \times n$ matrices A and B.

19. Let A and B be invertible $n \times n$ matrices. Show that det $A =$ det B if and only if $A = UB$, where U is a matrix with det $U = 1$.

20. For each of the matrices in Exercise 2, find the inverse for those values of c for which it exists.

21. If $A = \begin{bmatrix} 1 & a & b \\ -a & 1 & c \\ -b & -c & 1 \end{bmatrix}$, show that det $A = 1 + a^2 + b^2 + c^2$.

Hence find A^{-1} for any a, b, and c.

22. (a) Show that $A = \begin{bmatrix} a & p & q \\ 0 & b & r \\ 0 & 0 & c \end{bmatrix}$ has an inverse if and only if $abc \neq 0$, and find A^{-1} in that case.

  ◆ **(b)** Show that if an upper triangular matrix is invertible, the inverse is also upper triangular.

23. Let A be a matrix each of whose entries are integers. Show that each of the following conditions implies the other.

  (1) A is invertible and A^{-1} also has integer entries.

  (2) det $A = 1$ or -1

◆ **24.** If $A^{-1} = \begin{bmatrix} 3 & 0 & 1 \\ 0 & 2 & 3 \\ 3 & 1 & -1 \end{bmatrix}$, find adj A.

25. If A is 3×3 and det $A = 2$, find $\det(A^{-1} + 4$ adj $A)$.

26. If A and B are 2×2, show that $\det \begin{bmatrix} 0 & A \\ B & X \end{bmatrix} =$ det A det B.

What if A and B are 3×3? [*Hint:* Multiply by $\begin{bmatrix} 0 & I \\ I & 0 \end{bmatrix}$]

27. Let A be $n \times n$, $n \geq 2$, and assume one column of A consists of zeros. Find the possible values of rank (adj A).

28. If A is 3×3 and invertible, compute $\det(-A^2(\text{adj } A)^{-1})$.

29. Show that adj$(uA) = u^{n-1}$ adj A for all $n \times n$ matrices A.

30. Let A and B denote invertible $n \times n$ matrices. Show that:

  (a) adj$(\text{adj } A) = (\det A)^{n-2}A$ (here $n \geq 2$)

   [*Hint:* See Example 8.]

  ◆ **(b)** adj$(A^{-1}) = (\text{adj } A)^{-1}$

  (c) adj$(A^T) = (\text{adj } A)^T$

  ◆ **(d)** adj$(AB) = (\text{adj}B)(\text{adj}A)$ [*Hint:* Show that AB adj$(AB) = AB$ adj B adj A.]

Section 3.3   An Application to Polynomial Interpolation (Optional)

There often arise situations wherein two variables x and y are related but the actual functional form $y = f(x)$ of the relationship is unknown. Suppose that for certain values $x_1, x_2, \ldots, x_n$ of x, the corresponding values $y_1, y_2, \ldots, y_n$ are known (say from experimental measurements). One way to estimate the value of y corresponding to some other value of x is to find a polynomial $p(x)$ that "fits" the data, that is, $p(x_i) = y_i$ holds for each $i = 1, 2, \ldots, n$. Then the estimate for y is $p(x)$. Such a polynomial always exists if the x_i are distinct.

THEOREM 1

Let n data pairs (x_1, y_1), (x_2, y_2), . . . , (x_n, y_n) be given, and assume that the x_i are distinct. Then there exists a unique polynomial

$$p(x) = r_0 + r_1x + r_2x^2 + \cdots + r_{n-1}x^{n-1}$$

such that $p(x_i) = y_i$ for each $i = 1, 2, \ldots, n$.

Proof   The conditions that $p(x_i) = y_i$ are

$$r_0 + r_1x_1 + r_2x_1^2 + \cdots + r_{n-1}x_1^{n-1} = y_1$$
$$r_0 + r_1x_2 + r_2x_2^2 + \cdots + r_{n-1}x_2^{n-1} = y_2$$
$$\vdots \qquad \vdots \qquad \vdots \qquad\qquad \vdots \qquad \vdots$$
$$r_0 + r_1x_n + r_2x_n^2 + \cdots + r_{n-1}x_n^{n-1} = y_n$$

In matrix form, this is

$$\begin{bmatrix} 1 & x_1 & x_1^2 & \cdots & x_1^{n-1} \\ 1 & x_2 & x_2^2 & \cdots & x_2^{n-1} \\ \vdots & \vdots & \vdots & & \vdots \\ 1 & x_n & x_n^2 & \cdots & x_n^{n-1} \end{bmatrix} \begin{bmatrix} r_0 \\ r_1 \\ \vdots \\ r_{n-1} \end{bmatrix} = \begin{bmatrix} y_1 \\ y_2 \\ \vdots \\ y_n \end{bmatrix}$$

It can be shown (see Theorem 2) that the determinant of the coefficient matrix equals the product of all terms $(x_i - x_j)$ with $i > j$ and so is nonzero (because the x_i are distinct). Hence the equations have a unique solution $r_0, r_1, \ldots, r_{n-1}$ and the theorem follows.   ◆

The polynomial in Theorem 1 is called the **interpolating polynomial** for the data.

EXAMPLE 1

Find a polynomial $p(x)$ of degree 2 such that $p(0) = 1, p(1) = 3$, and $p(2) = 2$.

Solution   Write $p(x) = r_0 + r_1x + r_2x^2$. The conditions are

$$p(0) = r_0 \qquad\qquad = 1$$
$$p(1) = r_0 + r_1 + r_2 = 3$$
$$p(2) = r_0 + 2r_1 + 4r_2 = 2$$

The solution is $r_0 = 1, r_1 = \frac{7}{2}$, and $r_2 = -\frac{3}{2}$, so $p(x) = \frac{1}{2}(2 + 7x - 3x^2)$.   ◆◆◆

The next example shows how Theorem 1 is used in interpolation.

EXAMPLE 2

Given the data values

$$(0, 1.21), (1, 3.53), (2, 5.01), (3, 3.79)$$

use polynomial interpolation to estimate the value of y corresponding to $x = 1.5$.

Solution   We find the polynomial $p(x)$ of the degree 3 that fits these data. If $p(x) = r_0 + r_1 x + r_2 x^2 + r_3 x^3$, the conditions are

$$
\begin{aligned}
r_0 &= 1.21 \\
r_0 + r_1 + r_2 + r_3 &= 3.53 \\
r_0 + 2r_1 + 4r_2 + 8r_3 &= 5.01 \\
r_0 + 3r_1 + 9r_2 + 27r_3 &= 3.79
\end{aligned}
$$

The solution is $r_0 = 1.21$, $r_1 = 2.12$, $r_2 = 0.51$, and $r_3 = -0.31$, so the interpolating polynomial is $p(x) = 1.21 + 2.12x + 0.51x^2 - 0.31x^3$. Hence the estimated value of y corresponding to $x = 1.5$ is $p(1.5) = 4.49$.

◆◆◆

As a final example, we construct a polynomial that approximates a known function. This type of approximation is often useful in practical situations because polynomials are easy to compute.

EXAMPLE 3

Find a cubic polynomial $p(x)$ that approximates the function $\sin x$ on the interval $0 \le x \le \frac{\pi}{2}$ (x in radians). Use the following values of $\sin x$.

$$\sin 0 = 0, \quad \sin \left(\tfrac{\pi}{6}\right) = 0.5000, \quad \sin \left(\tfrac{\pi}{3}\right) = 0.8660, \quad \sin \left(\tfrac{\pi}{2}\right) = 1$$

Then use $p(x)$ to approximate $\sin(0.5)$.

Solution   If $p(x) = r_0 + r_1 x + r_2 x^2 + r_3 x^3$, use $x_0 = 0$, $x_1 = \frac{\pi}{6}$, $x_2 = \frac{\pi}{3}$, and $x_3 = \frac{\pi}{2}$, and $y_i = \sin(x_i)$ as given.

$$p(0) = r_0 = 0$$

$$p\left(\frac{\pi}{6}\right) = r_0 + r_1\left(\frac{\pi}{6}\right) + r_2\left(\frac{\pi}{6}\right)^2 + r_3\left(\frac{\pi}{6}\right)^3 = 0.5000$$

$$p\left(\frac{\pi}{3}\right) = r_0 + r_1\left(\frac{\pi}{3}\right) + r_2\left(\frac{\pi}{3}\right)^2 + r_3\left(\frac{\pi}{3}\right)^3 = 0.8660$$

$$p\left(\frac{\pi}{2}\right) = r_0 + r_1\left(\frac{\pi}{2}\right) + r_2\left(\frac{\pi}{2}\right)^2 + r_3\left(\frac{\pi}{2}\right)^3 = 1.0000$$

Clearly $r_0 = 0$, so multiplying the remaining equations by $\frac{6}{\pi}$, $\frac{3}{\pi}$, and $\frac{2}{\pi}$, respectively, gives

$$r_1 + \frac{r_2 \pi}{6} + \frac{r_3 \pi^2}{36} = 0.9549$$

$$r_1 + \frac{r_2 \pi}{3} + \frac{r_3 \pi^2}{9} = 0.8270$$

$$r_1 + \frac{r_2 \pi}{2} + \frac{r_3 \pi^2}{4} = 0.6366$$

If these are regarded as equations in r_1, $r_2 \pi$, and $r_3 \pi^2$, the coefficient matrix has an inverse:

$$\begin{bmatrix} 1 & \frac{1}{6} & \frac{1}{36} \\ 1 & \frac{1}{3} & \frac{1}{9} \\ 1 & \frac{1}{2} & \frac{1}{4} \end{bmatrix}^{-1} = \begin{bmatrix} 3 & -3 & 1 \\ -15 & 24 & -9 \\ 18 & -36 & 18 \end{bmatrix}$$

This leads to $r_1 = 1.0203$, $r_2 \pi = -0.2049$, and $r_3 \pi^2 = -1.1250$. Finally, then, $r_2 = -0.0652$ and $r_3 = -0.1140$, so

$$p(x) = 1.0203x - 0.0652x^2 - 0.1140x^3$$

This gives $p(0.5) = 0.4796$ as the approximation to $\sin(0.5)$. The true value is $\sin(0.5) = 0.4794$ to four decimal places. This is quite good, and even better approximations are achieved with polynomials of higher degree.

◆◆◆

We conclude this section by evaluating the determinant of the matrix that arose in the proof of Theorem 1. If $a_1, a_2, \ldots, a_n$ are numbers, the $n \times n$ matrix

$$\begin{bmatrix} 1 & 1 & 1 & \cdots & 1 & 1 \\ a_1 & a_2 & a_3 & \cdots & a_{n-1} & a_n \\ a_1^2 & a_2^2 & a_3^2 & \cdots & a_{n-1}^2 & a_n^2 \\ \vdots & \vdots & \vdots & & \vdots & \vdots \\ a_1^{n-1} & a_2^{n-1} & a_3^{n-1} & \cdots & a_{n-1}^{n-1} & a_n^{n-1} \end{bmatrix}$$

(or its transpose) is called a **Vandermonde matrix,** and its determinant is called a **Vandermonde determinant.** There is a simple formula for this determinant. If $n = 2$, the determinant is $(a_2 - a_1)$; if $n = 3$, it is $(a_3 - a_2)(a_3 - a_1)(a_2 - a_1)$ by Example 9§3.1. The general result is as follows:

THEOREM 2
Vandermonde Determinant

Let $a_1, a_2, \ldots, a_n$ be real numbers, $n \geq 2$. Then the corresponding Vandermonde determinant is given by

$$\det \begin{bmatrix} 1 & 1 & 1 & \cdots & 1 \\ a_1 & a_2 & a_3 & \cdots & a_n \\ a_1^2 & a_2^2 & a_3^2 & \cdots & a_n^2 \\ \vdots & \vdots & \vdots & & \vdots \\ a_1^{n-1} & a_2^{n-1} & a_3^{n-1} & \cdots & a_n^{n-1} \end{bmatrix} = \prod_{1 \le j < i \le n} (a_i - a_j)$$

where $\prod_{1 \le j < i \le n} (a_i - a_j)$ means the product of all factors $(a_i - a_j)$, where $i > j$ and both i and j are between 1 and n.

Proof   We may assume that the a_i are distinct because otherwise both sides would be zero. Proceed by induction on $n \ge 2$ and assume inductively that the theorem is true for $n - 1$. The trick is to replace a_n with a variable x and consider the determinant

$$p(x) = \det \begin{bmatrix} 1 & 1 & \cdots & 1 & 1 \\ a_1 & a_2 & \cdots & a_{n-1} & x \\ a_1^2 & a_2^2 & \cdots & a_{n-1}^2 & x^2 \\ \vdots & \vdots & & \vdots & \vdots \\ a_1^{n-1} & a_2^{n-1} & \cdots & a_{n-1}^{n-1} & x^{n-1} \end{bmatrix}$$

If this is expanded along the last column, it is clear that $p(x)$ is a polynominal in x of degree at most $n - 1$. Moreover, $p(a_1) = 0$ (because $x = a_1$ produces identical columns), so $x - a_1$ is a factor of $p(x)$ by the factor theorem — say, $p(x) = (x - a_1)p_1(x)$. Then the fact that $p(a_2) = 0$ and $a_2 \ne a_1$ means that $p_1(a_2) = 0$. So, again by the factor theorem $p_1(x) = (x - a_2)p_2(x)$. This gives $p(x) = (x - a_1)(x - a_2)p_2(x)$. The process continues (the a_i are distinct) to give

$$p(x) = (x - a_1)(x - a_2) \cdots (x - a_{n-1})d \qquad (*)$$

where d is a constant. In fact, d is the coefficient of x^{n-1} in $p(x)$ and so, by the Laplace expansion, d is the (n, n)-cofactor of the matrix:

$$d = (-1)^{n+n} \det \begin{bmatrix} 1 & 1 & \cdots & 1 \\ a_1 & a_2 & \cdots & a_{n-1} \\ a_1^2 & a_2^2 & \cdots & a_{n-1}^2 \\ \vdots & \vdots & & \vdots \\ a_1^{n-2} & a_2^{n-2} & \cdots & a_{n-1}^{n-2} \end{bmatrix}$$

Because $(-1)^{n+n} = 1$, the induction hypothesis shows that d is the product of all terms $(a_i - a_j)$ where $1 \le j < i \le n - 1$, and the result now follows from equation $(*)$ by substituting $x = a_n$ in $p(x)$.   ◆

EXERCISES 3.3

1. Find a polynomial $p(x)$ of degree 2 such that:
   (a) $p(0) = 2, p(1) = 3, p(3) = 8$
   ◆(b) $p(0) = 5, p(1) = 3, p(2) = 5$

2. Find a polynomial $p(x)$ of degree 3 such that:
   (a) $p(0) = p(1) = 1, p(-1) = 4, p(2) = -5$
   ◆(b) $p(0) = p(1) = 1, p(-1) = 2, p(-2) = -3$

3. Given the following data pairs, find the interpolating polynomial of degree 3 and estimate the value of y corresponding to $x = 1.5$.

   (a) $(0,1), (1,2), (2,5), (3,10)$
   ◆(b) $(0,1), (1,1.49), (2,-0.42), (3,-11.33)$
   (c) $(0,2), (1,2.03), (2,-0.40), (-1,0.89)$

4. Use the polynomial $p(x)$ in Example 3 to approximate
   (a) $\sin(0.3)$ and ◆(b) $\sin(0.7)$.

5. Find a quadratic polynomial $p(x)$ approximating e^{3x} on the range $0 \le x \le \frac{1}{2}$. (Use $x_0 = 0, x_1 = \frac{1}{4}$, and $x_2 = \frac{1}{2}$, so $y_0 = 1, y_1 = 2.117$, and $y_2 = 4.482$.) Use $p(x)$ to estimate $e^{9/8}$.

Section 3.4   Proof of the Laplace Expansion (Optional)

Recall that our definition of the term *determinant* is inductive: The determinant of any 1×1 matrix is defined first; then it is used to define the determinants of 2×2 matrices. Then that is used for the 3×3 case, and so on. The case of a 1×1 matrix $[a]$ poses no problem. We simply define

$$\det[a] = a$$

as in Section 3.1. Given an $n \times n$ matrix A, define A_{ij} to be the $(n-1) \times (n-1)$ matrix obtained from A by deleting row i and column j. Now assume that the determinant of any $(n-1) \times (n-1)$ matrix has been defined. Then the determinant of A is defined to be

$$\det A = a_{11} \det A_{11} - a_{21} \det A_{21} + \cdots \pm a_{n1} \det A_{n1}$$

$$= \sum_{i=1}^{n} (-1)^{i+1} a_{i1} \det A_{i1}$$

Observe that, in the terminology of Section 3.1, this is just the Laplace expansion of $\det A$ along the first column, that $\det A_{ij}$ is just the (i, j)-minor of A (previously denoted as $M_{ij}(A)$), and that $(-1)^{i+j} \det A_{ij}$ is the (i, j)-cofactor (previously denoted as $C_{ij}(A)$). To illustrate the definition consider the 2×2 matrix $A = \begin{bmatrix} a_{11} & a_{12} \\ a_{21} & a_{22} \end{bmatrix}$. Then the definition gives

$$\det \begin{bmatrix} a_{11} & a_{12} \\ a_{21} & a_{22} \end{bmatrix} = a_{11} \det[a_{22}] - a_{21} \det[a_{12}] = a_{11}a_{22} - a_{21}a_{12}$$

and this is the same as the definition in Section 3.1.

Of course, the task now is to use this definition to *prove* that the Laplace expansion along *any* row or column yields $\det A$ (this is Theorem 1§3.1). The proof proceeds by first establishing the properties of determinants stated in Theorem 2§3.1, but for *rows* only (see Lemma 2). This being done, the full proof of Theorem 1§3.1 is not difficult. The proof of Lemma 2 requires the following preliminary result.

LEMMA 1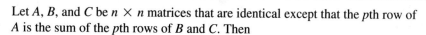

Let A, B, and C be $n \times n$ matrices that are identical except that the pth row of A is the sum of the pth rows of B and C. Then

$$\det A = \det B + \det C.$$

Proof We proceed by induction on n, the cases $n = 1$ and $n = 2$ being easily checked. Consider a_{i1} and A_{i1}:

Case 1: If $i \neq p$,

$$a_{i1} = b_{i1} = c_{i1} \quad \text{and} \quad \det A_{i1} = \det B_{i1} + \det C_{i1}$$

by induction because A_{i1}, B_{i1}, C_{i1} are identical except that one row of A_{i1} is the sum of the corresponding rows of B_{i1} and C_{i1}.

Case 2: If $i = p$

$$a_{p1} = b_{p1} + c_{p1} \quad \text{and} \quad A_{p1} = B_{p1} = C_{p1}$$

Now write out the defining sum for $\det A$, splitting off the pth term for special attention.

$$\det A = \sum_{i \neq p} a_{i1}(-1)^{i+1} \det A_{i1} + a_{p1}(-1)^{p+1} \det A_{p1}$$

$$= \sum_{i \neq p} a_{i1}(-1)^{i+1} [\det B_{i1} + \det C_{i1}] + (b_{p1} + c_{p1})(-1)^{p+1} \det A_{p1}$$

But the terms here involving B_{i1} and b_{p1} add up to $\det B$ because $a_{i1} = b_{i1}$ if $i \neq p$ and $A_{p1} = B_{p1}$. Similarly, the terms involving C_{i1} and c_{p1} add up to $\det C$. Hence $\det A = \det B + \det C$, as required.   ◆

LEMMA 2

Let $A = [a_{ij}]$ denote an $n \times n$ matrix.

1. If $B = [b_{ij}]$ is formed from A by multiplying a row of A by a number u, then $\det B = u \det A$.
2. If A contains a row of zeros, then $\det A = 0$.
3. If $B = [b_{ij}]$ is formed by interchanging two rows of A, then $\det B = -\det A$.
4. If A contains two identical rows, then $\det A = 0$.
5. If $B = [b_{ij}]$ is formed by adding a multiple of one row of A to a different row, then $\det B = \det A$.

Proof   For later reference the defining sums for det A and det B are as follows:

$$\det A = \sum_{i=1}^{n} a_{i1}(-1)^{i+1} \det A_{i1} \qquad\qquad (*)$$

$$\det B = \sum_{i=1}^{n} b_{i1}(-1)^{i+1} \det B_{i1}. \qquad\qquad (**)$$

Property 1.   The proof is by induction on n, the cases $n = 1$ and $n = 2$ being easily verified. Consider the ith term in the sum $(**)$ for det B where B is the result of multiplying row p of A by u.

    a. If $i \neq p$, then $b_{i1} = a_{i1}$ and det $B_{i1} = u$ det A_{i1} by induction because B_{i1} comes from A_{i1} by multiplying a row by u.

    b. If $i = p$, then $b_{p1} = ua_{p1}$ and $B_{p1} = A_{p1}$.

In either case, each term in equation $(**)$ is u times the corresponding term in equation $(*)$, so it is clear that det $B = u$ det A.

Property 2.   This is clear by property 1 because the row of zeros has a common factor $u = 0$.

Property 3.   Observe first that it suffices to prove property 3 for interchanges of adjacent rows. (Rows p and q $(q > p)$ can be interchanged by carrying out $2(q - p) - 1$ adjacent changes, which results in an *odd* number of sign changes in the determinant.) So suppose that rows p and $p + 1$ of A are interchanged to obtain B. Again consider the ith term in $(**)$.

    a. If $i \neq p$ and $i \neq p + 1$, then $b_{i1} = a_{i1}$ and det $B_{i1} = -\det A_{i1}$ by induction because B_{i1} results from interchanging adjacent rows in A_{i1}. Hence the ith term in $(**)$ is the negative of the ith term in $(*)$, and so det $B = -\det A$.

    b. If $i = p$ or $i = p + 1$, then $b_{p1} = a_{p+1\,1}$ and $B_{p1} = A_{p+1\,1}$, whereas $b_{p+1\,1} = a_{p1}$ and $B_{p+1\,1} = A_{p1}$. Hence terms p and $p + 1$ in $(**)$ are

$$b_{p1}(-1)^{p+1} \det B_{p1} = -a_{p+1\,1}(-1)^{(p+1)+1} \det(A_{p+1\,1})$$

$$b_{p+1\,1}(-1)^{(p+1)+1}\det(B_{p+1\,1}) = -a_{p1}(-1)^{p+1}\det A_{p1}$$

This means that terms p and $p + 1$ in $(**)$ are the same as these terms in $(*)$, except that the order is reversed and the signs are changed. Thus the sum $(**)$ is the negative of the sum $(*)$; that is, det $B = -\det A$.

Property 4.   If rows p and q in A are identical, let B be obtained from A by interchanging these rows. Then $B = A$ so det $A =$ det B. But det $B = -\det A$ by property 3 so det $A = -\det A$. This implies that det $A = 0$.

Property 5.   Suppose B results from adding u times row q of A to row p. Then Lemma 1 applies to B to show that det $B =$ det $A +$ det C, where C is obtained from

A by replacing row p by u times row q. It now follows from properties 1 and 4 that $\det C = 0$ so $\det B = \det A$, as asserted.   ◆

These facts are enough to enable us to prove Theorem 1 §3.1. For convenience, it is restated here in the notation of the foregoing lemmas. The only difference between the notations is that the (i, j)-cofactor of an $n \times n$ matrix A was denoted earlier by

$$C_{ij}(A) = (-1)^{i+j} \det A_{ij}$$

THEOREM 1

If $A = [a_{ij}]$ is an $n \times n$ matrix, then

1. $\det A = \displaystyle\sum_{i=1}^{n} a_{ij}(-1)^{i+j} \det A_{ij}$ (Laplace expansion along column j)

2. $\det A = \displaystyle\sum_{j=1}^{n} a_{ij}(-1)^{i+j} \det A_{ij}$ (Laplace expansion along row i).

Here A_{ij} denotes the matrix obtained from A by deleting row i and column j.

Proof Lemma 2 establishes the truth of Theorem 2§3.1 for *rows*. With this information, the arguments in Section 3.2 proceed exactly as written to establish that $\det A = \det A^T$ holds for any $n \times n$ matrix A. Now suppose B is obtained from A by interchanging two columns. Then B^T is obtained from A^T by interchanging two rows so, by property 3 of Lemma 2,

$$\det B = \det B^T = -\det A^T = -\det A$$

Hence property 3 of Lemma 2 holds for *columns* too.

This enables us to prove the Laplace expansion for columns. Given an $n \times n$ matrix $A = [a_{ij}]$, let $B = [b_{ij}]$ be obtained by moving column j to the left side, using $j - 1$ interchanges of adjacent columns. Then $\det B = (-1)^{j-1} \det A$ and, because $B_{i1} = A_{ij}$ and $b_{i1} = a_{ij}$ for all i, we obtain

$$\det A = (-1)^{j-1} \det B = (-1)^{j-1} \sum_{i=1}^{n} b_{i1}(-1)^{1+i} \det B_{i1}$$

$$= \sum_{i=1}^{n} a_{ij}(-1)^{i+j} \det A_{ij}$$

This is the Laplace expansion of $\det A$ along column j.

Finally, to prove the row expansion, write $B = A^T$. Then $B_{ij} = A_{ji}^T$ and $b_{ij} = a_{ji}$ for all i and j. Expanding $\det B$ along column j gives

$$\det A = \det A^T = \det B = \sum_{i=1}^{n} b_{ij}(-1)^{i+j}\det B_{ij}$$

$$= \sum_{i=1}^{n} a_{ji}(-1)^{j+i}\det(A_{ji}^T) = \sum_{i=1}^{n} a_{ji}(-1)^{j+i}\det A_{ji}$$

This is the required expansion of det A along row j.   ◆

EXERCISES 3.4

1. Prove Lemma 1 for columns.

◆**2.** Verify that interchanging rows p and q $(q > p)$ can be accomplished using $2(q - p) - 1$ adjacent interchanges.

3. If u is a number and A is an $n \times n$ matrix, prove that $\det(uA) = u^n \det A$ by induction on n, using only the definition of det A.

SUPPLEMENTARY EXERCISES FOR CHAPTER 3

1. Show that
$$\det\begin{bmatrix} a + px & b + qx & c + rx \\ p + ux & q + vx & r + wx \\ u + ax & v + bx & w + cx \end{bmatrix} = (1 + x^3)\det\begin{bmatrix} a & b & c \\ p & q & r \\ u & v & w \end{bmatrix}.$$

2. (a) Show that $M_{ij}(A)^T = M_{ji}(A^T)$ for all i, j and all square matrices A.

◆**(b)** Use **(a)** to prove that $\det A^T = \det A$. [*Hint:* Induction on n where A is $n \times n$.]

3. Show that $\det\begin{bmatrix} 0 & I_n \\ I_m & 0 \end{bmatrix} = (-1)^{nm}$ for all $n \geq 1$ and $m \geq 1$.

Vector Geometry[1]

◆◆◆

SECTION 4.1
Vectors and Lines

Many quantities in nature are completely specified by one number (called the *magnitude* of the quantity) and are usually referred to as scalar quantities. Some examples are temperature, time, length, and net assets. However, certain quantities require both a magnitude and a direction to specify them. Consider displacement: To say that a boat sailed 10 kilometers (km) does not specify where it went. It is necessary to give the direction too; perhaps it sailed 10 km northwest. Quantities that require both a magnitude and a direction to describe them are called **vector** quantities. Other examples include velocity and force. Vector quantities will be denoted by boldface type: **u**, **v**, **w**, and so on.

A vector **v** can be represented geometrically as a directed line segment or arrow (see Figure 4.1). The head and tail of the arrow are called, respectively, the **terminal point** and the **initial point** of the vector. The **magnitude** of a vector **v** will be denoted by ∥**v**∥, and is sometimes referred to as the **length** of **v** because it is represented by the length of the arrow.

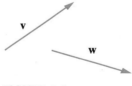

FIGURE 4.1

Two vectors **v** and **w** are **equal** (written **v** = **w**) if they have the same length and the same direction. Thus, for example, the two vectors in Figure 4.2 are equal even though the initial and terminal points are different.[2]

---

[1]Readers familiar with this material can proceed to Chapter 5.
[2]This is like the situation for rational fractions: We write $\frac{4}{7} = \frac{28}{49} = \frac{-8}{-14}$ even though the numerators and denominators are different.

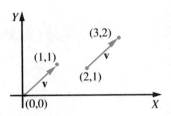

FIGURE 4.2

There is only one vector with zero length. It is called the **zero vector** and is denoted by **0**. In other words,

$$\mathbf{v} = \mathbf{0} \quad \text{if and only if} \quad \|\mathbf{v}\| = 0$$

No direction is assigned to the vector **0**. Two nonzero vectors are called **parallel** if they have the same or opposite directions (Figure 4.3). Given a vector **v**, the vector with the same magnitude as **v** but the opposite direction is called the **negative** of **v** and is denoted $-\mathbf{v}$ (Figure 4.4). Because **0** is the only vector with length 0, it follows that $-\mathbf{0} = \mathbf{0}$. Clearly **v** and $-\mathbf{v}$ are parallel if $\mathbf{v} \neq \mathbf{0}$.

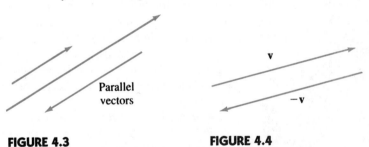

Parallel
vectors

FIGURE 4.3   **FIGURE 4.4**

Probably the most useful aspect of vectors is that they can be added and multiplied by a number in such a way that these operations reflect physical or geometrical facts. Consider the following example.

EXAMPLE 1

Find the displacement resulting from a 1-km walk northeast and a 1-km walk east.

Solution

Let $\mathbf{w}_1$ and $\mathbf{w}_2$ denote the northeast and east displacements, respectively. If $\mathbf{w}_1$ is done first, the resulting displacement **v** is given by the first diagram. The second diagram shows the resulting displacement when $\mathbf{w}_2$ is done first. The resulting displacement is the same in both cases, so it is designated **v** in both. Moreover, the magnitude $\|\mathbf{v}\|$ and the direction of **v** (given by θ) are determined (Exercise 1), so **v** is determined completely by $\mathbf{w}_1$ and $\mathbf{w}_2$.

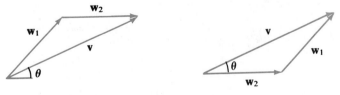

The situation in Example 1 is typical of many vector quantities other than displacements, and this leads to a general notion of vector addition.

DEFINITION

If **u** and **w** are two vectors, their **sum u + w** is defined as follows: Position **u** and **w** so that they emanate from a common point *P*. They determine a parallelogram, and the diagonal drawn from *P* represents **u + w**. This is called the **parallelogram rule.**

The situation is shown in Figure 4.5(a). We have **u + w = w + u** because the two vectors enter into the definition in a symmetric fashion. However, it is often convenient to regard **u + w** as "first **u** and then **w**" by placing the initial point of **w** at the terminal point of **u**, as in Figure 4.5(b). Similarly, **u + w = w + u** is the result of "first **w** and then **u**" (Figure 4.5(c)).

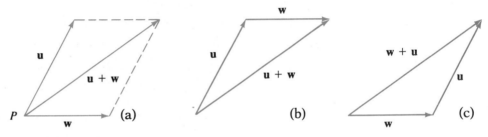

FIGURE 4.5

The basic properties of vector addition are as follows:

1. **u + v = v + u** for all vectors **u** and **v**
2. **u + (v + w) = (u + v) + w** for all vectors **u, v,** and **w**
3. **v + 0 = v** for all vectors **v**
4. **v + (−v) = 0** for all vectors **v**

The first of these has already been mentioned, and the last two follow from the parallelogram rule. The second is illustrated by Figure 4.6. The two diagrams have **u, v,** and **w** in the same position. The vector across the bottom is **u + (v + w)** in Figure 4.6(a) and **(u + v) + w** in Figure 4.6(b).

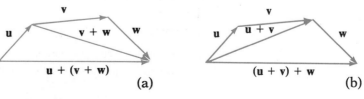

FIGURE 4.6

Because of the fact that $\mathbf{u} + (\mathbf{v} + \mathbf{w}) = (\mathbf{u} + \mathbf{v}) + \mathbf{w}$, we shall write this simply as $\mathbf{u} + \mathbf{v} + \mathbf{w}$. The foregoing discussion makes it clear that $\mathbf{u} + \mathbf{v} + \mathbf{w}$ can be regarded as the result of $\mathbf{u}$, $\mathbf{v}$, and $\mathbf{w}$ being placed end to end with the terminal point of each vector coinciding with the initial point of the next vector. The sum $\mathbf{u} + \mathbf{v} + \mathbf{w}$ then has the same initial point as $\mathbf{u}$ and the same terminal point as $\mathbf{w}$.

Vectors can be used effectively to give proofs of theorems in Euclidean geometry that make no use of coordinates. If A and B are two geometrical points, the vector from A to B is denoted as $\overrightarrow{AB}$.

EXAMPLE 2

Show that the diagonals of a parallelogram bisect each other.

Solution

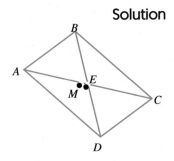

Let the parallelogram have vertices A, B, C, and D, as shown; let E denote the intersection of the two diagonals; and let M denote the midpoint of diagonal AC. We must show that $M = E$ and that this is the midpoint of diagonal BD. This is accomplished by showing that $\overrightarrow{BM} = \overrightarrow{MD}$. (The fact that $\overrightarrow{BM}$ and $\overrightarrow{MD}$ have the same direction means that $M = E$, and the fact that they have the same length means that $M = E$ is the midpoint of BD.) Now $\overrightarrow{AM} = \overrightarrow{MC}$ because M is the midpoint of AC, and $\overrightarrow{BA} = \overrightarrow{CD}$ because the figure is a parallelogram. Hence

$$\overrightarrow{BM} = \overrightarrow{BA} + \overrightarrow{AM} = \overrightarrow{CD} + \overrightarrow{MC} = \overrightarrow{MC} + \overrightarrow{CD} = \overrightarrow{MD}$$

where the first and last equalities use the parallelogram rule of vector addition.

◆◆◆

By analogy with numerical arithmetic, vector subtraction is defined as follows:

$$\mathbf{u} - \mathbf{v} = \mathbf{u} + (-\mathbf{v})$$

As for numerical subtraction, the solution $\mathbf{x}$ to a vector equation $\mathbf{x} + \mathbf{v} = \mathbf{u}$ is $\mathbf{x} = \mathbf{u} - \mathbf{v}$. In fact, adding $-\mathbf{v}$ to each side of $\mathbf{x} + \mathbf{v} = \mathbf{u}$ gives

$$\mathbf{x} + \mathbf{v} + (-\mathbf{v}) = \mathbf{u} + (-\mathbf{v})$$
$$\mathbf{x} + \mathbf{0} = \mathbf{u} - \mathbf{v}$$
$$\mathbf{x} = \mathbf{u} - \mathbf{v}$$

Like vector addition, subtraction has a geometric interpretation, as shown in Figure 4.7: We have $\mathbf{v} + \overrightarrow{AB} = \mathbf{u}$ by vector addition, so $\overrightarrow{AB} = \mathbf{u} - \mathbf{v}$ as shown. The situation in Figure 4.7 is best remembered by observing that

$$\mathbf{u} - \mathbf{v} \text{ is the vector that, when added to } \mathbf{v}, \text{ gives } \mathbf{u}$$

FIGURE 4.7

Scalar Multiplication

DEFINITION

Given any vector $\mathbf{v}$ and any number a, the **scalar multiple** of $\mathbf{v}$ by a is the vector $a\mathbf{v}$ defined by specifying its magnitude and direction as follows:

1. The magnitude of $a\mathbf{v}$ is $\|a\mathbf{v}\| = |a|\ \|\mathbf{v}\|$.

2. The direction of $a\mathbf{v}$ is $\begin{cases} \text{the same as that of } \mathbf{v} \text{ if } a > 0 \text{ and } \mathbf{v} \neq \mathbf{0}. \\ \text{unspecified if } a = 0 \text{ or } \mathbf{v} = \mathbf{0}. \\ \text{opposite to that of } \mathbf{v} \text{ if } a < 0 \text{ and } \mathbf{v} \neq \mathbf{0}. \end{cases}$

Some examples of scalar multiplication appear in Figure 4.8.

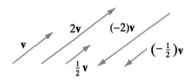

FIGURE 4.8

Taking $a = 0$ in the formula for $\|a\mathbf{v}\|$ yields $\|0\mathbf{v}\| = 0$. In other words, $0\mathbf{v}$ has magnitude zero, so it is the zero vector; that is, $0\mathbf{v} = \mathbf{0}$. Similarly, $a\mathbf{0} = \mathbf{0}$ for each number a. Observe also that $1\mathbf{v} = \mathbf{v}$ because $1\mathbf{v}$ has the same magnitude and direction as $\mathbf{v}$. However, $(-1)\mathbf{v}$ has the same magnitude as $\mathbf{v}$ but the *opposite* direction, so $(-1)\mathbf{v} = -\mathbf{v}$. These properties of scalar multiplication are collected below for reference:

1. $\|a\mathbf{v}\| = |a|\ \|\mathbf{v}\|$ for all scalars a and vectors $\mathbf{v}$

2. $1\mathbf{v} = \mathbf{v}$ for all vectors $\mathbf{v}$

3. $(-1)\mathbf{v} = -\mathbf{v}$ for all vectors $\mathbf{v}$

4. $0\mathbf{v} = \mathbf{0}$ for all vectors $\mathbf{v}$

5. $a\mathbf{0} = \mathbf{0}$ for all scalars a

A vector is called a **unit vector** if its magnitude is 1.

EXAMPLE 3

If $\mathbf{v} \neq \mathbf{0}$, show that $\dfrac{1}{\|\mathbf{v}\|}\mathbf{v}$ is a unit vector in the same direction as $\mathbf{v}$.

Solution   The vectors in the same direction as $\mathbf{v}$ are $a\mathbf{v}$, where $a > 0$. Because $\|a\mathbf{v}\| = a\|\mathbf{v}\|$ when $a > 0$, this is a unit vector when $a = \dfrac{1}{\|\mathbf{v}\|}$.

$\blacklozenge\blacklozenge\blacklozenge$

Many properties of vector addition and scalar multiplication have been mentioned (and utilized) so far. Several of these facts are consequences of the following eight fundamental properties.

THEOREM 1

Vector addition and scalar multiplication exhibit the following properties.

1. $\mathbf{u} + \mathbf{v} = \mathbf{v} + \mathbf{u}$ for all vectors $\mathbf{u}$ and $\mathbf{v}$
2. $\mathbf{u} + (\mathbf{v} + \mathbf{w}) = (\mathbf{u} + \mathbf{v}) + \mathbf{w}$ for all vectors $\mathbf{u}$, $\mathbf{v}$, and $\mathbf{w}$
3. $\mathbf{u} + \mathbf{0} = \mathbf{u}$ for all vectors $\mathbf{u}$
4. $\mathbf{u} + (-\mathbf{u}) = \mathbf{0}$ for all vectors $\mathbf{u}$
5. $1\mathbf{u} = \mathbf{u}$ for all vectors $\mathbf{u}$
6. $a(b\mathbf{u}) = (ab)\mathbf{u}$ for all vectors $\mathbf{u}$ and scalars a, b
7. $(a + b)\mathbf{u} = a\mathbf{u} + b\mathbf{u}$ for all vectors $\mathbf{u}$ and scalars a, b
8. $a(\mathbf{u} + \mathbf{v}) = a\mathbf{u} + a\mathbf{v}$ for all vectors $\mathbf{u}$, $\mathbf{v}$ and scalars a

Proof   Only the last three properties remain to be verified. They follow easily from the definitions when the scalars a and b are positive or zero. The general case is not difficult (though it is a bit tedious), and the details are left as Exercise 33.   ◆

As for matrices, these properties enable us to carry out algebraic manipulations of vectors as though the vectors were variables.

EXAMPLE 4

(a) Simplify $5(\mathbf{u} - 2\mathbf{v}) + 6(5\mathbf{u} + 2\mathbf{v}) - 2(\mathbf{v} - \mathbf{u})$.

(b) If $\mathbf{v} = 4\mathbf{w}$, show that $\mathbf{w} = \frac{1}{4}\mathbf{v}$.

Solution   (a) $5(\mathbf{u} - 2\mathbf{v}) + 6(5\mathbf{u} + 2\mathbf{v}) - 2(\mathbf{v} - \mathbf{u})$

$\qquad = 5\mathbf{u} - 10\mathbf{v} + 30\mathbf{u} + 12\mathbf{v} - 2\mathbf{v} + 2\mathbf{u}$

$\qquad = 37\mathbf{u}$

(b) $\frac{1}{4}\mathbf{v} = \frac{1}{4}(4\mathbf{w}) = (\frac{1}{4} \cdot 4)\mathbf{w} = 1\mathbf{w} = \mathbf{w}$ ◆◆◆

EXAMPLE 5

Show that the midpoints of the four sides of any quadrilateral are the vertices of a parallelogram. Here a quadrilateral is any figure with four vertices.

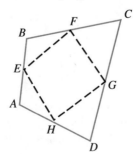

Solution   Suppose that the vertices of the quadrilateral are A, B, C, and D (in that order) and that E, F, G, and H are the midpoints of the sides as shown in the diagram. It suffices to show $\overrightarrow{EF} = \overrightarrow{HG}$ (because then these two sides are parallel and of equal length). Now the fact that E is the midpoint of AB means that $\overrightarrow{EB} = \frac{1}{2}\overrightarrow{AB}$. Similarly, $\overrightarrow{BF} = \frac{1}{2}\overrightarrow{BC}$, so

$$\overrightarrow{EF} = \overrightarrow{EB} + \overrightarrow{BF} = \tfrac{1}{2}\overrightarrow{AB} + \tfrac{1}{2}\overrightarrow{BC} = \tfrac{1}{2}(\overrightarrow{AB} + \overrightarrow{BC}) = \tfrac{1}{2}\overrightarrow{AC}$$

A similar argument shows that $\overrightarrow{HG} = \frac{1}{2}\overrightarrow{AC}$ too, so $\overrightarrow{EF} = \overrightarrow{HG}$ as required. ◆◆◆

Coordinates

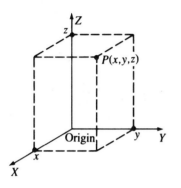

FIGURE 4.9

Examples 2 and 5 use vectors to verify geometrical propositions without the benefit of coordinates. On the other hand, the introduction of coordinates gives us a convenient way of doing vector calculations and enables us to use vectors to study lines and planes.

To introduce coordinates into space, choose three mutually perpendicular lines (called the **X** axis, the **Y** axis, and the **Z** axis) which meet at a point (called the **origin**). Each of these axes is a copy of the real line with 0 at the origin. The plane determined by the X and Y axes is called the **X-Y plane**; similarly we get the **X-Z plane** and the **Y-Z plane**. Each point P determines a unique triple (x, y, z) of numbers called the **coordinates** of P. For example, x is the point of intersection of the X axis and the plane through P parallel to the Y-Z plane (see Figure 4.9). When the coordinates of a point P are to be emphasized, we shall write $P = P(x, y, z)$.

DEFINITION

Given a point P, the **position vector** of P is defined to be the vector $\mathbf{p} = \overrightarrow{OP}$ from the origin to P. If $P = P(x, y, z)$ in space, the position vector is denoted

$$\mathbf{p} = (x, y, z)$$

and the numbers x, y, and z are called the X, Y, and Z **components** of $\mathbf{p}$. If $P = P(x, y)$ in the X-Y plane, the position vector is denoted

$$\mathbf{p} = (x, y)$$

Using vectors, we find that geometry in space is much like geometry in the plane. Consequently we shall emphasize the situation in space. In vector geometry the coordinates of a point $P(x, y, z)$ are computed by finding its position vector $\mathbf{p} = (x, y, z)$. This vector point of view is useful as we shall see.

Every vector $\mathbf{v}$ is the position vector of a unique point $P(x, y, z)$ (if $\mathbf{v}$ is positioned with its initial point at the origin, then the terminal point is P). In particular, the zero vector is the position vector of the origin itself:

$$\mathbf{0} = (0, 0, 0)$$

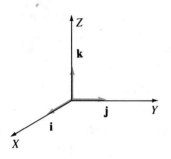

FIGURE 4.10

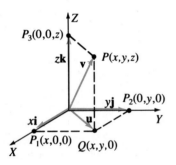

FIGURE 4.11

The **coordinate vectors i, j,** and **k** are defined by

$$\mathbf{i} = (1, 0, 0)$$
$$\mathbf{j} = (0, 1, 0)$$
$$\mathbf{k} = (0, 0, 1)$$

These are unit vectors pointing along the positive X, Y, and Z axes, respectively (see Figure 4.10), and as we shall see, every vector can be expressed in terms of **i, j,** and **k**.

Indeed, let $\mathbf{v} = (x, y, z)$ be *any* vector in component form. Consider the point $P_1(x, 0, 0)$ on the X axis. Because P_1 is at a distance $|x|$ from the origin, the position vector of P_1 is $x\mathbf{i}$. (In fact, the magnitude of $x\mathbf{i}$ is $\|x\mathbf{i}\| = |x|\,\|\mathbf{i}\| = |x|$, and the direction is along the positive X axis if $x > 0$ and along the negative X axis if $x < 0$.) This verifies the first of the following statements (the others are similar):

$$x\mathbf{i} = (x, \ 0, \ 0) \qquad \text{is the position vector of } P_1(x, \ 0, \ 0)$$
$$y\mathbf{j} = (0, \ y, \ 0) \qquad \text{is the position vector of } P_2(0, \ y, \ 0)$$
$$z\mathbf{k} = (0, \ 0, \ z) \qquad \text{is the position vector of } P_3(0, \ 0, \ z)$$

These vectors are shown in Figure 4.11. Vector addition gives $\mathbf{u} = x\mathbf{i} + y\mathbf{j}$, whence $\mathbf{v} = \mathbf{u} + z\mathbf{k} = (x\mathbf{i} + y\mathbf{j}) + z\mathbf{k}$. Thus

$$\mathbf{v} = (x, y, z) = x\mathbf{i} + y\mathbf{j} + z\mathbf{k} \qquad (*)$$

for all x, y, and z. This enables us to give the reason for writing vectors in component form: Vector addition and scalar multiplication correspond to matrix operations.

THEOREM 2

Let $\mathbf{u} = (x, y, z)$ and $\mathbf{u_1} = (x_1, y_1, z_1)$ be two vectors in component form. Then:

1. $\mathbf{u} = \mathbf{u_1}$ if and only if $x = x_1$, $y = y_1$, and $z = z_1$
2. $\mathbf{u} + \mathbf{u_1} = (x + x_1, y + y_1, z + z_1)$
3. $a\mathbf{u} = (ax, ay, az)$ for any scalar a
4. $\mathbf{u} - \mathbf{u_1} = (x - x_1, y - y_1, z - z_1)$

Proof   (1).   The vectors $\mathbf{u} = (x, y, z)$ and $\mathbf{u_1} = (x_1, y_1, z_1)$ are position vectors of points $P(x, y, z)$ and $P_1(x_1, y_1, z_1)$, respectively. Hence $\mathbf{u} = \mathbf{u_1}$ means $P = P_1$; that is, $x = x_1$, $y = y_1$, and $z = z_1$.

(2), (3), and (4).   Equation $(*)$ shows that $\mathbf{u} = x\mathbf{i} + y\mathbf{j} + z\mathbf{k}$ and $\mathbf{u_1} = x_1\mathbf{i} + y_1\mathbf{j} + z_1\mathbf{k}$. Hence Theorem 1 gives

$$\mathbf{u} + \mathbf{u_1} = (x\mathbf{i} + y\mathbf{j} + z\mathbf{k}) + (x_1\mathbf{i} + y_1\mathbf{j} + z_1\mathbf{k})$$
$$= (x + x_1)\mathbf{i} + (y + y_1)\mathbf{j} + (z + z_1)\mathbf{k}$$
$$= (x + x_1, \ y + y_1, \ z + z_1)$$

where the last step again uses Equation (∗). This gives (2); (3) and (4) are similar and are left as Exercise 22.   ◆

EXAMPLE 6

Given vectors $\mathbf{u} = (3, -1, 2)$ and $\mathbf{v} = (1, 0, -1)$, compute $\mathbf{u} + \mathbf{v}$, $3\mathbf{u}$, and $2\mathbf{u} - 3\mathbf{v}$.

Solution   These are just matrix calculations.

$$\mathbf{u} + \mathbf{v} = (3, -1, 2) + (1, 0, -1) = (4, -1, 1)$$
$$3\mathbf{u} = 3(3, -1, 2) = (9, -3, 6)$$
$$2\mathbf{u} - 3\mathbf{v} = 2(3, -1, 2) - 3(1, 0, -1) = (3, -2, 7)$$

◆◆◆

The following theorem is a particularly useful consequence of Theorem 2.

THEOREM 3

Given points $P_1(x_1, y_1, z_1)$ and $P_2(x_2, y_2, z_2)$, the vector from P_1 to P_2 is

$$\overrightarrow{P_1P_2} = (x_2 - x_1, y_2 - y_1, z_2 - z_1)$$

Proof   Let $\mathbf{p_1} = (x_1, y_1, z_1)$ and $\mathbf{p_2} = (x_2, y_2, z_2)$ denote the position vectors of P_1 and P_2. Then

$$\overrightarrow{P_1P_2} = \mathbf{p_2} - \mathbf{p_1} = (x_2 - x_1, y_2 - y_1, z_2 - z_1)$$

using Theorem 2. (See the accompanying figure.)   ◆

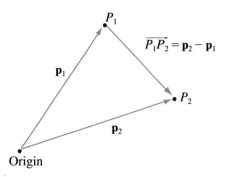

These results all have natural analogues for vectors in the plane. If $\mathbf{u} = (x, y)$ and $\mathbf{u_1} = (x_1, y_1)$, then:

1. $\mathbf{u} = \mathbf{u_1}$ if and only if $x = x_1$, and $y = y_1$
2. $\mathbf{u} + \mathbf{u_1} = (x + x_1, y + y_1)$

3. $a\mathbf{u} = (ax, ay)$ for any scalar a

4. $\mathbf{u} - \mathbf{u_1} = (x - x_1, y - y_1)$

5. The vector from $P_1(x_1, y_1)$ to $P_2(x_2, y_2)$ is $\overrightarrow{P_1P_2} = (x_2 - x_1, y_2 - y_1)$.

The verifications are entirely analogous to those in Theorems 2 and 3 and are omitted.

The **midpoint** of two points is the point on the line segment between them that is halfway from one to the other.

EXAMPLE 7

Show that the midpoint of $P_1(x_1, y_1, z_1)$ and $P_2(x_2, y_2, z_2)$ is

$$P\left(\frac{x_1 + x_2}{2}, \; \frac{y_1 + y_2}{2}, \; \frac{z_1 + z_2}{2}\right)$$

Solution

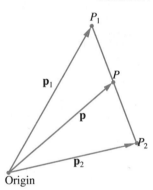

Let $\mathbf{p_1}$, $\mathbf{p_2}$, and $\mathbf{p}$ be the position vectors of P_1, P_2, and $P(x, y, z)$, respectively. We find x, y, and z by computing $\mathbf{p} = (x, y, z)$. We have $\overrightarrow{P_1P} = \frac{1}{2}\overrightarrow{P_1P_2}$ because P is the midpoint (see the diagram) and $\overrightarrow{P_1P_2} = \mathbf{p_2} - \mathbf{p_1}$. Hence

$$\mathbf{p} = \mathbf{p_1} + \overrightarrow{P_1P} = \mathbf{p_1} + \tfrac{1}{2}(\mathbf{p_2} - \mathbf{p_1}) = \tfrac{1}{2}(\mathbf{p_1} + \mathbf{p_2}) = \tfrac{1}{2}(x_1 + x_2, \; y_1 + y_2, \; z_1 + z_2)$$

This gives the result. Note that the position vector $\mathbf{p} = \frac{1}{2}(\mathbf{p_1} + \mathbf{p_2})$ of the midpoint is the "average" of the position vectors $\mathbf{p_1}$ and $\mathbf{p_2}$.

$\blacklozenge\blacklozenge\blacklozenge$

Lines in Space

These vector techniques can be used to give a very simple way of describing straight lines in space. We use the fact that there is exactly one line through a particular point in space that is parallel to a given nonzero vector.

> **DEFINITION**
>
> Given a straight line, any nonzero vector that is parallel to the line is called a **direction vector** for the line.

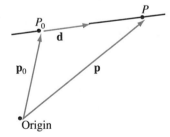

Every line has many direction vectors. In fact, any nonzero multiple of a direction vector also serves as a direction vector.

Suppose $P_0 = P_0(x_0, y_0, z_0)$ is any point and $\mathbf{d} = (a, b, c)$ is any vector (assumed to be nonzero). There is a unique line through P_0 with direction vector $\mathbf{d}$, and we want to give a condition that a point $P = P(x, y, z)$ lies on that line. Let $\mathbf{p_0} = (x_0, y_0, z_0)$ and $\mathbf{p} = (x, y, z)$ be the position vectors of P_0 and P, respectively (see Figure 4.12). Then

$$\mathbf{p} = \mathbf{p_0} + \overrightarrow{P_0P}$$

FIGURE 4.12

Hence P lies on the line if and only if $\overrightarrow{P_0P}$ is parallel to $\mathbf{d}$—that is, if and only if $\overrightarrow{P_0P} = t\mathbf{d}$ for some scalar t. Thus $\mathbf{p}$ is the position vector of a point on the line if and only if $\mathbf{p} = \mathbf{p}_0 + t\mathbf{d}$ for some scalar t.

VECTOR EQUATION OF A LINE

The line parallel to $\mathbf{d} \neq \mathbf{0}$ through the point with position vector $\mathbf{p}_0$ is given by

$$\mathbf{p} = \mathbf{p}_0 + t\mathbf{d} \qquad \text{for some scalar } t$$

In other words, the point with position vector $\mathbf{p}$ is on this line if and only if a real number t exists such that $\mathbf{p} = \mathbf{p}_0 + t\mathbf{d}.$

In component form the vector equation becomes

$$(x, y, z) = (x_0, y_0, z_0) + t(a, b, c)$$

Equating components gives a different description of the line.

PARAMETRIC EQUATIONS OF A LINE

The line through $P_0(x_0, y_0, z_0)$ with direction vector $\mathbf{d} = (a, b, c) \neq \mathbf{0}$ is given by

$$x = x_0 + ta$$
$$y = y_0 + tb \qquad t \text{ any scalar}$$
$$z = z_0 + tc$$

In other words, the point $P(x, y, z)$ is on this line if and only if a real number t exists such that $x = x_0 + ta$, $y = y_0 + tb$, and $z = z_0 + tc$.

EXAMPLE 8

Find the equations of the line through the points $P_0(2, 0, 1)$ and $P_1(4, -1, 1)$.

Solution Let $\mathbf{d} = \overrightarrow{P_0P_1} = (4 - 2, -1 - 0, 1 - 1) = (2, -1, 0)$ denote the vector from P_0 to P_1. Clearly $\mathbf{d}$ serves as a direction vector for the line. Using P_0 as the point on the line leads to the parametric equations

$$x = 2 + 2t$$
$$y = -t \qquad t \text{ a parameter}$$
$$z = 1$$

Note that if P_1 is used (rather than P_0), the equations are

$$x = 4 + 2s$$
$$y = -1 - s \qquad s \text{ a parameter}$$
$$z = 1$$

These are different from the preceding equations, but this is merely the result of a change in parameter. In fact, $s = t - 1$.   ◆◆◆

EXAMPLE 9

Find the equations of the line through $P_0(3, -1, 2)$ parallel to the line with equations

$$x = -1 + 2t$$
$$y = 1 + t$$
$$z = -3 + 4t$$

Solution   The coefficients of t give a direction vector $\mathbf{d} = (2, 1, 4)$ of the given line. Because the line we seek is parallel to this line, $\mathbf{d}$ serves as direction vector. Thus the parametric equations are

$$x = 3 + 2t$$
$$y = -1 + t$$
$$z = 2 + 4t$$
    ◆◆◆

EXAMPLE 10

Determine whether the following lines intersect and, if so, find the point of intersection.

$$x = 1 - 3t \qquad x = -1 + s$$
$$y = 2 + 5t \qquad y = 3 - 4s$$
$$z = 1 + t \qquad z = 1 - s$$

Solution   A typical point $P(x, y, z)$ on the first line has position vector $(x, y, z) = (1 - 3t, 2 + 5t, 1 + t)$ for some value of the parameter t. Similarly, a point on the second line has position vector $(x, y, z) = (-1 + s, 3 - 4s, 1 - s)$ for some value of s. Hence if $P(x, y, z)$ lies on *both* lines, there must exist t and s such that

$$(1 - 3t, 2 + 5t, 1 + t) = (x, y, z) = (-1 + s, 3 - 4s, 1 - s)$$

This means that the three equations

$$1 - 3t = -1 + s$$
$$2 + 5t = 3 - 4s$$
$$1 + t = 1 - s$$

must have a solution. Of course if there is *no* solution, the lines do not intersect. But in this case $t = 1$ and $s = -1$ satisfy all three equations. Thus the lines *do* intersect and the point of intersection is $(x, y, z) = (1 - 3t, 2 + 5t, 1 + t) = (-2, 7, 2)$

using $t = 1$. Of course this point can be found from $(x, y, z) = (-1 + s, 3 - 4s, 1 - s)$ using $s = -1$.

◆◆◆

The description of (nonvertical) lines in the plane is usually done using the notion of slope. The next example derives the point-slope formula using vector techniques.

EXAMPLE 11

Show that the line through $P_0(x_0, y_0)$ with slope m has direction vector $\mathbf{d} = (1, m)$ and equation $y - y_0 = m(x - x_0)$. This equation is called the point-slope formula.

Solution Let $P_1(x_1, y_1)$ be the point on the line one unit to the right of P_0—so that $x_1 = x_0 + 1$. Then $\mathbf{d} = \overrightarrow{P_0P_1}$ serves as direction vector of the line, so $\mathbf{d} = (x_1 - x_0, y_1 - y_0) = (1, y_1 - y_0)$. But the slope m can be computed as follows:

$$m = \frac{y_1 - y_0}{x_1 - x_0} = \frac{y_1 - y_0}{1} = y_1 - y_0$$

Hence $\mathbf{d} = (1, m)$ and the parametric equations are $x = x_0 + t$, $y = y_0 + mt$. Eliminating t gives $y - y_0 = mt = m(x - x_0)$, as asserted.

◆◆◆

Note that the vertical line through $P_0(x_0, y_0)$ has a direction vector $\mathbf{d} = (0, 1)$ that is not of the form $(1, m)$ for any m. This result confirms that the notion of slope makes no sense in this case. However, the vector method gives parametric equations for the line:

$$x = x_0$$
$$y = y_0 + t$$

Because y is arbitrary here (t is arbitrary), this is usually written simply as $x = x_0$.

EXERCISES 4.1

1. Find $\|\mathbf{v}\|$ and θ, in Example 1.

2. Use vectors to show that the line joining the midpoints of two sides of a triangle is parallel to the third side and half as long.

3. Let A, B, and C denote the three vertices of a triangle.
   (a) If E is the midpoint of side BC, show that
   $$\overrightarrow{AE} = \tfrac{1}{2}(\overrightarrow{AB} + \overrightarrow{AC}).$$
   ◆ (b) If F is the midpoint of side AC, show that
   $$\overrightarrow{FE} = \tfrac{1}{2}\overrightarrow{AB}.$$

4. Simplify the following where $\mathbf{u}$, $\mathbf{v}$, and $\mathbf{w}$ represent vectors.
   (a) $6(\mathbf{u} + 3\mathbf{v} - \mathbf{w}) - 12(\mathbf{v} - 3\mathbf{u}) + 3(2\mathbf{w} - 2\mathbf{u} - 2\mathbf{v})$
   ◆ (b) $8(2\mathbf{u} - \mathbf{v} + 3\mathbf{w}) + 3(5\mathbf{v} - 6\mathbf{w}) - 2(8\mathbf{u} + 3\mathbf{v} + 3\mathbf{w})$

5. Let $\mathbf{u} = (-1, 1, 2)$, $\mathbf{v} = (2, 0, 3)$, and $\mathbf{w} = (-1, 3, 9)$. Compute the following in component form.
   (a) $\mathbf{u} + 2\mathbf{v} - \mathbf{w}$ ◆ (b) $\tfrac{1}{3}(3\mathbf{u} - \mathbf{v} + 4\mathbf{w})$
   (c) $6\mathbf{u} + 2\mathbf{v} - 2\mathbf{w}$ ◆ (d) $2\mathbf{v} - 3(\mathbf{u} + \mathbf{w})$
   (e) $\tfrac{1}{3}(\mathbf{v} - 3\mathbf{u} + 2\mathbf{w})$ ◆ (f) $2(\mathbf{u} + \mathbf{v}) - (\mathbf{v} + \mathbf{w} - \mathbf{u})$

6. Determine whether $\mathbf{u}$ and $\mathbf{v}$ are parallel in each of the following cases.
   (a) $\mathbf{u} = (1, 2, -1)$; $\mathbf{v} = (2, 1, 0)$
   ◆ (b) $\mathbf{u} = (3, -6, 3)$; $\mathbf{v} = (-1, 2, -1)$
   (c) $\mathbf{u} = (1, 0, 1)$; $\mathbf{v} = (-1, 0, 1)$
   ◆(d) $\mathbf{u} = (2, 0, -1)$; $\mathbf{v} = (-8, 0, 4)$

7. Let $\mathbf{u}$ and $\mathbf{v}$ be the position vectors of points P and Q, respectively, and let R be the point whose position vector is $\mathbf{u} + \mathbf{v}$. Express the following in terms of $\mathbf{u}$ and $\mathbf{v}$.
   (a) $\overrightarrow{QP}$ ◆ (b) $\overrightarrow{QR}$
   (c) $\overrightarrow{RP}$ ◆ (d) $\overrightarrow{RO}$ where 0 is the origin

8. In each case, find $\overrightarrow{PQ}$ in component form.
   (a) $P(1, -1, 3)$, $Q(2, 1, 0)$
   ◆ (b) $P(2, 0, 1)$, $Q(1, -1, 6)$
   (c) $P(0, 0, 1)$, $Q(1, 0, -3)$
   ◆ (d) $P(1, -1, 2)$, $Q(1, -1, 2)$
   (e) $P(1, 0, -3)$, $Q(-1, 0, 3)$
   ◆ (f) $P(3, -1, 6)$, $Q(1, 1, 4)$

9. In each case, find a point Q such that $\overrightarrow{PQ}$ has (i) the same direction as $\mathbf{v}$; (ii) the opposite direction to $\mathbf{v}$.
   (a) $P(-1, 2, 2)$, $\mathbf{v} = (1, 2, -1)$ $Q(5, -1, 2)$
   ◆ (b) $P(3, 0, -1)$, $\mathbf{v} = (2, -1, 3)$ $Q(1, 1, -4)$

10. Let $\mathbf{u} = (3, -1, 0)$, $\mathbf{v} = (4, 0, 1)$, and $\mathbf{w} = (1, 1, 3)$. In each case, find $\mathbf{x}$ such that:
    (a) $3(2\mathbf{u} + \mathbf{x}) + \mathbf{w} = 2\mathbf{x} - \mathbf{v}$
    ◆ (b) $2(3\mathbf{v} - \mathbf{x}) = 5\mathbf{w} + \mathbf{u} - 3\mathbf{x}$

11. Let $\mathbf{u} = (1, 1, 2)$, $\mathbf{v} = (0, 1, 2)$, and $\mathbf{w} = (1, 0, -1)$. In each case, find numbers a, b, and c such that $\mathbf{x} = a\mathbf{u} + b\mathbf{v} + c\mathbf{w}$.
    (a) $\mathbf{x} = (2, -1, 6)$ ◆ (b) $\mathbf{x} = (1, 3, 0)$

12. Let $\mathbf{u} = (3, -1, 0)$, $\mathbf{v} = (4, 0, 1)$, and $\mathbf{z} = (1, 1, 1)$. In each case, show that there are no numbers a, b, and c such that:
    (a) $a\mathbf{u} + b\mathbf{v} + c\mathbf{z} = (1, 2, 1)$
    ◆ (b) $a\mathbf{u} + b\mathbf{v} + c\mathbf{z} = (5, 6, -1)$

13. Let $P_1 = P_1(2, 1, -2)$ and $P_2 = P_2(1, -2, 0)$. Find the coordinates of the point P:
    (a) $\frac{1}{5}$ the way from P_1 to P_2 $\frac{1}{4}(5, -5, -2)$
    ◆ (b) $\frac{1}{4}$ the way from P_2 to P_1

14. Find the two points trisecting the segment between $P(2, 3, 5)$ and $Q(8, -6, 2)$.

15. Let $P_1 = P_1(x_1, y_1, z_1)$ and $P_2 = P_2(x_2, y_2, z_2)$ be two points with position vectors $\mathbf{p_1}$ and $\mathbf{p_2}$, respectively. If r and s are positive integers, show that the point P lying $\dfrac{r}{r + s}$ the way from P_1 to P_2 has position vector
$$\mathbf{p} = \left[\frac{s}{r + s}\right]\mathbf{p_1} + \left[\frac{r}{r + s}\right]\mathbf{p_2}$$

16. Find the vector and parametric equations of the following lines.
    (a) The line parallel to $(2, -1, 0)$ and passing through $P(1, -1, 3)$
    ◆ (b) The line passing through $P(3, -1, 4)$ and $Q(1, 0, -1)$
    (c) The line passing through $P(3, -1, 4)$ and $Q(3, -1, 5)$
    ◆ (d) The line parallel to $(1, 1, 1)$ and passing through $P(1, 1, 1)$
    (e) The line passing through $P(1, 0, -3)$ and parallel to the line with parametric equations $x = -1 + 2t$, $y = 2 - t$, and $z = 3 + 3t$.
    ◆ (f) The line passing through $P(2, -1, 1)$ and parallel to the line with parametric equations $x = 2 - t$, $y = 1$, and $z = t$.
    (g) The lines through $P(1, 0, 1)$ that meet the line with vector equation $\mathbf{p} = (1, 2, 0) + t(2, -1, 2)$ at points at distance 3 from $P_0(1, 2, 0)$.

17. In each case, verify that the points P and Q lie on the line.
    (a) $x = 3 - 4t$ $P(-1, 3, 0)$, $Q(11, 0, 3)$
       $y = 2 + t$
       $z = 1 - t$
    ◆ (b) $x = 4 - t$ $P(2, 3, -3)$, $Q(-1, 3, -9)$
       $y = 3$
       $z = 1 - 2t$

18. Find the point of intersection (if any) of the following pairs of lines.
    (a) $x = 3 + t$ $x = 4 + 2s$
       $y = 1 - 2t$ $y = 6 + 3s$
       $z = 3 + 3t$ $z = 1 + s$
    ◆ (b) $x = 1 - t$ $x = 2s$
       $y = 2 + 2t$ $y = 1 + s$
       $z = -1 + 3t$ $z = 3$
    (c) $(x, y, z) = (3, -1, 2) + t(1, 1, -1)$
       $(x, y, z) = (1, 1, -2) + s(2, 0, 3)$
    ◆ (d) $(x, y, z) = (4, -1, 5) + t(1, 0, 1)$
       $(x, y, z) = (2, -7, 12) + s(0, -2, 3)$

19. Show that if a line passes through the origin, the position vectors of points on the line are all scalar multiples of some fixed nonzero vector.

20. Show that every line parallel to the Z axis has parametric equations $x = x_0$, $y = y_0$, $z = t$ for some fixed numbers x_0 and y_0.

21. Let $\mathbf{d} = (a, b, c)$ be a vector where a, b, and c are *all* nonzero. Show that the equations of the line through $P_0(x_0, y_0, z_0)$ with direction vector $\mathbf{d}$ can be written in the form

$$\frac{x - x_0}{a} = \frac{y - y_0}{b} = \frac{z - z_0}{c}$$

This is called the **symmetric form** of the equations.

22. **(a)** Prove (3) of Theorem 2.
   ◆**(b)** Prove (4) of Theorem 2.

23. A parallelogram has sides AB, BC, CD and DA. Given $A(1, -1, 2)$, $C(2, 1, 0)$, and the midpoint $M(1, 0, -3)$ of AB, find $\overrightarrow{BD}$.

◆**24.** Find all points C on the line through $A(1, -1, 2)$ and $B = (2, 0, 1)$ such that $\|\overrightarrow{AC}\| = 2\|\overrightarrow{BC}\|$.

25. Make a sketch like that in Figure 4.6, illustrating each of the following ways of adding four vectors.
   (a) $[(\mathbf{u} + \mathbf{v}) + \mathbf{w}] + \mathbf{z}$
   (b) $(\mathbf{u} + \mathbf{v}) + (\mathbf{w} + \mathbf{z})$
   (c) $\mathbf{u} + [\mathbf{v} + (\mathbf{w} + \mathbf{z})]$

26. Let A, B, C, D, E, and F be the vertices of a regular hexagon, taken in order. Show that $\overrightarrow{AB} + \overrightarrow{AC} + \overrightarrow{AD} + \overrightarrow{AE} + \overrightarrow{AF} = 3\overrightarrow{AD}$.

27. Let OAB be a right-angled triangle, as shown in the diagram, with the right angle at O. If C is the foot of the perpendicular from O to the hypotenuse, show that

$$\overrightarrow{AC} = \frac{\|\overrightarrow{OA}\|^2}{\|\overrightarrow{AB}\|^2}\overrightarrow{AB}$$

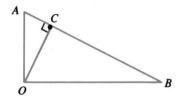

28. **(a)** Let P_1, P_2, P_3, P_4, P_5, and P_6 be six points equally spaced on a circle with center C. Show that $\overrightarrow{CP_1} + \overrightarrow{CP_2} + \overrightarrow{CP_3} + \overrightarrow{CP_4} + \overrightarrow{CP_5} + \overrightarrow{CP_6} = \mathbf{0}$.
   ◆**(b)** Show that the conclusion in part **(a)** holds for any *even* set of points evenly spaced on the circle.
   (c) Show that the conclusion in part **(a)** holds for *three* points.
   (d) Do you think it works for *any* finite set of points evenly spaced around the circle?

29. Consider a quadrilateral with vertices A, B, C, and D in order (as shown in the diagram). If the diagonals AC and

BD bisect each other, show that the quadrilateral is a parallelogram (this is the converse of Example 2). [*Hint:* Let E be the intersection of the diagonals. Show that $\overrightarrow{AB} = \overrightarrow{DC}$ by writing $\overrightarrow{AB} = \overrightarrow{AE} + \overrightarrow{EB}$.]

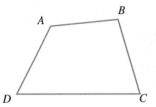

◆**30.** Consider the parallelogram $ABCD$ (see diagram), and let E be the midpoint of side AD. Show that BE and AC trisect each other; that is, show that the intersection point is one-third of the way from E to B and from A to C. [*Hint:* If F is one-third of the way from A to C, show that $\overrightarrow{FB} = 2\overrightarrow{EF}$ and argue as in Example 2.]

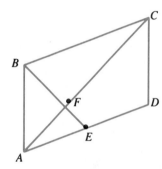

31. The intersection of the three medians of a triangle is called the **centroid** of the triangle. If $\mathbf{u}$, $\mathbf{v}$, and $\mathbf{w}$ are the position vectors of the vertices of a triangle, show that the centroid has position vector $\frac{1}{3}(\mathbf{u} + \mathbf{v} + \mathbf{w})$.

32. Given four noncoplanar points in space, the figure with these points as vertices is called a **tetrahedron.** The line from a vertex through the centroid (see previous exercise) of the triangle formed by the remaining vertices is called a **median** of the tetrahedron. If $\mathbf{u}$, $\mathbf{v}$, $\mathbf{w}$, and $\mathbf{x}$ are the position vectors of the four vertices, show that the point on a median one-fourth the way from the centroid to the vertex has position vector $\frac{1}{4}(\mathbf{u} + \mathbf{v} + \mathbf{w} + \mathbf{x})$. Conclude that the four medians are concurrent.

33. Prove the following parts of Theorem 1: **(a)** part (6); **(b)** part (7); **(c)** part (8).

FIGURE 4.13

Section 4.2　The Dot Product and Projections

Any student of geometry soon realizes that the notion of perpendicular lines is funda-mental. As an illustration, suppose a point P and a plane are given and it is desired to find the point Q that lies in the plane and is closest to P, as shown in Figure 4.13. Clearly, what is required is to find the line through P that is perpendicular to the plane and then to obtain Q as the point of intersection of this line with the plane. Finding the line *perpendicular* to the plane requires a way to determine when two vectors are perpendicular. Surprisingly enough, this can be done by using the follow-ing formula for the length of a vector in terms of its components.

THEOREM 1

Let $\mathbf{v} = (x, y, z)$ be a vector. Then

$$\|\mathbf{v}\| = \sqrt{x^2 + y^2 + z^2}$$

where the notation $\sqrt{}$ indicates the positive square root.

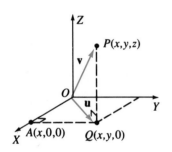

Proof　This is an application of the Pythagorean theorem.[3] Position $\mathbf{v}$ with its initial point at the origin and its terminal point at $P(x, y, z)$, as shown in the diagram, and let $\mathbf{u}$ be the position vector of $Q(x, y, 0)$. If $A = A(x, 0, 0)$, the lengths of the line segments OA and AQ are $|x|$ and $|y|$, respectively, so $\|\mathbf{u}\|^2 = x^2 + y^2$ by the Pythagorean theorem. However, the length of PQ is $|z|$, and so, again by the Pythagorean theorem, $\|\mathbf{v}\|^2 = \|\mathbf{u}\|^2 + z^2 = (x^2 + y^2) + z^2$, from which the desired conclusion follows.　◆

THEOREM 2
Distance Formula

The distance d between points $P_1(x_1, y_1, z_1)$ and $P_2(x_2, y_2, z_2)$ is given by

$$\sqrt{(x_2 - x_1)^2 + (y_2 - y_1)^2 + (z_2 - z_1)^2}$$

Proof　Theorem 3§4.1 gives $\overrightarrow{P_1P_2} = (x_2 - x_1, y_2 - y_1, z_2 - z_1)$. So the result follows by Theorem 1 because $d = \|\overrightarrow{P_1P_2}\|$.　◆

EXAMPLE 1

Compute the distance between $P_1(3, -1, 2)$ and $P_2(-1, 1, 0)$.

---

[3]The Pythagorean theorem is as follows: Given a right-angled triangle, the square of the length of the hypotenuse equals the sum of the squares of the lengths of the other two sides. A proof is sketched in Exercise 39.

Solution   The distance is

$$\sqrt{(-1-3)^2 + (1-(-1))^2 + (0-2)^2} = \sqrt{16+4+4} = \sqrt{24} = 2\sqrt{6}.$$

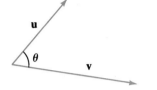

FIGURE 4.14

Let **u** and **v** be two nonzero vectors and suppose they are positioned with a common initial point. Then they determine a unique angle θ in the range

$$0 \le \theta \le \pi$$

(recall that π radians equals 180°). This angle θ will be referred to as **the angle between u and v** (see Figure 4.14). Clearly **u** and **v** are parallel if either $\theta = 0$ or $\theta = \pi$; they are said to be **orthogonal** if $\theta = \frac{\pi}{2}$—that is, if θ is a right angle. If one of **u** and **v** is **0**, the angle between them is not defined.

Among other uses, the following concept leads to a way of computing the angle between two nonzero vectors.

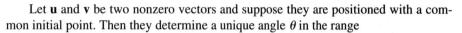

> **DEFINITION**
>
> The **dot product u · v** of two vectors **u** and **v** is defined as follows:
>
> $$\mathbf{u} \cdot \mathbf{v} = \begin{cases} \|\mathbf{u}\| \ \|\mathbf{v}\| \cos \theta & \text{if } \mathbf{u} \neq \mathbf{0} \text{ and } \mathbf{v} \neq \mathbf{0} \\ 0 & \text{if } \mathbf{u} = \mathbf{0} \text{ or } \mathbf{v} = \mathbf{0} \end{cases}$$
>
> where θ is the angle between **u** and **v**.

Note that **u · v** is a *number* (even though **u** and **v** are vectors). For this reason, **u · v** is sometimes called the **scalar product** of **u** and **v**.

The dot product of two vectors **u** and **v** is not easy to compute using the definition because the angle between the vectors is not usually known. However, the following fundamental theorem provides a very easy way of calculating **u · v**, provided that **u** and **v** are given in component form (and so leads to a method of computing the *angle*). The proof depends on the **law of cosines** from trigonometry: *If a triangle has sides of lengths a, b, and h, and if θ is the angle between the sides of lengths a and b (as in Figure 4.15), then*

$$h^2 = a^2 + b^2 - 2ab \cos \theta$$

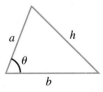

FIGURE 4.15

Observe that this is a generalization of the Pythagorean theorem: If $\theta = \frac{\pi}{2}$ (that is, $\theta = 90°$), then $\cos \theta = 0$ and the result becomes $h^2 = a^2 + b^2$.

THEOREM 3

Let $\mathbf{v_1} = (x_1, y_1, z_1)$ and $\mathbf{v_2} = (x_2, y_2, z_2)$ be two vectors given in component form. Then their dot product can be computed as follows:

$$\mathbf{v_1} \cdot \mathbf{v_2} = x_1 x_2 + y_1 y_2 + z_1 z_2$$

Proof   The result is true if $\mathbf{v_1} = \mathbf{0}$ or $\mathbf{v_2} = \mathbf{0}$. Otherwise consider the triangle in the diagram and apply the law of cosines.

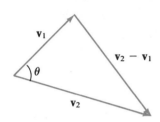

$$\|\mathbf{v_2} - \mathbf{v_1}\|^2 = \|\mathbf{v_1}\|^2 + \|\mathbf{v_2}\|^2 - 2\|\mathbf{v_1}\|\ \|\mathbf{v_2}\| \cos \theta$$
$$= \|\mathbf{v_1}\|^2 + \|\mathbf{v_2}\|^2 - 2\mathbf{v_1} \cdot \mathbf{v_2}$$

It follows that $\mathbf{v_1} \cdot \mathbf{v_2} = \frac{1}{2}\{\|\mathbf{v_1}\|^2 + \|\mathbf{v_2}\|^2 - \|\mathbf{v_2} - \mathbf{v_1}\|^2\}$. Because $\mathbf{v_2} - \mathbf{v_1} = (x_2 - x_1, y_2 - y_1, z_2 - z_1)$ in component form, Theorem 1 gives

$$\mathbf{v_1} \cdot \mathbf{v_2} = \frac{1}{2}\{(x_1^2 + y_1^2 + z_1^2) + (x_2^2 + y_2^2 + z_2^2)$$
$$-[(x_2 - x_1)^2 + (y_2 - y_1)^2 + (z_2 - z_1)^2]\}$$

The reader can verify that the right side reduces to $x_1 x_2 + y_1 y_2 + z_1 z_2$.   ◆

EXAMPLE 2

Compute $\mathbf{u} \cdot \mathbf{v}$ when $\mathbf{u} = (2, -1, 3)$ and $\mathbf{v} = (1, 4, -1)$.

Solution $\mathbf{u} \cdot \mathbf{v} = 2 \cdot 1 + (-1) \cdot 4 + 3(-1) = -5$ ◆◆◆

The next theorem lists several basic properties of the dot product that we will use repeatedly.

THEOREM 4

Let $\mathbf{u}, \mathbf{v},$ and $\mathbf{w}$ denote arbitrary vectors.

1. $\mathbf{u} \cdot \mathbf{v}$ is a real number
2. $\mathbf{u} \cdot \mathbf{v} = \mathbf{v} \cdot \mathbf{u}$
3. $\mathbf{u} \cdot \mathbf{0} = 0 = \mathbf{0} \cdot \mathbf{u}$
4. $\mathbf{u} \cdot \mathbf{u} = \|\mathbf{u}\|^2$
5. $(k\mathbf{u}) \cdot \mathbf{v} = k(\mathbf{u} \cdot \mathbf{v}) = \mathbf{u} \cdot (k\mathbf{v})$ for any scalar k
6. $\mathbf{u} \cdot (\mathbf{v} \pm \mathbf{w}) = \mathbf{u} \cdot \mathbf{v} \pm \mathbf{u} \cdot \mathbf{w}$

Proof   (1), (2), and (3)   These follow from the definition of $\mathbf{u} \cdot \mathbf{v}$.
(4) If $\mathbf{u} = (x, y, z)$, then $\mathbf{u} \cdot \mathbf{u} = x^2 + y^2 + z^2 = \|\mathbf{u}\|^2$ by Theorem 1.
(5) If $\mathbf{u} = (x, y, z)$ and $\mathbf{v} = (x_1, y_1, z_1)$, then Theorem 3 gives

$$
\begin{aligned}
(k\mathbf{u}) \cdot \mathbf{v} &= (kx, ky, kz) \cdot (x_1, y_1, z_1) \\
&= (kx)x_1 + (ky)y_1 + (kz)z_1 \\
&= k(xx_1 + yy_1 + zz_1) \\
&= k(\mathbf{u} \cdot \mathbf{v})
\end{aligned}
$$

The verification that $\mathbf{u} \cdot (k\mathbf{v}) = k(\mathbf{u} \cdot \mathbf{v})$ is analogous and is left to the reader.
(6) This is verified as in (5). It is left as Exercise 35.   ◆

Observe that (5) and (6) of Theorem 4 can be combined to enable us to make such calculations as the following:

$$
2\mathbf{u} \cdot (3\mathbf{v} - 2\mathbf{w} + 4\mathbf{z}) = 6(\mathbf{u} \cdot \mathbf{v}) - 4(\mathbf{u} \cdot \mathbf{w}) + 8(\mathbf{u} \cdot \mathbf{z})
$$

Such calculations are carried out without comment in what follows. Here is an example.

EXAMPLE 3

Verify that $\|\mathbf{u} + \mathbf{v}\|^2 = \|\mathbf{u}\|^2 + \|\mathbf{v}\|^2 + 2(\mathbf{u} \cdot \mathbf{v})$.

Solution   Apply Theorem 4.

$$
\begin{aligned}
\|\mathbf{u} + \mathbf{v}\|^2 &= (\mathbf{u} + \mathbf{v}) \cdot (\mathbf{u} + \mathbf{v}) = (\mathbf{u} + \mathbf{v}) \cdot \mathbf{u} + (\mathbf{u} + \mathbf{v}) \cdot \mathbf{v} \\
&= \mathbf{u} \cdot \mathbf{u} + \mathbf{v} \cdot \mathbf{u} + \mathbf{u} \cdot \mathbf{v} + \mathbf{v} \cdot \mathbf{v} \\
&= \|\mathbf{u}\|^2 + \|\mathbf{v}\|^2 + 2(\mathbf{u} \cdot \mathbf{v})
\end{aligned}
$$

◆◆◆

As we have mentioned, one important use of the dot product is in calculating the angle θ between two nonzero vectors $\mathbf{u}$ and $\mathbf{v}$. Because $\|\mathbf{u}\| \neq 0$ and $\|\mathbf{v}\| \neq 0$, the defining relationship $\mathbf{u} \cdot \mathbf{v} = \|\mathbf{u}\| \, \|\mathbf{v}\| \cos \theta$ can be solved for $\cos \theta$ to get

$$
\cos \theta = \frac{\mathbf{u} \cdot \mathbf{v}}{\|\mathbf{u}\| \, \|\mathbf{v}\|}
$$

This can be used to find θ. In this connection, it is worth noting that $\cos \theta$ has the same sign as $\mathbf{u} \cdot \mathbf{v}$, so

$$
\begin{array}{lll}
\mathbf{u} \cdot \mathbf{v} > 0 & \text{if and only if} & \theta \text{ is acute } \left(0 \leq \theta < \tfrac{\pi}{2}\right) \\
\mathbf{u} \cdot \mathbf{v} < 0 & \text{if and only if} & \theta \text{ is obtuse } \left(\tfrac{\pi}{2} < \theta \leq \pi\right) \\
\mathbf{u} \cdot \mathbf{v} = 0 & \text{if and only if} & \theta = \tfrac{\pi}{2}
\end{array}
$$

In this last case, the (nonzero) vectors are perpendicular. The following terminology is used in linear algebra.

DEFINITION

Two vectors **u** and **v** are said to be **orthogonal** if **u** = **0** or **v** = **0** or the angle between them is $\frac{\pi}{2}$.

Hence we have the following theorem:

THEOREM 5

Two nonzero vectors **u** and **v** are orthogonal if and only if $\mathbf{u} \cdot \mathbf{v} = 0$.

EXAMPLE 4

Compute the angle between **u** = $(-1, 1, 2)$ and **v** = $(2, 1, -1)$.

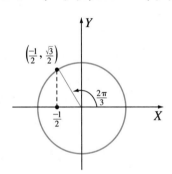

Solution   Compute $\cos\theta = \dfrac{\mathbf{u} \cdot \mathbf{v}}{\|\mathbf{u}\| \, \|\mathbf{v}\|} = \dfrac{-2 + 1 - 2}{\sqrt{6}\sqrt{6}} = \dfrac{-1}{2}$. Now recall that cos θ and sin θ

are defined so that (cos θ, sin θ) is the point on the unit circle determined by the angle θ (drawn counterclockwise, starting from the positive X axis). In the present case, we know that cos $\theta = -\frac{1}{2}$ and that $0 \le \theta \le \pi$. Because cos $\frac{\pi}{3} = \frac{1}{2}$, it follows that $\theta = \frac{2\pi}{3}$ (see the diagram).   ◆◆◆

EXAMPLE 5

Show that the points $P(3, -1, 1)$, $Q(4, 1, 4)$, and $R(6, 0, 4)$ are the vertices of a right triangle.

Solution   The vectors along the sides of the triangle are

$$\overrightarrow{PQ} = (1, 2, 3), \quad \overrightarrow{PR} = (3, 1, 3), \quad \text{and} \quad \overrightarrow{QR} = (2, -1, 0)$$

Evidently $\overrightarrow{PQ} \cdot \overrightarrow{QR} = (1, 2, 3) \cdot (2, -1, 0) = 2 - 2 + 0 = 0$, so $\overrightarrow{PQ}$ and $\overrightarrow{QR}$ are orthogonal vectors. This means sides PQ and QR are perpendicular — that is, the angle at Q is a right angle.

◆◆◆

Examples 6 and 7 demonstrate how the dot product can be used to verify geometrical theorems involving perpendicular lines.

EXAMPLE 6

A parallelogram with sides of equal length is called a **rhombus**. Show that the diagonals of a rhombus are perpendicular.

Solution

Let **u** and **v** denote vectors along two adjacent sides of a rhombus, as shown in the diagram. Then the diagonals are $\mathbf{u} - \mathbf{v}$ and $\mathbf{u} + \mathbf{v}$, and we compute

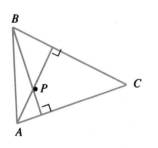

$$(\mathbf{u} - \mathbf{v}) \cdot (\mathbf{u} + \mathbf{v}) = \mathbf{u} \cdot (\mathbf{u} + \mathbf{v}) - \mathbf{v} \cdot (\mathbf{u} + \mathbf{v})$$
$$= \mathbf{u} \cdot \mathbf{u} + \mathbf{u} \cdot \mathbf{v} - \mathbf{v} \cdot \mathbf{u} - \mathbf{v} \cdot \mathbf{v}$$
$$= \|\mathbf{u}\|^2 - \|\mathbf{v}\|^2$$
$$= 0$$

because $\|\mathbf{u}\| = \|\mathbf{v}\|$ (it is a rhombus). Hence $\mathbf{u} - \mathbf{v}$ and $\mathbf{u} + \mathbf{v}$ are orthogonal.   ◆◆◆

EXAMPLE 7

The line through a vertex of a triangle, perpendicular to the opposite side is called an **altitude** of the triangle. Show that the three altitudes of any triangle are concurrent.

Solution

Let the vertices be A, B, and C, and let P be the point of intersection of the altitudes through A and B, as in the diagram. We must show that $\overrightarrow{PC}$ is orthogonal to $\overrightarrow{AB}$; that is, $\overrightarrow{PC} \cdot \overrightarrow{AB} = 0$. We have $\overrightarrow{AB} = \overrightarrow{AC} - \overrightarrow{BC}$, so

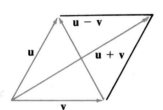

$$\overrightarrow{PC} \cdot \overrightarrow{AB} = \overrightarrow{PC} \cdot \overrightarrow{AC} - \overrightarrow{PC} \cdot \overrightarrow{BC}$$
$$= (\overrightarrow{PB} + \overrightarrow{BC}) \cdot \overrightarrow{AC} - (\overrightarrow{PA} + \overrightarrow{AC}) \cdot \overrightarrow{BC}$$
$$= \overrightarrow{PB} \cdot \overrightarrow{AC} + \overrightarrow{BC} \cdot \overrightarrow{AC} - \overrightarrow{PA} \cdot \overrightarrow{BC} - \overrightarrow{AC} \cdot \overrightarrow{BC}$$
$$= 0 + \overrightarrow{BC} \cdot \overrightarrow{AC} - 0 - \overrightarrow{AC} \cdot \overrightarrow{BC}$$
$$= 0$$

◆◆◆

Projections

If a nonzero vector **d** is given, it is often useful to be able to write an arbitrary vector **u** as a sum of two vectors,

$$\mathbf{u} = \mathbf{u}_1 + \mathbf{u}_2$$

where $\mathbf{u}_1$ is parallel to **d** and $\mathbf{u}_2 = \mathbf{u} - \mathbf{u}_1$ is orthogonal to **d**. Suppose that **u** and $\mathbf{d} \neq \mathbf{0}$ emanate from a common initial point Q (see Figure 4.16).

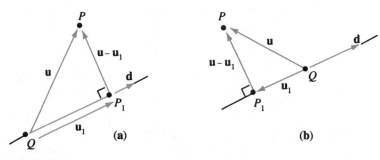

FIGURE 4.16

Let P be the terminal point of $\mathbf{u}$, and let P_1 denote the foot of the perpendicular dropped from P to the line through Q parallel to $\mathbf{d}$. Then $\mathbf{u}_1 = \overrightarrow{QP_1}$ has the required properties:

1. $\mathbf{u}_1$ is parallel to $\mathbf{d}$
2. $\mathbf{u}_2 = \mathbf{u} - \mathbf{u}_1$ is orthogonal to $\mathbf{d}$
3. $\mathbf{u} = \mathbf{u}_1 + \mathbf{u}_2$.

DEFINITION

The vector $\mathbf{u}_1 = \overrightarrow{QP_1}$ is called **the projection of u on d.** It is denoted

$$\mathbf{u}_1 = \operatorname{proj}_{\mathbf{d}} \mathbf{u}$$

In Figure 4.16(a) the vector $\mathbf{u}_1 = \operatorname{proj}_{\mathbf{d}} \mathbf{u}$ has the same direction as $\mathbf{d}$; however, it has the opposite direction from $\mathbf{d}$ if the angle between $\mathbf{u}$ and $\mathbf{d}$ is greater than $\frac{\pi}{2}$ (Figure 4.16(b)). Note that the projection $\mathbf{u}_1 = \operatorname{proj}_{\mathbf{d}} \mathbf{u}$ is zero if and only if $\mathbf{u}$ and $\mathbf{d}$ are orthogonal.

Calculating the projection of $\mathbf{u}$ on $\mathbf{d}$ is remarkably easy.

THEOREM 6

Let $\mathbf{u}$ and $\mathbf{d} \neq \mathbf{0}$ be vectors.

1. The projection $\mathbf{u}_1$ of $\mathbf{u}$ on $\mathbf{d}$ is given by $\operatorname{proj}_{\mathbf{d}} \mathbf{u} = \dfrac{\mathbf{u} \cdot \mathbf{d}}{\|\mathbf{d}\|^2} \mathbf{d}$.
2. The vector $\mathbf{u} - \operatorname{proj}_{\mathbf{d}} \mathbf{u}$ is orthogonal to $\mathbf{d}$.

Proof The vector $\mathbf{u}_1 = \operatorname{proj}_{\mathbf{d}} \mathbf{u}$ is parallel to $\mathbf{d}$ and so has the form $\mathbf{u}_1 = t\mathbf{d}$ for some scalar t. The requirement that $\mathbf{u} - \mathbf{u}_1$ and $\mathbf{d}$ are orthogonal determines t. In fact it means that $(\mathbf{u} - \mathbf{u}_1) \cdot \mathbf{d} = 0$ by Theorem 5; and if $\mathbf{u}_1 = t\mathbf{d}$ is substituted here, the condition is

$$0 = (\mathbf{u} - t\mathbf{d}) \cdot \mathbf{d} = \mathbf{u} \cdot \mathbf{d} - t(\mathbf{d} \cdot \mathbf{d}) = \mathbf{u} \cdot \mathbf{d} - t\|\mathbf{d}\|^2$$

It follows that $t = \dfrac{\mathbf{u} \cdot \mathbf{d}}{\|\mathbf{d}\|^2}$, where the assumption that $\mathbf{d} \neq \mathbf{0}$ guarantees that $\|\mathbf{d}\|^2 \neq 0$. ◆

EXAMPLE 8

Find the projection of $\mathbf{u} = (2, -3, 1)$ on $\mathbf{d} = (1, -1, 3)$ and express $\mathbf{u} = \mathbf{u}_1 + \mathbf{u}_2$ where $\mathbf{u}_1$ is parallel to $\mathbf{d}$ and $\mathbf{u}_2$ is orthogonal to $\mathbf{d}$.

Solution   The projection $\mathbf{u}_1$ of $\mathbf{u}$ on $\mathbf{d}$ is

$$\mathbf{u}_1 = \text{proj}_{\mathbf{d}}\,\mathbf{u} = \frac{\mathbf{u} \cdot \mathbf{d}}{\|\mathbf{d}\|^2}\,\mathbf{d} = \frac{2 + 3 + 3}{1^2 + (-1)^2 + 3^2}\,(1, -1, 3) = \tfrac{8}{11}\,(1, -1, 3)$$

The vector $\mathbf{u}_2 = \mathbf{u} - \mathbf{u}_1 = \tfrac{1}{11}(14, -25, -13)$ is orthogonal to $\mathbf{d}$ by Theorem 6 (alternatively, observe that $\mathbf{d} \cdot \mathbf{u}_2 = 0$), and $\mathbf{u} = \mathbf{u}_1 + \mathbf{u}_2$ as required.   ◆◆◆

EXAMPLE 9

Find the shortest distance (see diagram) from the point $P(1, 3, -2)$ to the line through $P_0(2, 0, -1)$ with direction vector $\mathbf{d} = (1, -1, 0)$. Also find the point P_1 that lies on the line and is closest to P.

Solution   Let $\mathbf{u} = (1, 3, -2) - (2, 0, -1) = (-1, 3, -1)$ denote the vector from P_0 to P, and let $\mathbf{u}_1$ denote the projection of $\mathbf{u}$ on $\mathbf{d}$. Thus

$$\mathbf{u}_1 = \frac{\mathbf{u} \cdot \mathbf{d}}{\|\mathbf{d}\|^2}\,\mathbf{d} = \frac{-1 - 3 + 0}{1^2 + (-1)^2 + 0^2}\,\mathbf{d} = -2\mathbf{d} = (-2, 2, 0)$$

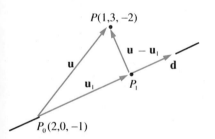

by Theorem 6. We see geometrically that the point P_1 on the line is closest to P, so the distance is

$$\|\overrightarrow{P_1P}\| = \|\mathbf{u} - \mathbf{u}_1\| = \|(1, 1, -1)\| = \sqrt{3}$$

To find the coordinates of P_1, let $\mathbf{p}_0$ and $\mathbf{p}_1$ denote the position vectors of P_0 and P_1, respectively. Then $\mathbf{p}_0 = (2, 0, -1)$ and $\mathbf{p}_1 = \mathbf{p}_0 + \mathbf{u}_1 = (0, 2, -1)$. Hence $P_1 = P_1(0, 2, -1)$ is the required point. It can be checked that the distance from P_1 to P is $\sqrt{3}$, as expected.   ◆◆◆

Exercises 4.2

1. Compute $\|\mathbf{v}\|$ if $\mathbf{v}$ equals:
   (a) $(2, -1, 1)$ ◆ (b) $(1, -1, 2)$
   (c) $(1, 0, -1)$ ◆ (d) $(-1, 0, 2)$
   (e) $2(1, -1, 2)$ ◆ (f) $-3(1, 1, 2)$

2. Find a unit vector in the direction of:
   (a) $(2, -2, 1)$ ◆ (b) $(-2, -1, 2)$

3. (a) Find a unit vector in the direction from $(3, -1, 4)$ to $(1, 3, 5)$.
   (b) If $\mathbf{u} \neq \mathbf{0}$, when is $a\mathbf{u}$ a unit vector?

4. Find the distance between the following pairs of points.
   (a) $(3, -1, 0)$ and $(2, 0, 1)$
   ◆ (b) $(2, -1, 2)$ and $(2, 0, 1)$

(c) $(-3, 5, 2)$ and $(-1, 3, 3)$

   ♦ (d) $(4, 0, -2)$ and $(3, 2, 0)$

5. Compute $\mathbf{u} \cdot \mathbf{v}$ where:

   (a) $\mathbf{u} = (2, -1, 3), \mathbf{v} = (1, 1, -2)$

   ♦ (b) $\mathbf{u} = (1, 2, -1), \mathbf{v} = \mathbf{u}$

   (c) $\mathbf{u} = (1, 1, -3), \mathbf{v} = 2(1, 1, 1)$

   ♦ (d) $\mathbf{u} = (3, -1, 5), \mathbf{v} = (6, -7, -5)$

   (e) $\mathbf{u} = (x, y, z), \mathbf{v} = (a, b, c)$

   ♦ (f) $\mathbf{u} = (a, b, c), \mathbf{v} = \mathbf{0}$

6. Find the angle between the following pairs of vectors.

   (a) $\mathbf{u} = (1, 0, 3), \mathbf{v} = (2, 0, 1)$

   ♦ (b) $\mathbf{u} = (3, -1, 0), \mathbf{v} = (-6, 2, 0)$

   (c) $\mathbf{u} = (7, -1, 3), \mathbf{v} = (1, 4, -1)$

   ♦ (d) $\mathbf{u} = (2, 1, -1), \mathbf{v} = (3, 6, 3)$

   (e) $\mathbf{u} = (1, -1, 0), \mathbf{v} = (0, 1, 1)$

   ♦ (f) $\mathbf{u} = (0, 3, 4), \mathbf{v} = (5\sqrt{2}, -7, -1)$

7. Find all real numbers x such that:

   (a) $(2, -1, 3)$ and $(x, -2, 1)$ are orthogonal

   ♦ (b) $(2, -1, 1)$ and $(1, x, 2)$ are at an angle of $\frac{\pi}{3}$

8. Find all vectors $\mathbf{v} = (x, y, z)$ orthogonal to both:

   (a) $\mathbf{u}_1 = (-1, -3, 2)$ and $\mathbf{u}_2 = (0, 1, 1)$

   ♦ (b) $\mathbf{u}_1 = (3, -1, 2)$ and $\mathbf{u}_2 = (2, 0, 1)$

   (c) $\mathbf{u}_1 = (2, 0, -1)$ and $\mathbf{u}_2 = (-4, 0, 2)$

   ♦ (d) $\mathbf{u}_1 = (2, -1, 3)$ and $\mathbf{u}_2 = (0, 0, 0)$

9. Find two vectors $\mathbf{x}$ and $\mathbf{y}$ that are both orthogonal to $\mathbf{v} = (1, 2, 0)$ and such that $\mathbf{x}$ is orthogonal to $\mathbf{y}$.

10. Consider the triangle with vertices $P(2, 0, -3)$, $Q(5, -2, 1)$, and $R(7, 5, 3)$.

   (a) Show that it is a right-angled triangle.

   ♦ (b) Find the lengths of the three sides and verify the Pythagorean theorem.

11. Show that the triangle with vertices $A(4, -7, 9), B(6, 4, 4)$, and $C(7, 10, -6)$ is not a right-angled triangle.

12. Find the three internal angles of the triangle with vertices:

   (a) $A(3, 1, -2), B(3, 0, -1)$, and $C(5, 2, -1)$

   ♦ (b) $A(3, 1, -2), B(5, 2, -1)$, and $C(4, 3, -3)$

13. Show that the line through $P_0(3, 1, 4)$ and $P_1(2, 1, 3)$ is perpendicular to the line through $P_2(1, -1, 2)$ and $P_3(0, 5, 3)$.

14. In each case, compute the projection of $\mathbf{u}$ on $\mathbf{v}$.

   (a) $\mathbf{u} = (5, 7, 1), \mathbf{v} = (2, -1, 3)$

   ♦ (b) $\mathbf{u} = (3, -2, 1), \mathbf{v} = (4, 1, 1)$

   (c) $\mathbf{u} = (1, -1, 2), \mathbf{v} = (3, -1, 1)$

   ♦ (d) $\mathbf{u} = (3, -2, -1), \mathbf{v} = (-6, 4, 2)$

15. In each case, write $\mathbf{u} = \mathbf{u}_1 + \mathbf{u}_2$, where $\mathbf{u}_1$ is parallel to $\mathbf{v}$ and $\mathbf{u}_2$ is orthogonal to $\mathbf{v}$.

   (a) $\mathbf{u} = (2, -1, 1), \mathbf{v} = (1, -1, 3)$

   ♦ (b) $\mathbf{u} = (3, 1, 0), \mathbf{v} = (-2, 1, 4)$

   (c) $\mathbf{u} = (2, -1, 0), \mathbf{v} = (3, 1, -1)$

   ♦ (d) $\mathbf{u} = (3, -2, 1), \mathbf{v} = (-6, 4, -1)$

16. Calculate the distance from the point P to the line in each case and find the point Q on the line closest to P.

   (a) $P(3, 2, -1)$ line: $(x, y, z) = (2, 1, 3) + t(3, -1, -2)$

   ♦ (b) $P(1, -1, 3)$ line: $(x, y, z) = (1, 0, -1) + t(3, 1, 4)$

17. Show that two lines in the plane with slopes m_1 and m_2 are perpendicular if and only if $m_1 m_2 = -1$. [*Hint:* Example 11§4.1.]

18. (a) Show that, of the four diagonals of a cube, no pair is perpendicular.

   (b) Show that each diagonal is perpendicular to the face diagonals it does not meet.

19. Given a rectangular solid with sides of lengths 1, 1, and $\sqrt{2}$, find the angle between a diagonal and one of the longest sides.

20. Consider a rectangular solid with sides of lengths $a, b,$ and c. Show that it has two orthogonal diagonals if and only if the sum of two of $a^2, b^2,$ and c^2 equals the third.

21. Let $A, B,$ and $C(2, -1, 1)$ be the vertices of a triangle where $\overrightarrow{AB}$ is parallel to $(1, -1, 1)$, $\overrightarrow{AC}$ is parallel to $(2, 0, -1)$, and angle C is a right angle. Find the equation of the line through B and C.

22. Given $\mathbf{v} = (x, y, z)$ in component form, show that the projections of $\mathbf{v}$ on $\mathbf{i}, \mathbf{j},$ and $\mathbf{k}$ are $x\mathbf{i}, y\mathbf{j},$ and $z\mathbf{k}$, respectively.

23. Can $\mathbf{u} \cdot \mathbf{v} = -7$ if $\|\mathbf{u}\| = 3$ and $\|\mathbf{v}\| = 2$? Defend your answer.

24. Show that $(\mathbf{u} + \mathbf{v}) \cdot (\mathbf{u} - \mathbf{v}) = \|\mathbf{u}\|^2 - \|\mathbf{v}\|^2$ for any vectors $\mathbf{u}$ and $\mathbf{v}$.

25. (a) Show that $\|\mathbf{u} + \mathbf{v}\|^2 + \|\mathbf{u} - \mathbf{v}\|^2 = 2(\|\mathbf{u}\|^2 + \|\mathbf{v}\|^2)$ for any vectors $\mathbf{u}$ and $\mathbf{v}$.

   ♦ (b) What does this say about parallelograms?

26. Show that if the diagonals of a parallelogram are perpendicular, it is necessarily a rhombus. [*Hint:* Example 6.]

27. Let A and B be the end points of a diameter of a circle (see diagram). If C is any point on the circle, show that $\overrightarrow{AC}$ and $\overrightarrow{BC}$ are perpendicular. [*Hint:* Express $\overrightarrow{AC}$ and $\overrightarrow{BC}$

in terms of $\mathbf{u} = \overrightarrow{OA}$ and $\mathbf{v} = \overrightarrow{OC}$, where O is the center.]

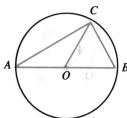

28. If $\mathbf{u}$ and $\mathbf{v}$ are orthogonal, show that $\|\mathbf{u} + \mathbf{v}\|^2 = \|\mathbf{u}\|^2 + \|\mathbf{v}\|^2$.

29. Let $\mathbf{u}$, $\mathbf{v}$, and $\mathbf{w}$ be pairwise orthogonal vectors.
   (a) Show that $\|\mathbf{u} + \mathbf{v} + \mathbf{w}\|^2 = \|\mathbf{u}\|^2 + \|\mathbf{v}\|^2 + \|\mathbf{w}\|^2$.
   ◆ **(b)** If $\mathbf{u}$, $\mathbf{v}$, and $\mathbf{w}$ are all the same length show that they all make the same angle with $\mathbf{u} + \mathbf{v} + \mathbf{w}$.

30. **(a)** Show that $\mathbf{n} = (a, b)$ is perpendicular to the line $ax + by + c = 0$.
   (b) Show that the shortest distance from $P_0(x_0, y_0)$ to the line is $\dfrac{|ax_0 + by_0 + c|}{\sqrt{a^2 + b^2}}$. [*Hint:* If P_1 is on the line, project $\mathbf{u} = \overrightarrow{P_1P_0}$ on $\mathbf{n}$.]

31. Assume $\mathbf{u}$ and $\mathbf{v}$ are nonzero vectors that are not parallel. Show that $\mathbf{w} = \|\mathbf{u}\|\mathbf{v} + \|\mathbf{v}\|\mathbf{u}$ is a nonzero vector that bisects the angle between $\mathbf{u}$ and $\mathbf{v}$.

32. Let α, β, and γ be the angles a vector $\mathbf{v} \neq \mathbf{0}$ makes with the positive X, Y, and Z axes, respectively. Then $\cos\alpha$, $\cos\beta$, and $\cos\gamma$ are called the **direction cosines** of the vector $\mathbf{v}$.
   (a) If $\mathbf{v} = (a, b, c)$, show that $\cos\alpha = \dfrac{a}{\|\mathbf{v}\|}$, $\cos\beta = \dfrac{b}{\|\mathbf{v}\|}$, and $\cos\gamma = \dfrac{c}{\|\mathbf{v}\|}$.
   ◆ **(b)** Show that $\cos^2\alpha + \cos^2\beta + \cos^2\gamma = 1$.

33. Let $\mathbf{v} \neq \mathbf{0}$ be any nonzero vector and suppose that a vector $\mathbf{u}$ can be written as $\mathbf{u} = \mathbf{p} + \mathbf{q}$, where $\mathbf{p}$ is parallel to $\mathbf{v}$ and $\mathbf{q}$ is orthogonal to $\mathbf{v}$. Show that $\mathbf{p}$ is necessarily the projection of $\mathbf{u}$ on $\mathbf{v}$. [*Hint:* Argue as in the proof of Theorem 6.]

34. Use Theorem 2§4.1 and Theorem 1 to verify the formula $\|k\mathbf{v}\| = |k| \|\mathbf{v}\|$ for all scalars k and vectors $\mathbf{v}$.

35. Prove (6) of Theorem 4.

36. Let $\mathbf{v} \neq \mathbf{0}$ be a nonzero vector and let $a \neq 0$ be a scalar. If $\mathbf{u}$ is any vector, show that the projection of $\mathbf{u}$ on $\mathbf{v}$ equals the projection of $\mathbf{u}$ on $a\mathbf{v}$.

37. **(a)** Show that the **Cauchy–Schwarz inequality** $|\mathbf{u} \cdot \mathbf{v}| \leq \|\mathbf{u}\| \|\mathbf{v}\|$ holds for all vectors $\mathbf{u}$ and $\mathbf{v}$. [*Hint:* $|\cos\theta| \leq 1$ for all angles θ.]
   (b) Show that $|\mathbf{u} \cdot \mathbf{v}| = \|\mathbf{u}\| \|\mathbf{v}\|$ if and only if $\mathbf{u}$ and $\mathbf{v}$ are parallel. [*Hint:* When is $\cos\theta = \pm 1$?]
   (c) Show that
   $$|x_1x_2 + y_1y_2 + z_1z_2| \leq \sqrt{x_1^2 + y_1^2 + z_1^2}\sqrt{x_2^2 + y_2^2 + z_2^2}$$
   holds for all numbers x_1, x_2, y_1, y_2, z_1, and z_2.
   (d) Show that $|xy + yz + zx| \leq x^2 + y^2 + z^2$ for all x, y, and z.
   (e) Show that $(x + y + z)^2 \leq 3(x^2 + y^2 + z^2)$ holds for all x, y, and z.

38. Prove that the **triangle inequality** $\|\mathbf{u} + \mathbf{v}\| \leq \|\mathbf{u}\| + \|\mathbf{v}\|$ holds for all vectors $\mathbf{u}$ and $\mathbf{v}$. [*Hint:* Consider the triangle with $\mathbf{u}$ and $\mathbf{v}$ as two sides.]

39. Prove the Pythagorean theorem. [*Hint:* In the diagram, let $h = p + q$. Then $\frac{a}{h} = \frac{p}{a}$ and $\frac{b}{h} = \frac{q}{b}$ using similar triangles.]

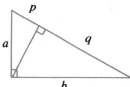

Section 4.3   Planes and the Cross Product

It is evident geometrically that among all planes that are perpendicular to a given straight line there is exactly one containing a given point. This fact can be used to give a very simple description of a plane. To do this, it is necessary to introduce the following notion:

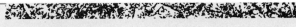

DEFINITION

A nonzero vector **n** is called a **normal** to a plane if it is orthogonal to every vector in the plane.

For example, the coordinate vector **k** is a normal to the *X-Y* plane.

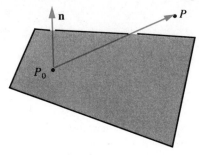

FIGURE 4.17

Given a point $P_0 = P_0(x_0, y_0, z_0)$ and a nonzero vector **n**, there is a unique plane through P_0 with normal **n**. By Figure 4.17 a point $P = P(x, y, z)$ lies on this plane if and only if the vector $\overrightarrow{P_0P}$ is orthogonal to **n** — that is, if and only if $\mathbf{n} \cdot \overrightarrow{P_0P} = 0$. Because $\overrightarrow{P_0P} = (x - x_0, y - y_0, z - z_0)$, this gives the following:

SCALAR EQUATION OF A PLANE

The plane through $P_0(x_0, y_0, z_0)$ with normal $\mathbf{n} = (a, b, c) \neq \mathbf{0}$ is given by

$$a(x - x_0) + b(y - y_0) + c(z - z_0) = 0$$

In other words, the point $P(x, y, z)$ is on this plane if and only if x, y, and z satisfy this equation.

EXAMPLE 1

Find an equation of the plane through $P_0(1, -1, 3)$ with normal $\mathbf{n} = (3, -1, 2)$.

Solution   Here the general scalar equation becomes

$$3(x - 1) - (y + 1) + 2(z - 3) = 0$$

This simplifies to $3x - y + 2z = 10$.

◆◆◆

If we write $d = ax_0 + by_0 + cz_0$, the scalar equation shows that every plane with normal $\mathbf{n} = (a, b, c)$ has a linear equation of the form

$$ax + by + cz = d \qquad (*)$$

for some constant d. Conversely, the graph of this equation is a plane with $\mathbf{n} = (a, b, c)$ as a normal vector (assuming that a, b, and c are not all zero).

EXAMPLE 2

Find an equation of the plane through $P_0(3, -1, 2)$ that is parallel to the plane with equation $2x - 3y = 6$.

Solution   The plane with equation $2x - 3y = 6$ has normal $\mathbf{n} = (2, -3, 0)$. Because the two planes are parallel, $\mathbf{n}$ serves as a normal to the plane we seek, so the equation is $2x - 3y = d$ for some d by equation $(*)$. Insisting that $P_0(3, -1, 2)$ lies on the plane determines d: $d = 2 \cdot 3 - 3(-1) = 9$. Hence the equation is $2x - 3y = 9$.

◆◆◆

Consider points $P_0(x_0, y_0, z_0)$ and $P(x, y, z)$ with position vectors $\mathbf{p}_0 = (x_0, y_0, z_0)$ and $\mathbf{p} = (x, y, z)$. Given a nonzero vector $\mathbf{n}$, the scalar equation of the plane through $P_0(x_0, y_0, z_0)$ with normal $\mathbf{n} = (a, b, c)$ takes the vector form:

> ### VECTOR EQUATION OF A PLANE
>
> The plane with normal $\mathbf{n} \neq \mathbf{0}$ through the point with position vector $\mathbf{p}_0$ is given by
>
> $$\mathbf{n} \cdot (\mathbf{p} - \mathbf{p}_0) = 0$$
>
> In other words, the point with position vector $\mathbf{p}$ is on the plane if and only if $\mathbf{p}$ satisfies this condition.

Moreover, equation $(*)$ translates as follows: *Every plane with normal* $\mathbf{n}$ *has vector equation*

$$\mathbf{n} \cdot \mathbf{p} = d$$

for some number d. This is useful in the second solution of Example 3.

EXAMPLE 3

Find the shortest distance from the point $P_1(2, 1, -3)$ to the plane with equation $3x - y + 4z = 1$. Also find the point on this plane closest to P_1.

Solution 1   The plane in question has normal $\mathbf{n} = (3, -1, 4)$. Choose any point P_0 on the plane — say $P_0(0, -1, 0)$ — and let $P(x, y, z)$ be the point on the plane closest to P_1 (see the diagram). The vector from P_0 to P_1 is $\mathbf{u} = (2, 2, -3)$. Now erect $\mathbf{n}$ with its initial point at P_0. Then $\overrightarrow{PP_1} = \mathbf{u}_1$ is the projection of $\mathbf{u}$ on $\mathbf{n}$:

$$\mathbf{u}_1 = \frac{\mathbf{n} \cdot \mathbf{u}}{\|\mathbf{n}\|^2} \mathbf{n} = \frac{-8}{26}(3, -1, 4) = \frac{-4}{13}(3, -1, 4)$$

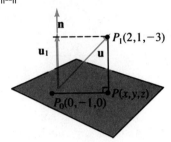

Hence the distance is $\|\overrightarrow{PP_1}\| = \|\mathbf{u}_1\| = \frac{4\sqrt{26}}{13}$. To calculate point P, let $\mathbf{p} = (x, y, z)$ and $\mathbf{p}_0 = (0, -1, 0)$ be the position vectors of P and P_0. Then $\mathbf{p} = \mathbf{p}_0 + \mathbf{u} - \mathbf{u}_1 = (0, -1, 0) + (2, 2, -3) + \frac{4}{13}(3, -1, 4) = \left(\frac{38}{13}, \frac{9}{13}, \frac{-23}{13}\right)$. This gives the coordinates of P.

Solution 2   Let $\mathbf{p} = (x, y, z)$ and $\mathbf{p}_1 = (2, 1, -3)$ be the position vectors of P and P_1. Then P is on the line through P_1 with direction vector $\mathbf{n}$, so $\mathbf{p} = \mathbf{p}_1 + t\mathbf{n}$ for some value of t. In addition, P lies on the plane, so $\mathbf{n} \cdot \mathbf{p} = 3x - y + 4z = 1$. This determines t:

$$1 = \mathbf{n} \cdot \mathbf{p} = \mathbf{n} \cdot (\mathbf{p}_1 + t\mathbf{n}) = \mathbf{n} \cdot \mathbf{p}_1 + t\|\mathbf{n}\|^2 = -7 + t(26)$$

This gives $t = \frac{8}{26} = \frac{4}{13}$, so

$$(x, y, z) = \mathbf{p} = \mathbf{p}_1 + t\mathbf{n} = (2, 1, -3) + \frac{4}{13}(3, -1, 4) = \frac{1}{13}(38, 9, -23)$$

as before. This determines P (in the diagram), and the reader can verify that the required distance is $\|\overrightarrow{PP_1}\| = \frac{4}{13}\sqrt{26}$, as before.

◆◆◆

It is clear geometrically that if three distinct points P, Q, and R are not all on some line, there is a unique plane that contains all three. Now the vectors $\overrightarrow{PQ}$ and $\overrightarrow{PR}$ lie in the plane, so any nonzero vector orthogonal to both $\overrightarrow{PQ}$ and $\overrightarrow{PR}$ serves as a normal. Hence we must find a systematic way to discover a vector orthogonal to two given vectors. The cross product provides such a vector.

The Cross Product

DEFINITION

Given vectors $\mathbf{v}_1 = (x_1, y_1, z_1)$ and $\mathbf{v}_2 = (x_2, y_2, z_2)$, the **cross product** $\mathbf{v}_1 \times \mathbf{v}_2$ is defined by

$$\mathbf{v}_1 \times \mathbf{v}_2 = ((y_1 z_2 - z_1 y_2), -(x_1 z_2 - z_1 x_2), (x_1 y_2 - y_1 x_2))$$

There is a useful way to remember this definition. Recall that any vector $\mathbf{u} = (x, y, z)$ can be written as $\mathbf{u} = x\mathbf{i} + y\mathbf{j} + z\mathbf{k}$, where $\mathbf{i} = (1, 0, 0)$, $\mathbf{j} = (0, 1, 0)$, and $\mathbf{k} = (0, 0, 1)$ are the coordinate vectors introduced in Section 4.1. Then the cross product can be described as follows:

DETERMINANT FORM OF THE CROSS PRODUCT

If $\mathbf{v}_1 = (x_1, y_1, z_1)$ and $\mathbf{v}_2 = (x_2, y_2, z_2)$ are two vectors, then

$$\mathbf{v}_1 \times \mathbf{v}_2 = \det \begin{bmatrix} \mathbf{i} & \mathbf{j} & \mathbf{k} \\ x_1 & y_1 & z_1 \\ x_2 & y_2 & z_2 \end{bmatrix}$$

$$= \begin{vmatrix} y_1 & z_1 \\ y_2 & z_2 \end{vmatrix} \mathbf{i} - \begin{vmatrix} x_1 & z_1 \\ x_2 & z_2 \end{vmatrix} \mathbf{j} + \begin{vmatrix} x_1 & y_1 \\ x_2 & y_2 \end{vmatrix} \mathbf{k}$$

where the determinant is expanded along the first row by cofactors.

EXAMPLE 4

Find $\mathbf{u} \times \mathbf{v}$ if $\mathbf{u} = (2, -1, 4)$ and $\mathbf{v} = (1, 3, 7)$.

Solution

$$\mathbf{u} \times \mathbf{v} = \det \begin{bmatrix} \mathbf{i} & \mathbf{j} & \mathbf{k} \\ 2 & -1 & 4 \\ 1 & 3 & 7 \end{bmatrix} = \begin{vmatrix} -1 & 4 \\ 3 & 7 \end{vmatrix} \mathbf{i} - \begin{vmatrix} 2 & 4 \\ 1 & 7 \end{vmatrix} \mathbf{j} + \begin{vmatrix} 2 & -1 \\ 1 & 3 \end{vmatrix} \mathbf{k}$$

$$= -19\mathbf{i} - 10\mathbf{j} + 7\mathbf{k}$$

$$= (-19, -10, 7)$$

It is easily verified that $\mathbf{u} \times \mathbf{v}$ is orthogonal to both $\mathbf{u}$ and $\mathbf{v}$ in Example 4. This holds in general by the following result.

THEOREM 1

Let $\mathbf{w} = (x_1, y_1, z_1)$, $\mathbf{u} = (x_2, y_2, z_2)$, and $\mathbf{v} = (x_3, y_3, z_3)$. Then

$$\mathbf{w} \cdot (\mathbf{u} \times \mathbf{v}) = \det \begin{bmatrix} x_1 & y_1 & z_1 \\ x_2 & y_2 & z_2 \\ x_3 & y_3 & z_3 \end{bmatrix}$$

Proof Recall that $\mathbf{w} \cdot (\mathbf{u} \times \mathbf{v})$ is computed by multiplying corresponding components of $\mathbf{w}$ and $\mathbf{u} \times \mathbf{v}$ and then adding. The result follows by expanding the determinant along row 1. ◆

Because of Theorem 1 and the determinant form of the cross product, several properties of the cross product follow easily from properties of determinants (they can also be verified directly).

THEOREM 2

Let $\mathbf{u}$, $\mathbf{v}$, and $\mathbf{w}$ denote arbitrary vectors.

1. $\mathbf{u} \times \mathbf{v}$ is a vector
2. $\mathbf{u} \times \mathbf{v}$ is orthogonal to both $\mathbf{u}$ and $\mathbf{v}$
3. $\mathbf{u} \times \mathbf{0} = \mathbf{0} = \mathbf{0} \times \mathbf{u}$
4. $\mathbf{u} \times \mathbf{u} = \mathbf{0}$
5. $\mathbf{u} \times \mathbf{v} = -(\mathbf{v} \times \mathbf{u})$
6. $(k\mathbf{u}) \times \mathbf{v} = k(\mathbf{u} \times \mathbf{v}) = \mathbf{u} \times (k\mathbf{v})$ for any scalar k
7. $\mathbf{u} \times (\mathbf{v} + \mathbf{w}) = (\mathbf{u} \times \mathbf{v}) + (\mathbf{u} \times \mathbf{w})$
8. $(\mathbf{v} + \mathbf{w}) \times \mathbf{u} = (\mathbf{v} \times \mathbf{u}) + (\mathbf{w} \times \mathbf{u})$

Proof (1) is clear; (2) follows from Theorem 1; and (3) and (4) follow because the determinant of a matrix is zero if one row is zero or two rows are identical. If two rows are interchanged, the determinant changes sign, and this proves (5). The proofs of (6), (7), and (8) are left as Exercise 27. ◆

EXAMPLE 5

Find an equation of the plane through $P(1, 3, -2)$, $Q(1, 1, 5)$, and $R(2, -2, 3)$.

Solution 1 The vectors $\overrightarrow{PQ} = (0, -2, 7)$ and $\overrightarrow{PR} = (1, -5, 5)$ lie in the plane, so

$$\overrightarrow{PQ} \times \overrightarrow{PR} = \det \begin{bmatrix} \mathbf{i} & \mathbf{j} & \mathbf{k} \\ 0 & -2 & 7 \\ 1 & -5 & 5 \end{bmatrix} = 25\mathbf{i} + 7\mathbf{j} + 2\mathbf{k} = (25,\ 7,\ 2)$$

is a normal to the plane (being orthogonal to both $\overrightarrow{PQ}$ and $\overrightarrow{PR}$). Because $P(1, 3, -2)$ lies in the plane, the equation is $25(x - 1) + 7(y - 3) + 2(z + 2) = 0$. This simplifies to $25x + 7y + 2z = 42$. Incidentally, the reader can verify that this same equation is obtained if either Q or R is used as the point lying in the plane.

Solution 2 The plane has equation $ax + by + cz - d = 0$ and substituting P, Q, and R gives three equations

$$a + 3b - 2c - d = 0$$
$$a + b + 5c - d = 0$$
$$2a - 2b + 3c - d = 0$$

Gaussian elimination gives $a = 25t$, $b = 7t$, $c = 2t$, $d = 42t$. Taking $t = 1$ gives the preceding solution. ◆◆◆

EXAMPLE 6

Find the shortest distance between the nonparallel lines

$$(x, y, z) = (1, 0, -1) + t(2, 0, 1)$$
$$(x, y, z) = (3, 1, 0) + s(1, 1, -1)$$

Then find the points A and B on the lines that are closest together.

Solution

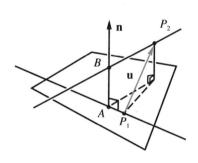

Direction vectors for the two lines are $\mathbf{d}_1 = (2, 0, 1)$ and $\mathbf{d}_2 = (1, 1, -1)$, so

$$\mathbf{n} = \mathbf{d}_1 \times \mathbf{d}_2 = \det \begin{bmatrix} \mathbf{i} & \mathbf{j} & \mathbf{k} \\ 2 & 0 & 1 \\ 1 & 1 & -1 \end{bmatrix} = (-1, 3, 2)$$

is perpendicular to both lines. Consider the plane containing the first line with $\mathbf{n}$ as normal. This plane contains $P_1(1, 0, -1)$ and is parallel to the second line. Because $P_2(3, 1, 0)$ is on the second line, the distance in question is just the shortest distance between $P_2(3, 1, 0)$ and this plane. The vector $\mathbf{u}$ from P_1 to P_2 is $\mathbf{u} = \overrightarrow{P_1P_2} = (2, 1, 1)$ and so, as in Example 3, the distance is the length of the projection of $\mathbf{u}$ on $\mathbf{n}$.

$$\text{distance} = \left\| \frac{\mathbf{u} \cdot \mathbf{n}}{\|\mathbf{n}\|^2} \, \mathbf{n} \right\| = \frac{|\mathbf{u} \cdot \mathbf{n}|}{\|\mathbf{n}\|} = \frac{3}{\sqrt{14}} = \frac{3\sqrt{14}}{14}$$

Note that it is necessary that $\mathbf{n} = \mathbf{d}_1 \times \mathbf{d}_2$ not be zero for this calculation to be possible. As is shown later (Theorem 4), this is guaranteed by the fact that $\mathbf{d}_1$ and $\mathbf{d}_2$ are *not* parallel.

The points A and B have coordinates $A(1 + 2t, 0, t - 1)$ and $B(3 + s, 1 + s, -s)$ for some s and t, so $\overrightarrow{AB} = (2 + s - 2t, 1 + s, 1 - s - t)$. This vector is orthogonal to $\mathbf{d}_1$ and $\mathbf{d}_2$, and the conditions $\overrightarrow{AB} \cdot \mathbf{d}_1 = 0$ and $\overrightarrow{AB} \cdot \mathbf{d}_2 = 0$ give equations $5t - s = 5$ and $t - 3s = 2$. The solution is $s = \frac{-5}{14}$ and $t = \frac{13}{14}$ so the points are $A\left(\frac{40}{14}, 0, \frac{-1}{14}\right)$ and $B\left(\frac{37}{14}, \frac{9}{14}, \frac{5}{14}\right)$. We have $\|\overrightarrow{AB}\| = \frac{3\sqrt{14}}{14}$, as before.

◆◆◆

Recall that the dot product of two vectors $\mathbf{u}$ and $\mathbf{v}$ was defined by $\mathbf{u} \cdot \mathbf{v} = \|\mathbf{u}\| \, \|\mathbf{v}\|$ cos θ, where θ is the angle between $\mathbf{u}$ and $\mathbf{v}$. One virtue of this definition is that it does not depend on any coordinate system (although $\mathbf{u} \cdot \mathbf{v}$ can be *computed* using Theorem 3§4.2 when the components of $\mathbf{u}$ and $\mathbf{v}$ are given in some coordinate system). However, the cross product $\mathbf{u} \times \mathbf{v}$ has been defined in terms of components of **u** and **v**, and the question naturally arises whether **u** $\times$ **v** can be defined in terms of the vectors **u** and **v** themselves without any reference to coordinates. In other words, can the length and direction of **u** $\times$ **v** be given in terms of the length and direction of **u** and **v**? The answer is affirmative and is based to some extent on the following identity relating the dot product and the cross product.

THEOREM 3
Lagrange Identity[4]

If **u** and **v** are any two vectors, then

$$\|\mathbf{u} \times \mathbf{v}\|^2 = \|\mathbf{u}\|^2 \|\mathbf{v}\|^2 - (\mathbf{u} \cdot \mathbf{v})^2$$

Proof   Given **u** and **v**, introduce a coordinate system and write $\mathbf{u} = (x_1, y_1, z_1)$ and $\mathbf{v} = (x_2, y_2, z_2)$ in component form. Then all the terms in the identity can be computed in terms of the components. The detailed proof is left as Exercise 26.   ◆

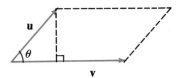

FIGURE 4.18

An expression for the magnitude of the vector **u** $\times$ **v** can be easily obtained from the Lagrange identity. If θ is the angle between **u** and **v**, substituting $\mathbf{u} \cdot \mathbf{v} = \|\mathbf{u}\| \|\mathbf{v}\| \cos \theta$ into the Lagrange identity gives

$$\|\mathbf{u} \times \mathbf{v}\|^2 = \|\mathbf{u}\|^2 \|\mathbf{v}\|^2 - \|\mathbf{u}\|^2 \|\mathbf{v}\|^2 \cos^2\theta = \|\mathbf{u}\|^2 \|\mathbf{v}\|^2 \sin^2 \theta$$

using the fact that $1 - \cos^2\theta = \sin^2\theta$. But $\sin \theta$ is nonnegative on the range $0 \le \theta \le \pi$, so taking the positive square root of both sides gives

$$\|\mathbf{u} \times \mathbf{v}\| = \|\mathbf{u}\| \|\mathbf{v}\| \sin \theta$$

This expression for $\|\mathbf{u} \times \mathbf{v}\|$ makes no reference to a coordinate system and, moreover, it has a nice geometrical interpretation. The parallelogram determined by the vectors **u** and **v** has base length $\|\mathbf{v}\|$ and altitude $\|\mathbf{u}\| \sin \theta$ (see Figure 4.18). Hence the area of the parallelogram formed by **u** and **v** is

$$(\|\mathbf{u}\| \sin \theta)\|\mathbf{v}\| = \|\mathbf{u} \times \mathbf{v}\|$$

This is also valid if **u** and **v** are parallel because then **u** $\times$ **v** = **0** by Theorem 2 and $\sin \theta = 0$ (as $\theta = 0$ or π). This proves the first part of Theorem 4.

---

[4]Joseph Louis Lagrange (1736–1813) was born in Italy and spent his early years in Turin. At the age of 19 he solved a famous problem by inventing an entirely new method, known today as the calculus of variations, and went on to become one of the greatest mathematicians of all time. His work brought a new level of rigor to analysis and his *Mécanique Analytique* is a masterpiece in which he introduced methods still in use. In 1766 he was appointed to the Berlin Academy by Frederick the Great who asserted that the "greatest mathematician in Europe" should be at the court of the "greatest king in Europe." After the death of Frederick, Lagrange went to Paris at the invitation of Louis XVI. He remained there throughout the revolution and was made a count by Napoleon, who called him the "lofty pyramid of the mathematical sciences."

> **THEOREM 4**
>
> If **u** and **v** are two vectors and θ is the angle between **u** and **v**, then
>
> 1. $\|\mathbf{u} \times \mathbf{v}\| = \|\mathbf{u}\|\, \|\mathbf{v}\|\, \sin\,\theta$ = area of the parallelogram determined by **u** and **v**
>
> 2. **u** and **v** are parallel if and only if $\mathbf{u} \times \mathbf{v} = \mathbf{0}$

Proof of (2)   By (1), $\mathbf{u} \times \mathbf{v} = \mathbf{0}$ if and only if the area of the parallelogram is zero. But the area vanishes if and only if **u** and **v** have the same or opposite direction — that is, if they are parallel.   ◆

EXAMPLE 7

Find the area of the triangle with vertices $P(2, 1, 0)$, $Q(3, -1, 1)$, and $R(1, 0, 1)$.

Solution   First compute $\overrightarrow{RP} = (1, 1, -1)$ and $\overrightarrow{RQ} = (2, -1, 0)$. The area of the triangle is half the area of the parallelogram determined by these vectors (see the diagram), so it equals $\frac{1}{2}\|\overrightarrow{RP} \times \overrightarrow{RQ}\|$. Now

$$\overrightarrow{RP} \times \overrightarrow{RQ} = \det\begin{bmatrix} \mathbf{i} & \mathbf{j} & \mathbf{k} \\ 1 & 1 & -1 \\ 2 & -1 & 0 \end{bmatrix} = (-1, -2, -3)$$

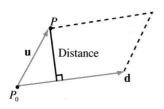

so the area of the triangle is $\frac{1}{2}\|(-1, -2, -3)\| = \frac{1}{2}\sqrt{1 + 4 + 9} = \frac{1}{2}\sqrt{14}$.   ◆◆◆

Theorem 4 can also be used to find the shortest distance from a point to a line. The next example illustrates this by resolving part of Example 9§4.2.

EXAMPLE 8

Find the shortest distance (see diagram) from the point $P(1, 3, -2)$ to the line through $P_0(2, 0, -1)$ with direction vector $\mathbf{d} = (1, -1, 0)$.

Solution   Compute $\mathbf{u} = \overrightarrow{P_0P} = (-1, 3, -1)$. Then

$$\mathbf{u} \times \mathbf{d} = \det\begin{bmatrix} \mathbf{i} & \mathbf{j} & \mathbf{k} \\ -1 & 3 & -1 \\ 1 & -1 & 0 \end{bmatrix} = (-1, -1, -2)$$

so the parallelogram determined by **u** and **d** has area $\|\mathbf{u} \times \mathbf{d}\| = \sqrt{1 + 1 + 4} = \sqrt{6}$. On the other hand, the area of this parallelogram equals $\|\mathbf{d}\|$ times the distance in question. Hence the distance from P to the line is

$$\frac{\|\mathbf{u} \times \mathbf{d}\|}{\|\mathbf{d}\|} = \frac{\sqrt{6}}{\sqrt{2}} = \sqrt{3}$$

Of course this agrees with Example 9§4.2, but the technique here is entirely different.

$\diamond\diamond\diamond$

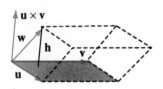

FIGURE 4.19

If three vectors **u**, **v**, and **w** are given, they determine a "squashed" rectangular solid called a **parallelepiped** (Figure 4.19), and it is often useful to be able to find the volume of such a solid. The base of the solid is the parallelogram determined by **u** and **v**, so it has area $A = \|\mathbf{u} \times \mathbf{v}\|$ by Theorem 4. The height of the solid is the length h of the projection of **w** on $\mathbf{u} \times \mathbf{v}$. Hence

$$h = \left|\frac{\mathbf{w} \cdot (\mathbf{u} \times \mathbf{v})}{\|\mathbf{u} \times \mathbf{v}\|^2}\right| \|\mathbf{u} \times \mathbf{v}\| = \frac{|\mathbf{w} \cdot (\mathbf{u} \times \mathbf{v})|}{\|\mathbf{u} \times \mathbf{v}\|} = \frac{|\mathbf{w} \cdot (\mathbf{u} \times \mathbf{v})|}{A}$$

Thus the volume of the parallelepiped is $hA = |\mathbf{w} \cdot (\mathbf{u} \times \mathbf{v})|$.

THEOREM 5

The volume of the parallelepiped determined by **w**, **u**, and **v** (Figure 4.19) is given by $|\mathbf{w} \cdot (\mathbf{u} \times \mathbf{v})|$.

EXAMPLE 9

Find the volume of the parallelepiped determined by the vectors $\mathbf{w} = (1, 2, -1)$, $\mathbf{u} = (1, 1, 0)$, and $\mathbf{v} = (-2, 0, 1)$.

Solution   We use Theorem 1.

$$\mathbf{w} \cdot (\mathbf{u} \times \mathbf{v}) = \det\begin{bmatrix} 1 & 2 & -1 \\ 1 & 1 & 0 \\ -2 & 0 & 1 \end{bmatrix} = -3$$

Hence the volume is $|\mathbf{w} \cdot (\mathbf{u} \times \mathbf{v})| = |-3| = 3$ by Theorem 5.

$\diamond\diamond\diamond$

We can now give a coordinate-free description of the cross product $\mathbf{u} \times \mathbf{v}$. Its magnitude is given by $\|\mathbf{u} \times \mathbf{v}\| = \|\mathbf{u}\| \|\mathbf{v}\| \sin \theta$, and if $\mathbf{u} \times \mathbf{v} \neq \mathbf{0}$, its direction is very nearly determined by the fact that it is orthogonal to both **u** and **v** and so points along the line normal to the plane determined by **u** and **v**. It remains only to decide which of the two possible directions is correct.

Before this can be done, the basic issue of how coordinates are assigned must be clarified. When coordinate axes are chosen in space, the procedure is as follows: An origin is selected and two perpendicular lines (the X and Y axes) are chosen through the origin, the positive direction on each of these axes being chosen quite arbitrarily. Then the line through the origin normal to this X-Y plane is called the Z axis, but there is a choice of which direction on this axis is the positive one. The two possibilities are shown in Figure 4.20, and it is a standard convention that Cartesian coordinates are always **right-hand coordinate systems.** The reason for this terminology is that, in such a system, if the Z axis is grasped in the right hand with the thumb pointing in the positive-Z direction, then the fingers curl around from the positive X axis to the positive Y axis (through a right angle).

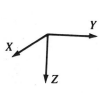 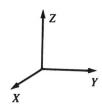

Left-hand system   Right-hand system

FIGURE 4.20

Suppose now that $\mathbf{u}$ and $\mathbf{v}$ are given and that θ is the angle between them (so $0 \leq \theta \leq \pi$). Then the direction of $\|\mathbf{u} \times \mathbf{v}\|$ is given by the right-hand rule.

RIGHT-HAND RULE

If the vector $\mathbf{u} \times \mathbf{v}$ is grasped in the right hand and the fingers curl around from $\mathbf{u}$ to $\mathbf{v}$ through the angle θ, the thumb points in the direction of $\mathbf{u} \times \mathbf{v}$.

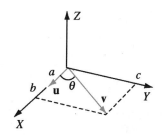

FIGURE 4.21

To indicate why this is true, introduce coordinates in such a way that the initial points of $\mathbf{u}$ and $\mathbf{v}$ are at the origin, $\mathbf{u}$ points along the positive X axis, $\mathbf{v}$ lies in the X-Y plane, and $\mathbf{v}$ and the positive Y axis are on the same side of the X axis. Then, in this system, $\mathbf{u}$ and $\mathbf{v}$ have component form $\mathbf{u} = (a, 0, 0)$ and $\mathbf{v} = (b, c, 0)$, where $a > 0$ and $c > 0$. The situation is depicted in Figure 4.21. The right-hand rule asserts that $\mathbf{u} \times \mathbf{v}$ should point in the positive Z direction. But our definition of $\mathbf{u} \times \mathbf{v}$ gives

$$\mathbf{u} \times \mathbf{v} = \det \begin{bmatrix} \mathbf{i} & \mathbf{j} & \mathbf{k} \\ a & 0 & 0 \\ b & c & 0 \end{bmatrix} = (0,\ 0,\ ac) = (ac)\mathbf{k}$$

and $(ac)\,\mathbf{k}$ has the positive Z direction because $ac > 0$.

Exercises 4.3

1. Compute $\mathbf{u} \times \mathbf{v}$ where:
   (a) $\mathbf{u} = (1, 2, 3), \mathbf{v} = (1, 0, -1)$
   ◆**(b)** $\mathbf{u} = (3, -1, 0), \mathbf{v} = (-6, 2, 0)$

   (c) $\mathbf{u} = (3, -2, 1), \mathbf{v} = (1, 1, 1)$
   ◆**(d)** $\mathbf{u} = (2, 0, -1), \mathbf{v} = (1, 4, 7)$

2. If **i**, **j**, and **k** are the coordinate vectors, verify that $\mathbf{i} \times \mathbf{j} = \mathbf{k}$, $\mathbf{j} \times \mathbf{k} = \mathbf{i}$, and $\mathbf{k} \times \mathbf{i} = \mathbf{j}$.

3. Show that $\mathbf{u} \times (\mathbf{v} \times \mathbf{w})$ need not equal $(\mathbf{u} \times \mathbf{v}) \times \mathbf{w}$ by calculating both when $\mathbf{u} = (1, 1, 1)$, $\mathbf{v} = (1, 1, 0)$, and $\mathbf{w} = (0, 0, 1)$.

4. Find two unit vectors orthogonal to both **u** and **v** if:
  (a) $\mathbf{u} = (1, 2, 2)$, $\mathbf{v} = (2, -1, 2)$
  ◆**(b)** $\mathbf{u} = (1, 2, -1)$, $\mathbf{v} = (3, 1, 2)$

5. Find the equation of each of the following planes.
  (a) Passing through $A(2, 1, 3)$, $B(3, -1, 5)$, and $C(0, 2, -4)$
  ◆**(b)** Passing through $A(1, -1, 6)$, $B(0, 0, 1)$, and $C(4, 7, -11)$
  (c) Passing through $P(2, -1, 4)$ and parallel to the plane with equation $3x - 2y - z = 0$
  ◆**(d)** Passing through $P(3, 0, -1)$ and parallel to the plane with equation $2x - y + z = 3$
  (e) Containing $P(3, 0, -1)$ and the line $(x, y, z) = (0, 0, 2) + t(1, 0, 1)$
  ◆**(f)** Containing $P(2, 1, 0)$ and the line $(x, y, z) = (3, -1, 2) + t(1, 0, -1)$
  (g) Containing the lines $(x, y, z) = (1, -1, 2) + t(1, 0, 1)$ and $(x, y, z) = (0, 0, 2) + t(1, -1, 0)$
  ◆**(h)** Containing the lines $(x, y, z) = (3, 1, 0) + t(1, -1, 3)$ and $(x, y, z) = (0, -2, 5) + t(2, 1, -1)$
  (i) Each point of which is equidistant from $P(2, -1, 3)$ and $Q(1, 1, -1)$
  ◆**(j)** Each point of which is equidistant from $P(0, 1, -1)$ and $Q(2, -1, -3)$

6. In each case, find the equation of the line.
  (a) Passing through $P(3, -1, 4)$ and perpendicular to the plane $3x - 2y - z = 0$
  ◆**(b)** Passing through $P(2, -1, 3)$ and perpendicular to the plane $2x + y = 1$
  (c) Passing through $P(0, 0, 0)$ and perpendicular to the lines $(x, y, z) = (1, 1, 0) + t(2, 0, -1)$ and $(x, y, z) = (2, 1, -3) + t(1, -1, 7)$
  ◆**(d)** Passing through $P(1, 1, -1)$ and perpendicular to the lines $(x, y, z) = (2, 0, 1) + t(1, 1, -2)$ and $(x, y, z) = (5, 5, -2) + t(1, 2, -3)$
  (e) Passing through $P(2, 1, -1)$, intersecting the line $(x, y, z) = (1, 2, -1) + t(3, 0, 1)$, and perpendicular to that line
  ◆**(f)** Passing through $P(1, 1, 2)$, intersecting the line $(x, y, z) = (2, 1, 0) + t(1, 1, 1)$, and perpendicular to that line

7. In each case, find the shortest distance from the point P to the plane and find the point Q on the plane closest to P.
  (a) $P(2, 3, 0)$; plane with equation $5x - y + z = 1$
  ◆**(b)** $P(3, 1, -1)$; plane with equation $2x + y - z = 6$

8. (a) Does the line through $P(1, 2, -3)$ with direction vector $\mathbf{d} = (1, 2, -3)$ lie in the plane $2x - y - z = 3$? Explain.
  ◆**(b)** Does the plane through $P(4, 0, 5)$, $Q(2, 2, 1)$, and $R(1, -1, 2)$ pass through the origin? Explain.

9. Show that every plane containing $P(1, 2, -1)$ and $Q(2, 0, 1)$ must also contain $R(-1, 6, -5)$.

10. Find the equations of the line of intersection of the following planes.
  (a) $2x - 3y + 2z = 5$ and $x + 2y - z = 4$
  ◆**(b)** $3x + y - 2z = 1$ and $x + y + z = 5$

11. In each case, find all points of intersection of the given plane and the line $(x, y, z) = (1, -2, 3) + t(2, 5, -1)$.
  (a) $x - 3y + 2z = 4$
  ◆**(b)** $2x - y - z = 5$
  (c) $3x - y + z = 8$
  ◆**(d)** $-x + 4y + 3z = 6$

12. Find the equations of *all* planes:
  (a) Perpendicular to the line $(x, y, z) = (2, -1, 3) + t(2, 1, 3)$
  ◆**(b)** Perpendicular to the line $(x, y, z) = (1, 0, -1) + t(3, 0, 2)$
  (c) Containing the origin
  ◆**(d)** Containing $P(3, 2, -4)$
  (e) Containing $P(1, 1, -1)$ and $Q(0, 1, 1)$
  ◆**(f)** Containing $P(2, -1, 1)$ and $Q(1, 0, 0)$
  (g) Containing the line $(x, y, z) = (2, 1, 0) + t(1, -1, 0)$
  ◆**(h)** Containing the line $(x, y, z) = (3, 0, 2) + t(1, 2, -1)$

13. If a plane contains two distinct points P_1 and P_2, show that it contains every point on the line through P_1 and P_2.

14. Find the shortest distance between the following pairs of parallel lines.
  (a) $(x, y, z) = (2, -1, 3) + t(1, -1, 4)$
    $(x, y, z) = (1, 0, 1) + t(1, -1, 4)$
  ◆**(b)** $(x, y, z) = (3, 0, 2) + t(3, 1, 0)$
    $(x, y, z) = (-1, 2, 2) + t(3, 1, 0)$

15. Find the shortest distance between the following pairs of nonparallel lines and find the points on the lines that are closest together.
  (a) $(x, y, z) = (3, 0, 1) + s(2, 1, -3)$
    $(x, y, z) = (1, 1, -1) + t(1, 0, 1)$
  ◆**(b)** $(x, y, z) = (1, -1, 0) + s(1, 1, 1)$
    $(x, y, z) = (2, -1, 3) + t(3, 1, 0)$
  (c) $(x, y, z) = (3, 1, -1) + s(1, 1, -1)$
    $(x, y, z) = (1, 2, 0) + t(1, 0, 2)$
  ◆**(d)** $(x, y, z) = (1, 2, 3) + s(2, 0, -1)$
    $(x, y, z) = (3, -1, 0) + t(1, 1, 0)$

16. Find the area of the triangle with the following vertices.
  (a) $A(3, -1, 2)$, $B(1, 1, 0)$, and $C(1, 2, -1)$
  ♦(b) $A(3, 0, 1)$, $B(5, 1, 0)$, and $C(7, 2, -1)$
  (c) $A(1, 1, -1)$, $B(2, 0, 1)$, and $C(1, -1, 3)$
  ♦(d) $A(3, -1, 1)$, $B(4, 1, 0)$, and $C(2, -3, 0)$

17. Find the volume of the parallelepiped determined by $\mathbf{w}$, $\mathbf{u}$, and $\mathbf{v}$ when:
  (a) $\mathbf{w} = (2, 1, 1)$, $\mathbf{v} = (1, 0, 2)$, and $\mathbf{u} = (2, 1, -1)$
  ♦(b) $\mathbf{w} = (1, 0, 3)$, $\mathbf{v} = (2, 1, -3)$, and $\mathbf{u} = (1, 1, 1)$

18. Let P_0 be a point with position vector $\mathbf{p}_0$, and let $ax + by + cz = d$ be the equation of a plane with normal $\mathbf{n} = (a, b, c)$.
  (a) Show that the point on the plane closest to P_0 has

  position vector $\mathbf{p} = \mathbf{p}_0 + \dfrac{d - (\mathbf{p}_0 \cdot \mathbf{n})}{\|\mathbf{n}\|^2} \, \mathbf{n}$. [*Hint:*

  $\mathbf{p} = \mathbf{p}_0 + t\mathbf{n}$ for some t, and $\mathbf{p} \cdot \mathbf{n} = d$.]
  ♦(b) Show that the shortest distance from P_0 to the plane

  is $\dfrac{|d - (\mathbf{p}_0 \cdot \mathbf{n})|}{\|\mathbf{n}\|}$.

  (c) Let P_0' denote the reflection of P_0 in the plane — that is, the point on the opposite side of the plane such that the line through P_0 and P_0' is perpendicular to

  the plane. Show that $\mathbf{p}_0 + 2 \dfrac{d - (\mathbf{p}_0 \cdot \mathbf{n})}{\|\mathbf{n}\|^2} \mathbf{n}$ is the

  position vector of P_0'.

19. Simplify $(a\mathbf{u} + b\mathbf{v}) \times (c\mathbf{u} + d\mathbf{v})$.

20. Show that the shortest distance from a point P to the line

  through P_0 with direction vector $\mathbf{d}$ is $\dfrac{\|\overrightarrow{P_0P} \times \mathbf{d}\|}{\|\mathbf{d}\|}$.

21. Let $\mathbf{u}$ and $\mathbf{v}$ be nonzero, nonorthogonal vectors. If θ is the

  angle between them, show that $\tan \theta = \dfrac{\|\mathbf{u} \times \mathbf{v}\|}{\mathbf{u} \cdot \mathbf{v}}$.

22. Show that points A, B, and C are all on one line if and only if $\overrightarrow{AB} \times \overrightarrow{AC} = \mathbf{0}$.

23. Show that points A, B, C, and D are all on one plane if and only if $\overrightarrow{AB} \cdot (\overrightarrow{AC} \times \overrightarrow{AD}) = 0$.

♦24. Use Theorem 5 to confirm that, if $\mathbf{u}$, $\mathbf{v}$, and $\mathbf{w}$ are mutually perpendicular, the (rectangular) parallelepiped they determine has volume $\|\mathbf{u}\| \, \|\mathbf{v}\| \, \|\mathbf{w}\|$.

25. Show that the volume of the parallelepiped determined by $\mathbf{u}$, $\mathbf{v}$, and $\mathbf{u} \times \mathbf{v}$ is $\|\mathbf{u} \times \mathbf{v}\|^2$.

26. Complete the proof of Theorem 3.

27. Prove the following properties in Theorem 2.
  (a) Property 6   **♦(b)** Property 7   **(c)** Property 8

28. **(a)** Show that $\mathbf{w} \cdot (\mathbf{u} \times \mathbf{v}) = \mathbf{u} \cdot (\mathbf{v} \times \mathbf{w}) = \mathbf{v} \cdot (\mathbf{w} \times \mathbf{u})$ holds for all vectors $\mathbf{w}$, $\mathbf{u}$, and $\mathbf{v}$. [*Hint:* Theorem 1.]

♦(b) Show that $\mathbf{v} - \mathbf{w}$ and $(\mathbf{u} \times \mathbf{v}) + (\mathbf{v} \times \mathbf{w}) + (\mathbf{w} \times \mathbf{u})$ are orthogonal.

29. Show that $\mathbf{u} \times (\mathbf{v} \times \mathbf{w}) = (\mathbf{u} \cdot \mathbf{w})\mathbf{v} - (\mathbf{u} \cdot \mathbf{v})\mathbf{w}$. [*Hint:* First do it for $\mathbf{u} = \mathbf{i}$, $\mathbf{j}$, and $\mathbf{k}$; then write $\mathbf{u} = x\mathbf{i} + y\mathbf{j} + z\mathbf{k}$ and use Theorem 2.]

30. Prove the **Jacobi identity:** $\mathbf{u} \times (\mathbf{v} \times \mathbf{w}) + \mathbf{v} \times (\mathbf{w} \times \mathbf{u}) + \mathbf{w} \times (\mathbf{u} \times \mathbf{v}) = \mathbf{0}$. [*Hint:* The previous exercise.]

31. Show that $(\mathbf{u} \times \mathbf{v}) \cdot (\mathbf{w} \times \mathbf{z}) = \det \begin{bmatrix} \mathbf{u} \cdot \mathbf{w} & \mathbf{u} \cdot \mathbf{z} \\ \mathbf{v} \cdot \mathbf{w} & \mathbf{v} \cdot \mathbf{z} \end{bmatrix}$
  [*Hint:* Exercises 28 and 29.]

32. Let P, Q, R, and S be four points, not all on one plane, as in the diagram. Show that the volume of the pyramid they determine is $\frac{1}{6}|\overrightarrow{PQ} \cdot (\overrightarrow{PR} \times \overrightarrow{PS})|$. [*Hint:* The volume of a cone with base area A and height h is $\frac{1}{3}Ah$.]

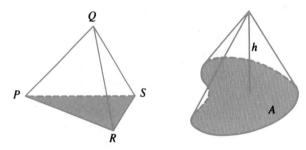

33. Consider a triangle with vertices A, B, and C, as in the diagram. Let α, β, and γ denote the angles at A, B, and C, respectively, and let a, b, and c denote the lengths of the sides opposite A, B, and C, respectively. Write $\mathbf{u} = \overrightarrow{AB}$, $\mathbf{v} = \overrightarrow{BC}$, and $\mathbf{w} = \overrightarrow{CA}$.
  (a) Deduce $\mathbf{u} + \mathbf{v} + \mathbf{w} = \mathbf{0}$.
  (b) Show that $\mathbf{u} \times \mathbf{v} = \mathbf{w} \times \mathbf{u} = \mathbf{v} \times \mathbf{w}$. [*Hint:* Compute $\mathbf{u} \times (\mathbf{u} + \mathbf{v} + \mathbf{w})$ and $\mathbf{v} \times (\mathbf{u} + \mathbf{v} + \mathbf{w})$.]
  (c) Deduce the **law of sines:**

  $$\frac{\sin \alpha}{a} = \frac{\sin \beta}{b} = \frac{\sin \gamma}{c}$$

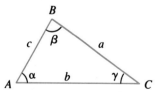

♦34. Let A be a 3×3 matrix. Given vectors $\mathbf{u}$, $\mathbf{v}$, and $\mathbf{w}$, show that the volume of the parallelepiped determined by $\mathbf{u}A$, $\mathbf{v}A$, and $\mathbf{w}A$ equals $|\det A|$ times the volume of the parallelepiped determined by $\mathbf{u}$, $\mathbf{v}$, and $\mathbf{w}$.

35. Show that the (shortest) distance between two planes $\mathbf{n} \cdot \mathbf{p} = d_1$ and $\mathbf{n} \cdot \mathbf{p} = d_2$ with $\mathbf{n}$ as normal is $\dfrac{|d_2 - d_1|}{\|\mathbf{n}\|}$.

36. Let A and B be points other than the origin, and let $\mathbf{a}$ and $\mathbf{b}$ be their position vectors. If $\mathbf{a}$ and $\mathbf{b}$ are not parallel, show that the plane through A, B, and the origin is given by $\{P(x, y, z) \mid (x, y, z) = s\mathbf{a} + t\mathbf{b}$ for some s and $t\}$.

37. Let A be a 2×3 matrix of rank 2 with rows $\mathbf{r}_1$ and $\mathbf{r}_2$. Show that $\mathscr{P} = \{XA \mid X = [x \quad y];\ x,\ y$ arbitrary$\}$ is the plane through the origin with normal $\mathbf{r}_1 \times \mathbf{r}_2$.

38. Given the cube with vertices $P(x, y, z)$, where each of x, y, and z is either 0 or 2, consider the plane perpendicular to the diagonal through $P(0, 0, 0)$ and $P(2, 2, 2)$ and bisecting it.

(a) Show that the plane meets six of the edges of the cube and bisects them.

(b) Show that the six points in **(a)** are the vertices of a regular hexagon.

⬛⬛⬛⬛ Section 4.4   An Application to Least Squares Approximation (Optional)

In many scientific investigations, data are collected that relate two variables. For example, if x is the number of dollars spent on advertising by a manufacturer and y is the value of sales in the region in question, the manufacturer could generate data by spending $x_1, x_2, \ldots, x_n$ dollars at different times and measuring the corresponding sales values $y_1, y_2, \ldots, y_n$.

Suppose it is known that a linear relationship exists between the variables x and y — in other words, that $y = a + bx$ for some constants a and b. If the data are plotted, the points $(x_1, y_1), (x_2, y_2), \ldots, (x_n, y_n)$ may appear to lie on a straight line and estimating a and b requires finding the "best-fitting" line through these data points. For example, if five data points occur as shown in Figure 4.22, line 1 is clearly a better fit than line 2. In general, the problem is to find the values of the constants a and b such that the line $y = a + bx$ best approximates the data in question. Note that an *exact* fit would be obtained if a and b were such that $y_i = a + bx_i$ were true for each data point (x_i, y_i). But this is too much to expect. Experimental errors in measurement are bound to occur, so the choice of a and b should be made in such a way that the errors between the observed values y_i and the corresponding fitted values $a + bx_i$ are in some sense minimized.

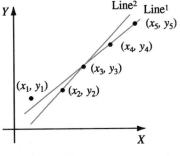

FIGURE 4.22

The first thing we must do is explain exactly what we mean by the *best fit* of a line $y = a + bx$ to an observed set of data points $(x_1, y_1), (x_2, y_2), \ldots, (x_n, y_n)$. For convenience, write the linear function $a + bx$ as

FIGURE 4.23

$$f(x) = a + bx$$

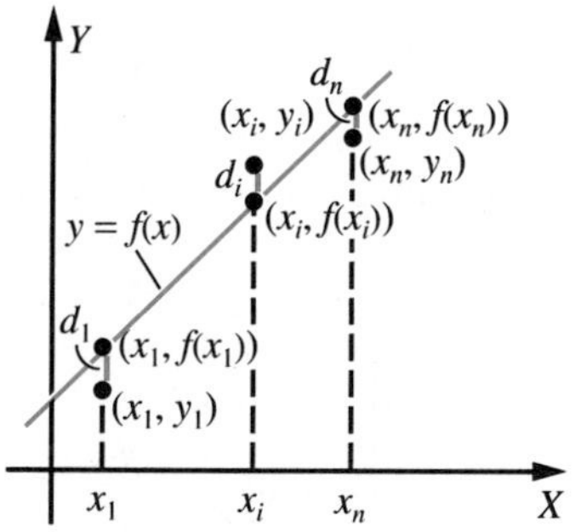

so that the fitted points (on the line) have coordinates $(x_1, f(x_1)), \ldots, (x_n, f(x_n))$. Figure 4.23 is a sketch of what the line $y = f(x)$ might look like. For each i the observed data point (x_i, y_i) and the fitted point $(x_i, f(x_i))$ need not be the same, and the distance d_i between them measures how far the line misses the observed point. For this reason d_i is often called the *error* at x_i, and a natural measure of how close the line $y = f(x)$ is to the observed data points is the sum $d_1 + d_2 + \cdots + d_n$ of all these errors. However, it turns out to be better to use the sum of squares

$$S = d_1^2 + d_2^2 + \cdots + d_n^2$$

as the measure of error, and the line $y = f(x)$ is to be chosen so as to make this sum as small as possible. This line is said to be the **least squares approximating line** for the data points $(x_1, y_1), (x_2, y_2), \ldots, (x_n, y_n)$.

The square of the distance d_i is given by $d_i^2 = [y_i - f(x_i)]^2$ for each i, so the quantity S to be minimized is the sum:

$$S = [y_1 - f(x_1)]^2 + [y_2 - f(x_2)]^2 + \cdots + [y_n - f(x_n)]^2$$

Note that all the numbers x_i and y_i are *given* here; what is required is that the *function* f be chosen in such a way as to minimize this expression. Because $f(x) = a + bx$, this amounts to choosing a and b so as to minimize S, and the problem can be solved using vector techniques. The following notations simplify the discussion.

$$Y = \begin{bmatrix} y_1 \\ y_2 \\ \vdots \\ y_n \end{bmatrix} \qquad M = \begin{bmatrix} 1 & x_1 \\ 1 & x_2 \\ \vdots & \vdots \\ 1 & x_n \end{bmatrix} \qquad Z = \begin{bmatrix} a \\ b \end{bmatrix}$$

Observe that

$$Y - MZ = \begin{bmatrix} y_1 - (a + bx_1) \\ y_2 - (a + bx_2) \\ \vdots \\ y_n - (a + bx_n) \end{bmatrix} = \begin{bmatrix} y_1 - f(x_1) \\ y_2 - f(x_2) \\ \vdots \\ y_n - f(x_n) \end{bmatrix}$$

so the quantity S that is to be minimized is just the sum of the squares of the entries of this column matrix.

Now suppose for a moment that $n = 3$. Then $Y - MZ$ is an ordered triple and so can be regarded as a vector (written as a column rather than as a row). Moreover, S is the square of the length of this vector.

$$S = \|Y - MZ\|^2$$

Here Y and M are given, and we are asked to choose Z such that the length of the vector $Y - MZ$ is as small as possible. To this end, consider the set P of all vectors MZ where $Z = \begin{bmatrix} a \\ b \end{bmatrix}$ varies. Then P takes the form

$$P = \left\{ MZ \;\middle|\; Z = \begin{bmatrix} a \\ b \end{bmatrix} \right\} = \left\{ \begin{bmatrix} a + bx_1 \\ a + bx_2 \\ a + bx_3 \end{bmatrix} \;\middle|\; a \text{ and } b \text{ are arbitrary} \right\}$$

If x_1, x_2, and x_3 are distinct, this is the plane through the origin with equation $(x_2 - x_3) x + (x_3 - x_1)y + (x_1 - x_2)z = 0$ (in fact, it contains

$$U = \begin{bmatrix} x_1 \\ x_2 \\ x_3 \end{bmatrix} \quad \text{and} \quad V = \begin{bmatrix} 1 \\ 1 \\ 1 \end{bmatrix}, \quad \text{so} \quad U \times V = \begin{bmatrix} x_2 - x_3 \\ x_3 - x_1 \\ x_1 - x_2 \end{bmatrix}$$

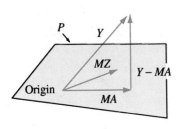

FIGURE 4.24

is a normal). Thus the task is to choose a point MA in P as close as possible to Y. It is clear geometrically (see Figure 4.24) that the vector $Y - MA$ is orthogonal to every vector MZ in the plane P. This means that

$$(MZ) \cdot (Y - MA) = 0$$

for all Z and this condition determines A.

To see this, observe that the dot product of column vectors U and V can be written as $U \cdot V = U^T V$, where U and V are regarded as 3×1 matrices. Hence, for each Z

$$0 = (MZ)^T(Y - MA) = Z^T M^T(Y - MA) = Z \cdot (M^T Y - M^T MA)$$

where the last dot product is in *two* dimensions. In other words, the vector $M^T Y - M^T MA$ is orthogonal to *every* two-dimensional vector Z and so must be zero (being orthogonal to itself!). This means that

$$(M^T M)A = M^T Y$$

These are called the **normal equations** for A and can be solved using Gaussian elimination. Moreover, $M^T M$ can be shown to be invertible when x_1, x_2, and x_3 are distinct (it is sufficient that at least two of x_1, x_2, and x_3 be distinct), so solving for A yields

$$A = (M^T M)^{-1} M^T Y$$

This solves our problem (at least when $n = 3$) because if $A = \begin{bmatrix} a_0 \\ a_1 \end{bmatrix}$, the best-fitting line is $y = a_0 + a_1 x$.

Of course this argument depends heavily on the fact that $n = 3$ so that we can avail ourselves of the theory of vectors in two and three dimensions and use such notions as the length of a vector, orthogonality, and the dot product. However, all these notions extend to vectors in higher dimensions than two or three, and the entire argument goes through almost unaltered in the general context. This argument is carried out in Chapter 6, and the result is the following useful theorem.

THEOREM 1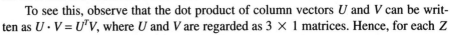

Suppose that n data points $(x_1, y_1), (x_2, y_2), \ldots, (x_n, y_n)$ are given, where at least two of $x_1, x_2, \ldots, x_n$ are distinct. Put

$$
Y = \begin{bmatrix} y_1 \\ y_2 \\ \vdots \\ y_n \end{bmatrix} \qquad M = \begin{bmatrix} 1 & x_1 \\ 1 & x_2 \\ \vdots & \vdots \\ 1 & x_n \end{bmatrix}
$$

Then the least squares approximating line for these data points has the equation

$$
y = a_0 + a_1 x
$$

where $A = \begin{bmatrix} a_0 \\ a_1 \end{bmatrix}$ is found by Gaussian elimination from the normal equations

$$
(M^T M)A = M^T Y
$$

The condition that at least two of $x_1, x_2, \ldots, x_n$ are distinct ensures that $M^T M$ is an invertible matrix, so A is unique:

$$
A = (M^T M)^{-1} M^T Y
$$

EXAMPLE 1

Let data points (x_1, y_1), (x_2, y_2), . . . , (x_5, y_5) be given as in the accompanying table. Find the least squares approximating line for these data.

Solution   In this case we have

x	y
1	1
3	2
4	3
6	4
7	5

$$
M^T M = \begin{bmatrix} 1 & 1 & \cdots & 1 \\ x_1 & x_2 & \cdots & x_5 \end{bmatrix} \begin{bmatrix} 1 & x_1 \\ 1 & x_2 \\ \vdots & \vdots \\ 1 & x_5 \end{bmatrix}
$$

$$
= \begin{bmatrix} 5 & x_1 + \cdots + x_5 \\ x_1 + \cdots + x_5 & x_1^2 + \cdots + x_5^2 \end{bmatrix} = \begin{bmatrix} 5 & 21 \\ 21 & 111 \end{bmatrix}
$$

$$
M^T Y = \begin{bmatrix} 1 & 1 & \cdots & 1 \\ x_1 & x_2 & \cdots & x_5 \end{bmatrix} \begin{bmatrix} y_1 \\ y_2 \\ \vdots \\ y_5 \end{bmatrix}
$$

$$
= \begin{bmatrix} y_1 + y_2 + \cdots + y_5 \\ x_1 y_1 + x_2 y_2 + \cdots + x_5 y_5 \end{bmatrix} = \begin{bmatrix} 15 \\ 78 \end{bmatrix}
$$

so the normal equations $(M^T M)A = M^T Y$ for $A = \begin{bmatrix} a_0 \\ a_1 \end{bmatrix}$ become

$$
\begin{bmatrix} 5 & 21 \\ 21 & 111 \end{bmatrix} \begin{bmatrix} a_0 \\ a_1 \end{bmatrix} = \begin{bmatrix} 15 \\ 78 \end{bmatrix}
$$

The solution (using Gaussian elimination) is $\begin{bmatrix} a_0 \\ a_1 \end{bmatrix} = \begin{bmatrix} 0.24 \\ 0.66 \end{bmatrix}$ to two decimal places, so
the least squares approximating line for these data is $y = 0.24 + 0.66x$. Note that $M^T M$ is indeed invertible here (the determinant is 114), and the exact solution is

$$A = (M^T M)^{-1} M^T Y = \frac{1}{114} \begin{bmatrix} 111 & -21 \\ -21 & 5 \end{bmatrix} \begin{bmatrix} 15 \\ 78 \end{bmatrix} = \frac{1}{114} \begin{bmatrix} 27 \\ 75 \end{bmatrix}$$

Suppose now that, rather than a straight line, we want to find the parabola $y = a_0 + a_1 x + a_2 x^2$ that is the least squares approximation to the data points $(x_1, y_1), \ldots, (x_n, y_n)$. In the function $f(x) = a_0 + a_1 x + a_2 x^2$, the *three* constants a_0, a_1, and a_2 must be chosen to minimize the sum of squares of the errors:

$$S = [y_1 - f(x_1)]^2 + [y_2 - f(x_2)]^2 + \cdots + [y_n - f(x_n)]^2$$

Choosing a_0, a_1, and a_2 amounts to choosing the (parabolic) function f that minimizes S.

In general, there is a relationship $y = f(x)$ between the variables, and the range of candidate functions is limited — say, to all lines or to all parabolas. The task is to find, among the suitable candidates, the function that makes the quantity S as small as possible. The function that does so is called the least squares approximating function (of that type) for the data points.

As might be imagined, this is not always an easy task. However, if the functions $f(x)$ are restricted to polynomials of degree m,

$$f(x) = a_0 + a_1 x + \cdots + a_m x^m$$

the analysis proceeds much as before (when $m = 1$). The problem is to choose the numbers $a_0, a_1, \ldots, a_m$ so as to minimize the sum

$$S = [y_1 - f(x_1)]^2 + [y_2 - f(x_2)]^2 + \cdots + [y_n - f(x_n)]^2$$

The resulting function $y = f(x) = a_0 + a_1 x + \cdots + a_m x^m$ is called the **least squares approximating polynomial of degree m** for the data $(x_1, y_1), \ldots, (x_n, y_n)$. By analogy with the preceding analysis, define

$$Y = \begin{bmatrix} y_1 \\ y_2 \\ \vdots \\ y_n \end{bmatrix} \qquad M = \begin{bmatrix} 1 & x_1 & x_1^2 & \cdots & x_1^m \\ 1 & x_2 & x_2^2 & \cdots & x_2^m \\ \vdots & \vdots & \vdots & & \vdots \\ 1 & x_n & x_n^2 & \cdots & x_n^m \end{bmatrix} \qquad A = \begin{bmatrix} a_0 \\ a_1 \\ \vdots \\ a_m \end{bmatrix}$$

Then

$$Y - MA = \begin{bmatrix} y_1 - (a_0 + a_1 x_1 + \cdots + a_m x_1^m) \\ y_2 - (a_0 + a_1 x_2 + \cdots + a_m x_2^m) \\ \vdots \\ y_n - (a_0 + a_1 x_n + \cdots + a_m x_n^m) \end{bmatrix} = \begin{bmatrix} y_1 - f(x_1) \\ y_2 - f(x_2) \\ \vdots \\ y_n - f(x_n) \end{bmatrix}$$

so S is the sum of the squares of the entries of $Y - MA$. An analysis similar to that for Theorem 1 can be used to prove Theorem 2.

THEOREM 2

Suppose n data points (x_1, y_1), (x_2, y_2), . . . , (x_n, y_n) are given, where at least $m + 1$ of $x_1, x_2, \ldots, x_n$ are distinct (in particular $n \geq m + 1$). Put

$$Y = \begin{bmatrix} y_1 \\ y_2 \\ \vdots \\ y_n \end{bmatrix} \qquad M = \begin{bmatrix} 1 & x_1 & x_1^2 & \cdots & x_1^m \\ 1 & x_2 & x_2^2 & \cdots & x_2^m \\ \vdots & \vdots & \vdots & & \vdots \\ 1 & x_n & x_n^2 & \cdots & x_n^m \end{bmatrix}$$

Then the least squares approximating polynomial of degree m for the data points has the equation

$$y = a_0 + a_1 x + \cdots + a_m x^m$$

where $A = \begin{bmatrix} a_0 \\ a_1 \\ \vdots \\ a_m \end{bmatrix}$ is found by Gaussian elimination from the normal equations

$$(M^T M)A = M^T Y$$

The condition that at least $m + 1$ of $x_1, x_2, \ldots, x_n$ be distinct ensures that the matrix MM^T is invertible, so A is unique:

$$A = (M^T M)^{-1} M^T Y$$

A proof of this theorem is given in Section 6.10.

EXAMPLE 2

Find the least squares approximating quadratic $y = a_0 + a_1 x + a_2 x^2$ for the following data points.

$$(-3,3), (-1,1), (0,1), (1,2), (3,4)$$

Solution   This is an instance of Theorem 2 with $m = 2$. Here

$$Y = \begin{bmatrix} 3 \\ 1 \\ 1 \\ 2 \\ 4 \end{bmatrix} \qquad M = \begin{bmatrix} 1 & -3 & 9 \\ 1 & -1 & 1 \\ 1 & 0 & 0 \\ 1 & 1 & 1 \\ 1 & 3 & 9 \end{bmatrix}$$

Hence,

$$M^T M = \begin{bmatrix} 1 & 1 & 1 & 1 & 1 \\ -3 & -1 & 0 & 1 & 3 \\ 9 & 1 & 0 & 1 & 9 \end{bmatrix} \begin{bmatrix} 1 & -3 & 9 \\ 1 & -1 & 1 \\ 1 & 0 & 0 \\ 1 & 1 & 1 \\ 1 & 3 & 9 \end{bmatrix} = \begin{bmatrix} 5 & 0 & 20 \\ 0 & 20 & 0 \\ 20 & 0 & 164 \end{bmatrix}$$

$$M^T Y = \begin{bmatrix} 1 & 1 & 1 & 1 & 1 \\ -3 & -1 & 0 & 1 & 3 \\ 9 & 1 & 0 & 1 & 9 \end{bmatrix} \begin{bmatrix} 3 \\ 1 \\ 1 \\ 2 \\ 4 \end{bmatrix} = \begin{bmatrix} 11 \\ 4 \\ 66 \end{bmatrix}$$

The normal equations for A are

$$\begin{bmatrix} 5 & 0 & 20 \\ 0 & 20 & 0 \\ 20 & 0 & 164 \end{bmatrix} A = \begin{bmatrix} 11 \\ 4 \\ 66 \end{bmatrix} \qquad \text{whence} \qquad A = \begin{bmatrix} 1.15 \\ 0.20 \\ 0.26 \end{bmatrix}$$

This means that the least squares approximating quadratic for these data is $y = 1.15 + 0.20x + 0.26x^2$. Again the matrix $M^T M$ is invertible, the inverse being

$$(M^T M)^{-1} = \frac{1}{420} \begin{bmatrix} 164 & 0 & -20 \\ 0 & 21 & 0 \\ -20 & 0 & 5 \end{bmatrix}, \text{ so } A \text{ can be calculated from } A = (M^T M)^{-1} M^T Y.$$

However, this takes much more computation than Gaussian elimination.

◆◆◆

Least squares approximation can be used to estimate physical constants, as is illustrated by the next example.

EXAMPLE 3

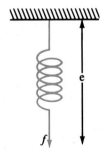

Hooke's law in mechanics asserts that the magnitude of the force f required to hold a spring is a linear function of the extension e of the spring (see the accompanying diagram). That is,

$$f = ke + e_0$$

where k and e_0 are constants depending only on the spring. The following data were collected for a particular spring.

e	9	11	12	16	19
f	33	38	43	54	61

Find the least squares approximating line $f = a_0 + a_1e$ to these data, and use it to estimate k.

Solution   Here f and e play the role of y and x in the general theory. We have

$$Y = \begin{bmatrix} 33 \\ 38 \\ 43 \\ 54 \\ 61 \end{bmatrix} \qquad M = \begin{bmatrix} 1 & 9 \\ 1 & 11 \\ 1 & 12 \\ 1 & 16 \\ 1 & 19 \end{bmatrix}$$

as in Theorem 1, so

$$M^TM = \begin{bmatrix} 5 & 67 \\ 67 & 963 \end{bmatrix} \qquad \text{and} \qquad M^TY = \begin{bmatrix} 229 \\ 3254 \end{bmatrix}$$

Hence the normal equations for A are

$$\begin{bmatrix} 5 & 67 \\ 67 & 963 \end{bmatrix} A = \begin{bmatrix} 229 \\ 3254 \end{bmatrix} \qquad \text{whence} \qquad A = \begin{bmatrix} 7.70 \\ 2.84 \end{bmatrix}$$

The least squares approximating line is $f = 7.70 + 2.84e$, so the estimate for k is $k = 2.84$.

◆◆◆

Exercises 4.4

1. Find the least squares approximating line $y = a_0 + a_1x$ for each of the following sets of data points.
  (a) $(1, 1), (3, 2), (4, 3), (6, 4)$
  ◆ **(b)** $(2, 4), (4, 3), (7, 2), (8, 1)$
  (c) $(-1, -1), (0, 1), (1, 2), (2, 4), (3, 6)$
  ◆ **(d)** $(-2, 3), (-1, 1), (0, 0), (1, -2), (2, -4)$

2. Find the least squares approximating quadratic $y = a_0 + a_1x + a_2x^2$ for each of the following sets of data points.
  (a) $(0, 1), (2, 2), (3, 3), (4, 5)$
  ◆ **(b)** $(-2, 1), (0, 0), (3, 2), (4, 3)$

3. If M is a square invertible matrix, show that $A = M^{-1}Y$ (in the notation of Theorem 2).

◆ **4.** Newton's laws of motion imply that an object dropped from rest at a height of 100 meters will be at a height $s = 100 - \frac{1}{2}gt^2$ meters t seconds later, where g is a constant called the acceleration of gravity. The values of s and t given in the table are observed. Write $x = t^2$, find the least squares approximating line $s = a + bx$ for these data and use b to estimate g. Then find the least squares approximating quadratic $s = a_0 + a_1t + a_2t^2$ and use the value of a_2 to estimate g.

t	1	2	3
s	95	80	56

5. A naturalist measured the heights y_i (in meters) of several spruce trees with trunk diameters x_i (in centimeters). The data are as given in the table. Find the least squares approximating line for these data and use it to estimate the height of a spruce tree with a trunk of diameter 10 cm.

x_i	5	7	8	12	13	16
y_i	2	3.3	4	7.3	7.9	10.1

6. (a) Use $m = 0$ in Theorem 2 to show that the best-fitting horizontal line $y = a_0$ through the data points (x_1, y_1), . . . , (x_n, y_n) is $y = \frac{1}{n}(y_1 + y_2 + \cdots + y_n)$, the average of the y coordinates.
  ◆ **(b)** Deduce the conclusion in **(a)** without using Theorem 2.

7. Assume $n = m + 1$ in Theorem 2 (so M is square). If the x_i are distinct, use Theorem 2§3.3 to show that M is invertible. Deduce that $A = M^{-1}Y$ and that the least squares polynomial is the interpolating polynomial (Section 3.3) and actually passes through the data points.

SUPPLEMENTARY EXERCISES FOR CHAPTER 4

1. Suppose that $\mathbf{u}$ and $\mathbf{v}$ are nonzero vectors. If $\mathbf{u}$ and $\mathbf{v}$ are not parallel, and $a\mathbf{u} + b\mathbf{v} = a_1\mathbf{u} + b_1\mathbf{v}$, show that $a = a_1$ and $b = b_1$.

2. Consider a triangle with vertices A, B, and C. Let E and F be the midpoints of sides AB and AC, respectively, and let the medians EC and FB meet at O. Write $\overrightarrow{EO} = s\overrightarrow{EC}$ and $\overrightarrow{FO} = t\overrightarrow{FB}$, where s and t are scalars. Show that $s = t = \frac{1}{3}$ by expressing $\overrightarrow{AO}$ two ways in the form $a\overrightarrow{AB} + b\overrightarrow{AC}$, and applying Exercise 1. Conclude that the medians of a triangle meet at the point on each that is one-third of the way from the midpoint to the vertex (and so are concurrent).

3. A river flows at 1 km per hour and a swimmer moves at 2 km per hour (relative to the water). At what angle must he swim to go straight across? What is his resulting speed?

4. A wind is blowing from the south at 75 knots, and an airplane flies heading east at 100 knots. Find the resulting velocity of the airplane.

5. An airplane pilot flies at 300 km/hr in a direction 30° south of east. The wind is blowing from the south at 150 km/hr.

(a) Find the resulting direction and speed of the airplane.
(b) Find the speed of the airplane if the wind is from the west (at 150 km/hr).

◆**6.** A rescue boat has a top speed of 13 knots. The captain wants to go due east as fast as possible in water with a current of 5 knots due south. Find the velocity vector $\mathbf{v} = (x, y)$ that she must achieve, assuming the X and Y axes point east and north, respectively, and find her resulting speed.

7. A boat goes 12 knots heading north. The current is 5 knots from the west. In what direction does the boat actually move and at what speed?

Vector Spaces

Section 5.1
Examples and Basic Properties

In Euclidean geometry, a point P in space is described by a triple (x, y, z) of real numbers called its Cartesian coordinates (the name honors René Descartes 1596–1650). This is called an *ordered* triple because the order of the coordinates is important. For example, $(1, 2, 3)$ represents a different point than $(1, 3, 2)$. Similarly, a point P in the plane is described by an ordered pair (x, y) of real numbers. Ordered sequences of n real numbers are important for values of n other than 2 or 3.

DEFINITION

An **ordered n–tuple** $\mathbf{v} = (v_1, v_2, \ldots, v_n)$ is an ordered sequence of n numbers $v_1, v_2, \ldots, v_n$ (called the **entries** of the n-tuple). Two such n-tuples are defined to be **equal** only when corresponding entries are equal:

$$(v_1, v_2, \ldots, v_n) = (w_1, w_2, \ldots, w_n) \text{ means } v_1 = w_1, v_2 = w_2, \ldots, v_n = w_n$$

Ordered 2- and 3-tuples are called **ordered pairs** and **ordered triples**, respectively. Our interest here is not so much in these n-tuples themselves as in the set of *all* n-tuples. This set is described as follows:

DEFINITION

Given an integer $n \geq 1$, the set of all n-tuples with real entries is called **Euclidean n-space** and is denoted $\mathbb{R}^n$. Here $\mathbb{R}$ denotes the set of real numbers.

The space $\mathbb{R}^3$ has a geometrical meaning when it is identified with the geometrical vectors discussed in Chapter 4. Thus a vector (a, b, c) in $\mathbb{R}^3$ is identified with the "arrow" from the origin to the point $P(a, b, c)$ with a, b, and c as coordinates. Then vector addition is given geometrically by the parallelogram law, and scalar multiplication takes on a geometric meaning as well. The geometry is useful in that it often provides a "picture" of some aspect of the space $\mathbb{R}^3$ that enhances our comprehension and even helps us understand the spaces $\mathbb{R}^n$ when $n > 3$. Of course $\mathbb{R}^2$ can be identified with the geometric plane. And $\mathbb{R}^1$ is $\mathbb{R}$ itself and so is identified with the points on a line.

These topics are discussed in Chapter 4, and the reader who is totally unfamiliar with them would do well to read Section 4.1. However, Chapter 4 is not required for understanding the present chapter because our treatment here is algebraic in nature.

DEFINITION

Let $\mathbf{v} = (v_1, v_2, \ldots, v_n)$, and $\mathbf{u} = (u_1, u_2, \ldots, u_n)$ be n-tuples in $\mathbb{R}^n$.

1. The **sum $\mathbf{u} + \mathbf{v}$** is defined by

$$\mathbf{u} + \mathbf{v} = (u_1 + v_1, u_2 + v_2, \ldots, u_n + v_n)$$

2. If a is any real number, the **scalar multiple** $a\mathbf{v}$ is defined by

$$a\mathbf{v} = (av_1, av_2, \ldots, av_n)$$

3. The **zero n-tuple 0** in $\mathbb{R}^n$ is defined by

$$\mathbf{0} = (0, 0, \ldots, 0)$$

4. The **negative** $-\mathbf{v}$ of the n-tuple $\mathbf{v}$ is defined by

$$-\mathbf{v} = (-v_1, -v_2, \ldots, -v_n)$$

5. The **difference $\mathbf{u} - \mathbf{v}$** is defined to be $\mathbf{u} + (-\mathbf{v})$. That is,

$$\mathbf{u} - \mathbf{v} = (u_1 - v_1, u_2 - v_2, \ldots, u_n - v_n)$$

Of course, these definitions are consistent with the corresponding matrix operations (regarding the n-tuples in $\mathbb{R}^n$ as $1 \times n$ matrices),[1] so all the computations with matrices are available in $\mathbb{R}^n$. These definitions also conform to the geometric operations in $\mathbb{R}^3$ introduced in Section 4.1.

EXAMPLE 1

Let $\mathbf{u} = (3, -1, 2, 4)$ and $\mathbf{v} = (5, 0, -6, 1)$ in $\mathbb{R}^4$. Then

$$\mathbf{u} + \mathbf{v} = (8, -1, -4, 5)$$
$$3\mathbf{u} - 5\mathbf{v} = (9, -3, 6, 12) - (25, 0, -30, 5) = (-16, -3, 36, 7)$$

◆◆◆

---

[1] The n-tuples in $\mathbb{R}^n$ will often be viewed as $1 \times n$ matrices, and will sometimes be written as columns.

The following fundamental properties of the n-tuples in $\mathbb{R}^n$ follow easily from the definitions (they also come from the corresponding properties of matrices).

PROPERTIES OF $\mathbb{R}^n$

Let $\mathbf{u}$, $\mathbf{v}$, and $\mathbf{w}$ denote n-tuples in $\mathbb{R}^n$, and let a and b be real numbers. Then

1. $\mathbf{u} + \mathbf{v} = \mathbf{v} + \mathbf{u}$
2. $\mathbf{u} + (\mathbf{v} + \mathbf{w}) = (\mathbf{u} + \mathbf{v}) + \mathbf{w}$
3. $\mathbf{v} + \mathbf{0} = \mathbf{v}$
4. $\mathbf{v} + (-\mathbf{v}) = \mathbf{0}$
5. $a(\mathbf{v} + \mathbf{w}) = a\mathbf{v} + a\mathbf{w}$
6. $(a + b)\mathbf{v} = a\mathbf{v} + b\mathbf{v}$
7. $a(b\mathbf{v}) = (ab)\mathbf{v}$
8. $1\mathbf{v} = \mathbf{v}$

The foregoing properties of $\mathbb{R}^n$ are not exclusive to $\mathbb{R}^n$. For example, the set of all $m \times n$ matrices also has an addition and a scalar multiplication that satisfy these conditions. It turns out that many other sets of mathematical objects have these properties, and the general study of such systems is the subject of this chapter. These systems are defined as follows.

DEFINITION

A **vector space** consists of a nonempty set V of objects (called **vectors**) that can be added, that can be multiplied by a real number (called a **scalar** in this context), and for which certain axioms hold. If $\mathbf{v}$ and $\mathbf{w}$ are two vectors in V, their sum is expressed as $\mathbf{v} + \mathbf{w}$, and the scalar product of $\mathbf{v}$ by a real number a is denoted as $a\mathbf{v}$. These operations are called **vector addition** and **scalar multiplication**, respectively, and the following axioms are assumed to hold.

Axioms for vector addition

A1. If $\mathbf{u}$ and $\mathbf{v}$ are in V, then $\mathbf{u} + \mathbf{v}$ is in V.

A2. $\mathbf{u} + \mathbf{v} = \mathbf{v} + \mathbf{u}$ for all $\mathbf{u}$ and $\mathbf{v}$ in V.

A3. $\mathbf{u} + (\mathbf{v} + \mathbf{w}) = (\mathbf{u} + \mathbf{v}) + \mathbf{w}$ for all $\mathbf{u}$, $\mathbf{v}$, and $\mathbf{w}$ in V.

A4. An element $\mathbf{0}$ in V exists such that $\mathbf{v} + \mathbf{0} = \mathbf{v} = \mathbf{0} + \mathbf{v}$ for every $\mathbf{v}$ in V.

A5. For each $\mathbf{v}$ in V, an element $-\mathbf{v}$ in V exists such that $-\mathbf{v} + \mathbf{v} = \mathbf{0}$ and $\mathbf{v} + (-\mathbf{v}) = \mathbf{0}$.

Axioms for scalar multiplication

S1. If $\mathbf{v}$ is in V, then $a\mathbf{v}$ is in V for all a in $\mathbb{R}$.

> **S2.** $a(\mathbf{v} + \mathbf{w}) = a\mathbf{v} + a\mathbf{w}$ for all $\mathbf{v}$ and $\mathbf{w}$ in V and all a in $\mathbb{R}$.
>
> **S3.** $(a + b)\mathbf{v} = a\mathbf{v} + b\mathbf{v}$ for all $\mathbf{v}$ in V and all a and b in $\mathbb{R}$.
>
> **S4.** $a(b\mathbf{v}) = (ab)\mathbf{v}$ for all $\mathbf{v}$ in V and all a and b in $\mathbb{R}$.
>
> **S5.** $1\mathbf{v} = \mathbf{v}$ for all $\mathbf{v}$ in V.

The content of axioms A1 and S1 is described by saying that V is **closed** under vector addition and scalar multiplication. The element $\mathbf{0}$ in axiom A4 is called the **zero vector**, and the vector $-\mathbf{v}$ in axiom A5 is called the **negative** of $\mathbf{v}$.

The properties of $\mathbb{R}^n$ that we have discussed give

EXAMPLE 2

$\mathbb{R}^n$ is a vector space using the preceding addition and scalar multiplication.

It is important to realize that, in a general vector space, the vectors need not be n-tuples. They can be any kind of object at all as long as the addition and scalar multiplication are defined and the axioms are satisfied. The following examples illustrate the diversity of the concept.

The space $\mathbb{R}^n$ consists of special types of matrices. More generally, let $\mathbf{M}_{mn}$ denote the set of all $m \times n$ matrices with real entries. Then Theorem 1§2.1 gives the following information.

EXAMPLE 3

The set $\mathbf{M}_{mn}$ of all $m \times n$ matrices is a vector space using matrix addition and scalar multiplication. The zero element in this vector space is the zero matrix of size $m \times n$, and the vector space negative of a matrix (required by axiom A5) is the same as that discussed in Section 2.1.

EXAMPLE 4

Show that[2]

$$V = \{(x, x, y) \,|\, x \text{ and } y \text{ in } \mathbb{R}\}$$

is a vector space using the operations of $\mathbb{R}^3$.

Solution   Axioms A2, A3, S2, S3, S4, and S5 all hold $\mathbb{R}^3$ and so are satisfied in V. Hence we check the remaining axioms. Given a in $\mathbb{R}$, and $\mathbf{u} = (x, x, y)$ and $\mathbf{v} = (x_1, x_1, y_1)$ in V, we have

---

[2]We use set-theoretic notation. If $p(x)$ is a condition on x, the notation $\{x \,|\, p(x)\}$ means "the set of all x such that the condition $p(x)$ is satisfied."

$$\mathbf{u} + \mathbf{v} = (x + x_1,\ x + x_1,\ y + y_1)$$
$$a\mathbf{v} = (ax_1,\ ax_1,\ ay_1)$$

so both lie in V (they have the correct form because the first components are equal). Hence axioms A1 and S1 are satisfied. To verify axiom A4, we must find a vector $\mathbf{0}$ in V such that $\mathbf{v} + \mathbf{0} = \mathbf{v} = \mathbf{0} + \mathbf{v}$ for every $\mathbf{v}$ in V. The vector $\mathbf{0} = (0, 0, 0)$ certainly has this property (it holds for all $\mathbf{v}$ in $\mathbb{R}^3$), so the fact that it lies in V means that it serves as the zero vector of V. Finally, if $\mathbf{v} = (x, x, y)$ is in V, take $-\mathbf{v} = (-x, -x, -y)$. Because this is also in V it serves as the negative of $\mathbf{v}$ in V. Hence axiom A5 is satisfied.   ◆◆◆

EXAMPLE 5

Let V denote the set of all ordered pairs $(x, y,)$ and define addition in V as in $\mathbb{R}^2$. However, define a new scalar multiplication in V by

$$a(x, y) = (ay, ax)$$

Determine if V is a vector space with these operations.

Solution   Axioms A1–A5 are valid for V because they hold in $\mathbb{R}^2$. Also $a(x, y) = (ay, ax)$ is again in V, so axiom S1 holds. To verify axiom S2, let $\mathbf{v} = (x, y)$ and $\mathbf{w} = (x_1, y_1)$ be typical elements in V and compute

$$a(\mathbf{v} + \mathbf{w}) = a(x + x_1,\ y + y_1) = (a(y + y_1),\ a(x + x_1))$$
$$a\mathbf{v} + a\mathbf{w} = (ay,\ ax) + (ay_1,\ ax_1) = (ay + ay_1,\ ax + ax_1)$$

Because these are equal, axiom S2 holds. Similarly, the reader can verify that axiom S3 holds. However, axiom S4 fails because

$$a(b(x, y)) = a(by, bx) = (abx, aby)$$

need not equal $ab(x, y) = (aby, abx)$. Hence V is not a vector space. (In fact axiom S5 also fails.)

◆◆◆

Sets of polynomials provide another important source of examples of vector spaces, so we review some basic facts. A **polynomial** in an indeterminate x is an expression

$$p(x) = a_0 + a_1 x + a_2 x^2 + \cdots + a_n x^n$$

where $a_0, a_1, a_2, \ldots, a_n$ are real numbers called **coefficients** of the polynomial. If all the coefficients are zero, the polynomial is called the **zero polynomial** and is denoted simply as 0. If $p(x) \neq 0$, the highest power of x with a nonzero coefficient is called the **degree** of $p(x)$ and is denoted as $\deg p(x)$. Hence $\deg(3 + 5x) = 1$, $\deg(1 + x + x^2) = 2$, and $\deg(4) = 0$. (The degree of the zero polynomial is not defined.)

Let **P** denote the set of all polynomials and suppose that

$$p(x) = a_0 + a_1x + a_2x^2 + \cdots$$
$$q(x) = b_0 + b_1x + b_2x^2 + \cdots$$

are two polynomials in **P** (possibly of different degrees). Then $p(x)$ and $q(x)$ are called **equal** [written $p(x) = q(x)$] if and only if all the corresponding coefficients agree—that is, $a_0 = b_0$, $a_1 = b_1$, $a_2 = b_2$, and so on. In particular, $a_0 + a_1x + a_2x^2 + \cdots = 0$ means that $a_0 = 0$, $a_1 = 0$, $a_2 = 0$, $\ldots$, and this is the reason for calling x an **indeterminate**. The set **P** has an addition and scalar multiplication defined on it as follows: If $p(x)$ and $q(x)$ are as before and a is a real number,

$$p(x) + q(x) = (a_0 + b_0) + (a_1 + b_1)x + (a_2 + b_2)x^2 + \cdots$$
$$ap(x) = aa_0 + (aa_1)x + (aa_2)x^2 + \cdots$$

Evidently these are again polynomials, so **P** is closed under these operations. The other vector space axioms are easily verified.

EXAMPLE 6

The set **P** of all polynomials is a vector space with the foregoing addition and scalar multiplication. The zero vector is the zero polynomial, and the negative of a polynomial $p(x) = a_0 + a_1x + a_2x^2 + \cdots$ is the polynomial $-p(x) = -a_0 - a_1x - a_2x^2 - \cdots$ obtained by negating all the coefficients.

◆◆◆

If a and b are real numbers and $a < b$, the **interval** $[a, b]$ is defined to be the set of all real numbers x such that $a \leq x \leq b$. A (real-valued) **function** f on $[a, b]$ is a rule that associates every number x in $[a, b]$ with a real number denoted $f(x)$. The rule is frequently specified by giving a formula for $f(x)$ in terms of x. For example, $f(x) = 2^x$, $f(x) = \sin x$, and $f(x) = x^2 + 1$ are familiar functions. In fact, every polynomial $p(x)$ can be regarded as the formula for a function p. The set of all functions on $[a, b]$ is denoted **F**$[a, b]$. Two functions f and g in **F**$[a, b]$ are **equal** if $f(x) = g(x)$ for every x in $[a, b]$, and we describe this by saying that f and g have the **same action**. Note that two polynomials are equal (defined prior to Example 6) if and only if they are equal as functions.

If f and g are two functions in **F**$[a, b]$, and r is a real number, define the sum $f + g$ and the scalar product rf by

$$(f + g)(x) = f(x) + g(x) \qquad \text{for each } x \text{ in } [a, b]$$
$$(rf)(x) = rf(x) \qquad \text{for each } x \text{ in } [a, b]$$

In other words, the action of $f + g$ upon x is to associated x with the number $f(x) + g(x)$, and rf associates x with $rf(x)$. These operations on **F**$[a, b]$ are called **pointwise** addition and scalar multiplication of functions.

EXAMPLE 7

The set **F**$[a, b]$ of all functions on the interval $[a, b]$ is a vector space if pointwise addition and scalar multiplication of functions are used. The zero function (in axiom A4) is denoted as 0 and has action defined by

$$0(x) = 0 \quad \text{for all } x \text{ in } [a, b]$$

The negative of a function f is denoted $-f$ and has action defined by

$$(-f)(x) = -f(x) \quad \text{for all } x \text{ in } [a, b]$$

Axioms A1 and S1 are clearly satisfied because, if f and g are functions on $[a, b]$, then $f + g$ and rf are again such functions. The verification of the remaining axioms is left as Exercise 21.

◆◆◆

Other examples of vector spaces will appear later, but these are sufficiently varied to indicate the scope of the concept and to illustrate the properties of vector spaces to be discussed. With such a variety of examples, it may come as a surprise that a well-developed *theory* of vector spaces exists. That is, many properties can be shown to hold for *all* vector spaces and hence hold in every example. Such properties are called *theorems* and can be deduced from the axioms. Here is an important example.

THEOREM 1
Cancellation

Let $\mathbf{u}$, $\mathbf{v}$, and $\mathbf{w}$ be vectors in a vector space V. If $\mathbf{v} + \mathbf{u} = \mathbf{v} + \mathbf{w}$, then $\mathbf{u} = \mathbf{w}$.

Proof   We are given $\mathbf{v} + \mathbf{u} = \mathbf{v} + \mathbf{w}$. If these were numbers instead of vectors, we would simply subtract $\mathbf{v}$ from both sides of the equation to obtain $\mathbf{u} = \mathbf{w}$. This can be accomplished with vectors by adding $-\mathbf{v}$ to both sides of the equation. The steps are as follows.

$$\mathbf{v} + \mathbf{u} = \mathbf{v} + \mathbf{w}$$
$$-\mathbf{v} + (\mathbf{v} + \mathbf{u}) = -\mathbf{v} + (\mathbf{v} + \mathbf{w}) \quad \text{(axiom A5)}$$
$$(-\mathbf{v} + \mathbf{v}) + \mathbf{u} = (-\mathbf{v} + \mathbf{v}) + \mathbf{w} \quad \text{(axiom A3)}$$
$$\mathbf{0} + \mathbf{u} = \mathbf{0} + \mathbf{w} \quad \text{(axiom A5)}$$
$$\mathbf{u} = \mathbf{w} \quad \text{(axiom A4)}$$

This is the desired conclusion.   ◆

As with many good mathematical theorems, the technique of the proof of Theorem 1 is at least as important as the theorem itself. The idea was to mimic the well-known process of numerical subtraction in a vector space V as follows: To subtract a vector $\mathbf{v}$ from both sides of a vector equation, we added $-\mathbf{v}$ to both sides. With this in mind, we define **difference $\mathbf{u} - \mathbf{v}$** of two vectors in V as

$$\mathbf{u} - \mathbf{v} = \mathbf{u} + (-\mathbf{v})$$

We shall say that this vector is the result of having **subtracted v** from **u**, and, as in arithmetic, this operation has the property given in Theorem 2.

THEOREM 2

If **u** and **v** are vectors in a vector space V, the equation

$$\mathbf{x} + \mathbf{v} = \mathbf{u}$$

has one and only one solution **x** in V given by

$$\mathbf{x} = \mathbf{u} - \mathbf{v}$$

Proof The difference $\mathbf{x} = \mathbf{u} - \mathbf{v}$ is a solution to the equation because (using several axioms)

$$\mathbf{x} + \mathbf{v} = (\mathbf{u} - \mathbf{v}) + \mathbf{v} = [\mathbf{u} + (-\mathbf{v})] + \mathbf{v} = \mathbf{u} + (-\mathbf{v} + \mathbf{v}) = \mathbf{u} + \mathbf{0} = \mathbf{u}$$

To see that this is the only solution, suppose $\mathbf{x}_1$ is another solution so that $\mathbf{x}_1 + \mathbf{v} = \mathbf{u}$. Then $\mathbf{x} + \mathbf{v} = \mathbf{x}_1 + \mathbf{v}$ (they both equal **u**), so $\mathbf{x} = \mathbf{x}_1$ by cancellation. ◆

Similarly, cancellation shows that there is only one zero vector in any vector space and only one negative of each vector. (Exercises 17 and 18). Hence we speak of *the* zero vector and *the* negative of a vector.

The next theorem introduces some basic facts that are used extensively later.

THEOREM 3

Let **v** denote a vector in a vector space V and let a denote a real number.

1. $0\mathbf{v} = \mathbf{0}$
2. $a\mathbf{0} = \mathbf{0}$
3. If $a\mathbf{v} = \mathbf{0}$, then either $a = 0$ or $\mathbf{v} = \mathbf{0}$
4. $(-1)\mathbf{v} = -\mathbf{v}$
5. $(-a)\mathbf{v} = -(a\mathbf{v}) = a(-\mathbf{v})$

Proof The proofs of (2) and (5) are left as Exercise 19.

1. Observe that $0\mathbf{v} + 0\mathbf{v} = (0 + 0)\mathbf{v} = 0\mathbf{v} = 0\mathbf{v} + \mathbf{0}$ where the first equality is by axiom S3. It follows that $0\mathbf{v} = \mathbf{0}$ by cancellation.

3. Assume $a\mathbf{v} = \mathbf{0}$; it suffices to show that if $a \neq 0$, then necessarily $\mathbf{v} = \mathbf{0}$. But $a \neq 0$ means we can scalar-multiply the given equation $a\mathbf{v} = \mathbf{0}$ by $\frac{1}{a}$ to obtain

$$\mathbf{v} = 1\mathbf{v} = \left(\tfrac{1}{a}\, a\right)\mathbf{v} = \tfrac{1}{a}\,(a\mathbf{v}) = \tfrac{1}{a}\mathbf{0} = \mathbf{0}$$

    using (2) and axioms S4 and S5.

4. We have $-\mathbf{v} + \mathbf{v} = \mathbf{0}$ by axiom A5. On the other hand,

$$(-1)\mathbf{v} + \mathbf{v} = (-1)\mathbf{v} + 1\mathbf{v} = (-1 + 1)\mathbf{v} = 0\mathbf{v} = \mathbf{0}$$

using (1) and axioms S5 and S3. Hence $(-1)\mathbf{v} + \mathbf{v} = -\mathbf{v} + \mathbf{v}$ (because both are equal to $\mathbf{0}$), so $(-1)\mathbf{v} = -\mathbf{v}$ by cancellation. ◆

Axioms S2 and S3 extend. For example, $a(\mathbf{u} + \mathbf{v} + \mathbf{w}) = a\mathbf{u} + a\mathbf{v} + a\mathbf{w}$ and $(a + b + c)\mathbf{v} = a\mathbf{v} + b\mathbf{v} + c\mathbf{v}$ hold for all values of the scalars and vectors involved. More generally,[3]

$$a(\mathbf{v}_1 + \mathbf{v}_2 + \cdots + \mathbf{v}_n) = a\mathbf{v}_1 + a\mathbf{v}_2 + \cdots + a\mathbf{v}_n$$
$$(a_1 + a_2 + \cdots + a_n)\mathbf{v} = a_1\mathbf{v} + a_2\mathbf{v} + \cdots + a_n\mathbf{v}$$

hold for all $n \geq 1$, all numbers $a, a_1, \ldots, a_n$, and all vectors, $\mathbf{v}, \mathbf{v}_1, \ldots, \mathbf{v}_n$. The verifications are by induction and are left to the reader (Exercise 20). These facts—together with the axioms, Theorem 3, and the definition of subtraction—enable us to simplify expressions involving sums of scalar multiples of vectors by collecting like terms, expanding, and taking out common factors. This has been discussed for the vector space of matrices in Section 2.1 (and for geometric vectors in Section 4.1); the manipulations in an arbitrary vector space are carried out in the same way. To illustrate, we rework Example 8§2.1 in the general context.

EXAMPLE 8

If $\mathbf{u}$, $\mathbf{v}$, and $\mathbf{w}$ are vectors in a vector space V, simplify

$$2(\mathbf{u} + 3\mathbf{w}) - 3(2\mathbf{w} - \mathbf{v}) - 3[2(2\mathbf{u} + \mathbf{v} - 4\mathbf{w}) - 4(\mathbf{u} - 2\mathbf{w})]$$

Solution   The reduction proceeds as though $\mathbf{u}$, $\mathbf{v}$, and $\mathbf{w}$ were matrices or variables.

$$\begin{aligned}
2(\mathbf{u} + 3\mathbf{w}) &- 3(2\mathbf{w} - \mathbf{v}) - 3[2(2\mathbf{u} + \mathbf{v} - 4\mathbf{w}) - 4(\mathbf{u} - 2\mathbf{w})] \\
&= 2\mathbf{u} + 6\mathbf{w} - 6\mathbf{w} + 3\mathbf{v} - 3[4\mathbf{u} + 2\mathbf{v} - 8\mathbf{w} - 4\mathbf{u} + 8\mathbf{w}] \\
&= 2\mathbf{u} + 3\mathbf{v} - 3[2\mathbf{v}] \\
&= 2\mathbf{u} + 3\mathbf{v} - 6\mathbf{v} \\
&= 2\mathbf{u} - 3\mathbf{v}
\end{aligned}$$

◆◆◆

The next example shows that the techniques for solving linear equations in Chapter 1 work for vector variables too.

EXAMPLE 9

Let $\mathbf{u}$ and $\mathbf{v}$ be vectors in a vector space V. Find vectors $\mathbf{x}$ and $\mathbf{y}$ in V such that

$$\mathbf{x} - 4\mathbf{y} = \mathbf{u}$$
$$2\mathbf{x} + 3\mathbf{y} = \mathbf{v}$$

---

[3]It is a consequence of axiom A3 that we can omit parentheses when writing a sum $\mathbf{v}_1 + \mathbf{v}_2 + \cdots + \mathbf{v}_n$ of vectors.

Solution 1   The usual row operations on equations work here. Subtract twice the first equation from the second to obtain $11\mathbf{y} = \mathbf{v} - 2\mathbf{u}$. This gives $\mathbf{y} = \frac{1}{11}\mathbf{v} - \frac{2}{11}\mathbf{u}$. Substituting this in the first equation gives $\mathbf{x} = \mathbf{u} + 4\mathbf{y} = \frac{3}{11}\mathbf{u} + \frac{4}{11}\mathbf{v}$.

Solution 2   Write the equations in matrix form

$$\begin{bmatrix} 1 & -4 \\ 2 & 3 \end{bmatrix}\begin{bmatrix} \mathbf{x} \\ \mathbf{y} \end{bmatrix} = \begin{bmatrix} \mathbf{u} \\ \mathbf{v} \end{bmatrix}$$

as in Section 2.2, where the product of a matrix and a column of vectors is defined in the obvious way. But $\begin{bmatrix} 1 & -4 \\ 2 & 3 \end{bmatrix}^{-1} = \frac{1}{11}\begin{bmatrix} 3 & 4 \\ -2 & 1 \end{bmatrix}$, so

$$\begin{bmatrix} \mathbf{x} \\ \mathbf{y} \end{bmatrix} = \begin{bmatrix} 1 & -4 \\ 2 & 3 \end{bmatrix}^{-1}\begin{bmatrix} \mathbf{u} \\ \mathbf{v} \end{bmatrix} = \frac{1}{11}\begin{bmatrix} 3 & 4 \\ -2 & 1 \end{bmatrix}\begin{bmatrix} \mathbf{u} \\ \mathbf{v} \end{bmatrix} = \frac{1}{11}\begin{bmatrix} 3\mathbf{u} + 4\mathbf{v} \\ -2\mathbf{u} + \mathbf{v} \end{bmatrix}$$

Hence $\mathbf{x} = \frac{1}{11}(3\mathbf{u} + 4\mathbf{v})$ and $\mathbf{y} = \frac{1}{11}(-2\mathbf{u} + \mathbf{v})$, as before.

◆◆◆

EXERCISES 5.1

1. Let $\mathbf{u} = (1, 2, -1, 0, 4)$ and $\mathbf{v} = (2, 7, 5, 3, -2)$. Compute:
   (a) $\mathbf{u} + \mathbf{v}$ ◆**(b)** $3\mathbf{u} - 2\mathbf{v}$
   (c) $-2\mathbf{u} + \mathbf{v}$ ◆**(d)** $-3(2\mathbf{u} - 3\mathbf{v})$

2. Vectors $\mathbf{u}$, $\mathbf{v}$, and $\mathbf{w}$ in $\mathbb{R}^n$ are called linearly independent if the only way $a\mathbf{u} + b\mathbf{v} + c\mathbf{w} = \mathbf{0}$ can hold is when $a = b = c = 0$. In each case, determine whether $\mathbf{u}$, $\mathbf{v}$, and $\mathbf{w}$ are linearly independent.
   (a) $\mathbf{u} = (1, 2, -1, 1)$ ◆**(b)** $\mathbf{u} = (1, 3, -1, 4)$
       $\mathbf{v} = (2, 1, 3, 0)$ $\mathbf{v} = (2, 1, 1, -2)$
       $\mathbf{w} = (1, 0, 1, 2)$ $\mathbf{w} = (4, -3, 5, -14)$
   (c) $\mathbf{u} = (1, -1, 3, 2, -4)$ ◆**(d)** $\mathbf{u} = (2, 1, 1, 3, 0)$
       $\mathbf{v} = (2, 0, 1, 3, -5)$ $\mathbf{v} = (1, 3, -1, 0, 4)$
       $\mathbf{w} = (0, -2, 5, 1, -3)$ $\mathbf{w} = (2, 1, 6, 8, 1)$

3. In each case, determine scalars, a, b, and c (if they exist) such that the condition is satisfied:
   (a) $a(1, 2, -1, 1) + b(2, 0, 1, 1) + c(1, 0, 2, 1) = (1, 4, -4, 1)$
   ◆**(b)** $a(1, 3, 0, 1) + b(2, -1, 1, 0) + c(3, 1, -1, 1) = (1, 4, -5, 2)$

4. In each case, show that V is a vector space using the operations of $\mathbb{R}^2$.
   (a) $V = \{(x, 0) \mid x \text{ in } \mathbb{R}\}$ ◆**(b)** $V = \{(x, -x) \mid x \text{ in } \mathbb{R}\}$
   (c) $V = \{(2x - y, x + y) \mid x \text{ and } y \text{ in } \mathbb{R}\}$
   ◆**(d)** $V = \{(3x - y, 2x + 5y) \mid x \text{ and } y \text{ in } \mathbb{R}\}$

5. Let V denote the set of ordered triples (x, y, z) and define addition on V as in $\mathbb{R}^3$. For each of the following definitions of scalar multiplication decide whether V is a vector space.
   (a) $a(x, y, z) = (ax, y, az)$ ◆**(b)** $a(x, y, z) = (ax, 0, az)$
   (c) $a(x, y, z) = (0, 0, 0)$ ◆**(d)** $a(x, y, z) = (2ax, 2ay, 2az)$

6. Are the following sets vector spaces with the indicated operations? If not, why not?
   (a) The set V of nonnegative real numbers; ordinary addition and scalar multiplication.
   ◆**(b)** The set V of all polynomials of degree ≥ 3, together with 0; operations of **P**.
   (c) The set V of all 2×2 matrices of the form $\begin{bmatrix} a & b \\ 0 & c \end{bmatrix}$; operations of $\mathbf{M}_{22}$.
   ◆**(d)** The set V of all 2×2 matrices with equal column sums; operations of $\mathbf{M}_{22}$.
   (e) The set V of 2×2 matrices with zero determinant; usual matrix operations.
   ◆**(f)** The set V of real numbers; usual operations.
   (g) The set V of complex numbers; usual addition and multiplication by a real number.
   ◆**(h)** A set $V = \{\mathbf{0}\}$ consisting of a single vector $\mathbf{0}$ where $\mathbf{0} + \mathbf{0} = \mathbf{0}$ and $a\mathbf{0} = \mathbf{0}$ for all a in $\mathbb{R}$.
   (i) The set V of all ordered pairs (x, y) with the addition of $\mathbb{R}^2$, but scalar multiplication $a(x, y) = (x, y)$ for all a in $\mathbb{R}$.
   ◆**(j)** The set V of all functions $f: \mathbb{R} \to \mathbb{R}$ with pointwise addition and scalar multiplication defined by $(af)(x) = f(ax)$.

(k) The set V of all 2×2 matrices whose entries sum to 0; operations of $\mathbf{M}_{22}$.

(l) The set V of all 2×2 matrices with the addition of $\mathbf{M}_{22}$ but scalar multiplication $*$ defined by $a * X = aX^T$.

7. Let V be the set of positive real numbers with vector addition being ordinary multiplication and scalar multiplication being $av = v^a$. Show that V is a vector space.

8. If V is the set of ordered pairs (x, y) of real numbers, show that it is a vector space if $(x, y) + (x_1, y_1) = (x + x_1, y + y_1 + 1)$ and $a(x, y) = (ax, ay + a - 1)$.

9. (a) The line through the origin with slope m has the equation $y = mx$ and so consists of points $P(x, mx)$ with x in $\mathbb{R}$. Show that $V = \{(x, mx) \mid x \text{ in } \mathbb{R}\}$ is a vector space using the operations of $\mathbb{R}^2$.

(b) The plane V through the origin with direction vector $(a, b, c) \neq \mathbf{0}$ has equation $ax + by + cz = 0$, and so $V = \{(x, y, z) \mid ax + by + cz = 0\}$. Show that V is a vector space using the operations of $\mathbb{R}^3$.

10. (a) Let V be the set of ordered pairs (x, y) of real numbers with the scalar multiplication of $\mathbb{R}^2$ but with addition defined by

$$(x_1, y_1) + (x_2, y_2) = \left(\sqrt[3]{x_1^3 + x_2^3}, \sqrt[3]{y_1^3 + y_2^3} \right)$$

Show that V is a vector space.

(b) What if $(x_1, y_1) + (x_2, y_2) = \left(\sqrt[2]{x_1^2 + x_2^2}, \sqrt[2]{y_1^2 + y_2^2} \right)$ where $\sqrt[2]{}$ indicates the positive square root?

11. Find $\mathbf{x}$ and $\mathbf{y}$ (in terms of $\mathbf{u}$ and $\mathbf{v}$) such that:

(a) $2\mathbf{x} + \mathbf{y} = \mathbf{u}$
    $5\mathbf{x} + 3\mathbf{y} = \mathbf{v}$

(b) $3\mathbf{x} - 2\mathbf{y} = \mathbf{u}$
    $4\mathbf{x} - 5\mathbf{y} = \mathbf{v}$

12. Find *all* vectors $\mathbf{x}$, $\mathbf{y}$, and $\mathbf{z}$ (in terms of $\mathbf{u}$, $\mathbf{v}$) such that:

(a) $\mathbf{x} - 2\mathbf{y} + \mathbf{z} = 2\mathbf{u} - \mathbf{v}$
     $2\mathbf{x} - 3\mathbf{y} - \mathbf{z} = \mathbf{v} - \mathbf{u}$
     $-\mathbf{x} + 3\mathbf{y} - 4\mathbf{z} = 4\mathbf{v} - 7\mathbf{u}$

(b) $\mathbf{x} - \mathbf{y} + 2\mathbf{z} = 3\mathbf{u} - \mathbf{v}$
     $-3\mathbf{x} + 4\mathbf{y} - \mathbf{z} = \mathbf{v} - \mathbf{u}$
     $5\mathbf{x} - 6\mathbf{y} + 5\mathbf{z} = 7\mathbf{u} - 3\mathbf{v}$

(c) $3\mathbf{x} + 2\mathbf{y} + \mathbf{z} = \mathbf{0}$
     $2\mathbf{x} + \mathbf{y} - 2\mathbf{z} = \mathbf{0}$
     $\mathbf{x} + \mathbf{y} + 3\mathbf{z} = \mathbf{0}$

(d) $3\mathbf{x} - \mathbf{y} + 4\mathbf{z} = \mathbf{0}$
     $\mathbf{x} + \mathbf{y} - 5\mathbf{z} = \mathbf{0}$
     $\mathbf{x} - 3\mathbf{y} + 14\mathbf{z} = \mathbf{0}$

13. In each case show that the condition $a\mathbf{u} + b\mathbf{v} + c\mathbf{w} = \mathbf{0}$ in V implies that $a = b = c = 0$.

(a) $V = \mathbb{R}^4$; $\mathbf{u} = (2, 1, 0, 2)$, $\mathbf{v} = (1, 1, -1, 0)$, $\mathbf{w} = (0, 1, 2, 1)$

(b) $V = \mathbf{M}_{22}$; $\mathbf{u} = \begin{bmatrix} 1 & 0 \\ 0 & 1 \end{bmatrix}$, $\mathbf{v} = \begin{bmatrix} 0 & 1 \\ 1 & 0 \end{bmatrix}$, $\mathbf{w} = \begin{bmatrix} 1 & 1 \\ 1 & -1 \end{bmatrix}$

(c) $V = \mathbf{P}$; $\mathbf{u} = x^3 + x$, $\mathbf{v} = x^2 + 1$, $\mathbf{w} = x^3 - x^2 + x + 1$

(d) $V = \mathbf{F}[0, \pi]$; $\mathbf{u} = \sin x$, $\mathbf{v} = \cos x$, $\mathbf{w} = 1$

14. Simplify each of the following.

(a) $3[2(\mathbf{u} - 2\mathbf{v} - \mathbf{w}) + 3(\mathbf{w} - \mathbf{v})] - 7(\mathbf{u} - 3\mathbf{v} - \mathbf{w})$

(b) $4(3\mathbf{u} - \mathbf{v} + \mathbf{w}) - 2[(3\mathbf{u} - 2\mathbf{v}) - 3(\mathbf{v} - \mathbf{w})] + 6(\mathbf{w} - \mathbf{u} - \mathbf{v})$

15. Show that $\mathbf{x} = \mathbf{v}$ is the only solution to the equation $\mathbf{x} + \mathbf{x} = 2\mathbf{v}$ in a vector space V. Cite all axioms used.

16. Show that $-\mathbf{0} = \mathbf{0}$ in any vector space. Cite all axioms used.

17. Show that the zero vector $\mathbf{0}$ is uniquely determined by the property in axiom A4.

18. Given a vector $\mathbf{v}$, show that its negative $-\mathbf{v}$ is uniquely determined by the property in axiom A5.

19. (a) Prove (2) of Theorem 3.

(b) Prove that $(-a)\mathbf{v} = -(a\mathbf{v})$ in Theorem 3 by first computing $(-a)\mathbf{v} + a\mathbf{v}$. Then do it using (4) of Theorem 3 and axiom S4.

(c) Prove that $a(-\mathbf{v}) = -(a\mathbf{v})$ in Theorem 3 in two ways, as in part (b).

20. Let $\mathbf{v}, \mathbf{v}_1, \ldots, \mathbf{v}_n$ denote vectors in a vector space V and let $a, a_1, \ldots, a_n$ denote numbers. Use induction on n to prove each of the following.

(a) $a(\mathbf{v}_1 + \mathbf{v}_2 + \cdots + \mathbf{v}_n) = a\mathbf{v}_1 + a\mathbf{v}_2 + \cdots + a\mathbf{v}_n$

(b) $(a_1 + a_2 + \cdots + a_n)\mathbf{v} = a_1\mathbf{v} + a_2\mathbf{v} + \cdots + a_n\mathbf{v}$

21. Verify axioms A2–A5 and S2–S5 for the space $\mathbf{F}[a, b]$ of functions on $[a, b]$ (Example 7).

22. Prove each of the following for vectors $\mathbf{u}$ and $\mathbf{v}$ and scalars a and b.

(a) If $a\mathbf{v} = b\mathbf{v}$ and $\mathbf{v} \neq \mathbf{0}$, then $a = b$.

(b) If $a\mathbf{v} = a\mathbf{w}$ and $a \neq 0$, then $\mathbf{v} = \mathbf{w}$.

23. By calculating $(1 + 1)(\mathbf{v} + \mathbf{w})$ in two ways (using axioms S2 and S3), show that axiom A2 follows from the other axioms.

24. Let V be a vector space, and define V^n to be the set of all n-tuples $(\mathbf{v}_1, \mathbf{v}_2, \ldots, \mathbf{v}_n)$ of n vectors $\mathbf{v}_i$, each belonging to V. Define addition and scalar multiplication in V^n as follows:

$$(\mathbf{u}_1, \mathbf{u}_2, \ldots, \mathbf{u}_n) + (\mathbf{v}_1, \mathbf{v}_2, \ldots, \mathbf{v}_n)$$
$$= (\mathbf{u}_1 + \mathbf{v}_1, \mathbf{u}_2 + \mathbf{v}_2, \ldots, \mathbf{u}_n + \mathbf{v}_n)$$
$$a(\mathbf{v}_1, \mathbf{v}_2, \ldots, \mathbf{v}_n) = (a\mathbf{v}_1, a\mathbf{v}_2, \ldots, a\mathbf{v}_n)$$

Show that V^n is a vector space.

25. Let V^n be the vector space of n-tuples from the preceding exercise, written as columns. If A is an $m \times n$ matrix, and X is in V^n, define AX in V^m by matrix multiplication. More precisely, if $A = [a_{ij}]$ and $X = \begin{bmatrix} \mathbf{v}_1 \\ \vdots \\ \mathbf{v}_n \end{bmatrix}$, let $AX = \begin{bmatrix} \mathbf{u}_1 \\ \vdots \\ \mathbf{u}_n \end{bmatrix}$, where $\mathbf{u}_i = a_{i1}\mathbf{v}_1 + a_{i2}\mathbf{v}_2 + \cdots + a_{in}\mathbf{v}_n$ for each i. Prove that:

(a) $B(AX) = (BA)X$

(b) $(A + A_1)X = AX + A_1X$

(c) $A(X + X_1) = AX + AX_1$

(d) $(kA)X = k(AX) = A(kX)$ if k is any number

(e) $IX = X$ if I is the $n \times n$ identity matrix

(f) Let E be an elementary matrix obtained by performing a row operation on (the rows of) I_n (see Section 2.4). Show that EX is the column resulting from performing that same row operation on the vectors (call them rows) of X. [*Hint*: Theorem 1 §2.4.]

Section 5.2   Subspaces and Spanning Sets

Very often the most interesting vector spaces arise as parts of larger vector spaces.

> **DEFINITION**
>
> If V is a vector space, a subset U of V is called a **subspace** of V if U is itself a vector space where U uses the vector addition and scalar multiplication of V.

If U is a subspace of V, it is clear (by axioms A1 and S1) that the sum of two vectors in U is again in U and that any scalar multiple of a vector in U is again in U—in short, that U is **closed** under the vector addition and scalar multiplication of V. The nice part is that the converse is also true: If U is closed under these operations, then all the other axioms are automatically satisfied. For example, axiom A2 asserts that $\mathbf{u} + \mathbf{u}_1 = \mathbf{u}_1 + \mathbf{u}$ holds for all vectors $\mathbf{u}$ and $\mathbf{u}_1$ in U. But this is clear because the equation is already true in V, and U uses the same addition as V. Similarly, axioms A3, S2, S3, S4, and S5 hold automatically in U because they are true in V. All that remains is to verify axioms A4 and A5.

> **THEOREM 1**
> **Subspace Test**
>
> Let U be a subset of a vector space V. Then U is a subspace of V if and only if it satisfies the following three conditions.[4]
>
> **1.** $\mathbf{0}$ lies in U where $\mathbf{0}$ is the zero vector of V.
>
> **2.** If $\mathbf{u}_1$ and $\mathbf{u}_2$ lie in U, then $\mathbf{u}_1 + \mathbf{u}_2$ lies in U.
>
> **3.** If $\mathbf{u}$ lies in U, then $a\mathbf{u}$ lies in U for all a in $\mathbb{R}$.

---

[4]Condition (1) can be replaced by the requirement that U is nonempty (Exercise 24).

Proof   If (1), (2) and (3) hold, then axiom A4 follows from (1), and axiom A5 follows from (3) (because $-\mathbf{u} = (-1)\mathbf{u}$ lies in U for all $\mathbf{u}$ in U). Hence U is a subspace by the discussion preceding the theorem. Conversely, if U is a subspace it is closed under addition and scalar multiplication, and this gives (2) and (3). If $\mathbf{z}$ denotes the zero vector of U, then $\mathbf{z} = 0\mathbf{z}$ in U by Theorem 3§5.1. But $0\mathbf{z} = \mathbf{0}$ in V by the same theorem, so $\mathbf{0} = \mathbf{z}$ lies in U. This proves (1).   ◆

If U is a subspace of V, the proof shows that U and V share the same zero vector. Also, if $\mathbf{u}$ lies in U, then $-\mathbf{u} = (-1)\mathbf{u}$ lies in U; that is, the negative of a vector in U is the same as its negative in V.

The subspace test provides an easy way of finding subspaces.

EXAMPLE 1

Show that $U = \{(x, -x) \mid x \text{ in } \mathbb{R}\}$ is a subspace of $\mathbb{R}^2$.

Solution   Clearly $\mathbf{0} = (0, 0)$ is in U, and the equations

$$(x, -x) + (y, -y) = ((x + y), -(x + y))$$
$$a(x, -x) = (ax, -(ax))$$

show that U is closed under addition and scalar multiplication. Hence the subspace test applies.

◆◆◆

EXAMPLE 2

If V is any vector space, show that $\{\mathbf{0}\}$ and V are subspaces of V.

Solution $U = V$ clearly satisfies the conditions of the test. As to $U = \{\mathbf{0}\}$, it satisfies the conditions because $\mathbf{0} + \mathbf{0} = \mathbf{0}$ and $a\mathbf{0} = \mathbf{0}$ for all a in $\mathbb{R}$.

◆◆◆

The vector space $\{\mathbf{0}\}$ is called the **zero subspace** of V. Because all zero subspaces look alike, we speak of the **zero vector space** and denote it by 0. It is the unique vector space containing just one vector.

EXAMPLE 3

Let $\mathbf{v}$ be a vector in a vector space V. Show that the set

$$\mathbb{R}\mathbf{v} = \{a\mathbf{v} \mid a \text{ in } \mathbb{R}\}$$

of all scalar multiples of $\mathbf{v}$ is a subspace of V.

Solution   Because $\mathbf{0} = 0\mathbf{v}$, it is clear that $\mathbf{0}$ lies in $\mathbb{R}\mathbf{v}$. Given two vectors $a\mathbf{v}$ and $a_1\mathbf{v}$ in $\mathbb{R}\mathbf{v}$, their sum $a\mathbf{v} + a_1\mathbf{v} = (a + a_1)\mathbf{v}$ is also a scalar multiple of $\mathbf{v}$ and so lies in $\mathbb{R}\mathbf{v}$. Hence $\mathbb{R}\mathbf{v}$ is closed under addition. Finally, given $a\mathbf{v}$, $r(a\mathbf{v}) = (ra)\mathbf{v}$ lies in $\mathbb{R}\mathbf{v}$, so $\mathbb{R}\mathbf{v}$ is closed under scalar multiplication. Hence the subspace test applies.

◆◆◆

EXAMPLE 4

Let A be an $m \times n$ matrix. Consider the set

$$U = \{AX \mid X \text{ lies in } \mathbb{R}^n, X \text{ written as a column}\}$$

Show that U is a subspace of $\mathbb{R}^m$ called the **range** of the matrix A.

Solution

Note first that U is in fact a subset of $\mathbb{R}^m$ because A is $m \times n$. Each vector in U is of the form AX for some vector X in $\mathbb{R}^n$. To apply the subspace test, note that $0 = A0$ has the required form, so 0 lies in U. Similarly, the equations $AX + AX_1 = A(X + X_1)$ and $r(AX) = A(rX)$ show that sums and scalar multiples of vectors in U again have the required form. Hence U is a subspace of $\mathbb{R}^m$.

◆◆◆

The next example gives another important subspace related to a matrix A. However, rather than specify the *form* of each vector in the subspace (as in Example 4), we describe it by specifying a *condition* that vectors must satisfy to be in the subspace.

EXAMPLE 5

Let A be an $m \times n$ matrix. Show that the set

$$U = \{X \text{ in } \mathbb{R}^n \mid AX = 0, X \text{ written as a column}\}$$

is a subspace of $\mathbb{R}^n$ called the **null space** of the matrix A and denoted null A. Note that U is the set of solutions to the homogeneous system of equations with A as coefficient matrix (and is also called the **solution space** of the system).

Solution

Here U consists of all columns X in $\mathbb{R}^n$ satisfying the condition that $AX = 0$. Because $A0 = 0$, it is clear that 0 lies in U. If X and X_1 both lie in U, then $A(X + X_1) = AX + AX_1 = 0 + 0 = 0$. This shows that $X + X_1$ qualifies for membership in U, so U is closed under addition. Similarly, $A(rX) = r(AX) = r0 = 0$, so rX lies in U. This means that U is closed under scalar multiplication and so is a subspace of $\mathbb{R}^n$.

◆◆◆

The next example describes a subset U of the space $\mathbf{M}_{22}$, first by giving a *condition* that each matrix of U must satisfy and second by giving the *form* of each matrix in U. Both characterizations of U are used to show that it is a subspace of $\mathbf{M}_{22}$.

EXAMPLE 6

Let $A = \begin{bmatrix} 1 & 1 \\ 0 & 0 \end{bmatrix}$ be a fixed matrix in $\mathbf{M}_{22}$, and let

$$U = \{X \text{ in } \mathbf{M}_{22} \mid AX = XA\}$$

Show that U is a subspace of $\mathbf{M}_{22}$.

Solution 1   If 0 is the 2 $\times$ 2 zero matrix, then $A0 = 0A$, so 0 satisfies the condition for membership in U. Next suppose that X and X_1 lie in U so that $AX = XA$ and $AX_1 = X_1A$. Then

$$A(X + X_1) = AX + AX_1 = XA + X_1A = (X + X_1)A$$
$$A(aX) = a(AX) = a(XA) = (aX)A$$

for all a in $\mathbb{R}$, so both $X + X_1$ and aX lie in U. Hence U is a subspace of $\mathbf{M}_{22}$.

Solution 2   If X lies in U, write $X = \begin{bmatrix} x & y \\ z & w \end{bmatrix}$. Then the condition $AX = XA$ becomes

$$\begin{bmatrix} 1 & 1 \\ 0 & 0 \end{bmatrix}\begin{bmatrix} x & y \\ z & w \end{bmatrix} = \begin{bmatrix} x & y \\ z & w \end{bmatrix}\begin{bmatrix} 1 & 1 \\ 0 & 0 \end{bmatrix}$$

Comparing entries gives $x + z = x$, $y + w = x$, and $z = 0$. Thus X has the form $X = \begin{bmatrix} y + w & y \\ 0 & w \end{bmatrix}$ where y and w are arbitrary real numbers. Hence

$$U = \left\{ \begin{bmatrix} y + w & y \\ 0 & w \end{bmatrix} \middle| \, y \text{ and } w \text{ in } \mathbb{R} \right\}$$

specifies U by giving the *form* of all matrices in U. Now 0 clearly lies in U (when y and w are zero), and it is easily verified that sums and scalar multiples of matrices of this form are again of the same form and so lie in U. This shows again that U is a subspace of $\mathbf{M}_{22}$.

$\blacklozenge\blacklozenge\blacklozenge$

The two solutions of Example 6 are quite different. The first has the advantage that it is brief and works in the same way for *any* square matrix A. The second is more tedious and would be different if another matrix A were used. However, the explicit form of the vectors (in this case, matrices) in a subspace is often needed in other contexts.

Suppose $p(x)$ is a polynomial and a is a number. Then the number $p(a)$ obtained by replacing x by a in the expression for $p(x)$ is called the **evaluation** of $p(x)$ at a. For example, if $p(x) = 5 - 6x + 2x^2$, then the evaluation of $p(x)$ at $a = 2$ is $p(2) = 1$. If $p(a) = 0$, the number a is called a **root** of $p(x)$.

EXAMPLE 7

Consider the set U of all polynomials in $\mathbf{P}$ that have 3 as a root:

$$U = \{p(x) \text{ in } \mathbf{P} \, | \, p(3) = 0\}$$

Show that U is a subspace of $\mathbf{P}$.

Solution 1   Clearly the zero polynomial lies in U. Now let $p(x)$ and $q(x)$ lie in U so $p(3) = 0$ and $q(3) = 0$. Then $(p + q)(x) = p(x) + q(x)$ for all x, so $(p + q)(3) =$

$p(3) + q(3) = 0 + 0 = 0$, and U is closed under addition. The verification that U is closed under scalar multiplication is similar.

Solution 2 The form of all the polynomials in U follows from the factor theorem:[5]

$$U = \{(x - 3)q(x) \mid q(x) \text{ in } \mathbf{P}\}$$

The verification of this, and of the fact that it shows U to be a subspace of $\mathbf{P}$, is left as Exercise 26.

◆◆◆

There are other important examples of vector spaces consisting of polynomials. Let $\mathbf{P}_n$ denote the set of all polynomials of degree at most n, together with the zero polynomial. In other words, $\mathbf{P}_n$ consists of all polynomials of the form

$$a_0 + a_1x + a_2x^2 + \cdots + a_nx^n$$

where $a_0, a_1, a_2, \ldots, a_n$ are real numbers and so is closed under the addition and scalar multiplication in $\mathbf{P}$. Moreover, the zero polynomial is included in $\mathbf{P}_n$. So the subspace test gives Example 8.

EXAMPLE 8

For each $n \geq 0$, $\mathbf{P}_n$ is a subspace of $\mathbf{P}$.

◆◆◆

The next example refers to material in Chapter 4.

EXAMPLE 9

Regard $\mathbb{R}^3$ as the set of points in space. Show that every plane through the origin is a subspace.

Solution As shown in Section 4.3, every plane P through the origin has equation $ax + by + cz = 0$ for some numbers a, b, and c, not all zero. In other words, P is the following subset of $\mathbb{R}^3$:

$$P = \{(x, y, z) \text{ in } \mathbb{R}^3 \mid ax + by + cz = 0\}$$

It is clear from this that $\mathbf{0} = (0, 0, 0)$ lies in P; the verification that P is closed under the addition and scalar multiplication of $\mathbb{R}^3$ is left as Exercise 27.

◆◆◆

The next example involves the notion of the derivative f' of a function f. (If the reader is not familiar with calculus, this example may be omitted.) A function f defined on the interval $[a, b]$ is called **differentiable** if the derivative function f' exists.

---

[5]The factor theorem is given in Section 5.6.

EXAMPLE 10

Show that the subset **D**[a, b] of all **differentiable functions** on [a, b] is a subspace of the vector space **F**[a, b] of all functions on [a, b].

Solution

The derivative of any constant function is the constant function 0; in particular, 0 itself is differentiable and so lies in **D**[a, b]. If f and g both lie in **D**[a, b] (so that f' and g' exist), then it is a theorem of calculus that f + g and af are both differentiable [in fact, (f + g)' = f' + g' and (af)' = af'], so both lie in **D**[a, b]. This shows that **D**[a, b] is a subspace of **F**[a, b].   ◆◆◆

EXAMPLE 11

Consider the two subsets P and Q of $\mathbb{R}^2$ defined by

$$P = \{(a, b) \text{ in } \mathbb{R}^2 \mid a \geq 0\} \qquad Q = \{(a, b) \text{ in } \mathbb{R}^2 \mid a^2 = b^2\}$$

Then P and Q both contain the zero vector (0, 0) of $\mathbb{R}^2$, but they are not subspaces. In fact, P is closed under addition but not scalar multiplication (for example (1, 0) lies in P but $(-1)(1, 0) = (-1, 0)$ is not in P), whereas Q is closed under scalar multiplication but not addition (for example (2, −2) and (1, 1) both lie in Q, but their sum $(2, -2) + (1, 1) = (3, -1)$ does not lie in Q).   ◆◆◆

Linear Combinations and Spanning Sets

The set of solutions to a system of m homogeneous linear equations in n variables is a subspace of $\mathbb{R}^n$ (Example 5), and the Gaussian algorithm gives a convenient way to describe this subspace. For example, consider the system

$$\begin{aligned}
x_1 - 2x_2 + x_3 + x_4 &= 0 \\
-x_1 + 2x_2 \qquad + x_4 &= 0 \\
2x_1 - 4x_2 + x_3 \qquad &= 0
\end{aligned}$$

The augmented matrix is reduced as follows:

$$\begin{bmatrix} 1 & -2 & 1 & 1 & 0 \\ -1 & 2 & 0 & 1 & 0 \\ 2 & -4 & 1 & 0 & 0 \end{bmatrix} \rightarrow \begin{bmatrix} 1 & -2 & 1 & 1 & 0 \\ 0 & 0 & 1 & 2 & 0 \\ 0 & 0 & -1 & -2 & 0 \end{bmatrix} \rightarrow \begin{bmatrix} 1 & -2 & 0 & -1 & 0 \\ 0 & 0 & 1 & 2 & 0 \\ 0 & 0 & 0 & 0 & 0 \end{bmatrix}$$

Taking $x_2 = s$ and $x_4 = t$, the solution takes the form

$$\begin{bmatrix} x_1 \\ x_2 \\ x_3 \\ x_4 \end{bmatrix} = \begin{bmatrix} 2s + t \\ s \\ -2t \\ t \end{bmatrix} = \begin{bmatrix} 2s \\ s \\ 0 \\ 0 \end{bmatrix} + \begin{bmatrix} t \\ 0 \\ -2t \\ t \end{bmatrix} = s \begin{bmatrix} 2 \\ 1 \\ 0 \\ 0 \end{bmatrix} + t \begin{bmatrix} 1 \\ 0 \\ -2 \\ 1 \end{bmatrix}$$

where s and t are arbitrary parameters. Hence every solution can be found as a sum of scalar multiples of the two basic solutions

$$\begin{bmatrix} 2 \\ 1 \\ 0 \\ 0 \end{bmatrix} \text{ and } \begin{bmatrix} 1 \\ 0 \\ -2 \\ 1 \end{bmatrix}$$

Such descriptions often turn out to be the most convenient way to describe a subspace.

DEFINITION

A vector $\mathbf{v}$ is called a **linear combination** of the vectors $\mathbf{v}_1, \mathbf{v}_2, \ldots, \mathbf{v}_n$ if it can be expressed in the form

$$\mathbf{v} = a_1\mathbf{v}_1 + a_2\mathbf{v}_2 + \cdots + a_n\mathbf{v}_n$$

where $a_1, a_2, \ldots, a_n$ are scalars called the **coefficients** of $\mathbf{v}_1, \mathbf{v}_2, \ldots, \mathbf{v}_n$.

EXAMPLE 12

Determine whether $(1, 1, 4)$ or $(1, 5, 1)$ is a linear combination of the vectors $\mathbf{v}_1 = (1, 2, -1)$ and $\mathbf{v}_2 = (3, 5, 2)$ in $\mathbb{R}^3$.

Solution   First, $(1, 1, 4)$ is a linear combination of $\mathbf{v}_1$ and $\mathbf{v}_2$; indeed,

$$(1, 1, 4) = s(1, 2, -1) + t(3, 5, 2)$$

where $s = -2$ and $t = 1$. Turning to $(1, 5, 1)$, the question is whether s and t can be found such that $(1, 5, 1) = s(1, 2, -1) + t(3, 5, 2)$. Equating components gives

$$1 = s + 3t$$
$$5 = 2s + 5t$$
$$1 = -s + 2t$$

These equations have no solution, so $(1, 5, 1)$ is *not* a linear combination of $\mathbf{v}_1$ and $\mathbf{v}_2$.

◆◆◆

In the system of three homogeneous linear equations considered prior to Example 12, the solutions turned out to be just the set of *all* linear combinations of two particular solutions. This prompts the following terminology:

DEFINITION

If $\{\mathbf{v}_1, \mathbf{v}_2, \ldots, \mathbf{v}_n\}$ is any set of vectors in a vector space V, the set of all linear combinations of these vectors is called their **span**, and is denoted by

$$\text{span}\{\mathbf{v}_1, \mathbf{v}_2, \ldots, \mathbf{v}_n\}$$

If it happens that $V = \text{span}\{\mathbf{v}_1, \mathbf{v}_2, \ldots, \mathbf{v}_n\}$, then these vectors are called a **spanning set** for V.

For example, the span of two vectors $\mathbf{v}$ and $\mathbf{w}$ is the set

$$\text{span}\{\mathbf{v}, \mathbf{w}\} = \{s\mathbf{v} + t\mathbf{w} \mid s \text{ and } t \text{ in } \mathbb{R}\}$$

of all sums of scalar multiples of the vectors.

EXAMPLE 13

Show that $\mathbb{R}^n = \text{span}\{\mathbf{e}_1, \mathbf{e}_2, \ldots, \mathbf{e}_n\}$, where

$$\mathbf{e}_1 = (1, 0, 0, \ldots, 0)$$
$$\mathbf{e}_2 = (0, 1, 0, \ldots, 0)$$
$$\vdots \qquad\qquad \vdots$$
$$\mathbf{e}_n = (0, 0, 0, \ldots, 1)$$

Solution   We must show that every vector in $\mathbb{R}^n$ lies in $\text{span}\{\mathbf{e}_1, \mathbf{e}_2, \ldots, \mathbf{e}_n\}$. But if $\mathbf{v} = (a_1, a_2, \ldots, a_n)$ is any vector in $\mathbb{R}^n$, then

$$\mathbf{v} = (a_1, a_2, \ldots, a_n) = a_1\mathbf{e}_1 + a_2\mathbf{e}_2 + \cdots + a_n\mathbf{e}_n$$

Hence $\mathbf{v}$ lies in $\text{span}\{\mathbf{e}_1, \mathbf{e}_2, \ldots, \mathbf{e}_n\}$.   ◆◆◆

EXAMPLE 14

Show that $\mathbf{P}_n = \text{span}\{1, x, x^2, \ldots, x^n\}$.

Solution   We need only show that each polynomial $p(x)$ in $\mathbf{P}_n$ is a linear combination of 1, $x, \ldots, x^n$. But this is clear because $p(x)$ has the form

$$p(x) = a_0 + a_1 x + a_2 x^2 + \cdots + a_n x^n.$$

◆◆◆

In the case of a single vector $\mathbf{v}$ in a vector space V, the span is

$$\text{span}\{\mathbf{v}\} = \{s\mathbf{v} \mid s \text{ in } \mathbb{R}\} = \mathbb{R}\mathbf{v}$$

The notation $\mathbb{R}\mathbf{v}$ was introduced in Example 3 where it was verified that $\mathbb{R}\mathbf{v}$ is a subspace of V. It turns out that the span of any set of vectors is a subspace.

THEOREM 2

Let $U = \text{span}\{\mathbf{v}_1, \mathbf{v}_2, \ldots, \mathbf{v}_n\}$ in a vector space V. Then

1. U is a subspace of V containing each of $\mathbf{v}_1, \mathbf{v}_2, \ldots, \mathbf{v}_n$.
2. U is the "smallest" subspace containing these vectors in the sense that any subspace of V that contains each of $\mathbf{v}_1, \mathbf{v}_2, \ldots, \mathbf{v}_n$ must contain U.

Proof

1. Clearly $\mathbf{0} = 0\mathbf{v}_1 + \cdots + 0\mathbf{v}_n$ belongs to U. If $\mathbf{v} = a_1\mathbf{v}_1 + \cdots + a_n\mathbf{v}_n$ and $\mathbf{w} = b_1\mathbf{v}_1 + \cdots + b_n\mathbf{v}_n$ are two members of U and a is in $\mathbb{R}$, then

$$\mathbf{v} + \mathbf{w} = (a_1 + b_1)\mathbf{v}_1 + \cdots + (a_n + b_n)\mathbf{v}_n$$
$$a\mathbf{v} = (aa_1)\mathbf{v}_1 + \cdots + (aa_n)\mathbf{v}_n$$

so both $\mathbf{v} + \mathbf{w}$ and $a\mathbf{v}$ lie in U. Hence U is a subspace of V. It contains each of $\mathbf{v}_1, \mathbf{v}_2, \ldots, \mathbf{v}_n$; for example, $\mathbf{v}_2 = 0\mathbf{v}_1 + 1\mathbf{v}_2 + 0\mathbf{v}_3 + \cdots + 0\mathbf{v}_n$. This proves (1).

2. Let W be a subspace of V that contains each of $\mathbf{v}_1, \mathbf{v}_2, \ldots, \mathbf{v}_n$. Because W is closed under scalar multiplication, each of $a_1\mathbf{v}_1, a_2\mathbf{v}_2, \ldots, a_n\mathbf{v}_n$ lies in W for any choice of $a_1, a_2, \ldots, a_n$ in $\mathbb{R}$. But then $a_1\mathbf{v}_1 + a_2\mathbf{v}_2 + \cdots + a_n\mathbf{v}_n$ lies in W because W is closed under addition. This means that W contains every member of U, which proves (2). ◆

Theorem 2 is useful for determining spanning sets, as the following examples show.

EXAMPLE 15

Show that $\mathbb{R}^3 = \text{span}\{(1, 1, 1), (1, 1, 0), (0, 1, 1)\}$.

Solution   Write $\mathbf{v}_1 = (1, 1, 1)$, $\mathbf{v}_2 = (1, 1, 0)$, $\mathbf{v}_3 = (0, 1, 1)$, and $U = \text{span}\{\mathbf{v}_1, \mathbf{v}_2, \mathbf{v}_3\}$. Clearly U is contained in $\mathbb{R}^3$. We have $\mathbb{R}^3 = \text{span}\{(1, 0, 0), (0, 1, 0), (0, 0, 1)\}$; so to prove that $\mathbb{R}^3$ is contained in U, it is enough by Theorem 2 to show that each of $(1, 0, 0)$, $(0, 1, 0)$, and $(0, 0, 1)$ lies in $\text{span}\{\mathbf{v}_1, \mathbf{v}_2, \mathbf{v}_3\}$. But they can be given explicitly as linear combinations of $\mathbf{v}_1$, $\mathbf{v}_2$, and $\mathbf{v}_3$:

$$(1, 0, 0) = (1, 1, 1) - (0, 1, 1) = \mathbf{v}_1 - \mathbf{v}_3$$
$$(0, 0, 1) = (1, 1, 1) - (1, 1, 0) = \mathbf{v}_1 - \mathbf{v}_2$$

and then, using the first of these, we have

$$(0, 1, 0) = (1, 1, 0) - (1, 0, 0) = \mathbf{v}_2 - (\mathbf{v}_1 - \mathbf{v}_3) = \mathbf{v}_2 - \mathbf{v}_1 + \mathbf{v}_3 \qquad ◆◆◆$$

EXAMPLE 16

Show that $\mathbf{P}_3 = \text{span}\{x^2 + x^3, x, 2x^2 + 1, 3\}$.

Solution   Write $U = \text{span}\{x^2 + x^3, x, 2x^2 + 1, 3\}$. Then U is contained in $\mathbf{P}_3$, and we use the fact that $\mathbf{P}_3 = \text{span}\{1, x, x^2, x^3\}$ to show that $\mathbf{P}_3$ is contained in U. In fact, x and $1 = \frac{1}{3} \cdot 3$ clearly lie in U. But then successively,

$$x^2 = \tfrac{1}{2}[(2x^2 + 1) - 1]$$
$$x^3 = (x^2 + x^3) - x^2$$

also lie in U. Hence $\mathbf{P}_3$ is contained in U by Theorem 2.

◆◆◆

The following notation is useful: If X and Y are two sets, then $X \subseteq Y$ means that X is **contained** in Y; that is, every member of X is a member of Y. Clearly $X = Y$ if and only if $X \subseteq Y$ and $Y \subseteq X$ are both true.

EXAMPLE 17

Let **u** and **v** be two vectors in a vector space V. Show that

$$\text{span}\{\mathbf{u}, \mathbf{v}\} = \text{span}\{\mathbf{u} + \mathbf{v}, \mathbf{u} - \mathbf{v}\}$$

Solution   We have $\text{span}\{\mathbf{u} + \mathbf{v}, \mathbf{u} - \mathbf{v}\} \subseteq \text{span}\{\mathbf{u}, \mathbf{v}\}$ because both $\mathbf{u} + \mathbf{v}$ and $\mathbf{u} - \mathbf{v}$ lie in $\text{span}\{\mathbf{u}, \mathbf{v}\}$. On the other hand,

$$\mathbf{u} = \tfrac{1}{2}(\mathbf{u} + \mathbf{v}) + \tfrac{1}{2}(\mathbf{u} - \mathbf{v})$$
$$\mathbf{v} = \tfrac{1}{2}(\mathbf{u} + \mathbf{v}) - \tfrac{1}{2}(\mathbf{u} - \mathbf{v})$$

so $\text{span}\{\mathbf{u}, \mathbf{v}\} \subseteq \text{span}\{\mathbf{u} + \mathbf{v}, \mathbf{u} - \mathbf{v}\}$ by Theorem 2.

$\blacklozenge\blacklozenge\blacklozenge$

EXERCISES 5.2

1. Which of the following are subspaces of $\mathbb{R}^3$? Support your answer.
   (a) $U = \{(a, b, 1) \mid a \text{ and } b \text{ in } \mathbb{R}\}$
   ◆ (b) $U = \{(a, b, c) \mid a + 2b - c = 0; a, b, \text{ and } c \text{ in } \mathbb{R}\}$
   (c) $U = \{(0, 0, c) \mid c \text{ in } \mathbb{R}\}$
   ◆ (d) $U = \{(a, b, 0) \mid a^2 = b^2, a \text{ and } b \text{ in } \mathbb{R}\}$
   (e) $U = \{(a, a - 1, c) \mid a \text{ and } c \text{ in } \mathbb{R}\}$
   ◆ (f) $U = \{(a, b, c) \mid a^2 + b^2 = c^2; a, b, \text{ and } c \text{ in } \mathbb{R}\}$
   (g) $U = \{(a, a, c) \mid a \text{ and } c \text{ in } \mathbb{R}\}$

2. Which of the following are subspaces of $\mathbf{P}_3$? Support your answer.
   (a) $U = \{f(x) \mid f(2) = 1\}$
   ◆ (b) $U = \{xg(x) \mid g(x) \text{ in } \mathbf{P}_2\}$
   (c) $U = \{xg(x) \mid g(x) \text{ in } \mathbf{P}_3\}$
   ◆ (d) $U = \{xg(x) + (1 - x)\,h(x) \mid g(x) \text{ and } h(x) \text{ in } \mathbf{P}_2\}$
   (e) $U = $ The set of all polynomials in $\mathbf{P}_3$ with constant term 0
   ◆ (f) $U = \{f(x) \mid \deg f(x) = 3\}$

3. Which of the following are subspaces of $\mathbf{M}_{22}$? Support your answer.
   (a) $U = \left\{ \begin{bmatrix} a & b \\ 0 & c \end{bmatrix} \middle| a, b, \text{ and } c, \text{ in } \mathbb{R} \right\}$
   ◆ (b) $U = \left\{ \begin{bmatrix} a & b \\ c & d \end{bmatrix} \middle| a + b = c + d; a, b, c, \text{ and } d \text{ in } \mathbb{R} \right\}$
   (c) $U = \{A \mid A \text{ in } \mathbf{M}_{22}, A = A^T\}$
   ◆ (d) $U = \{A \mid A \text{ in } \mathbf{M}_{22}, AB = 0\}$, B a fixed 2×2 matrix
   (e) $U = \{A \mid A \text{ in } \mathbf{M}_{22}, A^2 = A\}$
   ◆ (f) $U = \{A \mid A \text{ in } \mathbf{M}_{22}, A \text{ is not invertible}\}$

   (g) $U = \{A \mid A \text{ in } \mathbf{M}_{22}, BAC = CAB\}$, B and C fixed 2×2 matrices

4. Which of the following are subspaces of $\mathbf{F}[0, 1]$? Support your answer.
   (a) $U = \{f \mid f(0) = 0\}$
   ◆ (b) $U = \{f \mid f(0) = 1\}$
   (c) $U = \{f \mid f(0) = f(1)\}$
   ◆ (d) $U = \{f \mid f(x) \geq 0 \text{ for all } x \text{ in } [0, 1]\}$
   (e) $U = \{f \mid f(x) = f(y) \text{ for all } x \text{ and } y \text{ in } [0, 1]\}$
   ◆ (f) $U = \{f \mid f(x + y) = f(x) + f(y) \text{ for all } x \text{ and } y \text{ in } [0, 1]\}$
   (g) $U = \left\{ f \mid \int_0^1 f(x)dx = 0 \right\}$

5. Let A be an $m \times n$ matrix. For which columns B in $\mathbb{R}^m$ is $U = \{X \mid X \text{ in } \mathbb{R}^n, AX = B\}$ a subspace of $\mathbb{R}^n$?

6. Let X be a vector in $\mathbb{R}^n$ (written as a column), and define $U = \{AX \mid A \text{ in } \mathbf{M}_{mn}\}$.
   (a) Show that U is a subspace of $\mathbb{R}^m$.
   ◆ (b) Show that $U = \mathbb{R}^m$ if $X \neq \mathbf{0}$.

7. Write each of the following as a linear combination of $x + 1$, $x^2 + x$, and $x^2 + 2$.
   (a) $x^2 + 3x + 2$ ◆ (b) $2x^2 - 3x + 1$
   (c) $x^2 + 1$ ◆ (d) x

8. Write each of the following as a linear combination of $(1, -1, 1)$, $(1, 0, 1)$, and $(1, 1, 0)$.
   (a) $(2, 1, -1)$ ◆ (b) $(1, -7, 5)$
   (c) $\left(\tfrac{1}{2}, 2, \tfrac{1}{3}\right)$ ◆ (d) $(0, 0, 0)$

9. Determine whether **v** lies in $\text{span}\{\mathbf{u}, \mathbf{w}\}$ in each case.
   (a) $\mathbf{v} = (1, -1, 2)$; $\mathbf{u} = (1, 1, 1)$, $\mathbf{w} = (0, 1, 3)$
   ◆ (b) $\mathbf{v} = (3, 1, -3)$; $\mathbf{u} = (1, 1, 1)$, $\mathbf{w} = (0, 1, 3)$

(c) $\mathbf{v} = 3x^2 - 2x - 1; \mathbf{u} = x^2 + 1, \mathbf{w} = x + 2$

(d) $\mathbf{v} = x; \mathbf{u} = x^2 + 1, \mathbf{w} = x + 2$

(e) $\mathbf{v} = \begin{bmatrix} 1 & 3 \\ -1 & 1 \end{bmatrix}; \mathbf{u} = \begin{bmatrix} 1 & -1 \\ 2 & 1 \end{bmatrix}, \mathbf{w} = \begin{bmatrix} 2 & 1 \\ 1 & 0 \end{bmatrix}$

(f) $\mathbf{v} = \begin{bmatrix} 1 & -4 \\ 5 & 3 \end{bmatrix}; \mathbf{u} = \begin{bmatrix} 1 & -1 \\ 2 & 1 \end{bmatrix}, \mathbf{w} = \begin{bmatrix} 2 & 1 \\ 1 & 0 \end{bmatrix}$

10. Which of the following functions lie in span$\{\cos^2 x, \sin^2 x\}$? (Work in $\mathbf{F}[0, \pi]$.)

(a) $\cos 2x$ **(b)** 1

(c) x^2 **(d)** $1 + x^2$

11. **(a)** Show that $\mathbb{R}^3$ is spanned by $\{(1, 0, 1), (1, 1, 0), (0, 1, 1)\}$.

(b) Show that $\mathbf{P}_2$ is spanned by $\{1 + 2x^2, 3x, 1 + x\}$.

(c) Show that $\mathbf{M}_{22}$ is spanned by

$$\left\{ \begin{bmatrix} 1 & 0 \\ 0 & 0 \end{bmatrix}, \begin{bmatrix} 1 & 0 \\ 0 & 1 \end{bmatrix}, \begin{bmatrix} 0 & 1 \\ 1 & 0 \end{bmatrix}, \begin{bmatrix} 1 & 1 \\ 0 & 1 \end{bmatrix} \right\}.$$

12. If X and Y are two sets of vectors in a vector space V, and if $X \subseteq Y$, show that span $X \subseteq$ span Y.

13. Let $\mathbf{u}, \mathbf{v},$ and $\mathbf{w}$ denote vectors in a vector space V.

(a) Show that span$\{\mathbf{u}, \mathbf{v}, \mathbf{w}\} = $ span$\{\mathbf{u} + \mathbf{v}, \mathbf{u} + \mathbf{w}, \mathbf{v} + \mathbf{w}\}$

(b) Show that span$\{\mathbf{u}, \mathbf{v}, \mathbf{w}\} = $ span$\{\mathbf{u} - \mathbf{v}, \mathbf{u} + \mathbf{w}, \mathbf{w}\}$

14. Show that span$\{\mathbf{v}_1, \mathbf{v}_2, \ldots, \mathbf{v}_n, \mathbf{0}\} = $ span$\{\mathbf{v}_1, \mathbf{v}_2, \ldots, \mathbf{v}_n\}$ holds for any set of vectors $\{\mathbf{v}_1, \mathbf{v}_2, \ldots, \mathbf{v}_n\}$.

15. If X and Y are nonempty subsets of a vector space V such that span $X = $ span $Y = V$, must there be a vector common to both X and Y? Justify your answer.

16. Is it possible that $\{(1, 2, 0), (1, 1, 1)\}$ can span the subspace $U = \{(a, b, 0) \mid a \text{ and } b \text{ in } \mathbb{R}\}$?

17. Describe span$\{\mathbf{0}\}$.

18. Let $\mathbf{v}$ denote any vector in a vector space V. Show that span$\{\mathbf{v}\} = $ span$\{a\mathbf{v}\}$ for any $a \neq 0$.

19. Determine all subspaces of $\mathbb{R}\mathbf{v}$ where $\mathbf{v} \neq \mathbf{0}$ in some vector space V.

20. If $\mathbb{R}^3$ is regarded as the set of points in space and if $\mathbf{d} \neq \mathbf{0}$ in $\mathbb{R}^3$, show that span$\{\mathbf{d}\}$ is the line through the origin with direction vector $\mathbf{d}$.

21. If $M_{nn} = $ span$\{A_1, A_2, \ldots, A_k\}$, show that $M_{nn} = $ span $\{A_1^T, A_2^T, \ldots, A_k^T\}$.

22. If $P_n = $ span$\{p_1(x), p_2(x), \ldots, p_k(x)\}$ and a is in $\mathbb{R}$, show that $p_i(a) \neq 0$ for some i.

23. Let U be a subspace of a vector space V.

(a) If $a\mathbf{u}$ is in U where $a \neq 0$, show that $\mathbf{u}$ is in U.

(b) If $\mathbf{u}$ and $\mathbf{u} + \mathbf{v}$ are in U, show that $\mathbf{v}$ is in U.

24. Let U be a nonempty subset of a vector space V. Show that U is a subspace if and only if (2) and (3) in Theorem 1 hold.

25. Let U be a nonempty subset of a vector space V. Show that U is a subspace of V if and only if $\mathbf{u}_1 + a\mathbf{u}_2$ lies in U for all $\mathbf{u}_1$ and $\mathbf{u}_2$ in U and all a in $\mathbb{R}$.

26. Let U be the set in Example 7: $U = \{p(x) \text{ in } \mathbf{P} \mid p(3) = 0\}$. Use the factor theorem to show that U consists of multiples of $x - 3$; that is, show that $U = \{(x - 3)q(x) \mid q(x) \text{ in } \mathbf{P}\}$. Use this to show that U is a subspace of $\mathbf{P}$.

27. Let P denote the set in Example 9. Show that P is a subspace of $\mathbb{R}^3$ by:

(a) Using the description $P = \{(x, y, z) \text{ in } \mathbb{R}^3 \mid ax + by + cz = 0\}$.

(b) Using Theorem 4§4.2 and the description $P = \{\mathbf{v} \text{ in } \mathbb{R}^3 \mid \mathbf{v} \cdot \mathbf{n} = 0\}, \mathbf{n} = (a, b, c)$.

28. Let $A_1, A_2, \ldots, A_m$ denote $n \times n$ matrices. If Y is a nonzero column in $\mathbb{R}^n$ and $A_1 Y = A_2 Y = \cdots = A_m Y = 0$, show that $\{A_1, A_2, \ldots, A_m\}$ cannot span $\mathbf{M}_{nn}$.

29. Let $\{\mathbf{v}_1, \mathbf{v}_2, \ldots, \mathbf{v}_n\}$ and $\{\mathbf{u}_1, \mathbf{u}_2, \ldots, \mathbf{u}_n\}$ be sets of vectors in a vector space; and let

$$X = \begin{bmatrix} \mathbf{v}_1 \\ \vdots \\ \mathbf{v}_n \end{bmatrix} \qquad Y = \begin{bmatrix} \mathbf{u}_1 \\ \vdots \\ \mathbf{u}_n \end{bmatrix}$$

as in Exercise 25§5.1.

(a) Show that span$\{\mathbf{v}_1, \ldots, \mathbf{v}_n\} \subseteq $ span$\{\mathbf{u}_1, \ldots, \mathbf{u}_n\}$ if and only if $AY = X$ for some $n \times n$ matrix A.

(b) If $X = AY$ where A is invertible, show that span$\{\mathbf{v}_1, \ldots, \mathbf{v}_n\} = $ span$\{\mathbf{u}_1, \ldots, \mathbf{u}_n\}$.

30. If U and W are subspaces of a vector space V, let $U \cup W = \{\mathbf{v} \mid \mathbf{v} \text{ is in } U \text{ or } \mathbf{v} \text{ is in } W\}$. Show that $U \cup W$ is a subspace if and only if $U \subseteq W$ or $W \subseteq U$.

31. Show that $\mathbf{P}$ cannot be spanned by a finite set of polynomials.

Section 5.3 Linear Independence and Dimension

Some spanning sets in a vector space V are in some sense better than others. If $V = $ span$\{\mathbf{v}_1, \mathbf{v}_2, \ldots, \mathbf{v}_n\}$, then each vector $\mathbf{v}$ in V can be written as a linear combination of the vectors $\mathbf{v}_1, \mathbf{v}_2, \ldots, \mathbf{v}_n$. This section is devoted to the study of spanning sets having the property that each vector in V has *exactly one* representation as a linear combination of these vectors. It turns out to be enough to insist on this property for the zero vector only.

DEFINITION

A set of vectors $\{\mathbf{v}_1, \mathbf{v}_2, \ldots, \mathbf{v}_n\}$ is called **linearly independent** if it satisfies the following condition:

$$\text{If } s_1\mathbf{v}_1 + s_2\mathbf{v}_2 + \cdots + s_n\mathbf{v}_n = \mathbf{0}, \quad \text{then } s_1 = s_2 = \cdots = s_n = 0.$$

A set of vectors that is not linearly independent is said to be **linearly dependent**.

For example, $\{(1, -1), (1, 1), (2, 1)\}$ is linearly dependent because

$$(1, -1) + 3(1, 1) - 2(2, 1) = (0, 0)$$

Because we refer to linear independence frequently, it is worthwhile to formulate the definition slightly differently. The **trivial linear combination** of the vectors $\mathbf{v}_1, \mathbf{v}_2, \ldots, \mathbf{v}_n$ is the one with every coefficient zero.

$$0\mathbf{v}_1 + 0\mathbf{v}_2 + \cdots + 0\mathbf{v}_n$$

This is obviously one way of expressing $\mathbf{0}$ as a linear combination of these vectors, and they are linearly independent when it is the *only* way.

EXAMPLE 1

Show that $\{(1, 0, -1), (2, 1, 2), (3, -2, 0)\}$ is linearly independent in $\mathbb{R}^3$.

Solution   Suppose that a linear combination of these vectors gives zero.

$$s_1(1, 0, -1) + s_2(2, 1, 2), + s_3(3, -2, 0) = (0, 0, 0)$$

We must show that it is the trivial combination — that is, that $s_1 = s_2 = s_3 = 0$. Equating components gives

$$
\begin{aligned}
s_1 + 2s_2 + 3s_3 &= 0 \\
s_2 - 2s_3 &= 0 \\
-s_1 + 2s_2 \quad\quad &= 0
\end{aligned}
$$

These equations have only the trivial solution $s_1 = s_2 = s_3 = 0$.

◆◆◆

EXAMPLE 2

Show that $\{1 + x, \; 3x + x^2, \; 2 + x - x^2\}$ is linearly independent in $\mathbf{P}_2$.

Solution   Suppose a linear combination of these polynomials vanishes.

$$s_1(1 + x) + s_2(3x + x^2) + s_3(2 + x - x^2) = 0$$

Equating the coefficients of 1, x, and x^2 gives a set of linear equations.

$$s_1 + + 2s_3 = 0$$
$$s_1 + 3s_2 + s_3 = 0$$
$$s_2 - s_3 = 0$$

Here again the only solution is $s_1 = s_2 = s_3 = 0$.   ◆◆◆

EXAMPLE 3

Show that $\{\sin\ x, \cos\ x\}$ is linearly independent in the vector space $\mathbf{F}[0, 2\pi]$ of functions defined on the interval $[0, 2\pi]$.

Solution   Suppose that a linear combination of these functions vanishes.

$$s_1(\sin\ x) + s_2(\cos x) = 0$$

This must hold for *all* values of x (by the definition of equality in $\mathbf{F}[0, 2\pi]$). Taking $x = 0$ yields $s_2 = 0$ (because $\sin 0 = 0$ and $\cos 0 = 1$). Similarly, $s_1 = 0$ follows from taking $x = \frac{\pi}{2}$ (because $\sin \frac{\pi}{2} = 1$ and $\cos \frac{\pi}{2} = 0$).   ◆◆◆

EXAMPLE 4

Suppose that $\{\mathbf{u}, \mathbf{v}\}$ is a linearly independent set in a vector space V. Show that $\{\mathbf{u} + \mathbf{v}, \mathbf{u} - \mathbf{v}\}$ is also linearly independent.

Solution   Suppose a linear combination of $\mathbf{u} + \mathbf{v}$ and $\mathbf{u} - \mathbf{v}$ vanishes.

$$s(\mathbf{u} + \mathbf{v}) + t(\mathbf{u} - \mathbf{v}) = \mathbf{0}$$

We must deduce that $s = t = 0$. Collecting coefficients of $\mathbf{u}$ and $\mathbf{v}$ gives

$$(s + t)\mathbf{u} + (s - t)\mathbf{v} = \mathbf{0}$$

Now this is a linear combination of $\mathbf{u}$ and $\mathbf{v}$ that vanishes; so, because $\{\mathbf{u}, \mathbf{v}\}$ is linearly independent, all the coefficients must be zero. This yields linear equations $s + t = 0$ and $s - t = 0$, and the only solution is $s = t = 0$.   ◆◆◆

EXAMPLE 5

If $\mathbf{v} \neq \mathbf{0}$, the set $\{\mathbf{v}\}$ consisting of the single vector $\mathbf{v}$ is linearly independent.

Solution   The linear combinations in this case are just the scalar multiples $s\mathbf{v}$, s in $\mathbb{R}$. If $s\mathbf{v} = \mathbf{0}$, then $s = 0$ by Theorem 3§5.1 (because $\mathbf{v} \neq \mathbf{0}$). This shows that $\{\mathbf{v}\}$ is linearly independent.   ◆◆◆

EXAMPLE 6

If $\{\mathbf{v}_1, \mathbf{v}_2, \ldots, \mathbf{v}_n\}$ is linearly independent in a vector space V, show that $\{a_1\mathbf{v}_1, a_2\mathbf{v}_2, \ldots, a_n\mathbf{v}_n\}$ is also linearly independent, provided that the numbers $a_1, a_2, \ldots, a_n$ are all nonzero.

Solution   Suppose a linear combination of the new set vanishes;

$$s_1(a_1\mathbf{v}_1) + s_2(a_2\mathbf{v}_2) + \cdots + s_n(a_n\mathbf{v}_n) = \mathbf{0}$$

where $s_1, s_2, \ldots, s_n$ lie in $\mathbb{R}$. Then $s_1 a_1 = s_2 a_2 = \cdots = s_n a_n = 0$ by the linear independence of $\{\mathbf{v}_1, \ldots, \mathbf{v}_n\}$. The fact that each a_i is nonzero now implies that $s_1 = s_2 = \cdots = s_n = 0$. ◆◆◆

EXAMPLE 7

Show that no linearly independent set of vectors can contain the zero vector.

Solution   The set $\{\mathbf{0}, \mathbf{v}_1, \mathbf{v}_2, \ldots, \mathbf{v}_n\}$ cannot be linearly independent because $1\mathbf{0} + 0\mathbf{v}_1 + 0\mathbf{v}_2 + 0\mathbf{v}_3 + \cdots + 0\mathbf{v}_n = \mathbf{0}$ is a nontrivial linear combination that vanishes.

◆◆◆

The following is a convenient test for linear dependence.

THEOREM 1

A set $\{\mathbf{v}_1, \mathbf{v}_2, \ldots, \mathbf{v}_n\}$ of vectors in a vector space V is linearly dependent if and only if some $\mathbf{v}_i$ is a linear combination of the others.

Proof   Assume that $\{\mathbf{v}_1, \mathbf{v}_2, \ldots, \mathbf{v}_n\}$ is linearly dependent. Then some nontrivial linear combination vanishes—say, $a_1\mathbf{v}_1 + a_2\mathbf{v}_2 + \cdots + a_n\mathbf{v}_n = \mathbf{0}$ where some coefficient is not zero. Suppose $a_1 \neq 0$. Then $\mathbf{v}_1 = (-a_2/a_1)\mathbf{v}_2 + \cdots + (-a_n/a_1)\mathbf{v}_n$ gives $\mathbf{v}_1$ as a linear combination of the others. In general, if $a_i \neq 0$ then a similar argument expresses $\mathbf{v}_i$ as a linear combination of the others.

Conversely, suppose one of the vectors is a linear combination of the others—say, $\mathbf{v}_1 = a_2\mathbf{v}_2 + \cdots + a_n\mathbf{v}_n$. Then the nontrivial linear combination $1\mathbf{v}_1 - a_2\mathbf{v}_2 - \cdots - a_n\mathbf{v}_n$ equals zero, so the set $\{\mathbf{v}_1, \ldots, \mathbf{v}_n\}$ is not linearly independent; that is, it is linearly dependent. A similar argument works if *any* $\mathbf{v}_i$ is a linear combination of the others. ◆

Theorem 1 has a geometric interpretation in $\mathbb{R}^3$. Let $\mathbf{u}$ and $\mathbf{v}$ be nonzero vectors in $\mathbb{R}^3$ with their initial points at their origin (see Chapter 4). Theorem 1 shows that the set $\{\mathbf{u}, \mathbf{v}\}$ is linearly dependent if and only if one of them is a scalar multiple of the other; that is, if and only if they are parallel. Hence the set $\{\mathbf{u}, \mathbf{v}\}$ is linearly independent if and only if $\mathbf{u}$ and $\mathbf{v}$ are not parallel and, in this case, span $\{\mathbf{u}, \mathbf{v}\}$ is the plane through the origin containing $\mathbf{u}$ and $\mathbf{v}$.

Similarly, the set $\{\mathbf{u}, \mathbf{v}, \mathbf{w}\}$ is linearly dependent if and only if one of them is in the subspace (plane or line through the origin) spanned by the others. Hence the set $\{\mathbf{u}, \mathbf{v}, \mathbf{w}\}$ in Figure 5.1 is linearly independent because $\mathbf{w}$ is not in the plane containing $\mathbf{u}$ and $\mathbf{v}$.

The notion of independence was motivated by the insistence that the set of vectors in question be such that linear combinations have uniquely determined coefficients. However, the definition of linear independence requires only that linear

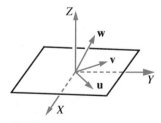

FIGURE 5.1

combinations *equaling zero* have uniquely determined coefficients (necessarily all zero). Theorem 2 asserts that linearly independent sets have the stronger uniqueness property.

THEOREM 2

Let $\{\mathbf{v}_1, \mathbf{v}_2, \ldots, \mathbf{v}_n\}$ be a linearly independent set of vectors in a vector space V. If a vector $\mathbf{v}$ has two (ostensibly different) representations

$$\mathbf{v} = s_1\mathbf{v}_1 + s_2\mathbf{v}_2 + \cdots + s_n\mathbf{v}_n$$
$$\mathbf{v} = t_1\mathbf{v}_1 + t_2\mathbf{v}_2 + \cdots + t_n\mathbf{v}_n$$

as linear combinations of these vectors then $s_1 = t_1,\ s_2 = t_2, \ldots, s_n = t_n$.

Proof We have $s_1\mathbf{v}_1 + s_2\mathbf{v}_2 + \cdots + s_n\mathbf{v}_n = t_1\mathbf{v}_1 + t_2\mathbf{v}_2 + \cdots + t_n\mathbf{v}_n$ (because both sides equal $\mathbf{v}$); so, taking everything to the left side, we get

$$(s_1 - t_1)\,\mathbf{v}_1 + (s_2 - t_2)\,\mathbf{v}_2 + \cdots + (s_n - t_n)\mathbf{v}_n = \mathbf{0}$$

All the coefficients are zero by the independence of the $\mathbf{v}_i$, so $s_1 = t_1,\ s_2 = t_2, \ldots,$ $s_n = t_n$, as required.   ◆

The following theorem uses a simple fact about linear equations (Theorem 1§3.1) to prove a basic result about linear independence: The number of vectors in an independent set can never exceed the number in a spanning set.

THEOREM 3

Fundamental Theorem

Suppose a vector space V can be spanned by n vectors. If any set of m vectors in V is linearly independent, then $m \leq n$.

Proof Let $V = \text{span}\{\mathbf{v}_1, \mathbf{v}_2, \ldots, \mathbf{v}_n\}$. We must show that every set $\{\mathbf{u}_1, \mathbf{u}_2, \ldots, \mathbf{u}_m\}$ of vectors in V with $m > n$ *fails* to be linearly independent. This is accomplished by showing that numbers $x_1, x_2, \ldots, x_m$ can be found, not all zero, such that

$$\sum_{j=1}^{m} x_j\,\mathbf{u}_j = x_1\mathbf{u}_1 + x_2\mathbf{u}_2 + \cdots + x_m\mathbf{u}_m = \mathbf{0}$$

Because V is spanned by the vectors $\mathbf{v}_1, \mathbf{v}_2, \ldots, \mathbf{v}_n$, each vector $\mathbf{u}_j$ can be expressed as a linear combination of the $\mathbf{v}_i$:

$$\mathbf{u}_j = a_{1j}\mathbf{v}_1 + a_{2j}\mathbf{v}_2 + \cdots + a_{nj}\mathbf{v}_n = \sum_{i=1}^{n} a_{ij}\mathbf{v}_i$$

Substituting these expressions into the preceding equation gives

$$0 = \sum_{j=1}^{m} x_j \left(\sum_{i=1}^{n} a_{ij}\mathbf{v}_i \right) = \sum_{i=1}^{n} \left(\sum_{j=1}^{m} a_{ij}x_j \right) \mathbf{v}_i$$

This is certainly the case if each coefficient of $\mathbf{v}_i$ is zero—that is, if

$$\sum_{j=1}^{m} a_{ij}x_j = 0 \qquad \text{for } i = 1, 2, \ldots, n$$

But this is a system of n equations in the m variables $x_1, x_2, \ldots, x_m$; so, because $m > n$, it has a nontrivial solution by Theorem 1§1.3. This is what we wanted.   ◆

We now come to a very important definition.

DEFINITION

A set $\{\mathbf{e}_1, \mathbf{e}_2, \ldots, \mathbf{e}_n\}$ of vectors in a vector space V is called a **basis** of V if it satisfies the following two conditions:

1. $\{\mathbf{e}_1, \mathbf{e}_2, \ldots, \mathbf{e}_n\}$ is linearly independent
2. $V = \text{span}\{\mathbf{e}_1, \mathbf{e}_2, \ldots, \mathbf{e}_n\}$

Thus if a set of vectors $\{\mathbf{e}_1, \mathbf{e}_2, \ldots, \mathbf{e}_n\}$ is a basis, then *every* vector in V can be written as a linear combination of these vectors in a *unique* way (Theorem 2). But even more is true: Any two (finite) bases of V contain the same number of vectors.

THEOREM 4

Let $\{\mathbf{e}_1, \mathbf{e}_2, \ldots, \mathbf{e}_n\}$ and $\{\mathbf{f}_1, \mathbf{f}_2, \ldots, \mathbf{f}_m\}$ be two bases of a vector space V. Then $n = m$.

Proof   Because $V = \text{span}\{\mathbf{e}_1, \mathbf{e}_2, \ldots, \mathbf{e}_n\}$, it follows from Theorem 3 that $m \leq n$. Similarly $n \leq m$, so $n = m$, as asserted.   ◆

Theorem 4 guarantees that no matter which basis of V is chosen it contains the same number of vectors as any other basis. Hence there is no ambiguity about the following definition.

> **DEFINITION**
>
> If $\{\mathbf{e}_1, \mathbf{e}_2, \ldots, \mathbf{e}_n\}$ is a basis of the nonzero vector space V, the number of n of vectors in the basis is called the **dimension** of V, and we write
>
> $$\dim V = n$$
>
> The zero vector space $V = 0$ is defined to have dimension 0:
>
> $$\dim 0 = 0$$
>
> A vector space V is called **finite dimensional** if $V = 0$ or V has a finite basis.

In the discussion of bases to this point, we have tacitly assumed that a basis is nonempty and hence that the dimension of the space is at least 1. On the other hand, the zero space 0, consisting of the zero vector alone, has *no* basis (Example 7), so our insistence that $\dim 0 = 0$ amounts to saying that the empty set of vectors is a basis of the zero space.

EXAMPLE 8

Show that $\dim \mathbb{R}^n = n$ and that $\{\mathbf{e}_1, \mathbf{e}_2, \ldots, \mathbf{e}_n\}$ is a basis where

$$\mathbf{e}_1 = (1, 0, 0, \ldots, 0), \mathbf{e}_2 = (0, 1, 0, \ldots, 0), \ldots, \mathbf{e}_n = (0, 0, 0, \ldots, 1)$$

This basis is called the **standard basis** for $\mathbb{R}^n$.

Solution $\mathbb{R}^n = \mathrm{span}\{\mathbf{e}_1, \mathbf{e}_2, \ldots, \mathbf{e}_n\}$ because

$$(a_1, a_2, \ldots, a_n) = a_1\mathbf{e}_1 + a_2\mathbf{e}_2 + \cdots + a_n\mathbf{e}_n$$

holds for all vectors $(a_1, a_2, \ldots, a_n)$ in $\mathbb{R}^n$. But this also shows that the vectors $\mathbf{e}_1, \mathbf{e}_2, \ldots, \mathbf{e}_n$ are linearly independent because $a_1\mathbf{e}_1 + \cdots + a_n\mathbf{e}_n = \mathbf{0}$ implies that $a_1 = \cdots = a_n = 0$. Hence $\{\mathbf{e}_1, \mathbf{e}_2, \ldots, \mathbf{e}_n\}$ is a basis of $\mathbb{R}^n$, so $\dim \mathbb{R}^n = n$. ◆◆◆

Similar considerations apply to the space of all $m \times n$ matrices.

EXAMPLE 9

The space $\mathbf{M}_{mn}$ has dimension mn, and one basis consists of all $m \times n$ matrices with exactly one entry equal to 1 and all other entries equal to 0. ◆◆◆

EXAMPLE 10

Show that $\dim \mathbf{P}_n = n + 1$ and that $\{1, x, x^2, \ldots, x^n\}$ is a basis.

Solution   Each polynomial $p(x) = a_0 + a_1x + \cdots + a_nx^n$ in $\mathbf{P}_n$ is clearly a linear combination of $1, x, \ldots, x^n$, so $\mathbf{P}_n = \mathrm{span}\{1, x, \ldots, x^n\}$. On the other hand, if a linear combination of these vectors vanishes, $a_0 1 + a_1 x + \cdots + a_n x^n = 0$, then $a_0 = a_1 = \cdots = a_n = 0$ because x is an indeterminate. So $\{1, x, \ldots, x^n\}$ is linearly independent and hence is a basis.

◆◆◆

Example 10 shows that the space **P** of *all* polynomials cannot be finite dimensional. In fact, if dim **P** $= n$, then **P** would be spanned by n polynomials. But the fact that $\{1, x, x^2, \ldots, x^n\}$ are $n + 1$ linearly independent vectors in **P** would then contradict the fundamental theorem.

EXAMPLE 11

If $\mathbf{v} \neq \mathbf{0}$ is any nonzero vector in a vector space V, show that span$\{\mathbf{v}\} = \mathbb{R}\mathbf{v}$ has dimension 1.

Solution $\{\mathbf{v}\}$ clearly spans $\mathbb{R}\mathbf{v}$, and it is linearly independent by Example 5. Hence $\{\mathbf{v}\}$ is a basis of $\mathbb{R}\mathbf{v}$, and so dim $\mathbb{R}\mathbf{v} = 1$.

◆◆◆

EXAMPLE 12

As in Example 6§5.2, let $A = \begin{bmatrix} 1 & 1 \\ 0 & 0 \end{bmatrix}$ and consider the subspace

$$U = \{X \text{ in } \mathbf{M}_{22} \mid AX = XA\}$$

of $\mathbf{M}_{22}$. Show that dim $U = 2$ and find a basis of U.

Solution   It was shown in Example 6§5.2 that

$$U = \left\{ \begin{bmatrix} y + w & y \\ 0 & w \end{bmatrix} \;\middle|\; y \text{ and } w \text{ in } \mathbb{R} \right\}.$$

Hence each matrix X in U can be written

$$X = \begin{bmatrix} y + w & y \\ 0 & w \end{bmatrix} = y \begin{bmatrix} 1 & 1 \\ 0 & 0 \end{bmatrix} + w \begin{bmatrix} 1 & 0 \\ 0 & 1 \end{bmatrix}$$

so $U = $ span B where $B = \left\{ \begin{bmatrix} 1 & 1 \\ 0 & 0 \end{bmatrix}, \begin{bmatrix} 1 & 0 \\ 0 & 1 \end{bmatrix} \right\}$. Moreover, the set B is linearly independent (verify this), so it is a basis of U and dim $U = 2$.

◆◆◆

EXAMPLE 13

Show that the set V of all symmetric 2×2 matrices is a vector space, and find the dimension of V.

Solution   A matrix A is symmetric if $A^T = A$. If A and B lie in V, then

$$(A + B)^T = A^T + B^T = A + B$$

$$(kA)^T = kA^T = kA$$

using Theorem 2§2.1; so $A + B$ and kA are also symmetric. This shows that V is a vector space (being a subspace of $\mathbf{M}_{22}$). Now a matrix A is symmetric when entries directly across the main diagonal are equal, so each 2×2 symmetric matrix has the form

$$\begin{bmatrix} a & c \\ c & b \end{bmatrix} = a \begin{bmatrix} 1 & 0 \\ 0 & 0 \end{bmatrix} + b \begin{bmatrix} 0 & 0 \\ 0 & 1 \end{bmatrix} + c \begin{bmatrix} 0 & 1 \\ 1 & 0 \end{bmatrix}$$

Hence the set $B = \left\{ \begin{bmatrix} 1 & 0 \\ 0 & 0 \end{bmatrix}, \begin{bmatrix} 0 & 0 \\ 0 & 1 \end{bmatrix}, \begin{bmatrix} 0 & 1 \\ 1 & 0 \end{bmatrix} \right\}$ spans V, and the reader can verify that

B is linearly independent. Thus B is a basis of V, so dim $V = 3$.

◆◆◆

The fundamental theorem takes the following useful form when stated for vector spaces of dimension n.

THEOREM 5

Let V be a vector space and assume that dim $V = n > 0$.

1. No set of more than n vectors in V can be linearly independent.

2. No set of fewer than n vectors can span V.

Proof V can be spanned by n vectors (any basis), so (1) restates the fundamental theorem. But the n basis vectors are also linearly independent, so no spanning set can have fewer than n vectors, again by Theorem 3. This gives (2).   ◆

Hence any set of more than n vectors in $\mathbb{R}^n$ must be linearly dependent by (1) of Theorem 5. Here is another application.

EXAMPLE 14

Let A denote an $n \times n$ matrix. Then there exist $n^2 + 1$ real numbers a_0, a_1, a_2, $\ldots$, a_{n^2}, not all zero, such that

$$a_0 I + a_1 A + a_2 A^2 + \cdots + a_{n^2} A^{n^2} = 0$$

where I denotes the $n \times n$ identity matrix.

Solution   The space $\mathbf{M}_{nn}$ of all $n \times n$ matrices has dimension n^2 by Example 9. Hence the $n^2 + 1$ matrices I, A, $\ldots$, A^{n^2} cannot be linearly independent by property 1 of

Theorem 5, so a nontrivial linear combination must vanish. This is the desired conclusion.

◆◆◆

We note in passing that the result in Example 14 can be written as $f(A) = 0$ where $f(x) = a_0 + a_1x + \cdots + a_{n^2}x^{n^2}$. In other words, A satisfies a nonzero polynomial of degree n^2. In fact, we know that A satisfies a nonzero polynomial of degree n (this is the Cayley–Hamilton theorem; see Theorem 2§7.7), but the brevity of the solution in Example 14 is an indication of the power of these methods.

EXERCISES 5.3

1. Show that each of the following sets of vectors is linearly independent.

(a) $\{(1, -1), (2, 0)\}$ in $\mathbb{R}^2$

◆ (b) $\{(1, 2), (-1, 1)\}$ in $\mathbb{R}^2$

(c) $\{(1, -1, 0), (0, -1, 2), (2, 1, 1)\}$ in $\mathbb{R}^3$

◆ (d) $\{(1, 1, 1), (0, 1, 1), (0, 0, 1)\}$ in $\mathbb{R}^3$

(e) $\{(1 + x, 1 - x, x + x^2)\}$ in $\mathbf{P}_2$

◆ (f) $\{x^2, x + 1, 1 - x - x^2\}$ in $\mathbf{P}_2$

(g) $\left\{ \begin{bmatrix} 1 & 1 \\ 0 & 0 \end{bmatrix}, \begin{bmatrix} 1 & 0 \\ 1 & 0 \end{bmatrix}, \begin{bmatrix} 0 & 0 \\ 1 & -1 \end{bmatrix}, \begin{bmatrix} 0 & 1 \\ 0 & 1 \end{bmatrix} \right\}$ in $\mathbf{M}_{22}$

◆ (h) $\left\{ \begin{bmatrix} 1 & 1 \\ 1 & 0 \end{bmatrix}, \begin{bmatrix} 0 & 1 \\ 1 & 1 \end{bmatrix}, \begin{bmatrix} 1 & 0 \\ 1 & 1 \end{bmatrix}, \begin{bmatrix} 1 & 1 \\ 0 & 1 \end{bmatrix} \right\}$ in $\mathbf{M}_{22}$

2. Which of the following subsets of V are linearly independent?

(a) $V = \mathbb{R}^3$; $\{(1, -1, 0), (3, 2, -1), (3, 5, -2)\}$

◆ (b) $V = \mathbb{R}^3$; $\{(1, 1, 1), (1, -1, 1), (0, 0, 1)\}$

(c) $V = \mathbb{R}^4$; $\{(1, -1, 1, -1), (2, 0, 1, 0), (0, -2, 1, -2)\}$

◆ (d) $V = \mathbb{R}^4$; $\{(1, 1, 0, 0), (1, 0, 1, 0), (0, 0, 1, 1), (0, 1, 0, 1)\}$

(e) $V = \mathbf{P}_2$; $\{x^2 + 1, x + 1, x\}$

◆ (f) $V = \mathbf{P}_2$; $\{x^2 - x + 3, 2x^2 + x + 5, x^2 + 5x + 1\}$

(g) $V = \mathbf{M}_{22}$; $\left\{ \begin{bmatrix} 1 & 1 \\ 0 & 1 \end{bmatrix}, \begin{bmatrix} 1 & 0 \\ 1 & 1 \end{bmatrix}, \begin{bmatrix} 1 & 0 \\ 0 & 1 \end{bmatrix} \right\}$

◆ (h) $V = \mathbf{M}_{22}$; $\left\{ \begin{bmatrix} -1 & 0 \\ 0 & -1 \end{bmatrix}, \begin{bmatrix} 1 & -1 \\ -1 & 1 \end{bmatrix}, \begin{bmatrix} 1 & 1 \\ 1 & 1 \end{bmatrix}, \begin{bmatrix} 0 & -1 \\ -1 & 0 \end{bmatrix} \right\}$

(i) $V = \mathbf{F}[1, 2]$; $\left\{ \dfrac{1}{x}, \dfrac{1}{x^2}, \dfrac{1}{x^3} \right\}$.

◆ (j) $V = \mathbf{F}[0, 1]$; $\left\{ \dfrac{1}{x^2 + x - 6}, \dfrac{1}{x^2 - 5x + 6}, \dfrac{1}{x^2 - 9} \right\}$

3. Which of the following are linearly independent in $\mathbf{F}[0, 2\pi]$?

(a) $\{\sin^2 x, \cos^2 x\}$

◆ (b) $\{1, \sin^2 x, \cos^2 x\}$

(c) $\{x, \sin^2 x, \cos^2 x\}$

4. Find all values of x such that the following are linearly independent in $\mathbb{R}^3$.

(a) $\{(1, -1, 0), (x, 1, 0), (0, 2, 3)\}$

◆ (b) $\{(2, x, 1), (1, 0, 1), (0, 1, 3)\}$

5. Show that the following are bases of the space V indicated.

(a) $\{(1, 1, 0), (1, 0, 1), (0, 1, 1)\}$; $V = \mathbb{R}^3$

◆ (b) $\{(-1, 1, 1), (1, -1, 1), (1, 1, -1)\}$; $V = \mathbb{R}^3$

(c) $\left\{ \begin{bmatrix} 1 & 0 \\ 0 & 1 \end{bmatrix}, \begin{bmatrix} 0 & 1 \\ 1 & 0 \end{bmatrix}, \begin{bmatrix} 1 & 1 \\ 0 & 1 \end{bmatrix}, \begin{bmatrix} 1 & 0 \\ 0 & 0 \end{bmatrix} \right\}$; $V = \mathbf{M}_{22}$

◆ (d) $\{1 + x, x + x^2, x^2 + x^3, x^3\}$; $V = \mathbf{P}_3$

6. Exhibit a basis and calculate the dimension of each of the following subspaces of $\mathbb{R}^4$.

(a) $\{(a, a + b, a - b, b) \mid a \text{ and } b \text{ in } \mathbb{R}\}$

◆ (b) $\{(a + b, a - b, a, b) \mid a \text{ and } b \text{ in } \mathbb{R}\}$

(c) $\{(a, b, c + a, c) \mid a, b, \text{ and } c \text{ in } \mathbb{R}\}$

◆ (d) $\{(a - b, b + c, a, b + c) \mid a, b, \text{ and } c \text{ in } \mathbb{R}\}$

(e) $\{(a, b, c, d) \mid a + b + c + d = 0\}$

◆ (f) $\{(a, b, c, d) \mid a + b = c + d\}$

(g) $\{X \text{ in } \mathbb{R}^4 \mid XA = 0\}$ where $A = \begin{bmatrix} 1 & 1 & 2 & 1 \\ 1 & -1 & 0 & 2 \end{bmatrix}^T$

◆ (h) $\{X \text{ in } \mathbb{R}^4 \mid XA = 0\}$ where $A = \begin{bmatrix} -1 & 2 & 1 & 2 \\ 1 & 1 & 0 & 3 \end{bmatrix}^T$

7. Exhibit a basis and calculate the dimension of each of the following subspaces of $\mathbf{P}_2$.

(a) $\{a(1 + x) + b(x + x^2) \mid a \text{ and } b \text{ in } \mathbb{R}\}$

◆ (b) $\{a + b(x + x^2) \mid a \text{ and } b \text{ in } \mathbb{R}\}$

(c) $\{p(x) \mid p(1) = 0\}$

◆ (d) $\{p(x) \mid p(x) = p(-x)\}$

8. Exhibit a basis and calculate the dimension of each of the following subspaces of $\mathbf{M}_{22}$.

(a) $\{A \mid A^T = -A\}$

◆ (b) $\left\{ A \mid A \begin{bmatrix} 1 & 1 \\ -1 & 0 \end{bmatrix} = \begin{bmatrix} 1 & 1 \\ -1 & 0 \end{bmatrix} A \right\}$

(c) $\left\{ A \mid A \begin{bmatrix} 1 & 0 \\ -1 & 0 \end{bmatrix} = \begin{bmatrix} 0 & 0 \\ 0 & 0 \end{bmatrix} \right\}$

◆ (d) $\left\{ A \mid A \begin{bmatrix} 1 & 1 \\ -1 & 0 \end{bmatrix} = \begin{bmatrix} 0 & 1 \\ -1 & 1 \end{bmatrix} A \right\}$

9. Let $A = \begin{bmatrix} 1 & 1 \\ 0 & 0 \end{bmatrix}$ and define $U = \{X \mid X$ is in $\mathbf{M}_{22}$ and $AX = X\}$.

(a) Find a basis of U containing A.

◆ (b) Find a basis of U not containing A.

10. In each case, find a basis of the solution space.

(a) $\begin{bmatrix} x \\ y \\ z \end{bmatrix}$ $\begin{aligned} 2x - y + z &= 0 \\ x + 2y - z &= 0 \\ x - 3y + 2z &= 0 \end{aligned}$

◆ (b) $\begin{bmatrix} x \\ y \\ z \end{bmatrix}$ $\begin{aligned} 3x - y + z &= 0 \\ x + 3y - 2z &= 0 \\ 2x - 4y + 3z &= 0 \end{aligned}$

(c) $\begin{bmatrix} r \\ s \\ t \\ u \end{bmatrix}$ $\begin{aligned} 3r - s \quad\;\; + 2u &= 0 \\ r + s + t \quad\;\; &= 0 \\ 2r - 2s - t + 2u &= 0 \\ r - 3s - 2t + 2u &= 0 \end{aligned}$

◆ (d) $\begin{bmatrix} r \\ s \\ t \\ u \end{bmatrix}$ $\begin{aligned} r + s - t + 2u &= 0 \\ 3r - s + 2t - u &= 0 \\ r - 3s + 4t - 5u &= 0 \\ 5r - 3s + 5t - 4u &= 0 \end{aligned}$

11. (a) Let V denote the set of all 2×2 matrices with equal column sums. Show that V is a subspace of $\mathbf{M}_{22}$, and compute dim V.

◆ (b) Repeat part (a) for 3×3 matrices.

◆ (c) Repeat part (a) for $n \times n$ matrices.

12. (a) Let $V = \{(x^2 + x + 1)p(x) \mid p(x)$ in $\mathbf{P}_2\}$. Show that V is a subspace of $\mathbf{P}_4$ and find dim V. [*Hint:* If $f(x) g(x) = 0$ in $\mathbf{P}$ then $f(x) = 0$ or $g(x) = 0$.]

◆ (b) Repeat with $V = \{(x^2 - x)p(x) \mid p(x)$ in $\mathbf{P}_3\}$, a subset of $\mathbf{P}_5$.

(c) Generalize.

13. In each case, either prove the assertion or give an example showing that it is false.

(a) Every set of four nonzero polynomials in $\mathbf{P}_3$ is a basis.

◆ (b) $\mathbf{P}_2$ has a basis of polynomials $f(x)$ such that $f(0) = 0$.

(c) $\mathbf{P}_2$ has a basis of polynomials $f(x)$ such that $f(0) = 1$.

◆ (d) Every basis of $\mathbf{M}_{22}$ contains a noninvertible matrix.

(e) No linearly independent subset of $\mathbf{M}_{22}$ contains a matrix A with $A^2 = 0$.

◆ (f) If $\{\mathbf{u}, \mathbf{v}, \mathbf{w}\}$ is linearly independent, then $a\mathbf{u} + b\mathbf{v} + c\mathbf{w} = \mathbf{0}$ for some a, b, c.

(g) $\{\mathbf{u}, \mathbf{v}, \mathbf{w}\}$ is linearly independent if $a\mathbf{u} + b\mathbf{v} + c\mathbf{w} = \mathbf{0}$ for some a, b, c.

◆ (h) If $\{\mathbf{u}, \mathbf{v}\}$ is linearly independent, so is $\{\mathbf{u}, \mathbf{u} + \mathbf{v}\}$.

(i) If $\{\mathbf{u}, \mathbf{v}\}$ is linearly independent, so is $\{\mathbf{u}, \mathbf{v}, \mathbf{u} + \mathbf{v}\}$.

◆ (j) If $\{\mathbf{u}, \mathbf{v}, \mathbf{w}\}$ is linearly independent, so is $\{\mathbf{u}, \mathbf{v}\}$.

(k) If $\{\mathbf{u}, \mathbf{v}, \mathbf{w}\}$ is linearly dependent, so is $\{\mathbf{u}, \mathbf{v}\}$.

14. Let $A \neq 0$ and $B \neq 0$ be $n \times n$ matrices, and assume that A is symmetric and B is skew-symmetric (that is, $B^T = -B$). Show that $\{A, B\}$ is linearly independent.

15. Show that every set of vectors containing a linearly dependent set is again linearly dependent.

◆ **16.** Show that every nonempty subset of a linearly independent set of vectors is again linearly independent.

17. Let f and g be functions on $[a, b]$, and assume that $f(a) = 1 = g(b)$ and $f(b) = 0 = g(a)$. Show that $\{f, g\}$ is linearly independent in $\mathbf{F}[a, b]$.

18. Let $\{A_1, A_2, \ldots, A_k\}$ be linearly independent in $\mathbf{M}_{mn}$, and suppose that U and V are invertible matrices of size $m \times m$ and $n \times n$, respectively. Show that $\{UA_1V, UA_2V, \ldots, UA_kV\}$ is linearly independent.

19. Let A be $m \times n$ and let $B_1, \ldots, B_n$ be columns in $\mathbb{R}^n$ such that $AX_i = B_i$ has a solution X_i for each i. If $\{B_1, \ldots, B_n\}$ is linearly independent, show that $\{X_1, \ldots, X_n\}$ is also linearly independent.

20. (a) Show that $\{\mathbf{v}, \mathbf{w}\}$ is linearly independent if and only if neither $\mathbf{v}$ nor $\mathbf{w}$ is a scalar multiple of the other.

◆ (b) If $\mathbf{v}, \mathbf{w}$ are nonzero vectors in $\mathbb{R}^3$, show that $\mathbf{v}$ and $\mathbf{w}$ are linearly independent if and only if the lines through the origin with $\mathbf{v}$ and $\mathbf{w}$ as direction vectors are not parallel [see Chapter 4].

21. If $\mathbf{u} = (a, b)$ and $\mathbf{v} = (c, d)$, show that $\{\mathbf{u}, \mathbf{v}\}$ is linearly independent if and only if $\begin{bmatrix} a & b \\ c & d \end{bmatrix}$ is invertible. [*Hint:* Theorem 4§2.4.]

◆ **22.** Assume that $\{\mathbf{u}, \mathbf{v}\}$ is linearly independent in a vector space V. Write $\mathbf{u}' = a\mathbf{u} + b\mathbf{v}$ and $\mathbf{v}' = c\mathbf{u} + d\mathbf{v}$, where $a, b, c,$ and d are numbers. Show that $\{\mathbf{u}', \mathbf{v}'\}$ is linearly independent if and only if the matrix $\begin{bmatrix} a & b \\ c & d \end{bmatrix}$ is invertible. [*Hint:* Theorem 4§2.4.]

23. If $\{\mathbf{v}_1, \mathbf{v}_2, \ldots, \mathbf{v}_k\}$ is linearly independent and $\mathbf{w}$ is not in span$\{\mathbf{v}_1, \ldots, \mathbf{v}_k\}$, show that:

(a) $\{\mathbf{w}, \mathbf{v}_1, \mathbf{v}_2, \ldots, \mathbf{v}_k\}$ is linearly independent.

(b) $\{\mathbf{v}_1 + \mathbf{w}, \ldots, \mathbf{v}_k + \mathbf{w}\}$ is linearly independent.

24. If $\{\mathbf{v}_1, \mathbf{v}_2, \ldots, \mathbf{v}_k\}$ is linearly independent, show that $\{\mathbf{v}_1, \mathbf{v}_1 + \mathbf{v}_2, \ldots, \mathbf{v}_1 + \mathbf{v}_2 + \cdots + \mathbf{v}_k\}$ is also linearly independent.

25. Let $\{\mathbf{u}, \mathbf{v}, \mathbf{w}, \mathbf{z}\}$ be linearly independent. Which of the following are linearly dependent?

(a) $\{\mathbf{u} - \mathbf{v}, \mathbf{v} - \mathbf{w}, \mathbf{w} - \mathbf{u}\}$

◆ (b) $\{\mathbf{u} + \mathbf{v}, \mathbf{v} + \mathbf{w}, \mathbf{w} + \mathbf{u}\}$

(c) $\{\mathbf{u} - \mathbf{v}, \mathbf{v} - \mathbf{w}, \mathbf{w} - \mathbf{z}, \mathbf{z} - \mathbf{u}\}$

◆ (d) $\{\mathbf{u} + \mathbf{v}, \mathbf{v} + \mathbf{w}, \mathbf{w} + \mathbf{z}, \mathbf{z} + \mathbf{u}\}$

26. Let U and W be subspaces of V with bases $\{\mathbf{u}_1, \mathbf{u}_2, \mathbf{u}_3\}$ and $\{\mathbf{w}_1, \mathbf{w}_2\}$ respectively. If U and W have only the zero vector in common, show that $\{\mathbf{u}_1, \mathbf{u}_2, \mathbf{u}_3, \mathbf{w}_1, \mathbf{w}_2\}$ is linearly independent.

27. Let $\{p, q\}$ be linearly independent polynomials. Show that $\{p, q, pq\}$ is linearly independent if and only if $\deg p \geq 1$ and $\deg q \geq 1$.

◆28. If z is a complex number, show that $\{z, z^2\}$ is linearly independent if and only if z is not real.

29. If $V = \mathbf{F}[a, b]$ as in Example 7§5.1, show that the set of constant functions is a subspace of dimension 1 (f is **constant** if there is a number c such that $f(x) = c$ for all x).

30. (a) If U is an invertible $n \times n$ matrix and $\{A_1, A_2, \ldots\}$ is a basis for $\mathbf{M}_{m\,n}$, show that $\{A_1 U, A_2 U, \ldots\}$ is also a basis.

   ◆(b) Show that part **(a)** fails if U is not invertible. [*Hint:* Theorem 6§2.4.]

31. Show that $\{(a, b), (a_1, b_1)\}$ is a basis of $\mathbb{R}^2$ if and only if $\{a + bx, a_1 + b_1 x\}$ is a basis of $\mathbf{P}_1$.

32. Find the dimension of the subspace $\mathrm{span}\{1, \sin^2 \theta, \cos 2\theta\}$ of $\mathbf{F}[0, 2\pi]$.

33. Show that $\mathbf{F}[0, 1]$ is not finite dimensional.

34. If U and W are subspaces of V, define their intersection $U \cap W$ as follows:

$$U \cap W = \{\mathbf{v} \mid \mathbf{v} \text{ in both } U \text{ and } W\}$$

   (a) Show that $U \cap W$ is a subspace contained in U and W.

   ◆(b) Show that $U \cap W = 0$ if and only if $\{\mathbf{u}, \mathbf{w}\}$ is linearly independent for any nonzero vectors $\mathbf{u}$ in U and $\mathbf{w}$ in W.

   (c) If B and D are bases of U and W, and if $U \cap W = \{\mathbf{0}\}$, show that $B \cup D = \{\mathbf{v} \mid \mathbf{v} \text{ is in } B \text{ or } D\}$ is linearly independent.

35. If U and W are vector spaces, let $V = \{(\mathbf{u}, \mathbf{w}) \mid \mathbf{u} \text{ in } U$ and $\mathbf{w} \text{ in } W\}$.

(a) Show that V is a vector space if $(\mathbf{u}, \mathbf{w}) + (\mathbf{u}_1, \mathbf{w}_1) = (\mathbf{u} + \mathbf{u}_1, \mathbf{w} + \mathbf{w}_1)$ and $a(\mathbf{u}, \mathbf{w}) = (a\mathbf{u}, a\mathbf{w})$.

(b) If $\dim U = m$ and $\dim V = n$, show that $\dim V = m + n$.

(c) If $V_1, \ldots, V_m$ are vector spaces, let $V = V_1 \times \cdots \times V_m = \{(\mathbf{v}_1, \ldots, \mathbf{v}_m) \mid \mathbf{v}_i \text{ in } V_i \text{ for each } i\}$ denote the space of n-tuples from the V_i with componentwise operations (see Exercise 24 §5.1). If $\dim V_i = n_i$ for each i, show that $\dim V = n_1 + \cdots + n_m$.

36. Let $\mathbf{D}_n$ denote the set of all functions f from the set $\{1, 2, \ldots, n\}$ to $\mathbb{R}$.

(a) Show that $\mathbf{D}_n$ is a vector space with pointwise addition and scalar multiplication.

(b) Show that $\{S_1, S_2, \ldots, S_n\}$ is a basis of $\mathbf{D}_n$ where, for each $k = 1, 2, \ldots, n$, the function S_k is defined by $S_k(k) = 1$, whereas $S_k(j) = 0$ if $j \neq k$.

37. A polynomial $p(x)$ is **even** if $p(-x) = p(x)$ and **odd** if $p(-x) = -p(x)$. Let E_n and O_n denote the sets of even and odd polynomials in $\mathbf{P}_n$.

(a) Show that E_n is a subspace of $\mathbf{P}_n$ and find $\dim E_n$.

◆(b) Show that O_n is a subspace of $\mathbf{P}_n$ and find $\dim O_n$.

38. Suppose that $\{\mathbf{v}_1, \mathbf{v}_2, \ldots, \mathbf{v}_n\}$ is a maximal linearly independent set in a vector space V. That is, $\{\mathbf{v}_1, \ldots, \mathbf{v}_n\}$ is linearly independent and no set of more than n vectors is linearly independent. Show that $\{\mathbf{v}_1, \ldots, \mathbf{v}_n\}$ is a basis of V.

39. Suppose that $\{\mathbf{v}_1, \mathbf{v}_2, \ldots, \mathbf{v}_n\}$ is a minimal spanning set for a vector space V. That is, $V = \mathrm{span}\{\mathbf{v}_1, \ldots, \mathbf{v}_n\}$ and V cannot be spanned by fewer than n vectors. Show that $\{\mathbf{v}_1, \ldots, \mathbf{v}_n\}$ is a basis of V.

40. Let $\{\mathbf{v}_1, \ldots, \mathbf{v}_n\}$ be linearly independent in a vector space V, and let A be an $n \times n$ matrix. Define $\mathbf{u}_1, \ldots, \mathbf{u}_n$ by

$$\begin{bmatrix} \mathbf{u}_1 \\ \vdots \\ \mathbf{u}_n \end{bmatrix} = A \begin{bmatrix} \mathbf{v}_1 \\ \vdots \\ \mathbf{v}_n \end{bmatrix}$$

(See Exercise 25§5.1.) Show that $\{\mathbf{u}_1, \ldots, \mathbf{u}_n\}$ is linearly independent if and only if A is invertible.

██████████ Section 5.4   **Existence of Bases**

Up to this point, we have had no guarantee that an arbitrary vector space *has* a basis—and hence no guarantee that one can speak *at all* of the dimension of V. However, Theorem 2 will show that any space that is spanned by a finite set of vectors has a (finite) basis. We first derive a theorem that provides a useful connection between linear independence and spanning sets.

> **THEOREM 1**
>
> Let $\{\mathbf{v}_1, \mathbf{v}_2, \ldots, \mathbf{v}_n\}$ be a linearly independent set of vectors in a vector space V. The following conditions are equivalent for a vector $\mathbf{v}$ in V:
>
> 1. $\{\mathbf{v}, \mathbf{v}_1, \mathbf{v}_2, \ldots, \mathbf{v}_n\}$ is linearly independent.
> 2. $\mathbf{v}$ does not lie in $\mathrm{span}\{\mathbf{v}_1, \mathbf{v}_2, \ldots, \mathbf{v}_n\}$.

Proof Assume (1) is true and suppose, if possible, that $\mathbf{v}$ lies in $\mathrm{span}\{\mathbf{v}_1, \ldots, \mathbf{v}_n\}$ — say, $\mathbf{v} = a_1\mathbf{v}_1 + \cdots + a_n\mathbf{v}_n$. Then $\mathbf{v} - a_1\mathbf{v}_1 - \cdots - a_n\mathbf{v}_n = \mathbf{0}$ is a nontrivial linear combination, contrary to (1). So (1) implies (2). Conversely, assume that (2) holds and suppose that $a\mathbf{v} + a_1\mathbf{v}_1 + \cdots + a_n\mathbf{v}_n = \mathbf{0}$. If $a \neq 0$, then $\mathbf{v} = (-a^{-1}a_1)\mathbf{v}_1 + \cdots + (-a^{-1}a_n)\mathbf{v}_n$, contrary to (2). So $a = 0$ and hence $a_1\mathbf{v}_1 + \cdots + a_n\mathbf{v}_n = \mathbf{0}$. This implies that $a_1 = \cdots = a_n = 0$ because the set $\{\mathbf{v}_1, \ldots, \mathbf{v}_n\}$ is linearly independent. This proves that (2) implies (1). ◆

Theorem 1 provides a method of creating larger and larger linearly independent sets.

EXAMPLE 1

Find a basis of $\mathbf{P}_3$ containing the linearly independent set $\{1 + x, 1 + x^2\}$.

Solution Observe that every polynomial in $\mathrm{span}\{1 + x, 1 + x^2\}$ has degree at most 2. In particular x^3 is not in $\mathrm{span}\{1 + x, 1 + x^2\}$, so $\{1 + x, 1 + x^2, x^3\}$ is linearly independent by Theorem 1. Next, 1 is not in $\mathrm{span}\{1 + x, 1 + x^2, x^3\}$ as the reader can verify, so

$$B = \{1, 1 + x, 1 + x^2, x^3\}$$

is linearly independent, again by Theorem 1. In fact B spans $\mathbf{P}_3$. For if not, B could be enlarged (by Theorem 1) to a linearly independent set of five polynomials, and this contradicts Theorem 5§5.3 because $\dim \mathbf{P}_3 = 4$.

◆◆◆

The technique in Example 1 applies much more generally. Indeed, if V is a vector space that is spanned by a finite number of vectors, we claim that *any* linearly independent subset

$$S = \{\mathbf{v}_1, \mathbf{v}_2, \ldots, \mathbf{v}_k\}$$

of V is contained in a basis of V. This is certainly true if $V = \mathrm{span}\, S$ because then S is *itself* a basis of V. Otherwise, choose $\mathbf{v}_{k+1}$ outside $\mathrm{span}\, S$. Then

$$S_1 = \{\mathbf{v}_1, \ldots, \mathbf{v}_k, \mathbf{v}_{k+1}\}$$

is linearly independent by Theorem 1. If $V = \text{span } S_1$ then S_1 is the desired basis containing S. If not, choose $\mathbf{v}_{k+2}$ outside $\text{span}\{\mathbf{v}_1, \ldots, \mathbf{v}_{k+1}\}$ so that

$$S_2 = \{\mathbf{v}_1, \ldots, \mathbf{v}_k, \mathbf{v}_{k+1}, \mathbf{v}_{k+2}\}$$

is linearly independent. Continue this process. Either a basis is reached at some stage or, if not, arbitrarily large independent sets are found in V. But this latter possibility cannot occur by the fundamental theorem (Theorem 3§5.3) because V is spanned by a finite number of vectors. This proves the first part of Theorem 2.

THEOREM 2

Let $V \neq 0$ be a vector space spanned by n vectors.

1. Each set of linearly independent vectors is part of a basis of V.
2. Each spanning set for V contains a basis of V.
3. V has a basis, and $\dim V \leq n$.

Proof

1. This part has just been proved.

2. Let $V = \text{span}\{\mathbf{v}_1, \mathbf{v}_2, \ldots, \mathbf{v}_m\}$, where (as $V \neq 0$) we may assume that each $\mathbf{v}_i \neq \mathbf{0}$. If $\{\mathbf{v}_1, \ldots, \mathbf{v}_m\}$ is linearly independent, it is itself a basis and we are finished. If not, one of these vectors lies in the span of the others (Theorem 1§5.3). Relabeling if necessary, assume that $\mathbf{v}_1$ lies in $\text{span}\{\mathbf{v}_2, \ldots, \mathbf{v}_m\}$ so that

$$V = \text{span}\{\mathbf{v}_2, \ldots, \mathbf{v}_m\}$$

Now repeat the argument. If $\{\mathbf{v}_2, \ldots, \mathbf{v}_m\}$ is linearly independent, we are finished. If not, we have (after possible relabeling) $V = \text{span}\{\mathbf{v}_3, \ldots, \mathbf{v}_m\}$. Continue this process. If a basis is encountered at some stage, we are finished. If not, we ultimately reach $V = \text{span}\{\mathbf{v}_m\}$. But then $\{\mathbf{v}_m\}$ is a basis because $\mathbf{v}_m \neq \mathbf{0}$ ($V \neq 0$).

3. V has a spanning set of n vectors, one of which is nonzero because $V \neq 0$. Hence (3) follows from (2).   ◆

Recall that a vector space is called finite dimensional if it has a finite basis. The following consequence of Theorem 2 is a stronger result.

COROLLARY

A nonzero vector space is finite dimensional if and only if it can be spanned by finitely many vectors.

Example 1 illustrates how an independent set can be enlarged to a basis as in (1) of Theorem 2. The following example shows how the Gaussian algorithm can be used to cut down a spanning set to a basis as in (2) of Theorem 2.

EXAMPLE 2

Find a basis of $U = \text{span}\{(1, -1, 3, 2), (0, -1, 2, 1), (2, 1, 0, 1)\}$.

Solution   Label these spanning vectors R_1, R_2, and R_3, respectively, and let A denote the matrix with these vectors as rows. Carry A to row-echelon form by the Gaussian algorithm.

$$A = \begin{bmatrix} 1 & -1 & 3 & 2 \\ 0 & -1 & 2 & 1 \\ 2 & 1 & 0 & 1 \end{bmatrix} = \begin{bmatrix} R_1 \\ R_2 \\ R_3 \end{bmatrix}$$

$$\begin{bmatrix} 1 & -1 & 3 & 2 \\ 0 & -1 & 2 & 1 \\ 0 & 3 & -6 & -3 \end{bmatrix} = \begin{bmatrix} R_1 \\ R_2 \\ R_3 - 2R_1 \end{bmatrix}$$

$$\begin{bmatrix} 1 & -1 & 3 & 2 \\ 0 & 1 & -2 & -1 \\ 0 & 0 & 0 & 0 \end{bmatrix} = \begin{bmatrix} R_1 \\ -R_2 \\ R_3 - 2R_1 + 3R_2 \end{bmatrix}$$

Hence $R_3 = 2R_1 - 3R_2$, so R_3 is in $\text{span}\{R_1, R_2\}$. It follows that

$$U = \text{span}\{R_1, R_2, R_3\} = \text{span}\{R_1, R_2\}$$

Because $\{R_1, R_2\}$ is linearly independent (verify), it is the desired basis. Observe that the nonzero rows $\{R_1, -R_2\}$ of the row-echelon form of A also form a basis of U. This turns out to be true in general and will be investigated in the following section.

$\blacklozenge\blacklozenge\blacklozenge$

Theorem 2 is very powerful, and two important consequences are collected in Theorem 3.

THEOREM 3

Let V be a vector space, and assume that $\dim V = n > 0$.

1. Any set of n linearly independent vectors in V is a basis (that is, it necessarily spans V).

2. Any spanning set of n nonzero vectors in V is a basis (that is, it is necessarily linearly independent).

Proof

1. If the n independent vectors do not span V, they are part of a basis of more than n vectors by property (1) of Theorem 2. This contradicts Theorem 5§5.3.

2. If the n vectors in a spanning set are not linearly independent, they contain a basis of fewer than n vectors by property (2) of Theorem 2, contradicting Theorem 5§5.3.   ◆

Theorem 3 saves time when we are verifying that a set of n vectors is a basis of a space known to have dimension n. Here are some examples.

EXAMPLE 3
_____

Show that $\{(1, 0, 0, 0), (1, 1, 0, 0), (1, 1, 1, 0), (1, 1, 1, 1)\}$ is a basis of $\mathbb{R}^4$.

Solution

Because dim $\mathbb{R}^4 = 4$, it suffices (by Theorem 3) to show either that these vectors are linearly independent or that they span $\mathbb{R}^4$. But subtracting successive vectors shows that the standard basis $\{(1, 0, 0, 0), (0, 1, 0, 0), (0, 0, 1, 0), (0, 0, 0, 1)\}$ is contained in the span of these vectors, so they span $\mathbb{R}^4$.   ◆◆◆

EXAMPLE 4
_____

Let V denote the space of all symmetric 2×2 matrices. Find a basis of V consisting of invertible matrices.

Solution

It was established in Example 13§5.3 that dim $V = 3$, so what is needed is a set of three invertible, symmetric matrices that are linearly independent (or, alternatively, span V). The set $\left\{ \begin{bmatrix} 1 & 0 \\ 0 & 1 \end{bmatrix}, \begin{bmatrix} 1 & 0 \\ 0 & -1 \end{bmatrix}, \begin{bmatrix} 0 & 1 \\ 1 & 0 \end{bmatrix} \right\}$ is independent (verify), and hence it is a basis of the required type.

_____◆◆◆

The next theorem collects some very useful information on the dimension of subspaces of finite dimensional spaces.

THEOREM 4

Let V be a vector space of dimension n and let U and W denote subspaces of V. Then:

1. U is finite dimensional and dim $U \leq n$.
2. Any basis of U is part of a basis for V.
3. If $U \subseteq W$ and dim $U = $ dim W, then $U = W$.

Proof

1. If $U = 0$, dim $U = 0$ by definition. So assume $U \neq 0$ and choose $\mathbf{u}_1 \neq \mathbf{0}$ in U. If $U = \text{span}\{\mathbf{u}_1\}$, then dim $U = 1$. If $U \neq \text{span}\{\mathbf{u}_1\}$, choose $\mathbf{u}_2$ in U outside $\text{span}\{\mathbf{u}_1\}$. Then $\{\mathbf{u}_1, \mathbf{u}_2\}$,is linearly independent by Theorem 1. If $U = \text{span}\ \{\mathbf{u}_1, \mathbf{u}_2\}$, then dim $U = 2$. If not, repeat the process to find $\mathbf{u}_3$ in U such that $\{\mathbf{u}_1, \mathbf{u}_2, \mathbf{u}_3\}$ is linearly independent. Continue in this way. The process must terminate because the space V (having dimension n) cannot contain more than n independent vectors. Hence U has a basis of at most n vectors, proving (1).

2. This follows from (1) and Theorem 2.

3. Let dim $U = $ dim $W = m$. Then any basis $\{\mathbf{u}_1, \ldots, \mathbf{u}_m\}$ of U is an independent set of m vectors in W and so is a basis of W by Theorem 3. In particular, $\{\mathbf{u}_1, \ldots, \mathbf{u}_m\}$ spans W so, because it also spans U, $W = \text{span}\{\mathbf{u}_1, \ldots, \mathbf{u}_m\} = U$. This proves (3). ◆

EXAMPLE 5

If a is a number, let W denote the subspace of all polynomials in $\mathbf{P}_n$ with a as a root.

$$W = \{p(x) \mid p(x) \text{ is in } \mathbf{P}_n \text{ and } p(a) = 0\}$$

Show that $\{(x - a), (x - a)^2, \ldots, (x - a)^n\}$ is a basis of W.

Solution   Observe first that $(x - a), (x - a)^2, (x - a)^3, \ldots, (x - a)^n$ are members of W and are linearly independent (Exercise 15). Write

$$U = \text{span}\{(x - a), (x - a)^2, (x - a)^3, \ldots, (x - a)^n\}$$

Then we have $U \subseteq W \subseteq \mathbf{P}_n$, dim $U = n$, and dim $\mathbf{P}_n = n + 1$. Hence $n \leq$ dim $W \leq n + 1$ by property (1) of Theorem 4, so dim $W = n$ or dim $W = n + 1$. But then $W = U$ or $W = \mathbf{P}_n$ by (3) of Theorem 4. Because $W \neq \mathbf{P}_n$, it follows that $W = U$.

◆◆◆

The next example uses ideas from Chapter 4.

EXAMPLE 6

Show that the only subspaces of three-dimensional Euclidean space $\mathbb{R}^3$ are 0, $\mathbb{R}^3$, lines through the origin, and planes through the origin.

Solution   Let U be such a subspace. Because dim $\mathbb{R}^3 = 3$, the only possibilities for dim U are $0, 1, 2,$ and 3. If dim $U = 0$ or 3, then $U = 0$ or $\mathbb{R}^3$. If dim $U = 1$, let $\{\mathbf{d}\}$ be a basis so that $U = \mathbb{R}\mathbf{d} = \{t\mathbf{d} \mid t \text{ in } \mathbb{R}\}$. This is just the line through the origin with direction vector $\mathbf{d}$. Finally, if dim $U = 2$, let $\{\mathbf{u}_1, \mathbf{u}_2\}$ be a basis of U. If $\mathbf{n} = \mathbf{u}_1 \times \mathbf{u}_2$, we claim that U is the plane P through the origin with normal $\mathbf{n}$. The plane P can be described by

$$P = \{\mathbf{v} \mid \mathbf{v} \cdot \mathbf{n} = 0\}$$

This is a subspace (as the reader can verify), and because $\mathbf{u}_1$ and $\mathbf{u}_2$ are orthogonal to $\mathbf{n} = \mathbf{u}_1 \times \mathbf{u}_2$, we have $U \subseteq P \subseteq \mathbb{R}^3$. Because dim $U = 2$ and dim $\mathbb{R}^3 = 3$, this means that dim $P = 2$ or 3, so $P = U$ or $P = \mathbb{R}^3$. But $P \neq \mathbb{R}^3$ because $\mathbf{n} \neq \mathbf{0}$ by Theorem 4§4.3, so $P = U$, as required.

◆◆◆

EXERCISES 5.4

1. In each case, find a basis for V that includes the vector $\mathbf{v}$.
  (a) $V = \mathbb{R}^3$, $\mathbf{v} = (1, -1, 1)$
  ◆**(b)** $V = \mathbb{R}^3$, $\mathbf{v} = (0, 1, 1)$
  (c) $V = \mathbf{M}_{22}$, $\mathbf{v} = \begin{bmatrix} 1 & 1 \\ 1 & 1 \end{bmatrix}$
  ◆**(d)** $V = \mathbf{P}_2$, $\mathbf{v} = x^2 - x + 1$

2. In each case, find a basis for V among the given vectors.
  (a) $V = \mathbb{R}^3$, $\{(1, 1, -1), (2, 0, 1), (-1, 1, -2), (1, 2, 1)\}$
  ◆**(b)** $V = \mathbf{P}_2$, $\{x^2 + 3, x + 2, x^2 - 2x - 1, x^2 + x\}$

3. In each case, find a basis of V containing $\mathbf{v}$ and $\mathbf{w}$.
  (a) $V = \mathbb{R}^4$, $\mathbf{v} = (1, -1, 1, -1)$, $\mathbf{w} = (0, 1, 0, 1)$
  ◆**(b)** $V = \mathbb{R}^4$, $\mathbf{v} = (0, 0, 1, 1)$, $\mathbf{w} = (1, 1, 1, 1)$
  (c) $V = \mathbf{M}_{22}$, $\mathbf{v} = \begin{bmatrix} 1 & 0 \\ 0 & 1 \end{bmatrix}$, $\mathbf{w} = \begin{bmatrix} 0 & 1 \\ 1 & 0 \end{bmatrix}$
  ◆**(d)** $V = \mathbf{P}_3$, $\mathbf{v} = x^2 + 1$, $\mathbf{w} = x^2 + x$

4. (a) If z is not a real number, show that $\{z, z^2\}$ is a basis of the space $\mathbb{C}$ of all complex numbers.
  ◆**(b)** If z is neither real nor pure imaginary, show that $\{z, \bar{z}\}$ is a basis of $\mathbb{C}$.

5. Find a basis of $\mathbf{M}_{22}$ consisting of matrices with the property that $A^2 = A$.

6. Find a basis of $\mathbf{P}_3$ consisting of polynomials whose coefficients sum to 4. What if they sum to 0?

7. If $\{\mathbf{u}, \mathbf{v}, \mathbf{w}\}$ is a basis of V, determine which of the following are bases.
  (a) $\{\mathbf{u} + \mathbf{v}, \mathbf{u} + \mathbf{w}, \mathbf{v} + \mathbf{w}\}$
  ◆**(b)** $\{2\mathbf{u} + \mathbf{v} + 3\mathbf{w}, 3\mathbf{u} + \mathbf{v} - \mathbf{w}, \mathbf{u} - 4\mathbf{w}\}$
  (c) $\{\mathbf{u}, \mathbf{u} + \mathbf{v} + \mathbf{w}\}$
  ◆**(d)** $\{\mathbf{u}, \mathbf{u} + \mathbf{w}, \mathbf{u} - \mathbf{w}, \mathbf{v} + \mathbf{w}\}$

8. (a) Can two vectors span $\mathbb{R}^3$? Can they be linearly independent? Explain.
  ◆**(b)** Can four vectors span $\mathbb{R}^3$? Can they be linearly independent? Explain.

9. Show that any nonzero vector in a finite dimensional vector space is part of a basis.

10. In each case, find a basis of the subspace U.
  (a) $U = \mathrm{span}\{(1, -1, 3, 2), (2, 1, 1, 3), (1, 5, -7, 0), (4, -1, 7, 7)\}$.

◆**(b)** $U = \mathrm{span}\{(1, 0, -1, 3), (2, 1, 0, -2), (-1, 1, 2, 1), (-3, 2, 4, 11)\}$.

11. If U and W are subspaces of V and dim $U = 2$, show that either $U \subseteq W$ or dim$(U \cap W) \leq 1$. [See Exercise 34§5.3]

12. Let A be a 2×2 matrix and write $U = \{X \text{ in } \mathbf{M}_{22} \mid XA = AX\}$. Show that dim $U \geq 2$. [*Hint: I* and A are in U.]

13. If $U \subseteq \mathbb{R}^2$ is a subspace, show that $U = 0$, $U = \mathbb{R}^2$, or U is a line through the origin.

14. Given $\mathbf{v}_1, \mathbf{v}_2, \mathbf{v}_3, \ldots, \mathbf{v}_k$, and $\mathbf{v}$, let $U = \mathrm{span}\{\mathbf{v}_1, \mathbf{v}_2, \ldots, \mathbf{v}_k\}$ and $W = \mathrm{span}\{\mathbf{v}_1, \ldots, \mathbf{v}_k, \mathbf{v}\}$. Show that either dim $W = $ dim U or dim $W = 1 + $ dim U.

15. (a) Show that any set of nonzero polynomials in $\mathbf{P}$ of distinct degrees form an independent set.
  ◆**(b)** Complete Example 5 by showing that $\{(x - a), (x - a)^2, \ldots, (x - a)^n\}$ is linearly independent in $\mathbf{P}_n$.

16. Suppose U is a subspace of $\mathbf{P}_1$ and $U \neq 0$, $U \neq \mathbf{P}_1$. Show that either $U = \mathbb{R}$ or $U = \mathbb{R}(a + x)$ for some a in $\mathbb{R}$.

17. Let U be a subspace of V and assume dim $V = 4$ and dim $U = 2$. Does every basis of V result from adding (two) vectors to some basis of U (as in Theorem 4)? Defend your answer.

18. Let U and W be subspaces of a vector space V.
  (a) If dim $V = 3$, dim $U = $ dim $W = 2$, and $U \neq W$, show that dim$(U \cap W) = 1$. [*Hint:* Exercise 34§5.3.]
  ◆**(b)** Interpret **(a)** geometrically if $V = \mathbb{R}^3$.

19. Let $U \subseteq W$ be subspaces of V with dim $U = k$ and dim $W = m$, where $k < m$. If $k < l < m$, show that a subspace X exists with $U \subseteq X \subseteq W$ and dim $X = l$.

20. (a) Let $p(x)$ and $q(x)$ lie in $\mathbf{P}_1$ and suppose that $p(1) \neq 0$, $q(2) \neq 0$, and $p(2) = 0 = q(1)$. Show that $\{p(x), q(x)\}$ is a basis of $\mathbf{P}_1$. [*Hint:* If $rp(x) + sq(x) = 0$, evaluate at $x = 1$, $x = 2$.]
  (b) Let $B = \{p_0(x), p_1(x), \ldots, p_n(x)\}$ be a set of polynomials in $\mathbf{P}_n$. Assume that there exist numbers a_0, a_1,

..., a_n such that $p_i(a_i) \neq 0$ for each i but $p_i(a_j) = 0$ if i is different from j. Show that B is a basis of $\mathbf{P}_n$.

21. Let V be the set of all infinite sequences $(a_0, a_1, a_2, \ldots)$ of real numbers. Define addition and scalar multiplication by $(a_0, a_1, \ldots) + (b_0, b_1, \ldots) = (a_0 + b_0, a_1 + b_1, \ldots)$ and $r(a_0, a_1, \ldots) = (ra_0, ra_1, \ldots)$.

(a) Show that V is a vector space.

◆(b) Show that V is not finite dimensional.

(c) [For those with some calculus.] Show that the set of convergent sequences (that is, $\lim_{n \to \infty} a_n$ exists) is a subspace, also of infinite dimension.

22. If U and W are subspaces of V, we define their **sum** $U + W$ as follows: $U + W = \{\mathbf{u} + \mathbf{w} \mid \mathbf{u} \text{ in } U \text{ and } \mathbf{w} \text{ in } W\}$.

(a) Show that $U + W$ is a subspace of V containing U and W.

◆(b) Show that $\text{span}\{\mathbf{u}, \mathbf{w}\} = \mathbb{R}\mathbf{u} + \mathbb{R}\mathbf{w}$ for any vectors $\mathbf{u}$ and $\mathbf{w}$.

(c) Show that $\text{span}\{\mathbf{u}_1, \ldots, \mathbf{u}_m, \mathbf{w}_1, \ldots, \mathbf{w}_n\} = \text{span}\{\mathbf{u}_1, \ldots, \mathbf{u}_m\} + \text{span}\{\mathbf{w}_1, \ldots, \mathbf{w}_n\}$ for any vectors $\mathbf{u}_i$ and $\mathbf{w}_j$.

(d) If U and W are finite dimensional, show that $U + W$ is finite dimensional and that $\dim(U + W) \leq \dim U + \dim W$. [*Hint:* A basis of U together with a basis of W spans $U + W$.]

(e) If U and W are finite dimensional, and if $U \cap W = \{\mathbf{0}\}$, show that $\dim(U + W) = \dim U + \dim W$. [*Hint:* See (d) and Exercise 34§5.3.]

Section 5.5   Rank of a Matrix

In this section we apply the notions of linear independence and spanning sets to matrices by using them to study invertibility and rank.

If A is an $m \times n$ matrix with rows $R_1, R_2, \ldots, R_m$ and columns $C_1, C_2, \ldots, C_n$, it is convenient to write A in block form as follows:

$$A = [C_1 \quad C_2 \cdots C_n] = \begin{bmatrix} R_1 \\ R_2 \\ \vdots \\ R_m \end{bmatrix}$$

Then block multiplication by a column or a row takes the form

$$A \begin{bmatrix} x_1 \\ \vdots \\ x_n \end{bmatrix} = [C_1 \quad \cdots \quad C_n] \begin{bmatrix} x_1 \\ \vdots \\ x_n \end{bmatrix} \quad \text{and} \quad [y_1 \quad \cdots \quad y_m]A = [y_1 \quad \cdots \quad y_m] \begin{bmatrix} R_1 \\ \vdots \\ R_m \end{bmatrix}$$

$$= x_1 C_1 + \cdots + x_n C_n \qquad\qquad\qquad = y_1 R_1 + \cdots + y_m R_m$$

These results will be used frequently. Here is an illustration.

EXAMPLE 1

If $A = \begin{bmatrix} 2 & -1 & 3 \\ 5 & 7 & -2 \end{bmatrix}$, the columns are $C_1 = \begin{bmatrix} 2 \\ 5 \end{bmatrix}$, $C_2 = \begin{bmatrix} -1 \\ 7 \end{bmatrix}$, and $C_3 = \begin{bmatrix} 3 \\ -2 \end{bmatrix}$.

Hence

$$A \begin{bmatrix} x_1 \\ x_2 \\ x_3 \end{bmatrix} = \begin{bmatrix} 2x_1 - x_2 + 3x_3 \\ 5x_1 + 7x_2 - 2x_3 \end{bmatrix} = x_1 \begin{bmatrix} 2 \\ 5 \end{bmatrix} + x_2 \begin{bmatrix} -1 \\ 7 \end{bmatrix} + x_3 \begin{bmatrix} 3 \\ -2 \end{bmatrix} = x_1 C_1 + x_2 C_2 + x_3 C_3$$

The reader can verify the analogous property for rows.

◆◆◆

Our first result is a useful condition that a matrix is invertible.

THEOREM 1

The following conditions are equivalent for an $n \times n$ matrix A.

1. The rows of A are linearly independent in $\mathbb{R}^n$.
2. The rows of A span $\mathbb{R}^n$.
3. The columns of A are linearly independent in $\mathbb{R}^n$.
4. The columns of A span $\mathbb{R}^n$.
5. A is invertible.

Proof   We show that (1), (2), and (5) are equivalent. The equivalence of (3), (4), and (5) is analogous and is left as Exercise 4.

(1) *implies* (2). This follows from Theorem 3§5.4 because dim $\mathbb{R}^n = n$.

(2) *implies* (5). By Theorem 9§2.4 it suffices to find a matrix B such that $BA = I$. Let $R_1, R_2, \ldots, R_n$ denote the rows of A, and write row i of B as a matrix $K = [k_1 \quad k_2 \cdots k_n]$. Then the condition $I = BA$ gives

$$\text{row } i \text{ of } I = \text{row } i \text{ of } BA = KA = [k_1 \quad k_2 \quad \cdots \quad k_n] \begin{bmatrix} R_1 \\ R_2 \\ \vdots \\ R_n \end{bmatrix} = k_1 R_1 + k_2 R_2 + \cdots + k_n R_n$$

Such k_i exist by (2), so each row of B can be found.

(5) *implies* (1). Let $k_1 R_1 + k_2 R_2 + \cdots + k_n R_n = 0$, where R_i denotes the ith row of A. If we write $K = [k_1 \quad k_2 \ldots k_n]$, then,

$$KA = k_1 R_1 + k_2 R_2 + \cdots + k_n R_n = 0$$

Right multiplication by A^{-1} [which exists by property (5)] gives $K = 0$, so $k_1 = k_2 = \cdots = k_n = 0$, as required.   ◆

Hence, for example, to verify that $A = \begin{bmatrix} 3 & -1 & 0 \\ 1 & 0 & 4 \\ 0 & 1 & 3 \end{bmatrix}$ is an invertible matrix, it suf-

fices to check that the rows $(3, -1, 0)$, $(1, 0, 4)$, and $(0, 1, 3)$ are linearly independent in $\mathbb{R}^3$ (or, equivalently, that they span $\mathbb{R}^3$). This holds in reverse too.

EXAMPLE 2

Show that $(1, 2, -1)$, $(3, 1, -4)$, and $(1, 1, 7)$ are a basis of $\mathbb{R}^3$.

Solution

Let $A = \begin{bmatrix} 1 & 2 & -1 \\ 3 & 1 & -4 \\ 1 & 1 & 7 \end{bmatrix}$ be the matrix with these vectors as rows. It suffices by Theorem 1 to check that this matrix is invertible. This is easily verified (for example, $\det A = -41$, so A is invertible by Theorem 2§3.2).

◆◆◆

Theorem 1 shows that linear independence and spanning sets are useful in discussing the invertibility of square matrices. The question arises whether these ideas shed any light on nonsquare matrices (which, of course, cannot be invertible). Before this can be discussed we must introduce two important subspaces related to a matrix.

DEFINITION

If A is an $m \times n$ matrix, the rows of A are vectors in $\mathbb{R}^n$, and the subspace of $\mathbb{R}^n$ spanned by these rows is called the **row space** of A and is denoted as row A. Similarly, the space spanned by the columns of A is called the **column space** of A and is denoted as col A.

The following properties of these spaces are needed.

THEOREM 2

Let A, U, and V be matrices of sizes $m \times n$, $p \times m$, and $n \times q$, respectively.

1. row$(UA) \subseteq$ row A with equality if U is (square and) invertible.
2. col$(AV) \subseteq$ col A with equality if V is (square and) invertible.

Proof Let $U_i = [u_{i1} \ u_{i2} \ \cdots \ u_{im}]$ denote row i of the matrix U. If $R_1, R_2, \ldots, R_m$ denote the rows of A, then block multiplication gives

$$\text{row } i \text{ of } UA = U_i A = [u_{i1} \ u_{i2} \ \cdots \ u_{im}] \begin{bmatrix} R_1 \\ R_2 \\ \vdots \\ R_m \end{bmatrix} = u_{i1}R_1 + u_{i2}R_2 + \cdots + u_{im}R_m$$

Thus each row of UA lies in the row space of A, so row$(UA) \subseteq$ row A, as asserted.

If U is invertible, then, using what we have just proved,

$$\text{row } A = \text{row}[U^{-1}(UA)] \subseteq \text{row}(UA)$$

Thus row $A = \text{row}(UA)$ if U is invertible. This proves (1), and (2) follows in the same way. [Alternatively, apply (1) to $(AV)^T = V^T A^T$.]   ◆

This theorem leads to important results in matrix theory. The first is the surprising fact that row A and col A have the *same dimension* for any matrix A.

To see this, consider first the case of a matrix R in row-echelon form (Section 1.2). Recall that in each nonzero row of R, the first nonzero entry (from the left) is a 1, called the leading 1 for that row. Moreover, these leading 1's occupy distinct columns, and these columns have only zero entries below the leading 1. This implies (Exercise 16) that

(a)   The columns of R containing leading 1's are a basis of col R.

(b)   The nonzero rows of R are a basis of row R.

These facts lead to the following fundamental theorem.

THEOREM 3

Rank Theorem

Let A denote any $m \times n$ matrix. Then

$$\dim(\text{row } A) = \dim(\text{col } A)$$

Moreover, suppose A can be carried to a matrix R in row-echelon form by a series of elementary row operations. If r denotes the number of nonzero rows in R, then

1.   The r nonzero rows of R are a basis of row A.

2.   If the leading 1's lie in columns $j_1, j_2, \ldots, j_r$ of R, then the corresponding columns $j_1, j_2, \ldots, j_r$ of A are a basis of col A.

Proof   Theorem 3§2.4 asserts that $R = UA$ for some invertible matrix U. Hence row $A = \text{row } R$ by Theorem 2, and (1) follows from (b) preceding this theorem.

To prove (2), let $C_1, C_2, \ldots, C_n$ denote the columns of A. Then $A = [C_1 \ C_2 \cdots C_n]$ in block form, and

$$R = UA = U[C_1 \ C_2 \cdots C_n] = [UC_1 \ UC_2 \cdots UC_n]$$

Hence, in the notation of (2), the set $B = \{UC_{j_1}, UC_{j_2}, \ldots, UC_{j_r}\}$ consists of the columns of R that contain a leading 1, so B is a basis of col R by (a) preceding this theorem. But then the fact that U is invertible implies that $\{C_{j_1}, C_{j_2}, \ldots, C_{j_r}\}$ is linearly independent. Furthermore, if C_j is any column of A, then UC_j is a linear combination of the columns in B. Again, the invertibility of U implies that C_j is a linear combination of $C_{j_1}, C_{j_2}, \ldots, C_{j_r}$. This proves (2).

Finally, $\dim(\text{row } A) = r = \dim(\text{col } A)$ by (1) and (2).   ◆

> **DEFINITION**
>
> The common dimension of the row and column spaces of an $m \times n$ matrix A is called the **rank** of A and is denoted rank A.

Recall that in Section 1.2 it was asserted (without proof) that, no matter how a matrix A is reduced (by row operations) to a matrix R in row-echelon form, the number r of nonzero rows of R is always the same (r was called the *rank* of A). The rank theorem shows that this assertion is true and that the two notions of rank agree. We record this for reference.

> **COROLLARY 1**
>
> Suppose a matrix A can be carried to a matrix R in row-echelon form by a series of elementary row operations. Then the rank of A is equal to the number of nonzero rows of R.

EXAMPLE 3

Compute the rank of matrix $A = \begin{bmatrix} 1 & 2 & 2 & -1 \\ 3 & 6 & 5 & 0 \\ 1 & 2 & 1 & 2 \end{bmatrix}$ and find bases for the row space and the column space of A.

Solution   The reduction of A to row-echelon form is as follows:

$$\begin{bmatrix} 1 & 2 & 2 & -1 \\ 3 & 6 & 5 & 0 \\ 1 & 2 & 1 & 2 \end{bmatrix} \rightarrow \begin{bmatrix} 1 & 2 & 2 & -1 \\ 0 & 0 & -1 & 3 \\ 0 & 0 & -1 & 3 \end{bmatrix} \rightarrow \begin{bmatrix} 1 & 2 & 2 & -1 \\ 0 & 0 & 1 & -3 \\ 0 & 0 & 0 & 0 \end{bmatrix}$$

Hence rank $A = 2$, and $\{(1, 2, 2, -1), (0, 0, 1, -3)\}$ is a basis of the row space of A. Moreover, the leading 1's are in columns 1 and 3 of the row-echelon matrix, so Theorem 3 shows that columns 1 and 3 of A are a basis $\left\{ \begin{bmatrix} 1 \\ 3 \\ 1 \end{bmatrix}, \begin{bmatrix} 2 \\ 5 \\ 1 \end{bmatrix} \right\}$ of col A.

◆◆◆

The rank theorem has several important consequences. Corollary 2 follows because the dimension of a vector space cannot exceed the number of vectors in any spanning set.

COROLLARY 2

If A is an $m \times n$ matrix, then rank $A \leq m$ and rank $A \leq n$.

The fact that the rows of A are just the columns of the transpose matrix A^T gives

COROLLARY 3

If A is any matrix, rank A = rank A^T.

Theorem 2 immediately yields

COROLLARY 4

Let A be an $m \times n$ matrix and let U and V be invertible matrices of size $m \times m$ and $n \times n$, respectively. Then

$$\text{rank } A = \text{rank}(UA) = \text{rank}(AV)$$

If A is an $n \times n$ matrix, then the row space is spanned by the n rows and so has dimension n if and only if those rows are linearly independent. Combining this with Theorem 1 gives the following:

COROLLARY 5

An $n \times n$ matrix A is invertible if and only if rank $A = n$.

The rank theorem can be used to find bases of subspaces in $\mathbb{R}^n$.

EXAMPLE 4

Find a basis for the following subspace of $\mathbb{R}^4$.

$$U = \text{span}\{(1, 1, 2, 3), (2, 4, 1, 0), (1, 5, -4, -9)\}$$

Solution U is just the row space of $\begin{bmatrix} 1 & 1 & 2 & 3 \\ 2 & 4 & 1 & 0 \\ 1 & 5 & -4 & -9 \end{bmatrix}$, so we reduce this to row-echelon form:

$$\begin{bmatrix} 1 & 1 & 2 & 3 \\ 2 & 4 & 1 & 0 \\ 1 & 5 & -4 & -9 \end{bmatrix} \rightarrow \begin{bmatrix} 1 & 1 & 2 & 3 \\ 0 & 2 & -3 & -6 \\ 0 & 4 & -6 & -12 \end{bmatrix} \rightarrow \begin{bmatrix} 1 & 1 & 2 & 3 \\ 0 & 1 & -\frac{3}{2} & -3 \\ 0 & 0 & 0 & 0 \end{bmatrix}$$

The required basis is $\{(1, 1, 2, 3), (0, 1, -\frac{3}{2}, -3)\}$. Thus $\{(1, 1, 2, 3), (0, 2, -3, -6)\}$ is also a basis and avoids fractions.

◆◆◆

Let A be an $m \times n$ matrix of rank r. Theorem 3§2.4 shows that, if R is the reduced row-echelon form of A, the row operations that carry A to R also carry I_m to an invertible matrix U such that

$$UA = R$$

Hence they carry the block matrix $[A \ I_m]$ to $[R \ U]$. Because rank $A = r$, only the first r rows of R are nonzero, and these rows contain every column of I_r. Hence elementary *column* operations will reduce R to the block $m \times n$ matrix $\begin{bmatrix} I_r & 0 \\ 0 & 0 \end{bmatrix}$. This can be accomplished by doing the corresponding *row* operations to R^T to obtain its $n \times m$ row-echelon form $U_1 R^T = \begin{bmatrix} I_r & 0 \\ 0 & 0 \end{bmatrix}$, where U_1 is $n \times n$ and invertible. If we write $V = U_1^T$, we get

$$UAV = RU_1^T = (U_1 R^T)^T = \begin{bmatrix} I_r & 0 \\ 0 & 0 \end{bmatrix}^T = \begin{bmatrix} I_r & 0 \\ 0 & 0 \end{bmatrix}$$

This proves half of the following theorem.

THEOREM 4

Let A denote an $m \times n$ matrix. Then A has rank r if and only if there exist invertible matrices U and V (of sizes $m \times m$ and $n \times n$, respectively) such that

$$UAV = \begin{bmatrix} I_r & 0 \\ 0 & 0 \end{bmatrix}$$

in block form, where I_r is the $r \times r$ identity matrix.

Proof It remains to show that rank $A = r$ if such U and V exist. But this follows from Corollary 4 of the rank theorem. ◆

The argument leading to Theorem 4 not only proves the existence of the matrices U and V but also gives us a way to find them. To summarize:

1. Find U from $[A \ I_m] \rightarrow [R \ U]$.

2. Find V from $[R^T \, I_n] \to \begin{bmatrix} \begin{bmatrix} I_r & 0 \\ 0 & 0 \end{bmatrix} V^T \end{bmatrix}$.

Here is an example.

EXAMPLE 5

Given $A = \begin{bmatrix} 1 & -1 & 1 & 2 \\ 2 & -2 & 1 & -1 \\ -1 & 1 & 0 & 3 \end{bmatrix}$, find invertible matrices U and V such that UAV

$= \begin{bmatrix} I_r & 0 \\ 0 & 0 \end{bmatrix}$, where $r = \text{rank } A$.

Solution

The matrix U and the reduced row-echelon form R of A are computed by the row reduction $[A \, I_3] \to [R \, U]$.

$$\begin{bmatrix} 1 & -1 & 1 & 2 & | & 1 & 0 & 0 \\ 2 & -2 & 1 & -1 & | & 0 & 1 & 0 \\ -1 & 1 & 0 & 3 & | & 0 & 0 & 1 \end{bmatrix} \to \begin{bmatrix} 1 & -1 & 0 & -3 & | & -1 & 1 & 0 \\ 0 & 0 & 1 & 5 & | & 2 & -1 & 0 \\ 0 & 0 & 0 & 0 & | & -1 & 1 & 1 \end{bmatrix}$$

Hence

$$R = \begin{bmatrix} 1 & -1 & 0 & -3 \\ 0 & 0 & 1 & 5 \\ 0 & 0 & 0 & 0 \end{bmatrix} \quad \text{and} \quad U = \begin{bmatrix} -1 & 1 & 0 \\ 2 & -1 & 0 \\ -1 & 1 & 1 \end{bmatrix}$$

In particular, $r = \text{rank } R = 2$. Now row-reduce $[R^T \, I_4] \to \begin{bmatrix} \begin{bmatrix} I_r & 0 \\ 0 & 0 \end{bmatrix} V^T \end{bmatrix}$:

$$\begin{bmatrix} 1 & 0 & 0 & | & 1 & 0 & 0 & 0 \\ -1 & 0 & 0 & | & 0 & 1 & 0 & 0 \\ 0 & 1 & 0 & | & 0 & 0 & 1 & 0 \\ -3 & 5 & 0 & | & 0 & 0 & 0 & 1 \end{bmatrix} \to \begin{bmatrix} 1 & 0 & 0 & | & 1 & 0 & 0 & 0 \\ 0 & 1 & 0 & | & 0 & 0 & 1 & 0 \\ 0 & 0 & 0 & | & 1 & 1 & 0 & 0 \\ 0 & 0 & 0 & | & 3 & 0 & -5 & 1 \end{bmatrix}$$

whence

$$V^T = \begin{bmatrix} 1 & 0 & 0 & 0 \\ 0 & 0 & 1 & 0 \\ 1 & 1 & 0 & 0 \\ 3 & 0 & -5 & 1 \end{bmatrix} \qquad V = \begin{bmatrix} 1 & 0 & 1 & 3 \\ 0 & 0 & 1 & 0 \\ 0 & 1 & 0 & -5 \\ 0 & 0 & 0 & 1 \end{bmatrix}$$

Then $UAV = \begin{bmatrix} I_2 & 0 \\ 0 & 0 \end{bmatrix}$, as is easily verified.

◆◆◆

Let A be an $m \times n$ matrix and consider the associated homogeneous system of linear equations

$$AX = 0$$

The set of solutions to this system is a subspace of $\mathbb{R}^n$ called the **null space** of the matrix A and is denoted null A. Theorem 5 determines dim (null A).

THEOREM 5

Let A be an $m \times n$ matrix and denote r = rank A. Then

$$\dim (\text{null } A) = n - r$$

Proof We have null $(A) = \{X \mid AX = 0\}$. By Theorem 4, let $UAV = \begin{bmatrix} I_r & 0 \\ 0 & 0 \end{bmatrix}$, where U and V are invertible. Then null(A) = null(UA) because U is invertible, and dim[null(UA)] = dim[null(UAV)] because V is invertible (Exercise 30). Hence dim[null(A)] = dim[null(UAV)], and this equals $n - r$ because $UAV = \begin{bmatrix} I_r & 0 \\ 0 & 0 \end{bmatrix}$ is $m \times n$. In fact, null (UAV) consists of all columns in $\mathbb{R}^n$ with the first r entries zero.

◆

Of course the Gaussian algorithm is designed to compute the null space. Here is an example.

EXAMPLE 6

Find a basis for the null space of $A = \begin{bmatrix} 1 & -2 & 1 & 1 \\ -1 & 2 & 0 & 1 \\ 2 & -4 & 1 & 0 \end{bmatrix}$.

Solution The corresponding system of linear equations is

$$\begin{aligned} x_1 - 2x_2 + x_3 + x_4 &= 0 \\ -x_1 + 2x_2 \qquad\quad + x_4 &= 0 \\ 2x_1 - 4x_2 + x_3 \qquad &= 0 \end{aligned}$$

If the augmented matrix is carried to reduced row-echelon form, the result is

$$\begin{bmatrix} 1 & -2 & 1 & 1 & | & 0 \\ -1 & 2 & 0 & 1 & | & 0 \\ 2 & -4 & 1 & 0 & | & 0 \end{bmatrix} \rightarrow \begin{bmatrix} 1 & -2 & 0 & -1 & | & 0 \\ 0 & 0 & 1 & 2 & | & 0 \\ 0 & 0 & 0 & 0 & | & 0 \end{bmatrix}$$

The leading variables are x_1 and x_3, so the nonleading variables are assigned parameters $x_2 = s$, $x_4 = t$. Then the equations are solved for the leading variables, and the result is

$$\begin{bmatrix} x_1 \\ x_2 \\ x_3 \\ x_4 \end{bmatrix} = s \begin{bmatrix} 2 \\ 1 \\ 0 \\ 0 \end{bmatrix} + t \begin{bmatrix} 1 \\ 0 \\ -2 \\ 1 \end{bmatrix}$$

It follows that null $A = \text{span} \left\{ \begin{bmatrix} 2 \\ 1 \\ 0 \\ 0 \end{bmatrix}, \begin{bmatrix} 1 \\ 0 \\ -2 \\ 1 \end{bmatrix} \right\}$. These vectors are in fact a basis because

the null space has dimension $n - r = 4 - 2 = 2$ by Theorem 5. (Of course the reader can check directly that these vectors are linearly independent.)

◆◆◆

In general, if A is any $m \times n$ matrix of rank r, the reduced row-echelon form R of A has exactly r nonzero rows. Hence there are r leading variables; so, because there are n variables in all, there are $n - r$ nonleading variables. As in Example 6, these variables are assigned parameter values. This gives every solution as a linear combination of $n - r$ particular solutions (the parameters are the coefficients), and so these $n - r$ solutions span null A. They are thus a basis by Theorem 5.

Corollary 5 of the rank theorem asserts that an $n \times n$ matrix A is invertible if and only if rank $A = n$. This can be rephrased as follows: A is invertible if and only if rank A is as large as possible. In this form, it makes sense for nonsquare matrices. If A is $m \times n$, then the largest possible rank is the smaller of m and n. The next theorem shows that $m \times n$ matrices with rank m do retain some of the properties of invertible matrices. (Of course, an analogous theorem holds if the rank is n.)

THEOREM 6

The following conditions are equivalent for an $m \times n$ matrix.

1. Rank $A = m$.
2. The rows of A are linearly independent.
3. If $YA = 0$ with Y in $\mathbb{R}^m$, then necessarily $Y = 0$.
4. AA^T is invertible.

Proof

(1) *implies* (2). This is by Theorem 3§5.4 because the rows span row A, and row A has dimension m by (1).

(2) *implies* (3). If $YA = 0$, write $Y = [y_1\ y_2\ \cdots\ y_m]$, and denote the rows of A by $R_1, R_2, \ldots, R_m$. Then block multiplication gives $y_1 R_1 + y_2 R_2 + \cdots + y_m R_m = YA = 0$, so (2) implies $y_1 = y_2 = \cdots = y_m = 0$. Hence $Y = 0$.

(3) *implies* (4). By Theorem 4§2.4 it suffices to show that, if $Y(AA^T) = 0$ with Y in $\mathbb{R}^m$, then $Y = 0$. Now YA is a row in $\mathbb{R}^n$, and we compute

$$(YA)(YA)^T = YAA^TY^T = 0Y^T = 0$$

This means that $YA = 0$ (if $XX^T = 0$ with X in $\mathbb{R}^n$, then $X = 0$). Hence $Y = 0$ by (3).

(4) *implies* (1). If R_i is the ith row of A, it suffices to show that $\{R_1, \ldots, R_m\}$ is linearly independent. Let $x_1 R_1 + x_2 R_2 + \cdots + x_m R_m = 0$, where the x_i lie in $\mathbb{R}$. If $X = [x_1\ x_2\ \cdots\ x_m]$, this means $XA = 0$, whence $XAA^T = 0$. But then $X = 0$ by (4), so $x_1 = x_2 = \cdots = x_m = 0$, as required. ◆

EXAMPLE 7

If at least two of the numbers x, y, and z are distinct, show that $S = \begin{bmatrix} 3 & x + y + z \\ x + y + z & x^2 + y^2 + z^2 \end{bmatrix}$ is invertible.

Solution If $A = \begin{bmatrix} 1 & 1 & 1 \\ x & y & z \end{bmatrix}$, then $S = AA^T$, so it is necessary only to verify that the rows of A are linearly independent. This is left to the reader. ◆◆◆

EXERCISES 5.5

1. Use Theorem 1 to test whether each of the following matrices is invertible.

(a) $\begin{bmatrix} 1 & -1 & 2 \\ 0 & 1 & 5 \\ 2 & 1 & 19 \end{bmatrix}$ ◆**(b)** $\begin{bmatrix} 1 & 2 & -1 \\ 3 & 1 & 1 \\ 5 & 3 & -1 \end{bmatrix}$

(c) $\begin{bmatrix} 2 & 1 & -2 \\ 1 & 0 & 3 \\ 4 & 1 & 4 \end{bmatrix}$

2. Use Theorem 1 to determine whether each of the following sets of vectors is a basis of $\mathbb{R}^4$.

(a) $\{(1, -1, 1, 0), (1, 0, 1, -1), (1, 1, 1, 1), (0, 1, 2, 3)\}$

◆**(b)** $\{(1, 1, 0, 0), (1, 0, 1, 0), (0, 1, 0, 1), (0, 0, 1, 1)\}$

3. Use Theorem 1 to show that a triangular matrix (Section 3.1) is invertible provided that the entries on the main diagonal are nonzero.

◆**4.** Complete the proof of Theorem 1.

5. If A is an $n \times n$ matrix, prove that $\det A = 0$ if and only if some row of A is a linear combination of the others.

6. If A is a square matrix, use Theorem 1 to show that A is invertible if and only if the transpose A^T is invertible.

7. In each case find bases for the row and column spaces of A and determine the rank of A.

(a) $A = \begin{bmatrix} 2 & -4 & 6 & 8 \\ 2 & -1 & 3 & 2 \\ 4 & -5 & 9 & 10 \\ 0 & -1 & 1 & 2 \end{bmatrix}$ ◆**(b)** $A = \begin{bmatrix} 2 & -1 & 1 \\ -2 & 1 & 1 \\ 4 & -2 & 3 \\ -6 & 3 & 0 \end{bmatrix}$

(c) $A = \begin{bmatrix} 1 & -1 & 5 & -2 & 2 \\ 2 & -2 & -2 & 5 & 1 \\ 0 & 0 & -12 & 9 & -3 \\ -1 & 1 & 7 & -7 & 1 \end{bmatrix}$

◆**(d)** $A = \begin{bmatrix} 1 & 2 & -1 & 3 \\ -3 & -6 & 3 & -2 \end{bmatrix}$

8. In each case find a basis of the subspace U.

(a) $U = \text{span}\{(1, -1, 0, 3), (2, 1, 5, 1), (4, -2, 5, 7)\}$

◆**(b)** $U = \text{span}\{(1, -1, 2, 5, 1), (3, 1, 4, 2, 7), (1, 1, 0, 0, 0), (5, 1, 6, 7, 8)\}$

(c) $U = \text{span}\left\{\begin{bmatrix} 1 \\ 1 \\ 0 \\ 0 \end{bmatrix}, \begin{bmatrix} 0 \\ 0 \\ 1 \\ 1 \end{bmatrix}, \begin{bmatrix} 1 \\ 0 \\ 1 \\ 0 \end{bmatrix}, \begin{bmatrix} 0 \\ 1 \\ 0 \\ 1 \end{bmatrix}\right\}$

◆ (d) $U = \text{span}\left\{\begin{bmatrix} 1 \\ 5 \\ -6 \end{bmatrix}, \begin{bmatrix} 2 \\ 6 \\ -8 \end{bmatrix}, \begin{bmatrix} 3 \\ 7 \\ -10 \end{bmatrix}, \begin{bmatrix} 4 \\ 8 \\ 12 \end{bmatrix}\right\}$

9. (a) Can a 3×4 matrix have independent columns? Independent rows? Explain.

◆ (b) If A is 4×3 and rank $A = 2$, can A have independent columns? Independent rows? Explain.

(c) If A is an $m \times n$ matrix and rank $A = m$, show that $m \leq n$.

◆ (d) Can a nonsquare matrix have its rows independent and its columns independent? Explain.

(e) Can the null space of a 3×6 matrix have dimension 2? Explain.

◆ (f) If A is not square, show that either the rows of A or the columns of A are linearly dependent.

10. (a) Show that rank $UA \leq$ rank A, with equality if U is invertible.

◆ (b) Show that rank $AV \leq$ rank A, with equality if V is invertible.

11. Show that rank$(AB) \leq$ rank$A$ and that rank$(AB) \leq$ rank B.

◆ 12. Prove (2) of Theorem 2 by applying (1) to $(AV)^T = V^T A^T$.

13. Show that the rank does not change when an elementary row or column operation is performed on a matrix.

14. Show that every $m \times n$ matrix A with m linearly independent rows can be obtained from some $n \times n$ invertible matrix by deleting the last $n - m$ rows. [*Hint:* Theorem 4§5.4.]

15. In each case find a basis of the null space of A. Then compute rank A and verify Theorem 5.

(a) $A = \begin{bmatrix} 3 & 1 & 1 \\ 2 & 0 & 1 \\ 4 & 2 & 1 \\ 1 & -1 & 1 \end{bmatrix}$

◆ (b) $A = \begin{bmatrix} 3 & 5 & 5 & 2 & 0 \\ 1 & 0 & 2 & 2 & 1 \\ 1 & 1 & 1 & -2 & -2 \\ -2 & 0 & -4 & -4 & -2 \end{bmatrix}$

16. Let R denote a matrix in row-echelon form.

(a) Show that the nonzero rows of R are linearly independent and so are a basis of row R.

◆ (b) Show that the columns of R containing leading 1's are a basis of col R.

17. In each case find invertible U and V such that $UAV = \begin{bmatrix} I_r & 0 \\ 0 & 0 \end{bmatrix}$, where $r =$ rank A.

(a) $A = \begin{bmatrix} 1 & 1 & -1 \\ -2 & -2 & 4 \end{bmatrix}$

◆ (b) $A = \begin{bmatrix} 3 & 2 \\ 2 & 1 \end{bmatrix}$

(c) $A = \begin{bmatrix} 1 & -1 & 2 & 1 \\ 2 & -1 & 0 & 3 \\ 0 & 1 & -4 & 1 \end{bmatrix}$

◆ (d) $A = \begin{bmatrix} 1 & 1 & 0 & -1 \\ 3 & 2 & 1 & 1 \\ 1 & 0 & 1 & 3 \end{bmatrix}$

18. Let $A = CR$ where $C \neq 0$ is a column in $\mathbb{R}^m$ and $R \neq 0$ is a row in $\mathbb{R}^n$.

(a) Show that col $A = \text{span}\{C\}$ and row $A = \text{span}\{R\}$.

◆ (b) Find dim(null A).

(c) Show that null A = null R.

19. Show that null $A = 0$ if and only if the columns of A are independent.

20. Let A be an $n \times n$ matrix.

(a) Show that $A^2 = 0$ if and only if col $A \subseteq$ null A.

◆ (b) Conclude that if $A^2 = 0$ then rank $A \leq \frac{n}{2}$.

(c) Find a matrix A for which col A = null A.

21. If A is $m \times n$ and B is $n \times m$, show that $AB = 0$ if and only if col $B \subseteq$ null A.

◆ 22. If A is an $m \times n$ matrix, show that col $A = \{AX \mid X$ in $\mathbb{R}^n\}$.

23. Let A be an $m \times n$ matrix. Show that $AX = B$ has a solution for every B in $\mathbb{R}^m$ if and only if rank $A = m$.

24. If A is $m \times n$ and B is $m \times 1$, show that B lies in the column space of A if and only if rank $[A\ B] =$ rank A.

25. (a) Show that $AX = B$ has a solution if and only if rank $A =$ rank $[A\ B]$. [*Hint:* Exercises 22 and 24.]

◆ (b) If $AX = B$ has no solution, show that rank $[A\ B] = 1 +$ rank A.

26. Formulate and prove the analogue of Theorem 6 for $m \times n$ matrices with independent columns.

27. Let X be an $m \times k$ matrix. If I is the $m \times m$ identity matrix, show that $I + XX^T$ is invertible. [*Hint:* $I + XX^T = [I\ X]\begin{bmatrix} I \\ X^T \end{bmatrix}$ in block form. Use Theorem 6.]

◆ 28. Show that an $m \times n$ matrix A has rank m if and only if there is an $n \times m$ matrix B such that $AB = I_m$.

29. If A and B are $m \times n$ matrices, show that rank $(A + B) \leq$ rank $A +$ rank B. [*Hint:* If U and V are the column spaces of A and B, respectively, show that the column

space of $A + B$ is contained in $U + V$ and that dim $(U + V) \leq$ dim $U +$ dim V. (See Exercise 22§5.4.)]

30. Consider the $m \times n$ matrix A.

   (a) Show that null$A =$ null(UA) for any invertible $m \times n$ matrix U.

   ◆**(b)** Show that dim [null A] = dim [null(AV)] for any invertible $n \times n$ matrix V. [*Hint:* If $\{X_1, \ldots, X_k\}$ is a basis of null A, consider $\{V^{-1}X_1, \ldots, V^{-1}X_k\}$.]

31. **(a)** Show that if A and B have independent rows, so does AB.

   (b) Show that if A and B have independent columns, so does AB.

32. Let A be an $n \times n$ matrix of rank r. If $U = \{X$ in $M_{n\,n}\,|\,AX = 0\}$, show that dim $U = n(n - r)$. [*Hint:* Theorem 5 and Exercise 35§5.3.]

33. A matrix obtained from A by deleting rows and columns is called a **submatrix** of A. If A has an invertible $k \times k$ submatrix, show that rank $A \geq k$. [*Hint:* Show that row and column operations carry $A \to \begin{bmatrix} I_k & P \\ 0 & Q \end{bmatrix}$ in block form.

See the proof of Theorem 4.] Remark: It can be shown that rank A is the largest integer r such that A has an invertible $r \times r$ submatrix.

34. Two $m \times n$ matrices A and B are called **equivalent** (written $A \overset{e}{\sim} B$) if there exist invertible matrices U and V (sizes $m \times m$ and $n \times n$) such that $A = UBV$.

   (a) Prove the following the properties of equivalence.

      (i) $A \overset{e}{\sim} A$ for all $m \times n$ matrices A.

      (ii) If $A \overset{e}{\sim} B$, then $B \overset{e}{\sim} A$.

      (iii) If $A \overset{e}{\sim} B$ and $B \overset{e}{\sim} C$, then $A \overset{e}{\sim} C$.

   (b) Prove that two $m \times n$ matrices are equivalent if and only if they have the same rank. [*Hint:* Use part **(a)** and Theorem 4.]

35. Let A be an $m \times n$ matrix of rank r.

   (a) Show that A can be factored as $A = PQ$, where P is $m \times r$ and has r independent columns and Q is $r \times m$ and has r independent rows. [*Hint:* Let $UAV = \begin{bmatrix} I_r & 0 \\ 0 & 0 \end{bmatrix}$ as in Theorem 4, and write U^{-1} and V^{-1} in block form as follows: $U^{-1} = \begin{bmatrix} U_1 & U_2 \\ U_3 & U_4 \end{bmatrix}$ and $V^{-1} = \begin{bmatrix} V_1 & V_2 \\ V_3 & V_4 \end{bmatrix}$, where U_1 and V_1 are $r \times r$.]

   ◆**(b)** For each matrix A in Exercise 17, write $A = PQ$ as in part **(a)**.

Section 5.6   An Application to Polynomials (Optional)[6]

The vector space of all polynomials of degree at most n is denoted $\mathbf{P}_n$, and it was established in Section 5.3 that $\mathbf{P}_n$ has dimension $n + 1$; in fact, $\{1, x, x^2, \ldots, x^n\}$ is a basis. The next theorem shows that *any* $n + 1$ polynomials of distinct degrees form a basis.

THEOREM 1

Let $p_0(x), p_1(x), p_2(x), \ldots, p_n(x)$ be polynomials in $\mathbf{P}_n$ of degrees $0, 1, 2, \ldots, n$, respectively. Then $\{p_0(x), \ldots, p_n(x)\}$ is a basis of $\mathbf{P}_n$.

Proof   Because dim $\mathbf{P}_n = n + 1$, it suffices by Theorem 3§5.4 to show that $\{p_0(x), p_1(x), \ldots, p_n(x)\}$ is linearly independent. So suppose some linear combination vanishes:

$$a_0 p_0(x) + a_1 p_1(x) + \cdots + a_n p_n(x) = 0$$

The aim is to show that $a_0 = a_1 = \cdots = a_n = 0$. Because $p_n(x)$ has degree n and

---

[6]The two applications in this chapter are independent and may be taken in any order.

every other polynomial $p_k(x)$ has degree *lower* than n, the term involving x^n on the left side is $a_n a x^n$, where $a \neq 0$ is the leading coefficient of $p_n(x)$. It follows that $a_n a = 0$, so $a_n = 0$. Hence the foregoing linear combination becomes

$$a_0 p_0(x) + a_1 p_1(x) + \cdots + a_{n-1} p_{n-1}(x) = 0$$

and a similar argument shows that $a_{n-1} = 0$. Continue in this way to obtain $a_k = 0$ for all k.   ◆

An immediate consequence is that $\{1, (x - a), (x - a)^2, \ldots, (x - a)^n\}$ is a basis of $\mathbf{P}_n$ for any number a. Hence we have the following:

COROLLARY 1

If a is any number, every polynomial $f(x)$ of degree at most n has an expansion in powers of $(x - a)$:

$$f(x) = a_0 + a_1(x - a) + a_2(x - a)^2 + \cdots + a_n(x - a)^n \qquad (*)$$

If $f(x)$ is evaluated at $x = a$, then equation $(*)$ becomes

$$f(a) = a_0 + a_1(a - a) + \cdots + a_n(a - a)^n = a_0$$

Hence $a_0 = f(a)$, and equation $(*)$ can be written $f(x) = f(a) + (x - a)g(x)$, where $g(x)$ is a polynomial of degree $n - 1$ (this assumes that $n \geq 1$). If it happens that $f(a) = 0$, then it is clear that $f(x)$ has the form $f(x) = (x - a)g(x)$. Conversely, every such polynomial certainly satisfies $f(a) = 0$, and we obtain:

COROLLARY 2

Let $f(x)$ be a polynomial of degree $n \geq 1$ and let a be any number. Then

Remainder Theorem

1. $f(x) = f(a) + (x - a)\, g(x)$ for some polynomial $g(x)$ of degree $n - 1$.

Factor Theorem

2. $f(a) = 0$ if and only if $f(x) = (x - a)g(x)$ for some polynomial $g(x)$.

The polynomial $g(x)$ can be easily computed by using "long division" to divide $f(x)$ by $(x - a)$.

All the coefficients in the expansion $(*)$ of $f(x)$ in powers of $(x - a)$ can be determined in terms of the derivatives of $f(x)$.[7] These will be familiar to students

---

[7] The discussion of Taylor's theorem can be omitted with no loss of continuity.

of calculus. Let $f^{(n)}(x)$ denote the nth derivative of the polynomial $f(x)$, and write $f^{(0)}(x) = f(x)$. Then, if

$$f(x) = a_0 + a_1(x - a) + a_2(x - a)^2 + \cdots + a_n(x - a)^n$$

it is clear that $a_0 = f(a) = f^{(0)}(a)$. Differentiation gives

$$f^{(1)}(x) = a_1 + 2a_2(x - a) + 3a_3(x - a)^2 + \cdots + na_n(x - a)^{n-1}$$

and substituting $x = a$ yields $a_1 = f^{(1)}(a)$. This process continues to give $a_2 = \dfrac{f^{(2)}(a)}{2!}$, $a_3 = \dfrac{f^{(3)}(a)}{3!}$, and so on, where $k! = k(k-1)\cdots 2 \cdot 1$. Hence we obtain the following:

COROLLARY 3

Taylor's Theorem

If $f(x)$ is a polynomial of degree n, then

$$f(x) = f(a) + \frac{f^{(1)}(a)}{1!}(x - a) + \frac{f^{(2)}(a)}{2!}(x - a)^2 + \cdots + \frac{f^{(n)}(a)}{n!}(x - a)^n$$

EXAMPLE 1

Expand $f(x) = 5x^3 + 10x + 2$ as a polynomial in powers of $x - 1$.

Solution   The derivatives are $f^{(1)}(x) = 15x^2 + 10$, $f^{(2)}(x) = 30x$, and $f^{(3)}(x) = 30$. Hence the Taylor expansion is

$$f(x) = f(1) + \frac{f^{(1)}(1)}{1!}(x - 1) + \frac{f^{(2)}(1)}{2!}(x - 1)^2 + \frac{f^{(3)}(1)}{3!}(x - 1)^3$$
$$= 17 + 25(x - 1) + 15(x - 1)^2 + 5(x - 1)^3$$

◆◆◆

Taylor's theorem is useful in that it provides a formula for the coefficients in the expansion. It is dealt with in calculus texts and will not be pursued here.

Theorem 1 produces bases of $\mathbf{P}_n$ consisting of polynomials of distinct degrees. A different criterion is involved in the next theorem.

THEOREM 2

Let $f_0(x), f_1(x), \ldots, f_n(x)$ be polynomials in $\mathbf{P}_n$. Assume that numbers $a_0, a_1, \ldots, a_n$ exist such that

$$f_i(a_i) \neq 0 \qquad \text{for each } i$$
$$f_i(a_j) = 0 \qquad \text{if } i \neq j$$

Then

1. $\{f_0(x), \ldots, f_n(x)\}$ is a basis of $\mathbf{P}_n$.

2. If $f(x)$ is any polynomial in $\mathbf{P}_n$, its expansion as a linear combination of these basis vectors is

$$f(x) = \frac{f(a_0)}{f_0(a_0)} f_0(x) + \frac{f(a_1)}{f_1(a_1)} f_1(x) + \cdots + \frac{f(a_n)}{f_n(a_n)} f_n(x)$$

Proof

1. It suffices to show that it is linearly independent (because dim $\mathbf{P}_n = n + 1$). Suppose that

$$r_0 \, f_0(x) + r_1 \, f_1(x) + \cdots + r_n \, f_n(x) = 0 \qquad r_i \text{ in } \mathbb{R}$$

Because $f_i(a_0) = 0$ for all $i > 0$, taking $x = a_0$ gives $r_0 \, f_0(a_0) = 0$. But then the fact that $f_0(a_0) \neq 0$ shows that $r_0 = 0$. The proof that $r_i = 0$ for $i > 0$ is analogous.

2. By (1), $f(x) = r_0 f_0(x) + \cdots + r_n f_n(x)$ for *some* numbers r_i. Again, evaluating at a_0 gives $f(a_0) = r_0 \, f_0(a_0)$, so $r_0 = f(a_0)/f_0(a_0)$. Similarly, $r_i = f(a_i)/f_i(a_i)$ for each i. ◆

EXAMPLE 2

Show that $\{x^2 - x, x^2 - 2x, x^2 - 3x + 2\}$ is a basis of $\mathbf{P}_2$.

Solution

Write $f_0(x) = x^2 - x = x(x - 1)$, $f_1(x) = x^2 - 2x = x(x - 2)$, and $f_2(x) = x^2 - 3x + 2 = (x - 1)(x - 2)$. Then the conditions of the theorem are satisfied with $a_0 = 2$, $a_1 = 1$, and $a_2 = 0$.   ◆◆◆

We investigate one natural choice of the polynomials $f_i(x)$ in Theorem 2. To illustrate, let a_0, a_1, and a_2 be distinct numbers and write

$$f_0(x) = \frac{(x - a_1)(x - a_2)}{(a_0 - a_1)(a_0 - a_2)} \qquad f_1(x) = \frac{(x - a_0)(x - a_2)}{(a_1 - a_0)(a_1 - a_2)}$$
$$f_2(x) = \frac{(x - a_0)(x - a_1)}{(a_2 - a_0)(a_2 - a_1)}$$

Then $f_0(a_0) = f_1(a_1) = f_2(a_2) = 1$, whereas $f_i(a_j) = 0$ for $i \neq j$. Hence Theorem 2 applies, and because $f_i(a_i) = 1$ for each i, the formula for expanding any polynomial is simplified.

In fact, this can be generalized with no extra effort. If $a_0, a_1, \ldots, a_n$ are distinct numbers, define the **Lagrange polynomials** $\delta_0(x), \delta_1(x), \ldots, \delta_n(x)$ relative to these numbers as follows:

$$\delta_k(x) = \frac{\prod_{i \neq k} (x - a_i)}{\prod_{i \neq k} (a_k - a_i)} \qquad k = 0, 1, 2, \ldots, n$$

Here the numerator is the product of all the terms $(x - a_0), (x - a_1), \ldots, (x - a_n)$ with $(x - a_k)$ omitted, and a similar remark applies to the denominator. If $n = 3$, these are just the polynomials in the preceding paragraph. If $n = 4$, the polynomial $\delta_1(x)$ takes the form

$$\delta_1(x) = \frac{(x - a_0)(x - a_2)(x - a_3)}{(a_1 - a_0)(a_1 - a_2)(a_1 - a_3)}$$

In the general case, it is clear that $\delta_i(a_i) = 1$ for each i and that $\delta_i(a_j) = 0$ if $i \neq j$. Hence Theorem 2 specializes as Theorem 3.

THEOREM 3
Lagrange Interpolation
Expansion

Let $a_0, a_1, \ldots, a_n$ be distinct numbers. The corresponding set

$$\{\delta_0(x), \delta_1(x), \ldots, \delta_n(x)\}$$

of Lagrange polynomials is a basis of $\mathbf{P}_n$, and any polynomial $f(x)$ in $\mathbf{P}_n$ has the following unique expansion as a linear combination of these polynomials.

$$f(x) = f(a_0)\delta_0(x) + f(a_1)\delta_1(x) + \cdots + f(a_n)\delta_n(x)$$

EXAMPLE 3

Find the Lagrange interpolation expansion for $f(x) = x^2 - 2x + 1$ relative to $a_0 = -1, a_1 = 0$, and $a_2 = 1$.

Solution   The Lagrange polynomials are

$$\delta_0(x) = \frac{(x - 0)(x - 1)}{(-1 - 0)(-1 - 1)} = \frac{1}{2}(x^2 - x)$$

$$\delta_1(x) = \frac{(x + 1)(x - 1)}{(0 + 1)(0 - 1)} = -(x^2 - 1)$$

$$\delta_2(x) = \frac{(x + 1)(x - 0)}{(1 + 1)(1 - 0)} = \frac{1}{2}(x^2 + x)$$

Because $f(-1) = 4, f(0) = 1$, and $f(1) = 0$, the expansion is

$$f(x) = 2(x^2 - x) - (x^2 - 1)$$

◆◆◆

The Lagrange interpolation expansion gives an easy proof of the following important fact.

> ### THEOREM 4
>
> Let $f(x)$ be a polynomial in $\mathbf{P}_n$, and let $a_0, a_1, \ldots, a_n$ denote distinct numbers. If $f(a_i) = 0$ for all i, then $f(x)$ is the zero polynomial (that is, all coefficients are zero).

Proof All the coefficients in the Lagrange expansion of $f(x)$ are zero.   ◆

EXERCISES 5.6

1. Show that any set of polynomials of distinct degrees is linearly independent.

⁸2. Expand each of the following as a polynomial in powers of $x - 1$.
  (a) $f(x) = x^3 - 2x^2 + x - 1$
  ◆(b) $f(x) = x^3 + x + 1$
  (c) $f(x) = x^4$
  ◆(d) $f(x) = x^3 - 3x^2 + 3x$

⁸3. Prove Taylor's theorem for polynomials.

⁸4. Use Taylor's theorem to derive the **binomial theorem:**

$$(1 + x)^n = \binom{n}{0} + \binom{n}{1}x + \binom{n}{2}x^2 + \cdots + \binom{n}{n}x^n$$

Here the **binomial coefficients** $\binom{n}{r}$ are defined by $\binom{n}{r}$

$$= \frac{n!}{r!\,(n-r)!} \quad \text{where } n! = n(n-1)\cdots 2 \cdot 1 \text{ if } n \geq 1,$$

and $0! = 1$.

⁸5. Let $f(x)$ be a polynomial of degree n. Show that, given any polynomial $g(x)$ in $\mathbf{P}_n$, there exist numbers $b_0, b_1, \ldots, b_n$ such that

$$g(x) = b_0\, f(x) + b_1\, f^{(1)}(x) + \cdots + b_n\, f^{(n)}(x)$$

where $f^{(k)}(x)$ denotes the kth derivative of $f(x)$.

6. Use Theorem 2 to show that the following are bases of $\mathbf{P}_2$.
  (a) $\{x^2 - 2x, x^2 + 2x, x^2 - 4\}$
  ◆(b) $\{x^2 - 3x + 2, x^2 - 4x + 3, x^2 - 5x + 6\}$

7. Find the Lagrange interpolation expansion of $f(x)$ relative to $a_0 = 1, a_1 = 2$, and $a_2 = 3$.

  (a) $f(x) = x^2 + 1$ ◆(b) $f(x) = x^2 + x + 1$

8. Let $a_0, a_1, \ldots, a_n$ be distinct numbers. If $f(x)$ and $g(x)$ in $\mathbf{P}_n$ satisfy $f(a_i) = g(a_i)$ for all i, show that $f(x) = g(x)$. [*Hint:* See Theorem 4.]

9. Let $a_0, a_1, \ldots, a_n$ be distinct numbers. If $f(x)$ in $\mathbf{P}_{n+1}$ satisfies $f(a_i) = 0$ for each $i = 0, 1, \ldots, n$, show that $f(x) = r(x - a_0)(x - a_1) \cdots (x - a_n)$ for some r in $\mathbb{R}$. [*Hint: r* is the coefficient of x^{n+1} in $f(x)$. Consider $f(x) - r(x - a_0) \cdots (x - a_n)$ and use Theorem 4.]

10. Let a and b denote distinct numbers.
  (a) Show that $\{(x - a), (x - b)\}$ is a basis of $\mathbf{P}_1$.
  ◆(b) Show that $\{(x - a)^2, (x - a)(x - b), (x - b)^2\}$ is a basis of $\mathbf{P}_2$.
  (c) Show that $\{(x - a)^n, (x - a)^{n-1}(x - b), \ldots, (x - a)(x - b)^{n-1}, (x - b)^n\}$ is a basis of $\mathbf{P}_n$. [*Hint* for part (c): If a linear combination vanishes, evaluate at $x = a$ and $x = b$. Then reduce to the case $n - 2$ by using the fact that if $p(x)q(x) = 0$ in $\mathbf{P}$, then either $p(x) = 0$ or $q(x) = 0$.]

11. Let a and b be two distinct numbers. Assume that $n \geq 2$ and let

$$U_n = \{f(x) \text{ in } \mathbf{P}_n \mid f(a) = 0 = f(b)\}$$

  (a) Show that $U_n = \{(x - a)(x - b)p(x) \mid p(x) \text{ in } \mathbf{P}_{n-2}\}$.
  ◆(b) Show that $\dim U_n = n - 1$. [*Hint:* If $p(x)q(x) = 0$ in $\mathbf{P}$, then either $p(x) = 0$ or $q(x) = 0$.]
  (c) Show that $\{(x - a)^{n-1}(x - b), (x - a)^{n-2}(x - b)^2, \ldots, (x - a)^2(x - b)^{n-2}, (x - a)(x - b)^{n-1}\}$ is a basis of U_n. [*Hint:* Exercise 10.]

---

⁸This exercise requires polynomial differentiation.

Section 5.7   An Application to Differential Equations (Optional)

Let f be a function of a real variable x, and let f' and f'' denote the first and second derivatives of f. Equations of the form

$$f' + 3f = 0$$
$$f'' + 2f' + f = 0$$

are called **differential equations**. Solving many practical problems comes down to finding functions f satisfying such an equation. The study of differential equations is a very large undertaking and the present book gives a short introduction to how linear algebra aids in the solution of these equations (we return to this subject in Section 6.11). Of course, an acquaintance with calculus is required.

The simplest example is the **first-order** equation

$$f' + af = 0$$

where a is a number. It is easily verified that $f(x) = e^{-ax}$ is one solution, and this equation is simple enough for us to find *all* solutions. In fact, suppose f is *any* solution so that $f'(x) + af(x) = 0$. Then consider the new function given by $g(x) = f(x)e^{ax}$. The product rule of differentiation gives

$$g'(x) = f(x)[ae^{ax}] + f'(x)e^{ax}$$
$$= af(x)e^{ax} - af(x)e^{ax}$$
$$= 0$$

Hence the function $g(x)$ has zero derivative and so must be a constant—say, $g(x) = c$. But then $f(x)e^{ax} = c$, so

$$f(x) = ce^{-ax}$$

In other words, every solution $f(x)$ is just a multiple of the "basic" solution e^{-ax}.

At this point we can see where linear algebra comes into play. The aim is to describe *all* solutions of the equation $f' + af = 0$ — that is, to describe the set

$$U = \{f \mid f' \text{ exists and } f' + af = 0\}$$

But this set U is a vector space. In fact, if f and f_1 both lie in U (so $f' + af = 0$ and $f_1' + af_1 = 0$), then given a number c, the basic theory of differentiation shows that $(f + f_1)' = f' + f_1'$ and $(cf)' = cf'$ both exist and that $f + f_1$ and cf lie in U:

$$(f + f_1)' + a(f + f_1) = (f' + af) + (f_1' + af_1) = 0$$
$$(cf)' + a(cf) = c(f' + af) = 0$$

Hence U is a vector space (in fact, it is a subspace of the space of all real-valued functions), and the previous paragraph shows that e^{-ax} lies in U and *every* member of U is a scalar multiple of e^{-ax}. This can be expressed as in Theorem 1.

> **THEOREM 1**
>
> The set of solutions of the first-order differential equation
>
> $$f' + af = 0$$
>
> is a one-dimensional vector space, and $\{e^{-ax}\}$ is a basis.

EXAMPLE 1

Assume that the number $n(t)$ of bacteria in a culture at time t has the property that the rate of change of n is proportional to n itself. If there are n_0 bacteria present when $t = 0$, find the number at time t.

Solution   Let k denote the proportionality constant. The rate of change of $n(t)$ is its time-derivative $n'(t)$, so the given relationship is $n'(t) = kn(t)$. Thus, $n' - kn = 0$, and Theorem 1 shows that all solutions n are given by $n(t) = ce^{kt}$, where c is a constant. In this case, the constant c is determined by the requirement that there be n_0 bacteria present when $t = 0$. Hence $n_0 = n(0) = ce^{k \cdot 0} = c$, so

$$n(t) = n_0 e^{kt}$$

gives the number at time t. Of course the constant k depends on the strain of bacteria.

◆◆◆

The condition that $n(0) = n_0$ in Example 1 is called an **initial condition** or a **boundary condition** and serves to select one solution from the available solutions. Only one initial condition is needed here because the space of solutions is one-dimensional.

Now consider **second-order** differential equations of the form

$$f'' + af' + bf = 0$$

where a and b are constants. Again the set

$$U = \{f \mid f'' + af' + bf = 0\}$$

is a vector space, and here dim $U = 2$ (we omit the proof). To find a basis for U, it is necessary to introduce the **characteristic polynomial**

$$x^2 + ax + b$$

of the differential equation. Suppose that λ is a real root of this polynomial — that is, $\lambda^2 + a\lambda + b = 0$. Then the function

$$g(x) = e^{\lambda x}$$

is a solution to the differential equation. Indeed,

$$g''(x) + ag'(x) + bg(x) = \lambda^2 e^{\lambda x} + a\lambda e^{\lambda x} + be^{\lambda x}$$
$$= (\lambda^2 + a\lambda + b)e^{\lambda x}$$
$$= 0$$

Hence if λ and μ are two distinct real roots of the characteristic polynomial, then $e^{\lambda x}$ and $e^{\mu x}$ are solutions to the differential equation and so lie in U. Moreover, they are linearly independent because, if $re^{\lambda x} + se^{\mu x} = 0$ for numbers r and s, and if $r \neq 0$, then $e^{(\lambda - \mu)x} = \dfrac{-s}{r}$, so $e^{(\lambda - \mu)x}$ is constant. This is not the case if $\lambda \neq \mu$, so the assumption that $r \neq 0$ is invalid. Thus $r = 0$, and similarly $s = 0$. Hence $\{e^{\lambda x}, e^{\mu x}\}$ is a linearly independent set in U and so, because dim $U = 2$, it is a basis. This establishes the first part of Theorem 2.

THEOREM 2

Let U denote the space of solutions of the second-order differential equation

$$f'' + af' + bf = 0$$

Assume that λ and μ are real roots of the characteristic polynomial $x^2 + ax + b$. Then

1. If $\lambda \neq \mu$, then $\{e^{\lambda x}, e^{\mu x}\}$ is a basis of U.
2. If $\lambda = \mu$, then $\{e^{\lambda x}, xe^{\lambda x}\}$ is a basis of U.

Proof   Except for the fact that dim $U = 2$, (1) was just proved. If $\lambda = \mu$, the verification that $xe^{\lambda x}$ is a solution and that $\{e^{\lambda x}, xe^{\lambda x}\}$ is linearly independent is left as Exercise 4. Then (2) follows because dim $U = 2$.   ◆

EXAMPLE 2

Find all solutions f of $f'' - f' - 6f = 0$.

Solution   The characteristic polynomial is $x^2 - x - 6 = (x - 3)(x + 2)$. The roots are 3 and -2, so $\{e^{3x}, e^{-2x}\}$ is a basis for the space of solutions. Hence every solution has the form

$$f(x) = ce^{3x} + de^{-2x}$$

where c and d are constants.

◆◆◆

The function $f(x) = ce^{3x} + de^{-2x}$ in Example 2 is sometimes referred to as the **general solution** of the differential equation. The constants c and d are determined by two boundary conditions.

EXAMPLE 3

Find the solution of $f'' + 4f' + 4f = 0$ that satisfies the boundary conditions $f(0) = 1$, $f(1) = -1$.

Solution

The characteristic polynomial is $x^2 + 4x + 4 = (x + 2)^2$, so -2 is a double root. Hence $\{e^{-2x}, xe^{-2x}\}$ is a basis for the space of solutions, and the general solution takes the form $f(x) = ce^{-2x} + dxe^{-2x}$. Applying the boundary conditions gives $1 = f(0) = c$ and $-1 = f(1) = (c + d)e^{-2}$. Hence $c = 1$ and $d = -(1 + e^2)$, so the required solution is

$$f(x) = e^{-2x} - (1 + e^2)xe^{-2x}$$

One other question remains: What happens if the roots of the characteristic polynomial are not real? To answer this, we must first state precisely what $e^{\lambda x}$ means when λ is not real. If q is a real number, define

$$e^{iq} = \cos q + i \sin q$$

where $i^2 = -1$. Then the relationship $e^{iq}e^{iq_1} = e^{i(q + q_1)}$ holds for all real q and q_1, as is easily verified. If $\lambda = p + iq$, where p and q are real numbers, we define

$$e^{\lambda} = e^p e^{iq} = e^p(\cos q + i \sin q)$$

Then it is a routine exercise to show that

1. $e^{\lambda}e^{\mu} = e^{\lambda + \mu}$
2. $e^{\lambda} = 1$ if and only if $\lambda = 0$
3. $(e^{\lambda x})' = \lambda e^{\lambda x}$

These imply easily that $f(x) = e^{\lambda x}$ is a solution to $f'' + af' + bf = 0$ if λ is a (possibly complex) root of the characteristic polynomial $x^2 + ax + b$. Now write $\lambda = p + iq$ so that

$$f(x) = e^{\lambda x} = e^{px}\cos(qx) + ie^{px}\sin(qx)$$

For convenience, denote the real and imaginary parts of $f(x)$ as $u(x) = e^{px}\cos(qx)$ and $v(x) = e^{px}\sin(qx)$. Then the fact that $f(x)$ satisfies the differential equation gives

$$0 = f'' + af' + bf = (u'' + au' + bu) + i(v'' + av' + bv)$$

Equating real and imaginary parts shows that $u(x)$ and $v(x)$ are both solutions to the differential equation. This proves part of Theorem 3.

> ### THEOREM 3
>
> Let U denote the space of solutions of the second-order differential equation
>
> $$f'' + af' + bf = 0$$
>
> where a and b are real. Suppose λ is a nonreal root of the characteristic polynomial $x^2 + ax + b$. If $\lambda = p + iq$, where p and q are real, then
>
> $$\{e^{px}\cos(qx),\ e^{px}\sin(qx)\}$$
>
> is a basis of U.

Proof   The foregoing discussion shows that these functions lie in U. Because dim U = 2 (a fact that we have not proved), it suffices to show that they are linearly independent. But if

$$re^{px}\cos(qx) + se^{px}\sin(qx) = 0$$

for all x, then $r\cos(qx) + s\sin(qx) = 0$ for all x (because $e^{px} \neq 0$). Taking $x = 0$ gives $r = 0$, and taking $x = \frac{\pi}{2q}$ gives $s = 0$ ($q \neq 0$ because λ is not real). This is what we wanted.   ◆

EXAMPLE 4

Find the solution $f(x)$ to $f'' - 2f' + 2f = 0$ that satisfies $f(0) = 2$ and $f(\frac{\pi}{2}) = 0$.

Solution   The characteristic polynomial $x^2 - 2x + 2$ has roots $1 + i$ and $1 - i$. Taking $\lambda = 1 + i$ (quite arbitrarily) gives $p = q = 1$ in the notation of Theorem 3, so $\{e^x\cos x, e^x\sin x\}$ is a basis for the space of solutions. The general solution is thus $f(x) = e^x(r\cos x + s\sin x)$. The boundary conditions yield $2 = f(0) = r$ and $0 = f(\frac{\pi}{2}) = e^{\pi/2}s$. Thus $r = 2$ and $s = 0$, and the required solution is $f(x) = 2e^x\cos x$.   ◆◆◆

The following theorem is an important special case of Theorem 3.

> ### THEOREM 4
>
> If $q \neq 0$ is a real number, the space of solutions to the differential equation $f'' + q^2f = 0$ has basis $\{\cos(qx), \sin(qx)\}$.

Proof   The characteristic polynomial $x^2 + q^2$ has roots qi and $-qi$, so Theorem 3 applies with $p = 0$.   ◆

In many situations, the displacement $s(t)$ of some object at time t turns out to have an oscillating form $s(t) = c\sin(at) + d\cos(at)$. These are called **simple harmonic motions**. An example follows.

EXAMPLE 5

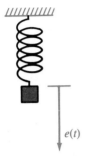

A weight is attached to an extension spring (see diagram). If it is pulled from the equilibrium position and released, it is observed to oscillate up and down. Let $e(t)$ denote the distance of the weight below the equilibrium position t seconds later. It is known (**Hooke's law**) that the acceleration $e''(t)$ of the weight is proportional to the displacement $e(t)$ and in the opposite direction. That is,

$$e''(t) = -ke(t)$$

where $k > 0$ is called the **spring constant**. Find $e(t)$ if the maximum extension is 10 cm below the equilibrium position and find the **period** of the oscillation (time taken for the weight to make a full oscillation).

Solution   It follows from Theorem 4 (with $q^2 = k$) that

$$e(t) = c\,\sin(\sqrt{k}t) + d\,\cos(\sqrt{k}t)$$

where c and d are constants. The condition $e(0) = 0$ gives $d = 0$, so $e(t) = c\,\sin(\sqrt{k}t)$. Now the maximum value of the function $\sin x$ is 1 (when $x = \frac{\pi}{2}$), so $c = 10$ $\left(\text{when } t = \frac{\pi}{2\sqrt{k}}\right)$. Hence

$$e(t) = 10\,\sin(\sqrt{k}t)$$

Finally the weight goes through a full oscillation as $\sqrt{k}t$ increases from 0 to 2π. The time taken is $t = \frac{2\pi}{\sqrt{k}}$, the period of the oscillation.

◆◆◆

EXERCISES 5.7

1. Find a solution f to each of the following differential equations satisfying the given boundary conditions.

   (a) $f' - 3f = 0; f(1) = 2$

   ◆**(b)** $f' + f = 0; f(1) = 1$

   (c) $f'' + 2f' - 15f = 0; f(1) = f(0) = 0$

   ◆**(d)** $f'' + f' - 6f = 0; f(0) = 0, f(1) = 1$

   (e) $f'' - 2f' + f = 0; f(1) = f(0) = 1$

   ◆**(f)** $f'' - 4f' + 4f = 0; f(0) = 2, f(-1) = 0$

   (g) $f'' - 3af' + 2a^2f = 0, a \neq 0; f(0) = 0,$
      $f(1) = 1 - e^a$

   ◆**(h)** $f'' - a^2f = 0, a \neq 0; f(0) = 1, f(1) = 0$

   (i) $f'' - 2f' + 5f = 0; f(0) = 1, f(\frac{\pi}{4}) = 0$

   ◆**(j)** $f'' + 4f' + 5f = 0; f(0) = 0, f(\frac{\pi}{2}) = 1$

2. Show that the solution to $f' + af = 0$ satisfying $f(x_0) = k$ is $f(x) = ke^{a(x_0 - x)}$.

3. If the characteristic polynomial of $f'' + af' + bf = 0$ has real roots, show that $f = 0$ is the only solution satisfying $f(0) = 0 = f(1)$.

4. Complete the proof of Theorem 2. [*Hint:* If λ is a double root of $x^2 + ax + b$, show that $a = -2\lambda$ and $b = \lambda^2$.]

5. (a) Given the equation $f' + af = b$, $(a \neq 0)$, make the substitution $f(x) = g(x) + b/a$ and obtain a differential equation for g. Then derive the general solution for $f' + af = b$.

   ◆**(b)** Find the general solution to $f' + f = 2$.

6. Consider the differential equation $f'' + af' + bf = g$, where g is some fixed function. Assume that f_0 is one solution of this equation.

   (a) Show that the general solution is $cf_1 + df_2 + f_0$, where c and d are constants and $\{f_1, f_2\}$ is any basis for the solutions to $f'' + af' + bf = 0$.

   ◆**(b)** Find a solution to $f'' + f' - 6f = 2x^3 - x^2 - 2x$.
      [*Hint:* Try $f(x) = \frac{-1}{3}x^3$.]

7. A radioactive element decays at a rate proportional to the amount present. Suppose an initial mass of 10 g decays to 8 g in 3 hours.

   (a) Find the mass t hours later.

   ◆**(b)** Find the *half-life* of the element — the time it takes to decay to half its mass.

8. The population $N(t)$ of a region at time t increases at a rate proportional to the population. If the population doubles in 5 years and is 3,000,000 initially, find $N(t)$.

◆**9.** Consider a spring, as in Example 5. If the period of the oscillation is 30 seconds, find the spring constant k.

10. As a pendulum swings (see the diagram), let t measure the time since it was vertical. The angle $\theta = \theta(t)$ from the vertical can be shown to satisfy the equation $\theta'' + k\theta = 0$, provided that θ is small. If the maximal angle is

$\theta = 0.05$ radians, find $\theta(t)$ in terms of k. If the period is 0.5 seconds, find k. [Assume that $\theta = 0$ when $t = 0$.]

SUPPLEMENTARY EXERCISES FOR CHAPTER 5

1. (Requires calculus) Let V denote the space of all functions $f\colon \mathbb{R} \to \mathbb{R}$ for which the derivatives f' and f'' exist. Show that f_1, f_2, and f_3 in V are linearly independent provided that their **Wronskian** $w(x)$ is nonzero for some x, where

$$w(x) = \det\begin{bmatrix} f_1(x) & f_2(x) & f_3(x) \\ f_1'(x) & f_2'(x) & f_3'(x) \\ f_1''(x) & f_2''(x) & f_3''(x) \end{bmatrix}$$

2. Let $\{v_1, v_2, \ldots, v_n\}$ be a basis of $\mathbb{R}^n$ (written as columns), and let A be an $n \times n$ matrix.

  (a) If A is invertible, show that $\{Av_1, Av_2, \ldots, Av_n\}$ is a basis of $\mathbb{R}^n$.

◆**(b)** If $\{Av_1, Av_2, \ldots, Av_n\}$ is a basis of $\mathbb{R}^n$, show that A is invertible.

3. If A is an $m \times n$ matrix, show that A has rank m if and only if col A contains every column of I_m.

◆**4.** Show that null $A = $ null$(A^T A)$ for any real matrix A.

5. If U and W are subspaces of V, consider the subspaces $U \cap W$ and $U + W$ (Exercises 34§5.3 and 22§5.4). Show that $\dim(U + W) = \dim U + \dim W - \dim(U \cap W)$. [*Hint:* If $\{z_1, \ldots, z_k\}$ is a basis of $U \cap W$, let $\{z_1, \ldots, z_k, u_{k+1}, \ldots, u_m\}$ and $\{z_1, \ldots, z_k, w_{k+1}, \ldots, w_n\}$ be bases of U and W, respectively. Show that $\{z_1, \ldots, z_k, u_{k+1}, \ldots, u_m, w_{k+1}, \ldots, w_n\}$ is a basis of $U + W$.]

6. Let A be an $m \times n$ matrix of rank r. Show that $\dim(\text{null } A) = n - r$ (Theorem 5§5.5) as follows. Choose a basis $\{X_1, \ldots, X_k\}$ of null A and extend it to a basis $\{X_1, \ldots, X_k, Z_1, \ldots, Z_m\}$ of $\mathbb{R}^n$. Show that $\{AZ_1, \ldots, AZ_m\}$ is a basis of col A.

7. Suppose that R and S are two $m \times n$ reduced row-echelon matrices that are row-equivalent. Prove that $R = S$. [*Hint:* Let rank $R = $ rank $S = r$. Use induction on m. If $m > 1$, write $R = US$, where U is an $m \times m$ invertible matrix. Show that the first column of U equals the first column of I_m by examining the first nonzero column of R and of US. Write $U = \begin{bmatrix} 1 & X \\ 0 & V \end{bmatrix}$, $R = \begin{bmatrix} r & P \\ 0 & R' \end{bmatrix}$, $S = \begin{bmatrix} s & Q \\ 0 & S' \end{bmatrix}$ in block form, and conclude $R' = S'$ by induction. Then show that the first r columns of U equal the first r columns of I_m by examining the columns in $R = US$ that contain leading 1s. Finally, write $U = \begin{bmatrix} I_r & Y \\ 0 & Z \end{bmatrix}$, $R = \begin{bmatrix} R_1 & R_2 \\ 0 & 0 \end{bmatrix}$, and $S = \begin{bmatrix} S_1 & S_2 \\ 0 & 0 \end{bmatrix}$, where R_1 and S_1 are $r \times r$. Conclude that $R = S$.]

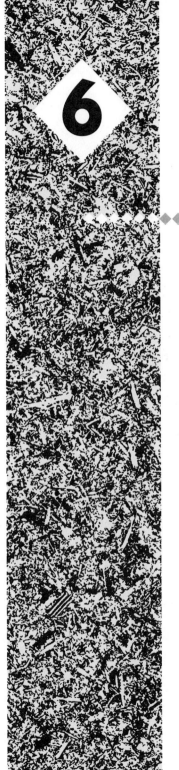

6

Eigenvalues and Diagonalization

If A is a square matrix, a number λ is called an *eigenvalue* of A if the system of linear equations

$$AX = \lambda X$$

has nonzero solutions X (called *eigenvectors*). Eigenvalues arise in many physical applications (such as natural frequencies and energy levels of electrons to name only two).[1] Applications often arise as follows: The eigenvalues and eigenvectors of A can be used to find an invertible matrix P such that $P^{-1} AP$ is a diagonal matrix (all entries off the main diagonal are zero). This *diagonalization* procedure is discussed in Section 6.2 and is applied to geometry in Section 6.9 and to differential equations in Section 6.11. Matrices whose eigenvalues are all positive are examined in Section 6.5.

Section 6.1
Eigenvalues and Similarity

DEFINITION

If A is an $n \times n$ matrix, a real number λ is called an **eigenvalue** of A if

$$AX = \lambda X$$

for some nonzero column X in $\mathbb{R}^n$.[2] In this case the (nonzero) vector X is called an **eigenvector** of A corresponding to λ. Eigenvalues and eigenvectors are often called **characteristic values** and **characteristic vectors,** respectively. If λ

---

[1] For example, soldiers break step when crossing a bridge to avoid inadvertently marching at the natural frequency of the bridge and so causing it to oscillate wildly.

[2] A real matrix A can have $AX = \lambda X$ where $X \neq 0$ and λ is a nonreal complex number (also called an eigenvalue of A). More on this later.

is an eigenvalue of A, the set

$$E_\lambda = E_\lambda(A) = \{X \mid X \text{ in } \mathbb{R}^n, AX = \lambda X\}$$

is a vector space (subspace of $\mathbb{R}^n$) called the **eigenspace** associated with λ.

Hence E_λ consists of all eigenvectors corresponding to λ, together with the zero vector. The condition $AX = \lambda X$ can be written

$$(\lambda I - A)X = 0$$

where I is the $n \times n$ identity matrix. Hence E_λ consists of all solutions to this system of linear equations, and λ is an eigenvalue if it happens that E_λ contains at least one nonzero vector.

EXAMPLE 1

Show that $\lambda = -3$ is an eigenvalue of $A = \begin{bmatrix} 5 & 8 & 16 \\ 4 & 1 & 8 \\ -4 & -4 & -11 \end{bmatrix}$, and find the corresponding eigenspace E_{-3}.

Solution   If $\lambda = -3$, the condition that $(\lambda I - A)X = 0$ becomes

$$\begin{aligned} -8x_1 - 8x_2 - 16x_3 &= 0 \\ -4x_1 - 4x_2 - 8x_3 &= 0 \qquad \text{where} \qquad X = \begin{bmatrix} x_1 \\ x_2 \\ x_3 \end{bmatrix} \\ 4x_1 + 4x_2 + 8x_3 &= 0 \end{aligned}$$

The solution is $X = sX_1 + tX_2$, where $X_1 = \begin{bmatrix} -1 \\ 1 \\ 0 \end{bmatrix}$ and $X_2 = \begin{bmatrix} -2 \\ 0 \\ 1 \end{bmatrix}$. Hence E_{-3} = span$\{X_1, X_2\}$.

The technique in Example 1 leads to a way of determining all the possible eigenvalues. Let A be any $n \times n$ matrix. A number λ is an eigenvalue of A if and only if

$$AX = \lambda X \text{ for some } X \neq 0$$

As before, this condition can be written

$$(\lambda I - A)X = 0 \text{ for some } X \neq 0$$

where I is the $n \times n$ identity matrix. Now recall (Theorem 6§2.4) that a matrix U is invertible if and only if $UX = 0$ implies that $X = 0$. Hence λ is an eigenvalue of A if and only if $\lambda I - A$ is *not* invertible, and this in turn means that

$$\det(\lambda I - A) = 0$$

by Theorem 2§3.2. This is a very useful test for the eigenvalues of A, and it suggests the following definition:

DEFINITION

The **characteristic polynomial** of the $n \times n$ matrix A is defined to be

$$c_A(x) = \det(xI - A)$$

Theorem 1 summarizes this discussion.

THEOREM 1

Let A be an $n \times n$ matrix. The eigenvalues of A are the real roots of the characteristic polynomial of A. That is, they are the real numbers λ satisfying

$$c_A(\lambda) = \det(\lambda I - A) = 0$$

where I is the $n \times n$ identity matrix. In this case, the eigenspace E_λ corresponding to λ is given by

$$E_\lambda = \{X \mid (\lambda I - A)X = 0\}$$

and so consists of all solutions to a system of n linear equations in n variables. The eigenvectors corresponding to λ are the nonzero vectors in the eigenspace E_λ.

Hence, determining the eigenspaces of a matrix is reduced to two problems: First find the eigenvalues, and then find the eigenspaces as sets of solutions to linear homogeneous equations. This latter problem has been treated in this book. The problem of determining the eigenvalues is more difficult, and we will not spend a great deal of time on it. Hence the examples and exercises will be so constructed that the roots of the characteristic polynomials encountered are relatively easy to find (usually integers). The reader should not be misled by this into thinking that eigenvalues are so easily obtained for the matrices that occur in practical applications! In fact, the eigenvalues are *not* usually found as roots of the characteristic polynomial, and two other techniques for finding them are briefly described in Section 6.7.

EXAMPLE 2

Find the characteristic polynomial of the matrix $A = \begin{bmatrix} 5 & 8 & 16 \\ 4 & 1 & 8 \\ -4 & -4 & -11 \end{bmatrix}$ mentioned in

Example 1 and use it to find all eigenvalues and eigenspaces of A.

Solution

The characteristic polynomial is

$$c_A(x) = \det(xI - A) = \det \begin{bmatrix} x - 5 & -8 & -16 \\ -4 & x - 1 & -8 \\ 4 & 4 & x + 11 \end{bmatrix}$$

Adding the last row to the second row gives

$$c_A(x) = \det \begin{bmatrix} x - 5 & -8 & -16 \\ 0 & x + 3 & x + 3 \\ 4 & 4 & x + 11 \end{bmatrix}$$

$$= (x + 3)(x^2 + 2x - 3) = (x + 3)^2(x - 1)$$

so the eigenvalues are $\lambda = -3$ and $\lambda = 1$. The case $\lambda = -3$ was dealt with in Example 1, where it was shown that the corresponding eigenspace is $E_{-3} = \text{span}\{X_1,$

$X_2\}$ where $X_1 = \begin{bmatrix} -1 \\ 1 \\ 0 \end{bmatrix}$ and $X_2 = \begin{bmatrix} -2 \\ 0 \\ 1 \end{bmatrix}$. However, the characteristic polynomial has

revealed a new eigenvalue, $\lambda = 1$. The corresponding eigenspace consists of all

$X = \begin{bmatrix} x_1 \\ x_2 \\ x_3 \end{bmatrix}$ satisfying $(I - A)X = 0$. This gives the equations

$$\begin{bmatrix} -4 & -8 & -16 \\ -4 & 0 & -8 \\ 4 & 4 & 12 \end{bmatrix} \begin{bmatrix} x_1 \\ x_2 \\ x_3 \end{bmatrix} = \begin{bmatrix} 0 \\ 0 \\ 0 \end{bmatrix} \quad \text{whence } X = tX_3, \text{ where } X_3 = \begin{bmatrix} 2 \\ 1 \\ -1 \end{bmatrix}$$

Hence the eigenspace corresponding to $\lambda = 1$ is $E_1 = \text{span}\{X_3\}$.

◆◆◆

The characteristic polynomial in Example 2 is $(x + 3)^2(x - 1)$, and the eigenspaces E_{-3} and E_1 have dimensions 2 and 1, respectively. These dimensions are equal to the multiplicities of $\lambda = -3$ and $\lambda = 1$ as roots of the characteristic polynomial, and the reader might be justified in asking whether this is always true. The next example shows that this inference cannot be drawn in general (it is true for symmetric matrices, however, as we shall see).

EXAMPLE 3

Find the characteristic polynomial, eigenvalues, and eigenspaces for A
$$= \begin{bmatrix} 2 & 1 & 1 \\ 2 & 1 & -2 \\ -1 & 0 & -2 \end{bmatrix}.$$

Solution   The characteristic polynomial is

$$c_A(x) = \det(xI - A) = \det \begin{bmatrix} x-2 & -1 & -1 \\ -2 & x-1 & 2 \\ 1 & 0 & x+2 \end{bmatrix}$$

$$= x^3 - x^2 - 5x - 3 = (x+1)^2(x-3)$$

so the eigenvalues are $\lambda = -1, 3$. For $\lambda = -1$, the eigenspace consists of all $X = \begin{bmatrix} x_1 \\ x_2 \\ x_3 \end{bmatrix}$ such that $(-I - A)X = 0$. This gives

$$\begin{bmatrix} -3 & -1 & -1 \\ -2 & -2 & 2 \\ 1 & 0 & 1 \end{bmatrix} \begin{bmatrix} x_1 \\ x_2 \\ x_3 \end{bmatrix} = \begin{bmatrix} 0 \\ 0 \\ 0 \end{bmatrix} \qquad \text{so that } X = tX_1, \text{ where } X_1 = \begin{bmatrix} -1 \\ 2 \\ 1 \end{bmatrix}$$

Hence the eigenspace is $E_{-1} = \mathbb{R}X_1$, and this has dimension 1 even though the eigenvalue $\lambda = -1$ is a *double* root of $c_A(x)$. For $\lambda = 3$, the eigenspace is given by

$$E_3 = \mathbb{R}X_2, \text{ where } X_2 = \begin{bmatrix} 5 \\ 6 \\ -1 \end{bmatrix}.$$

◆◆◆

The following example gives one situation wherein the characteristic polynomial and eigenvalues are easy to determine.

EXAMPLE 4

If A is a triangular matrix, show that the eigenvalues of A are the entries on the main diagonal.

Solution   If the main diagonal entries of A are $a_{11}, a_{22}, \ldots, a_{nn}$, then those of $xI - A$ are $x - a_{11}, x - a_{22}, \ldots, x - a_{nn}$. Hence Theorem 4§3.1 gives

$$c_A(x) = \det(xI - A) = (x - a_{11})(x - a_{22}) \cdots (x - a_{nn})$$

The result follows.   ◆◆◆

EXAMPLE 5

Show that A and A^T have the same characteristic polynomial and hence the same eigenvalues.

Solution   We use the fact (Theorem 3§3.2) that a matrix and its transpose have the same determinant. Hence

$$c_{A^T}(x) = \det(xI - A^T) = \det[(xI - A)^T] = \det(xI - A) = c_A(x)$$

The result follows from Theorem 1.

◆◆◆

There are many examples of matrices with real entries that have no (real) eigenvalues (for example, $\begin{bmatrix} 0 & 1 \\ -1 & 0 \end{bmatrix}$ has eigenvalues i and $-i$, where $i^2 = -1$). We will have more to say about such matrices in Section 6.8, where the following important result will be proved.

THEOREM 2

If A is a real symmetric matrix, each root of the characteristic polynomial $c_A(x)$ is real.

EXAMPLE 6

Confirm Theorem 2 for any symmetric 2×2 matrix A.

Solution   Write $A = \begin{bmatrix} a & b \\ b & c \end{bmatrix}$. The characteristic polynomial is

$$c_A(x) = \det\begin{bmatrix} x - a & -b \\ -b & x - c \end{bmatrix} = x^2 - (a + c)x + (ac - b^2)$$

This has real roots because the discriminant $(a + c)^2 - 4(ac - b^2) = (a - c)^2 + 4b^2$ is nonnegative.

◆◆◆

Similar Matrices

We now introduce a relationship for square matrices that will be fundamental for the rest of this chapter.

DEFINITION

Two $n \times n$ matrices A and B are called **similar** (denoted $A \sim B$) if

$$B = P^{-1}AP$$

holds for some invertible matrix P.

The matrix P is not unique (if $A = B = I$, *any* invertible P will do).

Similarity can be used to simplify a variety of matrix calculations, particularly when it can be shown that the matrix in question is similar to a diagonal matrix. Here is an example.

EXAMPLE 7

Let $A = \begin{bmatrix} 5 & 4 \\ -2 & -1 \end{bmatrix}$, $P = \begin{bmatrix} 1 & 2 \\ 1 & 1 \end{bmatrix}$, and $D = \begin{bmatrix} 1 & 0 \\ 0 & 3 \end{bmatrix}$. Show that $A = P^{-1}DP$ and use this to compute A^n for $n \geq 1$.

Solution   We leave it to the reader to verify that $A = P^{-1}DP$ (equivalently, $PA = DP$). Hence $A^2 = (P^{-1}DP)(P^{-1}DP) = P^{-1}D^2P$, and, similarly $A^3 = P^{-1}D^3P$. In general,

$$A^n = P^{-1}D^nP = \begin{bmatrix} -1 & 2 \\ 1 & -1 \end{bmatrix} \begin{bmatrix} 1 & 0 \\ 0 & 3^n \end{bmatrix} \begin{bmatrix} 1 & 2 \\ 1 & 1 \end{bmatrix}$$

$$= \begin{bmatrix} 2 \cdot 3^n - 1 & 2 \cdot 3^n - 2 \\ 1 - 3^n & 2 - 3^n \end{bmatrix}$$

◆◆◆

The work in Example 7 was simplified because calculations with a diagonal matrix are easy. Matrices that are similar to a diagonal matrix are discussed in the next section.

It is sometimes difficult to decide whether A and B are similar.

EXAMPLE 8

Let $A = \begin{bmatrix} 2 & 1 \\ -1 & -1 \end{bmatrix}$, $B = \begin{bmatrix} -2 & 5 \\ -1 & 3 \end{bmatrix}$, and $B_1 = \begin{bmatrix} 5 & 2 \\ 4 & 1 \end{bmatrix}$. Show that A and B are similar but A and B_1 are not.

Solution   If $P = \begin{bmatrix} -1 & 3 \\ 1 & -2 \end{bmatrix}$, the reader can verify that $P^{-1}AP = B$ (rather than computing P^{-1}, it is easier to check that P is invertible and that $AP = PB$). Hence A and B are similar. On the other hand, it is easy to verify that similar matrices have equal determinants, and so A and B_1 are not similar because $\det A = -1$ whereas $\det B_1 = -3$.

◆◆◆

The fact (utilized in Example 8) that similar matrices have equal determinants is sometimes expressed by saying that the determinant of a matrix is a **similarity invariant.** Other examples are given in Theorem 4; one of them is defined as follows.

DEFINITION

The **trace** of a square matrix A (denoted tr A) is defined to be the sum of the entries on the main diagonal of A.

We pause to record some properties of the trace.

THEOREM 3

Let A and B be $n \times n$ matrices, and let k be a scalar.

1. $\text{tr}(A + B) = \text{tr}A + \text{tr}B$ and $\text{tr}(kA) = k \text{ tr } A$
2. $\text{tr}(AB) = \text{tr}(BA)$

Proof We leave (1) to the reader. If $A = [a_{ij}]$ and $B = [b_{ij}]$, then the (i, j)-entry of AB is $\sum_{k=1}^{n} a_{ik}b_{kj}$. Hence

$$\text{tr}(AB) = \sum_{m=1}^{n} \left[\sum_{k=1}^{n} a_{mk} b_{km} \right] = \sum_{k=1}^{n} \left[\sum_{m=1}^{n} b_{km} a_{mk} \right] = \text{tr}(BA) \qquad \blacklozenge$$

THEOREM 4

If A and B are similar matrices, they have the same determinant, the same rank, the same trace, the same characteristic polynomial, and the same eigenvalues.

Proof Let $B = P^{-1}AP$, where P is invertible. Then Theorems 1§3.2 and 2§3.2 give

$$\det B = \det P^{-1} \det A \det P = \frac{1}{\det P} \det A \det P = \det A$$

Next, Corollary 4 of Theorem 3§5.5 gives

$$\text{rank } B = \text{rank}(P^{-1}AP) = \text{rank}(AP) = \text{rank } A$$

As to the trace, use Theorem 3 to get

$$\text{tr } B = \text{tr}(P^{-1}AP) = \text{tr}[(AP)P^{-1}] = \text{tr } A$$

Finally, the fact that $xI = P^{-1}(xI)P$ gives

$$c_B(x) = \det(xI - B) = \det[P^{-1}(xI - A)P] = \det(xI - A) = c_A(x)$$

Hence A and B share the same characteristic polynomial and so have the same eigenvalues by Theorem 1. $\qquad \blacklozenge$

EXAMPLE 9

The matrices $I = \begin{bmatrix} 1 & 0 \\ 0 & 1 \end{bmatrix}$ and $A = \begin{bmatrix} 1 & 2 \\ 0 & 1 \end{bmatrix}$ have the same determinant, the same rank, the same trace, the same characteristic polynomial, and the same eigenvalues.

However, they are *not* similar because $P^{-1}IP = I$ for all invertible P. Hence the invariants in Theorem 4 do not characterize similarity.

——————————————————————————————————————◆◆◆

One other comment on similarity is in order. Suppose $A \sim B$, say $B = P^{-1}AP$ for some invertible matrix P. This implies that $A = PBP^{-1}$, and this, in turn, can be written in the form $A = Q^{-1}BQ$, where $Q = P^{-1}$. Hence $B \sim A$, and we have verified the second of the following three properties of similarity (the others are left as an exercise).

> **THEOREM 5**
>
> Let A, B, and C denote $n \times n$ matrices. Then:
>
> 1. $A \sim A$ for all A.
> 2. If $A \sim B$, then $B \sim A$.
> 3. If $A \sim B$ and $B \sim C$, then $A \sim C$.

The properties of similarity in Theorem 5 are useful in the following way. If we want to prove that A and B are similar, it is sometimes convenient to show that both are similar to some "nice" matrix D (possibly diagonal): $A \sim D$ and $B \sim D$. Then $D \sim B$ by (2) of Theorem 5, so $A \sim D$ and $D \sim B$ imply that $A \sim B$ by (3).

EXERCISES 6.1

1. Find the characteristic polynomial, eigenvalues, and a basis for each eigenspace for A if:

   (a) $A = \begin{bmatrix} 1 & 2 \\ 3 & 2 \end{bmatrix}$

   ◆(b) $A = \begin{bmatrix} 2 & -4 \\ -1 & -1 \end{bmatrix}$

   (c) $A = \begin{bmatrix} 7 & 0 & -4 \\ 0 & 5 & 0 \\ 5 & 0 & -2 \end{bmatrix}$

   ◆(d) $A = \begin{bmatrix} 1 & 1 & -3 \\ 2 & 0 & 6 \\ 1 & -1 & 5 \end{bmatrix}$

   (e) $A = \begin{bmatrix} 1 & -2 & 3 \\ 2 & 6 & -6 \\ 1 & 2 & -1 \end{bmatrix}$

   ◆(f) $A = \begin{bmatrix} 0 & 1 & 1 \\ 1 & 0 & 1 \\ 1 & 1 & 0 \end{bmatrix}$

   (g) $A = \begin{bmatrix} 3 & 1 & 1 \\ -4 & -2 & -5 \\ 2 & 2 & 5 \end{bmatrix}$

   ◆(h) $A = \begin{bmatrix} 2 & 1 & 1 \\ 0 & 1 & 0 \\ 1 & -1 & 2 \end{bmatrix}$

   (i) $A = \begin{bmatrix} \lambda & 0 & 0 \\ 0 & \lambda & 0 \\ 0 & 0 & \mu \end{bmatrix}$, $\lambda \neq \mu$

2. If $A = \begin{bmatrix} \lambda & a & b \\ 0 & \lambda & c \\ 0 & 0 & \lambda \end{bmatrix}$, show that λ is the only eigenvalue of A and that

   $$E_\lambda = \left\{ \begin{bmatrix} r \\ s_1 \\ s_2 \end{bmatrix} \middle| \begin{bmatrix} a & b \\ 0 & c \end{bmatrix} \begin{bmatrix} s_1 \\ s_2 \end{bmatrix} = \begin{bmatrix} 0 \\ 0 \end{bmatrix} \right\}$$

3. Show that A has $\lambda = 0$ as an eigenvalue if and only if A is not invertible.

4. Let A denote $n \times n$ matrix and put $A_1 = A - \alpha I$, α in $\mathbb{R}$. Show that λ is an eigenvalue of A if and only if $\lambda - \alpha$ is an eigenvalue of A_1. How do the eigenvectors compare? (Hence the eigenvalues of A_1 are just those of A "shifted" by α.)

5. Show that the eigenvalues of $\begin{bmatrix} \cos\theta & \sin\theta \\ -\sin\theta & \cos\theta \end{bmatrix}$ are $e^{i\theta}$ and $e^{-i\theta}$. (See Appendix A.)

6. Find the characteristic polynomial of the $n \times n$ identity matrix I. Show that I has exactly one eigenvalue and find the eigenspace.

7. Given $A = \begin{bmatrix} a & b \\ c & d \end{bmatrix}$, show that:

(a) $c_A(x) = x^2 - \text{tr } A\, x + \det A$

(b) The eigenvalues are $\frac{1}{2}[(a + d) \pm \sqrt{(a - d)^2 + 4bc}]$

8. By computing the trace, determinant, and rank, show that A and B are *not* similar in each case.

(a) $A = \begin{bmatrix} 1 & 2 \\ 2 & 1 \end{bmatrix}$, $B = \begin{bmatrix} 1 & 1 \\ -1 & 1 \end{bmatrix}$

(b) $A = \begin{bmatrix} 3 & 1 \\ 2 & -1 \end{bmatrix}$, $B = \begin{bmatrix} 1 & 1 \\ 2 & 1 \end{bmatrix}$

(c) $A = \begin{bmatrix} 2 & 1 \\ 1 & -1 \end{bmatrix}$, $B = \begin{bmatrix} 3 & 0 \\ 1 & -1 \end{bmatrix}$

(d) $A = \begin{bmatrix} 3 & 1 \\ -1 & 2 \end{bmatrix}$, $B = \begin{bmatrix} 2 & -1 \\ 3 & 2 \end{bmatrix}$

(e) $A = \begin{bmatrix} 2 & 1 & 1 \\ 1 & 0 & 1 \\ 1 & 1 & 0 \end{bmatrix}$, $B = \begin{bmatrix} 1 & -2 & 1 \\ -2 & 4 & -2 \\ -3 & 6 & -3 \end{bmatrix}$

(f) $A = \begin{bmatrix} 1 & 2 & -3 \\ 1 & -1 & 2 \\ 0 & 3 & -5 \end{bmatrix}$, $B = \begin{bmatrix} -2 & 1 & 3 \\ 6 & -3 & -9 \\ 0 & 0 & 0 \end{bmatrix}$

9. Show that $\begin{bmatrix} 1 & 2 & -1 & 0 \\ 2 & 0 & 1 & 1 \\ 1 & 1 & 0 & -1 \\ 4 & 3 & 0 & 0 \end{bmatrix}$ and $\begin{bmatrix} 1 & -1 & 3 & 0 \\ -1 & 0 & 1 & 1 \\ 0 & -1 & 4 & 1 \\ 5 & -1 & -1 & -4 \end{bmatrix}$ are *not* similar.

10. If $A \sim B$, show that:

(a) $A^T \sim B^T$　　(b) $A^{-1} \sim B^{-1}$

(c) $rA \sim rB$ for r in $\mathbb{R}$　　(d) $A^n \sim B^n$ for $n \geq 1$

11. In each case, find $P^{-1}AP$ and then compute A^n.

(a) $A = \begin{bmatrix} 6 & -5 \\ 2 & -1 \end{bmatrix}$, $P = \begin{bmatrix} 1 & 5 \\ 1 & 2 \end{bmatrix}$

(b) $A = \begin{bmatrix} -7 & -12 \\ 6 & 10 \end{bmatrix}$, $P = \begin{bmatrix} -3 & 4 \\ 2 & -3 \end{bmatrix}$

[*Hint*: $(PDP^{-1})^n = PD^nP^{-1}$ for each $n = 1, 2, \dots$.]

12. Given a polynomial $p(x) = r_0 + r_1 x + \cdots + r_n x^n$ and a square matrix A, the matrix $p(A) = r_0 I + r_1 A + \cdots + r_n A^n$ is called the **evaluation** of $p(x)$ at A. Let $B = P^{-1}AP$.

(a) Show that $p(B) = P^{-1}p(A)P$ for all polynomials $p(x)$.

(b) Find $p(B)$ if $p(x) = x^3 + 3x^2 + x - 1$, $A = \begin{bmatrix} -1 & 0 \\ 0 & 2 \end{bmatrix}$, and $P = \begin{bmatrix} 3 & 2 \\ 2 & 1 \end{bmatrix}$.

13. If A is invertible, show that AB is similar to BA for all B.

14. (a) Show that the only matrix similar to a scalar matrix $A = rI$, r in $\mathbb{R}$, is A itself.

(b) If A has the property that the only matrix similar to it is A itself, show that $A = rI$ for some r in $\mathbb{R}$. [*Hint*: See Supplementary Exercise 6 of Chapter 2.]

15. Let λ be an eigenvalue of A with corresponding eigenvector X. If $B = P^{-1}AP$ is similar to A, show that $P^{-1}X$ is an eigenvector of B corresponding to λ.

16. Let A be any $n \times n$ matrix and $r \neq 0$ a real number.

(a) Show that the eigenvalues of rA are precisely the numbers $r\lambda$, where λ is an eigenvalue of A.

(b) Show that $c_{rA}(x) = r^n c_A\left(\frac{x}{r}\right)$.

17. (a) If all rows of A have the same sum s, show that s is an eigenvalue.

(b) If all columns of A have the same sum s, show that s is an eigenvalue.

18. Let A be an invertible $n \times n$ matrix.

(a) Show that the eigenvalues of A are nonzero.

(b) Show that the eigenvalues of A^{-1} are precisely the numbers $\frac{1}{\lambda}$, where λ is an eigenvalue of A.

(c) Show that $c_{A^{-1}}(x) = \frac{(-x)^n}{\det A} c_A\left(\frac{1}{x}\right)$.

19. Suppose λ is an eigenvalue of a square matrix A with eigenvector $X \neq 0$.

(a) Show that λ^2 is an eigenvalue of A^2 (with the same X).

(b) Show that $\lambda^3 - 2\lambda + 3$ is an eigenvalue of $A^3 - 2A + 3I$.

(c) Show that $p(\lambda)$ is an eigenvalue of $p(A)$ for any nonzero polynomial $p(x)$.

20. If A is an $n \times n$ matrix, show that $c_{A^2}(x^2) = (-1)^n c_A(x)c_A(-x)$.

21. An $n \times n$ matrix A is called **nilpotent** if $A^m = 0$ for some $m \geq 1$.

(a) Show that every triangular matrix with zeros on the main diagonal is nilpotent.

(b) Show that $\lambda = 0$ is the only eigenvalue (even complex) of A, if A is nilpotent.

(c) Deduce that $c_A(x) = x^n$, if A is $n \times n$ and nilpotent.

22. Given $p(x) = a_0 + a_1 x + a_2 x^2 + a_3 x^3 + x^4$, show that

$$A = \begin{bmatrix} 0 & 1 & 0 & 0 \\ 0 & 0 & 1 & 0 \\ 0 & 0 & 0 & 1 \\ -a_0 & -a_1 & -a_2 & -a_3 \end{bmatrix}$$

is a matrix whose characteristic polynomial equals $p(x)$. [A is called the **companion matrix** for $p(x)$.]

23. Let $A = \begin{bmatrix} 0 & a & b \\ a & 0 & c \\ b & c & 0 \end{bmatrix}$ and $B = \begin{bmatrix} c & a & b \\ a & b & c \\ b & c & a \end{bmatrix}$.

    (a) Show that $x^3 - (a^2 + b^2 + c^2)x - 2abc$ has real roots by considering A.

    ◆**(b)** Show that $a^2 + b^2 + c^2 \geq ab + ac + bc$ by considering B.

24. Let $A = \begin{bmatrix} B & 0 \\ 0 & C \end{bmatrix}$, where B and C are square matrices.

    (a) Show that $c_A(x) = c_B(x)\, c_C(x)$.

    (b) If X and Y are eigenvectors of B and C, respectively, show that $\begin{bmatrix} X \\ 0 \end{bmatrix}$ and $\begin{bmatrix} 0 \\ Y \end{bmatrix}$ are eigenvectors of A, and show how every eigenvector of A arises from such eigenvectors.

25. Suppose $\lambda_1, \lambda_2, \ldots, \lambda_n$ are the eigenvalues of an $n \times n$ matrix A, including repetitions. Show that $\det A =$ $\lambda_1 \lambda_2 \cdots \lambda_n$. [*Hint:* Look at the constant coefficient in $c_A(x)$. Use Theorem 1 to write $c_A(x) = (x - \lambda_1)(x - \lambda_2) \cdots (x - \lambda_n)$.]

26. Let P be an invertible $n \times n$ matrix. If A is any $n \times n$ matrix, write $T_P(A) = P^{-1}AP$. Verify that:

    (a) $T_P(I) = I$ ◆**(b)** $T_P(AB) = T_P(A)\, T_P(B)$

    (c) $T_P(A + B) = T_P(A) + T_P(B)$ **(d)** $T_P(rA) = rT_P(A)$

    (e) $T_P(A^k) = [T_P(A)]^k$ for $k \geq 1$

    (f) If A is invertible, $T_P(A^{-1}) = [T_P(A)]^{-1}$.

    (g) If Q is invertible, $T_Q[T_P(A)] = T_{PQ}(A)$.

27. Assume the 2×2 matrix A is similar to an upper triangular matrix. If tr $A = 0 =$ tr A^2, show that $A^2 = 0$.

28. Show that A is similar to A^T for all 2×2 matrices A. [*Hint:* Let $A = \begin{bmatrix} a & b \\ c & d \end{bmatrix}$. If $c = 0$, treat the cases $b = 0$ and $b \neq 0$ separately. If $c \neq 0$, reduce to the case $c = 1$ using Exercise 10(c).]

Section 6.2   Diagonalization

In this section we will be concerned with the following class of matrices.

DEFINITION

An $n \times n$ real matrix A is said to be **diagonalizable** if it is similar to a real diagonal matrix—that is, if $P^{-1}AP$ is diagonal for some invertible real matrix P.

Observe that A and $P^{-1}AP$ have the same eigenvalues by Theorem 4§6.1, and these are the diagonal entries of the diagonal matrix $P^{-1}AP$. Hence our assumption that P is real implies that the eigenvalues of A are real. The general case is discussed in Section 6.8.

Let A denote an $n \times n$ matrix, and suppose that an invertible matrix P exists that diagonalizes A, say

$$P^{-1}AP = D = \begin{bmatrix} \lambda_1 & 0 & 0 & \cdots & 0 \\ 0 & \lambda_2 & 0 & \cdots & 0 \\ 0 & 0 & \lambda_3 & \cdots & 0 \\ \vdots & \vdots & \vdots & & \vdots \\ 0 & 0 & 0 & \cdots & \lambda_n \end{bmatrix}$$

where $\lambda_1, \lambda_2, \ldots, \lambda_n$ are the (real) eigenvalues of A. Here is a way to find the diagonalizing matrix P. The idea is to determine P by finding each of its columns. So let $X_1, X_2, \ldots, X_n$ denote the columns of P and write P in block form as follows:

$$P = [X_1 \ X_2 \ \cdots \ X_n]$$

The condition $P^{-1}AP = D$ can be rewritten as $AP = PD$; that is

$$A[X_1 \ X_2 \ \cdots \ X_n] = [X_1 \ X_2 \ \cdots \ X_n] \begin{bmatrix} \lambda_1 & 0 & \cdots & 0 \\ 0 & \lambda_2 & \cdots & 0 \\ \vdots & \vdots & & \vdots \\ 0 & 0 & \cdots & \lambda_n \end{bmatrix}$$

If both sides are written in terms of their columns, the result is

$$[AX_1 \ AX_2 \ \cdots \ AX_n] = [\lambda_1 X_1 \ \lambda_2 X_2 \cdots \lambda_n X_n]$$

Equating columns gives

$$AX_i = \lambda_i X_i \qquad i = 1, 2, \ldots, n$$

so that the columns of P are eigenvectors of A corresponding (in order) to the eigenvalues $\lambda_1, \ldots, \lambda_n$ in D. The fact that P is invertible shows that the eigenvectors used must be linearly independent by Theorem 1 §5.5

   This proves that if A is diagonalizable, then it has n linearly independent eigenvectors. The converse of this statement is also true. If $X_1, X_2, \ldots, X_n$ are linearly independent eigenvectors, then $P = [X_1 \ X_2 \ \cdots \ X_n]$ is invertible and the preceding computation shows $AP = PD$ where D is diagonal. Hence $P^{-1}AP = D$, so the following theorem has been proved.

THEOREM 1

An $n \times n$ matrix is diagonalizable if and only if it has n linearly independent eigenvectors (which are then a basis of $\mathbb{R}^n$).

The foregoing argument actually gives the following method for finding the diagonalizing matrix P.

DIAGONALIZATION ALGORITHM

Given an $n \times n$ matrix A:

STEP (1) Find the eigenvalues of A.

STEP (2) Find (if possible) n linearly independent eigenvectors $X_1, X_2, \ldots, X_n$.

STEP (3) Form $P = [X_1 \ X_2 \ \cdots \ X_n]$ — the matrix with the X_i as columns.

STEP (4) Then $P^{-1}AP$ is diagonal, the diagonal entries being the eigenvalues corresponding (in order) to $X_1, X_2, \ldots, X_n$, respectively.

EXAMPLE 1

Diagonalize the matrix $A = \begin{bmatrix} 5 & 8 & 16 \\ 4 & 1 & 8 \\ -4 & -4 & -11 \end{bmatrix}$.

Solution   In Example 2§6.1 the eigenvalues were found to be $\lambda = -3$ and $\lambda = 1$, with eigenspaces $E_{-3} = \text{span}\{X_1, X_2\}$ and $E_1 = \text{span}\{X_3\}$, where

$$X_1 = \begin{bmatrix} -1 \\ 1 \\ 0 \end{bmatrix} \qquad X_2 = \begin{bmatrix} -2 \\ 0 \\ 1 \end{bmatrix} \qquad X_3 = \begin{bmatrix} 2 \\ 1 \\ -1 \end{bmatrix}$$

These vectors are linearly independent, so

$$P = [X_1 \quad X_2 \quad X_3] = \begin{bmatrix} -1 & -2 & 2 \\ 1 & 0 & 1 \\ 0 & 1 & -1 \end{bmatrix}$$

is invertible and will diagonalize A. In fact,

$$P^{-1} = \begin{bmatrix} -1 & 0 & -2 \\ 1 & 1 & 3 \\ 1 & 1 & 2 \end{bmatrix} \qquad \text{and} \qquad P^{-1}AP = \begin{bmatrix} -3 & 0 & 0 \\ 0 & -3 & 0 \\ 0 & 0 & 1 \end{bmatrix}$$

Note that the diagonal entries $-3, -3, 1$ of $P^{-1}AP$ are (in order) the eigenvalues corresponding to the columns X_1, X_2, and X_3 of P.

◆◆◆

The matrix P in Example 1 is not unique: The eigenvectors themselves can be *any* basis of the eigenspace and, further, the eigenvectors can be placed in any order as the columns of P.

Even if the eigenvalues are all real, it may happen that a matrix is *not* diagonalizable. Then the process will fail at step 2: It will be impossible to find n independent eigenvectors.

EXAMPLE 2

Show that the matrix $A = \begin{bmatrix} 2 & 1 & 1 \\ 2 & 1 & -2 \\ -1 & 0 & -2 \end{bmatrix}$ is not diagonalizable.

Solution   The eigenvalues were determined in Example 3§6.1 to be $\lambda = -1$ and $\lambda = 3$, with eigenspaces $E_{-1} = \text{span}\{X_1\}$ and $E_3 = \text{span}\{X_2\}$ where

$$X_1 = \begin{bmatrix} -1 \\ 2 \\ 1 \end{bmatrix} \qquad \text{and} \qquad X_2 = \begin{bmatrix} 5 \\ 6 \\ -1 \end{bmatrix}$$

These are independent, but there are not enough of them. All the eigenvectors corresponding to $\lambda = -1$ are multiples of X_1 so none can lie in an independent set containing X_1. Similarly, no eigenvector corresponding to $\lambda = 3$ is independent of X_2. Because these are the only eigenvalues of A it is impossible to find three independent eigenvectors. Hence A is not diagonalizable by Theorem 1.

◆◆◆

Observe that the matrices in Examples 1 and 2 were both of size 3×3 but each had only *two* eigenvalues. The reason the matrix in Example 1 was diagonalizable is that the eigenspaces had dimensions 2 and 1, respectively, so two independent eigenvectors could be found in the first space and one in the second, making up the required independent set of three. The situation in Example 2 was different: Both eigenspaces had dimension 1, so there was no possibility of finding *three* independent eigenvectors.

This discussion makes it clear that the dimensions of the eigenspaces play a vital role in determining whether there exist n linearly independent eigenvectors — and hence whether the matrix is diagonalizable. We can clearly choose independent eigenvectors in any eigenspace, and the question arises whether vectors from *different* eigenspaces (corresponding to *distinct* eigenvalues) will be linearly independent. The answer is yes.

THEOREM 2

Let $\lambda_1, \lambda_2, \ldots, \lambda_k$ be distinct eigenvalues of a matrix A. If $X_1, X_2, \ldots, X_k$ are eigenvectors corresponding to $\lambda_1, \lambda_2, \ldots, \lambda_k$, respectively, then $\{X_1, \ldots, X_k\}$ is a linearly independent set.

Proof   Use induction on k. If $k = 1$, then $\{X_1\}$ is linearly independent because $X_1 \neq 0$. Now assume that $k > 1$ and that the theorem holds for any $k - 1$ eigenvalues. Suppose a linear combination of the X_i vanishes:

$$r_1X_1 + r_2X_2 + r_3X_3 + \cdots + r_kX_k = 0 \qquad (*)$$

It is required to show that $r_1 = r_2 = \cdots = r_k = 0$. Left-multiplication by A together with the fact that $AX_i = \lambda_iX_i$ holds for each i yields

$$r_1\lambda_1X_1 + r_2\lambda_2X_2 + r_3\lambda_3X_3 + \cdots + r_k\lambda_kX_k = 0 \qquad (**)$$

Now multiply $(*)$ by λ_1 and subtract the result from $(**)$ to obtain

$$r_2(\lambda_2 - \lambda_1)X_2 + r_3(\lambda_3 - \lambda_1)X_3 + \cdots + r_k(\lambda_k - \lambda_1)X_k = 0$$

But $\{X_2, \ldots, X_k\}$ is linearly independent by the induction hypothesis, so $r_2(\lambda_2 - \lambda_1) = r_3(\lambda_3 - \lambda_1) = \cdots = r_k(\lambda_k - \lambda_1) = 0$. The fact that the λ_i are distinct now implies that $r_2 = r_3 = \cdots = r_k = 0$, so $(*)$ becomes $r_1X_1 = 0$. Because $X_1 \neq 0$, this implies that $r_1 = 0$ and completes the proof.   ◆

This theorem combines with Theorem 1 to yield the following useful fact.

> **THEOREM 3**
>
> An $n \times n$ matrix with n distinct eigenvalues is diagonalizable.

EXAMPLE 3

Show that $A = \begin{bmatrix} 1 & 0 & 0 \\ 1 & 2 & -3 \\ 1 & -1 & 0 \end{bmatrix}$ is diagonalizable.

Solution   The characteristic polynomial is

$$c_A(x) = \det(xI - A) = \det \begin{bmatrix} x-1 & 0 & 0 \\ -1 & x-2 & 3 \\ -1 & 1 & x \end{bmatrix} = (x-1)(x-3)(x+1)$$

Hence the eigenvalues are 1, 3, and -1, so A is diagonalizable by Theorem 3.

◆◆◆

It is important to note that the converse of Theorem 3 is false: There exist diagonalizable $n \times n$ matrices A that do not have n distinct eigenvalues. For example, the 3×3 matrix in Example 1 is diagonalizable, but $\lambda = -3$ and $\lambda = 1$ are the only eigenvalues.

An eigenvalue λ of a matrix A is said to have **multiplicity** m if it is repeated m times as a root of the characteristic polynomial — that is, if

$$c_A(x) = (x - \lambda)^m q(x) \qquad \text{where } q(\lambda) \neq 0$$

This is illustrated for the matrices in Examples 1, 2, and 3 in the accompanying table.

Matrix	Characteristic Polynomial	Eigenvalues	Multiplicity	Dimension of Eigenspace
$\begin{bmatrix} 5 & 8 & 16 \\ 4 & 1 & 8 \\ -4 & -4 & -11 \end{bmatrix}$	$(x+3)^2(x-1)$	-3 1	2 1	2 1
$\begin{bmatrix} 2 & 1 & 1 \\ 2 & 1 & -2 \\ -1 & 0 & -2 \end{bmatrix}$	$(x+1)^2(x-3)$	-1 3	2 1	1 1
$\begin{bmatrix} 1 & 0 & 0 \\ 1 & 2 & -3 \\ 1 & -1 & 0 \end{bmatrix}$	$(x-1)(x-3)(x+1)$	1 3 -1	1 1 1	1 1 1

The first and last matrices here are diagonalizable because the dimensions of the eigenspaces add up to $3 = n$ in both cases; however, this is not true for the second matrix, which is not diagonalizable (see the discussion following Example 2). Moreover, in both diagonalizable cases the multiplicity of each eigenvalue equals the dimension of the corresponding eigenspace, whereas this *fails* for the eigenvalue $\lambda = -1$ of the second (nondiagonalizable) matrix. These observations are all made precise in the following fundamental theorem.

THEOREM 4

Let A be an $n \times n$ matrix, and let

$$c_A(x) = (x - \lambda_1)^{m_1} (x - \lambda_2)^{m_2} \cdots (x - \lambda_k)^{m_k}$$

where $\lambda_1, \lambda_2, \ldots, \lambda_k$ denote the distinct real eigenvalues of A (with multiplicities $m_1, m_2, \ldots, m_k$). Write $d_i = \dim(E_{\lambda_i})$ for each i. Then the following statements are equivalent:

1. A is diagonalizable.
2. $d_1 + d_2 + \cdots + d_k = n$.
3. $d_i = m_i$ for each i.

The proof requires the following fact, which is proved in Theorem 5§7.7 where other methods are available.

THEOREM 5

If λ is an eigenvalue of A of multiplicity m, then $\dim(E_\lambda) \le m$.

Proof of Theorem 4   Observe first that $d_i \le m_i$ for each i by Theorem 5, and that $m_1 + m_2 + \cdots + m_k = \deg[c_A(x)] = n$.

(1) *implies* (2). Let B be a set of n independent eigenvectors, and let t_i of them lie in E_{λ_i}. Then $t_i \le d_i = \dim(E_{\lambda_i})$, so (2) follows from

$$n = t_1 + \cdots + t_k \le d_1 + \cdots + d_k \le m_1 + \cdots + m_k = n$$

(2) *implies* (3). By (2) $n = d_1 + \cdots + d_k \le m_1 + \cdots + m_k = n$. Hence $d_1 + \cdots + d_k = m_1 + \cdots + m_k$; so because $d_i \le m_i$ for each i, we must have $d_i = m_i$. This proves (3).

(3) *implies* (1). Let B_i denote a basis of E_{λ_i} for each i, and let B consist of all vectors belonging to at least one of the B_i. It suffices to show that B is linearly independent (it contains $d_1 + \cdots + d_k$ eigenvectors and $d_1 + \cdots + d_k = m_1 + \cdots + m_k = n$ by (3)). Suppose a linear combination of the vectors in B vanishes,

and let Y_i denote the sum of all terms that come from B_i. Then Y_i lies in E_{λ_i} for each i, so the nonzero Y_i are independent by Theorem 2. But the Y_i sum to zero, so each $Y_i = 0$. Hence all coefficients of terms in Y_i are zero (B_i is independent), proving that B is independent.   ◆

The proof that (3) implies (1) makes the process of diagonalization clear: Find the eigenspaces corresponding to the various eigenvectors of A and choose a basis of each. Then the collection of all vectors in these bases is an independent set of eigenvectors, and A is diagonalizable (by Theorem 1) if this set contains n vectors.

EXAMPLE 4

Show that $A = \begin{bmatrix} 1 & a & b & c \\ 0 & 1 & d & e \\ 0 & 0 & 2 & f \\ 0 & 0 & 0 & 3 \end{bmatrix}$ is not diagonalizable for any choice of $a \neq 0$, b, c, d, e, and f.

Solution   Because A is upper triangular, $c_A(x) = (x - 1)^2(x - 2)(x - 3)$. Hence $\lambda_1 = 1$ is an eigenvalue of multiplicity $m = 2$. But a routine calculation shows that $E_{\lambda_1} = \text{span}\{X\}$ where $X = [1\ 0\ 0\ 0]^T$, so $\dim E_{\lambda_1} = 1$ is not equal to the multiplicity of λ_1.   ◆◆◆

If A is diagonalizable, many of its properties are characterized by the eigenvalues. Here is an example. For convenience let

$$\text{diag}(\lambda_1, \lambda_2, \ldots, \lambda_n)$$

denote the $n \times n$ diagonal matrix with $\lambda_1, \lambda_2, \ldots, \lambda_n$ in order down the main diagonal.

EXAMPLE 5

Suppose $\lambda^2 = 2\lambda$ holds for every eigenvalue λ of the diagonalizable matrix A. Show that $A^2 = 2A$.

Solution   Let $P^{-1}AP = D = \text{diag}(\lambda_1, \ldots, \lambda_n)$. Because $\lambda_i^2 = 2\lambda_i$ holds for each i, we obtain

$$D^2 = \text{diag}(\lambda_1^2, \ldots, \lambda_n^2) = \text{diag}(2\lambda_1, \ldots, 2\lambda_n) = 2D$$

But then $A^2 = (PDP^{-1})^2 = PD^2P^{-1} = P(2D)P^{-1} = 2(PDP^{-1}) = 2A$, as required.   ◆◆◆

We conclude with an application of diagonalization to linear recurrence relations. This is treated in more generality in Section 7.8, but diagonalization techniques give a solution in certain cases. Here is an example.

EXAMPLE 6

Suppose a sequence of numbers $x_0, x_1, x_2, \ldots$ is determined by the condition that $x_0 = x_1 = 1$, and each successive x_n is given by

$$x_{n+2} = 6x_n + x_{n+1} \qquad n \geq 0$$

Find a formula for x_n in terms of n.

Solution

It is clear that the sequence is completely determined by these conditions. In fact, the next few numbers are

$$x_2 = 6x_0 + x_1 = 7$$
$$x_3 = 6x_1 + x_2 = 13$$
$$x_4 = 6x_2 + x_3 = 55$$
$$\vdots$$

A general pattern for these numbers is not apparent, but a clever device transforms the problem into a simpler recurrence involving matrices. The idea is to compute the column

$$V_n = \begin{bmatrix} x_n \\ x_{n+1} \end{bmatrix}$$

for each $n \geq 0$ rather than x_n itself. The recurrence $x_{n+2} = 6x_n + x_{n+1}$ gives

$$V_{n+1} = \begin{bmatrix} x_{n+1} \\ x_{n+2} \end{bmatrix} = \begin{bmatrix} 0 & 1 \\ 6 & 1 \end{bmatrix} \begin{bmatrix} x_n \\ x_{n+1} \end{bmatrix} = AV_n \qquad \text{where} \qquad A = \begin{bmatrix} 0 & 1 \\ 6 & 1 \end{bmatrix}$$

Hence $V_1 = AV_0$, $V_2 = AV_1 = A^2V_0$, $V_3 = AV_2 = A^3V_0, \ldots$, and clearly

$$V_n = A^nV_0 \qquad n = 0, 1, 2, \ldots$$

Moreover, $V_0 = \begin{bmatrix} 1 \\ 1 \end{bmatrix}$ is prescribed, so the problem comes down to computing A^n. This calculation is simplified if A is diagonalized.

The characteristic polynomial for A is $c_A(x) = (x - 3)(x + 2)$. Hence the eigenvectors are $X_1 = \begin{bmatrix} 1 \\ 3 \end{bmatrix}$ and $X_2 = \begin{bmatrix} 1 \\ -2 \end{bmatrix}$, so $P = \begin{bmatrix} 1 & 1 \\ 3 & -2 \end{bmatrix}$ is invertible and $P^{-1}AP$

$$= \begin{bmatrix} 3 & 0 \\ 0 & -2 \end{bmatrix} = D. \text{ Finally, } P^{-1} = \tfrac{1}{5} \begin{bmatrix} 2 & 1 \\ 3 & -1 \end{bmatrix}, \text{ so}$$

$$\begin{bmatrix} x_n \\ x_{n+1} \end{bmatrix} = V_n = A^nV_0 = PD^nP^{-1}V_0$$

$$= \tfrac{1}{5} \begin{bmatrix} 1 & 1 \\ 3 & -2 \end{bmatrix} \begin{bmatrix} 3^n & 0 \\ 0 & (-2)^n \end{bmatrix} \begin{bmatrix} 2 & 1 \\ 3 & -1 \end{bmatrix} \begin{bmatrix} 1 \\ 1 \end{bmatrix}$$

$$= \tfrac{1}{5} \begin{bmatrix} 3^{n+1} - (-2)^{n+1} \\ 3^{n+2} + (-2)^{n+2} \end{bmatrix}$$

It follows that $x_n = \frac{1}{5}(3^{n+1} - (-2)^{n+1})$ for each $n = 0, 1, 2, \ldots$. It is easy to verify that this formula gives $x_0 = 1$, $x_1 = 1$ and satisfies $x_{n+2} = 6x_n + x_{n+1}$ for any $n \geq 0$, as required.

◆◆◆

This approach works more generally — but only if the corresponding matrix is diagonalizable. This turns out to be the case precisely when the eigenvalues of A are distinct (Exercise 23). The general case is treated by other methods in Section 7.8.

EXERCISES 6.2

1. In each case, decide whether the matrix A is diagonalizable. If so, find P such that $P^{-1}AP$ is diagonal.

(a) $\begin{bmatrix} 1 & 4 \\ 3 & 2 \end{bmatrix}$ ◆**(b)** $\begin{bmatrix} 2 & -4 \\ -1 & 2 \end{bmatrix}$

(c) $\begin{bmatrix} 1 & 0 & 0 \\ 1 & 2 & 1 \\ 0 & 0 & 1 \end{bmatrix}$ ◆**(d)** $\begin{bmatrix} 3 & 0 & 6 \\ 0 & -3 & 0 \\ 5 & 0 & 2 \end{bmatrix}$

(e) $\begin{bmatrix} 3 & 1 & 6 \\ 2 & 1 & 0 \\ -1 & 0 & -3 \end{bmatrix}$ ◆**(f)** $\begin{bmatrix} 0 & 0 & 1 \\ 1 & 0 & -1 \\ 0 & 1 & 1 \end{bmatrix}$

◆**2.** If $A = \begin{bmatrix} 6 & 2 \\ 4 & -1 \end{bmatrix}$ and $B = \begin{bmatrix} 1 & 0 \\ 4 & -14 \end{bmatrix}$, verify that A and B are diagonalizable, but AB is not.

3. If A is an $n \times n$ matrix, show that A is diagonalizable if and only if A^T is diagonalizable.

4. If A and B are similar matrices, show that A is diagonalizable if and only if B is diagonalizable. (Hence, being diagonalizable is a similarity invariant.)

5. If A is diagonalizable, show that each of the following is also diagonalizable.

(a) A^n, $n \geq 1$ ◆**(b)** kA

(c) $p(A)$, $p(x)$ any polynomial (see Exercise 12§6.1)

◆**6.** Give an example of two diagonalizable matrices A and B whose sum $A + B$ is not diagonalizable.

7. If A is diagonalizable and 1 and -1 are the only eigenvalues, show that $A^{-1} = A$.

◆**8.** If A is diagonalizable and 0 and 1 are the only eigenvalues, show that $A^2 = A$.

9. If A is diagonalizable and $\lambda \geq 0$ for each eigenvalue of A, show that $A = B^2$ for some matrix B.

10. If $P^{-1}AP$ and $P^{-1}BP$ are both diagonal, show that $AB = BA$.

11. A square matrix A is called nilpotent if $A^n = 0$ for some $n \geq 1$. Find all nilpotent diagonalizable matrices.

◆**12.** Show that the only diagonalizable matrix A that has only one eigenvalue λ is the scalar matrix $A = \lambda I$.

13. Characterize the diagonalizable $n \times n$ matrices A such that $A^2 - 3A + 2I = 0$ in terms of their eigenvalues. [Hint: See Exercise 12§6.1.]

14. (a) Show that two diagonalizable matrices are similar if and only if they have the same eigenvalues with the same multiplicities.

(b) Show that $I = \begin{bmatrix} 1 & 0 \\ 0 & 1 \end{bmatrix}$ and $A = \begin{bmatrix} 1 & 1 \\ 0 & 1 \end{bmatrix}$ have the same characteristic polynomial (and hence the same eigenvalues) but are not similar.

15. Let A denote an $n \times n$ upper triangular matrix.

(a) If all the main diagonal entries of A are distinct, show that A is diagonalizable.

(b) If all the main diagonal entries of A are equal, show that A is diagonalizable only if it is *already* diagonal. [Hint: See Exercise 12.]

(c) Show that $\begin{bmatrix} 1 & 0 & 1 \\ 0 & 1 & 0 \\ 0 & 0 & 2 \end{bmatrix}$ is diagonalizable but that

$\begin{bmatrix} 1 & 1 & 0 \\ 0 & 1 & 0 \\ 0 & 0 & 2 \end{bmatrix}$ is not.

16. Let A be a diagonalizable $n \times n$ matrix with eigenvalues $\lambda_1, \lambda_2, \ldots, \lambda_n$ (including multiplicities). Show that:

(a) $\det A = \lambda_1 \lambda_2 \cdots \lambda_n$

◆**(b)** $\text{tr } A = \lambda_1 + \lambda_2 + \cdots + \lambda_n$

17. Let $A = \begin{bmatrix} B & 0 \\ 0 & C \end{bmatrix}$ where B and C are square matrices.

(a) If B and C are diagonalizable via Q and R (that is, $Q^{-1}BQ$ and $R^{-1}CR$ are diagonal), show that A is diagonalizable via $\begin{bmatrix} Q & 0 \\ 0 & R \end{bmatrix}$.

(b) Use **(a)** to diagonalize A if $B = \begin{bmatrix} 5 & 3 \\ 3 & 5 \end{bmatrix}$ and

$C = \begin{bmatrix} 7 & -1 \\ -1 & 7 \end{bmatrix}$.

(c) If A is diagonalizable, show that both B and C are diagonalizable.

18. If A is 2×2 and diagonalizable, show that $C(A)$ $= \{X \mid XA = AX\}$ has dimension 2 or 4.

19. If A is diagonalizable and $p(x)$ is a polynomial such that $p(\lambda) = 0$ for all eigenvalues λ of A, show that $p(A) = 0$ (see Exercise 12§6.1). In particular, show that $c_A(A) = 0$. [*Remark:* $c_A(A) = 0$ for *all* square matrices A — this is the Cayley–Hamilton theorem (see Theorem 2§7.7).]

◆20. Let A be $n \times n$ with n distinct real eigenvalues. If $AC = CA$, show that C is diagonalizable.

21. If A and B are diagonalizable $n \times n$ matrices, must AB be diagonalizable? Support your answer.

22. Solve each of the following linear recurrences by the method given in Example 6.

(a) $x_{n+2} = 3x_n + 2x_{n+1}, x_0 = 1, x_1 = 1$

◆(b) $x_{n+2} = 2x_n - x_{n+1}, x_0 = 1, x_1 = 2$

23. Generalize Example 6 as follows: Assume that the sequence $x_0, x_1, x_2, \ldots$ satisfies

$$x_{n+k} = r_0 x_n + r_1 x_{n+1} + \cdots + r_{k-1} x_{n+k-1}$$

for all $n \geq 0$. Define

$$A = \begin{bmatrix} 0 & 1 & 0 & \cdots & 0 \\ 0 & 0 & 1 & \cdots & 0 \\ \vdots & \vdots & \vdots & & \vdots \\ 0 & 0 & 0 & \cdots & 1 \\ r_0 & r_1 & r_2 & \cdots & r_{k-1} \end{bmatrix}, \quad V_n = \begin{bmatrix} x_n \\ x_{n+1} \\ \vdots \\ x_{n+k-1} \end{bmatrix}$$

Then show that:

(a) $V_n = A^n V_0$ for all n.

(b) $c_A(x) = x^k - r_{k-1}x^{k-1} - \cdots - r_1 x - r_0$.

(c) If λ is an eigenvalue of A, the eigenspace E_λ has dimension 1, and $X = (1, \lambda, \lambda^2, \ldots, \lambda^{k-1})^T$ is an eigenvector. [*Hint:* Use $c_A(\lambda) = 0$ to show that $E_\lambda = \mathbb{R}X$.]

(d) A is diagonalizable if and only if the eigenvalues of A are distinct. [*Hint:* See part **(c)** and Theorem 4.]

(e) If $\lambda_1, \lambda_2, \ldots, \lambda_k$ are distinct real eigenvalues, there exist constants $t_1, t_2, \ldots, t_k$ such that $x_n = t_1\lambda_1^n + \cdots + t_k\lambda_k^n$ holds for all n. [*Hint:* If D is diagonal with $\lambda_1, \lambda_2, \ldots, \lambda_k$ as the main diagonal entries, show that $A^n = PD^nP^{-1}$ has entries that are linear combinations of $\lambda_1^n, \lambda_2^n, \ldots, \lambda_k^n$.]

Section 6.3   Orthogonality in $\mathbb{R}^n$

The length of a vector has an intuitive meaning $\mathbb{R}^3$ as does the idea that two vectors are orthogonal (or perpendicular). These fundamental geometric concepts can both be defined using the dot product (see Chapter 4) and the purpose of this section is to extend these notions to $\mathbb{R}^n$. Applications to diagonalization are given in the next section.

It is convenient to regard n-tuples X and Y in $\mathbb{R}^n$ as row matrices. Then the matrix product XY^T is a 1×1 matrix that we regard as a real number.

DEFINITION

Given $X = [x_1 \quad x_2 \cdots x_n]$ and $Y = [y_1 \quad y_2 \cdots y_n]$ in $\mathbb{R}^n$, the **dot product** $X \cdot Y$ is defined by

$$X \cdot Y = XY^T = x_1y_1 + x_2y_2 + \cdots + x_ny_n$$

The **length** $\|X\|$ of X is defined by

$$\|X\| = \sqrt{X \cdot X} = \sqrt{x_1^2 + x_2^2 + \cdots + x_n^2}$$

This agrees with the usual notions in $\mathbb{R}^2$ and $\mathbb{R}^3$, and of course the dot product is fundamental in matrix multiplication.

EXAMPLE 1

If $X = [1 \ -1 \ 0 \ 3 \ 2]$ and $Y = [2 \ 1 \ 3 \ -2 \ 1]$ in $\mathbb{R}^5$, then

$$X \cdot Y = 2 - 1 + 0 - 6 + 2 = -3 \ \text{and} \ \|X\|^2 = 1 + 1 + 0 + 9 + 4 = 15$$

◆◆◆

The proofs of the following properties of the dot product are left to the reader.

THEOREM 1

The following hold for all X, Y, and Z in $\mathbb{R}^n$.

1. $X \cdot Y = Y \cdot X$
2. $X \cdot (Y + Z) = X \cdot Y + X \cdot Z$
3. $(aX) \cdot Y = X \cdot (aY) = a(X \cdot Y)$ for all a in $\mathbb{R}$
4. $\|X\| \geq 0$, and $\|X\| = 0$ if and only if $X = 0$

EXAMPLE 2

If $X \neq 0$ in $\mathbb{R}^2$, show that $\dfrac{1}{\|X\|} X$ in the unique positive multiple of X that is a unit vector — that is, it has length 1.

Solution

If $t > 0$, then $\|tX\|^2 = tX \cdot tX = t^2\|X\|^2$. Because $t > 0$, this equals 1 if and only if $t = \dfrac{1}{\|X\|}$.

◆◆◆

Thus, for example, the unit vector that is a positive multiple of $[1 \ -1 \ 2 \ 0]$ in $\mathbb{R}^4$ is $\frac{1}{\sqrt{6}}[1 \ -1 \ 2 \ 0]$.

Two vectors in $\mathbb{R}^3$ are orthogonal if and only if their dot product is zero (Theorem 5§4.2). Hence the following definition is natural.

DEFINITION

Two vectors X and Y in $\mathbb{R}^n$ are called **orthogonal** if $X \cdot Y = 0$. A set $\{X_1, \ldots, X_n\}$ in $\mathbb{R}^n$ is called an **orthogonal set** if

$$X_i \neq 0 \text{ for all } i \quad \text{and} \quad X_i \cdot X_j = 0 \text{ if } i \neq j$$

If also $\|X_i\| = 1$ for all i, then $\{X_1, \ldots, X_n\}$ is called an **orthonormal set**.

Clearly 0 is orthogonal to every vector, and 0 is the only vector in $\mathbb{R}^n$ that is orthogonal to itself. The reason for excluding the zero vector from orthogonal sets is that we are primarily concerned with orthogonal bases.

EXAMPLE 3

The standard basis of $\mathbb{R}^n$ is orthonormal.

◆◆◆

If $\{X_1, \ldots, X_m\}$ is orthogonal, so is $\{a_1 X_1, \ldots, a_m X_m\}$ for any $a_i \neq 0$. In particular, $\left\{ \dfrac{1}{\|X_1\|} X_1, \ldots, \dfrac{1}{\|X_m\|} X_m \right\}$ is orthonormal, and we say it is the result of **normalizing** the orthogonal set $\{X_1, \ldots, X_m\}$.

EXAMPLE 4

If $E_1 = [1\ 1\ 1\ -1]$, $E_2 = [1\ 0\ 1\ 2]$, $E_3 = [-1\ 0\ 1\ 0]$, and $E_4 = [-1\ 3\ -1\ 1]$, then $\{E_1, E_2, E_3, E_4\}$ is orthogonal in $\mathbb{R}^4$. After normalizing, the corresponding orthonormal set is $\left\{ \frac{1}{2}E_1,\ \frac{1}{\sqrt{6}}E_2,\ \frac{1}{\sqrt{2}}E_3,\ \frac{1}{2\sqrt{3}}E_4 \right\}$.

◆◆◆

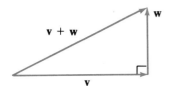

FIGURE 6.1

Given a right-angled triangle, let $\mathbf{v}$ and $\mathbf{w}$ be the geometric vectors along the two perpendicular sides as shown in Figure 6.1. Then the vector along the hypotenuse is $\mathbf{v} + \mathbf{w}$ by the parallelogram law, and the Pythagorean theorem can be expressed as $\|\mathbf{v} + \mathbf{w}\|^2 = \|\mathbf{v}\|^2 + \|\mathbf{w}\|^2$. The reason is that $\{\mathbf{v}, \mathbf{w}\}$ is an orthogonal set, and this suggests the following general form of the Pythagorean theorem.

THEOREM 2

Pythagorean Theorem

If X and Y are orthogonal in $\mathbb{R}^n$, then $\|X + Y\|^2 = \|X\|^2 + \|Y\|^2$.

Proof $\|X + Y\|^2 = (X + Y) \cdot (X + Y) = X \cdot X + X \cdot Y + Y \cdot X + Y \cdot Y$
$$= \|X\|^2 + 0 + 0 + \|Y\|^2 \qquad \blacklozenge$$

The next theorem gives a connection (perhaps unexpected) between orthogonality and independence. It is used extensively later.

THEOREM 3

Every orthogonal set of vectors in $\mathbb{R}^n$ is linearly independent.

Proof If $\{X_1, X_2, \ldots, X_m\}$ is orthogonal and $r_1X_1 + r_2X_2 + \cdots + r_mX_m = 0$, take the dot product with X_1:

$$0 = X_1 \cdot 0 = X_1 \cdot (r_1X_1 + r_2X_2 + \cdots + r_mX_m)$$
$$= r_1(X_1 \cdot X_1) + r_2(X_1 \cdot X_2) + \cdots + r_m(X_1 \cdot X_m)$$
$$= r_1\|X_1\|^2 + 0 + \cdots + 0$$

Hence $r_1 = 0$ because $X_1 \neq 0$ and, similarly, $r_2 = r_3 = \cdots = r_m = 0$. ◆

Orthogonal bases are convenient because, when we want to expand a vector as a linear combination of the basis vectors, explicit formulas exist for the coefficients.

THEOREM 4
Expansion Theorem

If $\{E_1\ E_2, \ldots, E_n\}$ is an orthogonal basis for $\mathbb{R}^n$, then

$$X = \frac{X \cdot E_1}{\|E_1\|^2} E_1 + \frac{X \cdot E_2}{\|E_2\|^2} E_2 + \cdots + \frac{X \cdot E_n}{\|E_n\|^2} E_n$$

for every X in $\mathbb{R}^n$.

Proof If $X = r_1E_1 + \cdots + r_iE_i + \cdots + r_nE_n$, then

$$X \cdot E_i = r(E_1 \cdot E_i) + \cdots + r_i\|E_i\|^2 + \cdots + r_n(E_n \cdot E_i) = r_i\|E_i\|^2$$

The result follows because $\|E_i\|^2 \neq 0$. ◆

EXAMPLE 5

The orthogonal set $\{E_1, E_2, E_3, E_4\}$ in Example 4 is a basis of $\mathbb{R}^4$ by Theorem 3. If $X = [a\ b\ c\ d]$, the expansion theorem gives $X = r_1E_1 + r_2E_2 + r_3E_3 + r_4E_4$ where

$$r_1 = \frac{X \cdot E_1}{\|E_1\|^2} = \tfrac{1}{4}(a + b + c - d) \qquad r_2 = \frac{X \cdot E_2}{\|E_2\|^2} = \tfrac{1}{6}(a + c + 2d)$$

$$r_3 = \frac{X \cdot E_3}{\|E_3\|^2} = \tfrac{1}{2}(-a + c) \qquad r_4 = \frac{X \cdot E_4}{\|E_4\|^2} = \tfrac{1}{12}(-a + 3b - c + d)$$

◆◆◆

If $\{v_1, \ldots, v_m\}$ is linearly independent in a vector space, and if v_{m+1} is not in $\text{span}\{v_1, \ldots, v_m\}$, then $\{v_1, \ldots, v_m, v_{m+1}\}$ is independent (Theorem 1§5.4). Here is the analogue for orthogonal sets in $\mathbb{R}^n$.

ORTHOGONAL LEMMA

Let $\{E_1, E_2, \ldots, E_m\}$ be an orthogonal set in $\mathbb{R}^n$. Given X in $\mathbb{R}^n$, write

$$E_{m+1} = X - \frac{X \cdot E_1}{\|E_1\|^2} E_1 - \frac{X \cdot E_2}{\|E_2\|^2} E_2 - \cdots - \frac{X \cdot E_m}{\|E_m\|^2} E_m$$

Then:

1. E_{m+1} is orthogonal to each of $E_1, E_2, \ldots, E_m$.
2. If X is not in span$\{E_1, \ldots, E_m\}$, then $E_{m+1} \neq 0$ and $\{E_1, \ldots, E_m, E_{m+1}\}$ is an orthogonal set.

Proof   For convenience write $t_i = (X \cdot E_i)/\|E_i\|^2$ for each i. Given $1 \leq k \leq m$:

$$\begin{aligned} E_{m+1} \cdot E_k &= (X - t_1 E_1 - \cdots - t_k E_k - \cdots - t_m E_m) \cdot E_k \\ &= X \cdot E_k - t_1(E_1 \cdot E_k) - \cdots - t_k(E_k \cdot E_k) - \cdots - t_m(E_m \cdot E_k) \\ &= X \cdot E_k - t_k\|E_k\|^2 \\ &= 0 \end{aligned}$$

This proves (1), and (2) follows because $E_{m+1} \neq 0$ if X is not in span$\{E_1, \ldots, E_m\}$. ◆

The orthogonal lemma has three important consequences for $\mathbb{R}^n$. The first is an extension for orthogonal sets of the fundamental fact that any independent set is part of a basis (Theorem 2§5.4).

THEOREM 5

Let U be a subspace of $\mathbb{R}^n$.

1. Every orthogonal subset $\{E_1, \ldots, E_m\}$ in U is part of an orthogonal basis of U.
2. If $U \neq 0$, it *has* an orthogonal basis.

Proof

1. Let dim$U = k$. If span$\{E_1, \ldots, E_m\} = U$ it is *already* a basis. Otherwise there exists X in U outside span$\{E_1, \ldots, E_m\}$. If E_{m+1} is as in the orthogonal lemma, then E_{m+1} is in U and $\{E_1, \ldots, E_m, E_{m+1}\}$ is orthogonal. If span$\{E_1, \ldots, E_m, E_{m+1}\} = U$, we are done. Otherwise the process continues to create larger and larger orthogonal subsets of U. They are all independent by Theorem 3, so we have a basis when we reach a subset containing k vectors.
2. If $E \neq 0$ is in U, then $\{E\}$ is orthogonal, so (2) follows from (1). ◆

We can improve upon (2) of Theorem 5. In fact, the second consequence of the orthogonal lemma is a procedure by which *any* basis $\{X_1, \ldots, X_m\}$ of a subspace U of $\mathbb{R}^n$ can be systematically modified to yield an orthogonal basis $\{E_1, \ldots, E_m\}$ of U. The E_i are constructed one at a time from the X_i. To start the process, take

$$E_1 = X_1$$

Then X_2 is not in span$\{E_1\}$ because $\{X_1, X_2\}$ is independent, so take

$$E_2 = X_2 - \frac{X \cdot E_1}{\|E_1\|^2} E_1$$

Thus $\{E_1, E_2\}$ is orthogonal by the orthogonal lemma. Moreover, span$\{E_1, E_2\}$ = span$\{X_1, X_2\}$ (verify), so E_3 is not in span$\{E_1, E_2\}$. Hence $\{E_1, E_2, E_3\}$ is orthogonal where

$$E_3 = X_3 - \frac{X_3 \cdot E_1}{\|E_1\|^2} E_1 - \frac{X_3 \cdot E_2}{\|E_2\|^2} E_2$$

Again, span$\{E_1, E_2, E_3\}$ = span$\{X_1, X_2, X_3\}$, so X_4 is not in span$\{E_1, E_2, E_3\}$ and the process continues. At the mth iteration we construct an orthogonal set $\{E_1, \ldots, E_m\}$ such that

$$\text{span}\{E_1, E_2, \ldots, E_m\} = \text{span}\{X_1, X_2, \ldots, X_m\} = U$$

Hence $\{E_1, E_2, \ldots, E_m\}$ is the desired orthogonal basis of U. The procedure can be summarized as follows.

THEOREM 6

Gram–Schmidt Orthogonalization Algorithm[3]

If $\{X_1, X_2, \ldots, X_m\}$ is any basis of a subspace U of $\mathbb{R}^n$, construct $E_1, E_2, \ldots,$ E_m in U successively as follows:

$$E_1 = X_1$$

$$E_2 = X_2 - \frac{X_2 \cdot E_1}{\|E_1\|^2} E_1$$

$$E_3 = X_3 - \frac{X_3 \cdot E_1}{\|E_1\|^2} E_1 - \frac{X_3 \cdot E_2}{\|E_2\|^2} E_2$$

$$\vdots$$

$$E_k = X_k - \frac{X_k \cdot E_1}{\|E_1\|^2} E_1 - \frac{X_k \cdot E_2}{\|E_2\|^2} E_2 - \cdots - \frac{X_k \cdot E_{k-1}}{\|E_{k-1}\|^2} E_{k-1}$$

---

[3]Erhardt Schmidt (1876–1959) was a German mathematician who studied under the great David Hilbert and later developed the theory of Hilbert spaces. He first described the present algorithm in 1907. Jörgen Pederson Gram (1850–1916) was a Danish actuary.

for each $k = 2, 3, \ldots, m$. Then

1. $\{E_1, E_2, \ldots, E_m\}$ is an orthogonal basis of U.
2. $\text{span}\{E_1, E_2, \ldots, E_k\} = \text{span}\{X_1, X_2, \ldots, X_k\}$ for $k = 1, 2, \ldots, m$.

Of course the algorithm converts any basis of $\mathbb{R}^n$ itself into an orthogonal basis.

EXAMPLE 6

Find an orthogonal basis of the row space of $A = \begin{bmatrix} 1 & 1 & -1 & -1 \\ 3 & 2 & 0 & 1 \\ 1 & 0 & 1 & 0 \end{bmatrix}$.

Solution Let X_1, X_2, X_3 denote the rows of A and observe that $\{X_1, X_2, X_3\}$ is linearly independent. Take $E_1 = X_1$. The algorithm gives

$$E_2 = X_2 - \frac{X_2 \cdot E_1}{\|E_1\|^2} E_1 = [3 \quad 2 \quad 0 \quad 1] - \tfrac{4}{4}[1 \quad 1 \quad -1 \quad -1] = [2 \quad 1 \quad 1 \quad 2]$$

$$E_3 = X_3 - \frac{X_3 \cdot E_1}{\|E_1\|^2} E_1 - \frac{X_3 \cdot E_2}{\|E_2\|^2} E_2 = X_3 - \tfrac{0}{4} E_1 - \tfrac{3}{10} E_2 = \tfrac{1}{10}[4 \quad -3 \quad 7 \quad -6]$$

Hence $\{[1 \ 1 \ -1 \ -1], [2 \ 1 \ 1 \ 2], \tfrac{1}{10}[4 \ -3 \ 7 \ -6]\}$ is the orthogonal basis provided by the algorithm. In hand calculations it may be convenient to eliminate fractions, so $\{[1 \ 1 \ -1 \ -1], [2 \ 1 \ 1 \ 2], [4 \ -3 \ 7 \ -6]\}$ is also an orthogonal basis for row A.

◆◆◆

Observe that the vector $\dfrac{X \cdot E_i}{\|E_i\|^2} E_i$ is unchanged if a nonzero scalar multiple of E_i is used in place of E_i. Hence if a newly constructed E_i is multiplied by a nonzero scalar at some stage of the Gram–Schmidt algorithm, the subsequent E's will be unchanged. This is useful in actual calculations.

Projections

Suppose a point P and a plane through the origin O are given, and we want to find the point Q in the plane that is closest to P. Our geometric intuition assures us that such a point Q exists. If we let $\mathbf{v}$ be the position vector of P (that is, the vector from O to P), then what is required is to find the position vector $\mathbf{p}$ of Q (see Figure 6.2). Again, our geometric insight assures us that, if $\mathbf{p}$ is chosen in such a way that $\mathbf{v} - \mathbf{p}$ is *perpendicular* to the plane, then $\mathbf{p}$ will be the vector we want.

Now we make two observations: first, that the set U of position vectors of points in the plane is a *subspace of* $\mathbb{R}^3$ (because the plane contains the origin) and, second,

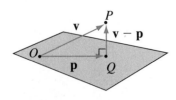

FIGURE 6.2

that the condition that $\mathbf{v} - \mathbf{p}$ is perpendicular to the plane U means that $\mathbf{v} - \mathbf{p}$ is *orthogonal* to every vector in U. In this form the whole discussion makes sense in $\mathbb{R}^n$. Furthermore the orthogonal lemma provides exactly what is needed to find $\mathbf{p}$ in this more general setting.

DEFINITION

If U is a subspace of $\mathbb{R}^n$, define the **orthogonal complement** $U^{\perp}$ of U (pronounced "U-perp") by

$$U^{\perp} = \{X \text{ in } \mathbb{R}^n \,|\, X \cdot Y = 0 \text{ for all } Y \text{ in } U\}$$

It is easily verified that $U^{\perp}$ is a subspace of $\mathbb{R}^n$ and, if $U = \text{span}\{X_1, \ldots, X_m\}$, that $U^{\perp} = \{X \text{ in } \mathbb{R}^n \,|\, X \cdot X_i = 0 \text{ for each } i = 1, 2, \ldots, m\}$. (See Exercise 19.)

EXAMPLE 7

Find $U^{\perp}$ if $U = \text{span}\{[1 \ -1 \ 2 \ 0], [1 \ 0 \ -2 \ 3]\}$ in $\mathbb{R}^4$.

Solution $X = [x \ y \ z \ w]$ is in $U^{\perp}$ if and only if it is orthogonal to both $[1 \ -1 \ 2 \ 0]$ and $[1 \ 0 \ -2 \ 3]$; that is,

$$\begin{aligned} x - y + 2z &= 0 \\ x - 2z + 3w &= 0 \end{aligned}$$

Gaussian elimination gives $U^{\perp} = \text{span}\{[2 \ 4 \ 1 \ 0], [3 \ 3 \ 0 \ -1]\}$.

◆◆◆

Now let $\{E_1, E_2, \ldots, E_m\}$ be an orthogonal basis of a subspace U of $\mathbb{R}^n$. Given X in $\mathbb{R}^n$, consider the vector

$$P = \frac{X \cdot E_1}{\|E_1\|^2} E_1 + \frac{X \cdot E_2}{\|E_2\|^2} E_2 + \cdots + \frac{X \cdot E_m}{\|E_m\|^2} E_m \qquad (*)$$

Then P is in U and $X - P$ is in $U^{\perp}$ (being orthogonal to each E_i by the orthogonal lemma). Moreover, P is independent of the choice of orthogonal basis in U. Indeed, suppose $\{E'_1, \ldots, E'_m\}$ is another orthogonal basis of U, and write

$$P' = \frac{X \cdot E'_1}{\|E'_1\|^2} E'_1 + \frac{X \cdot E'_2}{\|E'_2\|^2} E'_2 + \cdots + \frac{X \cdot E'_m}{\|E'_m\|^2} E'_m$$

Then, as before, P' is in U and $X - P'$ is in $U^{\perp}$. Write the vector $P - P'$ as follows:

$$P - P' = (X - P') - (X - P)$$

This vector is in U (because P and P' are in U) and also in $U^{\perp}$ (because $X - P$ and $X - P'$ are in $U^{\perp}$) and so must be zero (it is orthogonal to itself!). This means that $P = P'$, as asserted.

Hence the vector P in equation $(*)$ depends only on X and the subspace U, and *not* on the choice of orthogonal basis $\{E_1, \ldots, E_m\}$ of U used to compute it. Thus we are entitled to make the following definition.

DEFINITION

Let U be a subspace of $\mathbb{R}^n$ with orthogonal basis $\{E_1, E_2, \ldots, E_m\}$. If X is in $\mathbb{R}^n$, the vector

$$\text{proj}_U(X) = \frac{X \cdot E_1}{\|E_1\|^2} E_1 + \frac{X \cdot E_2}{\|E_2\|^2} E_2 + \cdots + \frac{X \cdot E_m}{\|E_m\|^2} E_m$$

is called the **orthogonal projection** of X on U. We define $\text{proj}_0(X) = 0$.

The preceding discussion proves (1) of the following theorem.

THEOREM 7
Projection Theorem

If U is a subspace of $\mathbb{R}^n$ and X is in $\mathbb{R}^n$, write $P = \text{proj}_U(X)$. Then

1. P is in U and $X - P$ is in $U^\perp$.

2. P is the vector in U closest to X in the sense that

$$\|X - P\| < \|X - Y\| \text{ for all } Y \text{ in } U,\ Y \neq P$$

3. $\dim U + \dim U^\perp = n$.

Proof

1. This is proved in the preceding discussion.

2. Write $X - Y = (X - P) + (P - Y)$. Then $P - Y$ is in U and so is orthogonal to $X - P$ by (1). Hence the Pythagorean theorem gives

$$\|X - Y\|^2 = \|X - P\|^2 + \|P - Y\|^2 > \|X - P\|^2$$

because $P - Y \neq 0$. This gives (2).

3. If $U = 0$, then $U^\perp = \mathbb{R}^n$ and (3) holds. If $U^\perp = 0$, then $U = \mathbb{R}^n$ by (1) and again (3) holds. So assume that U and $U^\perp$ are both nonzero and (by Theorem 5) let $\{E_1, \ldots, E_m\}$ and $\{F_1, \ldots, F_k\}$ be orthogonal bases of U and $U^\perp$, respectively. Then $\{E_1, \ldots, E_m, F_1, \ldots, F_k\}$ is orthogonal (by the definition of $U^\perp$), so it suffices to show that this set spans $\mathbb{R}^n$. But this follows from (1). ◆

EXAMPLE 8

Let $U = \text{span}\{X_1, X_2\}$ in $\mathbb{R}^4$ where $X_1 = [1\ 1\ 0\ 1]$ and $X_2 = [0\ 1\ 1\ 2]$. If $X = [3\ -1\ 0\ 2]$, find the vector in U closest to X and express X as the sum of a vector in U and a vector orthogonal to U.

Solution $\{X_1, X_2\}$ is independent but not orthogonal. The Gram–Schmidt process gives an orthogonal basis $\{E_1, E_2\}$ of U where $E_1 = X_1 = [1\ 1\ 0\ 1]$ and

$$E_2 = X_2 - \frac{X_2 \cdot E_1}{\|E_1\|^2} E_1 = X_2 - \tfrac{3}{3} E_1 = [-1\ \ 0\ \ 1\ \ 1]$$

Hence we can compute the projection using $\{E_1, E_2\}$:

$$P = \text{proj}_U(X) = \frac{X \cdot E_1}{\|E_1\|^2} E_1 + \frac{X \cdot E_2}{\|E_2\|^2} E_2 = \tfrac{4}{3} E_1 + \tfrac{-1}{3} E_2 = \tfrac{1}{3}[5\ \ 4\ \ -1\ \ 3].$$

Thus P is the vector in U closest to X, and $X - P = \tfrac{1}{3}[4\ -7\ 1\ 3]$ is orthogonal to every vector in U. (This can be verified by checking that it is orthogonal to the generators X_1 and X_2 of U.) The required decomposition of X is thus

$$X = P + (X - P) = \tfrac{1}{3}[5\ \ 4\ \ -1\ \ 3] + \tfrac{1}{3}[4\ \ -7\ \ 1\ \ 3] \qquad \blacklozenge\blacklozenge\blacklozenge$$

EXAMPLE 9

Find the point in the plane with equation $2x + y - z = 0$ that is closest to the point $P_0(2, -1, -3)$.

Solution   We revert to the notation of Chapter 4. The plane is the subspace U whose points (x,y,z) satisfy $z = 2x + y$. Hence

$$U = \{(s, t, 2s + t) \mid s, r \text{ in } \mathbb{R}\} = \text{span}\{(0, 1, 1), (1, 0, 2)\}$$

The Gram–Schmidt process produces an orthogonal basis $\{e_1, e_2\}$ of U where $e_1 = (0,1,1)$ and $e_2 = (1, -1,1)$. Hence the vector in U closest to $v = (2,-1,-3)$ is

$$\text{proj}_U(v) = \frac{v \cdot e_1}{\|e_1\|^2} e_1 + \frac{v \cdot e_2}{\|e_2\|^2} e_2 = -2e_1 + 0e_2 = (0, -2, -2)$$

Thus the point in U closest to $P_0(2,-1,-3)$ is $Q(0,-2,-2)$.

$\blacklozenge\blacklozenge\blacklozenge$

EXERCISES 6.3

1. Obtain an orthonormal basis of $\mathbb{R}^3$ by normalizing the following.
   (a) $\{[1\ -1\ 2], [0\ 2\ 1], [5\ 1\ -2]\}$
   ◆(b) $\{[1\ 1\ 1], [4\ 1\ -5], [2\ -3\ 1]\}$

2. In each case, show that the set of vectors is orthogonal in $\mathbb{R}^4$.

   (a) $\{[1\ -1\ 2\ 5], [4\ 1\ 1\ -1], [-7\ 28\ 5\ 5]\}$
   (b) $\{[2\ -1\ 4\ 5], [0\ -1\ 1\ -1], [0\ 3\ 2\ -1]\}$

3. In each case, show that B is an orthogonal basis of $\mathbb{R}^3$ and use Theorem 4 to expand $X = [a\ b\ c]$ as a linear combination of the basis vectors.
   (a) $\{[1\ -1\ 3], [-2\ 1\ 1], [4\ 7\ 1]\}$

◆ **(b)** $\{[1 \ 0 \ -1], [1 \ 4 \ 1], [2 \ -1 \ 2]\}$
  (c) $\{[1 \ 2 \ 3], [-1 \ -1 \ 1], [5 \ -4 \ 1]\}$
◆ **(d)** $\{[1 \ 1 \ 1], [1 \ -1 \ 0], [1 \ 1 \ -2]\}$

4. In each case, write X as a linear combination of the orthogonal basis of the subspace U.
  (a) $X = [13 \ -20 \ 15]$;
    $U = \text{span}\{[1 \ -2 \ 3], [-1 \ 1 \ 1]\}$
◆ **(b)** $X = [14 \ 1 \ -8 \ 5]$;
    $U = \text{span}\{[2 \ -1 \ 0 \ 3], [2 \ 1 \ -2 \ -1]\}$

5. In each case, find all $[a \ b \ c \ d]$ in $\mathbb{R}^3$ such that the given set is orthogonal.
  (a) $\{[1 \ 2 \ 1 \ 0], [1 \ -1 \ 1 \ 3], [2 \ -1 \ 0 \ -1], [a \ b \ c \ d]\}$
◆ **(b)** $\{[1 \ 0 \ -1 \ 1], [2 \ 1 \ 1 \ -1], [1 \ -3 \ 1 \ 0], [a \ b \ c \ d]\}$

6. Prove Theorem 1.

7. If $\|X\| = 3$, $\|Y\| = 1$, and $X \cdot Y = -2$, compute:
  (a) $\|3X - 5Y\|$;   ◆ **(b)** $\|2X + 7Y\|$;
  (c) $(3X - Y) \cdot (2Y - X)$; ◆ **(d)** $(X - 2Y) \cdot (3X + 5Y)$.

8. In each case, use the Gram–Schmidt algorithm to convert the given basis B of V into an orthogonal basis.
  (a) $V = \mathbb{R}^2$, $B = \{[1 \ -1], [2 \ 1]\}$
◆ **(b)** $V = \mathbb{R}^2$, $B = \{[2 \ 1], [1 \ 2]\}$
  (c) $V = \mathbb{R}^3$, $B = \{[1 \ -1 \ 1], [1 \ 0 \ 1], [1 \ 1 \ 2]\}$
◆ **(d)** $V = \mathbb{R}^3$, $B = \{[0 \ 1 \ 1], [1 \ 1 \ 1], [1 \ -2 \ 2]\}$

9. In each case, write X as the sum of a vector in U and a vector in $U^\perp$.
  (a) $X = [1 \ 5 \ 7]$, $U = \text{span}\{[1 \ -2 \ 3], [-1 \ 1 \ 1]\}$
◆ **(b)** $X = [2 \ 1 \ 6]$, $U = \text{span}\{[3 \ -1 \ 2], [2 \ 0 \ -3]\}$
  (c) $X = [3 \ 1 \ 5 \ 9]$,
    $U = \text{span}\{[1 \ 0 \ 1 \ 1], [0 \ 1 \ -1 \ 1], [-2 \ 0 \ 1 \ 1]\}$
◆ **(d)** $X = [2 \ 0 \ 1 \ 6]$,
    $U = \text{span}\{[1 \ 1 \ 1 \ 1], [1 \ 1 \ -1 \ -1], [1 \ -1 \ 1 \ -1]\}$
  (e) $X = [a \ b \ c \ d]$,
    $U = \text{span}\{[1 \ 0 \ 0 \ 0], [0 \ 1 \ 0 \ 0], [0 \ 0 \ 1 \ 0]\}$
◆ **(f)** $X = [a \ b \ c \ d]$,
    $U = \text{span}\{[1 \ -1 \ 2 \ 0], [-1 \ 1 \ 1 \ 1]\}$

10. Let $X = [1 \ -2 \ 1 \ 6]$ in $\mathbb{R}^4$, and let $U = \text{span}\{[2 \ 1 \ 3 \ -4], [1 \ 2 \ 0 \ 1]\}$.
◆ **(a)** Compute $\text{proj}_U(X)$.
  (b) Show that $\{[1 \ 0 \ 2 \ -3], [4 \ 7 \ 1 \ 2]\}$ is another orthogonal basis of U.
◆ **(c)** Use the basis in part **(b)** to compute $\text{proj}_U(X)$.

11. In each case, use the Gram–Schmidt algorithm to find an orthogonal basis of the subspace U and find the vector in U closest to X.
  (a) $U = \text{span}\{[1 \ 1 \ 1], [0 \ 1 \ 1]\}$; $X = [-1 \ 2 \ 1]$
◆ **(b)** $U = \text{span}\{[1 \ -1 \ 0], [-1 \ 0 \ 1]\}$; $X = [2 \ 1 \ 0]$

  (c) $U = \text{span}\{[1 \ 0 \ 1 \ 0], \quad [1 \ 1 \ 1 \ 0], \quad [1 \ 1 \ 0 \ 0]\}$;
    $X = [2 \ 0 \ -1 \ 3]$
◆ **(d)** $U = \text{span}\{[1 \ -1 \ 0 \ 1], \quad [1 \ 1 \ 0 \ 0], \quad [1 \ 1 \ 0 \ 1]\}$;
    $X = [2 \ 0 \ 3 \ 1]$

12. Let $U = \text{span}\{V_1, V_2, \ldots, V_k\}$, V_i in $\mathbb{R}^n$, and let A be the $k \times n$ matrix with the V_i as rows.
  (a) Show that $U^\perp = \{X \mid X \text{ in } \mathbb{R}^n, AX^T = 0\}$.
◆ **(b)** Use part **(a)** to find $U^\perp$ if $U = \text{span}\{[1 \ -1 \ 2 \ 1], [1 \ 0 \ -1 \ 1]\}$.

13. **(a)** Show that X and Y are orthogonal in $\mathbb{R}^n$ if and only if $\|X + Y\| = \|X - Y\|$.
◆ **(b)** Show that $X + Y$ and $X - Y$ are orthogonal in $\mathbb{R}^n$ if and only if $\|X\| = \|Y\|$.

14. Show that $\|X + Y\|^2 = \|X\|^2 + \|Y\|^2$ if and only if X is orthogonal to Y.

15. **(a)** Show that $X \cdot Y = \frac{1}{4}[\|X + Y\|^2 - \|X - Y\|^2]$ for all X, Y in $\mathbb{R}^n$.
  (b) Show that $\|X\|^2 + \|Y\|^2 = \frac{1}{2}[\|X + Y\|^2 + \|X - Y\|^2]$ for all X, Y in $\mathbb{R}^n$.

16. If A is $n \times n$, show that every eigenvalue of $A^T A$ is nonnegative. [*Hint*: Compute $\|AX\|^2$ where X is an eigenvector.]

17. If $\mathbb{R}^n = \text{span}\{X_1, \ldots, X_m\}$ and $X \cdot X_i = 0$ for all i, show that $X = 0$.

18. If $\mathbb{R}^n = \text{span}\{X_1, \ldots, X_m\}$ and $X \cdot X_i = Y \cdot X_i$ for all i, show that $X = Y$.

19. If $\{X_1, \ldots, X_m\}$ is orthogonal in $\mathbb{R}^n$, show that $\|X_1 + \cdots + X_m\|^2 = \|X_1\|^2 + \cdots + \|X_m\|^2$.

◆ **20.** Let $U = \text{span}\{X_1, \ldots, X_m\}$ be a subspace of $\mathbb{R}^n$. Show that $U^\perp = \{X \text{ in } \mathbb{R}^n \mid X \cdot X_i = 0 \text{ for each } i = 1, 2, \ldots, m\}$.

21. Let U be a subspace of $\mathbb{R}^n$. If X in $\mathbb{R}^n$ can be written in any at all as $X = P + Q$ with P in U and Q in $U^\perp$, show that necessarily $P = \text{proj}_U(X)$.

22. Let U be a subspace of $\mathbb{R}^n$ and let X be a vector in $\mathbb{R}^n$. Using Exercise 21, or otherwise, show that X is in U if and only if $X = \text{proj}_U(X)$.

23. Let U be a subspace of $\mathbb{R}^n$.
  (a) Show that $U^\perp = \mathbb{R}^n$ if and only if $U = 0$.
  (b) Show that $U^\perp = 0$ if and only if $U = \mathbb{R}^n$.

◆ **24.** If U is a subspace of $\mathbb{R}^n$, show that $U^\perp = \{X \text{ in } \mathbb{R}^n \mid \text{proj}_U(X) = 0\}$.

25. If U is a subspace of $\mathbb{R}^n$, show that $X = \text{proj}_U(X) + \text{proj}_{U^\perp}(X)$ for all X in $\mathbb{R}^n$.

26. If $\{E_1, \ldots, E_n\}$ is an orthogonal basis of $\mathbb{R}^n$ and $U = \text{span}\{E_1, \ldots, E_m\}$, show that $U^\perp = \text{span}\{E_{m+1}, \ldots, E_n\}$.

27. If U is a subspace of $\mathbb{R}^n$, show that $U^{\perp\perp} = U$. [*Hint:* Show that $U \subseteq U^{\perp\perp}$, then use Theorem 7(3) twice.]

28. If U is a subspace of $\mathbb{R}^n$, show how to find an $n \times n$ matrix A such that $U = \{X \mid AX = 0\}$. [*Hint:* Exercise 27.]

29. If A is an $n \times n$ matrix, write its null space as null $A = \{X \text{ in } \mathbb{R}^n \mid AX^T = 0\}$. Show that null $A = (\text{row } A)^\perp$.

30. Let $\{E_1, \ldots, E_n\}$ be an orthogonal basis of $\mathbb{R}^n$. Given X and Y in $\mathbb{R}^n$, show that

$$X \cdot Y = \frac{(X \cdot E_1)(Y \cdot E_1)}{\|E_1\|^2} + \cdots + \frac{(X \cdot E_n)(Y \cdot E_n)}{\|E_n\|^2}$$

31. **(a)** Let E be an $n \times n$ matrix, and let $U = \{XE \mid X \text{ in } \mathbb{R}^n\}$. Show that the following are equivalent.
  (i) $E^2 = E = E^T$ (E is a **projection matrix**)
  (ii) $(X - XE) \cdot (YE) = 0$ for all X and Y in $\mathbb{R}^n$
  (iii) $\text{proj}_U(X) = XE$ for all X in $\mathbb{R}^n$
  [*Hint:* For **(ii)** implies **(iii)**: Write $X = XE + (X - XE)$ and use the uniqueness argument preceding the definition of $\text{proj}_U(X)$. For **(iii)** implies **(ii)**: $X - XE$ is in $U^\perp$ for all X in $\mathbb{R}^n$.]
  (b) If E is a projection matrix, show that $I - E$ is also a projection matrix.
  (c) If $EF = 0 = FE$ and E and F are projection matrices, show that $E + F$ is also a projection matrix.
  ◆ **(d)** If A is $m \times n$ and AA^T is invertible, show that $E = A^T(AA^T)^{-1}A$ is a projection matrix.

32. Let A be an $n \times n$ matrix of rank r. Show that there is an invertible $n \times n$ matrix U such that UA is a row-echelon matrix with the property that the first r rows are orthogonal. [*Hint:* Let R be the row-echelon form of A, and use the Gram–Schmidt process on the nonzero rows of R from the bottom up. Use Theorem 1§2.4.]

33. Let A be an $(n - 1) \times n$ matrix with rows $X_1, X_2, \ldots, X_{n-1}$ and let A_i denote the $(n - 1) \times (n - 1)$ matrix obtained from A by deleting column i. Define the vector Y in $\mathbb{R}^n$ by

$$Y = [\det A_1 \quad -\det A_2 \quad \det A_3 \quad \cdots \quad (-1)^{n+1}\det A_n]$$

Show that:
  (a) $X_i \cdot Y = 0$ for all $i = 1, 2, \ldots, n - 1$.
  (b) $Y \neq 0$ if and only if $\{X_1, X_2, \ldots, X_{n-1}\}$ is linearly independent.
  (c) If $\{X_1, X_2, \ldots, X_{n-1}\}$ is linearly independent, use Theorem 7(3) to show that all solutions to the system of $n - 1$ homogeneous equations

$$AX^T = 0$$

are given by tY, t a parameter.
  [*Hints:* **(a)** Write $B_i = \begin{bmatrix} X_i \\ A \end{bmatrix}$ and show that $\det B_i = 0$.
  (b) If some $\det A_i \neq 0$, the rows of A_i are linearly independent. Conversely, if the X_i are independent, consider $A = UR$ where R is in reduced row-echelon form.]

Section 6.4   **Orthogonal Diagonalization**

Recall (Theorem 1§6.2) that an $n \times n$ matrix A is diagonalizable if and only if it has n linearly independent eigenvectors. Moreover, the matrix P with these eigenvectors as columns is a diagonalizing matrix for A, that is,

$$P^{-1}AP \text{ is diagonal}$$

As we have seen, the really nice bases of $\mathbb{R}^n$ are the orthogonal ones,[4] so a natural question is which $n \times n$ matrices have an orthogonal basis of eigenvectors. These turn out to be precisely the symmetric matrices, and this is the main result of this section.

If a matrix A has n orthogonal eigenvectors, they can (by normalizing) be taken to be orthonormal. The corresponding diagonalizing matrix P has orthonormal columns. Such matrices are very easy to invert.

---

[4] In Section 6.3, n-tuples in $\mathbb{R}^n$ are viewed as rows for convenience. If they are viewed as columns, the dot product is $X \cdot Y = X^TY$ and everything in Section 6.3 remains valid.

THEOREM 1

The following conditions are equivalent for an $n \times n$ matrix P.

1. P is invertible and $P^{-1} = P^T$.
2. The rows of P are orthonormal.
3. The columns of P are orthonormal.

Proof First recall that condition (1) is equivalent to $PP^T = I$ by Theorem 5§2.4. Let $X_1, X_2, \ldots, X_n$ denote the rows of P. Then X_j^T is the jth column of P^T, so the (i, j)-entry of PP^T is $X_i \cdot X_j$. Thus $PP^T = I$ means that $X_i \cdot X_j = 0$ if $i \neq j$ and $X_i \cdot X_j = 1$ if $i = j$, so condition (1) is equivalent to (2). The equivalence of (1) and (3) is similar. ◆

DEFINITION

An $n \times n$ matrix P is called an **orthogonal matrix** if it satisfies one (and hence all) of the conditions in Theorem 1.

EXAMPLE 1

The matrix $\begin{bmatrix} \cos\theta & \sin\theta \\ -\sin\theta & \cos\theta \end{bmatrix}$ is orthogonal for any angle θ. ◆◆◆

It is not enough that the rows of a matrix A are merely orthogonal for A to be an orthogonal matrix.

EXAMPLE 2

The matrix $\begin{bmatrix} 2 & 1 & 1 \\ -1 & 1 & 1 \\ 0 & -1 & 1 \end{bmatrix}$ has orthogonal rows but is not an orthogonal matrix.

However, if the rows are normalized, the resulting matrix $\begin{bmatrix} \frac{2}{\sqrt{6}} & \frac{1}{\sqrt{6}} & \frac{1}{\sqrt{6}} \\ \frac{-1}{\sqrt{3}} & \frac{1}{\sqrt{3}} & \frac{1}{\sqrt{3}} \\ 0 & \frac{-1}{\sqrt{2}} & \frac{1}{\sqrt{2}} \end{bmatrix}$ is

orthogonal (so the columns are also orthonormal). ◆◆◆

EXAMPLE 3

If P and Q are orthogonal matrices, then PQ is also orthogonal, as is $P^{-1} = P^T$.

Solution P and Q are invertible, so PQ is also invertible and $(PQ)^{-1} = Q^{-1}P^{-1} = Q^T P^T$ $= (PQ)^T$. Hence PQ is orthogonal. Similarly, $(P^{-1})^{-1} = P = (P^T)^T = (P^{-1})^T$ shows that P^{-1} is orthogonal.

◆◆◆

DEFINITION

An $n \times n$ matrix A is said to be **orthogonally diagonalizable** when an orthogonal matrix P can be found such that $P^{-1}AP = P^T AP$ is diagonal.

This condition turns out to characterize the symmetric matrices.

THEOREM 2
Principal Axes Theorem

The following conditions are equivalent for an $n \times n$ matrix A.

1. A has an orthonormal set of n eigenvectors.

2. A is orthogonally diagonalizable.

3. A is symmetric.

Proof

(1) *is equivalent to* (2). This follows from the following observations: If $P = [X_1 \; \cdots \; X_n]$ is an $n \times n$ matrix, then P is orthogonal if and only if $\{X_1, \ldots, X_n\}$ is an orthonormal set in $\mathbb{R}^n$; and $P^{-1}AP$ is diagonal if and only if $\{X_1, \ldots, X_n\}$ consists of eigenvectors of A (see the proof of Theorem 1§6.2).

(2) *implies* (3). If $P^T AP = D$ is diagonal, where $P^{-1} = P^T$, then $A = PDP^T$. Because $D^T = D$, this gives $A^T = P^{TT}D^T P^T = PD^T P^T = A$.

(3) *implies* (2). If A is an $n \times n$ symmetric matrix, we proceed by induction on n. If $n = 1$, A is already diagonal. If $n > 1$, assume that (3) implies (2) for $(n - 1) \times (n - 1)$ symmetric matrices. By Theorem 2§6.1 let λ_1 be a (real) eigenvalue of A, and let $AX_1 = \lambda_1 X_1$, where $\|X_1\| = 1$. Use the Gram–Schmidt algorithm to find an orthonormal basis $\{X_1, X_2, \ldots, X_n\}$ for $\mathbb{R}^n$, and let $P_1 = [X_1 \; X_2 \; \cdots \; X_n]$. Then

P_1 is an orthogonal matrix and $P_1^T AP_1 = \begin{bmatrix} \lambda_1 & X \\ 0 & A_1 \end{bmatrix}$ in block form. But $P_1^T AP_1$ is sym-

metric (A is), so it follows that $X = 0$ and A_1 is symmetric. Then, by induction, there exists an $(n - 1) \times (n - 1)$ orthogonal matrix Q such that $Q^T A_1 Q = D_1$ is

diagonal. Hence $P_2 = \begin{bmatrix} 1 & 0 \\ 0 & Q \end{bmatrix}$ is orthogonal and

$$(P_1P_2)^T A(P_1P_2) = P_2^T(P_1^T AP_1)P_2$$

$$= \begin{bmatrix} 1 & 0 \\ 0 & Q^T \end{bmatrix}\begin{bmatrix} \lambda_1 & 0 \\ 0 & A_1 \end{bmatrix}\begin{bmatrix} 1 & 0 \\ 0 & Q \end{bmatrix}$$

$$= \begin{bmatrix} \lambda_1 & 0 \\ 0 & D_1 \end{bmatrix}$$

is diagonal. Because P_1P_2 is orthogonal, this proves (2).   ◆

A set of orthonormal eigenvectors of a symmetric matrix A is called a set of **principal axes** for A. The name comes from geometry, and this is discussed in Section 6.9. Theorem 2 is also called the **real spectral theorem,** and the set of distinct eigenvalues is called the **spectrum** of the matrix. In full generality, the spectral theorem is a similar result for matrices with complex entries (Theorem 8§6.8).

EXAMPLE 4

Find an orthogonal matrix P such that $P^{-1}AP$ is diagonal, where $A = \begin{bmatrix} 1 & 0 & -1 \\ 0 & 1 & 2 \\ -1 & 2 & 5 \end{bmatrix}$.

Solution   The characteristic polynomial of A is

$$c_A(x) = \det\begin{bmatrix} x-1 & 0 & 1 \\ 0 & x-1 & -2 \\ 1 & -2 & x-5 \end{bmatrix} = x(x-1)(x-6)$$

Thus the eigenvalues are $\lambda = 0, 1$, and 6, and corresponding eigenvectors are

$$X_1 = \begin{bmatrix} 1 \\ -2 \\ 1 \end{bmatrix} \quad X_2 = \begin{bmatrix} 2 \\ 1 \\ 0 \end{bmatrix} \quad X_3 = \begin{bmatrix} -1 \\ 2 \\ 5 \end{bmatrix}$$

respectively. Moreover, by what appears to be remarkably good luck, these eigenvectors are *orthogonal*. We have $\|X_1\|^2 = 6$, $\|X_2\|^2 = 5$, and $\|X_3\|^2 = 30$, so

$$P = \begin{bmatrix} \frac{1}{\sqrt{6}}X_1 & \frac{1}{\sqrt{5}}X_2 & \frac{1}{\sqrt{30}}X_3 \end{bmatrix} = \frac{1}{\sqrt{30}}\begin{bmatrix} \sqrt{5} & 2\sqrt{6} & -1 \\ -2\sqrt{5} & \sqrt{6} & 2 \\ \sqrt{5} & 0 & 5 \end{bmatrix}$$

is orthogonal (thus $P^{-1} = P^T$), and

$$P^T AP = \begin{bmatrix} 0 & 0 & 0 \\ 0 & 1 & 0 \\ 0 & 0 & 6 \end{bmatrix}$$

by the diagonalization algorithm.   ◆◆◆

Actually, the fact that the eigenvectors in Example 4 are orthogonal is no coincidence. Theorem 2§6.2 guarantees they are linearly independent (they correspond to distinct eigenvalues); the fact that the matrix is symmetric implies that they are orthogonal. To prove this we need the following useful fact about symmetric matrices.

THEOREM 3

If A is an $n \times n$ symmetric matrix, then

$$(AX) \cdot Y = X \cdot (AY)$$

for all columns X and Y in $\mathbb{R}^n$.

Proof Recall that $X \cdot Y = X^T Y$ for all columns X and Y. Because $A^T = A$, we get

$$(AX) \cdot Y = (AX)^T Y = X^T A^T Y = X^T A Y = X \cdot (AY) \qquad \blacklozenge$$

THEOREM 4

If A is a symmetric matrix, then eigenvectors of A corresponding to distinct eigenvalues are orthogonal.

Proof Let $AX = \lambda X$ and $AY = \mu Y$, where $\lambda \neq \mu$. Using Theorem 3, we compute

$$\lambda(X \cdot Y) = (\lambda X) \cdot Y = (AX) \cdot Y = X \cdot (AY) = X \cdot (\mu Y) = \mu(X \cdot Y)$$

Hence $(\lambda - \mu)(X \cdot Y) = 0$, and so $X \cdot Y = 0$ because $\lambda \neq \mu$. $\qquad \blacklozenge$

Now the procedure for diagonalizing a symmetric $n \times n$ matrix is clear. Find the distinct eigenvalues (all real by Theorem 2§6.1) and find orthonormal bases for each eigenspace (the Gram–Schmidt algorithm may be needed). Then the set of all these basis vectors is orthonormal (by Theorem 4) and contains n vectors.

EXAMPLE 5

Diagonalize $A = \begin{bmatrix} 8 & -2 & 2 \\ -2 & 5 & 4 \\ 2 & 4 & 5 \end{bmatrix}$.

Solution The characteristic polynomial is

$$c_A(x) = \det \begin{bmatrix} x - 8 & 2 & -2 \\ 2 & x - 5 & -4 \\ -2 & -4 & x - 5 \end{bmatrix} = x(x - 9)^2$$

Hence the distinct eigenvalues are $\lambda = 0$ and 9 of multiplicities 1 and 2, respectively, so $\dim(E_0) = 1$ and $\dim(E_9) = 2$ by Theorem 4§6.2 (A is diagonalizable, being symmetric). One eigenvector for $\lambda = 0$ is $X_1 = \begin{bmatrix} 1 \\ 2 \\ -2 \end{bmatrix}$, so $E_0 = \text{span}\{X_1\}$. Gaussian elimination gives

$$E_9 = \text{span}\left\{ \begin{bmatrix} -2 \\ 1 \\ 0 \end{bmatrix}, \begin{bmatrix} 2 \\ 0 \\ 1 \end{bmatrix} \right\}$$

and these eigenvectors are both orthogonal to X_1 (as Theorem 4 guarantees) but not to each other. The Gram–Schmidt process gives an orthogonal basis $\{X_2, X_3\}$ of E_9 where $X_2 = \begin{bmatrix} -2 \\ 1 \\ 0 \end{bmatrix}$, $X_3 = \begin{bmatrix} 2 \\ 4 \\ 5 \end{bmatrix}$. Normalizing gives orthonormal eigenvectors

$$\left\{ \tfrac{1}{3} X_1, \ \tfrac{1}{\sqrt{5}} X_2, \ \tfrac{1}{3\sqrt{5}} X_3 \right\}$$

so

$$P = \begin{bmatrix} \tfrac{1}{3} X_1 & \tfrac{1}{\sqrt{5}} X_2 & \tfrac{1}{3\sqrt{5}} X_3 \end{bmatrix} = \tfrac{1}{3\sqrt{5}} \begin{bmatrix} \sqrt{5} & -6 & 2 \\ 2\sqrt{5} & 3 & 4 \\ -2\sqrt{5} & 0 & 5 \end{bmatrix}$$

is an orthogonal matrix such that $P^{-1}AP$ is diagonal.

It is worth noting that other, more convenient matrices P exist. For example, $Y_2 = \begin{bmatrix} 2 \\ 1 \\ 2 \end{bmatrix}$ and $Y_3 = \begin{bmatrix} -2 \\ 2 \\ 1 \end{bmatrix}$ lie in E_9 and they are orthogonal. Moreover, they both have norm 3 (as does X_1), so

$$Q = \begin{bmatrix} \tfrac{1}{3} X_1 & \tfrac{1}{3} Y_2 & \tfrac{1}{3} Y_3 \end{bmatrix} = \tfrac{1}{3} \begin{bmatrix} 1 & 2 & -2 \\ 2 & 1 & 2 \\ -2 & 2 & 1 \end{bmatrix}$$

is a nicer orthogonal matrix with the property that $Q^{-1}AQ$ is diagonal.

◆◆◆

If we are willing to replace *diagonal* by *upper triangular* in the principal axes theorem, we can weaken the requirement that A is symmetric to insisting only that A has real eigenvalues.

> **THEOREM 5**
> **Triangulation Theorem**
>
> If A is an $n \times n$ matrix with real eigenvalues, an orthogonal matrix P exists such that $P^T A P$ is upper triangular.[5]

Proof   We modify the proof of Theorem 2. If $AX_1 = \lambda_1 X_1$ where $\|X_1\| = 1$, let $\{X_1, X_2, \ldots, X_n\}$ be an orthonormal basis of $\mathbb{R}^n$, and let $P_1 = [X_1 \cdots X_n]$. Then P_1 is orthogonal and $P_1^T A P_1 = \begin{bmatrix} \lambda_1 & X \\ 0 & A_1 \end{bmatrix}$ in block form. By induction, let $Q^T A_1 Q = T_1$ be upper triangular where Q is orthogonal of size $(n-1) \times (n-1)$. Then $P_2 = \begin{bmatrix} 1 & 0 \\ 0 & Q \end{bmatrix}$ is orthogonal, so $P = P_1 P_2$ is also orthogonal and $P^T A P = \begin{bmatrix} \lambda_1 & XQ \\ 0 & T_1 \end{bmatrix}$ is upper triangular. ◆

The proof of Theorem 5 gives no way to construct the matrix P. However, an algorithm will be given in Section 7.7 where an improved version of Theorem 5 is presented.

As for a diagonal matrix, the eigenvalues of an upper triangular matrix are displayed along the main diagonal. Because, in Theorem 5, A and $P^T A P$ have the same determinant and trace, we obtain the following:

> **COROLLARY**
>
> If A is an $n \times n$ matrix with real eigenvalues $\lambda_1, \lambda_2, \ldots, \lambda_n$ (possibly not all distinct), then $\det A = \lambda_1 \lambda_2 \cdots \lambda_n$ and $\operatorname{tr} A = \lambda_1 + \lambda_2 + \cdots + \lambda_n$.

This corollary remains true even if the eigenvalues are not real. In fact a version of Theorem 5 holds for any $n \times n$ complex matrix (Schur's theorem). This is given in Section 6.8.

EXERCISES 6.4

1. Normalize the rows to make each of the following matrices orthogonal.

(a) $\begin{bmatrix} 1 & 1 \\ -1 & 1 \end{bmatrix}$

◆ **(b)** $\begin{bmatrix} 3 & -4 \\ 4 & 3 \end{bmatrix}$

(c) $\begin{bmatrix} 1 & 2 \\ -4 & 2 \end{bmatrix}$

◆ **(d)** $\begin{bmatrix} a & b \\ -b & a \end{bmatrix}$, $(a, b) \neq (0, 0)$

(e) $\begin{bmatrix} \cos\theta & \sin\theta & 0 \\ -\sin\theta & \cos\theta & 0 \\ 0 & 0 & 2 \end{bmatrix}$

◆ **(f)** $\begin{bmatrix} 2 & 1 & -1 \\ 1 & -1 & 1 \\ 0 & 1 & 1 \end{bmatrix}$

(g) $\begin{bmatrix} -1 & 2 & 2 \\ 2 & -1 & 2 \\ 2 & 2 & -1 \end{bmatrix}$

◆ **(h)** $\begin{bmatrix} 2 & 6 & -3 \\ 3 & 2 & 6 \\ -6 & 3 & 2 \end{bmatrix}$

---

[5]There is also a lower triangular version.

◆2. If P is a triangular orthogonal matrix, show that P is diagonal and that all diagonal entries are 1 or -1.

3. If P is orthogonal, show that kP is orthogonal if and only if $k = 1$ or $k = -1$.

4. If the first two rows of an orthogonal matrix are $\left[\frac{1}{3}, \frac{2}{3}, \frac{2}{3}\right]$ and $\left[\frac{2}{3}, \frac{1}{3}, \frac{-2}{3}\right]$, find all possible third rows.

5. For each matrix A, find an orthogonal matrix P such that $P^{-1}AP$ is diagonal.

(a) $A = \begin{bmatrix} 3 & 0 & 0 \\ 0 & 2 & 2 \\ 0 & 2 & 5 \end{bmatrix}$　**◆(b)** $A = \begin{bmatrix} 3 & 0 & 7 \\ 0 & 5 & 0 \\ 7 & 0 & 3 \end{bmatrix}$

(c) $A = \begin{bmatrix} 1 & 1 & 0 \\ 1 & 1 & 0 \\ 0 & 0 & 2 \end{bmatrix}$　**◆(d)** $A = \begin{bmatrix} 5 & -2 & -4 \\ -2 & 8 & -2 \\ -4 & -2 & 5 \end{bmatrix}$

(e) $A = \begin{bmatrix} 5 & 3 & 0 & 0 \\ 3 & 5 & 0 & 0 \\ 0 & 0 & 7 & 1 \\ 0 & 0 & 1 & 7 \end{bmatrix}$　**◆(f)** $A = \begin{bmatrix} 3 & 5 & -1 & 1 \\ 5 & 3 & 1 & -1 \\ -1 & 1 & 3 & 5 \\ 1 & -1 & 5 & 3 \end{bmatrix}$

◆6. Consider $A = \begin{bmatrix} 0 & a & 0 \\ a & 0 & c \\ 0 & c & 0 \end{bmatrix}$ where one of $a, c \neq 0$. Show that $c_A(x) = x(x - k)(x + k)$, where $k = \sqrt{a^2 + c^2}$ and find an orthogonal matrix P such that $P^{-1}AP$ is diagonal.

7. Consider $A = \begin{bmatrix} 0 & 0 & a \\ 0 & b & 0 \\ a & 0 & 0 \end{bmatrix}$. Show that $c_A(x) = (x - b)(x - a)(x + a)$ and find an orthogonal matrix P such that $P^{-1}AP$ is diagonal.

8. Given $A = \begin{bmatrix} b & a \\ a & b \end{bmatrix}$, show that $c_A(x) = (x - a - b)(x + a - b)$ and find an orthogonal matrix P such that $P^{-1}AP$ is diagonal.

9. Consider $A = \begin{bmatrix} b & 0 & a \\ 0 & b & 0 \\ a & 0 & b \end{bmatrix}$. Show that $c_A(x) = (x - b)(x - b - a)(x - b + a)$ and find an orthogonal matrix P such that $P^{-1}AP$ is diagonal.

10. Show that the following are equivalent for a symmetric matrix A.
(a) A is orthogonal.　　**(b)** $A^2 = I$
(c) All eigenvalues of A are ± 1. [*Hint:* For part **(b)** if and only if part **(c)**, use Theorem 2.]

11. We call matrices A and B **orthogonally similar** (and write $A \stackrel{\circ}{=} B$) if $B = P^TAP$ for an orthogonal matrix P.
(a) Show that $A \stackrel{\circ}{=} A$ for all A; $A \stackrel{\circ}{=} B$ implies that $B \stackrel{\circ}{=} A$; and $A \stackrel{\circ}{=} B$ and $B \stackrel{\circ}{=} C$ imply that $A \stackrel{\circ}{=} C$.

(b) Show that the following are equivalent for two symmetric matrices A and B.
(i) A and B are similar.
(ii) A and B are orthogonally similar.
(iii) A and B have the same eigenvalues.

12. Assume that A and B are orthogonally similar (Exercise 11).
(a) If A and B are invertible, show that A^{-1} and B^{-1} are orthogonally similar.
◆(b) Show that A^2 and B^2 are orthogonally similar.
(c) Show that, if A is symmetric, so is B.

13. If A is symmetric, show that every eigenvalue of A is nonnegative if and only if $A = B^2$ for some symmetric matrix B.

◆14. Prove the converse of Theorem 3: If $(AX) \cdot Y = X \cdot (AY)$ for all n-columns X and Y, then A is symmetric.

15. Show that every eigenvalue of A is zero if and only if A is nilpotent ($A^k = 0$ for some $k \geq 1$).

16. If A has real eigenvalues, show that $A = B + C$ where B is symmetric and C is nilpotent. [*Hint:* Theorem 5.]

17. Let P be an orthogonal matrix.
(a) Show that $\det P = 1$ or $\det P = -1$.
◆(b) Give 2×2 examples of P such that $\det P = 1$ and $\det P = -1$.
(c) If $\det P = -1$, show that $I + P$ has no inverse. [*Hint:* $P^T(I + P) = (I + P)^T$.]
(d) If P is $n \times n$ and $\det P \neq (-1)^n$, show that $I - P$ has no inverse. [*Hint:* $P^T(I - P) = -(I - P)^T$.]

18. We call a square matrix E a **projection matrix** if $E^2 = E = E^T$.
(a) If E is a projection matrix, show that $P = I - 2E$ is orthogonal and symmetric.
(b) If P is orthogonal and symmetric, show that $E = \frac{1}{2}(I - P)$ is a projection matrix.
(c) If U is $m \times n$ and $U^TU = I$ (for example, a unit column in $\mathbb{R}^n$), show that $E = UU^T$ is a projection matrix.

19. A matrix that we obtain from the identity matrix by writing its rows in a different order is called a **permutation matrix**. Show that every permutation matrix is orthogonal.

◆20. If the rows $R_1, \ldots, R_n$ of the $n \times n$ matrix $A = [a_{ij}]$ are orthogonal, show that the (i, j)-entry of A^{-1} is $\dfrac{a_{ji}}{\|R_j\|^2}$.

21. (a) Let A be an $m \times n$ matrix. Show that the following are equivalent.

     (i) A has orthogonal rows.

     (ii) A can be factored as $A = DP$, where D is invertible and diagonal and P has orthonormal rows.

     (iii) AA^T is an invertible, diagonal matrix.

  (b) Show that an $n \times n$ matrix A has orthogonal rows if and only if A can be factored as $A = DP$, where P is orthogonal and D is diagonal and invertible.

22. Let A be a skew-symmetric matrix; that is, $A^T = -A$. Assume that A is an $n \times n$ matrix.

  (a) Show that $I + A$ is invertible. [*Hint:* By Theorem 4§2.4, it suffices to show that $X(I + A) = 0$, X in $\mathbb{R}^n$, implies $X = 0$. Compute $X \cdot X = XX^T$, and use the fact that $XA = -X$ and $XA^2 = X$.]

  ◆ **(b)** Show that $P = (I - A)(I + A)^{-1}$ is orthogonal.

  (c) Show that every orthogonal matrix P such that $I + P$ is invertible arises as in part **(b)** from some skew-symmetric matrix A. [*Hint:* Solve $P = (I - A)(I + A)^{-1}$ for A.]

23. Show that the following are equivalent for an $n \times n$ matrix P.

  (a) P is orthogonal.

  (b) $\|XP\| = \|X\|$ for all rows X in $\mathbb{R}^n$.

  (c) $\|XP - YP\| = \|X - Y\|$ for all rows X and Y in $\mathbb{R}^n$.

  (d) $(XP) \cdot (YP) = X \cdot Y$ for all rows X and Y in $\mathbb{R}^n$.

     [*Hints:* For part **(c)** implies part **(d)**, see Exercise 15§6.3. For part **(d)** implies part **(a)**, show that row i of P equals E_iP, where E_i is row i of the identity matrix.]

24. Show that every 2×2 orthogonal matrix has the form
$$\begin{bmatrix} \cos\theta & \sin\theta \\ -\sin\theta & \cos\theta \end{bmatrix} \text{ or } \begin{bmatrix} \cos\theta & \sin\theta \\ \sin\theta & -\cos\theta \end{bmatrix} \text{ for some angle } \theta.$$
[*Hint:* If $a^2 + b^2 = 1$, then $a = \cos\theta$ and $b = \sin\theta$ for some angle θ.]

Section 6.5   Positive Definite Matrices

All the eigenvalues of any symmetric matrix are real; this section is about the case in which the eigenvalues are positive. These matrices, which arise whenever optimization (maximum and minimum) problems are encountered, have countless applications throughout science and engineering. They also arise in statistics (for example, in factor analysis used in the social sciences) and in geometry (see Section 6.9). We will encounter them again in Chapter 8 when describing all inner products in $\mathbb{R}^n$.

DEFINITION

A square matrix is called **positive definite** if it is symmetric and all its eigenvalues are positive.

Because these matrices are symmetric, the principal axes theorem plays a central role in the theory.

THEOREM 1

If A is positive definite then it is invertible and $\det A > 0$.

Proof   If A is $n \times n$ and the eigenvalues are $\lambda_1, \lambda_2, \ldots, \lambda_n$, then $\det A = \lambda_1\lambda_2\cdots\lambda_n > 0$ by the principal axes theorem (or the corollary to Theorem 5§6.4). ◆

If X is a column in $\mathbb{R}^n$ and A is any real $n \times n$ matrix, we view the 1×1 matrix X^TAX as a real number. With this convention, we have the following characterization of positive definite matrices.

> **THEOREM 2**
>
> A symmetric matrix A is positive definite if and only if $X^TAX > 0$ for every column $X \neq 0$ in $\mathbb{R}^n$.

Proof By the principal axes theorem, let $P^TAP = D = \text{diag}(\lambda_1, \lambda_2, \ldots, \lambda_n)$ where $P^{-1} = P^T$ and the λ_i are the eigenvalues of A. Given X in $\mathbb{R}^n$, write $Y = P^TX = [y_1\, y_2\, \cdots\, y_n]^T$. Then

$$X^TAX = Y^TDY = \lambda_1 y_1^2 + \lambda_2 y_2^2 + \cdots + \lambda_n y_n^2 \qquad (*)$$

If A is positive definite and $X \neq 0$, then $X^TAX > 0$ by $(*)$ because some $y_j \neq 0$. Conversely, if $X^TAX > 0$ whenever $X \neq 0$, let $X = PE_j \neq 0$ where E_j is column j of I_n. Then $Y = E_j$, so $(*)$ reads $\lambda_j = X^TAX > 0$.   ◆

EXAMPLE 1

If $A = C^TC$ where C is invertible, show that A is positive definite.

Solution If $X \neq 0$, $X^TAX = (CX)^T (CX) = \|CX\|^2 > 0$ because $CX \neq 0$ (C is invertible).   ◆◆◆

We are going to show that *every* positive definite matrix A can be factored as $A = U^TU$, where U is upper triangular with positive entries on the main diagonal. This factorization is unique (Exercise 11) and is called the **Cholesky decomposition** of the positive definite matrix A.[6] It can be found easily using row operations. Here is an example.

EXAMPLE 2

Find the Cholesky decomposition of $A = \begin{bmatrix} 10 & 5 & 2 \\ 5 & 3 & 2 \\ 2 & 2 & 3 \end{bmatrix}$.

Solution Proceed as in Gaussian elimination except that the only row operations used are to add multiples of a row to rows *below* it. If A is positive definite, the (proof of the) next lemma guarantees that A can be carried in this way to an upper triangular matrix U_1 with positive entries on the main diagonal. In the present case,

---

[6]The decomposition is sometimes defined as $A = LL^T$, where $L = U^T$ is lower triangular with positive diagonal entries.

$$A = \begin{bmatrix} 10 & 5 & 2 \\ 5 & 3 & 2 \\ 2 & 2 & 3 \end{bmatrix} \rightarrow \begin{bmatrix} 10 & 5 & 2 \\ 0 & \frac{1}{2} & 1 \\ 0 & 1 & \frac{13}{5} \end{bmatrix} \rightarrow \begin{bmatrix} 10 & 5 & 2 \\ 0 & \frac{1}{2} & 1 \\ 0 & 0 & \frac{3}{5} \end{bmatrix} = U_1$$

Now divide each row of U_1 by the square root of the diagonal entry in that row to produce U:

$$U = \begin{bmatrix} \sqrt{10} & \frac{5}{\sqrt{10}} & \frac{2}{\sqrt{10}} \\ 0 & \frac{1}{\sqrt{2}} & \sqrt{2} \\ 0 & 0 & \frac{\sqrt{3}}{\sqrt{5}} \end{bmatrix}$$

Then $A = U^T U$, as the reader can verify. This is the Cholesky decomposition.

◆◆◆

The method in Example 2 works for any positive definite matrix. Here is a formal statement of the procedure.

**ALGORITHM FOR
THE CHOLESKY
DECOMPOSITION**

If A is positive definite, a Cholesky decomposition $A = U^T U$ can be obtained as follows:

STEP 1 Carry A to an upper triangular matrix U_1 with positive diagonal entries using row operations, each of which adds a multiple of a row to a lower row.

STEP 2 Obtain U by dividing each row of U_1 by the square root of the diagonal entry in that row.

The key to the algorithm is the fact that Step 1 is possible for every positive definite matrix A. The following definition plays a crucial role in the proof.[7]

DEFINITION

If A is an $n \times n$ matrix and $1 \leq r \leq n$, let $^r\!A$ denote the $r \times r$ matrix obtained from A by deleting the last $n - r$ rows and columns. The matrices $^1\!A$, $^2\!A$, ..., $^n\!A = A$ are called the **principal submatrices** of A.

---

[7]The remainder of this section is not needed elsewhere in the book.

Hence if $A = [a_{ij}]$, then

$$^1A = [a_{11}], \quad {}^2A = \begin{bmatrix} a_{11} & a_{12} \\ a_{21} & a_{22} \end{bmatrix}, \quad {}^3A = \begin{bmatrix} a_{11} & a_{12} & a_{13} \\ a_{21} & a_{22} & a_{23} \\ a_{31} & a_{32} & a_{33} \end{bmatrix}, \dots$$

THEOREM 3

If A is positive definite, so also is ${}^rA$ for each $r = 1, 2, \dots, n$.

Proof   Write $A = \begin{bmatrix} {}^rA & P \\ Q & R \end{bmatrix}$. If $Y \neq 0$ in $\mathbb{R}^r$, let $X = \begin{bmatrix} Y \\ 0 \end{bmatrix}$ in $\mathbb{R}^n$. Then

$$0 < X^T A X = [Y^T 0] \begin{bmatrix} {}^rA & P \\ Q & R \end{bmatrix} \begin{bmatrix} Y \\ 0 \end{bmatrix} = Y^T {}^rA Y$$

This shows that ${}^rA$ is positive definite.   ◆

It follows from Theorems 1 and 3 that if A is positive definite, then $\det {}^rA > 0$ for each $r = 1, 2, \dots, n$. This condition actually characterizes the positive definite matrices; to prove this we first show that it implies that A can be carried to upper triangular form by row operations that add multiples of rows to *lower* rows. The precise result is as follows.

LEMMA

Let A be an $n \times n$ matrix for which $\det {}^rA > 0$ for each $r = 1, 2, \dots, n$. Then $LA = U$ where L is $n \times n$ lower triangular with 1's on the diagonal, and U is $n \times n$ upper triangular with positive entries on the diagonal.

Proof   Proceed by induction on n. If $n = 1$, then $A = [a]$, $a > 0$, so take $L = [1]$ and $U = A$. If $n > 1$, write $A = [a_{ij}]$ and observe that $a_{11} = \det {}^1A > 0$. Hence $A \rightarrow \begin{bmatrix} a_{11} & X \\ 0 & B \end{bmatrix}$ by row operations that add multiples of row 1 to lower rows. Each such row operation is achieved by left-multiplication by an elementary matrix that is lower triangular with 1's on the diagonal. The product L_1 of these elementary matrices is of the same form, so

$$L_1 A = \begin{bmatrix} a_{11} & X \\ 0 & B \end{bmatrix}$$

On the other hand, writing L_1 and A in block form gives matrices P, Q, and R such that

$$L_1 A = \begin{bmatrix} {}^r L_1 & 0 \\ M & N \end{bmatrix} \begin{bmatrix} {}^r A & C \\ D & E \end{bmatrix} = \begin{bmatrix} {}^r L_1 {}^r A & P \\ Q & R \end{bmatrix}$$

Comparing these, we conclude that, if $r \geq 2$, ${}^r L_1 {}^r A$ has the form

$$ {}^r L_1 {}^r A = {}^r (L_1 A) = \begin{bmatrix} a_{11} & X \\ 0 & B \end{bmatrix} = \begin{bmatrix} a_{11} & X' \\ 0 & {}^{r-1} B \end{bmatrix}$$

In particular, the fact that $\det {}^r L_1 = 1$ gives

$$a_{11} \det({}^{r-1} B) = \det({}^r L_1 {}^r A) = \det {}^r A$$

Hence $\det({}^{r-1} B) > 0$ for each $r \geq 2$; so, by induction, let $L_0 B = U_0$ be as in the lemma. If $L_2 = \begin{bmatrix} 1 & 0 \\ 0 & L_0 \end{bmatrix}$, then

$$L_2 L_1 A = \begin{bmatrix} 1 & 0 \\ 0 & L_0 \end{bmatrix} \begin{bmatrix} a_{11} & X \\ 0 & B \end{bmatrix} = \begin{bmatrix} a_{11} & X \\ 0 & U_0 \end{bmatrix}$$

is upper triangular with positive diagonal entries. Because $L_2 L_1$ is lower triangular with 1's on the diagonal, take $L = L_2 L_1$ and $U = L_2 L_1 A$. ◆

Note that the proof of the lemma shows that L is the product of the elementary matrices corresponding to row operations that add multiples of a row to rows below it. This proves step 1 of the algorithm. The rest of the algorithm is verified in the proof of the following theorem.

THEOREM 4

The following conditions are equivalent for an $n \times n$ symmetric matrix A.

1. A is positive definite.
2. $\det {}^r A > 0$ for each $r = 1, 2, \ldots, n$
3. $A = U^T U$ where U is upper triangular with positive diagonal entries.

Proof

(1) *implies* (2). This is by Theorems 1 and 3.

(2) *implies* (3). Let $LA = U_0$ as in the lemma and let $D = \text{diag}(d_1, \ldots, d_n)$ be the diagonal matrix with the same diagonal as U_0. Then $U_1 = D^{-1} U_0$ is upper triangular with 1's on the diagonal and

$$L^{-1} D U_1 = A = A^T = U_1^T D (L^T)^{-1}$$

Hence $DU_1L^TD^{-1} = LU_1^T$ is both upper triangular (left side) and lower triangular with 1's on the diagonal (right side), and so equals I_n. It follows that $L^{-1} = U_1^T$. Now write $D_0 = \text{diag}(\sqrt{d_1}, \ldots, \sqrt{d_n})$, so $D_0^2 = D$. Then $U = D_0^{-1}U_0$ is upper triangular with positive diagonal entries, and

$$U^TU = U_0^TD^{-1}U_0 = U_1^TU_0 = L^{-1}U_0 = A$$

(3) *implies* (1). This is by Example 1.   ◆

Finally we note that the proof of Theorem 4 actually completes the proof of the algorithm. We have $LA = U_0$ as in the proof by step 1 of the algorithm, and $U^TU = A$ where $U = D_0^{-1}U_0$. This is just as in step 2 of the algorithm because $D_0^{-1}U_0$ is the matrix obtained from U_0 by multiplying each row of U_0 by the inverse of the diagonal entry of that row.

EXERCISES 6.5

1. Find the Cholesky decomposition of each of the following matrices.

(a) $\begin{bmatrix} 4 & 3 \\ 3 & 5 \end{bmatrix}$ ◆(b) $\begin{bmatrix} 2 & -1 \\ -1 & 1 \end{bmatrix}$

(c) $\begin{bmatrix} 12 & 4 & 3 \\ 4 & 2 & -1 \\ 3 & -1 & 7 \end{bmatrix}$ ◆(d) $\begin{bmatrix} 20 & 4 & 5 \\ 4 & 2 & 3 \\ 5 & 3 & 5 \end{bmatrix}$

2. If A is positive definite, show that A^k is positive definite for all $k \geq 1$. Prove the converse if k is odd. What if k is even?

3. Find a symmetric matrix A such that A^2 is positive definite, but A is not.

◆4. If A and B are positive definite, show that $A + B$ is positive definite.

5. If A and B are positive definite, show that $\begin{bmatrix} A & 0 \\ 0 & B \end{bmatrix}$ is positive definite.

6. If A is an $n \times n$ positive definite matrix and U is an $n \times m$ matrix of rank m, show that U^TAU is positive definite.

7. If A is positive definite, show that each diagonal entry is positive.

8. Let A_0 be formed from A by deleting rows 2 and 4 and deleting columns 2 and 4. If A is positive definite, show that A_0 is positive definite.

9. If A is positive definite, show that $A = CC^T$ where C has orthogonal columns.

◆10. If A is positive definite, show that $A = C^2$ where C is positive definite.

11. Show that the Cholesky decomposition is unique; that is, if $U^TU = U_1^TU_1$ where U and U_1 are upper triangular with positive diagonal entries, show that $U = U_1$. [*Hint:* Stare at $UU_1^{-1} = (U^{-1})^TU_1^T$.]

12. (a) Suppose an invertible matrix A can be factored in $\mathbf{M}_{nn}$ as $A = LDU$ where L is lower triangular with 1's on the diagonal, U is upper triangular with 1's on the diagonal, and D is diagonal with positive diagonal entries. Show that the factorization is unique: If $A = L_1D_1U_1$ is another such factorization, show that $L_1 = L$, $D_1 = D$, and $U_1 = U$.

   ◆(b) Show that a matrix A is positive definite if and only if A is symmetric and admits a factorization $A = LDU$ as in (a).

13. Let A be positive definite and write $d_r = \det {}'A$ for each $r = 1, 2, \ldots, n$. If U is the upper triangular matrix obtained in step 1 of the algorithm, show that the diagonal elements $u_{11}, u_{22}, \ldots, u_{nn}$ of U are given by $u_{11} = d_1$, $u_{jj} = d_j/d_{j-1}$ if $j > 1$. [*Hint:* If $LA = U$ where L is lower triangular with 1's on the diagonal, use block multiplication to show that $\det {}'A = \det {}'U$ for each r.]

░▒▓█ Section 6.6   LP-Factorization (Optional)

The main virtue of orthogonal matrices is that they can be inverted easily: simply take the transpose. This fact combines with the following theorem to give a useful way of simplifying many matrix calculations (for example, in least squares approximation; see Section 6.10). The result concerns matrices with independent rows (or columns), and the idea is to factor such a matrix as the product of an invertible lower triangular matrix and a matrix with *orthonormal* rows (respectively, columns).

Suppose A is an $m \times n$ matrix with linearly independent rows $R_1, R_2, \ldots, R_m$. Then the Gram–Schmidt process can be applied to these rows to produce orthogonal rows $E_1, E_2, \ldots, E_m$, where $E_1 = R_1$ and

$$E_k = R_k - \frac{R_k \cdot E_1}{\|E_1\|^2} E_1 - \frac{R_k \cdot E_2}{\|E_2\|^2} E_2 - \cdots - \frac{R_k \cdot E_{k-1}}{\|E_{k-1}\|^2} E_{k-1}$$

for each $k = 2, 3, \ldots, m$. Now let $P_i = \dfrac{1}{\|E_i\|} E_i$ for each i. Then $P_1, P_2, \ldots, P_m$ are orthonormal rows, and the foregoing equation becomes

$$\|E_k\| P_k = R_k - (R_k \cdot P_1)P_1 - (R_k \cdot P_2)P_2 - \cdots - (R_k \cdot P_{k-1})P_{k-1}$$

If, for each k, this is solved for R_k in terms of $P_1, P_2, \ldots, P_k$, we get

$$R_i = \|E_1\|P_1$$
$$R_2 = (R_2 \cdot P_1)P_1 + \|E_2\|P_2$$
$$R_3 = (R_3 \cdot P_1)P_1 + (R_3 \cdot P_2)P_2 + \|E_3\|P_3$$
$$\vdots \qquad \vdots$$
$$R_m = (R_m \cdot P_1)P_1 + (R_m \cdot P_2)P_2 + (R_m \cdot P_3)P_3 + \cdots + \|E_m\|P_m$$

In matrix form, these equations give the required factorization of A:

$$A = \begin{bmatrix} R_1 \\ R_2 \\ R_3 \\ \vdots \\ R_m \end{bmatrix} = \begin{bmatrix} \|E_1\| & 0 & 0 & \cdots & 0 \\ R_2 \cdot P_1 & \|E_2\| & 0 & \cdots & 0 \\ R_3 \cdot P_1 & R_3 \cdot P_2 & \|E_3\| & \cdots & 0 \\ \vdots & \vdots & \vdots & & \vdots \\ R_m \cdot P_1 & R_m \cdot P_2 & R_m \cdot P_3 & \cdots & \|E_m\| \end{bmatrix} \cdot \begin{bmatrix} P_1 \\ P_2 \\ P_3 \\ \vdots \\ P_m \end{bmatrix} = LP \qquad (*)$$

Note that the first factor L is invertible because $\|E_i\| \neq 0$ for each i, and the second factor P has orthonormal rows $P_1, P_2, \ldots, P_m$. Moreover, an analogous procedure works if A has linearly independent columns (alternatively, apply the foregoing to A^T). The result is Theorem 1.

THEOREM 1   ░▒▓█▓▒░

Let A denote an $m \times n$ matrix.

1. If the rows of A are linearly independent, then A can be factored as

$$A = LP$$

where L is an $m \times m$ lower triangular invertible matrix and P is an $m \times n$ matrix with orthonormal rows. In particular, $PP^T = I_m$.

2. If the columns of A are linearly independent, then A can be factored as

$$A = QR$$

where R is an $n \times n$ invertible upper triangular matrix and Q is an $m \times n$ matrix with orthonormal columns. In particular, $Q^TQ = I_n$.

For a square matrix, having independent rows (or columns) is equivalent to being invertible, whereas having orthonormal rows (or columns) is equivalent to being orthogonal. Hence:

THEOREM 2

If A is square and invertible, then A has factorizations

$$A = LP \quad \text{and} \quad A = QR$$

where P and Q are orthogonal matrices, L is invertible and lower triangular, and R is invertible and upper triangular.

The argument leading to Theorem 1 gives formulas (in equation $(*)$) for L and P. Here is an example.

EXAMPLE 1

Express $A = \begin{bmatrix} 1 & -1 & 0 & 0 \\ 1 & 0 & 1 & 0 \\ 0 & 1 & 1 & 1 \end{bmatrix}$ in the form $A = LP$, where L is lower triangular and invertible and P has orthonormal rows.

Solution   Write $R_1 = [1 \ -1 \ 0 \ 0]$, $R_2 = [1 \ 0 \ 1 \ 0]$, $R_3 = [0 \ 1 \ 1 \ 1]$ for the rows of A. Apply the Gram–Schmidt algorithm to obtain orthogonal rows E_1, E_2, and E_3.

$$
\begin{aligned}
E_1 &= R_1 & &= [1 \ -1 \ 0 \ 0] \\
E_2 &= R_2 - \tfrac{1}{2} E_1 & &= [\tfrac{1}{2} \ \tfrac{1}{2} \ 1 \ 0] \\
E_3 &= R_3 + \tfrac{1}{2} E_1 - E_2 & &= [0 \ 0 \ 0 \ 1]
\end{aligned}
$$

Hence, using the notation preceding Theorem 1, we get

$$P = \begin{bmatrix} P_1 \\ P_2 \\ P_3 \end{bmatrix} = \begin{bmatrix} \frac{1}{\sqrt{2}} & \frac{-1}{\sqrt{2}} & 0 & 0 \\ \frac{1}{\sqrt{6}} & \frac{1}{\sqrt{6}} & \frac{2}{\sqrt{6}} & 0 \\ 0 & 0 & 0 & 1 \end{bmatrix} = \frac{1}{\sqrt{6}} \begin{bmatrix} \sqrt{3} & -\sqrt{3} & 0 & 0 \\ 1 & 1 & 2 & 0 \\ 0 & 0 & 0 & \sqrt{6} \end{bmatrix}$$

$$L = \begin{bmatrix} \|E_1\| & 0 & 0 \\ R_2 \cdot P_1 & \|E_2\| & 0 \\ R_3 \cdot P_1 & R_3 \cdot P_2 & \|E_3\| \end{bmatrix} = \frac{1}{2} \begin{bmatrix} 2\sqrt{2} & 0 & 0 \\ \sqrt{2} & \sqrt{6} & 0 \\ -\sqrt{2} & \sqrt{6} & 2 \end{bmatrix}$$

The reader can confirm that $A = LP$.

◆◆◆

Theorem 1 is useful in computations in the following way: If A has independent rows, then AA^T is invertible (Theorem 6§5.5); and in practice it is often necessary to invert AA^T (see, for example, Section 6.9). But if $A = LP$ as in (1) of Theorem 1, then $AA^T = LPP^TL^T = LL^T$, so

$$(AA^T)^{-1} = (L^{-1})^T L^{-1}$$

The inverse of L is particularly easy to find because L is triangular. In fact, the main reason for the importance of Theorem 1 is that AA^T may be difficult to invert. This is because the inversion process can be numerically unstable so that errors in the computation becomes too large. The processes of factoring $A = LP$ and finding L^{-1} are essentially back-substitutions and so are less unstable.

EXAMPLE 2

Use Theorem 1 to invert AA^T, where A is the matrix in Example 1.

Solution   The factorization $A = LP$ was found in Example 1. We have

$$L = \frac{1}{2} \begin{bmatrix} 2\sqrt{2} & 0 & 0 \\ \sqrt{2} & \sqrt{6} & 0 \\ -\sqrt{2} & \sqrt{6} & 2 \end{bmatrix}, \quad L^{-1} = \frac{1}{6} \begin{bmatrix} 3\sqrt{2} & 0 & 0 \\ -\sqrt{6} & 2\sqrt{6} & 0 \\ 6 & -6 & 6 \end{bmatrix}$$

Hence

$$(AA^T)^{-1} = (L^{-1})^T L^{-1} = \frac{1}{3} \begin{bmatrix} 5 & -4 & 3 \\ -4 & 5 & -3 \\ 3 & -3 & 3 \end{bmatrix}$$

The reader can verify that this is the inverse of AA^T.

◆◆◆

EXERCISES 6.6

1. In each case, factor A as $A = LP$, where L is invertible and lower triangular and P has orthonormal rows. Then compute $(AA^T)^{-1}$ as in Example 2.

(a) $A = \begin{bmatrix} 1 & -1 \\ -1 & 0 \end{bmatrix}$ ◀**(b)** $A = \begin{bmatrix} 2 & 1 \\ 1 & 1 \end{bmatrix}$

(c) $A = \begin{bmatrix} 1 & 1 & 1 & 0 \\ 1 & 1 & 0 & 0 \\ 1 & 0 & 0 & 0 \end{bmatrix}$ ◀**(d)** $A = \begin{bmatrix} 1 & -1 & 0 & 1 \\ 1 & 0 & 1 & -1 \\ 0 & 1 & 1 & 0 \end{bmatrix}$

2. If $A = LP = L_1 P_1$ are two LP-factorizations of the invertible matrix A, shows that $L_1 = LD$ and $P_1 = DP$ for some diagonal matrix D with diagonal entries ± 1. [*Hint:* Consider $L_1^{-1}L = P_1 P^{-1}$.]

Section 6.7   Computing Eigenvalues (Optional)

In practice, the problem of finding eigenvalues and eigenvectors of a matrix is virtually never solved by finding the roots of the characteristic polynomial. Iterative methods are much better. Two of these will be described briefly in this section.

An eigenvalue λ of an $n \times n$ matrix A is said to be a **dominant eigenvalue** if

$$|\lambda| > |\mu| \qquad \text{for all eigenvalues } \mu \neq \lambda$$

Any corresponding eigenvector is called a **dominant eigenvector** of A. When such an eigenvalue exists, one technique for finding it is as follows: Let X_0 in $\mathbb{R}^n$ be a first approximation to a dominant eigenvector, and compute successive approximations $X_1, X_2, \ldots$ by

$$X_1 = AX_0 \qquad X_2 = AX_1 \qquad X_3 = AX_2 \qquad \cdots$$

In general, we define

$$X_{k+1} = AX_k \qquad \text{for each } k \geq 0$$

If the first estimate X_0 is good enough (see later), these vectors X_n will approximate dominant eigenvectors of A. This technique is called the **power method**. Moreover, it can be used to approximate the dominant eigenvalue λ. Observe that if Z is any dominant eigenvector, then

$$\frac{Z \cdot (AZ)}{\|Z\|^2} = \frac{Z \cdot (\lambda Z)}{\|Z\|^2} = \lambda$$

Because the vectors $X_1, X_2, \ldots, X_n, \ldots$ approximate dominant eigenvectors, we define the **Rayleigh quotients** as follows:

$$r_k = \frac{X_k \cdot AX_k}{\|X_k\|^2} = \frac{X_k \cdot X_{k+1}}{\|X_k\|^2} \qquad \text{for } k \geq 1$$

Then the numbers r_k approximate the dominant eigenvalue λ.

EXAMPLE 1

Use the power method to approximate a dominant eigenvector and eigenvalue of
$$A = \begin{bmatrix} 1 & 1 \\ 2 & 0 \end{bmatrix}.$$

Solution
The eigenvalues of A are 2 and -1, with eigenvectors $\begin{bmatrix} 1 \\ 1 \end{bmatrix}$ and $\begin{bmatrix} 1 \\ -2 \end{bmatrix}$. Take $X_0 = \begin{bmatrix} 1 \\ 0 \end{bmatrix}$ as the first approximation and compute $X_1, X_2, \ldots$, successively, from $X_1 = AX_0$, $X_2 = AX_1, \ldots$. The result is

$$X_1 = \begin{bmatrix} 1 \\ 2 \end{bmatrix}, \; X_2 = \begin{bmatrix} 3 \\ 2 \end{bmatrix}, \; X_3 = \begin{bmatrix} 5 \\ 6 \end{bmatrix}, \; X_4 = \begin{bmatrix} 11 \\ 10 \end{bmatrix}, \; X_5 = \begin{bmatrix} 21 \\ 22 \end{bmatrix}, \; \ldots$$

These vectors are approaching scalar multiples of the dominant eigenvector $\begin{bmatrix} 1 \\ 1 \end{bmatrix}$. Moreover, the Rayleigh quotients are

$$r_1 = \tfrac{7}{5}, \; r_2 = \tfrac{27}{13}, \; r_3 = \tfrac{115}{61}, \; r_4 = \tfrac{451}{221}, \; \ldots$$

and these are approaching the dominant eigenvalue 2.

◆◆◆

To see why the power method works, let $\lambda_1, \lambda_2, \ldots, \lambda_m$ be eigenvalues of A with λ_1 dominant and let $Y_1, Y_2, \ldots, Y_m$ be corresponding eigenvectors. What is required is that the first approximation X_0 be a linear combination of these eigenvectors:

$$X_0 = a_1 Y_1 + a_2 Y_2 + \cdots + a_m Y_m \qquad \text{with } a_1 \neq 0$$

If $k \geq 1$, the fact that $A^k Y_i = \lambda_i^k Y_i$ for each i gives

$$X_k = a_1 \lambda_1^k Y_1 + a_2 \lambda_2^k Y_2 + \cdots + a_m \lambda_m^k Y_m \qquad \text{for } k \geq 1$$

Hence

$$\frac{1}{\lambda_1^k} X_k = a_1 Y_1 + a_2 \left(\frac{\lambda_2}{\lambda_1} \right)^k Y_2 + \cdots + a_m \left(\frac{\lambda_m}{\lambda_1} \right)^k Y_m$$

The right side approaches $a_1 Y_1$ as k increases because λ_1 is dominant $\left(\left| \frac{\lambda_i}{\lambda_1} \right| < 1 \text{ for each } i > 1 \right)$. Because $a_1 \neq 0$, this means that X_k approximates the dominant eigenvector $a_1 \lambda_1^k Y_1$.

The power method requires that the first approximation X_0 be a linear combination of eigenvalues. (In Example 1 the eigenvectors form a basis of $\mathbb{R}^2$.) But even in this case the method fails if $a_1 = 0$, where a_1 is the coefficient of the dominant eigenvector (try $X_0 = \begin{bmatrix} -1 \\ 2 \end{bmatrix}$ in Example 1). In general, the rate of convergence is quite slow if any of the ratios $\left|\dfrac{\lambda_i}{\lambda_1}\right|$ is near 1. Also, because the method requires repeated multiplications by A, it is not recommended unless these multiplications are easy to carry out (for example, if most of the entries of A are zero).

QR-Algorithm

A much better method depends on the factorization (using the Gram–Schmidt algorithm) of an invertible matrix A in the form

$$A = QR$$

where Q is orthogonal and R is invertible and upper triangular (see Theorem 2§6.6). The **QR-algorithm** uses this repeatedly to create a sequence of matrices $A_1 = A$, A_2, $A_3, \ldots$, as follows:

1. Define $A_1 = A$ and factor it as $A_1 = Q_1R_1$.
2. Define $A_2 = R_1Q_1$ and factor it as $A_2 = Q_2R_2$.
3. Define $A_3 = R_2Q_2$ and factor it as $A_3 = Q_3R_3$.
$$\vdots$$

In general, A_k is factored as $A_k = Q_kR_k$ and we define $A_{k+1} = R_kQ_k$. Then A_{k+1} is similar to A_k (in fact, $A_{k+1} = R_kQ_k = (Q_k^{-1}A_kQ_k)$), and hence each A_k has the same eigenvalues as A. If the eigenvalues of A are real and have distinct absolute values, the remarkable thing is that the sequence of matrices $A_1, A_2, A_3, \ldots$ converges to an upper triangular matrix with these eigenvalues on the main diagonal. [See later for the case of complex eigenvalues.]

EXAMPLE 2

If $A = \begin{bmatrix} 1 & 1 \\ 2 & 0 \end{bmatrix}$ as in Example 1, use the QR-algorithm to approximate the eigenvalues.

Solution   The matrices A_1, A_2, and A_3 are as follows:

$$A_1 = \begin{bmatrix} 1 & 1 \\ 2 & 0 \end{bmatrix} = Q_1R_1 \text{ where } Q_1 = \tfrac{1}{\sqrt{5}}\begin{bmatrix} 1 & -2 \\ 2 & 1 \end{bmatrix} \text{ and } R_1 = \tfrac{1}{\sqrt{5}}\begin{bmatrix} 5 & 1 \\ 0 & -2 \end{bmatrix}$$

$$A_2 = \tfrac{1}{5}\begin{bmatrix} 7 & -9 \\ -4 & -2 \end{bmatrix} = \begin{bmatrix} 1.4 & -1.8 \\ -0.8 & -0.4 \end{bmatrix} = Q_2R_2 \text{ where } Q_2 = \tfrac{1}{\sqrt{65}}\begin{bmatrix} 7 & 4 \\ -4 & 7 \end{bmatrix}$$

$$\text{and } R_2 = \tfrac{1}{\sqrt{65}}\begin{bmatrix} 13 & -11 \\ 0 & -10 \end{bmatrix}$$

$$A_3 = \tfrac{1}{13}\begin{bmatrix} 27 & -5 \\ 8 & -14 \end{bmatrix} = \begin{bmatrix} 2.08 & -.038 \\ 0.62 & -1.08 \end{bmatrix}$$

This is converging to $\begin{bmatrix} 2 & * \\ 0 & -1 \end{bmatrix}$ and so is approximating the eigenvalues 2 and -1 on the main diagonal.

◆◆◆

It is beyond the scope of this book to pursue a detailed discussion of these methods, and the reader is referred to J. M. Wilkinson, *The Algebraic Eigenvalue Problem* (Oxford, England: Oxford University Press, 1965) or G. W. Stewart, *Introduction to Matrix Computations* (New York: Academic Press, 1973). We conclude with some remarks on the QR-algorithm.

Shifting   Convergence is accelerated if, at stage k of the algorithm, a number s_k is chosen and $A_k - s_k I$ is factored in the form $Q_k R_k$ rather than A_k itself. Then

$$Q_k^{-1} A_k Q_k = Q_k^{-1}(Q_k R_k + s_k I) Q_k = R_k Q_k + s_k I$$

so we take $A_{k+1} = R_k Q_k + s_k I$. If the shifts s_k are carefully chosen, convergence can be greatly improved.

Preliminary Preparation   A matrix such as

$$\begin{bmatrix} * & * & * & * & * \\ * & * & * & * & * \\ 0 & * & * & * & * \\ 0 & 0 & * & * & * \\ 0 & 0 & 0 & * & * \end{bmatrix}$$

is said to be in **upper Hessenberg** form, and the QR-factorizations of such matrices are greatly simplified. A series of orthogonal matrices $H_1, H_2, \ldots, H_m$ (called **Householder matrices**) can be easily constructed such that

$$B = H_m^T \cdots H_1^T A H_1 \cdots H_m$$

is in upper Hessenberg form. Then the QR-algorithm can be efficiently applied to B and, because B is similar to A, it produces the eigenvalues of A.

Complex Eigenvalues   If some of the eigenvalues of a real matrix A are not real, the QR-algorithm converges to a block upper triangular matrix where the diagonal blocks are either 1×1 (the real eigenvalues) or 2×2 (each providing a pair of conjugate complex eigenvalues of A).

EXERCISES 6.7

1. In each case, find the exact eigenvalues and determine corresponding eigenvectors. Then start with $X_0 = \begin{bmatrix} 1 \\ 1 \end{bmatrix}$ and compute X_4 and r_3 using the power method.

(a) $\begin{bmatrix} 2 & -4 \\ -3 & 3 \end{bmatrix}$

◆(b) $\begin{bmatrix} 5 & 2 \\ -3 & -2 \end{bmatrix}$

(c) $\begin{bmatrix} 1 & 2 \\ 2 & 1 \end{bmatrix}$

◆(d) $\begin{bmatrix} 3 & 1 \\ 1 & 0 \end{bmatrix}$

2. In each case, find the exact eigenvalues and then approximate them using the QR-algorithm.

(a) $\begin{bmatrix} 1 & 1 \\ 1 & 0 \end{bmatrix}$ ◀(b) $\begin{bmatrix} 3 & 1 \\ 1 & 0 \end{bmatrix}$

3. Apply the power method to $A = \begin{bmatrix} 0 & 1 \\ -1 & 0 \end{bmatrix}$, starting at $X_0 = \begin{bmatrix} 1 \\ 1 \end{bmatrix}$. Does it converge? Explain.

◀**4.** If A is symmetric, show that each matrix A_k in the QR-algorithm is also symmetric. Deduce that they converge to a diagonal matrix.

5. Apply the QR-algorithm to $A = \begin{bmatrix} 2 & -3 \\ 1 & -2 \end{bmatrix}$. Explain.

6. Given a matrix A, let A_k, Q_k, and R_k, $k \geq 1$, be the matrices constructed in the QR-algorithm. Show that $A_k = (Q_1 Q_2 \cdots Q_k)(R_k \cdots R_2 R_1)$ for each $k \geq 1$ and hence that this is a QR-factorization of A_k. [*Hint:* Show that $Q_k R_k = R_{k-1} Q_{k-1}$ for each $k \geq 2$, and use this equality to compute $(Q_1 Q_2 \cdots Q_k)(R_k \cdots R_2 R_1)$ "from the center out." Use the fact that $(AB)^{n+1} = A(BA)^n B$ for any square matrices A and B.]

■■■■■■ **Section 6.8**   ## Complex Matrices (Optional)

If A is an $n \times n$ matrix, the characteristic polynomial $c_A(x)$ is a polynomial of degree n and the eigenvalues of A are just the roots of $c_A(x)$. In each of our examples these roots have been *real* numbers (in fact, the examples have been carefully chosen so this will be the case!); but it need not happen, even when the characteristic polynomial has real coefficients. For example, if $A = \begin{bmatrix} 0 & 1 \\ -1 & 0 \end{bmatrix}$, then $c_A(x) = x^2 + 1$ has roots i and $-i$, where i is the complex number satisfying $i^2 = -1$. Therefore, we have to deal with the possibility that the eigenvalues of a (real) square matrix might be complex numbers.

In fact, nearly everything in this book would remain true if the phrase *real number* were replaced by *complex number* wherever it occurs. Then we would deal with matrices with complex entries, systems of linear equations with complex coefficients (and complex solutions), determinants of complex matrices, and vector spaces with scalar multiplication by any complex number allowed. Moreover, the proofs of most theorems about (the real version of) these concepts extend easily to the complex case. It is not our intention here to give a full treatment of complex linear algebra. However, we will carry the theory far enough to prove that the eigenvalues of a real symmetric matrix A are real (Theorem 2§6.1) and to prove the spectral theorem (an extension of the principal axes theorem (Theorem 2§6.4).

The set of complex numbers is denoted $\mathbb{C}$. We require only the most basic properties of these numbers (mainly conjugation and absolute values), and the reader can find this material in Appendix A.

If $n \geq 1$, we denote the set of all n-tuples of complex numbers by $\mathbb{C}^n$. As for $\mathbb{R}^n$, these n-tuples will be written either as row or column matrices and will be referred to as vectors. We define vector operations on $\mathbb{C}^n$ as follows:

$$[v_1 \quad v_2 \quad \cdots \quad v_n] + [w_1 \quad w_2 \quad \cdots \quad w_n] = [v_1 + w_1 \quad v_2 + w_2 \quad \cdots \quad v_n + w_n]$$
$$u[v_1 \quad v_2 \quad \cdots \quad v_n] = [uv_1 \quad uv_2 \quad \cdots \quad uv_n] \quad \text{for } u \text{ in } \mathbb{C}$$

With these definitions, $\mathbb{C}^n$ satisfies the axioms for a vector space (with complex scalars) given in Chapter 5. Thus we can speak of spanning sets for $\mathbb{C}^n$, of linearly

independent subsets, and of bases. In all cases, the definitions are identical to the real case, except that the scalars are allowed to be complex numbers. In particular, the standard basis of $\mathbb{R}^n$ remains a basis of $\mathbb{C}^n$, called the **standard basis** of $\mathbb{C}^n$.

There is a natural generalization to $\mathbb{C}^n$ of the dot product in $\mathbb{R}^n$. Given $Z = [z_1\ z_2 \cdots z_n]$ and $W = [w_1\ w_2 \cdots w_n]$ in $\mathbb{C}^n$, define their **standard inner product** $\langle Z, W \rangle$ by

$$\langle Z, W \rangle = z_1\bar{w}_1 + z_2\bar{w}_2 + \cdots + z_n\bar{w}_n$$

where $\bar{w}$ is the conjugate of the complex number w. Clearly, if z and w actually lie in $\mathbb{R}^n$, then $\langle Z, W \rangle = Z \cdot W$ is the usual dot product.

EXAMPLE 1

If $Z = [2\quad 1-i\quad 2i\quad 3-i]$ and $W = [1-i\quad -1\quad -i\quad 3+2i]$, then

$$\langle Z,\ W \rangle = 2(1 + i) + (1 - i)(-1) + (2i)(i) + (3 - i)(3 - 2i) = 6 - 6i$$
$$\langle Z,\ Z \rangle = 2 \cdot 2 + (1 - i)(1 + i) + (2i)(-2i) + (3 - i)(3 + i) = 20$$

◆◆◆

Note that $\langle Z,\ W \rangle$ is a complex number in general. However, if $W = Z = [z_1\ z_2 \cdots z_n]$, the definition gives $\langle Z,\ Z \rangle = |z_1|^2 + \cdots + |z_n|^2$, which is a nonnegative real number and equals 0 if and only if $Z = 0$. This explains the conjugation in the definition of $\langle Z,\ W \rangle$, and it gives (4) of the following theorem.

THEOREM 1

Let $Z, Z_1, W,$ and W_1 denote vectors in $\mathbb{C}^n$, and let λ denote a complex number.

1. $\langle Z + Z_1, W \rangle = \langle Z, W \rangle + \langle Z_1, W \rangle$ and $\langle Z, W + W_1 \rangle\ \langle Z, W \rangle + \langle Z, W_1 \rangle$
2. $\langle \lambda Z, W \rangle = \lambda\langle Z, W \rangle$ and $\langle Z, \lambda W \rangle = \bar{\lambda}\langle Z, W \rangle$
3. $\langle Z, W \rangle = \overline{\langle W, Z \rangle}$
4. $\langle Z, Z \rangle \geq 0$, and $\langle Z, Z \rangle = 0$ if and only if $Z = 0$

Proof   We leave (1) and (2) to the reader (Exercise 8), and (4) has already been proved. To prove (3), write $Z = [z_1\ z_2 \cdots z_n]$ and $W = [w_1\ w_2 \cdots w_n]$. Then

$$\overline{\langle W, Z \rangle} = \overline{(w_1\bar{z}_1 + \cdots + w_n\bar{z}_n)} = \bar{w}_1\bar{\bar{z}}_1 + \cdots + \bar{w}_n\bar{\bar{z}}_n$$
$$= z_1\bar{w}_1 + \cdots + z_n\bar{w}_n = \langle Z,\ W \rangle \qquad ◆$$

As for the dot product on $\mathbb{R}^n$, property (4) enables us to define the **norm** or **length** $\|Z\|$ of a vector $Z = [z_1\ z_2 \cdots z_n]$ in $\mathbb{C}^n$:

$$\|Z\| = \sqrt{\langle Z, Z \rangle} = \sqrt{|z_1|^2 + |z_2|^2 + \cdots + |z_n|^2}$$

The only properties of the norm function we shall need are the following (the proof is left to the reader):

THEOREM 2

If Z is any vector in $\mathbb{C}^n$, then

1. $\|Z\| \geq 0$, and $\|Z\| = 0$ if and only if $Z = 0$
2. $\|\lambda Z\| = |\lambda|\,\|Z\|$ for all complex numbers λ

A vector U in $\mathbb{C}^n$ is called a **unit vector** if $\|U\| = 1$. Property (2) then shows that if $Z \neq 0$ is any nonzero vector in $\mathbb{C}^n$, then

$$U = \frac{1}{\|Z\|}\, Z$$

is a unit vector.

EXAMPLE 2

In $\mathbb{C}^4$, find a unit vector U that is a positive real multiple of $Z = [1-i \quad i \quad 2 \quad 3+4i]$.

Solution $\|Z\| = \sqrt{2 + 1 + 4 + 25} = \sqrt{32} = 4\sqrt{2}$, so take $U = \frac{1}{4\sqrt{2}} Z$.

◆◆◆

A matrix $Z = [z_{ij}]$ is called a **complex matrix** if every entry z_{ij} is a complex number. The notion of conjugation for complex numbers extends to matrices as follows: Define the **conjugate** of $Z = [z_{ij}]$ to be the matrix

$$\overline{Z} = [\overline{z}_{ij}]$$

obtained from Z by conjugating every entry. Then

$$\overline{Z + W} = \overline{Z} + \overline{W} \qquad \text{and} \qquad \overline{ZW} = \overline{Z}\ \overline{W}$$

holds for all (complex) matrices of appropriate size (using properties C_1 and C_2 of Appendix A).

Transposition of complex matrices is defined just as in the real case. The following notion is fundamental in the study of complex matrices.

DEFINITION

The **conjugate transpose** Z^* of a complex matrix Z is defined by

$$Z^* = (\overline{Z})^T = (\overline{Z^T})$$

Observe that $Z^* = Z^T$ when Z is real.

EXAMPLE 3

$$\begin{bmatrix} 3 & 1-i & 2+i \\ 2i & 5+2i & -i \end{bmatrix}^* = \begin{bmatrix} 3 & -2i \\ 1+i & 5-2i \\ 2-i & i \end{bmatrix}$$

The following properties of Z^* follow easily from the rules for transposition of real matrices (Exercise 8) and extend these rules to complex matrices. Note the conjugate in property (3).

THEOREM 3

Let Z and W denote the complex matrices, and let λ be a complex number.

1. $(Z^*)^* = Z$
2. $(Z + W)^* = Z^* + W^*$
3. $(\lambda Z)^* = \bar{\lambda} Z^*$
4. $(ZW)^* = W^*Z^*$

If A is a real symmetric matrix, it is clear that $A^* = A$. The complex matrices that satisfy this condition turn out to be the most natural generalization of the real symmetric matrices.

DEFINITION

A square complex matrix H is called **Hermitian** if $H^* = H$, equivalently $\bar{H} = H^T$. [8]

Hermitian matrices are easy to recognize because the entries on the main diagonal must be real, and the "reflection" of each nondiagonal entry in the main diagonal must be the conjugate of that entry.

EXAMPLE 4

$\begin{bmatrix} 3 & i & 2+i \\ -i & -2 & -7 \\ 2-i & -7 & 1 \end{bmatrix}$ is Hermitian, whereas $\begin{bmatrix} 1 & i \\ i & -2 \end{bmatrix}$ and $\begin{bmatrix} 1 & i \\ -i & i \end{bmatrix}$ are not.

---

[8] The name Hermitian honors Charles Hermite (1822–1901), a French mathematician who worked primarily in analysis and is remembered as the first to show that the number e from calculus is transcendental— that is, e is not the root of any polynomial with integer coefficients.

The following gives a very useful characterization of Hermitian matrices in terms of the standard inner product in $\mathbb{C}^n$.

THEOREM 4

An $n \times n$ complex matrix H is Hermitian if and only if

$$\langle HZ, W \rangle = \langle Z, HW \rangle$$

for all columns Z and W in $\mathbb{C}^n$.

Proof If H is Hermitian, we have $H^T = \bar{H}$. If Z and W are columns in $\mathbb{C}^n$, then $\langle Z, W \rangle = Z^T \bar{W}$, so

$$\langle HZ, W \rangle = (HZ)^T \bar{W} = Z^T H^T \bar{W} = Z^T \bar{H} \bar{W} = Z^T (\overline{HW}) = \langle Z, HW \rangle$$

To prove the converse, let E_j denote column j of the identity matrix. If $H = [h_{ij}]$ we have

$$\bar{h}_{ij} = \langle E_i, HE_j \rangle = \langle HE_i, E_j \rangle = h_{ji}$$

Hence $\bar{H} = H^T$, so H is Hermitian.   ◆

Let Z be an $n \times n$ complex matrix. As in the real case, a complex number λ is called an **eigenvalue** of Z if $ZX = \lambda X$ holds for some column $X \neq 0$ in $\mathbb{C}^n$. In this case X is called an **eigenvector** of Z corresponding to λ. The **characteristic polynomial** $c_Z(x)$ is defined by

$$c_Z(x) = \det(xI - A)$$

This polynomial has complex coefficients (possibly nonreal). However, the proof of Theorem 1§6.1 goes through to show that the eigenvalues of Z are the roots (possibly complex) of $c_Z(x)$. It is at this point that the advantage of working with complex numbers becomes apparent. The real numbers are incomplete in the sense that the characteristic polynomial of a real matrix may fail to have all its roots real. However, this difficulty does not occur for the complex numbers. The so-called fundamental theorem of algebra ensures that *every* polynomial of positive degree with complex coefficients has a complex root. Hence every square complex matrix has a (complex) eigenvalue. Indeed (Appendix A), $c_Z(x)$ factors completely as follows:

$$c_Z(x) = (x - \lambda_1)(x - \lambda_2) \cdots (x - \lambda_n)$$

where $\lambda_1, \lambda_2, \ldots, \lambda_n$ are the eigenvalues of Z (with possible repetitions due to multiple roots).

The next result shows that, for Hermitian matrices, the eigenvalues are actually real. Because symmetric real matrices are Hermitian, it gives a proof of Theorem 2§6.1. It also extends Theorem 4§6.4, which asserts that eigenvectors of a symmetric real matrix corresponding to distinct eigenvalues are actually orthogonal.

In the complex context, two columns Z and W in $\mathbb{C}^n$ are said to be **orthogonal** if $\langle Z, W \rangle = 0$.

THEOREM 5

Let H denote a Hermitian matrix.

1. The eigenvalues of H are real.
2. Eigenvectors of H corresponding to distinct eigenvalues are orthogonal.

Proof Let λ and μ be the eigenvalues of H with (nonzero) eigenvectors Z and W. Then $HZ = \lambda Z$ and $HW = \mu W$, so Theorem 4 gives

$$\lambda\langle Z, W \rangle = \langle \lambda Z, W \rangle = \langle HZ, W \rangle = \langle Z, HW \rangle = \langle Z, \mu W \rangle = \bar{\mu}\langle Z, W \rangle \qquad (*)$$

If $\mu = \lambda$ and $W = Z$, this becomes $\lambda\langle Z, Z \rangle = \bar{\lambda}\langle Z, Z \rangle$. Because $\langle Z, Z \rangle = \|Z\|^2 \neq 0$, this implies $\lambda = \bar{\lambda}$. Thus λ is real, proving (1). Similarly, μ is real, so equation $(*)$ gives $\lambda\langle Z, W \rangle = \mu\langle Z, W \rangle$. If $\lambda \neq \mu$, this implies $\langle Z, W \rangle = 0$, proving (2). ◆

The principal axes theorem (Theorem 2§6.4) asserts that every real symmetric matrix A is orthogonally diagonalizable — that is, $P^T A P$ is diagonal where P is an orthogonal matrix ($P^{-1} = P^T$). The next theorem identifies the complex analogues of these orthogonal real matrices. As in the real case, a set of nonzero vectors $\{Z_1, Z_2, \ldots, Z_m\}$ in $\mathbb{C}^n$ is called **orthogonal** if $\langle Z_i, Z_j \rangle = 0$ whenever $i \neq j$, and it is **orthonormal** if, in addition, $\|Z_i\| = 1$ for each i.

THEOREM 6

The following are equivalent for an $n \times n$ complex matrix U.

1. $U^{-1} = U^*$
2. The rows of U are an orthonormal set in $\mathbb{C}^n$.
3. The columns of U are an orthonormal set in $\mathbb{C}^n$.

The proof is a direct adaptation of the proof of Theorem 1§6.4.

DEFINITION

A square complex matrix U is called **unitary** if it satisfies the conditions in Theorem 6.

Thus a real matrix is unitary if and only if it is orthogonal.

EXAMPLE 5

The matrix $Z = \begin{bmatrix} 1+i & 1 \\ 1-i & i \end{bmatrix}$ has orthogonal columns, but the rows are not orthogonal.

Normalizing gives the unitary matrix $\frac{1}{2}\begin{bmatrix} 1+i & \sqrt{2} \\ 1-i & \sqrt{2}i \end{bmatrix}$.

◆◆◆

Given a real symmetric matrix A, the diagonalization algorithm in Section 6.2 is a procedure for finding an orthogonal matrix P such that $P^T A P$ is diagonal. The following example illustrates Theorem 5 and shows that the algorithm works for complex matrices.

EXAMPLE 6

Consider the Hermitian matrix $H = \begin{bmatrix} 3 & 2+i \\ 2-i & 7 \end{bmatrix}$. Find the eigenvalues of H, find two orthonormal eigenvectors, and so find a unitary matrix U such that U^*HU is diagonal.

Solution   The characteristic polynomial of H is

$$c_H(x) = \det(xI - H) = \det\begin{bmatrix} x-3 & -2-i \\ -2+i & x-7 \end{bmatrix} = (x-2)(x-8)$$

Hence the eigenvalues are 2 and 8 (both real as expected), and corresponding eigenvectors are $\begin{bmatrix} 2+i \\ -1 \end{bmatrix}$ and $\begin{bmatrix} 1 \\ 2-i \end{bmatrix}$ (orthogonal as expected). Each has length $\sqrt{6}$, so, as in the (real) diagonalization algorithm, let $U = \frac{1}{\sqrt{6}}\begin{bmatrix} 2+i & 1 \\ -1 & 2-i \end{bmatrix}$ be the unitary matrix with the normalized eigenvectors as columns. Then $U^*HU = \begin{bmatrix} 2 & 0 \\ 0 & 8 \end{bmatrix}$ is diagonal.

◆◆◆

An $n \times n$ complex matrix Z is called **unitarily diagonalizable** if U^*ZU is diagonal for some unitary matrix U. As Example 6 suggests, we are going to prove that every Hermitian matrix is unitarily diagonalizable. However, with only a little extra effort, we can get a very important theorem that has this result as an easy consequence.

A complex matrix is called **upper triangular** if every entry below the main diagonal is zero. We owe the following theorem to Issai Schur.[9]

---

[9]Issai Schur (1875–1941) was a German mathematician who did fundamental work in the theory of representations of groups as matrices.

THEOREM 7

Schur's Theorem

If Z is any $n \times n$ complex matrix, there exists a unitary matrix U such that

$$U^*ZU = T$$

is upper triangular. Moreover, the entries on the main diagonal of T are the eigenvalues $\lambda_1, \lambda_2, \ldots, \lambda_n$ of Z (including multiplicities).

Proof   We use induction on n. If $n = 1$, Z is already upper triangular. If $n > 1$, assume the theorem is valid for $(n - 1) \times (n - 1)$ complex matrices. Let λ_1 be an eigenvalue of Z, and let Y_1 be an eigenvector with $\|Y_1\| = 1$. Then Y_1 is part of a basis of $\mathbb{C}^n$ (by the analogue of Theorem 2§5.4), so the (complex analogue of the) Gram–Schmidt process provides $Y_2, \ldots, Y_n$ such that $\{Y_1, \ldots, Y_n\}$ is an orthonormal basis of $\mathbb{C}^n$. If $U_1 = [Y_1 \ Y_2 \ \cdots \ Y_n]$ is the matrix with these vectors as its columns, then

$$U_1^*ZU_1 = \begin{bmatrix} \lambda_1 & X_1 \\ 0 & Z_1 \end{bmatrix}$$

in block form. Now apply induction to find a unitary $(n - 1) \times (n - 1)$ matrix W_1 such that $W_1^*Z_1W_1 = T_1$ is upper triangular. Then $U_2 = \begin{bmatrix} 1 & 0 \\ 0 & W_1 \end{bmatrix}$ is a unitary $n \times n$ matrix. Hence $U = U_1U_2$ is unitary (using Theorem 6), and

$$U * ZU = U_2^*(U_1^*ZU_1)U_2$$

$$= \begin{bmatrix} 1 & 0 \\ 0 & W_1^* \end{bmatrix}\begin{bmatrix} \lambda_1 & X_1 \\ 0 & Z_1 \end{bmatrix}\begin{bmatrix} 1 & 0 \\ 0 & W_1 \end{bmatrix} = \begin{bmatrix} \lambda_1 & X_1W_1 \\ 0 & T_1 \end{bmatrix}$$

is upper triangular. Finally, Z and $U^*ZU = T$ have the same eigenvalues by (the complex version of) Theorem 4§6.1, and they are the diagonal entries of T because T is upper triangular.   ◆

The fact that similar matrices have the same traces and determinants gives the following consequence of Schur's theorem.

COROLLARY

Let Z be an $n \times n$ complex matrix, and let $\lambda_1, \lambda_2, \ldots, \lambda_n$ denote the eigenvalues of Z, including multiplicities. Then

$$\det Z = \lambda_1\lambda_2 \cdots \lambda_n \qquad \text{and} \qquad \text{tr } Z = \lambda_1 + \lambda_2 + \cdots + \lambda_n.$$

Schur's theorem asserts that every complex matrix can be "unitarily triangularized." However, we cannot substitute "unitarily diagonalized" here. In fact, if $Z = \begin{bmatrix} 1 & 1 \\ 0 & 1 \end{bmatrix}$, there is no invertible complex matrix U at all such that $U^{-1}ZU$ is diagonal. However, the situation is much better for Hermitian matrices.

THEOREM 8

Spectral Theorem

If H is Hermitian, there is a unitary matrix U such that U^*HU is diagonal.

Proof   By Schur's theorem, let $U^*HU = T$ be upper triangular where U is unitary. Since H is Hermitian, this gives

$$T^* = (U^*HU)^* = U^*H^*U^{**} = U^*HU = T$$

This means that T is both upper and lower triangular. Hence T is actually diagonal. ◆

The principal axes theorem asserts that a real matrix A is symmetric if and only if it is orthogonally diagonalizable (that is, P^TAP is diagonal for some real orthogonal matrix P). Theorem 8 is the complex analogue of half of this result. However, the converse is false for complex matrices: There exist unitarily diagonalizable matrices that are not Hermitian.

EXAMPLE 7

Show that the non-Hermitian matrix $Z = \begin{bmatrix} 0 & 1 \\ -1 & 0 \end{bmatrix}$ is unitarily diagonalizable.

Solution   The characteristic polynomial is $c_Z(x) = x^2 + 1$. Hence the eigenvalues are i and $-i$, and it is easy to verify that $\begin{bmatrix} i \\ -1 \end{bmatrix}$ and $\begin{bmatrix} -1 \\ i \end{bmatrix}$ are corresponding eigenvectors. Moreover, these eigenvectors are orthogonal and both have length $\sqrt{2}$, so $U = \frac{1}{\sqrt{2}}\begin{bmatrix} i & -1 \\ -1 & i \end{bmatrix}$ is a unitary matrix such that $U^*ZU = \begin{bmatrix} i & 0 \\ 0 & -i \end{bmatrix}$ is diagonal.

◆◆◆

There is a very simple way to characterize those complex matrices that are unitarily diagonalizable.

DEFINITION

An $n \times n$ complex matrix N is called **normal** if $NN^* = N^*N$.

It is clear that every Hermitian or unitary matrix is normal, as is the matrix $\begin{bmatrix} 0 & 1 \\ -1 & 0 \end{bmatrix}$

in Example 7. We conclude with the following result.

THEOREM 9

An $n \times n$ complex matrix Z is unitarily diagonalizable if and only if Z is normal.

Proof Assume first that $U^*ZU = D$, where U is unitary and D is diagonal. Then $DD^* = D^*D$ as is easily verified. Because $DD^* = U^*(ZZ^*)U$ and $D^*D = U^*(Z^*Z)U$, it follows by cancellation that $ZZ^* = Z^*Z$. Conversely, assume Z is normal — that is, $ZZ^* = Z^*Z$. By Schur's theorem, let $U^*ZU = T$, where T is upper triangular and U is unitary. Then T is normal too:

$$TT^* = U^*(ZZ^*)U = U^*(Z^*Z)U = T^*T$$

Hence it suffices to show that a normal $n \times n$ upper triangular matrix T must be diagonal. We induct on n; it is clear if $n = 1$. If $n > 1$ and $T = [t_{ij}]$, then equating $(1, 1)$-entries in TT^* and T^*T gives

$$|t_{11}|^2 + |t_{12}|^2 + \cdots + |t_{1n}|^2 = |t_{11}|^2$$

This implies $t_{12} = t_{13} = \cdots = t_{1n} = 0$, so $T = \begin{bmatrix} t_{11} & 0 \\ 0 & T_1 \end{bmatrix}$ in block form. Hence

$T^* = \begin{bmatrix} \bar{t}_{11} & 0 \\ 0 & T_1^* \end{bmatrix}$ so $TT^* = T^*T$ implies $T_1T_1^* = T_1^*T_1$. Thus T_1 is diagonal by induc-

tion, and the proof is complete. ◆

EXERCISES 6.8

1. In each case, compute the norm of the complex vector.
   (a) $(1, 1 - i, -2, i)$ ◆(b) $(1 - i, 1 + i, 1, -1)$
   (c) $(2 + i, 1 - i, 2, 0, -i)$
   ◆(d) $(-2, -i, 1 + i, 1 - i, 2i)$

2. In each case, determine whether the two vectors are orthogonal.
   (a) $(4, -3i, 2 + i), (i, 2, 2 - 4i)$
   ◆(b) $(i, -i, 2 + i), (i, i, 2 - i)$
   (c) $(1, 1, i, i), (1, i, -i, 1)$
   ◆(d) $(4 + 4i, 2 + i, 2i), (-1 + i, 2, 3 - 2i)$

3. A subset U of $\mathbb{C}^n$ is called a **complex subspace** of $\mathbb{C}^n$ if it contains 0 and if, given Z and W in U, both $Z + W$ and zZ lie in U (z any complex number). In each case, determine whether U is a complex subspace of $\mathbb{C}^3$.
   (a) $U = \{(w, \bar{w}, 0) \mid w \text{ in } \mathbb{C}\}$
   ◆(b) $U = \{(w, 2w, a) \mid w \text{ in } \mathbb{C}, a \text{ in } \mathbb{R}\}$
   (c) $U = \mathbb{R}^3$
   ◆(d) $U = \{(v + w, v - 2w, v) \mid v, w \text{ in } \mathbb{C}\}$

4. In each case, find a basis over $\mathbb{C}$, and determine the dimension of the complex subspace U of $\mathbb{C}^3$ (see the previous exercise).
   (a) $U = \{(w, v + w, v - iw) \mid v, w \text{ in } \mathbb{C}\}$
   ◆(b) $U = \{(iv + w, 0, 2v - w) \mid v, w \text{ in } \mathbb{C}\}$
   (c) $U = \{(u, v, w) \mid iu - 3v + (1 - i)w = 0; u, v, w \text{ in } \mathbb{C}\}$
   ◆(d) $U = \{(u, v, w) \mid 2u + (1 + i)v - iw = 0; u, v, w \text{ in } \mathbb{C}\}$

5. In each case, determine whether the given matrix is Hermitian, unitary, or normal.

(a) $\begin{bmatrix} 1 & -i \\ i & i \end{bmatrix}$

◆(b) $\begin{bmatrix} 2 & 3 \\ -3 & 2 \end{bmatrix}$

(c) $\begin{bmatrix} 1 & i \\ -i & 2 \end{bmatrix}$

◆(d) $\begin{bmatrix} 1 & -i \\ i & -1 \end{bmatrix}$

(e) $\frac{1}{\sqrt{2}} \cdot \begin{bmatrix} 1 & -1 \\ 1 & 1 \end{bmatrix}$

◆(f) $\begin{bmatrix} 1 & 1+i \\ 1+i & i \end{bmatrix}$

(g) $\begin{bmatrix} 2i & -1+i \\ 1+i & i \end{bmatrix}$

◆(h) $\frac{1}{\sqrt{2}|z|} \cdot \begin{bmatrix} z & z \\ \bar{z} & -\bar{z} \end{bmatrix}$, $z \neq 0$

6. In each case, find a unitary matrix U such that U^*ZU is diagonal.

(a) $Z = \begin{bmatrix} 1 & i \\ -i & 1 \end{bmatrix}$

◆(b) $Z = \begin{bmatrix} 4 & 3-i \\ 3+i & 1 \end{bmatrix}$

(c) $Z = \begin{bmatrix} a & b \\ -b & a \end{bmatrix}$; a, b real

◆(d) $Z = \begin{bmatrix} 2 & 1+i \\ 1-i & 3 \end{bmatrix}$

(e) $Z = \begin{bmatrix} 1 & 0 & 1+i \\ 0 & 2 & 0 \\ 1-i & 0 & 0 \end{bmatrix}$

◆(f) $Z = \begin{bmatrix} 1 & 0 & 0 \\ 0 & 1 & 1+i \\ 0 & 1-i & 2 \end{bmatrix}$

7. Show that $\langle ZX, Y \rangle = \langle X, Z^*Y \rangle$ holds for all $n \times n$ matrices Z and for all columns X and Y in $\mathbb{C}^n$.

8. (a) Prove (1) and (2) of Theorem 1.
◆(b) Prove Theorem 2.
(c) Prove Theorem 3.

9. (a) Show that Z is Hermitian if and only if $\bar{Z} = Z^T$.
◆(b) Show that the diagonal entries of any Hermitian matrix are real.

10. (a) Show that every complex matrix Z can be written uniquely in the form $Z = A + iB$, where A and B are real matrices.
(b) If $Z = A + iB$ as in (a), show that Z is Hermitian if and only if A is symmetric and B is skew-symmetric (that is, $B^T = -B$).

11. If Z is any complex $n \times n$ matrix, show that ZZ^* and $Z + Z^*$ are Hermitian.

12. A complex matrix S is called **skew-Hermitian** if $S^* = -S$.
(a) Show that $Z - Z^*$ is skew-Hermitian for any square complex matrix Z.
◆(b) If S is skew-Hermitian, show that S^2 and iS are Hermitian.
(c) If S is skew-Hermitian, show that the eigenvalues of S are pure imaginary ($i\lambda$ for real λ).

◆(d) Show that every $n \times n$ complex matrix Z can be written uniquely as $Z = H + S$, where H is Hermitian and S is skew-Hermitian.

13. Let U be a unitary matrix. Show that:
(a) $\|UX\| = \|X\|$ for all columns X in $\mathbb{C}^n$.
(b) $|\lambda| = 1$ for every eigenvalue λ of U.

14. (a) If Z is an invertible complex matrix, show that Z^* is invertible and that $(Z^*)^{-1} = (Z^{-1})^*$.
◆(b) Show that the inverse of a unitary matrix is again unitary.
(c) If U is unitary, show that U^* is unitary.

15. Let Z be an $m \times n$ matrix such that $Z^*Z = I_n$ (for example, Z is a unit column in $\mathbb{C}^n$).
(a) Show that $H = ZZ^*$ is Hermitian and satisfies $H^2 = H$.
(b) Show that $U = I - 2ZZ^*$ is both unitary and Hermitian (so $U^{-1} = U^* = U$).

16. (a) If N is normal, show that zN is also normal for all complex numbers z.
(b) Show that (a) fails if *normal* is replaced by *Hermitian*.

17. Show that a real 2×2 normal matrix is either symmetric or has the form $\begin{bmatrix} a & b \\ -b & a \end{bmatrix}$.

◆**18.** If H is Hermitian, show that all the coefficients of $c_H(x)$ are real numbers.

19. (a) If $A = \begin{bmatrix} 1 & 1 \\ 0 & 1 \end{bmatrix}$, show that $U^{-1}AU$ is not diagonal for any invertible complex matrix U.
◆(b) If $A = \begin{bmatrix} 0 & 1 \\ -1 & 0 \end{bmatrix}$, show that $U^{-1}AU$ is not upper triangular for any *real* invertible matrix U.

20. If Z is any $n \times n$ matrix, show that U^*ZU is lower triangular for some unitary matrix U.

21. If Z is a 3×3 matrix, show that $Z^2 = 0$ if and only if there exists a unitary matrix U such that U^*ZU has the form $\begin{bmatrix} 0 & 0 & u \\ 0 & 0 & v \\ 0 & 0 & 0 \end{bmatrix}$ or the form $\begin{bmatrix} 0 & u & v \\ 0 & 0 & 0 \\ 0 & 0 & 0 \end{bmatrix}$.

22. If $Z^2 = Z$, show that rank $Z = $ tr Z. [*Hint:* Use Schur's theorem.]

Section 6.9   An Application to Quadratic Forms (Optional)[10]

An expression like $x_1^2 + x_2^2 + x_3^2 - 2x_1x_3 + x_2x_3$ is called a quadratic form in the variables x_1, x_2, and x_3. In this section we show that a change of variables can always be made so that the quadratic form, when expressed in terms of the new variables y_1, y_2, and y_3, has no cross terms y_1y_2, y_1y_3, or y_2y_3. Moreover, we do this for forms involving any finite number of variables using orthogonal diagonalization. This has far-reaching applications; quadratic forms arise in such diverse areas as statistics, physics, theory of functions of several variables, number theory, and geometry.

A **quadratic form** q in the n variables x_1, x_2, . . . , x_n is a linear combination of terms x_1^2, x_2^2, . . . , x_n^2 and cross terms x_1x_2, x_1x_3, x_2x_3 . . . . If $n = 3$, q has the form

$$q = a_{11}x_1^2 + a_{22}x_2^2 + a_{33}x_3^2 + a_{12}\,x_1\,x_2 + a_{21}\,x_2\,x_1$$
$$+ a_{13}\,x_1\,x_3 + a_{31}\,x_3\,x_1 + a_{23}\,x_2\,x_3 + a_{32}\,x_3\,x_2$$

In general

$$q = a_{11}x_1^2 + a_{22}x_2^2 + \cdots + a_{nn}x_n^2 + a_{12}x_1x_2 + a_{13}x_1x_3 + \cdots$$

This sum can be written compactly as a matrix product

$$q = q(X) = X^TAX$$

where $X = \begin{bmatrix} x_1 \\ \vdots \\ x_n \end{bmatrix}$ and $A = [a_{ij}]$ is a real $n \times n$ matrix. Note that if $i \neq j$, two separate terms $a_{ij}x_ix_j$ and $a_{ji}x_jx_i$ are listed, each of which involves x_ix_j, and they can (rather cleverly) be replaced by

$$\tfrac{1}{2}(a_{ij} + a_{ji})x_ix_j \qquad \text{and} \qquad \tfrac{1}{2}(a_{ij} + a_{ji})x_jx_i$$

respectively, *without altering the sum*. Hence there is no loss of generality in assuming that x_ix_j and x_jx_i have the same coefficient in the sum for q. In other words, *we may assume that A is symmetric*.

EXAMPLE 1

Write $q = x_1^2 + 3x_3^2 + 2x_1x_2 - x_1x_3$ in the form $q(X) = X^TAX$, where A is a symmetric 3×3 matrix.

Solution

The cross terms are $2x_1x_2 = x_1x_2 + x_2x_1$ and $-x_1x_3 = -\tfrac{1}{2}x_1x_3 - \tfrac{1}{2}x_3x_1$. Of course, x_2x_3 and x_3x_2 both have coefficient zero, as does x_2^2. Hence

$$q(X) = \begin{bmatrix} x_1 & x_2 & x_3 \end{bmatrix} \begin{bmatrix} 1 & 1 & \frac{-1}{2} \\ 1 & 0 & 0 \\ \frac{-1}{2} & 0 & 3 \end{bmatrix} \begin{bmatrix} x_1 \\ x_2 \\ x_3 \end{bmatrix}$$

is the required form.

◆◆◆

---

[10]This section requires only Section 6.4.

We shall assume from now on that all quadratic forms are given by

$$q(X) = X^T A X,$$

where A is symmetric. Given such a form, the problem is to find new variables y_1, y_2, ..., y_n, related to $x_1, x_2, \ldots, x_n$, with the property that, when q is expressed in terms of $y_1, y_2, \ldots, y_n$, there are no cross terms. If we write

$$Y = \begin{bmatrix} y_1 \\ y_2 \\ \vdots \\ y_n \end{bmatrix}$$

this amounts to asking that $q = Y^T D Y$ where D is diagonal. It turns out that this can always be accomplished and, not surprisingly, that D is the matrix obtained when the symmetric matrix A is diagonalized. In fact, as Theorem 2§6.4 shows, a matrix P can be found that is orthogonal (that is, $P^{-1} = P^T$) and diagonalizes A:

$$P^T A P = D = \begin{bmatrix} \lambda_1 & 0 & \cdots & 0 \\ 0 & \lambda_2 & \cdots & 0 \\ \vdots & \vdots & & \vdots \\ 0 & 0 & \cdots & \lambda_n \end{bmatrix}$$

The diagonal entries $\lambda_1, \lambda_2, \ldots, \lambda_n$ are the (not necessarily distinct) eigenvalues of A, repeated according to their multiplicities in $c_A(x)$, and the columns of P are the corresponding (orthonormal) eigenvectors of A. As A is symmetric, the λ_i are real.

Now define new variables Y by the equations

$$X = PY \qquad \text{equivalently} \qquad Y = P^T X$$

Then substitution in $q(X) = X^T A X$ gives

$$q = (PY)^T A (PY) = Y^T (P^T A P) Y = Y^T D Y = \lambda_1 y_1^2 + \lambda_2 y_2^2 + \cdots + \lambda_n y_n^2$$

Hence this change of variables produces the desired simplification in q.

THEOREM 1
Diagonalization Theorem

Let $q = X^T A X$ be a quadratic form in the variables $x_1, x_2, \ldots, x_n$, where

$$X = \begin{bmatrix} x_1 \\ x_2 \\ \vdots \\ x_n \end{bmatrix}$$ and A is a symmetric $n \times n$ matrix. Let P be an orthogonal matrix

such that P^TAP is diagonal, and define new variables $Y = \begin{bmatrix} y_1 \\ y_2 \\ \vdots \\ y_n \end{bmatrix}$ by

$$X = PY \qquad \text{equivalently} \qquad Y = P^TX$$

If q is expressed in terms of these new variables $y_1, y_2, \ldots, y_n$, the result is

$$q = \lambda_1 y_1^2 + \lambda_2 y_2^2 + \cdots + \lambda_n y_n^2$$

where $\lambda_1, \lambda_2, \ldots, \lambda_n$ are the eigenvalues of A repeated according to their multiplicities.

Let $q = X^TAX$ be a quadratic form where A is a symmetric matrix and let $\lambda_1, \ldots, \lambda_n$ be the (real) eigenvalues of A repeated according to their multiplicities. A corresponding set $\{E_1, \ldots, E_n\}$ of orthonormal eigenvectors for A is called a set **principal axes** for the quadratic form q. (The reason for the name will be clear later.) The orthogonal matrix P in Theorem 1 is given as $P = [E_1 \cdots E_n]$, so the variables X and Y are related by

$$X = PY = [E_1 \quad E_2 \cdots E_n] \begin{bmatrix} y_1 \\ y_2 \\ \vdots \\ y_n \end{bmatrix} = y_1 E_1 + y_2 E_2 + \cdots + y_n E_n$$

Thus the new variables y_i are the coefficients when X is expanded in terms of the orthonormal basis $\{E_1, \ldots, E_n\}$ of $\mathbb{R}^n$. In particular, the coefficients y_i are given by $y_i = X \cdot E_i$ by the expansion theorem (Theorem 4§6.3). Hence q itself is easily computed from the eigenvalues λ_i and the principal axes E_i:

$$q = q(X) = \lambda_1(X \cdot E_1)^2 + \cdots + \lambda_n(X \cdot E_n)^2$$

EXAMPLE 2

Find new variables $y_1, y_2\, y_3$, and y_4 such that

$$q = 3(x_1^2 + x_2^2 + x_3^2 + x_4^2) + 2x_1x_2 - 10x_1x_3 + 10x_1x_4 + 10x_2x_3$$
$$-10x_2x_4 + 2x_3x_4$$

has diagonal form, and find the corresponding principal axes.

Solution　The form can be written as $q = X^T A X$, where

$$X = \begin{bmatrix} x_1 \\ x_2 \\ x_3 \\ x_4 \end{bmatrix} \qquad A = \begin{bmatrix} 3 & 1 & -5 & 5 \\ 1 & 3 & 5 & -5 \\ -5 & 5 & 3 & 1 \\ 5 & -5 & 1 & 3 \end{bmatrix}$$

A routine calculation yields

$$c_A(x) = \det(xI - A) = (x - 12)(x + 8)(x - 4)^2$$

so the eigenvalues are $\lambda_1 = 12$, $\lambda_2 = -8$, and $\lambda_3 = \lambda_4 = 4$. The corresponding orthonormal eigenvectors are the principal axes:

$$E_1 = \tfrac{1}{2}\begin{bmatrix} 1 \\ -1 \\ -1 \\ 1 \end{bmatrix} \qquad E_2 = \tfrac{1}{2}\begin{bmatrix} 1 \\ -1 \\ 1 \\ -1 \end{bmatrix} \qquad E_3 = \tfrac{1}{2}\begin{bmatrix} 1 \\ 1 \\ 1 \\ 1 \end{bmatrix} \qquad E_4 = \tfrac{1}{2}\begin{bmatrix} 1 \\ 1 \\ -1 \\ -1 \end{bmatrix}$$

The matrix

$$P = [E_1 \; E_2 \; E_3 \; E_4] = \tfrac{1}{2}\begin{bmatrix} 1 & 1 & 1 & 1 \\ -1 & -1 & 1 & 1 \\ -1 & 1 & 1 & -1 \\ 1 & -1 & 1 & -1 \end{bmatrix}$$

is thus orthogonal, and $P^{-1}AP = P^T A P$ is diagonal. Hence the new variables Y and the old variables X are related by $Y = P^T X$ and $X = PY$. Explicitly,

$$\begin{aligned}
y_1 &= \tfrac{1}{2}(x_1 - x_2 - x_3 + x_4) & x_1 &= \tfrac{1}{2}(y_1 + y_2 + y_3 + y_4) \\
y_2 &= \tfrac{1}{2}(x_1 - x_2 + x_3 - x_4) & x_2 &= \tfrac{1}{2}(-y_1 - y_2 + y_3 + y_4) \\
y_3 &= \tfrac{1}{2}(x_1 + x_2 + x_3 + x_4) & x_3 &= \tfrac{1}{2}(-y_1 + y_2 + y_3 - y_4) \\
y_4 &= \tfrac{1}{2}(x_1 + x_2 - x_3 - x_4) & x_4 &= \tfrac{1}{2}(y_1 - y_2 + y_3 - y_4)
\end{aligned}$$

If these x_i are substituted in the original expression for q, the result is

$$q = 12y_1^2 - 8y_2^2 + 4y_3^2 + 4y_4^2$$

This is the required diagonal form.

◆◆◆

It is instructive to look at the case of quadratic forms in two variables x_1 and x_2. Then the principal axes can always be found by rotating the X_1 and X_2 axes counterclockwise about the origin through an angle θ to produce the Y_1 and Y_2 axes (see Figure 6.3). To conform with Theorem 1, we write ordered pairs as columns. Then the rotation carries the X_1 and X_2 coordinate vectors $B_1 = \begin{bmatrix} 1 \\ 0 \end{bmatrix}$ and $B_2 = \begin{bmatrix} 0 \\ 1 \end{bmatrix}$ to the Y_1 and Y_2 coordinate vectors E_1 and E_2. Simple trigonometry gives (see Figure 6.3)

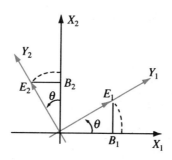

FIGURE 6.3

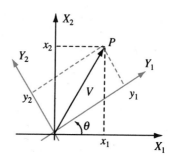

FIGURE 6.4

$$E_1 = \begin{bmatrix} \cos\theta \\ \sin\theta \end{bmatrix} \quad \text{and} \quad E_2 = \begin{bmatrix} -\sin\theta \\ \cos\theta \end{bmatrix} \tag{$*$}$$

These can be used to determine the Y_1 and Y_2 coordinates y_1, y_2 of a point P from the X_1 and X_2 coordinates x_1, x_2 of P. If V is the position vector of P as in Figure 6.4, then $V = x_1B_1 + x_2B_2$ in the original X_1 and X_2 coordinate system, and $V = y_1E_1 + y_2E_2$ in the new Y_1 and Y_2 system. Hence

$$\begin{bmatrix} x_1 \\ x_2 \end{bmatrix} = V = y_1E_1 + y_2E_2 = y_1\begin{bmatrix} \cos\theta \\ \sin\theta \end{bmatrix} + y_2\begin{bmatrix} -\sin\theta \\ \cos\theta \end{bmatrix} = \begin{bmatrix} \cos\theta & -\sin\theta \\ \sin\theta & \cos\theta \end{bmatrix}\begin{bmatrix} y_1 \\ y_2 \end{bmatrix} \tag{$**$}$$

is the change-of-variables formula for the rotation. Note that the matrix $P = \begin{bmatrix} \cos\theta & -\sin\theta \\ \sin\theta & \cos\theta \end{bmatrix}$ is orthogonal, so equation ($**$) takes the form $X = PY$ as in Theorem 1.

We can now completely analyze quadratic forms in two variables. Consider the graph of the equation $rx^2 + sy^2 = t$. Call the graph an **ellipse** if $rs > 0$ and a **hyperbola** if $rs < 0$. (We regard the point $x^2 + y^2 = 0$ as a degenerate ellipse and the lines $x^2 - y^2 = 0$ as a degenerate hyperbola.) Theorem 2 asserts that every equation $ax^2 + bxy + cy^2 = d$ can be transformed into such a diagonal form by a rotation, and also gives a simple way of deciding which conic it is.

THEOREM 2

Consider the quadratic form $q = ax^2 + bxy + cy^2$ where a, b, and c are not all zero.

1. There is a counterclockwise rotation of the coordinate axes about the origin such that, in the new coordinate system, q has no cross term.

2. The graph of the equation $ax^2 + bxy + cy^2 = d$ is an ellipse if $b^2 - 4ac < 0$ and a hyperbola if $b^2 - 4ac > 0$.

Proof   If $b = 0$, q *already* has no cross term and (1) and (2) are clear. So assume $b \neq 0$. Write $x_1 = x$ and $x_2 = y$ as before. The matrix $A = \begin{bmatrix} a & \frac{1}{2}b \\ \frac{1}{2}b & c \end{bmatrix}$ of q has characteristic polynomial $c_A(x) = x^2 - (a + c)x - \frac{1}{4}(b^2 - 4ac)$. We write $d = \sqrt{b^2 + (a - c)^2}$ for convenience; then the quadratic formula gives the eigenvalues

$$\lambda_1 = \tfrac{1}{2}[a + c - d] \quad \text{and} \quad \lambda_2 = \tfrac{1}{2}[a + c + d]$$

with corresponding principal axes

$$E_1 = \frac{1}{\sqrt{b^2 + (a + c - d)^2}} \begin{bmatrix} a + c - d \\ b \end{bmatrix} \quad \text{and}$$

$$E_2 = \frac{1}{\sqrt{b^2 + (a + c - d)^2}} \begin{bmatrix} -b \\ a + c - d \end{bmatrix}$$

as the reader can verify. These agree with Equation (∗) if θ is an angle such that

$$\cos \theta = \frac{a + c - d}{\sqrt{b^2 + (a + c - d)^2}} \quad \text{and} \quad \sin \theta = \frac{b}{\sqrt{b^2 + (a + c - d)^2}}$$

Then $P = [E_1 \;\; E_2] = \begin{bmatrix} \cos \theta & -\sin \theta \\ \sin \theta & \cos \theta \end{bmatrix}$ diagonalizes A and Equation (∗∗) becomes the formula $X = PY$ in Theorem 1. This proves (1).

Finally, A is similar to $\begin{bmatrix} \lambda_1 & 0 \\ 0 & \lambda_2 \end{bmatrix}$, so $\lambda_1 \lambda_2 = \det A = \frac{1}{4}(4ac - b^2)$. Hence the graph of $\lambda_1 y_1^2 + \lambda_2 y_2^2 = d$ is an ellipse if $b^2 < 4ac$ and a hyperbola if $b^2 > 4ac$. This proves (2). ◆

EXAMPLE 3

Consider the equation $x^2 + xy + y^2 = 1$. Find a rotation so that the equation has no cross term.

Solution   Here $a = b = c = 1$ in the notation of Theorem 2, so $\cos \theta = \frac{1}{\sqrt{2}} = \sin \theta$. Hence $\theta = \frac{\pi}{4}$ will do it. If the new variables are x_1 and y_1, then $x = \frac{1}{\sqrt{2}}(x_1 - y_1)$ and $y = \frac{1}{\sqrt{2}}(x_1 + y_1)$ by (∗∗), and the equation becomes $3x_1^2 + y_1^2 = 2$. The graph is an ellipse, and the angle θ has been chosen such that the new X_1 and Y_1 axes are the axes of symmetry of the ellipse (see the diagram). The eigenvectors $E_1 = \frac{1}{\sqrt{2}}\begin{bmatrix} 1 \\ 1 \end{bmatrix}$ and $E_2 = \frac{1}{\sqrt{2}}\begin{bmatrix} -1 \\ 1 \end{bmatrix}$ point along these axes of symmetry, and this is the reason for the name *principal axes*.

●●●

The determinant of any orthogonal matrix P is either 1 or -1 (because $PP^T = I$). The orthogonal matrices $\begin{bmatrix} \cos\theta & -\sin\theta \\ \sin\theta & \cos\theta \end{bmatrix}$ arising from rotations all have determinant 1. More generally, given any quadratic form $q = X^TAX$, the orthogonal matrix P such that P^TAP is diagonal can always be chosen so that $\det P = 1$ by interchanging two eigenvalues (and hence the corresponding columns of P). It is shown in Section 8.4 that orthogonal 2×2 matrices with determinant 1 correspond to rotations. Similarly, orthogonal 3×3 matrices with determinant 1 correspond to rotations about a line through the origin. This extends Theorem 2: Every quadratic form in two or three variables can be diagonalized by a rotation of the coordinate system.

Congruence

We return to the study of quadratic forms in general.

THEOREM 3

If $q(x) = X^TAX$ is a quadratic form given by a symmetric matrix A, then A is uniquely determined by q.

Proof Let $q(x) = X^TBX$ for all X where $B^T = B$. If $C = A - B$, then $X^TCX = 0$ for all X, and we must show that $C = 0$. Given Y in $\mathbb{R}^n$,

$$0 = (X + Y)^TC(X + Y) = X^TCX + X^TCY + Y^TCX + Y^TCY = X^TCY + Y^TCX$$

But $Y^TCX = (X^TCY)^T = X^TCY$ (it is 1×1). Hence $X^TCY = 0$ for all X and Y in $\mathbb{R}^n$. If X_j is column j of I_n, then $0 = X_i^TCX_j$ is the (i, j)-entry of C. Hence $C = 0$. ◆

Hence we can speak of *the* symmetric matrix of a quadratic form.

A quadratic form q can be diagonalized in several ways. For example, if $q = 2x_1^2 - 4x_1x_2 + 5x_2^2$, the diagonalization theorem gives

$$q = 6y_1^2 + y_2^2 \quad \text{where} \quad Y = P^TX, \ P = \tfrac{1}{\sqrt{5}}\begin{bmatrix} 1 & 2 \\ -2 & 1 \end{bmatrix}$$

On the other hand, $q = 2(x_1 - x_2)^2 + 3x_2^2$, so

$$q = 2z_1^2 + 3z_2^2 \quad \text{where} \quad Z = QX, \ Q = \begin{bmatrix} 1 & -1 \\ 0 & 1 \end{bmatrix}$$

The question arises: How are these changes of variables related and what properties do they share?

Given a quadratic form $q = q(X) = X^TAX$, suppose new variables Y are given by an invertible matrix U:

$$Y = U^{-1}X \qquad \text{equivalently} \qquad X = UY$$

In terms of these new variables, q takes the form

$$q = q(Y) = (UY)^TA(UY) = Y^T(U^TAU)Y$$

That is, q has matrix U^TAU with respect to the new variables Y. Hence, to study changes of variables in quadratic forms, we study the following relationship on matrices.

DEFINITION

Two $n \times n$ matrices A and B are called **congruent**, written $A \overset{c}{\sim} B$, if $B = U^TAU$ for some invertible matrix U.

Here are some properties of congruence:

1. $A \overset{c}{\sim} A$ for all A.
2. If $A \overset{c}{\sim} B$, then $B \overset{c}{\sim} A$.
3. If $A \overset{c}{\sim} B$ and $B \overset{c}{\sim} C$, then $A \overset{c}{\sim} C$.
4. If $A \overset{c}{\sim} B$, then A is symmetric if and only if B is symmetric.
5. If $A \overset{c}{\sim} B$, then rank A = rank B.

The converse to (5) can fail even for symmetric matrices.

EXAMPLE 4

The symmetric matrices $A = \begin{bmatrix} 1 & 0 \\ 0 & 1 \end{bmatrix}$ and $B = \begin{bmatrix} 1 & 0 \\ 0 & -1 \end{bmatrix}$ have the same rank but are not congruent. Indeed, if $A \overset{c}{\sim} B$, an invertible matrix U exists such that $B = U^TAU = U^TU$. But then $-1 = \det B = (\det U)^2$, a contradiction because we are working with real matrices.

♦♦♦

The key distinction between A and B in Example 4 is that A has two positive eigenvalues (counting multiplicities) whereas B has only one.

THEOREM 4

Sylvester's Law of Inertia

If $A \overset{c}{\sim} B$, then A and B have the same number of positive eigenvalues.

The proof is given at the end of this section.

DEFINITION

The **index** of a symmetric matrix A is the number of positive eigenvalues of A. If $q = q(X) = X^TAX$ is a quadratic form, the **index** and **rank** of q are defined to be, respectively, the index and rank of A.

As we saw before, if the variables expressing a quadratic form q are changed, the new matrix is congruent to the old one. Hence the index and rank depend only on q and not on the way it is expressed.

Now let $q = q(X) = X^TAX$ be any quadratic form in n variables, of index k and rank r, where A is symmetric. We claim that new variables Z can be found so that q is **completely diagonalized** — that is,

$$q(Z) = z_1^2 + \cdots + z_k^2 - z_{k+1}^2 - \cdots - z_r^2$$

If $k \leq r \leq n$, let $D_n(k, r)$ denote the $n \times n$ diagonal matrix whose main diagonal consists of k ones, followed by $r - k$ minus ones, followed by $n - r$ zeros. Then we seek new variables Z such that

$$q(Z) = Z^TD_n(k,r)Z$$

To determine Z, first diagonalize A as follows: Find an orthogonal matrix P_0 such that

$$P_0^TAP_0 = D = \text{diag}(\lambda_1, \lambda_2, \ldots, \lambda_r, 0, \ldots, 0)$$

is diagonal with the nonzero eigenvalues $\lambda_1, \lambda_2, \ldots, \lambda_r$ of A on the main diagonal (followed by $n - r$ zeros). By reordering the columns of P_0, if necessary, we may assume that $\lambda_1, \ldots, \lambda_k$ are positive and $\lambda_{k+1}, \ldots, \lambda_r$ are negative. This being the case, let D_0 be the $n \times n$ diagonal matrix

$$D_0 = \text{diag}\left(\frac{1}{\sqrt{\lambda_1}}, \ldots, \frac{1}{\sqrt{\lambda_k}}, \frac{1}{\sqrt{-\lambda_{k+1}}}, \ldots, \frac{1}{\sqrt{-\lambda_r}}, 1, \ldots, 1 \right)$$

Then $D_0^TDD_0 = D_n(k, r)$, so if new variables Z are given by $X = (P_0D_0)Z$, we obtain

$$q(Z) = Z^TD_n(k,r)Z = z_1^2 + \cdots + z_k^2 - z_{k+1}^2 - \cdots - z_r^2$$

as required. Note that the change-of-variables matrix P_0D_0 from X to Z has orthogonal columns (in fact scalar multiples of the columns of P_0).

EXAMPLE 5

Completely diagonalize the quadratic form q in Example 2 and find the index and rank.

Solution   In the notation of Example 2, the eigenvalues of the matrix A of q are 12, -8, 4, 4; so the index is 3 and the rank is 4. Moreover, the corresponding orthogonal eigenvalues are E_1, E_2, E_3, and E_4. Hence $P_0 = [E_1\ E_3\ E_4\ E_2]$ is orthogonal and

$$P_0^T A P_0 = \text{diag}(12, 4, 4, -8)$$

As before, take $D_0 = \text{diag}\left(\frac{1}{\sqrt{12}}, \frac{1}{2}, \frac{1}{2}, \frac{1}{\sqrt{8}}\right)$ and define the new variables Z by $X = (P_0 D_0)Z$. Hence the new variables are given by $Z = D_0^{-1} P_0^T X$. The result is

$$z_1 = \sqrt{3}(x_1 - x_2 - x_3 + x_4)$$
$$z_2 = x_1 + x_2 + x_3 + x_4$$
$$z_3 = x_1 + x_2 - x_3 - x_4$$
$$z_4 = \sqrt{2}(x_1 - x_2 + x_3 - x_4)$$

◆◆◆

This discussion gives the following information about symmetric matrices.

THEOREM 5

Let A and B be symmetric $n \times n$ matrices, and let $0 \le k \le r \le n$.

1. A has index k and rank r if and only if $A \overset{c}{\sim} D_n(k, r)$
2. $A \overset{c}{\sim} B$ if and only if they have the same rank and index.

Proof

1. If A has index k and rank r, take $U = P_0 D_0$ where P_0 and D_0 are as described prior to Example 5. Then $U^T A U = D_n(k, r)$. The converse is because $D_n(k, r)$ has index k and rank r (using Theorem 4).

2. If A and B both have index k and rank r, then $A \overset{c}{\sim} D_n[k, r] \overset{c}{\sim} B$ by (1). The converse was given earlier.   ◆

Proof of Theorem 4

By Theorem 1, $A \overset{c}{\sim} D_1$ and $B \overset{c}{\sim} D_2$ where D_1 and D_2 are diagonal and have the same eigenvalues as A and B, respectively. We have $D_1 \overset{c}{\sim} D_2$, (because $A \overset{c}{\sim} B$), so we may assume that A and B are both diagonal. Consider the quadratic form $q(X) = X^T A X$. If A has k positive eigenvalues, q has the form

$$q(X) = a_1 x_1^2 + \cdots + a_k x_k^2 - a_{k+1} x_{k+1}^2 - \cdots - a_r x_r^2 \qquad a_i > 0$$

where $r = \text{rank}\ A = \text{rank}\ B$. The subspace $W_1 = \{X \mid x_{k+1} = \cdots = x_r = 0\}$ of $\mathbb{R}^n$ has dimension $n - r + k$ and satisfies $q(X) > 0$ for all $X \ne 0$ in W_1.

On the other hand, if $B = U^T A U$, define new variables Y by $X = UY$. If B has k' positive eigenvalues, q has the form

$$q(X) = b_1 y_1^2 + \cdots + b_k y_{k'}^2 - b_{k'+1}\ y_{k'+1}^2 - \cdots - b_r y_r^2 \qquad b_i > 0$$

Let $E_1, \ldots, E_n$ denote the columns of U. They are a basis of $\mathbb{R}^n$ and

$$X = UY = [E_1 \ \cdots \ E_n]\begin{bmatrix} y_1 \\ \vdots \\ y_n \end{bmatrix} = y_1 E_1 + \cdots + y_n E_n$$

Hence the subspace $W_2 = \text{span}\{E_{k'+1}, \ldots, E_r\}$ satisfies $q(X) < 0$ for all $X \neq 0$ in W_2. Note that $\dim W_2 = r - k'$. It follows that W_1 and W_2 have only the zero vector in common. Hence, if B_1 and B_2 are bases of W_1 and W_2, respectively, then (Exercise 34§5.3) $B_1 \cup B_2$ is an independent set of $(n - r + k) + (r - k') = n + k - k'$ vectors in $\mathbb{R}^n$. This implies that $k \leq k'$, and a similar argument shows $k' \leq k$. ◆

EXERCISES 6.9

1. In each case, find a symmetric matrix A such that $q = X^T BX$ takes the form $q = X^T AX$.

(a) $B = \begin{bmatrix} 1 & 1 \\ 0 & 1 \end{bmatrix}$

◆**(b)** $B = \begin{bmatrix} 1 & 1 \\ -1 & 2 \end{bmatrix}$

(c) $B = \begin{bmatrix} 1 & 0 & 1 \\ 1 & 1 & 0 \\ 0 & 1 & 1 \end{bmatrix}$

◆**(d)** $B = \begin{bmatrix} 1 & 2 & -1 \\ 4 & 1 & 0 \\ 5 & -2 & 3 \end{bmatrix}$

2. In each case, find a change of variables that will diagonalize the quadratic form q. Determine the index and rank of q.

(a) $q = x_1^2 + 2x_1 x_2 + x_2^2$

◆**(b)** $q = x_1^2 + 4x_1 x_2 + x_2^2$

(c) $q = x_1^2 + x_2^2 + x_3^2 - 4(x_1 x_2 + x_1 x_3 + x_2 x_3)$

◆**(d)** $q = 7x_1^2 + x_2^2 + x_3^2 + 8x_1 x_2 + 8x_1 x_3 - 16x_2 x_3$

(e) $q = 2(x_1^2 + x_2^2 + x_3^2 - x_1 x_2 + x_1 x_3 - x_2 x_3)$

◆**(f)** $q = 5x_1^2 + 8x_2^2 + 5x_3^2 - 4(x_1 x_2 + 2x_1 x_3 + x_2 x_3)$

(g) $q = x_1^2 - x_3^2 - 4x_1 x_2 + 4x_2 x_3$

◆**(h)** $q = x_1^2 + x_3^2 - 2x_1 x_2 + 2x_2 x_3$

3. For each of the following, write the equation in terms of new variables so that it is in standard position, and identify the curve.

(a) $xy = 1$

◆**(b)** $3x^2 - 4xy = 2$

(c) $6x^2 + 6xy - 2y^2 = 5$

◆**(d)** $2x^2 + 4xy + 5y^2 = 1$

◆ **4.** Consider the equation $ax^2 + bxy + cy^2 = d$, where $b \neq 0$. Introduce new variables x_1 and y_1 by rotating the axes counterclockwise through an angle θ. Show that the resulting equation has no $x_1 y_1$-term if θ is given by

$$\cos 2\theta = \frac{a - c}{\sqrt{b^2 + (a - c)^2}}, \quad \sin 2\theta = \frac{b}{\sqrt{b^2 + (a - c)^2}}$$

[*Hint:* Use equation (∗∗) to get x and y in terms of x_1 and y_1, and substitute.]

5. Prove properties (1)–(5) preceding Example 4.

6. If $A \stackrel{c}{\sim} B$ show that A is invertible if and only if B is invertible.

7. If $X = [x_1 \ \cdots \ x_n]^T$ is a column of variables, $A = A^T$ is $n \times n$, B is $1 \times n$, and c is a constant, $X^T AX + BX = c$ is called a **quadratic equation** in the variables x_i.

(a) Show that new variables $y_1, \ldots, y_n$ can be found such that the equation takes the form $\lambda_1 y_1^2 + \cdots + \lambda_n y_n^2 + k_1 y_1 + \cdots + k_n y_n = c$.

◆**(b)** Write $x_1^2 + 3x_2^2 + 3x_3^2 + 4x_1 x_2 - 4x_1 x_3 + 5x_1 - 6x_3 = 7$ in this form and find variables y_1, y_2, y_3 as in **(a)**.

8. Given a symmetric matrix A, define $q_A(X) = X^T AX$. Show that $B \stackrel{c}{\sim} A$ if and only if B is symmetric and there is an invertible matrix U such that $q_B(X) = q_A(UX)$ for all X. [*Hint:* Theorem 3.]

9. Let $q(X) = X^T AX$ be a quadratic form, $A = A^T$.

(a) Show that $q(X) > 0$ for all $X \neq 0$, if and only if A is positive definite (all eigenvalues are positive). In this case, q is called **positive definite.**

(b) Show that new variables Y can be found such that $q = \|Y\|^2$ and $Y = UX$ where U is upper triangular with positive diagonal entries. [*Hint:* Theorem 4§6.5.]

10. A **bilinear form** β on $\mathbb{R}^n$ is a function that assigns to every pair X, Y of columns in $\mathbb{R}^n$ a number $\beta(X, Y)$ in such a way that

$$\beta(rX + sY, Z) = r\beta(X, Z) + s\beta(Y, Z)$$
$$\beta(X, rY + sZ) = r\beta(X, Y) + s\beta(X, Z)$$

for all X, Y, Z in $\mathbb{R}^n$ and r, s in $\mathbb{R}$. If $\beta(X, Y) = \beta(Y, X)$ for all X, Y, β is called **symmetric.**

(a) If β is a bilinear form, show that an $n \times n$ matrix A exists such that $\beta(X, Y) = X^T AY$ for all X, Y.

(b) Show that A is uniquely determined by β.

(c) Show that β is symmetric if and only if $A = A^T$.

Section 6.10   ## An Application to Best Approximation and Least Squares (Optional)[11]

A system of linear equations need not have a solution. However, even when no solution exists, it is often desirable to find a "best approximation" to a solution. In this section one definition of best approximation is given. Then it is shown that such an approximation always exists, and a method for finding it is described. The result is then applied to least squares approximation of data, a subject introduced in Section 4.4.

Suppose A is an $m \times n$ matrix and B is a column in $\mathbb{R}^m$, and consider the system

$$AX = B$$

of m linear equations in n variables. This need not have a solution. However, given any column Z in $\mathbb{R}^n$, the distance $\|B - AZ\|$ is a measure of how far AZ is from B. Hence it is natural to ask whether there is a column Z in $\mathbb{R}^n$ that is as close as possible to a solution in the sense that

$$\|B - AZ\|$$

is the minimum value of $\|B - AX\|$ as X ranges over all columns in $\mathbb{R}^n$. Theorem 7§6.3 (the projection theorem) answers this question in the affirmative. To see how, define

$$U = \{AX \mid X \text{ lies in } \mathbb{R}^n\}$$

Then U is a subspace of $\mathbb{R}^m$, so we are to find AX in U as close as possible to B. Theorem 7§6.3 guarantees a solution — call it AZ — and in fact

$$AZ = \text{proj}_U (B)$$

However, two computational problems are involved here. First we need an orthogonal basis of U to compute $\text{proj}_U(B)$. Second we end up with AZ rather than Z itself. So it is useful to find a way to compute Z directly. The key observation is that $B - AZ$ is in $U^\perp$ (Theorem 7§6.3) and so is orthogonal to every vector AX in U. Thus,

$$0 = (AX) \cdot (B - AZ) = (AX)^T(B - AZ) = X^T A^T(B - AZ) = X \cdot [A^T(B - AZ)]$$

for all X in $\mathbb{R}^n$. In other words, $A^T(B - AZ)$ is orthogonal to *every* vector in $\mathbb{R}^n$ and so must be zero. Hence Z satisfies

$$(A^TA)Z = A^TB$$

---

[11]This section requires only Section 6.3.

This is a system of linear equations called the **normal equations** for Z. Note that this system can have more than one solution (see Exercise 5). However the $n \times n$ matrix A^TA is invertible if (and only if) the columns of A are linearly independent (Theorem 6§5.5); so in this case Z is uniquely determined and is given explicitly by

$$Z = (A^TA)^{-1}A^TB$$

However, the most efficient way to find Z is to apply Gaussian elimination to the normal equations.

This discussion is summarized in the following theorem.

THEOREM 1

Best Approximation Theorem

Let A be an $m \times n$ matrix, let B be any column in $\mathbb{R}^m$, and consider the system

$$AX = B$$

of m equations in n variables. Any solution Z to the normal equations

$$(A^TA)Z = A^TB$$

is a best approximation to a solution to $AX = B$ in the sense that $\|B - AZ\|$ is the minimum value of $\|B - AX\|$ as X ranges over all columns in $\mathbb{R}^n$.

If the columns of A are linearly independent, then A^TA is invertible and Z is given uniquely by $Z = (A^TA)^{-1}A^TB$.

Note that if A is $n \times n$ and invertible, then

$$Z = (A^TA)^{-1}A^TB = A^{-1}B$$

is the solution to the system of equations, and $\|B - AZ\| = 0$. Hence $(A^TA)^{-1}A^T$ is playing the role of the inverse of the nonsquare matrix A in the case in which A has linearly independent columns. The matrix $A^T(AA^T)^{-1}$ plays a similar role when the rows of A are linearly independent. These are both special cases of the **generalized inverse** of a matrix A. However, we shall not pursue this topic here.

EXAMPLE 1

The equations

$$3x - y = 4$$
$$x + 2y = 0$$
$$2x + y = 1$$

have no solution. Find the vector $Z = \begin{bmatrix} x_0 \\ y_0 \end{bmatrix}$ that best approximates a solution.

Solution   In this case,

$$A = \begin{bmatrix} 3 & -1 \\ 1 & 2 \\ 2 & 1 \end{bmatrix}, \quad \text{so} \quad A^T A = \begin{bmatrix} 3 & 1 & 2 \\ -1 & 2 & 1 \end{bmatrix} \begin{bmatrix} 3 & -1 \\ 1 & 2 \\ 2 & 1 \end{bmatrix} = \begin{bmatrix} 14 & 1 \\ 1 & 6 \end{bmatrix}$$

is invertible. The normal equations $(A^T A)Z = A^T B$ are

$$\begin{bmatrix} 14 & 1 \\ 1 & 6 \end{bmatrix} Z = \begin{bmatrix} 14 \\ -3 \end{bmatrix}, \quad \text{so} \quad Z = \frac{1}{83} \begin{bmatrix} 87 \\ -56 \end{bmatrix}$$

Thus $x_0 = \frac{87}{83}$ and $y_0 = \frac{-56}{83}$. With these values of x and y, the left sides of the equations are

$$3x_0 - y_0 = \tfrac{317}{83} = 3.82$$
$$x_0 + 2y_0 = \tfrac{-25}{83} = -0.30$$
$$2x_0 + y_0 = \tfrac{118}{83} = 1.42$$

This is as close as possible to a solution.   ◆◆◆

EXAMPLE 2

The average number g of goals per game scored by a hockey player seems to be related linearly to two factors: the number x_1 of years of experience and the number x_2 of goals in the preceding 10 games. The accompanying data were collected on four players. Find the linear function $g = a_0 + a_1 x_1 + a_2 x_2$ that best fits these data.

Solution   If the relationship is given by $g = r_0 + r_1 x_1 + r_2 x_2$, then the data can be described as follows:

g	x_1	x_2
0.8	5	3
0.8	3	4
0.6	1	5
0.4	2	1

$$\begin{bmatrix} 1 & 5 & 3 \\ 1 & 3 & 4 \\ 1 & 1 & 5 \\ 1 & 2 & 1 \end{bmatrix} \begin{bmatrix} r_0 \\ r_1 \\ r_2 \end{bmatrix} = \begin{bmatrix} 0.8 \\ 0.8 \\ 0.6 \\ 0.4 \end{bmatrix}$$

Using the notation in Theorem 1, we get

$$Z = (A^T A)^{-1} A^T B$$

$$= \frac{1}{294} \begin{bmatrix} 833 & -119 & -133 \\ -119 & 35 & 7 \\ -133 & 7 & 35 \end{bmatrix} \begin{bmatrix} 1 & 1 & 1 & 1 \\ 5 & 3 & 1 & 2 \\ 3 & 4 & 5 & 1 \end{bmatrix} \begin{bmatrix} 0.8 \\ 0.8 \\ 0.6 \\ 0.4 \end{bmatrix} = \begin{bmatrix} 0.14 \\ 0.09 \\ 0.08 \end{bmatrix}$$

Hence the best-fitting function is $g = 0.14 + 0.09x_1 + 0.08x_2$. The computation would have been reduced if the normal equations had been constructed and then solved by Gaussian elimination.

◆◆◆

Least Squares Approximation

Theorem 1 applies directly to least squares approximation. This was treated in Section 4.4, though it was not possible to prove the main theorem (Theorem 2§4.4) with the techniques then available.

Suppose that data are available giving pairs of corresponding values of the two variables x and y:

$$(x_1, y_1), (x_2, y_2), \ldots, (x_n, y_n)$$

Given such data pairs, assume for the moment that the variables x and y are related by a polynomial of degree m.

$$y = p(x) = r_0 + r_1 x + \cdots + r_m x^m$$

Then for each x_i we have *two* values of the variable y, the observed value y_i and the computed value $p(x_i)$. The question now is this: Is it possible to choose the coefficients $r_0, r_1, \ldots, r_m$ in such a way that the $p(x_i)$ are as close as possible to the corresponding y_i? To apply Theorem 1, the following notation is convenient:

$$Y = \begin{bmatrix} y_1 \\ y_2 \\ \vdots \\ y_n \end{bmatrix} \qquad p(X) = \begin{bmatrix} p(x_1) \\ p(x_2) \\ \vdots \\ p(x_n) \end{bmatrix}$$

Then the problem takes the following form: Choose $r_0, r_1, \ldots, r_n$ such that

$$\|Y - p(X)\|^2 = [y_1 - p(x_1)]^2 + [y_2 - p(x_2)]^2 + \cdots + [y_n - p(x_n)]^2$$

is as small as possible. A polynomial $p(x)$ satisfying this condition is called a **least squares approximating polynomial** of degree m for the data pairs given. Now write

$$R = \begin{bmatrix} r_0 \\ r_1 \\ \vdots \\ r_m \end{bmatrix} \qquad M = \begin{bmatrix} 1 & x_1 & x_1^2 & \cdots & x_1^m \\ 1 & x_2 & x_2^2 & \cdots & x_2^m \\ \vdots & \vdots & \vdots & & \vdots \\ 1 & x_n & x_n^2 & \cdots & x_n^m \end{bmatrix}$$

Then $p(X)$ can be written

$$p(X) = \begin{bmatrix} p(x_1) \\ p(x_2) \\ \vdots \\ p(x_n) \end{bmatrix} = \begin{bmatrix} r_0 + r_1 x_1 + \cdots + r_m x_1^m \\ r_0 + r_1 x_2 + \cdots + r_m x_2^m \\ \vdots \\ r_0 + r_1 x_n + \cdots + r_m x_n^m \end{bmatrix} = MR$$

so we are to find R in $\mathbb{R}^{m+1}$ such that $\|Y - MR\|^2$ is as small as possible. In this form, Theorem 1 applies directly and gives the first part of Theorem 2.

THEOREM 2

Let n data pairs (x_1, y_1), (x_2, y_2), ..., (x_n, y_n) be given, and write

$$Y = \begin{bmatrix} y_1 \\ y_2 \\ \vdots \\ y_n \end{bmatrix} \qquad M = \begin{bmatrix} 1 & x_1 & x_1^2 & \cdots & x_1^m \\ 1 & x_2 & x_2^2 & \cdots & x_2^m \\ \vdots & \vdots & \vdots & & \vdots \\ 1 & x_n & x_n^2 & \cdots & x_n^m \end{bmatrix}$$

1. If $Z = \begin{bmatrix} z_0 \\ z_1 \\ \vdots \\ z_m \end{bmatrix}$ is any solution to the normal equations

$$(M^T M)Z = M^T Y$$

then the polynomial

$$\bar{p}(x) = z_0 + z_1 x + z_2 x^2 + \cdots + z_m x^m$$

is a least squares approximating polynomial of degree m for the given data pairs.

2. If at least $m + 1$ of the numbers $x_1, x_2, \ldots, x_n$ are distinct (so $n \geq m + 1$), the matrix $M^T M$ is invertible and Z is uniquely determined by

$$Z = (M^T M)^{-1} M^T Y$$

Proof   It remains to prove that the columns of M are linearly independent in (2). Suppose a linear combination of the columns vanishes:

$$r_0 \begin{bmatrix} 1 \\ 1 \\ \vdots \\ 1 \end{bmatrix} + r_1 \begin{bmatrix} x_1 \\ x_2 \\ \vdots \\ x_n \end{bmatrix} + \cdots + r_m \begin{bmatrix} x_1^m \\ x_2^m \\ \vdots \\ x_n^m \end{bmatrix} = \begin{bmatrix} 0 \\ 0 \\ \vdots \\ 0 \end{bmatrix}$$

If we write $p(x) = r_0 + r_1 x + \cdots + r_m x^m$, equating coefficients shows that $p(x_1) = p(x_2) = \cdots = p(x_n) = 0$. Because $p(x)$ is a polynomial of degree m with at least $m + 1$ distinct roots, $p(x)$ must be the zero polynomial. Thus $r_0 = r_1 = \cdots = r_m = 0$, as required. ◆

Several examples illustrating the use of this theorem were given in Section 4.4. The interested reader is referred to them.

There is an extension of Theorem 2 that should be mentioned. Given the data pairs (x_1, y_1), ..., (x_n, y_n), that theorem shows how to find a polynomial

$$p(x) = r_0 + r_1 x + \cdots + r_m x^m$$

such that $\|Y - p(X)\|^2$ is as small as possible, where X and $p(X)$ are as before.

Choosing the appropriate polynomial $p(x)$ amounts to choosing the coefficients $r_0, r_1, \ldots, r_m$, and the theorem gives a formula for the optimal choices. Now $p(x)$ is a linear combination of the functions $1, x, \ldots, x^m$, where the r_i are the coefficients, and this suggests applying the method to linear combinations of other functions. If $f_0(x), f_1(x), \ldots, f_m(x)$ are given functions, write

$$f(x) = r_0 f_0(x) + r_1 f_1(x) + \cdots + r_m f_m(x)$$

where $r_0, r_1, \ldots, r_m$ are real numbers. Then the more general question is whether $r_0, r_1, \ldots, r_m$ can be found such that $\|Y - f(X)\|^2$ is as small as possible, where now we write

$$f(X) = \begin{bmatrix} f(x_1) \\ f(x_2) \\ \vdots \\ f(x_n) \end{bmatrix}$$

The theorem follows.

THEOREM 3

Let n data pairs $(x_1, y_1), (x_2, y_2), \ldots, (x_n, y_n)$ be given and suppose that $m + 1$ functions $f_0(x), f_1(x), \ldots, f_m(x)$ are specified. Write

$$Y = \begin{bmatrix} y_1 \\ y_2 \\ \vdots \\ y_n \end{bmatrix} \qquad M = \begin{bmatrix} f_0(x_1) & f_1(x_1) & \cdots & f_m(x_1) \\ f_0(x_2) & f_1(x_2) & \cdots & f_m(x_2) \\ \vdots & \vdots & & \vdots \\ f_0(x_n) & f_1(x_n) & \cdots & f_m(x_n) \end{bmatrix}$$

1. If $Z = \begin{bmatrix} z_0 \\ z_1 \\ \vdots \\ z_m \end{bmatrix}$ is any solution to the normal equations

$$(M^T M)Z = M^T Y$$

then

$$\bar{f}(x) = z_0 f_0(x) + z_1 f_1(x) + \cdots + z_m f_m(x)$$

is the best approximation for these data among all functions $f(x)$ of the form

$$f(x) = r_0 f_0(x) + r_1 f_1(x) + \cdots + r_m f_m(x) \qquad r_i \text{ in } \mathbb{R}$$

in the sense that $\|Y - \bar{f}(X)\| \leq \|Y - f(X)\|$ holds for all choices of the r_i.

2. If $M^T M$ is invertible (that is, rank $M = m + 1$), then Z is uniquely determined by

$$Z = (M^T M)^{-1} M^T Y$$

Proof Observe that $f(X) = MR$, where $R = \begin{bmatrix} r_0 \\ r_1 \\ \vdots \\ r_m \end{bmatrix}$, so we are asked to choose R to

minimize $\|Y - MR\|^2$. Theorem 1 applies as before. ◆

The function $\bar{f}(x) = z_0 f_0(x) + z_1 f_1(x) + \cdots + z_m f_m(x)$ in Theorem 3 is called a **least squares approximating function** of the form $r_0 f_0(x) + \cdots + r_m f_m(x)$. This theorem contains Theorem 2 as a special case ($f_i(x) = x^i$ for each i), but there is no guarantee that $M^T M$ will be invertible in the general case if $m + 1$ of the x_i are distinct. Conditions for this to hold depend on the choice of the functions $f_0(x)$, $f_1(x)$, . . . , $f_m(x)$.

EXAMPLE 3

Given the data pairs $(-1, 0)$, $(0, 1)$, and $(1, 4)$, find the least squares approximating function of the form $r_0 x + r_1 2^x$.

Solution The functions are $f_0(x) = x$ and $f_1(x) = 2^x$, so the matrix M is

$$M = \begin{bmatrix} f_0(x_1) & f_1(x_1) \\ f_0(x_2) & f_1(x_2) \\ f_0(x_3) & f_1(x_3) \end{bmatrix} = \begin{bmatrix} -1 & 2^{-1} \\ 0 & 2^0 \\ 1 & 2^1 \end{bmatrix} = \tfrac{1}{2} \begin{bmatrix} -2 & 1 \\ 0 & 2 \\ 2 & 4 \end{bmatrix}$$

In this case $M^T M = \tfrac{1}{4} \begin{bmatrix} 8 & 6 \\ 6 & 21 \end{bmatrix}$ is invertible, so the normal equations

$$\tfrac{1}{4} \begin{bmatrix} 8 & 6 \\ 6 & 21 \end{bmatrix} Z = \begin{bmatrix} 4 \\ 9 \end{bmatrix} \qquad \text{have solution} \qquad Z = \tfrac{1}{11} \begin{bmatrix} 10 \\ 16 \end{bmatrix}$$

Hence the best-fitting function of the form $r_0 x + r_1 2^x$ is $\bar{f}(x) = \tfrac{10}{11} x + \tfrac{16}{11} 2^x$. Note that

$$\bar{f}(X) = \begin{bmatrix} \bar{f}(-1) \\ \bar{f}(0) \\ \bar{f}(1) \end{bmatrix} = \begin{bmatrix} \tfrac{-2}{11} \\ \tfrac{16}{11} \\ \tfrac{42}{11} \end{bmatrix}, \text{ compared with } Y = \begin{bmatrix} 0 \\ 1 \\ 4 \end{bmatrix}.$$

◆◆◆

EXERCISES 6.10

1. Find the best approximation to a solution of each of the following systems of equations.

(a)
$$\begin{aligned} x + y - z &= 5 \\ 2x - y + 6z &= 1 \\ 3x + 2y - z &= 6 \\ -x + 4y + z &= 0 \end{aligned}$$

◆(b)
$$\begin{aligned} 3x + y + z &= 6 \\ 2x + 3y - z &= 1 \\ 2x - y + z &= 0 \\ 3x - 3y + 3z &= 8 \end{aligned}$$

2. Find a least squares approximating function of the form $r_0 x + r_1 x^2 + r_2(-1)^x$ for each of the following sets of data pairs.

(a) $(-1, 1)$, $(0, 3)$, $(1, 1)$, $(2, 0)$

◆(b) $(0, 1)$, $(1, 1)$, $(2, 5)$, $(3, 10)$

3. Find the least squares approximating function of the form $r_0 + r_1x^2 + r_2 \sin \frac{\pi x}{2}$ for each of the following sets of data pairs.

    (a) $(0, 3), (1, 0), (1, -1), (-1, 2)$

    ◆**(b)** $\left(-1, \frac{1}{2}\right), (0, 1), (2, 5), (3, 9)$

◆**4.** The yield y of wheat in bushels per acre appears to be a linear function of the number of days x_1 of sunshine, the number of inches x_2 of rain, and the number of pounds x_3 of fertilizer applied per acre. Find the best fit to the data in the table by an equation of the form $y = r_0 + r_1x_1 + r_2x_2 + r_3x_3$. [*Hint:* If a calculator for inverting A^TA is not available, the inverse is given in the answer.]

y	x_1	x_2	x_3
28	50	18	10
30	40	20	16
21	35	14	10
23	40	12	12
23	30	16	14

5. Let A be any $m \times n$ matrix and write $K = \{X \mid A^TAX = 0\}$. Let B be an m-column. Show that, if Z is an n-column such that $\|B - AZ\|$ is minimal, then *all* such vectors have the form $Z + X$ for some X in K. [*Hint:* $\|B - AY\|$ is minimal if and only if $A^TAY = A^TB$.]

6. Given the situation in Theorem 3, write

$$f(x) = r_0f_0(x) + r_1f_1(x) + \cdots + r_mf_m(x)$$

Suppose that $f(x)$ has at most k roots for any choice of the coefficients $r_0, r_1, \ldots, r_m$, not all zero.

    (a) Show that M^TM is invertible if at least $k + 1$ of the x_i are distinct.

    ◆**(b)** If at least two of the x_i are distinct, show that there is always a best approximation of the form $r_0 + r_1e^x$.

    (c) If at least three of the x_i are distinct, show that there is always a best approximation of the form $r_0 + r_1x + r_2e^x$. [Calculus is needed.]

7. If A is an $m \times n$ matrix, it can be proved that there exists a unique $n \times m$ matrix $A^\#$ satisfying the following four conditions: $AA^\#A = A$; $A^\#AA^\# = A^\#$; $AA^\#$ and $A^\#A$ are symmetric. The matrix $A^\#$ is called the **generalized inverse** of A, or the **Moore–Penrose** inverse.

    (a) If A is square and invertible, show that $A^\# = A^{-1}$.

    (b) If rank $A = m$, show that $A^\# = A^T(AA^T)^{-1}$.

    (c) If rank $A = n$, show that $A^\# = (A^TA)^{-1}A^T$.

Section 6.11   An Application to Systems of Differential Equations (Optional)[12]

Solving a variety of problems, particularly in science and engineering, comes down to solving a differential equation or a system of such equations. In this section, vector spaces and matrix multiplication will be used to describe systems of differential equations, and diagonalization will be used to solve such systems. Of course, our methods really are only a first step into the vast theory of differential equations, but, at least for linear systems of first-order differential equations, the techniques do solve the problem and provide a basis for further work.

If f is a function of a real variable x, and if f' and f'' denote the first and second derivatives of f, then equations of the form

$$f' + af = 0 \qquad \text{or} \qquad f'' + af' + bf = 0 \qquad (a \text{ and } b \text{ numbers})$$

are called **differential equations of order 1 and 2,** respectively. One approach to such equations is to reduce those of order greater than one to *systems* of first-order equations. Then matrix diagonalization techniques can be applied. We treat such systems in this section.

The general problem is to find differentiable functions $f_1, f_2, \ldots, f_n$ that satisfy a system of equations of the form

---

[12]This section requires only Section 6.2.

$$f_1' = a_{11}f_1 + a_{12}f_2 + \cdots + a_{1n}f_n$$
$$f_2' = a_{21}f_1 + a_{22}f_2 + \cdots + a_{2n}f_n$$
$$\vdots \qquad \vdots$$
$$f_n' = a_{n1}f_1 + a_{n2}f_2 + \cdots + a_{nn}f_n$$

where the a_{ij} are constants. This is called a **linear system of differential equations.** The first step is to put it in matrix form. Write

$$\mathbf{f} = \begin{bmatrix} f_1 \\ f_2 \\ \vdots \\ f_n \end{bmatrix} \qquad \mathbf{f}' = \begin{bmatrix} f_1' \\ f_2' \\ \vdots \\ f_n' \end{bmatrix} \qquad A = \begin{bmatrix} a_{11} & a_{12} & \cdots & a_{1n} \\ a_{21} & a_{22} & \cdots & a_{2n} \\ \vdots & \vdots & & \vdots \\ a_{n1} & a_{n2} & \cdots & a_{nn} \end{bmatrix}$$

Then the system can be written compactly as

$$\mathbf{f}' = A\mathbf{f}$$

and, given the matrix A, the problem is to find a column $\mathbf{f}$ of differentiable functions that satisfies this condition.

Linear algebra enters into this as follows: These columns of functions become a vector space if matrix addition and scalar multiplication are used:

$$\begin{bmatrix} f_1 \\ f_2 \\ \vdots \\ f_n \end{bmatrix} + \begin{bmatrix} g_1 \\ g_2 \\ \vdots \\ g_n \end{bmatrix} = \begin{bmatrix} f_1 + g_1 \\ f_2 + g_2 \\ \vdots \\ f_n + g_n \end{bmatrix} \qquad a \begin{bmatrix} f_1 \\ f_2 \\ \vdots \\ f_n \end{bmatrix} = \begin{bmatrix} af_1 \\ af_2 \\ \vdots \\ af_n \end{bmatrix}$$

Of course, addition $f_i + g_i$ and scalar multiplication af_i of the individual functions are defined pointwise (as in Example 7§5.1). That is, the actions of $f_i + g_i$ and af_i are given by

$$(f_i + g_i)(x) = f_i(x) + g_i(x) \qquad \text{and} \qquad (af_i)(x) = af_i(x)$$

for all x. With these definitions, the set of all n-columns of functions becomes a vector space, as the reader can verify. The zero vector and the negative of a vector are

$$\mathbf{0} = \begin{bmatrix} 0 \\ 0 \\ \vdots \\ 0 \end{bmatrix} \qquad - \begin{bmatrix} f_1 \\ f_2 \\ \vdots \\ f_n \end{bmatrix} = \begin{bmatrix} -f_1 \\ -f_2 \\ \vdots \\ -f_n \end{bmatrix}$$

just as for matrices. This vector space will be denoted F^n.

Our concern here is not for F^n but for those columns of functions in it that satisfy the linear system:

$$U = \{\mathbf{f} \mid \mathbf{f} \text{ lies in } F^n \text{ and } \mathbf{f}' = A\mathbf{f}\}$$

Now recall the following three basic facts about differentiation:

$$\mathbf{0}' = \mathbf{0} \qquad (\mathbf{f} + \mathbf{g})' = \mathbf{f}' + \mathbf{g}' \qquad (c\mathbf{f})' = c\mathbf{f}'$$

Hence U is a subspace of F^n. The problem now is to find a convenient basis for U.

The case $n = 1$ has been discussed earlier. Here the problem is to find all functions f satisfying

$$f' = af \qquad a \text{ a constant}$$

and it follows from Theorem 1§5.7 that the space of solutions has dimension 1 and that $\{e^{ax}\}$ is a basis. If the matrix A is diagonalizable, this case can be used in the general situation. The following example provides an illustration.

EXAMPLE 1
<hr>

Find a solution to the system

$$f_1' = f_1 + 3f_2$$
$$f_2' = 2f_1 + 2f_2$$

that satisfies $f_1(0) = 0, f_2(0) = 5$.

Solution   This is $\mathbf{f}' = A\mathbf{f}$, where $\mathbf{f} = \begin{bmatrix} f_1 \\ f_2 \end{bmatrix}$ and $A = \begin{bmatrix} 1 & 3 \\ 2 & 2 \end{bmatrix}$. The reader can verify that $c_A(x)$
$= (x - 4)(x + 1)$, and that $X_1 = \begin{bmatrix} 1 \\ 1 \end{bmatrix}$ and $X_2 = \begin{bmatrix} 3 \\ -2 \end{bmatrix}$ are eigenvectors corresponding
to the eigenvalues 4 and -1, respectively. Hence the diagonalization algorithm gives
$P^{-1}AP = \begin{bmatrix} 4 & 0 \\ 0 & -1 \end{bmatrix}$, where $P = [X_1 \ X_2] = \begin{bmatrix} 1 & 3 \\ 1 & -2 \end{bmatrix}$. Now consider new functions
g_1 and g_2 given by $\mathbf{f} = P\mathbf{g}$ (equivalently, $\mathbf{g} = P^{-1}\mathbf{f}$):

$$\begin{bmatrix} f_1 \\ f_2 \end{bmatrix} = \begin{bmatrix} 1 & 3 \\ 1 & -2 \end{bmatrix} \begin{bmatrix} g_1 \\ g_2 \end{bmatrix} \qquad \text{that is,} \qquad \begin{aligned} f_1 &= g_1 + 3g_2 \\ f_2 &= g_1 - 2g_2 \end{aligned}$$

Then $f_1' = g_1' + 3g_2'$ and $f_2' = g_1' - 2g_2'$ so that

$$\mathbf{f}' = \begin{bmatrix} f_1' \\ f_2' \end{bmatrix} = \begin{bmatrix} 1 & 3 \\ 1 & -2 \end{bmatrix} \begin{bmatrix} g_1' \\ g_2' \end{bmatrix} = P\mathbf{g}'$$

If this is substituted in $\mathbf{f}' = A\mathbf{f}$, the result is $P\mathbf{g}' = AP\mathbf{g}$, whence

$$\mathbf{g}' = P^{-1}AP\mathbf{g}$$

But this means that

$$\begin{bmatrix} g_1' \\ g_2' \end{bmatrix} = \begin{bmatrix} 4 & 0 \\ 0 & -1 \end{bmatrix} \begin{bmatrix} g_1 \\ g_2 \end{bmatrix}, \qquad \text{so} \qquad \begin{aligned} g_1' &= 4g_1 \\ g_2' &= -g_2 \end{aligned}$$

Then the case $n = 1$ gives $g_1(x) = ce^{4x}$, $g_2(x) = de^{-x}$, where c and d are constants. Finally, then,

$$\begin{bmatrix} f_1(x) \\ f_2(x) \end{bmatrix} = P\begin{bmatrix} g_1(x) \\ g_2(x) \end{bmatrix} = \begin{bmatrix} 1 & 3 \\ 1 & -2 \end{bmatrix}\begin{bmatrix} ce^{4x} \\ de^{-x} \end{bmatrix} = \begin{bmatrix} ce^{4x} + 3de^{-x} \\ ce^{4x} - 2de^{-x} \end{bmatrix}$$

so the *general solution* is

$$f_1(x) = ce^{4x} + 3de^{-x}$$
$$f_2(x) = ce^{4x} - 2de^{-x}$$ c and d constants

It is worth observing that this can be written in matrix form as

$$\begin{bmatrix} f_1(x) \\ f_2(x) \end{bmatrix} = c\begin{bmatrix} 1 \\ 1 \end{bmatrix}e^{4x} + d\begin{bmatrix} 3 \\ -2 \end{bmatrix}e^{-x}$$

That is,

$$\mathbf{f}(x) = cX_1e^{4x} + dX_2e^{-x}$$

This form of the solution works more generally, as will be shown.

Finally, the requirement in this example that $f_1(0) = 0$ and $f_2(0) = 5$ determines the constants c and d:

$$0 = f_1(0) = ce^0 + 3de^0 = c + 3d$$
$$5 = f_2(0) = ce^0 - 2de^0 = c - 2d$$

These equations give $c = 3$ and $d = -1$, so

$$f_1(x) = 3e^{4x} - 3e^{-x}$$
$$f_2(x) = 3e^{4x} + 2e^{-x}$$

satisfy all the requirements.

◆◆◆

The technique of this example works in general.

THEOREM 1

Consider a linear system

$$\mathbf{f}' = A\mathbf{f}$$

of differential equations, where A is an $n \times n$ diagonalizable matrix. Let $P^{-1}AP$ be diagonal, where P is given in terms of its columns

$$P = [X_1 \ X_2 \ \cdots \ X_n]$$

and $\{X_1, X_2, \ldots, X_n\}$ are independent eigenvectors of A. If X_i corresponds to the eigenvalue λ_i for each i, then

$$\{X_1 e^{\lambda_1 x}, X_2 e^{\lambda_2 x}, \ldots, X_n e^{\lambda_n x}\}$$

is a basis for the space of solutions of $\mathbf{f}' = A\mathbf{f}$.

Proof Such X_i exist by virtue of the assumption that A is diagonalizable, and their independence guarantees that P is invertible. Let

$$P^{-1}AP = \begin{bmatrix} \lambda_1 & 0 & \cdots & 0 \\ 0 & \lambda_2 & \cdots & 0 \\ \vdots & \vdots & & \vdots \\ 0 & 0 & \cdots & \lambda_n \end{bmatrix}$$

where $\lambda_1, \lambda_2, \ldots, \lambda_n$ are the (not necessarily distinct) eigenvalues of A. As in the example, define a new column of functions $\mathbf{g} = \begin{bmatrix} g_1 \\ g_2 \\ \vdots \\ g_n \end{bmatrix}$ by $\mathbf{g} = P^{-1}\mathbf{f}$; equivalently,

$\mathbf{f} = P\mathbf{g}$. If f_i is the ith component of $\mathbf{f}$ and $P = [p_{ij}]$, this gives

$$f_i = p_{i1}g_1 + p_{i2}g_2 + \cdots + p_{in}g_n$$

Differentiation preserves this relationship:

$$f_i' = p_{i1}g_1' + p_{i2}g_2' + \cdots + p_{in}g_n'$$

so $\mathbf{f}' = P\mathbf{g}'$. Substituting this into $\mathbf{f}' = A\mathbf{f}$ gives $P\mathbf{g}' = AP\mathbf{g}$. But then multiplication by P^{-1} gives $\mathbf{g}' = P^{-1}AP\mathbf{g}$, so the original system of equations for $\mathbf{f}$ becomes much simpler in terms of $\mathbf{g}$:

$$\begin{bmatrix} g_1' \\ g_2' \\ \vdots \\ g_n' \end{bmatrix} = \begin{bmatrix} \lambda_1 & 0 & \cdots & 0 \\ 0 & \lambda_2 & \cdots & 0 \\ \vdots & \vdots & & \vdots \\ 0 & 0 & \cdots & \lambda_n \end{bmatrix} \begin{bmatrix} g_1 \\ g_2 \\ \vdots \\ g_n \end{bmatrix}$$

Hence $g_i' = \lambda_i g_i$ holds for each i, and Theorem 1§5.7 implies that the only solutions are

$$g_i(x) = c_i e^{\lambda_i x} \qquad c_i \text{ some constant}$$

Then the relationship $\mathbf{f} = P\mathbf{g}$ gives the functions $f_1, f_2, \ldots, f_n$ as follows:

$$\mathbf{f}(x) = [X_1 \ X_2, \ldots, \ X_n] \begin{bmatrix} c_1 e^{\lambda_1 x} \\ c_2 e^{\lambda_2 x} \\ \vdots \\ c_n e^{\lambda_n x} \end{bmatrix} = c_1 X_1 e^{\lambda_1 x} + c_2 X_2 e^{\lambda_2 x} + \cdots + c_n X_n e^{\lambda_n x}$$

Hence the columns $\{X_1 e^{\lambda_1 x}, X_2 e^{\lambda_2 x}, \ldots, X_n e^{\lambda_n x}\}$ span the space U of solutions. They are independent because, if

$$c_1 e^{\lambda_1 x} X_1 + c_2 e^{\lambda_2 x} X_2 + \cdots + c_n e^{\lambda_n x} X_n = 0$$

in F^n, then the left side vanishes for all x. In particular, taking $x = 0$ gives $c_1 X_1 + \cdots + c_n X_1 = 0$, so the independence of the X_i gives $c_1 = \cdots = c_n = 0$.   ◆

The theorem shows that *every* solution to $\mathbf{f}' = A\mathbf{f}$ is a linear combination

$$\mathbf{f}(x) = c_1 X_1 e^{\lambda_1 x} + c_2 X_2 e^{\lambda_2 x} + \cdots + c_n X_n e^{\lambda_n x}$$

where the coefficients c_i are arbitrary. Hence this is called the **general solution** to the system of differential equations. In most cases the solution functions $f_i(x)$ are required to satisfy boundary conditions, often of the form $f_i(a) = b_i$, where $a, b_1, \ldots, b_n$ are prescribed numbers. These conditions determine the constants c_i. The following example illustrates this and displays a situation where one eigenvalue has multiplicity greater than 1.

EXAMPLE 2

Find the general solution to the system

$$\begin{aligned} f_1' &= 5f_1 + 8f_2 + 16f_3 \\ f_2' &= 4f_1 + f_2 + 8f_3 \\ f_3' &= -4f_1 - 4f_2 - 11f_3 \end{aligned}$$

Then find a solution satisfying the boundary conditions $f_1(0) = f_2(0) = f_3(0) = 1$.

Solution   The system has the form $\mathbf{f}' = A\mathbf{f}$, where $A = \begin{bmatrix} 5 & 8 & 16 \\ 4 & 1 & 8 \\ -4 & -4 & -11 \end{bmatrix}$. This matrix was considered in Example 2§6.1 where it was found that $c_A(x) = (x + 3)^2(x - 1)$ and that independent eigenvectors corresponding to the eigenvalues -3, -3, and 1 are, respectively,

$$X_1 = \begin{bmatrix} -1 \\ 1 \\ 0 \end{bmatrix} \qquad X_2 = \begin{bmatrix} -2 \\ 0 \\ 1 \end{bmatrix} \qquad X_3 = \begin{bmatrix} 2 \\ 1 \\ -1 \end{bmatrix}$$

Hence $\{X_1 e^{-3x}, X_2 e^{-3x}, X_3 e^x\}$ spans the space of solutions, so the general solution is

$$\mathbf{f}(x) = c_1 \begin{bmatrix} -1 \\ 1 \\ 0 \end{bmatrix} e^{-3x} + c_2 \begin{bmatrix} -2 \\ 0 \\ 1 \end{bmatrix} e^{-3x} + c_3 \begin{bmatrix} 2 \\ 1 \\ -1 \end{bmatrix} e^x \qquad c_i \text{ constants}$$

The boundary conditions $f_1(0) = f_2(0) = f_3(0) = 1$ determine the constants c_i.

$$\begin{bmatrix} 1 \\ 1 \\ 1 \end{bmatrix} = \mathbf{f}(0) = c_1 \begin{bmatrix} -1 \\ 1 \\ 0 \end{bmatrix} + c_2 \begin{bmatrix} -2 \\ 0 \\ 1 \end{bmatrix} + c_3 \begin{bmatrix} 2 \\ 1 \\ -1 \end{bmatrix}$$

$$= \begin{bmatrix} -1 & -2 & 2 \\ 1 & 0 & 1 \\ 0 & 1 & -1 \end{bmatrix} \begin{bmatrix} c_1 \\ c_2 \\ c_3 \end{bmatrix}$$

The solution is $c_1 = -3$, $c_2 = 5$, $c_3 = 4$, so the required specific solution is

$$f_1(x) = -7e^{-3x} + 8e^x$$
$$f_2(x) = -3e^{-3x} + 4e^x$$
$$f_3(x) = 5e^{-3x} - 4e^x$$

The foregoing analysis fails if A is not diagonalizable, a situation that will not be treated in this book.

EXERCISES 6.11

1. Use Theorem 1 to find the general solution to each of the following systems. Then find a specific solution satisfying the given boundary condition.

   (a) $f_1' = 2f_1 + 4f_2,\ f_1(0) = 0$
        $f_2' = 3f_1 + 3f_2,\ f_2(0) = 1$

   ◆(b) $f_1' = -f_1 + 5f_2,\ f_1(0) = 1$
        $f_2' = f_1 + 3f_2,\ f_2(0) = -1$

   (c) $f_1' = \qquad 4f_2 + 4f_3$
        $f_2' = f_1 + f_2 - 2f_3$
        $f_3' = -f_1 + f_2 + 4f_3$
        $f_1(0) = f_2(0) = f_3(0) = 1$

   ◆(d) $f_1' = 2f_1 + f_2 + 2f_3$
        $f_2' = 2f_1 + 2f_2 - 2f_3$
        $f_3' = 3f_1 + f_2 + f_3$
        $f_1(0) = f_2(0) = f_3(0) = 1$

2. (a) Show that $e^{\lambda x}$, $e^{\mu x}$, and $e^{\delta x}$ are linearly independent functions if λ, μ, and δ are distinct. [*Hint:* If $re^{\lambda x} + se^{\mu x} + te^{\delta x} = 0$, differentiate twice and use Theorem 2§3.3.]

   (b) Generalize part (a) to $\{e^{\lambda_1 x}, e^{\lambda_2 x}, \ldots, e^{\lambda_n x}\}$.

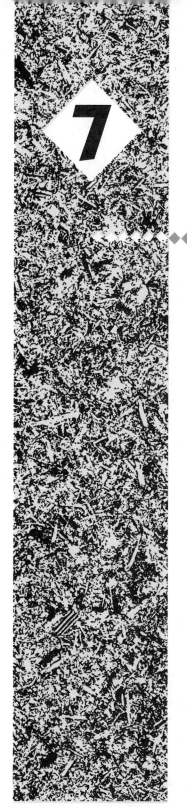

7 Linear Transformations

Section 7.1
Examples and Elementary Properties

Much of mathematics is concerned with the study of functions. Polynomial functions such as $p(x) = 3x^2 - 5x + 1$ come up in a wide variety of situations. Functions such as the exponential function e^x, the logarithm $\ln x$, and the trigonometric functions $\sin x$ and $\cos x$ play a fundamental role in calculus as well as in other areas of mathematics. If X and Y are sets, a **function** f from X to Y (written $f : X \to Y$ or $X \overset{f}{\to} Y$) is a rule that associates with every element x of X a uniquely determined element $f(x)$ of Y. In all these examples $X = \mathbb{R}$ and $Y = \mathbb{R}$, but we shall be considering functions where X and Y are both vector spaces.

DEFINITION

If V and W are two vector spaces, a function $T : V \to W$ is called a **linear transformation** if it satisfies the following axioms.

> **T1.** $T(\mathbf{v} + \mathbf{v}_1) = T(\mathbf{v}) + T(\mathbf{v}_1)$ for all $\mathbf{v}$ and $\mathbf{v}_1$ in V
>
> **T2.** $T(r\mathbf{v}) = rT(\mathbf{v})$ for all $\mathbf{v}$ in V and all r in $\mathbb{R}$

A linear transformation $T : V \to V$ is called a **linear operator** on V.

Axiom T1 is just the requirement that T preserves vector addition. It asserts that the result $T(\mathbf{v} + \mathbf{v}_1)$ of adding $\mathbf{v}$ and $\mathbf{v}_1$ first and then applying T is the same as applying T first to get $T(\mathbf{v})$ and $T(\mathbf{v}_1)$ and then adding. Similarly, axiom T2 means that T preserves scalar multiplication. Note that, even though the additions in axiom T1 are both denoted by the same symbol $+$, the addition on the left forming $\mathbf{v} + \mathbf{v}_1$ is carried out in V, whereas the addition $T(\mathbf{v}) + T(\mathbf{v}_1)$ is done in W. Similarly, the scalar multiplications $r\mathbf{v}$ and $rT(\mathbf{v})$ in axiom T2 refer to the spaces V and W, respectively.

The situation can be visualized as in Figure 7.1.

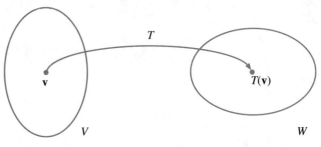

FIGURE 7.1

EXAMPLE 1

Define a function $T : \mathbb{R}^2 \to \mathbb{R}^3$ by $T\begin{bmatrix} x \\ y \end{bmatrix} = \begin{bmatrix} x + y \\ x - 2y \\ 3x \end{bmatrix}$ for all $\begin{bmatrix} x \\ y \end{bmatrix}$ in $\mathbb{R}^2$.

Show that T is a linear transformation.

Solution   We verify the axioms. Given $\begin{bmatrix} x \\ y \end{bmatrix}$ and $\begin{bmatrix} x_1 \\ y_1 \end{bmatrix}$ in $\mathbb{R}^2$, compute

$$T\left(\begin{bmatrix} x \\ y \end{bmatrix} + \begin{bmatrix} x_1 \\ y_1 \end{bmatrix} \right) = T\begin{bmatrix} x + x_1 \\ y + y_1 \end{bmatrix} = \begin{bmatrix} (x + x_1) + (y + y_1) \\ (x + x_1) - 2(y + y_1) \\ 3(x + x_1) \end{bmatrix}$$

$$= \begin{bmatrix} x + y \\ x - 2y \\ 3x \end{bmatrix} + \begin{bmatrix} x_1 + y_1 \\ x_1 - 2y_1 \\ 3x_1 \end{bmatrix} = T\begin{bmatrix} x \\ y \end{bmatrix} + T\begin{bmatrix} x_1 \\ y_1 \end{bmatrix}$$

This proves axiom T1, and axiom T2 is proved as follows:

$$T\left(r\begin{bmatrix} x \\ y \end{bmatrix} \right) = T\begin{bmatrix} rx \\ ry \end{bmatrix} = \begin{bmatrix} rx + ry \\ rx - 2ry \\ 3rx \end{bmatrix} = r\begin{bmatrix} x + y \\ x - 2y \\ 3x \end{bmatrix} = rT\begin{bmatrix} x \\ y \end{bmatrix}$$

Hence T preserves addition and scalar multiplication and so is a linear transformation.

◆◆◆

The linear transformation in Example 1 can be described using matrix multiplication:

$$T\begin{bmatrix} x \\ y \end{bmatrix} = \begin{bmatrix} 1 & 1 \\ 1 & -2 \\ 3 & 0 \end{bmatrix} \begin{bmatrix} x \\ y \end{bmatrix} \qquad \text{for all } \begin{bmatrix} x \\ y \end{bmatrix} \text{ in } \mathbb{R}^2$$

This suggests the following important class of linear transformations.

DEFINITION

If A is any $m \times n$ matrix, the **matrix transformation** $T_A : \mathbb{R}^n \to \mathbb{R}^m$ is defined by $T_A(X) = AX$ for all columns X in $\mathbb{R}^n$.

THEOREM 1

$T_A : \mathbb{R}^n \to \mathbb{R}^m$ is a linear transformation for each $m \times n$ matrix A.

Proof   Given X and X_1 in $\mathbb{R}^n$ and r in $\mathbb{R}$, matrix arithmetic yields

$$T_A(X + X_1) = A(X + X_1) = AX + AX_1 = T_A(X) + T_A(X_1)$$
$$T_A(rX) = A(rX) = r(AX) = rT_A(X)$$

Hence T_A is a linear transformation.   ◆

These matrix transformations are important for two reasons. First, *every* transformation from $\mathbb{R}^n$ to $\mathbb{R}^m$ arises in this way from an $m \times n$ matrix (this is Theorem 5, given later.) Second, we achieve a useful perspective on matrices when we view them as linear transformations in this way.

For example, consider the matrix operators on $\mathbb{R}^2$ given by the three types of elementary matrices

$$E = \begin{bmatrix} 0 & 1 \\ 1 & 0 \end{bmatrix} \qquad F = \begin{bmatrix} 2 & 0 \\ 0 & 1 \end{bmatrix} \qquad G = \begin{bmatrix} 1 & 1 \\ 0 & 1 \end{bmatrix}$$

obtained from I_2 by elementary operations. Thus

$$T_E\begin{bmatrix} x \\ y \end{bmatrix} = \begin{bmatrix} y \\ x \end{bmatrix} \qquad T_F\begin{bmatrix} x \\ y \end{bmatrix} = \begin{bmatrix} 2x \\ y \end{bmatrix} \qquad T_G\begin{bmatrix} x \\ y \end{bmatrix} = \begin{bmatrix} x + y \\ y \end{bmatrix}$$

Then T_E is **reflection** in the line $y = x$ because T carries every point to its mirror image with the line acting as the mirror. This is illustrated in Figure 7.2, which displays the effect of T_E on a rectangle.

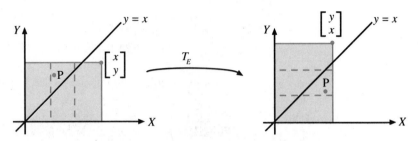

FIGURE 7.2   The reflection T_E in the line $y = x$, $E = \begin{bmatrix} 0 & 1 \\ 1 & 0 \end{bmatrix}$

Since $T_F \begin{bmatrix} x \\ y \end{bmatrix} = \begin{bmatrix} 2x \\ y \end{bmatrix}$, the operator T_F is called an **X expansion** because the effect is to expand the rectangle in Figure 7.3 in the X direction. If, instead, the matrix is $\begin{bmatrix} \frac{1}{2} & 0 \\ 0 & 1 \end{bmatrix}$, the operator is called an **X compression**.

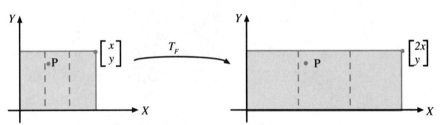

FIGURE 7.3   The X expansion T_F, $F = \begin{bmatrix} 2 & 0 \\ 0 & 1 \end{bmatrix}$

The operator T_G is called an **X shear.** The action is $T_G \begin{bmatrix} x \\ y \end{bmatrix} = \begin{bmatrix} x + y \\ y \end{bmatrix}$, so each horizontal line $y = b$ is carried to itself by moving each point b units along the line (to the right if $b > 0$). The action is illustrated in Figure 7.4.

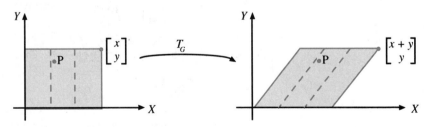

FIGURE 7.4   The X shear T_G, $G = \begin{bmatrix} 1 & 1 \\ 0 & 1 \end{bmatrix}$

Other familiar geometric transformations are linear. As in Chapter 4, identify the point $\begin{bmatrix} x \\ y \end{bmatrix}$ in $\mathbb{R}^2$ with the geometric vector from the origin to the point. Then the usual vector addition in $\mathbb{R}^2$ corresponds to the parallelogram law of addition for the geometric vectors, and a similar correspondence holds for scalar multiplication.

EXAMPLE 2

Let $R_\theta : \mathbb{R}^2 \to \mathbb{R}^2$ denote a counterclockwise rotation of the plane about the origin through an angle θ, and let $S : \mathbb{R}^2 \to \mathbb{R}^2$ denote a reflection in a line through the origin. Show that R_θ and S are linear operators on $\mathbb{R}^2$.

Solution

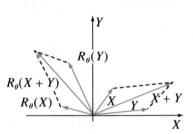

Two points X and Y are represented in the diagram as geometric vectors (arrows) emanating from the origin, and their sum $X + Y$ is the diagonal of the parallelogram they determine. The effect of R_θ is as follows: $R_\theta(X)$ is the vector obtained by rotating X counterclockwise about the origin through an angle θ. Hence the vectors $R_\theta(X)$, $R_\theta(Y)$, and $R_\theta(X + Y)$ can be obtained simultaneously by rotating the *entire parallelogram*. The result is the parallelogram determined by $R_\theta(X)$ and $R_\theta(Y)$, and the diagonal is $R_\theta(X + Y)$. But the diagonal is also $R_\theta(X) + R_\theta(Y)$ by the parallelogram law, so it follows that $R_\theta(X + Y) = R_\theta(X) + R_\theta(Y)$. A similar argument shows that $R_\theta(rX) = rR_\theta(X)$, so R_θ is linear.

The proof for S is analogous, because S flips the entire parallelogram over the line of reflection.

◆◆◆

As in Example 2, rotations about lines through the origin and reflections in planes through the origin can be seen to be linear operators from $\mathbb{R}^3$ to $\mathbb{R}^3$. Other geometric linear transformations are given in Exercise 18.

The next example lists three important linear transformations that will be referred to later. The verification of axioms T1 and T2 is left to the reader.

EXAMPLE 3

If V and W are vector spaces, the following are linear transformations:

Identity operator on V	$1_V : V \to V$ where $1_V(\mathbf{v}) = \mathbf{v}$ for all $\mathbf{v}$ in V
Zero transformation V $\to$ W	$0 : V \to W$ where $0(\mathbf{v}) = \mathbf{0}$ for all $\mathbf{v}$ in V
Scalar operator V $\to$ V	$a : V \to V$ where $a(\mathbf{v}) = a\mathbf{v}$ for all $\mathbf{v}$ in V

(Here a is any real number.)

◆◆◆

The symbol 0 will be used to denote the zero transformation from V to W for *any* spaces V and W. It was also used earlier (Example 7§5.1) to denote the zero function $[a, b] \to \mathbb{R}$.

The next example gives two important transformations of matrices. Recall the the trace tr A of an $n \times n$ matrix A is the sum of the entries on the main diagonal.

EXAMPLE 4

Transposition and trace are linear transformations. More precisely,

$$T : \mathbf{M}_{mn} \to \mathbf{M}_{nm} \qquad \text{where } T(A) = A^T \text{ for all } A \text{ in } \mathbf{M}_{mn}$$
$$T : \mathbf{M}_{nn} \to \mathbb{R} \qquad \text{where } T(A) = \text{tr } A \text{ for all } A \text{ in } \mathbf{M}_{nn}$$

are both linear transformations.

Solution   Axioms T1 and T2 for transposition are $(A + B)^T = A^T + B^T$ and $(rA)^T = r(A^T)$, respectively. The verifications for trace are left to the reader.

The projections we studied in Chapter 4 are also linear transformations, and the next example shows that this is true for $\mathbb{R}^n$.

EXAMPLE 5

Let U be a subspace of $\mathbb{R}^n$. Then projection on U is a linear transformation. More precisely, T is a linear transformation, where

$$T : \mathbb{R}^n \to U \text{ is defined by } T(X) = \text{proj}_U(X) \text{ for all } X \text{ in } \mathbb{R}^n$$

Solution   Let $\{E_1, E_2, \ldots, E_m\}$ be an orthonormal basis of U. Then

$$T(X) = (X \cdot E_1)E_1 + (X \cdot E_2)E_2 + \cdots + (X \cdot E_m)E_m$$

Axioms T1 and T2 hold because $(X + Y) \cdot E_i = X \cdot E_i + Y \cdot E_i$ and $(rX) \cdot E_i = r(X \cdot E_i)$ are valid for each i.

The next example involves some calculus.

EXAMPLE 6

The differentiation and integration operations are linear transformations. More precisely,

$$D : \mathbf{P}_n \to \mathbf{P}_{n-1} \quad \text{where } D[p(x)] = p'(x) \text{ for all } p(x) \text{ in } \mathbf{P}_n$$

$$I : \mathbf{P}_n \to \mathbf{P}_{n+1} \quad \text{where } I[p(x)] = \int_0^x p(t) \, dt \text{ for all } p(x) \text{ in } \mathbf{P}_n$$

are linear transformations.

Solution   These restate the following fundamental properties of differentiation and integration.

$$[p(x) + q(x)]' = p'(x) + q'(x) \quad \text{and} \quad [rp(x)]' = rp'(x)$$

$$\int_0^x [p(t) + q(t)]\, dt = \int_0^x p(t)\, dt + \int_0^x q(t)\, dt \quad \text{and} \quad \int_0^x rp(t)\, dt = r \int_0^x p(t)\, dt$$

◆◆◆

The next theorem collects three useful properties of *all* linear transformations. They can be described by saying that, in addition to preserving addition and scalar multiplication (these are the axioms), linear transformations preserve the zero vector, negatives, and linear combinations.

THEOREM 2

Let $T : V \to W$ be a linear transformation.

1. $T(\mathbf{0}) = \mathbf{0}$
2. $T(-\mathbf{v}) = -T(\mathbf{v})$ for all $\mathbf{v}$ in V
3. $T(r_1\mathbf{v}_1 + r_2\mathbf{v}_2 + \cdots + r_k\mathbf{v}_k) = r_1T(\mathbf{v}_1) + r_2T(\mathbf{v}_2) + \cdots + r_kT(\mathbf{v}_k)$ for all $\mathbf{v}_i$ in V and all r_i in $\mathbb{R}$

Proof

1. $T(\mathbf{0}) = T(0\mathbf{v}) = 0T(\mathbf{v}) = \mathbf{0}$ for any $\mathbf{v}$ in V.
2. $T(-\mathbf{v}) = T[(-1)\mathbf{v}] = (-1)T(\mathbf{v}) = -T(\mathbf{v})$ for any $\mathbf{v}$ in V.
3. If $k = 1$, this is $T(r_1\mathbf{v}_1) = r_1T(\mathbf{v}_1)$ by axiom T2. The general result is proved by induction on k. If it holds for a particular $k \geq 1$ then, using axiom T1 and the induction assumption, we have

$$T(r_1\mathbf{v}_1 + \cdots + r_k\mathbf{v}_k + r_{k+1}\mathbf{v}_{k+1}) = T(r_1\mathbf{v}_1 + \cdots + r_k\mathbf{v}_k) + T(r_{k+1}\mathbf{v}_{k+1})$$
$$= r_1T(\mathbf{v}_1) + \cdots + r_kT(\mathbf{v}_k) + r_{k+1}T(\mathbf{v}_{k+1})$$

This completes the induction and so proves property (3). ◆

The ability to use the last part of Theorem 2 effectively is vital to achieving any facility with linear transformations. The next two examples provide illustrations.

EXAMPLE 7

Let $T : V \to W$ be a linear transformation. If $T(\mathbf{v} - 3\mathbf{v}_1) = \mathbf{w}$ and $T(2\mathbf{v} - \mathbf{v}_1) = \mathbf{w}_1$, find $T(\mathbf{v})$ and $T(\mathbf{v}_1)$ in terms of $\mathbf{w}$ and $\mathbf{w}_1$.

Solution   The given relations imply that

$$T(\mathbf{v}) - 3T(\mathbf{v}_1) = \mathbf{w}$$
$$2T(\mathbf{v}) - T(\mathbf{v}_1) = \mathbf{w}_1$$

by Theorem 2. Subtracting twice the first from the second gives $T(\mathbf{v}_1) = \frac{1}{5}(\mathbf{w}_1 - 2\mathbf{w})$. Then substitution gives $T(\mathbf{v}) = \frac{1}{5}(3\mathbf{w}_1 - \mathbf{w})$. ◆◆◆

EXAMPLE 8

If $T : \mathbb{R}^3 \to \mathbb{R}$ is a linear transformation with $T(3, -1, 2) = 5$ and $T(1, 0, 1) = 2$, compute $T(-1, 1, 0)$.

Solution This can be done by Theorem 2, provided that $(-1, 1, 0)$ can be expressed as a linear combination of $(3, -1, 2)$ and $(1, 0, 1)$. This is indeed possible: $(-1, 1, 0) = -(3, -1, 2) + 2(1, 0, 1)$, so

$$T(-1, 1, 0) = -T(3, -1, 2) + 2T(1, 0, 1) = -5 + 4 = -1$$

◆◆◆

The full effect of property (3) in Theorem 2 is this: If $T : V \to W$ is a linear transformation and $T(\mathbf{v}_1)$, $T(\mathbf{v}_2)$, . . . , $T(\mathbf{v}_n)$ are known, then $T(\mathbf{v})$ can be computed for *every* $\mathbf{v}$ in span$\{\mathbf{v}_1, \mathbf{v}_2, \ldots, \mathbf{v}_n\}$. In particular, if $\{\mathbf{v}_1, \mathbf{v}_2, \ldots, \mathbf{v}_n\}$ spans V, then $T(\mathbf{v})$ is determined for all $\mathbf{v}$ in V by the choice of $T(\mathbf{v}_1)$, $T(\mathbf{v}_2)$, . . . , $T(\mathbf{v}_n)$. The next theorem states this somewhat differently. As for functions in general, two linear transformations $T : V \to W$ and $S : V \to W$ are called **equal** (written $T = S$) if they have the same **action**; that is, if $T(\mathbf{v}) = S(\mathbf{v})$ for all $\mathbf{v}$ in V.

THEOREM 3

Let $T : V \to W$ and $S : V \to W$ be two linear transformations. Suppose that $V = $ span$\{\mathbf{v}_1, \mathbf{v}_2, \ldots, \mathbf{v}_n\}$. If $T(\mathbf{v}_i) = S(\mathbf{v}_i)$ for each i, then $S = T$.

Proof Given $\mathbf{v}$ in V, write $\mathbf{v} = t_1\mathbf{v}_1 + t_2\mathbf{v}_2 + \cdots + t_n\mathbf{v}_n$, t_i in $\mathbb{R}$. Then Theorem 2 gives

$$\begin{aligned} T(\mathbf{v}) &= t_1T(\mathbf{v}_1) + t_2T(\mathbf{v}_2) + \cdots + t_nT(\mathbf{v}_n) \\ &= t_1S(\mathbf{v}_1) + t_2S(\mathbf{v}_2) + \cdots + t_nS(\mathbf{v}_n) \\ &= S(\mathbf{v}) \end{aligned}$$

Hence S and T have the same action, so $S = T$. ◆

EXAMPLE 9

Let $V = $ span$\{\mathbf{v}_1, \ldots, \mathbf{v}_n\}$. If $T : V \to W$ is a linear transformation and $T(\mathbf{v}_1) = \cdots = T(\mathbf{v}_n) = \mathbf{0}$, then $T = 0$, the zero transformation from V to W.

Solution The zero transformation $0 : V \to W$ is defined by $0(\mathbf{v}) = \mathbf{0}$ for all $\mathbf{v}$ in V (Example 3), so $T(\mathbf{v}_i) = 0(\mathbf{v}_i)$ holds for each i. Hence $T = 0$ by Theorem 3.

◆◆◆

Theorem 3 can be expressed as follows: If we know what a linear transformation $T : V \to W$ does to every vector in a spanning set for V, then we know what T does to *every* vector in V. If the spanning set is a basis, we can say more.

THEOREM 4

Let V and W be vector spaces and let $\{\mathbf{e}_1, \mathbf{e}_2, \ldots, \mathbf{e}_n\}$ be a basis of V. Given any vectors $\mathbf{w}_1, \mathbf{w}_2, \ldots, \mathbf{w}_n$ in W (they needn't be distinct), there exists a unique linear transformation $T : V \to W$ satisfying $T(\mathbf{e}_i) = \mathbf{w}_i$ for each $i = 1, 2, \ldots, n$. In fact, the action of T is as follows: Given $\mathbf{v} = v_1\mathbf{e}_1 + v_2\mathbf{e}_2 + \cdots + v_n\mathbf{e}_n$ in V, then

$$T(\mathbf{v}) = T(v_1\mathbf{e}_1 + v_2\mathbf{e}_2 + \cdots + v_n\mathbf{e}_n) = v_1\mathbf{w}_1 + v_2\mathbf{w}_2 + \cdots + v_n\mathbf{w}_n$$

Proof   If such a transformation T *does* exist, and if S is any other such transformation, then $T(\mathbf{e}_i) = \mathbf{w}_i = S(\mathbf{e}_i)$ holds for each i, so $S = T$ by Theorem 3. Hence T is unique if it exists, and it remains to show that there really is such a linear transformation. Given $\mathbf{v}$ in V, we must specify $T(\mathbf{v})$ in W. Because $\{\mathbf{e}_1, \ldots, \mathbf{e}_n\}$ is a basis of V, we have $\mathbf{v} = v_1\mathbf{e}_1 + \cdots + v_n\mathbf{e}_n$, where $v_1, \ldots, v_n$ are *uniquely* determined by $\mathbf{v}$ (this is Theorem 2§5.3). Hence we can define $T : V \to W$ by

$$T(\mathbf{v}) = T(v_1\mathbf{e}_1 + \cdots + v_n\mathbf{e}_n) = v_1\mathbf{w}_1 + v_2\mathbf{w}_2 + \cdots + v_n\mathbf{w}_n$$

for all $\mathbf{v} = v_1\mathbf{e}_1 + \cdots + v_n\mathbf{e}_n$ in V. This satisfies $T(\mathbf{e}_i) = \mathbf{w}_i$ for each i; the verification that T is linear is left to the reader. ◆

This theorem shows that linear transformations can be defined almost at will: Simply specify where the basis vectors are to be taken, and the rest of the action is dictated by the linearity. Moreover, Theorem 3 shows that deciding whether two linear transformations are equal comes down to determining whether they have the same effect on the basis vectors. So, given a basis $\{\mathbf{e}_1, \ldots, \mathbf{e}_n\}$ of a vector space V, there is a different linear transformation $V \to W$ for every ordered selection $\mathbf{w}_1, \mathbf{w}_2, \ldots, \mathbf{w}_n$ of vectors in W (not necessarily distinct).

EXAMPLE 10

Find a linear transformation $T : \mathbb{R}^3 \to \mathbb{R}^2$ such that

$$T\begin{bmatrix} 1 \\ 1 \\ 0 \end{bmatrix} = \begin{bmatrix} 2 \\ 1 \end{bmatrix} \qquad T\begin{bmatrix} 1 \\ 0 \\ 1 \end{bmatrix} = \begin{bmatrix} 1 \\ -1 \end{bmatrix} \qquad T\begin{bmatrix} 0 \\ 1 \\ 1 \end{bmatrix} = \begin{bmatrix} 0 \\ 0 \end{bmatrix}$$

Solution   The set $\left\{ \begin{bmatrix} 1 \\ 1 \\ 1 \end{bmatrix}, \begin{bmatrix} 1 \\ 0 \\ 1 \end{bmatrix}, \begin{bmatrix} 0 \\ 1 \\ 1 \end{bmatrix} \right\}$ is a basis of $\mathbb{R}^3$, so Theorem 4 applies. The expansion of an arbitrary vector in $\mathbb{R}^3$ as a linear combination of these vectors is

$$\begin{bmatrix} x \\ y \\ z \end{bmatrix} = \tfrac{1}{2}(x + y - z)\begin{bmatrix} 1 \\ 1 \\ 0 \end{bmatrix} + \tfrac{1}{2}(x - y + z)\begin{bmatrix} 1 \\ 0 \\ 1 \end{bmatrix} + \tfrac{1}{2}(-x + y + z)\begin{bmatrix} 0 \\ 1 \\ 1 \end{bmatrix}$$

Hence the transformation T must be given by

$$T\begin{bmatrix} x \\ y \\ z \end{bmatrix} = \tfrac{1}{2}(x + y - z)T\begin{bmatrix} 1 \\ 1 \\ 0 \end{bmatrix} + \tfrac{1}{2}(x - y + z)T\begin{bmatrix} 1 \\ 0 \\ 1 \end{bmatrix} + \tfrac{1}{2}(-x + y + z)T\begin{bmatrix} 0 \\ 1 \\ 1 \end{bmatrix}$$

$$= \tfrac{1}{2}(x + y - z)\begin{bmatrix} 2 \\ 1 \end{bmatrix} + \tfrac{1}{2}(x - y + z)\begin{bmatrix} 1 \\ -1 \end{bmatrix} + \tfrac{1}{2}(-x + y + z)\begin{bmatrix} 0 \\ 0 \end{bmatrix}$$

$$= \tfrac{1}{2}\begin{bmatrix} 3x + y - z \\ 2(y - z) \end{bmatrix}$$

Recall (Theorem 1) that every $m \times n$ matrix A gives rise to the matrix transformation $T_A : \mathbb{R}^n \to \mathbb{R}^m$ given by $T_A(X) = AX$ for all columns X in $\mathbb{R}^n$. In fact *all* linear transformations $\mathbb{R}^n \to \mathbb{R}^m$ arise in this way.

THEOREM 5

Let $T : \mathbb{R}^n \to \mathbb{R}^m$ be a linear transformation. Write vectors in $\mathbb{R}^n$ as columns.

1. There exists an $m \times n$ matrix A such that $T(X) = AX$ for all columns X in $\mathbb{R}^n$; that is, $T = T_A$.

2. The columns of A are respectively $T(E_1)$, $T(E_2)$, ... , $T(E_n)$, where $\{E_1, E_2, \ldots, E_n\}$ is the standard basis of $\mathbb{R}^n$. Hence A can be written in terms of its columns as

$$A = [T(E_1) \quad T(E_2) \cdots T(E_n)]$$

Proof Write

$$T(E_1) = \begin{bmatrix} a_{11} \\ a_{21} \\ \vdots \\ a_{m1} \end{bmatrix}, \; T(E_2) = \begin{bmatrix} a_{12} \\ a_{22} \\ \vdots \\ a_{m2} \end{bmatrix}, \ldots, \; T(E_n) = \begin{bmatrix} a_{1n} \\ a_{2n} \\ \vdots \\ a_{mn} \end{bmatrix}$$

Then $A = [a_{ij}]$ is an $m \times n$ matrix whose jth column is $T(E_j)$. Given X in $\mathbb{R}^n$, write

$$X = x_1 E_1 + x_2 E_2 + \cdots + x_n E_n \qquad x_i \text{ in } \mathbb{R}$$

Now compute $T(X)$, using Theorem 2.

$$T(X) = x_1 T(E_1) + x_2 T(E_2) + \cdots + x_n T(E_n)$$

$$= x_1 \begin{bmatrix} a_{11} \\ a_{21} \\ \vdots \\ a_{m1} \end{bmatrix} + x_2 \begin{bmatrix} a_{12} \\ a_{22} \\ \vdots \\ a_{m2} \end{bmatrix} + \cdots + x_n \begin{bmatrix} a_{1n} \\ a_{2n} \\ \vdots \\ a_{mn} \end{bmatrix}$$

$$= \begin{bmatrix} a_{11}x_1 + a_{12}x_2 + \cdots + a_{1n}x_n \\ a_{21}x_1 + a_{22}x_2 + \cdots + a_{2n}x_n \\ \vdots \qquad \vdots \qquad \qquad \vdots \\ a_{m1}x_1 + a_{m2}x_2 + \cdots + a_{mn}x_n \end{bmatrix}$$

$$= AX$$

$$= T_A(X).$$

Because this holds for all X in $\mathbb{R}^n$, it follows that $T = T_A$. ◆

The matrix A in Theorem 5 is called the **standard matrix** of T.

EXAMPLE 11

Find the standard matrix of $T : \mathbb{R}^3 \to \mathbb{R}^2$ when $T \begin{bmatrix} x \\ y \\ z \end{bmatrix} = \begin{bmatrix} x - 2y + z \\ x - z \end{bmatrix}$.

Solution   The desired matrix can be observed directly:

$$T \begin{bmatrix} x \\ y \\ z \end{bmatrix} = \begin{bmatrix} 1 & -2 & 1 \\ 1 & 0 & -1 \end{bmatrix} \begin{bmatrix} x \\ y \\ z \end{bmatrix}$$

However, the second part of Theorem 5 also gives the matrix. If $\{E_1, E_2, E_3\}$ is the standard basis of $\mathbb{R}^3$, then the columns are indeed $T(E_1)$, $T(E_2)$, and $T(E_3)$, as the reader can verify. ◆◆◆

EXAMPLE 12

Find the standard matrix of the following linear transformations $\mathbb{R}^2 \to \mathbb{R}^2$.

1. **Rotation** R_θ about the origin through the angle θ.
2. **Projection** P_m on the line $y = mx$.
3. **Reflection** S_m in the line $y = mx$.

Solution

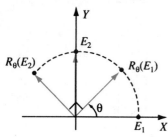

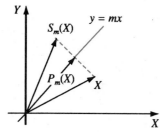

1. If $\{E_1, E_2\}$ is the standard basis, the diagram gives $R_\theta(E_1) = \begin{bmatrix} \cos \theta \\ \sin \theta \end{bmatrix}$ and $R_\theta(E_2) = \begin{bmatrix} -\sin \theta \\ \cos \theta \end{bmatrix}$, so the matrix is $\begin{bmatrix} \cos \theta & -\sin \theta \\ \sin \theta & \cos \theta \end{bmatrix}$

2. Write $X = \begin{bmatrix} x \\ y \end{bmatrix}$ and $D = \begin{bmatrix} 1 \\ m \end{bmatrix}$. Then D is a direction vector for the line, so Theorem 6§4.2 gives

$$P_m(X) = \frac{X \cdot D}{\|D\|^2} D = \frac{x + ym}{1 + m^2} \begin{bmatrix} 1 \\ m \end{bmatrix} = \frac{1}{1 + m^2} \begin{bmatrix} x + ym \\ xm + ym^2 \end{bmatrix}$$

$$= \frac{1}{1 + m^2} \begin{bmatrix} 1 & m \\ m & m^2 \end{bmatrix} \begin{bmatrix} x \\ y \end{bmatrix}$$

Hence the matrix is $\dfrac{1}{1 + m^2} \begin{bmatrix} 1 & m \\ m & m^2 \end{bmatrix}$.

3. The second diagram gives $S_m(X) = X + 2[P_m(X) - X] = 2P_m(X) - X$. Then (2) gives

$$S_m(X) = \frac{2}{1 + m^2} \begin{bmatrix} 1 & m \\ m & m^2 \end{bmatrix} \begin{bmatrix} x \\ y \end{bmatrix} - \begin{bmatrix} x \\ y \end{bmatrix}$$

$$= \frac{1}{1 + m^2} \begin{bmatrix} 1 - m^2 & 2m \\ 2m & m^2 - 1 \end{bmatrix} \begin{bmatrix} x \\ y \end{bmatrix}$$

Hence the matrix is $\dfrac{1}{1 + m^2} \begin{bmatrix} 1 - m^2 & 2m \\ 2m & m^2 - 1 \end{bmatrix}$.

◆◆◆

EXERCISES 7.1

1. Show that each of the following functions is a linear transformation.

   (a) $T : \mathbb{R}^2 \to \mathbb{R}^3$; $T\begin{bmatrix} x \\ y \end{bmatrix} = \begin{bmatrix} x - y \\ x + 2y \\ 3y \end{bmatrix}$

   ◆ (b) $T : \mathbb{R}^3 \to \mathbb{R}^2$; $T\begin{bmatrix} x \\ y \\ z \end{bmatrix} = \begin{bmatrix} 2x - 3y + 5z \\ 0 \end{bmatrix}$

   (c) $T : \mathbb{R}^2 \to \mathbb{R}^2$; $T(x, y) = (x, -y)$ (reflection in the X axis)

   ◆ (d) $T : \mathbb{R}^3 \to \mathbb{R}^3$; $T(x, y, z) = (x, y, -z)$ (reflection in the X-Y plane)

   (e) $T : \mathbb{C} \to \mathbb{C}$; $T(z) = \bar{z}$ (conjugation)

   ◆ (f) $T : \mathbf{M}_{mn} \to \mathbf{M}_{ki}$; $T(A) = PAQ$, P a $k \times m$ matrix, Q an $n \times l$ matrix

   (g) $T : \mathbf{M}_{nn} \to \mathbf{M}_{nn}$; $T(A) = A^T + A$

   ◆ (h) $T : \mathbf{P}_n \to \mathbb{R}$; $T[p(x)] = p(0)$

   (i) $T : \mathbf{P}_n \to \mathbb{R}$; $T(r_0 + r_1 x + \cdots + r_n x^n) = r_n$

   ◆ (j) $T : \mathbb{R}^n \to \mathbb{R}$; $T(X) = X \cdot Z$, Z a fixed vector in $\mathbb{R}^n$.

   (k) $T : \mathbf{P}_n \to \mathbf{P}_n$; $T[p(x)] = p(x + 1)$

   ◆ (l) $T : V \to \mathbb{R}$; $T(r_1 e_1 + \cdots + r_n e_n) = r_1$, where $\{e_1, \ldots, e_n\}$ is a fixed basis of V.

2. In each case, show that T is *not* a linear transformation.

   (a) $T : \mathbf{M}_{nn} \to \mathbb{R}$; $T(A) = \det A$

   ◆ (b) $T : \mathbf{M}_{nm} \to \mathbb{R}$; $T(A) = \text{rank } A$

(c) $T : \mathbb{R} \to \mathbb{R}; T(x) = x^2$

(d) $T : V \to V; T(\mathbf{v}) = \mathbf{v} + \mathbf{u}$ where $\mathbf{u} \neq \mathbf{0}$ is a fixed vector in V (T is called the **translation** by $\mathbf{u}$)

3. In each case, assume that T is a linear transformation.
   (a) If $T : V \to \mathbb{R}$ and $T(\mathbf{v}_1) = 1$, $T(\mathbf{v}_2) = -1$, find $T(3\mathbf{v}_1 - 5\mathbf{v}_2)$.
   (b) If $T : V \to \mathbb{R}$ and $T(\mathbf{v}_1) = 2$, $T(\mathbf{v}_2) = -3$, find $T(3\mathbf{v}_1 + 2\mathbf{v}_2)$.
   (c) If $T : \mathbb{R}^2 \to \mathbb{R}^2$ and $T\begin{bmatrix} 1 \\ 3 \end{bmatrix} = \begin{bmatrix} 1 \\ 1 \end{bmatrix}$, $T\begin{bmatrix} 1 \\ 1 \end{bmatrix} = \begin{bmatrix} 0 \\ 1 \end{bmatrix}$, find $T\begin{bmatrix} -1 \\ 3 \end{bmatrix}$.
   (d) If $T : \mathbb{R}^2 \to \mathbb{R}^2$ and $T\begin{bmatrix} 1 \\ -1 \end{bmatrix} = \begin{bmatrix} 0 \\ 1 \end{bmatrix}$, $T\begin{bmatrix} 1 \\ 1 \end{bmatrix} = \begin{bmatrix} 1 \\ 0 \end{bmatrix}$, find $T\begin{bmatrix} 1 \\ -7 \end{bmatrix}$.
   (e) If $T : \mathbf{P}_2 \to \mathbf{P}_2$ and $T(x + 1) = x$, $T(x - 1) = 1$, $T(x^2) = 0$, find $T(2 + 3x - x^2)$.
   (f) If $T : \mathbf{P}_2 \to \mathbb{R}$ and $T(x + 2) = 1$, $T(1) = 5$, $T(x^2 + x) = 0$, find $T(2 - x + 3x^2)$.

4. In each case, find a linear transformation with the given properties and compute $T(\mathbf{v})$.
   (a) $T : \mathbb{R}^2 \to \mathbb{R}^3; T(1, 2) = (1, 0, 1), T(-1, 0) = (0, 1, 1); \mathbf{v} = (2, 1)$
   (b) $T : \mathbb{R}^2 \to \mathbb{R}^3; T(2, -1) = (1, -1, 1), T(1, 1) = (0, 1, 0); \mathbf{v} = (-1, 2)$
   (c) $T : \mathbf{P}_2 \to \mathbf{P}_3; T(x^2) = x^3, T(x + 1) = 0, T(x - 1) = x; \mathbf{v} = x^2 + x + 1$
   (d) $T : \mathbf{M}_{22} \to \mathbb{R}; T\begin{bmatrix} 1 & 0 \\ 0 & 0 \end{bmatrix} = 3, T\begin{bmatrix} 0 & 1 \\ 1 & 0 \end{bmatrix} = -1,$

$T\begin{bmatrix} 1 & 0 \\ 1 & 0 \end{bmatrix} = 0 = T\begin{bmatrix} 0 & 0 \\ 0 & 1 \end{bmatrix}; \mathbf{v} = \begin{bmatrix} a & b \\ c & d \end{bmatrix}$

5. If $T : V \to V$ is a linear transformation, find $T(\mathbf{v})$ and $T(\mathbf{w})$ if:
   (a) $T(\mathbf{v} + \mathbf{w}) = \mathbf{v} - 2\mathbf{w}$ and $T(2\mathbf{v} - \mathbf{w}) = 2\mathbf{v}$
   (b) $T(\mathbf{v} + 2\mathbf{w}) = 3\mathbf{v} - \mathbf{w}$ and $T(\mathbf{v} - \mathbf{w}) = 2\mathbf{v} - 4\mathbf{w}$

6. If $T : V \to W$ is a linear transformation, show that $T(\mathbf{v} - \mathbf{v}_1) = T(\mathbf{v}) - T(\mathbf{v}_1)$ for all $\mathbf{v}$ and $\mathbf{v}_1$ in V.

7. Let $\{\mathbf{e}_1, \mathbf{e}_2\}$ be the standard basis of $\mathbb{R}^2$. Is it possible to have a linear transformation T such that $T(\mathbf{e}_1)$ lies in $\mathbb{R}$ while $T(\mathbf{e}_2)$ lies in $\mathbb{R}^2$? Explain your answer.

8. Let $\{\mathbf{v}_1, \ldots, \mathbf{v}_n\}$ be a basis of V and let $T : V \to V$ be a linear transformation.
   (a) If $T(\mathbf{v}_i) = \mathbf{v}_i$ for each i, show that $T = 1_V$.
   (b) If $T(\mathbf{v}_i) = \mathbf{0}$ for each i, show that $T = 0$ is the zero transformation.

9. If A is an $m \times n$ matrix, let $C_k(A)$ denote column k of A. Show that $C_k : \mathbf{M}_{m,n} \to \mathbb{R}^m$ is a linear transformation for each $k = 1, \ldots, n$.

10. Let $\{E_1, \ldots, E_n\}$ be an orthogonal basis of $\mathbb{R}^n$. Given k, $1 \leq k \leq n$, define $P_k : \mathbb{R}^n \to \mathbb{R}^n$ by $P_k(r_1E_1 + \cdots + r_nE_n) = r_kE_k$. Show that $P_k = \text{proj}_U(\ )$ where $U = \text{span}\{E_k\}$.

11. Let $S : V \to W$ and $T : V \to W$ be linear transformations. Given a in $\mathbb{R}$, define functions $(S + T) : V \to W$ and $(aT) : V \to W$ by $(S + T)(\mathbf{v}) = S(\mathbf{v}) + T(\mathbf{v})$ and $(aT)(\mathbf{v}) = aT(\mathbf{v})$ for all $\mathbf{v}$ in V. Show that $S + T$ and aT are linear transformations.

12. Describe all linear transformations $T : \mathbb{R} \to V$.

13. Let V and W be vector spaces, let V be finite dimensional, and let $\mathbf{v} \neq \mathbf{0}$ in V. Given any $\mathbf{w}$ in W, show that there exists a linear transformation $T : V \to W$ with $T(\mathbf{v}) = \mathbf{w}$. [*Hint:* Theorem 2§5.4 and Theorem 4.]

14. Show that every linear transformation $T : \mathbb{R}^n \to \mathbb{R}$ has the form $T(x_1, x_2, \ldots, x_n) = a_1x_1 + a_2x_2 + \cdots + a_nx_n$ for fixed $a_1, a_2, \ldots, a_n$ in $\mathbb{R}$.

15. Let $T : V \to W$ be a linear transformation. Show that:
   (a) If U is a subspace of V, then $T(U) = \{T(\mathbf{u}) \mid \mathbf{u} \text{ in } U\}$ is a subspace of W (called the **image** of U under T).
   (b) If P is a subspace of W, then $T^{-1}(P) = \{\mathbf{v} \text{ in } V \mid T(\mathbf{v}) \text{ in } P\}$ is a subspace of V (called the **preimage** of P under T).

16. If $T_A = T_B$ where A and B are $m \times n$ matrices, show that $A = B$.

17. Let $T : \mathbb{R}^m \to \mathbb{R}^n$ be a linear transformation where the vectors are written as rows.
   (a) Show that there is an $m \times n$ matrix A such that $T(X) = XA$ for all X in $\mathbb{R}^m$.
   (b) Show that the rows of A are $T(E_1), T(E_2), \ldots, T(E_m)$, respectively, where $\{E_1, \ldots, E_m\}$ is the standard basis of $\mathbb{R}^m$.

18. Find the standard matrix of the operator $T : \mathbb{R}^2 \to \mathbb{R}^2$ with the given geometric action.
   (a) Given $a > 0$, each point at distance r from the origin is moved radially out if $a > 1$ (or in if $a < 1$) from the origin to the point at distance ar from the origin. (This is a **contraction** if $a < 1$ and a **dilation** if $a > 1$.)
   (b) Each point is moved to its reflection in the Y axis.
   (c) Each point is moved to its reflection in the X axis.
   (d) Each point is moved to its reflection in the origin.
   (e) Each point is moved to the point on the X axis with the same X coordinate. (This is **projection** on the X axis.)

◆ **(f)** The plane is rotated clockwise through a right angle.

(g) The plane is rotated counterclockwise through a right angle.

◆ **(h)** Each point is moved to the point on the Y axis with the same Y coordinate (**projection** on the Y axis).

19. Find the linear transformation $T : \mathbb{R}^3 \to \mathbb{R}^3$ that has the given geometric action, and find the standard matrix of T.

(a) Rotation of θ about the Z axis counterclockwise in the X-Y plane

◆ **(b)** Reflection in the X-Y plane

20. Let $\mathbf{n} = \begin{bmatrix} a \\ b \\ c \end{bmatrix} \neq \mathbf{0}$ in $\mathbb{R}^3$, and let U denote the plane through the origin with normal $\mathbf{n}$ (as in Chapter 4). In each case, find the standard matrix of the linear transformation $\mathbb{R}^3 \to \mathbb{R}^3$.

(a) The projection on U

◆ **(b)** The reflection in U [*Hint:* Example 12]

21. Let $\mathbf{u}$, $\mathbf{v}$, and $\mathbf{w}$ denote vectors in $\mathbb{R}^3$ (as in Chapter 4). Show that $\mathbf{u} \times (\mathbf{v} \times \mathbf{w}) = (\mathbf{u} \cdot \mathbf{w})\mathbf{v} - (\mathbf{u} \cdot \mathbf{v})\mathbf{w}$ by applying Theorem 3 to $S : \mathbb{R}^3 \to \mathbb{R}^3$ and $T : \mathbb{R}^3 \to \mathbb{R}^3$, where $S(\mathbf{x}) = \mathbf{x} \times (\mathbf{v} \times \mathbf{w})$ and $T(\mathbf{x}) = (\mathbf{x} \cdot \mathbf{w})\,\mathbf{v} - (\mathbf{x} \cdot \mathbf{v})\mathbf{w}$.

22. Show that differentiation is the only linear transformation $\mathbf{P}_n \to \mathbf{P}_n$ that satisfies $T(x^k) = kx^{k-1}$ for each $k = 0, 1, 2, \ldots, n$.

23. Let $T : V \to W$ be a linear transformation and let $\mathbf{v}_1, \ldots, \mathbf{v}_n$ denote vectors in V.

(a) If $\{T(\mathbf{v}_1), \ldots, T(\mathbf{v}_n)\}$ is linearly independent, so also is $\{\mathbf{v}_1, \ldots, \mathbf{v}_n\}$

(b) Find $T : \mathbb{R}^2 \to \mathbb{R}^2$ for which the converse of part **(a)** is false.

◆**24.** Suppose $T : V \to V$ is a linear operator with the property that $T[T(\mathbf{v})] = \mathbf{v}$ for all $\mathbf{v}$ in V. (For example, transposi-

tion in $\mathbf{M}_{nn}$ or conjugation in $\mathbb{C}$.) If $\mathbf{v} \neq \mathbf{0}$ in V, show that $\{\mathbf{v}, T(\mathbf{v})\}$ is linearly independent if and only if $T(\mathbf{v}) \neq \mathbf{v}$ and $T(\mathbf{v}) \neq -\mathbf{v}$.

25. If a and b are real numbers, define $T_{a,b} : \mathbb{C} \to \mathbb{C}$ by $T_{a,b}(r + si) = ra + sbi$.

(a) Show that $T_{a,b}$ is linear and $T_{a,b}(\overline{z}) = \overline{T_{a,b}(z)}$ for all z in $\mathbb{C}$. (Here $\overline{z}$ denotes the conjugate of z.)

(b) If $T : \mathbb{C} \to \mathbb{C}$ is linear and $T(\overline{z}) = \overline{T(z)}$ for all z in $\mathbb{C}$, show that $T = T_{a,b}$ for some real a and b.

26. Show that the following conditions are equivalent for a linear transformation $T : \mathbf{M}_{22} \to \mathbf{M}_{22}$.

(1) $\text{tr}[T(A)] = \text{tr} A$ for all A in $\mathbf{M}_{22}$

(2) $T\begin{bmatrix} r_{11} & r_{12} \\ r_{21} & r_{22} \end{bmatrix} = r_{11}B_{11} + r_{12}B_{12} + r_{21}B_{21} + r_{22}B_{22}$

for matrices B_{ij} such that $\text{tr}\, B_{11} = 1 = \text{tr}\, B_{22}$ and $\text{tr}\, B_{12} = 0 = \text{tr}\, B_{21}$

27. Given a in $\mathbb{R}$, define the **evaluation** map $E_a : \mathbf{P}_n \to \mathbb{R}$ by $E_a[p(x)] = p(a)$ for all $p(x)$ in $\mathbf{P}_n$.

(a) Show that E_a is a linear transformation satisfying the additional condition that $E_a(x^k) = [E_a(x)]^k$ holds for all $k = 0, 1, 2, \ldots$. [*Note:* $x^0 = 1$.]

(b) If $T : \mathbf{P}_n \to R$ is a linear transformation satisfying $T(x^k) = [T(x)]^k$ for all $k = 0, 1, 2, \ldots$, show that $T = E_a$ for some a in R.

28. If $T : \mathbf{M}_{nn} \to \mathbb{R}$ is any linear transformation satisfying $T(AB) = T(BA)$ for all A and B in $\mathbf{M}_{nn}$, show that there exists a number k such that $T(A) = k\, \text{tr}\, A$ for all A. (See Theorem 3§6.1.)

[*Hint:* Let E_{ij} denote the $n \times n$ matrix with 1 in the (i, j) position and zeros elsewhere. Show that $E_{ik}E_{lj} = \begin{cases} 0 & \text{if } k \neq l \\ E_{ij} & \text{if } k = l \end{cases}$. Use this to show that $T(E_{ij}) = 0$ if $i \neq j$ and $T(E_{11}) = T(E_{22}) = \cdots = T(E_{nn})$. Put $k = T(E_{11})$ and use the fact the $\{E_{ij} \mid 1 \leq i, j \leq n\}$ is a basis of $\mathbf{M}_{nn}$.]

▓▓▓▓▓ Section 7.2   Kernel and Image of a Linear Transformation

This section is devoted to two important subspaces associated with a linear transformation.

> **DEFINITION**
>
> Let $T : V \to W$ denote a linear transformation. The **kernel** of T (denoted $\ker T$) and the **image** of T (denoted $\text{im}\, T$ or $T(V)$) are defined by
>
> $$\ker T = \{\mathbf{v} \text{ in } V \mid T(\mathbf{v}) = \mathbf{0}\}$$
> $$\text{im}\, T = \{T(\mathbf{v}) \mid \mathbf{v} \text{ in } V\}$$

The kernel of T is often called the **nullspace** of T. It consists of all vectors $\mathbf{v}$ in V satisfying the *condition* that $T(\mathbf{v}) = \mathbf{0}$. The image of T is often called the **range** of T and consists of all vectors $\mathbf{w}$ in W of the *form* $\mathbf{w} = T(\mathbf{v})$ for some $\mathbf{v}$ in V. These subspaces are depicted in Figure 7.5.

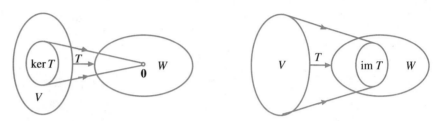

FIGURE 7.5

THEOREM 1

If $T : V \to W$ is a linear transformation, ker T is a subspace of V, and im T is a subspace of W.

Proof   The fact that $T(\mathbf{0}) = \mathbf{0}$ shows that both ker T and im T contain the zero vector. If $\mathbf{v}$ and $\mathbf{v}_1$ lie in ker T, then $T(\mathbf{v}) = \mathbf{0} = T(\mathbf{v}_1)$, so

$$T(\mathbf{v} + \mathbf{v}_1) = T(\mathbf{v}) + T(\mathbf{v}_1) = \mathbf{0} + \mathbf{0} = \mathbf{0}$$
$$T(r\mathbf{v}) = rT(\mathbf{v}) = r\mathbf{0} = \mathbf{0} \qquad \text{for all } r \text{ in } \mathbb{R}$$

Hence $\mathbf{v} + \mathbf{v}_1$ and $r\mathbf{v}$ lie in ker T (they satisfy the required condition), so ker T is a subspace of V. If $\mathbf{w}$ and $\mathbf{w}_1$ lie in im T, write $\mathbf{w} = T(\mathbf{v})$ and $\mathbf{w}_1 = T(\mathbf{v}_1)$ where $\mathbf{v}$ and $\mathbf{v}_1$ lie in V. Then

$$\mathbf{w} + \mathbf{w}_1 = T(\mathbf{v}) + T(\mathbf{v}_1) = T(\mathbf{v} + \mathbf{v}_1)$$
$$r\mathbf{w} = rT(\mathbf{v}) = T(r\mathbf{v}) \qquad \text{for all } r \text{ in } \mathbb{R}$$

Hence $\mathbf{w} + \mathbf{w}_1$ and $r\mathbf{w}$ both lie in im T (they have the required form), so im T is a subspace of W.   ◆

EXAMPLE 1

If $T : \mathbb{R}^3 \to \mathbb{R}^3$ is defined by $T(x, y, z) = (x - y, z, y - x)$, find ker T and im T, and compute their dimensions.

Solution   We use the definitions:

$$\ker T = \{(x, y, z) \mid (x - y, z, y - x) = (0, 0, 0)\} = \{(t, t, 0) \mid t \text{ in } \mathbb{R}\}$$
$$\operatorname{im} T = \{(x - y, z, y - x) \mid x, y, z \text{ in } \mathbb{R}\} = \{(s, t, -s) \mid s, t \text{ in } \mathbb{R}\}$$

Hence $\dim(\ker T) = 1$ and $\dim(\operatorname{im} T) = 2$.

◆◆◆

EXAMPLE 2

If A is an $m \times n$ matrix and $T_A : \mathbb{R}^n \to \mathbb{R}^m$ is defined by $T_A(X) = AX$ for every column X in $\mathbb{R}^n$, then

$$\ker T_A = \{X \mid AX = 0\} \qquad \text{and} \qquad \operatorname{im} T_A = \{AX \mid X \text{ in } \mathbb{R}^n\}$$

are, respectively, the null space and range of A (see Examples 5§5.2 and 4§5.2).

> **DEFINITION**
>
> Given a linear transformation $T : V \to W$,
>
> $\dim(\ker T)$ is called the **nullity** of T and denoted as nullity(T).
>
> $\dim(\operatorname{im} T)$ is called the **rank** of T and denoted as rank(T).

The rank of a matrix A was defined earlier to be the dimension of col A, the column space of A. The two usages of the word *rank* are consistent in the following sense:

EXAMPLE 3

Given an $m \times n$ matrix A, show that im $T_A = \operatorname{col} A$, so rank $T_A = \operatorname{rank} A$.

Solution   Write $A = [C_1 \cdots C_n]$ in terms of its columns. Then

$$\operatorname{im} T_A = \{AX \mid X \text{ in } \mathbb{R}^n\} = \left\{ [C_1 \ \cdots \ C_n] \begin{bmatrix} x_1 \\ \vdots \\ x_n \end{bmatrix} \ \middle| \ x_i \text{ in } \mathbb{R} \right\}$$

$$= \{x_1 C_1 + \cdots + x_n C_n \mid x_i \text{ in } \mathbb{R}\}$$

Hence im$T_A = \operatorname{span}\{C_1, \ldots, C_n\}$ is the column space of A.

EXAMPLE 4

Given the 4×3 matrix $A = \begin{bmatrix} 1 & -1 & 2 \\ 3 & 0 & 1 \\ 1 & 2 & -3 \\ -2 & -1 & 1 \end{bmatrix}$, compute the kernel and image of the

corresponding matrix transformation $T_A : \mathbb{R}^3 \to \mathbb{R}^4$, and determine the rank and nullity of T_A.

Solution   Bring A to reduced row-echelon form:

$$\begin{bmatrix} 1 & -1 & 2 \\ 3 & 0 & 1 \\ 1 & 2 & -3 \\ -2 & -1 & 1 \end{bmatrix} \rightarrow \begin{bmatrix} 1 & -1 & 2 \\ 0 & 3 & -5 \\ 0 & 3 & -5 \\ 0 & -3 & 5 \end{bmatrix} \rightarrow \begin{bmatrix} 1 & 0 & \frac{1}{3} \\ 0 & 1 & -\frac{5}{3} \\ 0 & 0 & 0 \\ 0 & 0 & 0 \end{bmatrix}$$

so rank T_A = rank A = 2. Moreover, the solutions to $AX = 0$ are $[-t \quad 5t \quad 3t]^T$, where t is a parameter. Because ker T_A = $\{X \text{ in } \mathbb{R}^3 \mid AX = 0\}$, this means that nullity T_A = dim(ker T_A) = 1.

◆◆◆

A useful way to study a subspace of a vector space is often to exhibit it as the kernel or image of a linear transformation. Here is an example.

EXAMPLE 5

Define a transformation $T : \mathbf{M}_{nn} \rightarrow \mathbf{M}_{nn}$ by $T(A) = A - A^T$ for all A in $\mathbf{M}_{nn}$. Show that T is linear and that:

(a)   ker T consists of all symmetric matrices.

(b)   im T consists of all skew-symmetric matrices.

Solution   The verification that T is linear is left to the reader. To prove part (a), note that a matrix A lies in ker T just when $0 = T(A) = A - A^T$, and this occurs if and only if $A = A^T$ — that is, A is symmetric. Turning to part (b), the space im T consists of all matrices $T(A)$, A in $\mathbf{M}_{nn}$. Every such matrix is skew-symmetric because

$$T(A)^T = (A - A^T)T = A^T - A = -T(A)$$

On the other hand, if S is skew-symmetric (that is, $S^T = -S$), then S lies in im T. In fact,

$$T\left[\tfrac{1}{2}S\right] = \tfrac{1}{2}S - \left[\tfrac{1}{2}S\right]^T = \tfrac{1}{2}(S + S) = S$$

◆◆◆

DEFINITION

Let $T : V \rightarrow W$ be a linear transformation.

1. T is said to be **onto** if im $T = W$.

2. T is said to be **one-to-one** if $T(\mathbf{v}) = T(\mathbf{v}_1)$ implies $\mathbf{v} = \mathbf{v}_1$.

Thus T is onto if *every* vector $\mathbf{w}$ in W has the form $\mathbf{w} = T(\mathbf{v})$ for some (not necessarily unique) vector $\mathbf{v}$ in V, whereas T is one-to-one if *no* two distinct vectors $\mathbf{v} \neq \mathbf{v}_1$ in V are carried to the same image $T(\mathbf{v}) = T(\mathbf{v}_1)$ in W. The onto transformations T are those for which im T is as large a subspace of W as possible. By contrast,

Theorem 2 shows that the one-to-one transformations T are the ones with ker T as *small* as possible.

THEOREM 2

If $T : V \to W$ is a linear transformation, then T is one-to-one if and only if ker $T = 0$.

Proof   If T is one-to-one, let $\mathbf{v}$ be any vector in ker T. Then $T(\mathbf{v}) = \mathbf{0}$, so $T(\mathbf{v}) = T(\mathbf{0})$. Hence $\mathbf{v} = \mathbf{0}$ because T is one-to-one. Conversely, assume that ker $T = 0$ and let $T(\mathbf{v}) = T(\mathbf{v}_1)$ with $\mathbf{v}$ and $\mathbf{v}_1$ in V. Then $T(\mathbf{v} - \mathbf{v}_1) = T(\mathbf{v}) - T(\mathbf{v}_1) = \mathbf{0}$, so $\mathbf{v} - \mathbf{v}_1$ lies in ker $T = 0$. This means that $\mathbf{v} - \mathbf{v}_1 = \mathbf{0}$, so $\mathbf{v} = \mathbf{v}_1$. This proves that T is one-to-one.   ◆

EXAMPLE 6

The identity transformation $1_V : V \to V$ is both one-to-one and onto for any vector space V.   ◆◆◆

EXAMPLE 7

Consider the linear transformations

$$S : \mathbb{R}^3 \to \mathbb{R}^2 \qquad \text{given by } S(x, y, z) = (x + y, x - y)$$
$$T : \mathbb{R}^2 \to \mathbb{R}^3 \qquad \text{given by } T(x, y) = (x + y, x - y, x)$$

Show that T is one-to-one but not onto, whereas S is onto but not one-to-one.

Solution   The verification that they are linear is omitted. T is one-to-one because

$$\ker T = \{(x, y) \mid x + y = x - y = x = 0\} = \{(0, 0)\} = 0$$

However, it is not onto. For example $(0, 0, 1)$ does not lie in im T because if $(0, 0, 1) = (x + y, x - y, x)$ for some x and y, then $x + y = 0 = x - y$ and $x = 1$, an impossibility. Turning to S, it is not one-to-one by Theorem 2 because $(0, 0, 1)$ lies in ker S. But every element (s, t) in $\mathbb{R}^2$ lies in im S because $(s, t) = (x + y, x - y)$ for some x and y (in fact $x = \frac{1}{2}(s + t)$ and $y = \frac{1}{2}(s - t)$). Hence S is onto.   ◆◆◆

EXAMPLE 8

Let U be an invertible $m \times m$ matrix and define

$$T : \mathbf{M}_{mn} \to \mathbf{M}_{mn} \qquad \text{by} \qquad T(X) = UX \text{ for all } X \text{ in } \mathbf{M}_{mn}$$

Show that T is a linear transformation that is both one-to-one and onto.

Solution   The verification that T is linear is left to the reader. To see that T is one-to-one, let $T(X) = 0$. Then $UX = 0$, so left-multiplication by U^{-1} gives $X = 0$. Hence ker $T = 0$, so T is one-to-one. Finally, if Y is any member of $\mathbf{M}_{mn}$, then $U^{-1}Y$ lies in $\mathbf{M}_{mn}$ too, and $T(U^{-1}Y) = U(U^{-1}Y) = Y$. This shows that T is onto.

                                                         ◆◆◆

The linear transformations $\mathbb{R}^n \rightarrow \mathbb{R}^m$ all have the form T_A for some $m \times n$ matrix A (Theorem 5§7.1). The next theorem gives conditions under which they are onto or one-to-one.

THEOREM 3

Let A be an $m \times n$ matrix, and let $T_A : \mathbb{R}^n \rightarrow \mathbb{R}^m$ be the matrix transformation defined by $T_A(X) = AX$.

1. T_A is onto if and only if rank $A = m$ (linearly independent rows).

2. T_A is one-to-one if and only if rank $A = n$ (linearly independent columns).

Proof

   1. We have that im T_A is the column space of A (see Example 3), so T_A is onto if and only if the column space of A is $\mathbb{R}^m$. Because the rank of A is the dimension of the column space, this holds if and only if rank $A = m$; this is equivalent to A having independent rows by Theorem 6§5.5.

   2. ker $T_A = \{X \text{ in } \mathbb{R}^n \mid AX = 0\}$, so (using Theorem 2) T_A is one-to-one if and only if $AX = 0$ implies $X = 0$. But AX is a linear combination of the columns of A, the coefficients being the components of X, so this holds if and only if A has independent columns. This is equivalent to rank $A = n$ by Theorem 6§5.5 (applied to A^T).   ◆

The following theorem is the main result of this section.

THEOREM 4
Dimension Theorem

Let $T : V \rightarrow W$ be any linear transformation and assume that ker T and im T are both finite dimensional. Then V is also finite dimensional and

$$\dim V = \dim(\ker T) + \dim(\operatorname{im} T)$$

In other words, $\dim V = \text{nullity}(T) + \text{rank}(T)$.

Proof   Every vector in im $T = T(V)$ has the form $T(\mathbf{v})$ for some $\mathbf{v}$ in V. Hence let $\{T(\mathbf{e}_1), T(\mathbf{e}_2), \ldots, T(\mathbf{e}_r)\}$ be a basis of im T, where the $\mathbf{e}_i$ lie in V. Let $\{\mathbf{f}_1, \mathbf{f}_2, \ldots, \mathbf{f}_k\}$

be any basis of ker T. Then dim(im T) $= r$ and dim(ker T) $= k$, so it suffices to show that $B = \{\mathbf{e}_1, \ldots, \mathbf{e}_r, \mathbf{f}_1, \ldots, \mathbf{f}_k\}$ is a basis of V.

1. *B spans V.* If $\mathbf{v}$ lies in V, then $T(\mathbf{v})$ lies in im V, so

$$T(\mathbf{v}) = t_1 T(\mathbf{e}_1) + t_2 T(\mathbf{e}_2) + \cdots + t_r T(\mathbf{e}_r) \qquad t_i \text{ in } \mathbb{R}$$

This implies that $\mathbf{v} - t_1\mathbf{e}_1 - t_2\mathbf{e}_2 - \cdots - t_r\mathbf{e}_r$ lies in ker T and so is a linear combination of $\mathbf{f}_1, \ldots, \mathbf{f}_k$. Hence $\mathbf{v}$ is a linear combination of the vectors in B.

2. *B is linearly independent.* Suppose that t_i and s_j in $\mathbb{R}$ satisfy

$$t_1\mathbf{e}_1 + \cdots + t_r\mathbf{e}_r + s_1\mathbf{f}_1 + \cdots + s_k\mathbf{f}_k = \mathbf{0} \qquad (*)$$

Applying T gives $t_1 T(\mathbf{e}_1) + \cdots + t_r T(\mathbf{e}_r) = \mathbf{0}$ (because $T(\mathbf{f}_i) = \mathbf{0}$ for each i), so the independence of $\{T(\mathbf{e}_1), \ldots, T(\mathbf{e}_r)\}$ yields $t_1 = \cdots = t_r = 0$. Hence $(*)$ becomes

$$s_1\mathbf{f}_1 + \cdots + s_k\mathbf{f}_k = \mathbf{0}$$

so $s_1 = \cdots = s_k = 0$ by the independence of $\{\mathbf{f}_1, \ldots, \mathbf{f}_k\}$. This proves that B is linearly independent.   ◆

Note that $r + k = n$ in the proof, so after relabeling, we end up with a basis

$$B = \{\mathbf{e}_1, \mathbf{e}_2, \ldots, \mathbf{e}_r, \mathbf{e}_{r+1}, \ldots, \mathbf{e}_n\}$$

of V with the property that $\{\mathbf{e}_{r+1} \ldots, \mathbf{e}_n\}$ is a basis of ker T and $\{T(\mathbf{e}_1), \ldots, T(\mathbf{e}_r)\}$ is a basis of im T. In fact, if V is known in advance to be finite dimensional, then *any* basis $\{\mathbf{e}_{r+1}, \ldots, \mathbf{e}_n\}$ of ker T can be extended to a basis $\{\mathbf{e}_1, \mathbf{e}_2, \ldots, \mathbf{e}_r, \mathbf{e}_{r+1}, \ldots, \mathbf{e}_n\}$ of V by Theorem 2§5.4. Moreover, it turns out that, no matter how this is done, the vectors $\{T(\mathbf{e}_1), \ldots, T(\mathbf{e}_r)\}$ will be a basis of im T. This result is useful, and we record it for reference. The proof is much like that of Theorem 4 and is left as Exercise 26.

THEOREM 5

Let $T : V \rightarrow W$ be a linear transformation, and let $\{\mathbf{e}_1, \ldots, \mathbf{e}_r, \mathbf{e}_{r+1}, \ldots, \mathbf{e}_n\}$ be a basis of V such that $\{\mathbf{e}_{r+1}, \ldots, \mathbf{e}_n\}$ is a basis of ker T. Then $\{T(\mathbf{e}_1), \ldots, T(\mathbf{e}_r)\}$ is a basis of im T, and hence $r = $ rank T.

The dimension theorem is one of the most useful results in all of linear algebra. It shows that if either dim(ker T) or dim(im T) can be found, then the other is automatically known. In many cases it is easier to compute one than the other, so the theorem is a real asset. The rest of this section is devoted to illustrations of this fact. The next example uses the dimension theorem to give a different proof of Theorem 5§5.5.

EXAMPLE 9

Let A be an $m \times n$ matrix of rank r. Show that the space of all solutions of the system $AX = 0$ of m homogeneous equations in n variables has dimension $n - r$.

Solution   The space in question is just $\ker T_A$, where $T_A : \mathbb{R}^n \to \mathbb{R}^m$ is defined by $T_A(X) = AX$ for all columns X in $\mathbb{R}^n$. But $\dim(\operatorname{im} T_A) = \operatorname{rank} T_A = \operatorname{rank} A = r$ by Example 3, so $\dim(\ker T_A) = n - r$ by the dimension theorem.   ◆◆◆

EXAMPLE 10

If $T : V \to W$ is a linear transformation where V is finite dimensional, then

$$\dim(\ker T) \leq \dim V \quad \text{and} \quad \dim(\operatorname{im} T) \leq \dim V$$

because $\dim V = \dim(\ker T) + \dim(\operatorname{im} T)$. Of course the first inequality also follows because $\ker T$ is a subspace of V.   ◆◆◆

EXAMPLE 11[1]

Let $D : \mathbf{P}_n \to \mathbf{P}_{n-1}$ be the differentiation map defined by $D[p(x)] = p'(x)$. Compute $\ker D$ and hence conclude that D is onto.

Solution   Because $p'(x) = 0$ means $p(x)$ is constant, we have $\dim(\ker D) = 1$. Because $\dim \mathbf{P}_n = n + 1$, the dimension theorem gives

$$\dim(\operatorname{im} D) = (n + 1) - \dim(\ker D) = n = \dim(\mathbf{P}_{n-1})$$

This implies that $\operatorname{im} D = \mathbf{P}_{n-1}$, so D is onto.   ◆◆◆

Of course it is not difficult to verify directly that each polynomial $q(x)$ in $\mathbf{P}_{n-1}$ is the derivative of some polynomial in $\mathbf{P}_n$ (simply integrate $q(x)$!), so the dimension theorem is not needed in this case. However, in some situations it is difficult to see directly that a linear transformation is onto, and the method used in Example 11 may be the easiest way by far to proceed. Here is another illustration.

EXAMPLE 12

Given a in $\mathbb{R}$, define the **evaluation map** $E_a : \mathbf{P}_n \to \mathbb{R}$ by $E_a[p(x)] = p(a)$. Show that E_a is linear and onto, and hence conclude that $\{(x - a), (x - a)^2, \ldots, (x - a)^n\}$ is a basis of $\ker E_a$, the subspace of all polynomials $p(x)$ for which $p(a) = 0$.

Solution   The verification that E_a is linear and onto is left to the reader. Hence $\dim(\operatorname{im} E_a) = \dim(\mathbb{R}) = 1$, so $\dim(\ker E_a) = (n + 1) - 1 = n$ by the dimension theorem. Now

---

[1]This example uses calculus and can be omitted with no loss of continuity.

each of the n polynomials $(x - a), (x - a)^2, \ldots, (x - a)^n$ clearly lies in ker E_a, so they are a basis because they are linearly independent (they have distinct degrees).

◆◆◆

We conclude by applying the dimension theorem to the rank of a matrix.

EXAMPLE 13

If A is any $m \times n$ matrix, show that rank A = rank $A^T A$ = rank AA^T.

Solution   It suffices to show that rank A = rank $A^T A$ (the rest follows by replacing A with A^T). Write $B = A^T A$, and consider the associated matrix transformations

$$T_A : \mathbb{R}^n \to \mathbb{R}^m \quad \text{and} \quad T_B : \mathbb{R}^n \to \mathbb{R}^n$$

The dimension theorem and Example 3 give

$$\text{rank } A = \text{rank } T_A = \dim(\text{im } T_A) = n - \dim(\text{ker } T_A)$$
$$\text{rank } B = \text{rank } T_B = \dim(\text{im } T_B) = n - \dim(\text{ker } T_B)$$

so it suffices to show that ker T_A = ker T_B. Now $AX = 0$ implies that $BX = A^T AX = 0$, so ker T_A is contained in ker T_B. On the other hand, if $BX = 0$, then $A^T AX = 0$, so

$$\|AX\|^2 = (AX)^T(AX) = X^T A^T AX = X^T 0 = 0$$

This implies that $AX = 0$, so ker T_B is contained in ker T_A.

◆◆◆

EXERCISES 7.2

1. For each matrix A, find a basis for the kernel and image of T_A, and find the rank and nullity of T_A.

(a) $\begin{bmatrix} 1 & 2 & -1 & 1 \\ 3 & 1 & 0 & 2 \\ 1 & -3 & 2 & 0 \end{bmatrix}$ ◆(b) $\begin{bmatrix} 2 & 1 & -1 & 3 \\ 1 & 0 & 3 & 1 \\ 1 & 1 & -4 & 2 \end{bmatrix}$

(c) $\begin{bmatrix} 1 & 2 & -1 \\ 3 & 1 & 2 \\ 4 & -1 & 5 \\ 0 & 2 & -2 \end{bmatrix}$ ◆(d) $\begin{bmatrix} 2 & 1 & 0 \\ 1 & -1 & 3 \\ 1 & 2 & -3 \\ 0 & 3 & -6 \end{bmatrix}$

2. In each case, (i) find a basis of ker T, and (ii) find a basis of im T.

(a) $T : \mathbf{P}_2 \to \mathbb{R}^2$; $T(a + bx + cx^2) = (a, b)$

◆(b) $T : \mathbf{P}_2 \to \mathbb{R}^2$; $T(p(x)) = (p(0), p(1))$

(c) $T : \mathbb{R}^3 \to \mathbb{R}^3$; $T(x, y, z) = (x + y, x + y, 0)$

◆(d) $T : \mathbb{R}^3 \to \mathbb{R}^4$; $T(x, y, z) = (x, x, y, y)$

(e) $T : \mathbf{M}_{22} \to \mathbf{M}_{22}$; $T\begin{bmatrix} a & b \\ c & d \end{bmatrix} = \begin{bmatrix} a + b & b + c \\ c + d & d + a \end{bmatrix}$

◆(f) $T : \mathbf{M}_{22} \to \mathbb{R}$; $T\begin{bmatrix} a & b \\ c & d \end{bmatrix} = a + d$

(g) $T : \mathbf{P}_n \to \mathbb{R}$; $T(r_0 + r_1 x + \cdots + r_n x^n) = r_n$

◆(h) $T : \mathbb{R}^n \to \mathbb{R}$; $T(r_1, r_2, \ldots, r_n) = r_1 + r_2 + \cdots + r_n$

(i) $T : \mathbf{M}_{22} \to \mathbf{M}_{22}$; $T(X) = XA - AX$, where $A = \begin{bmatrix} 0 & 1 \\ 1 & 0 \end{bmatrix}$

◆(j) $T : \mathbf{M}_{22} \to \mathbf{M}_{22}$; $T(X) = XA$, where $A = \begin{bmatrix} 1 & 1 \\ 0 & 0 \end{bmatrix}$

3. Let $P : V \to \mathbb{R}$ and $Q : V \to \mathbb{R}$ be linear transformations, where V is a vector space. Define $T : V \to \mathbb{R}^2$ by $T(\mathbf{v}) = (P(\mathbf{v}), Q(\mathbf{v}))$

(a) Show that T is a linear transformation.

◆(b) Show that ker T = ker $P \cap$ ker Q, the set of vectors in both ker P and ker Q.

4. In each case, find a basis $B = \{\mathbf{e}_1, \ldots, \mathbf{e}_r, \mathbf{e}_{r+1}, \ldots, \mathbf{e}_n\}$ of V such that $\{\mathbf{e}_{r+1}, \ldots, \mathbf{e}_n\}$ is a basis of ker T, and verify Theorem 5.

(a) $T : \mathbb{R}^3 \to \mathbb{R}^4$; $T(x, y, z) = (x - y + 2z, x + y - z, 2x + z, 2y - 3z)$

◆**(b)** $T : \mathbb{R}^3 \to \mathbb{R}^4$; $T(x, y, z) = (x + y + z, 2x - y + 3z, z - 3y, 3x + 4z)$

5. Show that every matrix X in $\mathbf{M}_{nn}$ has the form $X = A^T - 2A$ for some matrix A in $\mathbf{M}_{nn}$. [*Hint:* The dimension theorem.]

6. In each case either prove the statement or give an example in which it is false. Throughout, let $T : V \to W$ be a linear transformation where V and W are finite dimensional.
   (a) If $V = W$, then $\ker T \subseteq \operatorname{im} T$.
   ◆**(b)** If $\dim V = 5$, $\dim W = 3$, and $\dim(\ker T) = 2$, then T is onto.
   (c) If $\dim V = 5$ and $\dim W = 4$, then $\ker T \neq 0$.
   ◆**(d)** If $\ker T = V$, then $W = 0$.
   (e) If $W = 0$, then $\ker T = V$.
   ◆**(f)** If $W = V$, and $\operatorname{im} T \subseteq \ker T$, then $T = 0$.
   (g) If $\{e_1, e_2, e_3\}$ is a basis of V and $T(e_1) = \mathbf{0} = T(e_2)$, then $\dim(\operatorname{im} T) \leq 1$.
   ◆**(h)** If $\dim(\ker T) \leq \dim W$, then $\dim W \geq \frac{1}{2}\dim V$.
   (i) If T is one-to-one, then $\dim V \leq \dim W$.
   ◆**(j)** If $\dim V \leq \dim W$, then T is one-to-one.
   (k) If T is onto, then $\dim V \geq \dim W$.
   ◆**(l)** If $\dim V \geq \dim W$, then T is onto.

7. Show that linear independence is preserved by one-to-one transformations and that spanning sets are preserved by onto transformations. More precisely, if $T : V \to W$ is a linear transformation, show that:
   (a) If T is one-to-one and $\{v_1, \ldots, v_n\}$ is linearly independent in V, then $\{T(v_1), \ldots, T(v_n)\}$ is linearly independent in W.
   ◆**(b)** If T is onto and $V = \operatorname{span}\{v_1, \ldots, v_n\}$, then $W = \operatorname{span}\{T(v_1), \ldots, T(v_n)\}$.

8. Given $\{v_1, \ldots, v_n\}$ in a vector space V, define $T : \mathbb{R}^n \to V$ by $T(r_1, \ldots, r_n) = r_1 v_1 + \cdots + r_n v_n$. Show that T is linear, and that:
   (a) T is one-to-one if and only if $\{v_1, \ldots, v_n\}$ is linearly independent.
   ◆**(b)** T is onto if and only if $V = \operatorname{span}\{v_1, \ldots, v_n\}$.

9. Let $T : V \to V$ be a linear transformation in which V is finite dimensional. Show that exactly one of **(i)** and **(ii)** holds: **(i)** $T(v) = \mathbf{0}$ for some $v \neq \mathbf{0}$ in V; **(ii)** $T(x) = v$ has a solution x in V for every v in V.

◆**10.** Let $T : \mathbf{M}_{nn} \to \mathbb{R}$ denote the trace map: $T(A) = \operatorname{tr} A$ for all A in $\mathbf{M}_{nn}$. Show that $\dim(\ker T) = n^2 - 1$.

11. Show that the following are equivalent for a linear transformation $T : V \to W$.
   (a) $\ker T = V$ **(b)** $\operatorname{im} T = 0$ **(c)** $T = 0$

12. Let A and B be $m \times n$ and $k \times n$ matrices, respectively. Assume that $AX = 0$ implies $BX = 0$ for every n-column X. Show that rank $A \geq$ rank B.

13. Let A be an $m \times n$ matrix of rank r. Define $V = \{X$ in $\mathbb{R}^m \mid XA = 0\}$. Show that $\dim V = m - r$.

14. Consider $V = \left\{ \begin{bmatrix} a & b \\ c & d \end{bmatrix} \,\middle|\, a + c = b + d \right\}$.
   (a) Consider $S : \mathbf{M}_{22} \to \mathbb{R}$ with $S\begin{bmatrix} a & b \\ c & d \end{bmatrix} = a + c - b - d$. Show that S is linear and onto and that V is a subspace of $\mathbf{M}_{22}$, and compute $\dim V$.
   (b) Consider $T : V \to \mathbb{R}$ with $T\begin{bmatrix} a & b \\ c & d \end{bmatrix} = a + c$. Show that T is linear and onto, and use this information to compute $\dim(\ker T)$.

15. Define $T : \mathbf{P}_n \to \mathbb{R}$ by $T[p(x)] = $ the sum of all the coefficients of $p(x)$.
   (a) Use the dimension theorem to show that $\dim(\ker T) = n$.
   ◆**(b)** Conclude that $\{x - 1, x^2 - 1, \ldots, x^n - 1\}$ is a basis of $\ker T$.

16. Use the dimension theorem to prove Theorem 1§1.3: If A is an $m \times n$ matrix with $m < n$, the system $AX = 0$ of m homogeneous equations in n variables always has a non-trivial solution.

17. Let B be an $n \times n$ matrix, and consider the subspaces $U = \{A \mid A$ in $\mathbf{M}_{mn}, AB = 0\}$ and $V = \{AB \mid A$ in $\mathbf{M}_{mn}\}$. Show that $\dim U + \dim V = mn$.

18. Let U and V denote, respectively, the spaces of even and odd polynomials in $\mathbf{P}_n$. Show that $\dim U + \dim V = n + 1$. [*Hint:* Consider $T : \mathbf{P}_n \to \mathbf{P}_n$ where $T(p) = p(x) - p(-x)$.]

19. Show that every polynomial $f(x)$ in $\mathbf{P}_{n-1}$ can be written as $f(x) = p(x + 1) - p(x)$ for some polynomial $p(x)$ in $\mathbf{P}_n$. [*Hint:* Define $T : \mathbf{P}_n \to \mathbf{P}_{n-1}$ by $T[p(x)] = p(x + 1) - p(x)$.]

20. Let U and V denote the spaces of symmetric and skew-symmetric $n \times n$ matrices. Show that $\dim U + \dim V = n^2$.

21. Assume that B in $\mathbf{M}_{nn}$ satisfies $B^k = 0$ for some $k \geq 1$. Show that every matrix in $\mathbf{M}_{nn}$ has the form $BA - A$ for some A in $\mathbf{M}_{nn}$. [*Hint:* Show that $T : \mathbf{M}_{nn} \to \mathbf{M}_{nn}$ is linear and one-to-one where $T(A) = BA - A$ for each A.]

◆**22.** Fix a column $Y \neq 0$ in $\mathbb{R}^n$ and let $U = \{A$ in $\mathbf{M}_{nn} \mid AY = 0\}$. Show that $\dim U = m(n - 1)$.

23. If B in $\mathbf{M}_{mn}$ has rank r, let $U = \{A \text{ in } \mathbf{M}_{nn} \mid BA = 0\}$ and $W = \{BA \mid A \text{ in } \mathbf{M}_{nn}\}$. Show that dim $U = n(n - r)$ and dim $W = nr$. [*Hint:* Show that U consists of all matrices A whose columns are in the null space of B. Use Example 9.]

24. Let $T : V \to V$ be a linear transformation where dim $V = n$. If $\mathbf{0}$ is the only vector in both ker T and im T, show that every vector $\mathbf{v}$ in V can be written $\mathbf{v} = \mathbf{u} + \mathbf{w}$ for some $\mathbf{u}$ in ker T and $\mathbf{w}$ in im T. [*Hint:* Exercise 34§5.3.]

25. If $T : \mathbb{R}^n \to \mathbb{R}^n$ is a linear transformation of rank 1, show that there exist numbers $a_1, a_2, \ldots, a_n$ and $b_1, b_2, \ldots, b_n$ such that $T(X) = XA$ for all rows X in $\mathbb{R}^n$, where

$$A = \begin{bmatrix} a_1b_1 & a_1b_2 & \cdots & a_1b_n \\ a_2b_1 & a_2b_2 & \cdots & a_2b_n \\ \vdots & \vdots & & \vdots \\ a_nb_1 & a_nb_2 & \cdots & a_nb_n \end{bmatrix}$$

[*Hint:* im $T = \mathbb{R}\mathbf{w}$ for some $\mathbf{w} = (b_1, \ldots, b_n)$ in $\mathbb{R}^n$.]

26. Prove Theorem 5.

27. Let $T : V \to \mathbb{R}$ be a nonzero linear transformation, where dim $V = n$. Show that there is a basis $\{\mathbf{e}_1, \ldots, \mathbf{e}_n\}$ of V such that $T(r_1\mathbf{e}_1 + r_2\mathbf{e}_2 + \cdots + r_n\mathbf{e}_n) = r_1$.

28. Let U be a subspace of a finite dimensional vector space V.
  (a) Show that $U = \ker T$ for some linear transformation $T : V \to V$.
  (b) Show that $U = \operatorname{im} S$ for some linear transformation $S : V \to V$. [*Hint:* Theorems 4§5.4 and 4§7.1.]

29. Let V and W be finite dimensional vector spaces.
  (a) Show that dim $W \le$ dim V if and only if there exists an onto linear transformation $T : V \to W$. [*Hint:* Theorems 4§5.4 and 4§7.1.]
  (b) Show that dim $W \ge$ dim V if and only if there exists a one-to-one linear transformation $T : V \to W$. [*Hint:* Theorems 4§5.4 and 4§7.1.]

Section 7.3   Isomorphisms and Composition

Often two vector spaces can look quite different but, on closer examination, turn out to be the same vector space displayed in different symbols. The notion of isomorphism clarifies this.

> **DEFINITION**
>
> A linear transformation $T : V \to W$ is called an **isomorphism** if it is both onto and one-to-one. The vector spaces V and W are called **isomorphic** if there exists an isomorphism $T : V \to W$.

EXAMPLE 1

The identity transformation $1_V : V \to V$ is an isomorphism for any vector space V.

◆◆◆

EXAMPLE 2

If $T : \mathbf{M}_{mn} \to \mathbf{M}_{nm}$ is defined by $T(A) = A^T$ for all A in $\mathbf{M}_{mn}$, then T is an isomorphism.

◆◆◆

EXAMPLE 3

If U is any invertible $m \times m$ matrix, the map $T : \mathbf{M}_{mn} \to \mathbf{M}_{mn}$ given by $T(X) = UX$ is an isomorphism by Example 8§7.2.

◆◆◆

The word *isomorphism* comes from two Greek roots: *iso*, meaning "same," and *morphos*, meaning "form." The isomorphism T induces a pairing

$$\mathbf{v} \leftrightarrow T(\mathbf{v})$$

between vectors $\mathbf{v}$ in V and vectors $T(\mathbf{v})$ in W that preserves vector addition and scalar multiplication. Hence, *as far as their vector space properties are concerned*, the spaces V and W are identical except for notation. Because addition and scalar multiplication in either space are completely determined by the same operations in the other space, all *vector space* properties of either space are completely determined by those of the other.

We considered one of the most important examples of isomorphic spaces in Chapter 4. There the space $\mathbb{R}^3$ was *identified* with the space of geometric vectors by pairing each 3-tuple (x, y, z) with the "arrow" $\mathbf{v}$ from the origin to the point $P(x, y, z)$ — called the *position vector* of the point P. These arrows form a vector space using the parallelogram law of vector addition and the scalar multiplication described in Section 4.1. The remarkable thing is that the function T from $\mathbb{R}^3$ to this space given by $T(x, y, z) = \mathbf{v}$ is an isomorphism (this is Theorem 2§4.1). This fact justifies the *identification*

$$\mathbf{v} = (x, y, z)$$

of the geometric arrows with the algebraic 3-tuples that was made in Chapter 4. This identification is very useful. The arrows give a picture of the vectors and so bring geometric intuition into $\mathbb{R}^3$; the 3-tuples are useful for doing detailed calculations and so bring analytic power into geometry. This is one of the best examples of the power of an isomorphism to shed light on *both* spaces being considered.

The following theorem gives a very useful characterization of isomorphisms: They are the linear transformations that preserve bases.

THEOREM 1

If V and W are finite dimensional spaces, the following conditions are equivalent for a linear transformation $T : V \to W$.

1. T is an isomorphism.

2. If $\{\mathbf{e}_1, \mathbf{e}_2, \ldots, \mathbf{e}_n\}$ is any basis of V, then $\{T(\mathbf{e}_1), T(\mathbf{e}_2), \ldots, T(\mathbf{e}_n)\}$ is a basis of W.

3. There exists a basis $\{\mathbf{e}_1, \mathbf{e}_2, \ldots, \mathbf{e}_n\}$ of V such that $\{T(\mathbf{e}_1), T(\mathbf{e}_2), \ldots, T(\mathbf{e}_n)\}$ is a basis of W.

Proof

(1) *implies* (2). Let $\{\mathbf{e}_1, \ldots, \mathbf{e}_n\}$ be a basis of V. If $t_1 T(\mathbf{e}_1) + \cdots + t_n T(\mathbf{e}_n) = \mathbf{0}$ with t_i in $\mathbb{R}$, then $T[t_1 \mathbf{e}_1 + \cdots + t_n \mathbf{e}_n] = \mathbf{0}$, so $t_1 \mathbf{e}_1 + \cdots + t_n \mathbf{e}_n = \mathbf{0}$ (because ker

$T = 0$). But then each $t_i = 0$ by the independence of the $\mathbf{e}_i$, so $\{T(\mathbf{e}_1), \ldots, T(\mathbf{e}_n)\}$ is linearly independent. To show that it spans W, choose $\mathbf{w}$ in W. Because T is onto, $\mathbf{w} = T(\mathbf{v})$ for some $\mathbf{v}$ in V, so write $\mathbf{v} = t_1\mathbf{e}_1 + \cdots + t_n\mathbf{e}_n$. Then $\mathbf{w} = T(\mathbf{v}) = t_1T(\mathbf{e}_1) + \cdots + t_nT(\mathbf{e}_n)$, so $\{T(\mathbf{e}_1), \ldots, T(\mathbf{e}_n)\}$ spans W.

(2) *implies* (3). This is because V *has* a basis.

(3) *implies* (1). If $T(\mathbf{v}) = \mathbf{0}$, write $\mathbf{v} = v_1\mathbf{e}_1 + \cdots + v_n\mathbf{e}_n$. Then $\mathbf{0} = T(\mathbf{v}) = v_1T(\mathbf{e}_1) + \cdots + v_nT(\mathbf{e}_n)$, so $v_1 = \cdots = v_n = 0$ by (3). Hence $\mathbf{v} = \mathbf{0}$, so $\ker T = 0$ and T is one-to-one. To show that T is onto, let $\mathbf{w}$ be any vector in W. By (3) there exist $w_1, \ldots, w_n$ in $\mathbb{R}$ such that $\mathbf{w} = w_1T(\mathbf{e}_1) + \cdots + w_nT(\mathbf{e}_n) = T(w_1\mathbf{e}_1 + \cdots + w_n\mathbf{e}_n)$. Thus T is onto. ◆

Theorem 1 dovetails nicely with Theorem 4§7.1 as follows. Let V and W be vector spaces of dimension n, and suppose that $\{\mathbf{e}_1, \mathbf{e}_2, \ldots, \mathbf{e}_n\}$ and $\{\mathbf{f}_1, \mathbf{f}_2, \ldots, \mathbf{f}_n\}$ are bases of V and W, respectively. Theorem 4§7.1 asserts that there exists a linear transformation $T : V \to W$ such that

$$T(\mathbf{e}_i) = \mathbf{f}_i \quad \text{for each } i = 1, 2, \ldots, n$$

Then $\{T(\mathbf{e}_1), \ldots, T(\mathbf{e}_n)\}$ is evidently a basis of W, so T is an isomorphism by Theorem 1. Furthermore, the action of T is prescribed by

$$T(r_1\mathbf{e}_1 + \cdots + r_n\mathbf{e}_n) = r_1\mathbf{f}_1 + \cdots + r_n\mathbf{f}_n$$

so isomorphisms between spaces of equal dimension can be easily defined as soon as bases are known. In particular, we have proved half of the following theorem.

THEOREM 2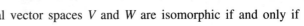

Two finite dimensional vector spaces V and W are isomorphic if and only if $\dim V = \dim W$.

Proof   It remains to show that if $T : V \to W$ is an isomorphism, then $\dim V = \dim W$. But if $\{\mathbf{e}_1, \ldots, \mathbf{e}_n\}$ is a basis of V, then $\{T(\mathbf{e}_1), \ldots, T(\mathbf{e}_n)\}$ is a basis of W by Theorem 1, so $\dim W = n = \dim V$. ◆

In particular, every vector space of dimension n is isomorphic to $\mathbb{R}^n$.

EXAMPLE 4

Let V denote the space of all 2×2 symmetric matrices. Find an isomorphism $T : \mathbf{P}_2 \to V$ such that $T(1) = I$.

Solution $\{1, x, x^2\}$ is a basis of $\mathbf{P}_2$, and we want a basis of V containing I. The set $\left\{ \begin{bmatrix} 1 & 0 \\ 0 & 1 \end{bmatrix}, \begin{bmatrix} 0 & 1 \\ 1 & 0 \end{bmatrix}, \begin{bmatrix} 0 & 0 \\ 0 & 1 \end{bmatrix} \right\}$ is independent in V, so it is a basis because $\dim V = 3$

(by Example 13§5.3). Hence define $T: \mathbf{P}_2 \to V$ by taking $T(1) = \begin{bmatrix} 1 & 0 \\ 0 & 1 \end{bmatrix}$, $T(x)$ $= \begin{bmatrix} 0 & 1 \\ 1 & 0 \end{bmatrix}$, and $T(x^2) = \begin{bmatrix} 0 & 0 \\ 0 & 1 \end{bmatrix}$ and extending linearly as in Theorem 4§7.1. Then T is an isomorphism by Theorem 1, and its action is given by $T(a + bx + cx^2) = aT(1)$ $+ bT(x) + cT(x^2) = \begin{bmatrix} a & b \\ b & a + c \end{bmatrix}$.

$\blacklozenge \blacklozenge \blacklozenge$

The dimension theorem (Theorem 4§7.2) gives the following useful fact about isomorphisms.

THEOREM 3

If dim V = dim $W = n$, a linear transformation $T : V \to W$ is an isomorphism if it is either one-to-one or onto.

Proof The dimension theorem asserts that $\dim(\ker T) + \dim(\operatorname{im} T) = n$, so $\dim(\ker T) = 0$ if and only if $\dim(\operatorname{im} T) = n$. Thus T is one-to-one if and only if T is onto, and the result follows. $\blacklozenge$

Composition

Suppose that $T : V \to W$ and $S : W \to U$ are linear transformations. They link together as in Figure 7.6, so it is possible to define a new function $V \to U$ by first applying T and then S.

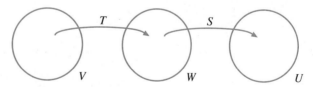

FIGURE 7.6

DEFINITION

Given linear transformations $V \xrightarrow{T} W \xrightarrow{S} U$, the **composite** $ST : V \to U$ of T and S is defined by

$$ST(\mathbf{v}) = S[T(\mathbf{v})] \qquad \text{for all } \mathbf{v} \text{ in } V$$

The operation of forming the new function ST is called **composition.**

The action of ST can be described compactly as follows: ST means first T then S. (Incidentally, some authors write $S \circ T$ in place of ST, but we shall stick to the simpler notation.)

EXAMPLE 5

Let $T : \mathbb{R}^3 \to \mathbb{R}^2$ and $S : \mathbb{R}^2 \to \mathbb{R}^4$ be defined by $T(x, y, z) = (x + y, y + z)$ and $S(x, y) = (x - y, x + y, y, x)$. Describe the action of ST.

Solution   Given (x, y, z) in $\mathbb{R}^3$, the definition yields

$$ST(x, y, z) = S[T(x, y, z)] = S(x + y, y + z)$$
$$= (x - z, x + 2y + z, y + z, x + y)$$

This describes $ST(x, y, z)$ for all (x, y, z) in $\mathbb{R}^3$.

◆◆◆

Not all pairs of linear transformations can be composed. In fact, if S and T are the transformations in Example 5, then ST is defined as was shown, but TS *cannot* be formed because

$$\mathbb{R}^2 \xrightarrow{S} \mathbb{R}^4 \quad \text{and} \quad \mathbb{R}^3 \xrightarrow{T} \mathbb{R}^2$$

do not "link" in this order. Moreover, even if ST and TS *can* both be formed, the new functions ST and TS need not be equal.

EXAMPLE 6

Define: $S : \mathbf{M}_{22} \to \mathbf{M}_{22}$ and $T : \mathbf{M}_{22} \to \mathbf{M}_{22}$ by $S\begin{bmatrix} a & b \\ c & d \end{bmatrix} = \begin{bmatrix} c & d \\ a & b \end{bmatrix}$ and $T(A) = A^T$.

Describe the action of ST and TS, and show that $ST \neq TS$.

Solution $ST\begin{bmatrix} a & b \\ c & d \end{bmatrix} = S\begin{bmatrix} a & c \\ b & d \end{bmatrix} = \begin{bmatrix} b & d \\ a & c \end{bmatrix}$, whereas $TS\begin{bmatrix} a & b \\ c & d \end{bmatrix} = T\begin{bmatrix} c & d \\ a & b \end{bmatrix} = \begin{bmatrix} c & a \\ d & b \end{bmatrix}$. It is

clear that $TS\begin{bmatrix} a & b \\ c & d \end{bmatrix}$ need not equal $ST\begin{bmatrix} a & b \\ c & d \end{bmatrix}$, so $TS \neq ST$.

◆◆◆

The next theorem collects some basic properties of the composition operation.

THEOREM 4

Let $V \xrightarrow{T} W \xrightarrow{S} U \xrightarrow{R} Z$ be linear transformations.

1. The composite ST is again a linear transformation.
2. $T1_V = T$ and $1_W T = T$
3. $(RS)T = R(ST)$

Proof   To prove (3), observe that, for all **v** in V:

$$\{(RS)T\}(\mathbf{v}) = (RS)\{T(\mathbf{v})\} + R\{S[T(\mathbf{v})]\} = R\{(ST)(\mathbf{v})\} = \{R(ST)\}(\mathbf{v})$$

The proofs of (1) and (2) are left as Exercise 28.   ◆

Up to this point composition seems to have no connection with isomorphisms. In fact the two notions are closely related.

THEOREM 5

Let V and W be finite dimensional vector spaces. The following conditions are equivalent for a linear transformation $T : V \to W$.

1. T is an isomorphism.
2. There exists a linear transformation $S : W \to V$ such that $ST = 1_V$ and $TS = 1_W$.

Moreover, in this case S is also an isomorphism and is uniquely determined by T: If **w** in W is written as $\mathbf{w} = T(\mathbf{v})$, then $S(\mathbf{w}) = \mathbf{v}$.

Proof

(1) *implies* (2). If $B = \{\mathbf{e}_1, \ldots, \mathbf{e}_n\}$ is a basis of V, then $D = \{T(\mathbf{e}_1), \ldots, T(\mathbf{e}_n)\}$ is a basis of W by Theorem 1. Hence (using Theorem 4§7.1), define $S : W \to V$ by

$$S[T(\mathbf{e}_i)] = \mathbf{e}_i \qquad \text{for each } i \qquad\qquad (*)$$

This gives $ST = 1_V$ by Theorem 3§7.1. But applying T gives $T[S[T(\mathbf{e}_i)]] = T(\mathbf{e}_i)$ for each i, so $TS = 1_W$ (again by Theorem 3§7.1, using the basis D of W).

(2) *implies* (1). If $T(\mathbf{v}) = T(\mathbf{v}_1)$, then $S[T(\mathbf{v})] = S[T(\mathbf{v}_1)]$. Because $ST = 1_V$, this reads $\mathbf{v} = \mathbf{v}_1$; that is, T is one-to-one. Given **w** in W, the fact that $TS = 1_W$ means that $\mathbf{w} = T[S(\mathbf{w})]$, and T is onto.

S is uniquely determined by the condition $ST = 1_V$ because this condition implies $(*)$. S is an isomorphism because it carries the basis D to B. Finally, given **w** in W, write $\mathbf{w} = r_1 T(\mathbf{e}_1) + \cdots + r_n T(\mathbf{e}_n) = T(\mathbf{v})$, where $\mathbf{v} = r_1 \mathbf{e}_1 + \cdots + r_n \mathbf{e}_n$. Then $S(\mathbf{w}) = \mathbf{v}$ by $(*)$.   ◆

DEFINITION

Given an isomorphism $T : V \to W$, the isomorphism $S : W \to V$ satisfying condition (2) of Theorem 5 is called the **inverse** of T and is denoted by T^{-1}.

Hence $T : V \to W$ and $T^{-1} : W \to V$ are related by the **fundamental identities**:

$$T^{-1}[T(\mathbf{v})] = \mathbf{v} \text{ for all } \mathbf{v} \text{ in } V \qquad \text{and} \qquad T[T^{-1}(\mathbf{w})] = \mathbf{w} \text{ for all } \mathbf{w} \text{ in } W$$

In other words, each of T and T^{-1} reverses the action of the other. In particular, equation (∗) in the proof of Theorem 5 shows how to define T^{-1} using the image of a basis under the isomorphism T. Here is an example.

EXAMPLE 7

Define $T : P_1 \to P_1$ by $T(a + bx) = (a - b) + ax$. Show that T has an inverse, and find the action of T^{-1}.

Solution Because $T(1) = 1 + x$ and $T(x) = -1$, T carries the basis $B = \{1, x\}$ to the basis $D = \{1 + x, -1\}$. Hence T is an isomorphism, and T^{-1} is defined by $T^{-1}(1 + x) = 1$ and $T^{-1}(-1) = x$. Because $a + bx = b(1 + x) + (b - a)(-1)$, we obtain

$$T^{-1}(a + bx) = bT^{-1}(1 + x) + (b - a)T^{-1}(x) = b + (b - a)x.$$

Condition (2) in Theorem 5 characterizes the inverse of a linear transformation $T : V \to W$ as the (unique) transformation $S : W \to V$ that satisfies $ST = 1_V$ and $TS = 1_W$. This often determines the inverse.

EXAMPLE 8

Define $T : \mathbb{R}^3 \to \mathbb{R}^3$ by $T(x, y, z) = (z, x, y)$. Show that $T^3 = 1_{\mathbb{R}^3}$, and hence find T^{-1}.

Solution $T^2(x, y, z) = T[T(x, y, z)] = T(z, x, y) = (y, z, x)$. Hence

$$T^3(x, y, z) = T[T^2(x, y, z)] = T(y, z, x) = (x, y, z)$$

This shows that $T^3 = 1_{\mathbb{R}^3}$, so $T(T^2) = 1_{\mathbb{R}^3} = (T^2)T$. Thus $T^{-1} = T^2$ by (2) of Theorem 5.

EXAMPLE 9

Define $T : P_n \to \mathbb{R}^{n+1}$ by $T(p) = (p(0), p(1), \ldots, p(n))$ for all p in P_n. Show that T^{-1} exists.

Solution The verification that T is linear is left to the reader. If $T(p) = 0$, then $p(k) = 0$ for $k = 0, 1, \ldots, n$, so p has $n + 1$ distinct roots. Because p has degree at most n, this implies that $p = 0$ is the zero polynomial (corollary to Theorem 3§5.6) and hence that T is one-to-one. But $\dim P_n = n + 1 = \dim \mathbb{R}^{n+1}$, so this means that T is also onto and hence is an isomorphism. Thus T^{-1} exists by Theorem 5. Note that we have not given an explicit description of the action of T^{-1}. We have merely shown that such a description exists. To give it requires some ingenuity; one method involves the Lagrange interpolation formula (Theorem 3§5.6).

Of course all the general results in this section apply to matrix transformations. Several such results are collected in Theorem 6 for later reference.

THEOREM 6

Let A and B denote matrices.

1. If $T_A = T_B$, then $A = B$.
2. $T_{I_n} = 1_{\mathbb{R}^n}$
3. $T_A T_B = T_{AB}$
4. T_A has an inverse if and only if A is invertible, and then $(T_A)^{-1} = T_{A^{-1}}$.
5. $\operatorname{im} T_A = \operatorname{col} A$, so $\operatorname{rank} T_A = \operatorname{rank} A$.

Proof If E_j is column j of I_n, then $T_A(E_j) = AE_j$ is column j of A. This implies (1), (2) is obvious, and (5) is Example 3§7.2. Property (3) follows from the fact that

$$T_A[T_B(X)] = A[BX] = (AB)X = T_{AB}(X)$$

holds for all columns X. To prove (4), assume first that A^{-1} exists. Then $T_A T_{A^{-1}} = T_{AA^{-1}} = T_I$ is the identity map, using (2) and (3). Similarly $T_{A^{-1}}T_A = T_I$, so $T_{A^{-1}}$ is the inverse of T_A by Theorem 5. Conversely, assume that A is $n \times n$ and that $(T_A)^{-1} : \mathbb{R}^n \to \mathbb{R}^n$ exists. Then $(T_A)^{-1} = T_B$ for some $n \times n$ matrix B (Theorem 5§7.1), and so (2) and (3) give

$$T_{AB} = T_A T_B = T_A(T_A)^{-1} = 1_{\mathbb{R}^n} = T_I$$

Hence $AB = I$ by (1). Similarly, $BA = I$, so $B = A^{-1}$.   ◆

EXERCISES 7.3

1. Verify that each of the following is an isomorphism (Theorem 3 is useful).
   (a) $T : \mathbb{R}^3 \to \mathbb{R}^3$; $T(x, y, z) = (x + y, y + z, z + x)$
   ◆(b) $T : \mathbb{R}^3 \to \mathbb{R}^3$; $T(x, y, z) = (x, x + y, x + y + z)$
   (c) $T : \mathbb{C} \to \mathbb{C}$; $T(z) = \overline{z}$
   ◆(d) $T : \mathbf{M}_{mn} \to \mathbf{M}_{mn}$; $T(X) = UXV$, U and V invertible
   (e) $T : \mathbf{P}_1 \to \mathbb{R}^2$; $T[p(x)] = [p(0), p(1)]$
   ◆(f) $T : V \to V$; $T(\mathbf{v}) = k\mathbf{v}$, $k \neq 0$ a fixed number, V any vector space
   (g) $T : \mathbf{M}_{22} \to \mathbb{R}^4$; $T\begin{bmatrix} a & b \\ c & d \end{bmatrix} = (a + b, d, c, a - b)$
   ◆(h) $T : \mathbf{M}_{mn} \to \mathbf{M}_{nm}$; $T(A) = A^T$

2. Show that $\{a + bx + cx^2, a_1 + b_1 x + c_1 x^2, a_2 + b_2 x + c_2 x^2\}$ is a basis of $\mathbf{P}_2$ if an only if $\{(a, b, c), (a_1, b_1, c_1), (a_2, b_2, c_2)\}$ is a basis of $\mathbb{R}^3$.

3. If V is any vector space, let V^n denote the space of all n-tuples $(\mathbf{v}_1, \mathbf{v}_2, \ldots, \mathbf{v}_n)$, where each $\mathbf{v}_i$ lies in V. (This is a vector space with componentwise operations; see Exercise 24§5.1.) If $C_j(A)$ denotes the jth column of the $m \times n$ matrix A, show that $T : \mathbf{M}_{mn} \to (\mathbb{R}^m)^n$ is an isomorphism if $T(A) = [C_1(A) \quad C_2(A) \quad \cdots \quad C_n(A)]$. (Here $\mathbb{R}^m$ consists of columns.)

4. In each case, compute the action of ST and TS, and show that $ST \neq TS$.
   (a) $S : \mathbb{R}^2 \to \mathbb{R}^2$ with $S(x, y) = (y, x)$; $T : \mathbb{R}^2 \to \mathbb{R}^2$ with $T(x, y) = (x, 0)$
   ◆(b) $S : \mathbb{R}^3 \to \mathbb{R}^3$ with $S(x, y, z) = (x, 0, z)$; $T : \mathbb{R}^3 \to \mathbb{R}^3$ with $T(x, y, z) = (x + y, 0, y + z)$
   (c) $S : \mathbf{P}_2 \to \mathbf{P}_2$ with $S(p) = p(0) + p(1)x + p(2)x^2$; $T : \mathbf{P}_2 \to \mathbf{P}_2$ with $T(a + bx + cx^2) = b + cx + ax^2$
   ◆(d) $S : \mathbf{M}_{22} \to \mathbf{M}_{22}$ with $S\begin{bmatrix} a & b \\ c & d \end{bmatrix} = \begin{bmatrix} a & 0 \\ 0 & d \end{bmatrix}$; $T : \mathbf{M}_{22} \to \mathbf{M}_{22}$ with $T\begin{bmatrix} a & b \\ c & d \end{bmatrix} = \begin{bmatrix} c & a \\ d & b \end{bmatrix}$

5. In each case, show that the linear transformation T satisfies $T^2 = T$.
   (a) $T : \mathbb{R}^4 \to \mathbb{R}^4$; $T(x, y, z, w) = (x, 0, z, 0)$
   ◆(b) $T : \mathbb{R}^2 \to \mathbb{R}^2$; $T(x, y) = (x + y, 0)$

(c) $T : P_2 \rightarrow P_2$; $T(a + bx + cx^2) = (a + b - c) + cx + cx^2$

◆**(d)** $T : M_{22} \rightarrow M_{22}$; $T\begin{bmatrix} a & b \\ c & d \end{bmatrix} = \frac{1}{2}\begin{bmatrix} a + c & b + d \\ a + c & b + d \end{bmatrix}$

6. Determine whether each of the following transformations T has an inverse and, if so, determine the action of T^{-1}.

(a) $T : \mathbb{R}^3 \rightarrow \mathbb{R}^3$; $T(x, y, z) = (x + y, y + z, z + x)$

◆**(b)** $T : \mathbb{R}^4 \rightarrow \mathbb{R}^4$; $T(x, y, z, t) = (x + y, y + z, z + t, t + x)$

(c) $T : M_{22} \rightarrow M_{22}$; $T\begin{bmatrix} a & b \\ c & d \end{bmatrix} = \begin{bmatrix} a - c & b - d \\ 2a - c & 2b - d \end{bmatrix}$

◆**(d)** $T : M_{22} \rightarrow M_{22}$; $T\begin{bmatrix} a & b \\ c & d \end{bmatrix} = \begin{bmatrix} a + 2c & b + 2d \\ 3c - a & 3d - b \end{bmatrix}$

(e) $T : P_2 \rightarrow \mathbb{R}^3$; $T(a + bx + cx^2) = (a - c, 2b, a + c)$

◆**(f)** $T : P_2 \rightarrow \mathbb{R}^3$; $T(p) = [p(0), p(1), p(-1)]$

7. In each case, show that T is self-inverse: $T^{-1} = T$.

(a) $T : \mathbb{R}^4 \rightarrow \mathbb{R}^4$; $T(x, y, z, w) = (x, -y, -z, w)$

◆**(b)** $T : \mathbb{R}^2 \rightarrow \mathbb{R}^2$; $T(x, y) = (ky - x, y)$, k any fixed number

(c) $T : P_n \rightarrow P_n$; $T(p(x)) = p(3 - x)$

◆**(d)** $T : M_{22} \rightarrow M_{22}$; $T(X) = AX$ where $A = \frac{1}{4}\begin{bmatrix} 5 & -3 \\ 3 & -5 \end{bmatrix}$

8. Let $R_{\frac{\pi}{2}}$ be as in Example 12§7.1, and let S be reflection of $\mathbb{R}^2$ in the Y axis.

(a) Show that $SR_{\frac{\pi}{2}}$ is reflection in the line $y = x$.

◆**(b)** Show that $R_{\frac{\pi}{2}}S$ is reflection in the line $y = -x$.

9. Using the notation of Example 12§7.1, show that $R_\theta R_\phi = R_{\theta + \phi}$ for all angles θ and ϕ.

10. In each case, show that $T^6 = 1_{\mathbb{R}^4}$ and so determine T^{-1}.

(a) $T : \mathbb{R}^4 \rightarrow \mathbb{R}^4$; $T(x, y, z, w) = (-x, z, w, y)$

◆**(b)** $T : \mathbb{R}^4 \rightarrow \mathbb{R}^4$; $T(x, y, z, w) = (-y, x - y, z, -w)$

11. In each case, show that T is an isomorphism by defining T^{-1} explicitly.

(a) $T : P_n \rightarrow P_n$ is given by $T[p(x)] = p(x + 1)$.

◆**(b)** $T : M_{nn} \rightarrow M_{nn}$ is given by $T(A) = UA$ where U is invertible in M_{nn}.

12. Given linear transformations $V \xrightarrow{T} W \xrightarrow{S} U$:

(a) If S and T are both one-to-one, show that ST is one-to-one.

◆**(b)** If S and T are both onto, show that ST is onto.

13. Let $T : V \rightarrow W$ be a linear transformation.

(a) If T is one-to-one and $TR = TR_1$ for transformations R and $R_1 : U \rightarrow V$, show that $R = R_1$.

(b) If T is onto and $ST = S_1T$ for transformations S and $S_1 : W \rightarrow U$, show that $S = S_1$.

14. Consider the linear transformations $V \xrightarrow{T} W \xrightarrow{R} U$.

(a) Show that $\ker T \subseteq \ker RT$.

(b) Show that $\operatorname{im} RT \subseteq \operatorname{im} R$.

15. Let $V \xrightarrow{T} U \xrightarrow{S} W$ be linear transformations.

(a) If ST is one-to-one, show that T is one-to-one and that $\dim V \leq \dim U$.

◆**(b)** If ST is onto, show that S is onto and that $\dim W \leq \dim U$.

◆**16.** Let $T : V \rightarrow V$ be a linear transformation. Show that $T^2 = 1_V$ if and only if T is invertible and $T = T^{-1}$.

17. Let N be a nilpotent $n \times n$ matrix (that is, $N^k = 0$ for some k). Show that $T : M_{nm} \rightarrow M_{nm}$ is an isomorphism if $T(X) = X - NX$. [*Hint:* If X is in $\ker T$, show that $X = NX = N^2X = \cdots$. Then use Theorem 3.]

◆**18.** Let $T : V \rightarrow W$ be a linear transformation, and let $\{e_1, \ldots, e_r, e_{r+1}, \ldots, e_n\}$ be any basis of V such that $\{e_{r+1}, \ldots, e_n\}$ is a basis of $\ker T$. Show that $\operatorname{im} T \cong \operatorname{span}\{e_1, \ldots, e_r\}$. [*Hint:* See Theorem 5§7.2.]

19. Is every isomorphism $T : M_{22} \rightarrow M_{22}$ given (as in Example 3) by an invertible matrix U such that $T(X) = UX$ for all X in M_{22}? Prove your answer.

20. Let D_n denote the space of all functions f from $\{1, 2, \ldots, n\}$ to $\mathbb{R}$ (see Exercise 36§5.3). If $T : D_n \rightarrow \mathbb{R}^n$ is defined by $T(f) = (f(1), f(2), \ldots, f(n))$, show that T is an isomorphism.

21. **(a)** Let V be the vector space of Exercise 7§5.1. Find an isomorphism $T : V \rightarrow \mathbb{R}^1$.

◆**(b)** Let V be the vector space of Exercise 8§5.1. Find an isomorphism $T : V \rightarrow \mathbb{R}^2$.

22. Let $V \xrightarrow{T} W \xrightarrow{S} V$ be linear transformations such that $ST = 1_V$. If $\dim V = \dim W = n$, show that $S = T^{-1}$ and $T = S^{-1}$. [*Hint:* Exercise 15 and Theorems 3, 4, and 5.]

23. Let $V \xrightarrow{T} W \xrightarrow{S} V$ be functions such that $TS = 1_W$ and $ST = 1_V$. If T is linear, show that S is also linear.

24. Use Theorems 4 and 6 to prove that $(AB)C = A(BC)$ for matrices of sizes $m \times n$, $n \times k$, and $k \times p$, respectively.

25. Let A and B be matrices of size $p \times m$ and $n \times q$. Assume that $mn = pq$. Define $R : M_{mn} \rightarrow M_{pq}$ by $R(X) = AXB$.

(a) Show that $M_{mn} \cong M_{pq}$ by comparing dimensions.

(b) Show that R is a linear transformation.

(c) Show that if R is an isomorphism, then $m = p$ and $n = q$. [*Hint:* Show that $T : M_{mn} \rightarrow M_{pn}$ given by $T(X) = AX$ and $S : M_{mn} \rightarrow M_{mq}$ given by $S(X) = XB$ are both one-to-one, and use the dimension theorem.]

26. Let $T : V \rightarrow V$ be a linear transformation such that $T^2 = 0$ is the zero transformation.

(a) If $V \neq 0$, show that T cannot be invertible.

(b) If $R : V \rightarrow V$ is defined by $R(v) = v + T(v)$ for all v in V, show that R is linear and invertible.

27. Let V consist of all sequences $[x_0, x_1, x_2, \ldots)$ of numbers, and define vector operations

$$[x_0, x_1, \ldots) + [y_0, y_1, \ldots) = [x_0 + y_0, x_1 + y_1, \ldots)$$
$$r[x_0, x_1, \ldots) = [rx_0, rx_1, \ldots)$$

   (a) Show that V is a vector space of infinite dimension.
   (b) Define $T : V \rightarrow V$ and $S : V \rightarrow V$ by $T[x_0, x_1, \ldots) = [x_1, x_2, \ldots)$ and $S[x_0, x_1, \ldots) = [0, x_0, x_1, \ldots)$. Show that $TS = 1_V$, so TS is one-to-one and onto, but that T is not one-to-one and S is not onto.

28. Prove (1) and (2) of Theorem 4.

29. Define $T : \mathbf{P}_n \rightarrow \mathbf{P}_n$ by $T(p) = p(x) + xp'(x)$ for all p in $\mathbf{P}_n$.
   (a) Show that T is linear.
   ◆**(b)** Show that $\ker T = 0$ and conclude that T is an isomorphism. [*Hint:* Write $p(x) = a_0 + a_1 x + \cdots + a_n x^n$ and compare coefficients if $p(x) = -xp'(x)$.]
   (c) Conclude that each $q(x)$ in $\mathbf{P}_n$ has the form $q(x) = p(x) + xp'(x)$ for some unique polynomial $p(x)$.
   (d) Does this remain valid if T is defined by $T[p(x)] = p(x) - xp'(x)$? Explain.

30. Let $T : V \rightarrow W$ be a linear transformation, where V and W are finite dimensional.
   (a) Show that T is one-to-one if and only if there exists a linear transformation $S : W \rightarrow V$ with $ST = 1_V$. [*Hint:* If $\{\mathbf{e}_1, \ldots, \mathbf{e}_n\}$ is a basis of V and T is one-to-one, show that W has a basis $\{T(\mathbf{e}_1), \ldots, T(\mathbf{e}_n), \mathbf{f}_{n+1}, \ldots, \mathbf{f}_{n+k}\}$ and use Theorems 3 and 4§7.1.]
   ◆**(b)** Show that T is onto if and only if there exists a linear transformation $S : W \rightarrow V$ with $TS = 1_W$. [*Hint:* Let $\{\mathbf{e}_1, \ldots, \mathbf{e}_r, \ldots, \mathbf{e}_n\}$ be a basis of V such that

$\{\mathbf{e}_{r+1}, \ldots, \mathbf{e}_n\}$ is a basis of $\ker T$. Use Theorem 5§7.2 and Theorems 3 and 4§7.1.]

31. Let S and T be linear transformations $V \rightarrow W$, where $\dim V = n$ and $\dim W = m$.
   (a) Show that $\ker S = \ker T$ if and only if $T = RS$ for some isomorphism $R : W \rightarrow W$. [*Hint:* Let $\{\mathbf{e}_1, \ldots, \mathbf{e}_r, \ldots, \mathbf{e}_n\}$ be a basis of V such that $\{\mathbf{e}_{r+1}, \ldots, \mathbf{e}_n\}$ is a basis of $\ker S = \ker T$. Use Theorem 5§7.2 to extend $\{S(\mathbf{e}_1), \ldots, S(\mathbf{e}_r)\}$ and $\{T(\mathbf{e}_1), \ldots, T(\mathbf{e}_r)\}$ to bases of W.]
   ◆**(b)** Show that $\operatorname{im} S = \operatorname{im} T$ if and only if $T = SR$ for some isomorphism $R : V \rightarrow V$. [*Hint:* Show that $\dim(\ker S) = \dim(\ker T)$ and choose bases $\{\mathbf{e}_1, \ldots, \mathbf{e}_r, \ldots, \mathbf{e}_n\}$ and $\{\mathbf{f}_1, \ldots, \mathbf{f}_r, \ldots, \mathbf{f}_n\}$ of V where $\{\mathbf{e}_{r+1}, \ldots, \mathbf{e}_n\}$ and $\{\mathbf{f}_{r+1}, \ldots, \mathbf{f}_n\}$ are bases of $\ker S$ and $\ker T$ respectively. If $1 \leq i \leq r$, show that $S(\mathbf{e}_i) = T(\mathbf{g}_i)$ for some $\mathbf{g}_i$ in V, and prove that $\{\mathbf{g}_1, \ldots, \mathbf{g}_r, \mathbf{f}_{r+1}, \ldots, \mathbf{f}_n\}$ is a basis of V.]

◆**32.** If $T : V \rightarrow V$ is a linear transformation where $\dim V = n$, show that $TST = T$ for some isomorphism $S : V \rightarrow V$. [*Hint:* Let $\{\mathbf{e}_1, \ldots, \mathbf{e}_r, \mathbf{e}_{r+1}, \ldots, \mathbf{e}_n\}$ be as in Theorem 5§7.2. Extend $\{T(\mathbf{e}_1), \ldots, T(\mathbf{e}_r)\}$ to a basis of V, and use Theorem 1 and Theorems 3 and 4§7.1.]

33. Let A and B denote $m \times n$ matrices. In each case show that (1) and (2) are equivalent.
   (a) (1) A and B have the same null space.
        (2) $B = PA$ for some invertible $m \times m$ matrix P.
   (b) (1) A and B have the same range.
        (2) $B = AQ$ for some invertible $n \times n$ matrix Q.
   [*Hint:* Use Exercise 31 and Theorem 6.]

▨▨▨▨▨ Section 7.4   The Matrix of a Linear Transformation

It was shown in Theorem 5§7.1 that every linear transformation from $\mathbb{R}^n$ to $\mathbb{R}^m$ is a matrix transformation

$$T_A : \mathbb{R}^n \rightarrow \mathbb{R}^m \text{ given by } T_A(X) = AX \text{ for all columns } X \text{ in } \mathbb{R}^n$$

Furthermore, the $m \times n$ matrix A is uniquely determined by T; in fact, its columns (in order) are $T(E_1)$, $T(E_2)$, $\ldots$, $T(E_n)$, where $\{E_1, \ldots, E_n\}$ is the standard basis of $\mathbb{R}^n$. In general, let $T : V \rightarrow W$ be *any* linear transformation where $\dim V = n$ and $\dim W = m$. The aim in this section is to describe the action of T as multiplication by an $m \times n$ matrix. The idea is to convert the vectors in V and W into columns in $\mathbb{R}^n$ and $\mathbb{R}^m$, respectively, and then represent the corresponding transformation $\mathbb{R}^n \rightarrow \mathbb{R}^m$ by a matrix.

Converting vectors to columns is a simple matter, but one small change is needed. Up to now the *order* of the vectors in a basis has been of no importance. However, in this section, we shall speak of an **ordered basis** $\{\mathbf{e}_1, \mathbf{e}_2, \ldots, \mathbf{e}_n\}$, which is just a basis where the order in which the vectors are listed is taken into account. Hence $\{\mathbf{e}_2, \mathbf{e}_1, \mathbf{e}_3\}$ is a different *ordered* basis from $\{\mathbf{e}_1, \mathbf{e}_2, \mathbf{e}_3\}$.

DEFINITION

If $B = \{\mathbf{e}_1, \mathbf{e}_2, \ldots, \mathbf{e}_n\}$ is an ordered basis in a vector space V, and if

$$\mathbf{v} = v_1\mathbf{e}_1 + v_2\mathbf{e}_2 + \cdots + v_n\mathbf{e}_n$$

is a vector in V, then the (uniquely determined) numbers $v_1, v_2, \ldots, v_n$ are called the **coordinates** of $\mathbf{v}$ with respect to the basis B. The **coordinate vector** of $\mathbf{v}$ with respect to B is defined to be

$$C_B(\mathbf{v}) = \begin{bmatrix} v_1 \\ v_2 \\ \vdots \\ v_n \end{bmatrix}$$

The reason for writing $C_B(\mathbf{v})$ as a column instead of a row will become clear later.

EXAMPLE 1

The coordinate vector for $\mathbf{v} = (2, 1, 3)$ with respect to the ordered basis $B = \{(1, 1, 0), (1, 0, 1), (0, 1, 1)\}$ of $\mathbb{R}^3$ is

$$C_B(\mathbf{v}) = \begin{bmatrix} 0 \\ 2 \\ 1 \end{bmatrix}$$

because $\mathbf{v} = (2, 1, 3) = 0(1, 1, 0) + 2(1, 0, 1) + 1(0, 1, 1)$.

THEOREM 1

If V has dimension n and B is any ordered basis of V, the coordinate transformation $C_B : V \to \mathbb{R}^n$ is an isomorphism.

Proof The verification that C_B is linear is Exercise 13. If $B = \{\mathbf{e}_1, \ldots, \mathbf{e}_n\}$, then $C_B(\mathbf{e}_j)$ is column j of the identity matrix. Hence C_B carries B to the standard basis of $\mathbb{R}^n$, so it is an isomorphism by Theorem 1§7.3. ◆

Now let $T : V \to W$ be a linear transformation where $\dim V = n$ and $\dim W = m$, and let $B = \{\mathbf{e}_1, \mathbf{e}_2, \ldots, \mathbf{e}_n\}$ and D be ordered bases of V and W, respectively.

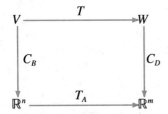

FIGURE 7.7

Then $C_B : V \to \mathbb{R}^n$ and $C_D : W \to \mathbb{R}^m$ are isomorphisms and we have the situation shown in Figure 7.7 where A is an $m \times n$ matrix (to be determined). Indeed, the composite $C_D T C_B^{-1} : \mathbb{R}^n \to \mathbb{R}^m$ is a linear transformation, so Theorem 5§7.1 shows that an $m \times n$ matrix A exists such that

$$C_D T C_B^{-1} = T_A, \qquad \text{equivalently} \qquad C_D T = T_A C_B$$

T_A acts by left multiplication by A, so this condition is

$$C_D[T(\mathbf{v})] = A C_B(\mathbf{v}) \qquad \text{for all } \mathbf{v} \text{ in } V$$

This requirement completely determines A. Indeed, the fact that $C_B(\mathbf{e}_j)$ is column j of the identity matrix gives

$$\text{Column } j \text{ of } A = A C_B(\mathbf{e}_j) = C_D[T(\mathbf{e}_j)]$$

for all j. Hence

$$A = [C_D[T(\mathbf{e}_1)] \quad C_D[T(\mathbf{e}_2)] \quad \cdots \quad C_D[T(\mathbf{e}_n)]]]$$

This prompts the following definition:

DEFINITION

Given $T : V \to W$, $B = \{\mathbf{e}_1, \ldots, \mathbf{e}_n\}$, and D as before, write

$$M_{DB}(T) = [C_D[T(\mathbf{e}_1)] \quad C_D[T(\mathbf{e}_2)] \quad \cdots \quad C_D[T(\mathbf{e}_n)]]$$

This is called the **matrix of T corresponding to the ordered bases B and D**.

The discussion is summarized in the following important theorem.

THEOREM 2

Let $T : V \to W$ be a linear transformation where $\dim V = n$ and $\dim W = m$, and let $B = \{\mathbf{e}_1, \ldots, \mathbf{e}_n\}$ and D be ordered bases of V and W, respectively. Then the matrix $M_{DB}(T)$ just given is the unique $m \times n$ matrix A that satisfies

$$C_D T = T_A C_B$$

Hence the defining property of $M_{DB}(T)$ is

$$C_D[T(\mathbf{v})] = M_{DB}(T) C_B(\mathbf{v}) \quad \text{for all } \mathbf{v} \text{ in } V$$

The fact that $T = C_D^{-1} T_A C_B$ means that the action of T on a vector $\mathbf{v}$ in V can be performed by first taking coordinates (that is, applying C_B to $\mathbf{v}$), then multiplying by A (applying T_A), and finally converting the resulting m-tuple back to a vector in W (applying C_D^{-1}).

EXAMPLE 2

Define $T: \mathbf{P}_2 \rightarrow \mathbb{R}^2$ by $T(a + bx + cx^2) = (a + c, b - a - c)$ for all polynomials $a + bx + cx^2$. If $B = \{\mathbf{e}_1, \mathbf{e}_2, \mathbf{e}_3\}$ and $D = \{\mathbf{f}_1, \mathbf{f}_2\}$ where

$$\mathbf{e}_1 = 1, \mathbf{e}_2 = x, \mathbf{e}_3 = x^2 \quad \text{and} \quad \mathbf{f}_1 = (1, 0), \mathbf{f}_2 = (0, 1)$$

compute $M_{DB}(T)$ and verify Theorem 2.

Solution   We have $T(\mathbf{e}_1) = \mathbf{f}_1 - \mathbf{f}_2$, $T(\mathbf{e}_2) = \mathbf{f}_2$, and $T(\mathbf{e}_3) = \mathbf{f}_1 - \mathbf{f}_2$. Hence

$$M_{DB}(T) = \begin{bmatrix} C_D[T(\mathbf{e}_1)] & C_D[T(\mathbf{e}_2)] & C_D[T(\mathbf{e}_3)] \end{bmatrix} = \begin{bmatrix} 1 & 0 & 1 \\ -1 & 1 & -1 \end{bmatrix}$$

If $\mathbf{v} = a + bx + cx^2 = a\mathbf{e}_1 + b\mathbf{e}_2 + c\mathbf{e}_3$, then $T(\mathbf{v}) = (a + c)\mathbf{f}_1 + (b - a - c)\mathbf{f}_2$, so

$$C_D[T(\mathbf{v})] = \begin{bmatrix} a + c \\ b - a - c \end{bmatrix} = \begin{bmatrix} 1 & 0 & 1 \\ -1 & 1 & -1 \end{bmatrix} \begin{bmatrix} a \\ b \\ c \end{bmatrix} = M_{DB}(T)C_B(\mathbf{v})$$

as Theorem 2 asserts.   ◆◆◆

EXAMPLE 3

Suppose that $T: \mathbf{P}_2 \rightarrow \mathbb{R}^2$ has matrix $M_{DB}(T) = \begin{bmatrix} 5 & 2 & -1 \\ 3 & 0 & 4 \end{bmatrix}$ where $B = \{1, x, x^2\}$ and $D = \{(1, 0), (1, 1)\}$. Find $T(\mathbf{v})$ where $\mathbf{v} = a + bx + cx^2$.

Solution   The idea is to compute $C_D[T(\mathbf{v})] = M_{DB}(T)C_B(\mathbf{v})$ first and then get $T(\mathbf{v})$:

$$C_D[T(\mathbf{v})] = \begin{bmatrix} 5 & 2 & -1 \\ 3 & 0 & 4 \end{bmatrix} \begin{bmatrix} a \\ b \\ c \end{bmatrix} = \begin{bmatrix} 5a + 2b - c \\ 3a + 4c \end{bmatrix}$$

Hence $T(\mathbf{v}) = (5a + 2b - c)(1, 0) + (3a + 4c)(1, 1) = (8a + 2b + 3c, 3a + 4c)$.   ◆◆◆

The next two examples will be referred to later.

EXAMPLE 4

Let A be an $m \times n$ matrix, and let $T_A: \mathbb{R}^n \rightarrow \mathbb{R}^m$ be the matrix transformation induced by $A: T_A(X) = AX$ for all columns X in $\mathbb{R}^n$. If B and D are the standard bases of $\mathbb{R}^n$ and $\mathbb{R}^m$, respectively (in the usual order), then

$$M_{DB}(T_A) = A$$

In other words, the matrix of T_A corresponding to the standard bases is A itself.

Solution   Write $B = \{E_1, \ldots, E_n\}$. Because D is the standard basis of $\mathbb{R}^m$, it is easy to verify that $C_D(Y) = Y$ for all columns Y in $\mathbb{R}^m$. Hence

$$M_{DB}(T_A) = [T_A(E_1)\ \ T_A(E_2)\ \ \cdots\ \ T_A(E_n)] = [AE_1\ \ AE_2\ \ \cdots\ \ AE_n] = A$$

because AE_j is the jth column of A. ◆◆◆

EXAMPLE 5

Let V and W have ordered bases B and D, respectively. Let dim $V = n$.

1. The identity transformation $1_V : V \to V$ has matrix $M_{BB}(1_V) = I_n$.
2. The zero transformation $0 : V \to W$ has matrix $M_{DB}(0) = 0$.

◆◆◆

The first result in Example 5 is false if the two bases of V are not equal. In fact, if B is the standard basis of $\mathbb{R}^n$, then the basis D of $\mathbb{R}^n$ can be chosen so that $M_{BD}(1_{\mathbb{R}^n})$ turns out to be any invertible matrix at all (Exercise 14).

The next two theorems show that composition of linear transformations is compatible with multiplication of the corresponding matrices.

THEOREM 3

Let $V \xrightarrow{T} W \xrightarrow{S} U$ be linear transformations and let B, D, and E be finite ordered bases of V, W, and U, respectively. Then

$$M_{EB}(ST) = M_{ED}(S) \cdot M_{DB}(T)$$

Proof   We use the property in Theorem 2 three times. If **v** is in V,

$$M_{ED}(S)M_{DB}(T)C_B(\mathbf{v}) = M_{ED}(S)C_D[T(\mathbf{v})] = C_E[ST(\mathbf{v})] = M_{EB}(ST)C_B(\mathbf{v})$$

If $B = \{\mathbf{e}_1, \ldots, \mathbf{e}_n\}$, then $C_B(\mathbf{e}_j)$ is column j of I_n. Hence taking $\mathbf{v} = \mathbf{e}_j$ shows that $M_{ED}(S)M_{DB}(T)$ and $M_{EB}(ST)$ have equal jth columns. ◆

THEOREM 4

Let $T : V \to W$ be a linear transformation, where dim $V =$ dim $W = n$. The following are equivalent.

1. T is an isomorphism.
2. $M_{DB}(T)$ is invertible for all ordered bases B and D of V and W.
3. $M_{DB}(T)$ is invertible for some pair of ordered bases B and D of V and W.

When this is the case, $M_{DB}(T)^{-1} = M_{BD}(T^{-1})$

Proof

(1) *implies* (2). We have $V \xrightarrow{T} W \xrightarrow{T^{-1}} V$, so Theorem 3 and Example 5 give

$$M_{BD}(T^{-1})M_{DB}(T) = M_{BB}(T^{-1}T) = M_{BB}(1_V) = I_n$$

Similarly, $M_{DB}(T)M_{BD}(T^{-1}) = I_n$, proving (2) (and the last statement in the theorem).

(2) *implies* (3). This is clear.

(3) *implies* (1). Let $M_{DB}(T)$ be invertible for some B and D and for convenience write $A = M_{DB}(T)$. Then $C_D T = T_A C_B$ by Theorem 2. Hence $T = C_D^{-1} T_A C_B$ is a composite of isomorphisms (Theorem 1 and Theorem 6§7.3) and so is itself an isomorphism. ◆

EXAMPLE 6

Let $p_0, p_1, \ldots, p_n$ be polynomials in $\mathbf{P}_n$ and assume that p_k has degree k for each $k = 0, 1, 2, \ldots, n$. Let $T : \mathbf{P}_n \to \mathbf{P}_n$ be the linear transformation such that $T(x^k) = p_k$ for each $k = 0, 1, 2, \ldots, n$. Show that T is an isomorphism and hence that $\{p_0, p_1, \ldots, p_n\}$ is a basis of $\mathbf{P}_n$ (Theorem 1§5.6).

Solution   Let $B = \{1, x, \ldots, x^n\}$ and let a_k denote the coefficient of x^k in p_k. The fact that $\deg(p_k) = k$ shows that $a_k \neq 0$ and that the coefficients of higher powers of x in p_k are zero. Hence

$$M_{BB}(T) = \begin{bmatrix} C_B[p_0] & C_B[p_1] & \cdots & C_B[p_n] \end{bmatrix} = \begin{bmatrix} a_0 & * & * & \cdots & * \\ 0 & a_1 & * & \cdots & * \\ 0 & 0 & a_2 & \cdots & * \\ \vdots & \vdots & \vdots & & \vdots \\ 0 & 0 & 0 & \cdots & a_n \end{bmatrix}$$

where the $*$ entries may be nonzero. This matrix has determinant $a_0 a_1 \ldots a_n \neq 0$, so T is an isomorphism by Theorem 4. The rest follows by Theorem 1§7.3.

◆◆◆

In view of Theorem 4, it is not surprising that there is a connection between the rank of a linear transformation and the rank of the corresponding matrices.

THEOREM 5

Let $T : V \to W$ be a linear transformation where $\dim V = n$ and $\dim W = m$. If B and D are any ordered bases of V and W, then $\text{rank } T = \text{rank}[M_{DB}(T)]$.

Proof   Write $A = M_{DB}(T)$ for convenience. The column space of A is $U = \{AX \mid X$ in $\mathbb{R}^n\}$. Hence $\text{rank } A = \dim U$ and so, because $\text{rank } T = \dim(\text{im } T)$, it suffices to find an isomorphism $S : \text{im } T \to U$. Now every vector in $\text{im } T$ has the form $T(\mathbf{v})$, $\mathbf{v}$ in V, and by Theorem 2, $C_D[T(\mathbf{v})] = AC_B(\mathbf{v})$ lies in U. So define $S : \text{im } T \to U$ by

$$S[T(\mathbf{v})] = C_D[T(\mathbf{v})] \qquad \text{for all vectors } T(\mathbf{v}) \text{ in im } T$$

The fact that C_D is linear and one-to-one implies immediately that S is linear and one-to-one. To see that S is onto, let AX be any member of U, X in $\mathbb{R}^n$. Then $X = C_B(\mathbf{v})$ for some $\mathbf{v}$ in V because C_B is onto. Hence $AX = AC_B(\mathbf{v}) = C_D[T(\mathbf{v})] = S[T(\mathbf{v})]$, so S is onto. This means that S is an isomorphism. ◆

We conclude with an example showing that the matrix of a linear transformation can be made very simple by a careful choice of the two bases.

EXAMPLE 7

Let $T : V \to W$ be a linear transformation where dim $V = n$ and dim $W = m$. Choose an ordered basis $B = \{\mathbf{e}_1, \ldots, \mathbf{e}_r, \mathbf{e}_{r+1}, \ldots, \mathbf{e}_n\}$ of V in which $\{\mathbf{e}_{r+1}, \ldots, \mathbf{e}_n\}$ is a basis of ker T. Then $\{T(\mathbf{e}_1), \ldots, T(\mathbf{e}_r)\}$ is a basis of im T by Theorem 5§7.2 so extend it to an ordered basis $D = \{T(\mathbf{e}_1), \ldots, T(\mathbf{e}_r), \mathbf{f}_{r+1}, \ldots, \mathbf{f}_m\}$ of W. Show that

$$M_{DB}(T) = \begin{bmatrix} I_r & 0 \\ 0 & 0 \end{bmatrix}$$

in block form. Moreover $r = $ rank T.

Solution

Because $T(\mathbf{e}_{r+1}) = \cdots = T(\mathbf{e}_n) = \mathbf{0}$, we have

$$M_{DB}(T) = \begin{bmatrix} C_D[T(\mathbf{e}_1)] \cdots C_D[T(\mathbf{e}_r)] & C_D[T(\mathbf{e}_{r+1})] \cdots C_D[T(\mathbf{e}_n)] \end{bmatrix} = \begin{bmatrix} I_r & 0 \\ 0 & 0 \end{bmatrix}$$

Then rank $T = r$ by Theorem 5. ◆◆◆

EXERCISES 7.4

1. In each case, find the coordinates of $\mathbf{v}$ with respect to the basis B of the vector space V.

   (a) $V = \mathbf{P}_2$, $\mathbf{v} = 2x^2 + x - 1$, $B = \{x + 1, x^2, 3\}$
   ◆ (b) $V = \mathbf{P}_2$, $\mathbf{v} = ax^2 + bx + c$, $B = \{x^2, x + 1, x + 2\}$
   (c) $V = \mathbb{R}^3$, $\mathbf{v} = (1, -1, 2)$, $B = \{(1, -1, 0), (1, 1, 1), (0, 1, 1)\}$
   ◆ (d) $V = \mathbb{R}^3$, $\mathbf{v} = (a, b, c)$, $B = \{(1, -1, 2), (1, 1, -1), (0, 0, 1)\}$

   (e) $V = \mathbf{M}_{22}$, $\mathbf{v} = \begin{bmatrix} 1 & 2 \\ -1 & 0 \end{bmatrix}$,

   $B = \left\{ \begin{bmatrix} 1 & 1 \\ 0 & 0 \end{bmatrix}, \begin{bmatrix} 1 & 0 \\ 1 & 0 \end{bmatrix}, \begin{bmatrix} 0 & 0 \\ 1 & 1 \end{bmatrix}, \begin{bmatrix} 1 & 0 \\ 0 & 1 \end{bmatrix} \right\}$

2. Suppose $T : \mathbf{P}_2 \to \mathbb{R}^2$ is a linear transformation. If $B = \{1, x, x^2\}$ and $D = \{(1, 1), (0, 1)\}$, find the action of T given:

   (a) $M_{DB}(T) = \begin{bmatrix} 1 & 2 & -1 \\ -1 & 0 & 1 \end{bmatrix}$

   ◆ (b) $M_{DB}(T) = \begin{bmatrix} 2 & 1 & 3 \\ -1 & 0 & -2 \end{bmatrix}$

3. In each case, find the matrix of $T : V \to W$ corresponding to the bases B and D of V and W, respectively.

   (a) $T : \mathbf{M}_{22} \to \mathbb{R}$, $T(A) = $ tr A;

   $B = \left\{ \begin{bmatrix} 1 & 0 \\ 0 & 0 \end{bmatrix}, \begin{bmatrix} 0 & 1 \\ 0 & 0 \end{bmatrix}, \begin{bmatrix} 0 & 0 \\ 1 & 0 \end{bmatrix}, \begin{bmatrix} 0 & 0 \\ 0 & 1 \end{bmatrix} \right\}$, $D = \{1\}$

   ◆ (b) $T : \mathbf{M}_{22} \to \mathbf{M}_{22}$, $T(A) = A^T$;

   $B = D = \left\{ \begin{bmatrix} 1 & 0 \\ 0 & 0 \end{bmatrix}, \begin{bmatrix} 0 & 1 \\ 0 & 0 \end{bmatrix}, \begin{bmatrix} 0 & 0 \\ 1 & 0 \end{bmatrix}, \begin{bmatrix} 0 & 0 \\ 0 & 1 \end{bmatrix} \right\}$

   (c) $T : \mathbf{P}_2 \to \mathbf{P}_3$, $T[p(x)] = xp(x)$; $B = \{1, x, x^2\}$, $D = \{1, x, x^2, x^3\}$

   ◆ (d) $T : \mathbf{P}_2 \to \mathbf{P}_2$, $T[p(x)] = p(x + 1)$; $B = D = \{1, x, x^2\}$

4. In each case, find the matrix of $T : V \to W$ corresponding to the bases B and D, respectively, and use it to compute $C_D[T(\mathbf{v})]$, and hence $T(\mathbf{v})$.

(a) $T: \mathbb{R}^3 \to \mathbb{R}^4$, $T(x, y, z) = (x + z, 2z, y - z, x + 2y)$; B and D standard; $\mathbf{v} = (1, -1, 3)$

◆(b) $T: \mathbb{R}^2 \to \mathbb{R}^4$, $T(x, y) = (2x - y, 3x + 2y, 4y, x)$; $B = \{(1, 1), (1, 0)\}$, D standard; $\mathbf{v} = (a, b)$

(c) $T: \mathbf{P}_2 \to \mathbb{R}^2$, $T(a + bx + cx^2) = (a + c, 2b)$; $B = \{1, x, x^2\}$, $D = \{(1, 0),(1,-1)\}$; $\mathbf{v} = a + bx + cx^2$

◆(d) $T: \mathbf{P}_2 \to \mathbb{R}^2$, $T(a + bx + cx^2) = (a + b, c)$; $B = \{1, x, x^2\}$, $D = \{(1,-1), (1, 1)\}$; $\mathbf{v} = a + bx + cx^2$

(e) $T: \mathbf{M}_{22} \to \mathbb{R}$, $T\begin{bmatrix} a & b \\ c & d \end{bmatrix} = a + b + c + d$;

$B = \left\{ \begin{bmatrix} 1 & 0 \\ 0 & 0 \end{bmatrix}, \begin{bmatrix} 0 & 1 \\ 0 & 0 \end{bmatrix}, \begin{bmatrix} 0 & 0 \\ 1 & 0 \end{bmatrix}, \begin{bmatrix} 0 & 0 \\ 0 & 1 \end{bmatrix} \right\}$,

$D = \{1\}$; $\mathbf{v} = \begin{bmatrix} a & b \\ c & d \end{bmatrix}$

◆(f) $T: \mathbf{M}_{22} \to \mathbf{M}_{22}$, $T\begin{bmatrix} a & b \\ c & d \end{bmatrix} = \begin{bmatrix} a & b + c \\ b + c & d \end{bmatrix}$;

$B = D = \left\{ \begin{bmatrix} 1 & 0 \\ 0 & 0 \end{bmatrix}, \begin{bmatrix} 0 & 1 \\ 0 & 0 \end{bmatrix}, \begin{bmatrix} 0 & 0 \\ 1 & 0 \end{bmatrix}, \begin{bmatrix} 0 & 0 \\ 0 & 1 \end{bmatrix} \right\}$;

$\mathbf{v} = \begin{bmatrix} a & b \\ c & d \end{bmatrix}$

5. In each case, verify Theorem 3. Use the standard basis in $\mathbb{R}^n$ and $\{1, x, x^2\}$ in $\mathbf{P}_2$.

(a) $\mathbb{R}^3 \xrightarrow{T} \mathbb{R}^2 \xrightarrow{S} \mathbb{R}^4$; $T(a, b, c) = (a + b, b - c)$, $S(a, b) = (a, b - 2a, 3b, a + b)$

◆(b) $\mathbb{R}^3 \xrightarrow{T} \mathbb{R}^4 \xrightarrow{S} \mathbb{R}^2$; $T(a, b, c) = (a + b, c + b, a + c, b - a)$, $S(a, b, c, d) = (a + b, c - d)$

(c) $\mathbf{P}_2 \xrightarrow{T} \mathbb{R}^3 \xrightarrow{S} \mathbf{P}_2$; $T(a + bx + cx^2) = (a, b - c, c - a)$, $S(a, b, c) = b + cx + (a - c)x^2$

◆(d) $\mathbb{R}^3 \xrightarrow{T} \mathbf{P}_2 \xrightarrow{S} \mathbb{R}^2$; $T(a, b, c) = (a - b) + (c - a)x + bx^2$, $S(a + bx + cx^2) = (a - b, c)$

◆6. Verify Theorem 3 for $\mathbf{M}_{22} \xrightarrow{T} \mathbf{M}_{22} \xrightarrow{S} \mathbf{P}_2$, where $T(A) = A^T$ and $S\begin{bmatrix} a & b \\ c & d \end{bmatrix} = b + (a + d)x + cx^2$. Use the bases $B = D$

$= \left\{ \begin{bmatrix} 1 & 0 \\ 0 & 0 \end{bmatrix}, \begin{bmatrix} 0 & 1 \\ 0 & 0 \end{bmatrix}, \begin{bmatrix} 0 & 0 \\ 1 & 0 \end{bmatrix}, \begin{bmatrix} 0 & 0 \\ 0 & 1 \end{bmatrix} \right\}$ and $E = \{1, x, x^2\}$.

7. In each case, find T^{-1} and verify $M_{DB}(T)^{-1} = M_{BD}(T^{-1})$.

(a) $T: \mathbb{R}^2 \to \mathbb{R}^2$, $T(a, b) = (a + 2b, 2a + 5b)$; $B = D$ = standard

◆(b) $T: \mathbb{R}^3 \to \mathbb{R}^3$, $T(a, b, c) = (b + c, a + c, a + b)$; $B = D =$ standard

(c) $T: \mathbf{P}_2 \to \mathbb{R}^3$, $T(a + bx + cx^2) = (a - c, b, 2a - c)$; $B = \{1, x, x^2\}$, $D =$ standard

◆(d) $T: \mathbf{P}_2 \to \mathbb{R}^3$, $T(a + bx + cx^2) = (a + b + c, b + c, c)$; $B = \{1, x, x^2\}$, $D =$ standard

8. In each case, show that $M_{DB}(T)$ is invertible and use the fact that $M_{BD}(T^{-1}) = [M_{BD}(T)]^{-1}$ to determine the action of T^{-1}.

(a) $T: \mathbf{P}_2 \to \mathbb{R}^3$, $T(a + bx + cx^2) = (a + c, c, b - c)$; $B = \{1, x, x^2\}$, $D =$ standard

◆(b) $T: \mathbf{M}_{22} \to \mathbb{R}^4$, $T\begin{bmatrix} a & b \\ c & d \end{bmatrix} = (a + b + c, b + c, c, d)$;

$B = \left\{ \begin{bmatrix} 1 & 0 \\ 0 & 0 \end{bmatrix}, \begin{bmatrix} 0 & 1 \\ 0 & 0 \end{bmatrix}, \begin{bmatrix} 0 & 0 \\ 1 & 0 \end{bmatrix}, \begin{bmatrix} 0 & 0 \\ 0 & 1 \end{bmatrix} \right\}$,

$D =$ standard

9. Let $D : \mathbf{P}_3 \to \mathbf{P}_2$ be the differentiation map given by $D[p(x)] = p'(x)$. Find the matrix of D corresponding to the bases $B = \{1, x, x^2, x^3\}$ and $E = \{1, x, x^2\}$, and use it to compute $D(a + bx + cx^2 + dx^3)$.

10. Use Theorem 4 to show that $T : V \to V$ is not an isomorphism if ker $T \neq 0$ (assume dim $V = n$). [*Hint:* Choose any ordered basis B containing a vector in ker T.]

11. Let $T : V \to \mathbb{R}$ be a linear transformation, and let $D = \{1\}$ be the basis of $\mathbb{R}$. Given any ordered basis $B = \{\mathbf{e}_1, \ldots, \mathbf{e}_n\}$ of V, show that $M_{DB}(T) = [T(\mathbf{e}_1) \quad \cdots \quad T(\mathbf{e}_n)]$.

◆12. Let $T : V \to W$ be an isomorphism, let $B = \{\mathbf{e}_1, \ldots, \mathbf{e}_n\}$ be an ordered basis of V, and let $D = \{T(\mathbf{e}_1), \ldots, T(\mathbf{e}_n)\}$. Show that $M_{DB}(T) = I_n$—the $n \times n$ identity matrix.

13. Complete the proof of Theorem 1.

14. Let U be any invertible $n \times n$ matrix, and let $D = \{\mathbf{f}_1, \mathbf{f}_2, \ldots, \mathbf{f}_n\}$ where $\mathbf{f}_j$ is column j of U. Show that $M_{BD}(1_{\mathbb{R}^n}) = U$ when B is the standard basis of $\mathbb{R}^n$.

15. Let B be an ordered basis of the n-dimensional space V and let $C_B : V \to \mathbb{R}^n$ be the coordinate transformation. If D is the standard basis of $\mathbb{R}^n$, show that $M_{DB}(C_B) = I_n$.

16. Let $T : \mathbf{P}_2 \to \mathbb{R}^3$ be defined by $T(p) = (p(0), p(1), p(2))$ for all p in $\mathbf{P}_2$. Let $B = \{1, x, x^2\}$ and $D = \{(1, 0, 0), (0, 1, 0), (0, 0, 1)\}$.

(a) Show that $M_{DB}(T) = \begin{bmatrix} 1 & 0 & 0 \\ 1 & 1 & 1 \\ 1 & 2 & 4 \end{bmatrix}$ and conclude that T is an isomorphism.

◆(b) Generalize to $T : \mathbf{P}_n \to \mathbb{R}^{n+1}$ where $T(p) = (p(a_0), p(a_1), \cdots, p(a_n))$ and $a_0, a_1, \ldots, a_n$ are distinct real numbers. [*Hint:* Theorem 2§3.3.]

17. Let $T : \mathbf{P}_n \to \mathbf{P}_n$ be defined by $T[p(x)] = p(x) + xp'(x)$, where $p'(x)$ denotes the derivative. Show that T is an isomorphism by finding $M_{BB}(T)$ when $B = \{1, x, x^2, \ldots, x^n\}$.

18. If k is any number, define $T_k : \mathbf{M}_{22} \to \mathbf{M}_{22}$ by $T_k(A) = A + kA^T$.

(a) If $B = \left\{ \begin{bmatrix} 1 & 0 \\ 0 & 0 \end{bmatrix}, \begin{bmatrix} 0 & 0 \\ 0 & 1 \end{bmatrix}, \begin{bmatrix} 0 & 1 \\ 1 & 0 \end{bmatrix}, \begin{bmatrix} 0 & 1 \\ -1 & 0 \end{bmatrix} \right\}$ find $M_{BB}(T_k)$, and conclude that T_k is invertible if $k \neq 1$ and $k \neq -1$.

(b) Repeat for $T_k : \mathbf{M}_{33} \to \mathbf{M}_{33}$. Can you generalize?

The remaining exercises require the following definitions. If V and W are vector spaces, the set of all linear transformations from V to W will be denoted by

$$\mathbf{L}(V, W) = \{T \mid T : V \to W \text{ is a linear transformation}\}$$

Given S and T in $\mathbf{L}(V, W)$ and a in $\mathbb{R}$, define $S + T : V \to W$ and $aT : V \to W$ by

$$(S + T)(\mathbf{v}) = S(\mathbf{v}) + T(\mathbf{v}) \quad \text{for all } \mathbf{v} \text{ in } V$$
$$(aT)(\mathbf{v}) = aT(\mathbf{v}) \quad \text{for all } \mathbf{v} \text{ in } V$$

19. Show that $\mathbf{L}(V, W)$ is a vector space.

20. Show that the following properties hold provided that the transformations link together in such a way that all the operations are defined.
   (a) $R(ST) = (RS)T$
   (b) $1_W T = T = T 1_V$
   (c) $R(S + T) = RS + RT$
   ◆**(d)** $(S + T)R = SR + TR$
   (e) $(aS)T = a(ST) = S(aT)$

21. Given S and T in $\mathbf{L}(V, W)$, show that:
   (a) $\ker S \cap \ker T \subseteq \ker(S + T)$ [See Exercise 34§5.3.]
   ◆**(b)** $\text{im}(S + T) \subseteq \text{im } S + \text{im } T$ [See Exercise 22§5.4.]

22. Let V and W be vector spaces. If X is a subset of V, define
$$X^0 = \{T \text{ in } \mathbf{L}(V, W) \mid T(\mathbf{v}) = \mathbf{0} \text{ for all } \mathbf{v} \text{ in } X\}$$
   (a) Show that X^0 is a subspace of $\mathbf{L}(V, W)$.
   ◆**(b)** If $X \subseteq X_1$, show that $X_1^0 \subseteq X^0$.
   (c) If U and U_1 are subspaces of V, show that $(U + U_1)^0 = U^0 \cap U_1^0$. [See Exercises 34§5.3 and 22§5.4.]

23. Define $R : \mathbf{M}_{mn} \to \mathbf{L}(\mathbb{R}^n, \mathbb{R}^m)$ by $R(A) = T_A$ for each $m \times n$ matrix A, where $T_A : \mathbb{R}^n \to \mathbb{R}^m$ is the matrix transformation given by $T_A(X) = AX$ for all columns X in $\mathbb{R}^n$. Show that R is an isomorphism.

24. Let V be any vector space (we do not assume it is finite dimensional). Given $\mathbf{v}$ in V, define $S_\mathbf{v} : \mathbb{R} \to V$ by $S_\mathbf{v}(r) = r\mathbf{v}$ for all r in $\mathbb{R}$.
   (a) Show that $S_\mathbf{v}$ lies in $\mathbf{L}(\mathbb{R}, V)$ for each $\mathbf{v}$ in V.
   ◆**(b)** Show that the map $R : V \to \mathbf{L}(\mathbb{R}, V)$ given by $R(\mathbf{v}) = $

$S_\mathbf{v}$ is an isomorphism. [*Hint:* To show that R is onto, if T lies in $\mathbf{L}(\mathbb{R}, V)$, show that $T = S_\mathbf{v}$ where $\mathbf{v} = T(1)$.]

25. Let V be a vector space with ordered basis $B = \{\mathbf{e}_1, \mathbf{e}_2, \ldots, \mathbf{e}_n\}$. For each $i = 1, 2, \ldots, m$, define $S_i ; \mathbb{R} \to V$ by $S_i(r) = r\mathbf{e}_i$ for all r in $\mathbb{R}$.
   (a) Show that each S_i lies in $\mathbf{L}(\mathbb{R}, V)$ and $S_i(1) = \mathbf{e}_i$.
   ◆**(b)** Given T in $\mathbf{L}(\mathbb{R}, V)$, let $T(1) = a_1\mathbf{e}_1 + a_2\mathbf{e}_2 + \cdots + a_n\mathbf{e}_n$, a_i in $\mathbb{R}$. Show that $T = a_1 S_1 + a_2 S_2 + \cdots + a_n S_n$.
   (c) Show that $\{S_1, S_2, \ldots, S_n\}$ is a basis of $\mathbf{L}(\mathbb{R}, V)$.

26. Let $\dim V = n$, $\dim W = m$, and let B and D be ordered bases of V and W, respectively. Show that $M_{DB} : \mathbf{L}(V, W) \to \mathbf{M}_{mn}$ is an isomorphism of vector spaces. [*Hint:* Let $B = \{\mathbf{e}_1, \ldots, \mathbf{e}_n\}$ and $D = \{\mathbf{f}_1, \ldots, \mathbf{f}_m\}$. Given $A = [a_{ij}]$ in $\mathbf{M}_{mn}$, show that $A = \mathbf{M}_{mn}(T)$ where $T : V \to W$ is defined by $T(\mathbf{e}_j) = a_{1j}\mathbf{f}_1 + a_{2j}\mathbf{f}_2 + \cdots + a_{mj}\mathbf{f}_m$ for each j.]

27. If V is a vector space, the space $V^* = \mathbf{L}(V, \mathbb{R})$ is called the **dual** of V. Given a basis $B = \{\mathbf{e}_1, \mathbf{e}_2, \ldots, \mathbf{e}_n\}$ of V, let $E_i : V \to \mathbb{R}$ for each $i = 1, 2, \ldots, m$ be the linear transformation satisfying

$$E_i(\mathbf{e}_j) = \begin{cases} 0 & \text{if } i \neq j \\ 1 & \text{if } i = j \end{cases}$$

(each E_i exists by Theorem 4§7.1). Prove the following:
   (a) $E_i(r_1\mathbf{e}_1 + \cdots + r_n\mathbf{e}_n) = r_i$ for each $i = 1, 2, \ldots, n$
   ◆**(b)** $\mathbf{v} = E_1(\mathbf{v})\mathbf{e}_1 + E_2(\mathbf{v})\mathbf{e}_2 + \cdots + E_n(\mathbf{v})\mathbf{e}_n$ for all $\mathbf{v}$ in V
   (c) $T = T(\mathbf{e}_1)E_1 + T(\mathbf{e}_2)E_2 + \cdots + T(\mathbf{e}_n)E_n$ for all T in V^*
   (d) $\{E_1, E_2, \ldots, E_n\}$ is a basis of V^* (called the **dual basis** of B).

   Given $\mathbf{v}$ in V, define $\mathbf{v}^* : V \to \mathbb{R}$ by

$$\mathbf{v}^*(\mathbf{w}) = E_1(\mathbf{v})E_1(\mathbf{w}) + E_2(\mathbf{v})E_2(\mathbf{w}) + \cdots + E_n(\mathbf{v})E_n(\mathbf{w})$$

   for all $\mathbf{w}$ in V. Show that:
   (e) $\mathbf{v}^* : V \to \mathbb{R}$ is linear, so $\mathbf{v}^*$ lies in V^*.
   (f) $\mathbf{e}_i^* = E_i$ for each $i = 1, 2, \ldots, n$
   (g) The map $R : V \to V^*$ with $R(\mathbf{v}) = \mathbf{v}^*$ is an isomorphism. [*Hint:* Show that R is linear and one-to-one and use Theorem 3§7.3. Alternatively, show that $R^{-1}(T) = T(\mathbf{e}_1)\mathbf{e}_1 + \cdots + T(\mathbf{e}_n)\mathbf{e}_n$.]

Section 7.5   **Change of Basis**

When working in a vector space V we often find that one basis of V is more convenient than another. For example, changing coordinate axes in $\mathbb{R}^2$ or $\mathbb{R}^3$ amounts to choosing a different basis. In this section we investigate how the coordinates of a vector change when the basis is changed and then use this to describe how the matrix of a linear transformation changes when the basis changes.

We begin by showing how to compute the coordinate vector $C_D(\mathbf{v})$ of a vector $\mathbf{v}$ in V from its coordinates $C_B(\mathbf{v})$ with respect to a different basis B. It turns out that B and D determine an invertible matrix P with the property that

$$C_D(\mathbf{v}) = PC_B(\mathbf{v})$$

holds for every vector $\mathbf{v}$ in V. To see this, consider first the case in which V has dimension $n = 2$. If $B = \{\mathbf{b}_1, \mathbf{b}_2\}$ and $D = \{\mathbf{d}_1, \mathbf{d}_2\}$, express the vectors $\mathbf{b}_1$ and $\mathbf{b}_2$ in terms of $\mathbf{d}_1$ and $\mathbf{d}_2$:

$$\mathbf{b}_1 = r\mathbf{d}_1 + s\mathbf{d}_2 \qquad \text{that is,} \quad C_D(\mathbf{b}_1) = \begin{bmatrix} r \\ s \end{bmatrix}$$

$$\mathbf{b}_2 = t\mathbf{d}_1 + u\mathbf{d}_2 \qquad \text{that is,} \quad C_D(\mathbf{b}_2) = \begin{bmatrix} t \\ u \end{bmatrix}$$

Given any vector $\mathbf{v}$ in V, write it as a linear combination of $\mathbf{b}_1$ and $\mathbf{b}_2$.

$$\mathbf{v} = v_1\mathbf{b}_1 + v_2\mathbf{b}_2 \qquad \text{that is,} \quad C_B(\mathbf{v}) = \begin{bmatrix} v_1 \\ v_2 \end{bmatrix}$$

Substitution gives

$$\mathbf{v} = v_1(r\mathbf{d}_1 + s\mathbf{d}_2) + v_2(t\mathbf{d}_1 + u\mathbf{d}_2) = (rv_1 + tv_2)\mathbf{d}_1 + (sv_1 + uv_2)\mathbf{d}_2$$

Hence

$$C_D(\mathbf{v}) = \begin{bmatrix} rv_1 + tv_2 \\ sv_1 + uv_2 \end{bmatrix} = \begin{bmatrix} r & t \\ s & u \end{bmatrix}\begin{bmatrix} v_1 \\ v_2 \end{bmatrix} = PC_B(\mathbf{v})$$

where $P = \begin{bmatrix} r & t \\ s & u \end{bmatrix}$ is a matrix depending only on B and D and not on $\mathbf{v}$. Indeed, the

columns of P are just $C_D(\mathbf{b}_1) = \begin{bmatrix} r \\ s \end{bmatrix}$ and $C_D(\mathbf{b}_2) = \begin{bmatrix} t \\ u \end{bmatrix}$, so we can represent P in

block form by listing its columns in order:

$$P = [C_D(\mathbf{b}_1) \quad C_D(\mathbf{b}_2)]$$

P is called the change matrix from B to D; and when it is necessary to emphasize the bases, we write $P = P_{D \leftarrow B}$. This process works in general.

DEFINITION

If $B = \{\mathbf{b}_1, \mathbf{b}_2, \ldots, \mathbf{b}_n\}$ and D are any two ordered bases of V, define the **change of basis matrix** $P_{D \leftarrow B}$ from B to D in block form by listing its columns:

$$P_{D \leftarrow B} = [C_D(\mathbf{b}_1) \quad C_D(\mathbf{b}_2) \quad \cdots \quad C_D(\mathbf{b}_n)]$$

We refer to $P_{D \leftarrow B}$ as the **change** matrix for short.

This means that the first column of $P_{D\leftarrow B}$ is $C_D(\mathbf{b}_1)$, the second column is $C_D(\mathbf{b}_2)$, and so on. With this, the preceding argument proves the case $n = 2$ of the following theorem. The general case is analogous and is left as Exercise 16.

THEOREM 1

Let $B = \{\mathbf{b}_1, \mathbf{b}_2, \ldots, \mathbf{b}_n\}$ and $D = \{\mathbf{d}_1, \mathbf{d}_2, \ldots, \mathbf{d}_n\}$ be two ordered bases of a vector space V. If $P_{D\leftarrow B}$ is defined as just shown, then

$$C_D(\mathbf{v}) = P_{D\leftarrow B}C_B(\mathbf{v})$$

holds for every vector $\mathbf{v}$ in V.

EXAMPLE 1

In $\mathbf{P}_2$ find $P_{D\leftarrow B}$ if $B = \{1, x, x^2\}$ and $D = \{1, (1 - x), (1 - x)^2\}$. Then use this to express $p = p(x) = a + bx + cx^2$ as a polynomial in powers of $(1 - x)$.

Solution   The change matrix $P_{D\leftarrow B}$ is computed as follows:

$$1 = 1 + 0(1 - x) + 0(1 - x)^2$$

$$x = 1 - 1(1 - x) + 0(1 - x)^2$$

$$x^2 = 1 - 2(1 - x) + 1(1 - x)^2$$

so $P_{D\leftarrow B} = [C_D(1) \quad C_D(x) \quad C_D(x^2)] = \begin{bmatrix} 1 & 1 & 1 \\ 0 & -1 & -2 \\ 0 & 0 & 1 \end{bmatrix}$. We have $C_B(p) = \begin{bmatrix} a \\ b \\ c \end{bmatrix}$, so

$$C_D(p) = P_{D\leftarrow B}C_B(p) = \begin{bmatrix} 1 & 1 & 1 \\ 0 & -1 & -2 \\ 0 & 0 & 1 \end{bmatrix}\begin{bmatrix} a \\ b \\ c \end{bmatrix} = \begin{bmatrix} a + b + c \\ -b - 2c \\ c \end{bmatrix}$$

Hence $p(x) = (a + b + c) - (b + 2c)(1 - x) + c(1 - x)^2$.

◆◆◆

Part (3) of the next theorem explains the notation $P_{D\leftarrow B}$.

THEOREM 2

Let B, D, and E be three ordered bases of an n-dimensional vector space V.

1. $P_{B\leftarrow B} = I$
2. $P_{D\leftarrow B}$ is invertible and $(P_{D\leftarrow B})^{-1} = P_{B\leftarrow D}$
3. $P_{E\leftarrow D}P_{D\leftarrow B} = P_{E\leftarrow B}$

Proof   The proof of (1) is left as an exercise, and (1) and (3) imply (2). In fact, taking $E = B$ in (3) gives $P_{B \leftarrow D} P_{D \leftarrow B} = P_{B \leftarrow B} = I$. Similarly, $P_{D \leftarrow B} P_{B \leftarrow D} = I$, and (2) follows.

It remains to verify (3). Given $\mathbf{v}$ in V, apply Theorem 1 thrice:

$$P_{E \leftarrow D} P_{D \leftarrow B} C_B(\mathbf{v}) = P_{E \leftarrow D} C_D(\mathbf{v}) = C_E(\mathbf{v}) = P_{E \leftarrow B} C_B(\mathbf{v})$$

Now (3) follows by taking $\mathbf{v} = \mathbf{b}_i$ for each i. [For any $n \times n$ matrix A, $A[C_B(\mathbf{b}_i)]$ is the ith column of A, because $C_B(\mathbf{b}_i)$ is the ith column of I_n.]   ◆

Now consider a linear operator $T : V \rightarrow V$ on the vector space V. If *different* bases B and D are admitted in the two copies of V, then they can be chosen such that the matrix $M_{DB}(T)$ has a very simple form. Indeed, we can make $M_{DB}(T) = \begin{bmatrix} I_r & 0 \\ 0 & 0 \end{bmatrix}$ in block form (Example 7§7.4). However, if we insist that $B = D$, the following question arises:

How simple can the matrix $M_{BB}(T)$ be made by an appropriate choice of basis B?

This question will occupy our attention for much of the rest of this chapter. With this in mind, we introduce some terminology.

DEFINITION

Given a linear operator $T : V \rightarrow V$ and an ordered basis $B = \{\mathbf{e}_1, \mathbf{e}_2, \ldots, \mathbf{e}_n\}$ of V, the matrix $M_{BB}(T)$ will be called **the matrix of T with respect to B** and will be denoted by $M_B(T)$. Hence

$$M_B(T) = [C_B[T(\mathbf{e}_1)] \quad C_B[T(\mathbf{e}_2)] \quad \cdots \quad C_B[T(\mathbf{e}_n)]]$$

Now let $B = \{\mathbf{e}_1, \mathbf{e}_2, \ldots, \mathbf{e}_n\}$ and B_0 be two ordered bases of a vector space V. Theorem 1 asserts that

$$C_{B_0}(\mathbf{v}) = P_{B_0 \leftarrow B} \, C_B(\mathbf{v}) \qquad \text{for all } \mathbf{v} \text{ in } V$$

On the other hand, Theorem 2§7.4 gives

$$C_B[T(\mathbf{v})] = M_B(T) \, C_B(\mathbf{v}) \qquad \text{for all } \mathbf{v} \text{ in } V$$

Combining these (and writing $P = P_{B_0 \leftarrow B}$ for convenience) gives

$$
\begin{aligned}
PM_B(T)C_B(\mathbf{v}) &= PC_B[T(\mathbf{v})] \\
&= C_{B_0}[T(\mathbf{v})] \\
&= M_{B_0}(T)C_{B_0}(\mathbf{v}) \\
&= M_{B_0}(T)PC_B(\mathbf{v})
\end{aligned}
$$

This holds for all $\mathbf{v}$ in V; and, because $C_B(\mathbf{e}_j)$ is the jth column of the identity matrix, it follows that

$$PM_B(T) = M_{B_0}(T)P$$

Now P is invertible (in fact, $P^{-1} = P_{B \leftarrow B_0}$ by Theorem 2), so this gives

$$M_B(T) = P^{-1}M_{B_0}(T)P$$

This asserts that $M_{B_0}(T)$ and $M_B(T)$ are similar matrices.

THEOREM 3

Let B_0 and B be two ordered bases of a finite dimensional vector space V. If $T : V \to V$ is any linear operator, the matrices $M_B(T)$ and $M_{B_0}(T)$ of T with respect to these bases are similar. More precisely,

$$M_B(T) = P^{-1}M_{B_0}(T)P$$

where $P = P_{B_0 \leftarrow B}$ is the change matrix from B to B_0:

$$P_{B_0 \leftarrow B} = [C_{B_0}(\mathbf{e}_1) \quad \cdots \quad C_{B_0}(\mathbf{e}_n)]$$

where $B = \{\mathbf{e}_1, \ldots, \mathbf{e}_n\}$.

EXAMPLE 2

Let $T : \mathbb{R}^3 \to \mathbb{R}^3$ be defined by $T(a, b, c) = (2a - b, b + c, c - 3a)$. If B_0 denotes the standard basis of $\mathbb{R}^3$ and $B = \{(1, 1, 0), (1, 0, 1), (0, 1, 0)\}$, find an invertible matrix P such that $P^{-1}M_{B_0}(T)P = M_B(T)$.

Solution   We have

$$M_{B_0}(T) = [C_{B_0}(2, 0, -3) \quad C_{B_0}(-1, 1, 0) \quad C_{B_0}(0, 1, 1)] = \begin{bmatrix} 2 & -1 & 0 \\ 0 & 1 & 1 \\ -3 & 0 & 1 \end{bmatrix}$$

$$M_B(T) = [C_B(1, 1, -3) \quad C_B(2, 1, -2) \quad C_B(-1, 1, 0)] = \begin{bmatrix} 4 & 4 & -1 \\ -3 & -2 & 0 \\ -3 & -3 & 2 \end{bmatrix}$$

$$P = P_{B_0 \leftarrow B} = [C_{B_0}(1, 1, 0) \quad C_{B_0}(1, 0, 1) \quad C_{B_0}(0, 1, 0)] = \begin{bmatrix} 1 & 1 & 0 \\ 1 & 0 & 1 \\ 0 & 1 & 0 \end{bmatrix}$$

The reader can verify that $P^{-1}M_{B_0}(T)P = M_B(T)$; equivalently, $M_{B_0}(T)P = PM_B(T)$.

◆◆◆

Similar matrices were studied in Section 6.1 in preparation for the discussion of diagonalization in Section 6.2. Theorem 3 comes into this as follows: Suppose an $n \times n$ matrix $A = M_{B_0}(T)$ is the matrix of some operator $T : V \to V$ with respect to an ordered basis B_0. If another ordered basis B of V can be found such that $M_B(T) = D$ is diagonal, then Theorem 3 shows how to find an invertible P such that $P^{-1}AP = D$. In other words, the "algebraic" problem of choosing P such that $P^{-1}AP$ is diagonal comes down to the "geometric" problem of finding a basis B such that $M_B(T)$ is diagonal. This shift of emphasis is one of the most important techniques in linear algebra.

Each $n \times n$ matrix A can easily be realized as the matrix of an operator. In fact (Example 4§7.4),

$$M_{B_0}(T_A) = A$$

where $T_A : \mathbb{R}^n \to \mathbb{R}^n$ is the matrix operator given by $T_A(X) = AX$, and B_0 is the standard basis of $\mathbb{R}^n$. The first part of the next theorem gives the converse of Theorem 3: Any pair of similar matrices can be realized as the matrices of the same linear transformation with respect to different bases.

THEOREM 4

Let A be an $n \times n$ matrix and let B_0 be the standard basis of $\mathbb{R}^n$.

1. If $A' = P^{-1}AP$, let B be the ordered basis of $\mathbb{R}^n$ consisting of the columns of P in order. Then

$$M_{B_0}(T_A) = A \qquad \text{and} \qquad M_B(T_A) = A'$$

2. If B is any ordered basis of $\mathbb{R}^n$, let P be the (invertible) matrix whose columns are the vectors in B in order. Then

$$M_B(T_A) = P^{-1}AP$$

Proof   Let B and P be as in (1), and let $X_1, \ldots, X_n$ be the columns of P. Then

$$P_{B_0 \leftarrow B} = [C_{B_0}(X_1) \quad \cdots \quad C_{B_0}(X_n)] = [X_1 \quad \cdots \quad X_n] = P$$

Now (1) follows from Theorem 3. Similarly, $P = P_{B_0 \leftarrow B}$ in (2) and again Theorem 3 applies.   ◆

EXAMPLE 3

Given $A = \begin{bmatrix} 10 & 6 \\ -18 & -11 \end{bmatrix}$, $P = \begin{bmatrix} 2 & -1 \\ -3 & 2 \end{bmatrix}$, and $D = \begin{bmatrix} 1 & 0 \\ 0 & -2 \end{bmatrix}$, verify that $P^{-1}AP = D$

and use this fact to find a basis B of $\mathbb{R}^2$ such that $M_B(T_A) = D$.

Solution $P^{-1}AP = D$ holds if $AP = PD$; this verification is left to the reader. Let B consist of the columns of P : $B = \left\{ \begin{bmatrix} 2 \\ -3 \end{bmatrix}, \begin{bmatrix} -1 \\ 2 \end{bmatrix} \right\}$. Then Theorem 4 gives $M_B(T_A) = P^{-1}AP = D$. More explicitly,

$$M_B(T_A) = \left[C_B\left(T_A\begin{bmatrix} 2 \\ -3 \end{bmatrix}\right) \ \ C_B\left(T_A\begin{bmatrix} -1 \\ 2 \end{bmatrix}\right)\right] = \left[C_B\begin{bmatrix} 2 \\ -3 \end{bmatrix} \ \ C_B\begin{bmatrix} 2 \\ -4 \end{bmatrix}\right] = \begin{bmatrix} 1 & 0 \\ 0 & -2 \end{bmatrix} = D$$

◆◆◆

Given an $n \times n$ matrix A, Theorem 4 provides a new way to find an invertible matrix P such that $P^{-1}AP$ is "nice," say, diagonal. The idea is to find a basis $B = \{E_1, E_2, \ldots, E_n\}$ of $\mathbb{R}^n$ such that $M_B(T_A) = D$ is diagonal and take $P = [E_1 \ \ E_2 \ \cdots \ E_n]$ to be the matrix with the E_j as columns. Then

$$P^{-1}AP = M_B(T_A) = D$$

by Theorem 4. This converts the algebraic problem of finding P into the geometric problem of finding B. This new point of view is very powerful and will be explored in the next two sections.

Theorem 4 enables facts about matrices to be deduced from the corresponding properties of operators. Here is an example.

EXAMPLE 4

1. If $T : V \to V$ is an operator where V is finite dimensional, show that $TST = T$ for some invertible operator $S : V \to V$.

2. If A is an $n \times n$ matrix, show that $AUA = A$ for some invertible matrix U.

Solution

1. Let $B = \{\mathbf{e}_1, \ldots, \mathbf{e}_r, \mathbf{e}_{r+1}, \ldots, \mathbf{e}_n\}$ be a basis of V such that $\ker T = \text{span} \{\mathbf{e}_{r+1}, \ldots, \mathbf{e}_n\}$. Then $\{T(\mathbf{e}_1), \ldots, T(\mathbf{e}_r)\}$ is independent (Theorem 5§7.2), so complete it to a basis $\{T(\mathbf{e}_1), \ldots, T(\mathbf{e}_r), \mathbf{f}_{r+1}, \ldots, \mathbf{f}_n\}$ of V. Define $S : V \to V$ by

$$S[T(\mathbf{e}_i)] = \mathbf{e}_i \qquad \text{for } 1 \leq i \leq r$$
$$S(\mathbf{f}_j) = \mathbf{e}_j \qquad \text{for } r < j \leq n$$

Then S is an isomorphism by Theorem 2§7.3, and $TST = T$ because these operators agree on the basis B. In fact, $(TST)(\mathbf{e}_i) = T[ST(\mathbf{e}_i)] = T(\mathbf{e}_i)$ if $1 \leq i \leq r$, and $(TST)(\mathbf{e}_j) = TS(\mathbf{0}) = \mathbf{0} = T(\mathbf{e}_j)$ for $r < j \leq n$.

2. Given A, let $T = T_A : \mathbb{R}^n \to \mathbb{R}^n$. By (1) let $TST = T$ where $S : \mathbb{R}^n \to \mathbb{R}^n$ is an isomorphism. If B_0 is the standard basis of $\mathbb{R}^n$, then $A = M_{B_0}(T)$ by Theorem 4. If $U = M_{B_0}(S)$, then U is invertible by Theorem 4§7.4, and

$$AUA = M_{B_0}(T)M_{B_0}(S)M_{B_0}(T) = M_{B_0}(TST) = M_{B_0}(T) = A$$

as required.

◆◆◆

Recall that a property of $n \times n$ matrices is called a similarity invariant if, whenever a given $n \times n$ matrix A has the property, every matrix similar to A also has the property. Theorem 4§6.1 shows that rank, determinant, and trace are all similarity invariants.

To illustrate how such similarity invariants are related to linear operators, consider the case of rank. If $T : V \to V$ is a linear operator, the matrices of T with respect to various bases of V all have the same rank (being similar), so it is natural to regard the common rank of all these matrices as a property of T itself and not of the particular matrix used to describe T. Hence the rank of T could be *defined* to be the rank of A, where A is *any* matrix of T. This would be unambiguous by the above discussion. Of course, this is unnecessary in the case of rank because rank T was defined earlier to be the dimension of im T, and this turned out to equal the rank of every matrix representing T (Theorem 5§7.4). This definition of rank T is said to be *intrinsic* because it makes no reference to the matrices representing T. However, the technique serves to identify a property of T with *every* similarity invariant, and some of these properties are not so easily defined intrinsically.

In particular, if $T : V \to V$ is a linear operator on a finite dimensional space V, define the **determinant** of T (denoted det T) by

$$\det T = \det M_B(T), \quad B \text{ any basis of } V$$

This is independent of the choice of basis B, because if B_0 is any other basis of V, the matrices $M_B(T)$ and $M_{B_0}(T)$ are similar and so have the same determinant. In the same way, the **trace** of T (denoted tr T) can be defined by

$$\text{tr } T = \text{tr } M_B(T), \quad B \text{ any basis of } V$$

This is unambiguous for the same reason.

Properties of matrices can often be translated to properties of linear operators. Here is an example.

EXAMPLE 5

Let S and T denote linear operators on the finite dimensional space V. Show that

$$\det(ST) = \det S \det T$$

Solution   Choose a basis B of V and use Theorem 3§7.4.

$$\det(ST) = \det [M_B(ST)] = \det[M_B(S) \, M_B(T)]$$
$$= \det[M_B(S)] \det[M_B(T)] = \det S \det T$$

◆◆◆

Recall next that the characteristic polynomial of a matrix is another similarity invariant: If A and A' are similar matrices, then $c_A(x) = c_{A'}(x)$ (Theorem 4§6.1). As discussed above, the discovery of a similarity invariant means the discovery of a property of linear operators. In this case, if $T : V \to V$ is a linear operator on the finite dimensional space V, define the **characteristic polynomial** of T by

$$c_T(x) = c_A(x) \quad \text{where } A = M_B(T), \quad B \text{ any basis of } V$$

In other words, the characteristic polynomial of an operator T is the characteristic polynomial of *any* matrix representing T. This is unambiguous because any two such matrices are similar by Theorem 3.

EXAMPLE 6

Compute the characteristic polynomial $c_T(x)$ of the operator $T : \mathbf{P}_2 \to \mathbf{P}_2$ given by
$T(a + bx + cx^2) = (b + c) + (a + c)x + (a + b)x^2$.

Solution   If $B = \{1, x, x^2\}$, the corresponding matrix of T is

$$M_B(T) = [C_B[T(1)] \quad C_B[T(x)] \quad C_B[T(x^2)]] = \begin{bmatrix} 0 & 1 & 1 \\ 1 & 0 & 1 \\ 1 & 1 & 0 \end{bmatrix}$$

Hence the characteristic polynomial of T is

$$c_T(x) = \det[xI - M_B(T)] = \det \begin{bmatrix} x & -1 & -1 \\ -1 & x & -1 \\ -1 & -1 & x \end{bmatrix}$$

$$= (x + 1)^2(x - 2) = x^3 - 3x - 2$$

EXERCISES 7.5

1. In each case find $P_{D \leftarrow B}$, where B and D are ordered bases of V. Then verify $C_D(\mathbf{v}) = P_{D \leftarrow B}C_B(\mathbf{v})$.
  (a) $V = \mathbb{R}^2$, $B = \{(0, -1), (2, 1)\}$, $D = \{(0,1), (1, 1)\}$, $\mathbf{v} = (3, -5)$
  ◆ **(b)** $V = \mathbf{P}_2$, $B = \{x, 1 + x, x^2\}$, $D = \{2, x + 3, x^2 - 1\}$, $\mathbf{v} = 1 + x + x^2$
  (c) $V = \mathbf{M}_{22}$, $B = \left\{ \begin{bmatrix} 1 & 0 \\ 0 & 0 \end{bmatrix}, \begin{bmatrix} 0 & 1 \\ 0 & 0 \end{bmatrix}, \begin{bmatrix} 0 & 0 \\ 0 & 1 \end{bmatrix}, \begin{bmatrix} 0 & 0 \\ 1 & 0 \end{bmatrix} \right\}$,
  $D = \left\{ \begin{bmatrix} 1 & 1 \\ 0 & 0 \end{bmatrix}, \begin{bmatrix} 1 & 0 \\ 1 & 0 \end{bmatrix}, \begin{bmatrix} 1 & 0 \\ 0 & 1 \end{bmatrix}, \begin{bmatrix} 0 & 1 \\ 1 & 0 \end{bmatrix} \right\}$,
  $\mathbf{v} = \begin{bmatrix} 3 & -1 \\ 1 & 4 \end{bmatrix}$

2. In $\mathbb{R}^3$ find $P_{D \leftarrow B}$, where $B = \{(1, 0, 0), (1, 1, 0), (1, 1, 1)\}$ and $D = \{(1, 0, 1), (1, 0, -1), (0, 1, 0)\}$. If $\mathbf{v} = (a, b, c)$, show that $C_D(\mathbf{v}) = \frac{1}{2} \begin{bmatrix} a + c \\ a - c \\ 2b \end{bmatrix}$ and $C_B(\mathbf{v}) = \begin{bmatrix} a - b \\ b - c \\ c \end{bmatrix}$, and verify Theorem 1.

3. In $\mathbf{P}_3$ find $P_{D \leftarrow B}$ if $B = \{1, x, x^2, x^3\}$ and $D = \{1, (1 - x), (1 - x)^2, (1 - x)^3\}$. Then express $p = a + bx + cx^2 + dx^3$ as a polynomial in powers of $(1 - x)$.

4. In each case verify that $P_{D \leftarrow B}$ is the inverse of $P_{B \leftarrow D}$ and that $P_{E \leftarrow D}P_{D \leftarrow B} = P_{E \leftarrow B}$, where B, D, and E are ordered bases of V.
  (a) $V = \mathbb{R}^3$, $B = \{(1, 1, 1), (1, -2, 1), (1, 0, -1)\}$, $D =$ standard basis, $E = \{(1, 1, 1), (1, -1, 0), (-1, 0, 1)\}$
  ◆ **(b)** $V = \mathbf{P}_2$, $B = \{1, x, x^2\}$, $D = \{1 + x + x^2, 1 - x, -1 + x^2\}$, $E = \{x^2, x, 1\}$

5. Use property (2) of Theorem 2, with D the standard basis of $\mathbb{R}^n$, to find the inverse of:
  (a) $A = \begin{bmatrix} 1 & 1 & 0 \\ 1 & 0 & 1 \\ 0 & 1 & 1 \end{bmatrix}$
  ◆ **(b)** $A = \begin{bmatrix} 1 & 2 & 1 \\ 2 & 3 & 0 \\ -1 & 0 & 2 \end{bmatrix}$

6. Find $P_{D \leftarrow B}$ if $B = \{\mathbf{b}_1, \mathbf{b}_2, \mathbf{b}_3, \mathbf{b}_4\}$ and $D = \{\mathbf{b}_2, \mathbf{b}_3, \mathbf{b}_1, \mathbf{b}_4\}$. Matrices arising when the bases differ only in the *order* of the vectors are called **permutation matrices**.

7. In each case, find $P = P_{B_0 \leftarrow B}$ and verify that $P^{-1}M_{B_0}(T)P = M_B(T)$ for the given operator T.

  (a) $T : \mathbb{R}^3 \to \mathbb{R}^3$, $T(a, b, c) = (2a - b, b + c, c - 3a)$; $B_0 = \{(1, 1, 0), (1, 0, 1), (0, 1, 0)\}$ and B is the standard basis.

  ◆**(b)** $T : \mathbf{P}_2 \to \mathbf{P}_2$, $T(a + bx + cx^2) = (a + b) + (b + c)x + (c + a)x^2$; $B_0 = \{1, x, x^2\}$ and $B = \{1 - x^2, 1 + x, 2x + x^2\}$

  (c) $T : \mathbf{M}_{22} \to \mathbf{M}_{22}$, $T = \begin{bmatrix} a & b \\ c & d \end{bmatrix} = \begin{bmatrix} a + d & b + c \\ a + c & b + d \end{bmatrix}$

  $B_0 = \left\{ \begin{bmatrix} 1 & 0 \\ 0 & 0 \end{bmatrix}, \begin{bmatrix} 0 & 1 \\ 0 & 0 \end{bmatrix}, \begin{bmatrix} 0 & 0 \\ 1 & 0 \end{bmatrix}, \begin{bmatrix} 0 & 0 \\ 0 & 1 \end{bmatrix} \right\}$

  and $B = \left\{ \begin{bmatrix} 1 & 1 \\ 0 & 0 \end{bmatrix}, \begin{bmatrix} 0 & 0 \\ 1 & 1 \end{bmatrix}, \begin{bmatrix} 1 & 0 \\ 0 & 1 \end{bmatrix}, \begin{bmatrix} 0 & 1 \\ 1 & 1 \end{bmatrix} \right\}$

8. In each case, verify that $P^{-1}AP = D$ and find a basis B of $\mathbb{R}^2$ such that $M_B(T_A) = D$.

  (a) $A = \begin{bmatrix} 11 & -6 \\ 12 & -6 \end{bmatrix}$ $P = \begin{bmatrix} 2 & 3 \\ 3 & 4 \end{bmatrix}$ $D = \begin{bmatrix} 2 & 0 \\ 0 & 3 \end{bmatrix}$

  ◆**(b)** $A = \begin{bmatrix} 29 & -12 \\ 70 & -29 \end{bmatrix}$ $P = \begin{bmatrix} 3 & 2 \\ 7 & 5 \end{bmatrix}$ $D = \begin{bmatrix} 1 & 0 \\ 0 & -1 \end{bmatrix}$

9. In each case, compute the characteristic polynomial $c_T(x)$.

  (a) $T : \mathbb{R}^2 \to \mathbb{R}^2$, $T(a, b) = (a - b, 2b - a)$

  ◆**(b)** $T : \mathbb{R}^2 \to \mathbb{R}^2$, $T(a, b) = (3a + 5b, 2a + 3b)$

  (c) $T : \mathbf{P}_2 \to \mathbf{P}_2$, $T(a + bx + cx^2) = (a - 2c) + (2a + b + c)x + (c - a)x^2$

  ◆**(d)** $T : \mathbf{P}_2 \to \mathbf{P}_2$, $T(a + bx + cx^2) = (a + b - 2c) + (a - 2b + c)x + (b - 2a)x^2$

  (e) $T : \mathbb{R}^3 \to \mathbb{R}^3$, $T(a, b, c) = (b, c, a)$

◆**(f)** $T : \mathbf{M}_{22} \to \mathbf{M}_{22}$, $T\begin{bmatrix} a & b \\ c & d \end{bmatrix} = \begin{bmatrix} a - c & b - d \\ a - c & b - d \end{bmatrix}$

10. If V is finite dimensional, show that a linear operator T on V has an inverse if and only if $\det T \neq 0$.

11. Let S and T be linear operators on V where V is finite dimensional.

  (a) Show that $\text{tr}(ST) = \text{tr}(TS)$. [*Hint:* Theorem 3§6.1.]

  (b) [See Exercise 19§7.4.] Show that $\text{tr}(S + T) = \text{tr}S + \text{tr}T$, and $\text{tr}(aT) = a\,\text{tr}(T)$.

12. If A and B are $n \times n$ matrices, show that they have the same null space if and only if $A = UB$ for some invertible matrix U. [*Hint:* Exercise 31§7.3.]

13. If A and B are $n \times n$ matrices, show that they have the same column space if and only if $A = BU$ for some invertible matrix U [*Hint:* Exercise 31§7.3.]

14. Let $B = \{\mathbf{e}_1, \ldots, \mathbf{e}_n\}$ be the standard ordered basis of $\mathbb{R}^n$, written as columns. If $D = \{\mathbf{d}_1, \ldots, \mathbf{d}_n\}$ is any ordered basis, show that $P_{B \leftarrow D} = [\mathbf{d}_1 \cdots \mathbf{d}_n]$.

15. Let $B = \{\mathbf{b}_1, \mathbf{b}_2, \ldots, \mathbf{b}_n\}$ be any ordered basis of $\mathbb{R}^n$, written as columns. If $Q = [\mathbf{b}_1 \quad \mathbf{b}_2 \quad \cdots \quad \mathbf{b}_n]$ is the matrix with the $\mathbf{b}_i$ as columns, show that $QC_B(\mathbf{v}) = \mathbf{v}$ for all $\mathbf{v}$ in $\mathbb{R}^n$.

◆**16.** Prove Theorem 1 by expressing each $\mathbf{b}_j$ as follows:

  $$\mathbf{b}_j = p_{1j}\mathbf{d}_1 + p_{2j}\mathbf{d}_2 + \cdots + p_{nj}\mathbf{d}_n = \sum_{i=1}^{n} p_{ij}\mathbf{d}_i$$

  Show that, if $P = [p_{ij}]$, then $P = [C_D(\mathbf{b}_1) \cdots C_D(\mathbf{b}_n)]$ and $C_D(\mathbf{v}) = PC_B(\mathbf{v})$ for all $\mathbf{v}$ in V.

Section 7.6   Invariant Subspaces and Direct Sums

A fundamental question in linear algebra is the following: If $T : V \to V$ is a linear operator, how can a basis B of V be chosen so the matrix $M_B(T)$ is as simple as possible? A basic technique for answering such questions will be explained in this section. If U is a subspace of V, write $T(U) = \{T(\mathbf{u}) \mid \mathbf{u} \text{ in } U\}$.

> **DEFINITION**
>
> If $T : V \to V$ is a linear operator, a subspace U of V is said to be T-**invariant** if $T(\mathbf{u})$ lies in U for every vector $\mathbf{u}$ in U; that is $T(U) \subseteq U$.

EXAMPLE 1

If $T : V \rightarrow V$ is any linear operator, then V and 0 are T-invariant subspaces.   ◆◆◆

EXAMPLE 2

Define $T : \mathbb{R}^3 \rightarrow \mathbb{R}^3$ by $T(a, b, c) = (3a + 2b, b - c, 4a + 2b - c)$. Then $U = \{(a, b, a \mid a, b \text{ in } \mathbb{R}\}$ is T-invariant because

$$T(a, b, a) = (3a + 2b, b - a, 3a + 2b)$$

is in U for all a and b (the first and last entries are equal).

─── ◆◆◆

If a spanning set for a subspace U is known, it is easy to check whether U is invariant.

EXAMPLE 3

Let $T : V \rightarrow V$ be a linear operator, and suppose that $U = \text{span}\{\mathbf{u}_1, \mathbf{u}_2, \ldots, \mathbf{u}_k\}$ is a subspace of V. Show that U is T-invariant if and only if $T(\mathbf{u}_i)$ lies in U for each $i = 1, 2, \ldots, k$.

Solution   Given $\mathbf{u}$ in U, write it as $\mathbf{u} = r_1\mathbf{u}_1 + \cdots + r_k\mathbf{u}_k$, r_i in $\mathbb{R}$. Then

$$T(\mathbf{u}) = r_1 T(\mathbf{u}_1) + \cdots + r_k T(\mathbf{u}_k)$$

and this lies in U if each $T(\mathbf{u}_i)$ lies in U. This shows that U is T-invariant if each $T(\mathbf{u}_i)$ lies in U; the converse is clear.   ◆◆◆

EXAMPLE 4

Define $T : \mathbb{R}^2 \rightarrow \mathbb{R}^2$ by $T(a, b) = (b, -a)$. Show that $\mathbb{R}^2$ contains no T-invariant subspace except 0 and $\mathbb{R}^2$.

Solution   Suppose, if possible, that U is T-invariant, but $U \neq 0$, $U \neq \mathbb{R}^2$. Then U has dimension 1 — say, $U = \mathbb{R}X$ where $X \neq 0$. Now $T(X)$ lies in U — say, $T(X) = rX$, r in $\mathbb{R}$. If we write $X = (a, b)$, this is $(b, -a) = r(a, b)$, which gives $b = ra$ and $-a = rb$. Eliminating b gives $r^2 a = rb = -a$, whence $a = 0$. Then $b = ra = 0$ too, contrary to the assumption that $X \neq 0$. Hence no one-dimensional T-invariant subspace exists.

─── ◆◆◆

Let $T : V \rightarrow V$ be a linear operator. If U is any T-invariant subspace of V, then

$$T : U \rightarrow U$$

is a linear operator *on the subspace* U, called the **restriction** of T to U. This is the reason for the importance of T-invariant subspaces and is the first step toward finding a basis that simplifies the matrix of T.

THEOREM 1

Let $T : V \to V$ be a linear operator where V has dimension n and suppose that U is any T-invariant subspace of V. Let $B_1 = \{e_1, \ldots, e_k\}$ be any basis of U and extend it to a basis $B = \{e_1, \ldots, e_k, e_{k+1}, \ldots, e_n\}$ of V in any way. Then $M_B(T)$ has the block triangular form

$$M_B(T) = \begin{bmatrix} M_{B_1}(T) & Y \\ 0 & Z \end{bmatrix}$$

where Z is $(n - k) \times (n - k)$ and $M_{B_1}(T)$ is the matrix of the restriction of T to U.

Proof The matrix of (the restriction of) $T : U \to U$ with respect to the basis B_1 is the $k \times k$ matrix

$$M_{B_1}(T) = [\,C_{B_1}[T(e_1)]\ [C_{B_1}[T(e_2)]\ \cdots\ C_{B_1}[T(e_k)]\,]$$

Now compare the first column $C_{B_1}[T(e_1)]$ here with the first column $C_B[T(e_1)]$ of $M_B(T)$. The fact that $T(e_1)$ lies in U (because U is T-invariant) means that $T(e_1)$ has the form

$$T(e_1) = t_1 e_1 + t_2 e_2 + \cdots + t_k e_k + 0 e_{k+1} + \cdots + 0 e_n$$

Consequently,

$$C_{B_1}[T(e_1)] = \begin{bmatrix} t_1 \\ t_2 \\ \vdots \\ t_k \end{bmatrix} \text{ in } \mathbb{R}^k \qquad \text{whereas} \qquad C_B[T(e_1)] = \begin{bmatrix} t_1 \\ t_2 \\ \vdots \\ t_k \\ 0 \\ \vdots \\ 0 \end{bmatrix} \text{ in } \mathbb{R}^n$$

This shows that $M_B(T)$ and $\begin{bmatrix} M_{B_1}(T) & Y \\ 0 & Z \end{bmatrix}$ have identical first columns. Similar statements apply to columns $2, 3, \ldots, k$, and this proves the theorem.   ◆

The block upper triangular form for the matrix $M_B(T)$ in Theorem 1 is very useful because the determinant of such a matrix equals the product of the determinants of each of the diagonal blocks. This is recorded in Theorem 2 for reference, together with an important application to characteristic polynomials.

THEOREM 2

Let A be a block upper triangular matrix, say

$$A = \begin{bmatrix} A_{11} & A_{12} & A_{13} & \cdots & A_{1n} \\ 0 & A_{22} & A_{23} & \cdots & A_{2n} \\ 0 & 0 & A_{33} & \cdots & A_{3n} \\ \vdots & \vdots & \vdots & & \vdots \\ 0 & 0 & 0 & \cdots & A_{nn} \end{bmatrix}$$

where the diagonal blocks are square. Then:

1. $\det A = (\det A_{11})(\det A_{22})(\det A_{33}) \cdots (\det A_{nn})$

2. $c_A(x) = c_{A_{11}}(x)\, c_{A_{22}}(x)\, c_{A_{33}}(x) \cdots c_{A_{nn}}(x)$

Proof   If $n = 2$, (1) is Theorem 5§3.1; the general case is by induction on n. Then (2) follows from (1) because

$$xI - A = \begin{bmatrix} xI - A_{11} & -A_{12} & -A_{13} & \cdots & -A_{1n} \\ 0 & xI - A_{22} & -A_{23} & \cdots & -A_{2n} \\ 0 & 0 & xI - A_{33} & \cdots & -A_{3n} \\ \vdots & \vdots & \vdots & & \vdots \\ 0 & 0 & 0 & \cdots & xI - A_{nn} \end{bmatrix}$$

where, in each diagonal block, the symbol I stands for the identity matrix of the appropriate size.   ♦

EXAMPLE 5

Consider the linear operator $T : \mathbf{P}_2 \rightarrow \mathbf{P}_2$ given by

$$T(a + bx + cx^2) = (-2a - b + 2c) + (a + b)x + (-6a - 2b + 5c)x^2$$

Show that $U = \text{span}\{x, 1 + 2x^2\}$ is T-invariant, use it to find a block upper triangular matrix for T, and use that to compute $c_T(x)$.

Solution U is T-invariant because $T(x)$ and $T(1 + 2x^2)$ both lie in U:

$$T(x) = -1 + x - 2x^2 = x - (1 + 2x^2)$$
$$T(1 + 2x^2) = 2 + x + 4x^2 = x + 2(1 + 2x^2)$$

Extend the basis $B_1 = \{x, 1 + 2x^2\}$ of U to a basis B of $\mathbf{P}_2$ in any way at all — say, $B = \{x, 1 + 2x^2, x^2\}$. Then

$$M_B(T) = [C_B[T(x)] \quad C_B[T(1 + 2x^2)] \quad C_B[T(x^2)]]$$
$$= [C_B(-1 + x - 2x^2) \quad C_B(2 + x + 4x^2) \quad C_B(2 + 5x^2)]$$
$$= \left[\begin{array}{cc|c} 1 & 1 & 0 \\ -1 & 2 & 2 \\ \hline 0 & 0 & 1 \end{array}\right]$$

is in block upper triangular form. Finally,

$$c_T(x) = \det \left[\begin{array}{cc|c} x-1 & -1 & 0 \\ 1 & x-2 & -2 \\ \hline 0 & 0 & x-1 \end{array}\right] = (x^2 - 3x + 3)(x - 1)$$

◆◆◆

Eigenvalues

Let $T : V \rightarrow V$ be a linear operator. A one-dimensional subspace $\mathbb{R}\mathbf{v}$, $\mathbf{v} \neq \mathbf{0}$, is T-invariant if and only if $T(r\mathbf{v}) = rT(\mathbf{v})$ lies in $\mathbb{R}\mathbf{v}$ for all r in $\mathbb{R}$. This holds if and only if $T(\mathbf{v})$ lies in $\mathbb{R}\mathbf{v}$; that is, $T(\mathbf{v}) = \lambda\mathbf{v}$ for some λ in $\mathbb{R}$.

DEFINITION

A real number λ is called an **eigenvalue** of an operator $T : V \rightarrow V$ if

$$T(\mathbf{v}) = \lambda\mathbf{v}$$

holds for some nonzero vector $\mathbf{v}$ in V. In this case, $\mathbf{v}$ is called an **eigenvector** of T corresponding to λ. The subspace

$$E_\lambda(T) = \{\mathbf{v} \text{ in } V \mid T(\mathbf{v}) = \lambda\mathbf{v}\}$$

is called the **eigenspace** of T corresponding to λ.

These terms are consistent with those used in Section 6.1 for matrices. If A is an $n \times n$ matrix, a real number λ is an eigenvalue of the operator $T_A : \mathbb{R}^n \rightarrow \mathbb{R}^n$ if and only if λ is an eigenvalue of the matrix A. Moreover, the eigenspaces agree:

$$E_\lambda(T_A) = \{X \text{ in } \mathbb{R}^n \mid AX = \lambda X\} = E_\lambda(A)$$

The following theorem reveals the connection between the eigenspaces of an operator T and those of the matrices representing T.

THEOREM 3

Let $T : V \rightarrow V$ be a linear operator where dim $V = n$, let B denote any ordered basis of V, and let $C_B : V \rightarrow \mathbb{R}^n$ denote the coordinate isomorphism. Then:

1. The eigenvalues λ of T are precisely the eigenvalues of the matrix $M_B(T)$ and thus are the roots of the characteristic polynomial $c_T(x)$.

2. In this case the eigenspaces $E_\lambda(T)$ and $E_\lambda[M_B(T)]$ are isomorphic via $C_B : E_\lambda(T) \to E_\lambda[M_B(T)]$.

Proof   Write $A = M_B(T)$. If $T(\mathbf{v}) = \lambda\mathbf{v}$, then applying C_B gives $\lambda C_B(\mathbf{v}) = C_B[T(\mathbf{v})]$ $= AC_B(\mathbf{v})$, using Theorem 1§7.5. Hence $C_B(\mathbf{v})$ lies in $E_\lambda(A)$, so we do indeed have a function $C_B : E_\lambda(T) \to E_\lambda(A)$. It is clearly linear and one-to-one; we claim it is onto. If X is in $E_\lambda(A)$, write $X = C_B(\mathbf{v})$ for some $\mathbf{v}$ in V (C_B is onto). This $\mathbf{v}$ actually lies in $E_\lambda(T)$:

$$C_B[T(\mathbf{v})] = AC_B(\mathbf{v}) = AX = \lambda X = \lambda C_B(\mathbf{v}) = C_B(\lambda\mathbf{v})$$

so $T(\mathbf{v}) = \lambda\mathbf{v}$ (C_B is one-to-one). This proves (2). As to (1), we have already shown that eigenvalues of T are eigenvalues of A. The converse follows, as in the foregoing proof that C_B is onto.   ◆

Theorem 3 shows how to pass back and forth between the eigenvectors of an operator T and the eigenvectors of any matrix $M_B(T)$ of T:

$$\mathbf{v} \text{ lies in } E_\lambda(T) \qquad \text{if and only if} \qquad C_B(\mathbf{v}) \text{ lies in } E_\lambda[M_B(T)]$$

EXAMPLE 6

Find the eigenvalues and eigenspaces for $T : \mathbf{P}_3 \to \mathbf{P}_3$ given by

$$T(a + bx + cx^2) = (2a + b + c) + (2a + b - 2c)x - (a + 2c)x^2$$

Solution   If $B = \{1, x, x^2\}$, then

$$M_B(T) = [C_B[T(1)] \quad C_B[T(x)] \quad C_B[T(x^2)]] = \begin{bmatrix} 2 & 1 & 1 \\ 2 & 1 & -2 \\ -1 & 0 & -2 \end{bmatrix}$$

Hence $c_T(x) = \det[xI - M_B(T)] = (x + 1)^2(x - 3)$ by Example 3§6.1. Moreover,

$$E_{-1}[M_B(T)] = \mathbb{R}\begin{bmatrix} -1 \\ 2 \\ 1 \end{bmatrix} \text{ and } E_3[M_B(T)] = \mathbb{R}\begin{bmatrix} 5 \\ 6 \\ -1 \end{bmatrix}, \text{ so Theorem 3 gives } E_{-1}(T) =$$

$\mathbb{R}(-1 + 2x + x^2)$ and $E_3(T) = \mathbb{R}(5 + 6x - x^2)$.

◆◆◆

THEOREM 4

Each eigenspace of a linear operator $T : V \to V$ is a T-invariant subspace of V.

Proof If $\mathbf{v}$ lies in the eigenspace $E_\lambda(T)$, then $T(\mathbf{v}) = \lambda\mathbf{v}$, so $T[T(\mathbf{v})] = T(\lambda\mathbf{v}) = \lambda T(\mathbf{v})$. This shows that $T(\mathbf{v})$ lies in $E_\lambda(T)$ too. ◆

If T is an operator on an n-dimensional space V, and if λ is an eigenvalue of T, then λ is a root of the characteristic polynomial $c_T(x)$ by Theorem 2. As for matrices, λ is said to be an eigenvalue of **multiplicity** m if

$$c_T(x) = (x - \lambda)^m q(x), \qquad q(\lambda) \neq 0$$

The multiplicity m is the highest power of $x - \lambda$ that is a factor of $c_T(x)$. The next theorem shows that m is an upper bound on the dimension of the eigenspace $E_\lambda(T)$ determined by T. The matrix version of this result was stated as Theorem 5§6.2 but the matrix techniques then available could not easily provide a proof.

THEOREM 5

Let $T : V \to V$ be a linear operator where dim $V = n$. Let λ be an eigenvalue of T of multiplicity m. Then $m \geq \dim[E_\lambda(T)]$.

Proof Write $d = \dim[E_\lambda(T)]$ and let $B = \{\mathbf{e}_1, \ldots, \mathbf{e}_d, \ldots, \mathbf{e}_n\}$ be an ordered basis of V such that $B_1 = \{\mathbf{e}_1, \ldots, \mathbf{e}_d\}$ is a basis of $E_\lambda(T)$. Theorem 4 shows that $E_\lambda(T)$ is T-invariant, so $T : E_\lambda(T) \to E_\lambda(T)$ is a linear operator and $M_{B_1}(T) = \lambda I_d$ because each vector in B_1 is an eigenvector corresponding to λ. Hence Theorem 1 shows that $M_B(T)$ has the block form

$$M_B(T) = \begin{bmatrix} \lambda I_d & Y \\ 0 & Z \end{bmatrix}$$

But then $c_T(x)$ has the form

$$c_T(x) = \det[xI - M_B(T)] = (x - \lambda)^d p(x)$$

The multiplicity m is the *highest* power of $x - \lambda$ that is a factor of $c_T(x)$, so it follows that $m \geq d$, as required. ◆

Direct Sums

If U and W are subspaces of V, their **sum** $U + W$ and their **intersection** $U \cap W$ are defined by

$$U + W = \{\mathbf{u} + \mathbf{w} \mid \mathbf{u} \text{ in } U \text{ and } \mathbf{w} \text{ in } W\}$$

$$U \cap W = \{\mathbf{v} \mid \mathbf{v} \text{ lies in both } U \text{ and } W\}$$

These are subspaces of V, the sum containing both U and W and the intersection contained in both U and W. It turns out that the most interesting pairs U and W of subspaces are those for which $U \cap W$ is as small as possible and $U + W$ is as large as possible.

DEFINITION

A vector space V is said to be the **direct sum** of subspaces U and W if

$$U \cap W = 0 \qquad \text{and} \qquad U + W = V$$

In this case we write $V = U \oplus W$. Given a subspace U, any subspace W such that $V = U \oplus W$ is called a **complement** of U in V.

EXAMPLE 7

In the space $\mathbb{R}^5$, consider the subspaces $U = \{(a, b, c, 0, 0) \mid a, b, \text{ and } c \text{ in } \mathbb{R}\}$ and $W = \{(0, 0, 0, d, e) \mid d \text{ and } e \text{ in } \mathbb{R}\}$. Show that $\mathbb{R}^5 = U \oplus W$.

Solution

If $X = (a, b, c, d, e)$ is any vector in $\mathbb{R}^5$, then $X = (a, b, c, 0, 0) + (0, 0, 0, d, e)$, so X lies in $U + W$. Hence $\mathbb{R}^5 = U + W$. To show that $U \cap W = 0$, let $X = (a, b, c, d, e)$ lie in $U \cap W$. Then $d = e = 0$ because X lies in U, and $a = b = c = 0$ because X lies in W. Thus $X = (0, 0, 0, 0, 0) = 0$, so 0 is the only vector in $U \cap W$. Hence $U \cap W = 0$.

$\blacklozenge\blacklozenge\blacklozenge$

The next example shows that the projection theorem (Theorem 7§6.3) is really about a direct sum decomposition.

EXAMPLE 8

If U is a subspace of $\mathbb{R}^n$, show that $\mathbb{R}^n = U \oplus U^\perp$.

Solution

The equation $\mathbb{R}^n = U + U^\perp$ holds because, given X in $\mathbb{R}^n$, the vector $\text{proj}_U(X)$ lies in U and $X - \text{proj}_U(X)$ lies in $U^\perp$. To see that $U \cap U^\perp = 0$, observe that any vector in $U \cap U^\perp$ is orthogonal to itself and hence must be zero.

$\blacklozenge\blacklozenge\blacklozenge$

EXAMPLE 9

Let $\{\mathbf{e}_1, \mathbf{e}_2, \ldots, \mathbf{e}_n\}$ be a basis of a vector space V, and partition it into two parts: $\{\mathbf{e}_1, \ldots, \mathbf{e}_k\}$ and $\{\mathbf{e}_{k+1}, \ldots, \mathbf{e}_n\}$. If $U = \text{span}\{\mathbf{e}_1, \ldots, \mathbf{e}_k\}$ and $W = \text{span}\{\mathbf{e}_{k+1}, \ldots, \mathbf{e}_n\}$, show that $V = U \oplus W$.

Solution

If $\mathbf{v}$ lies in $U \cap W$, then $\mathbf{v} = a_1\mathbf{e}_1 + \cdots + a_k\mathbf{e}_k$ and $\mathbf{v} = b_{k+1}\mathbf{e}_{k+1} + \cdots + b_n\mathbf{e}_n$ hold for some a_i and b_j in $\mathbb{R}$. The fact that the $\mathbf{e}_i$ are linearly independent forces all $a_i = b_j = 0$, so $\mathbf{v} = \mathbf{0}$. Hence $U \cap W = 0$. Now, given $\mathbf{v}$ in V, write $\mathbf{v} = v_1\mathbf{e}_1 + \cdots + v_n\mathbf{e}_n$ where the v_i are in $\mathbb{R}$. Then $\mathbf{v} = \mathbf{u} + \mathbf{w}$, where $\mathbf{u} = v_1\mathbf{e}_1 + \cdots + v_k\mathbf{e}_k$ lies in U and $\mathbf{w} = v_{k+1}\mathbf{e}_{k+1} + \cdots + v_n\mathbf{e}_n$ lies in W. This proves that $V = U + W$.

$\blacklozenge\blacklozenge\blacklozenge$

Example 9 is typical of all direct sum decompositions.

THEOREM 6

Let U and W be subspaces of a finite dimensional vector space V. The following conditions are equivalent:

1. $V = U \oplus W$.

2. Each vector $\mathbf{v}$ in V can be written uniquely in the form

$$\mathbf{v} = \mathbf{u} + \mathbf{w} \qquad \mathbf{u} \text{ in } U, \mathbf{w} \text{ in } W$$

   (The uniqueness means that if $\mathbf{v} = \mathbf{u}_1 + \mathbf{w}_1$ is another such representation, then $\mathbf{u}_1 = \mathbf{u}$ and $\mathbf{w}_1 = \mathbf{w}$.)

3. If $\{\mathbf{u}_1, \ldots, \mathbf{u}_k\}$ and $\{\mathbf{w}_1, \ldots, \mathbf{w}_m\}$ are bases of U and W, respectively, then $B = \{\mathbf{u}_1, \ldots, \mathbf{u}_k, \mathbf{w}_1, \ldots, \mathbf{w}_m\}$ is a basis of V.

Proof

(1) *implies* (2). Given $\mathbf{v}$ in V, we have $\mathbf{v} = \mathbf{u} + \mathbf{w}$, $\mathbf{u}$ in U, $\mathbf{w}$ in W, because $V = U + W$. If also $\mathbf{v} = \mathbf{u}_1 + \mathbf{w}_1$, then $\mathbf{u} - \mathbf{u}_1 = \mathbf{w}_1 - \mathbf{w}$ lies in $U \cap W = 0$, so $\mathbf{u} = \mathbf{u}_1$ and $\mathbf{w} = \mathbf{w}_1$.

   (2) *implies* (3). Given $\mathbf{v}$ in V, we have $\mathbf{v} = \mathbf{u} + \mathbf{w}$, $\mathbf{u}$ in U, $\mathbf{w}$ in W. Hence $\mathbf{v}$ lies in span B; that is, B spans V. To see that B is independent, let $a_1\mathbf{u}_1 + \cdots + a_k\mathbf{u}_k + b_1\mathbf{w}_1 + \cdots + b_m\mathbf{w}_m = \mathbf{0}$. Write $\mathbf{u} = a_1\mathbf{u}_1 + \cdots + a_k\mathbf{u}_k$ and $\mathbf{w} = b_1\mathbf{w}_1 + \cdots + b_m\mathbf{w}_m$. Then $\mathbf{u} + \mathbf{w} = \mathbf{0}$, and so $\mathbf{u} = \mathbf{0}$ and $\mathbf{w} = \mathbf{0}$ by the uniqueness in (2). Hence $a_i = 0$ and $b_j = 0$ for all i and j.

   (3) *implies* (1). This is by Example 9.   ◆

Condition (3) gives the following useful result.

THEOREM 7

If a finite dimensional vector space V is the direct sum $V = U \oplus W$ of subspaces U and W, then

$$\dim V = \dim U + \dim W$$

These direct sum decompositions of V play an important role in any discussion of invariant subspaces. If $T : V \rightarrow V$ is a linear operator and if U_1 is a T-invariant subspace, the block upper triangular matrix

$$M_B(T) = \begin{bmatrix} M_{B_1}(T) & Y \\ 0 & Z \end{bmatrix}$$

in Theorem 1 is achieved by choosing any basis $B_1 = \{\mathbf{e}_1, \ldots, \mathbf{e}_k\}$ of U_1 and completing it to a basis $B = \{\mathbf{e}_1, \ldots, \mathbf{e}_k, \mathbf{e}_{k+1}, \ldots, \mathbf{e}_n\}$ of V in any way at all. The fact that U_1 is T-invariant ensures that the first k columns of $M_B(T)$ have the given form, and the question arises whether the additional basis vectors $\mathbf{e}_{k+1}, \ldots, \mathbf{e}_n$ can be chosen such that

$$U_2 = \text{span}\{\mathbf{e}_{k+1}, \ldots, \mathbf{e}_n\}$$

is *also* T-invariant. In other words, does each T-invariant subspace of V have a T-invariant complement? Unfortunately the answer is no in general (see Example 11); but when it is possible, the matrix $M_B(T)$ simplifies further. The fact that the complement $U_2 = \text{span}\{\mathbf{e}_{k+1}, \ldots, \mathbf{e}_n\}$ is T-invariant too means that $Y = 0$ in the preceding notation and that $Z = M_{B_2}(T)$ is the matrix of the restriction of T to U_2 (where $B_2 = \{\mathbf{e}_{k+1}, \ldots, \mathbf{e}_n\}$). The verification is the same as in the proof of Theorem 1.

THEOREM 8

Let $T : V \to V$ be a linear operator where V has dimension n. Suppose $V = U_1 \oplus U_2$ where both U_1 and U_2 are T-invariant. If $B_1 = \{\mathbf{e}_1, \ldots, \mathbf{e}_k\}$ and $B_2 = \{\mathbf{e}_{k+1}, \ldots, \mathbf{e}_n\}$ are bases of U_1 and U_2 respectively, then

$$B = \{\mathbf{e}_1, \ldots, \mathbf{e}_k, \mathbf{e}_{r+1}, \ldots, \mathbf{e}_n\}$$

is a basis of V, and $M_B(T)$ has the block diagonal form

$$M_B(T) = \begin{bmatrix} M_{B_1}(T) & 0 \\ 0 & M_{B_2}(T) \end{bmatrix}$$

where $M_{B_1}(T)$ and $M_{B_2}(T)$ are the matrices of the restrictions of T to U_1 and to U_2, respectively.

The linear operator $T : V \to V$ is said to be **reducible** if nonzero T-invariant subspaces U_1 and U_2 can be found such that $V = U_1 \oplus U_2$. Then T has a matrix in block diagonal form as in Theorem 8, and the study of T is reduced to studying its restrictions to the lower-dimensional spaces U_1 and U_2. If these can be determined, so can T. Here is an example in which the action of T on the invariant subspaces U_1 and U_2 is very simple indeed. The result for operators is used to derive the corresponding similarity theorem for matrices.

EXAMPLE 10

Let $T : V \to V$ be a linear operator satisfying $T^2 = 1$ (such operators are called **involutions**). Define

$$U_1 = \{\mathbf{v} \mid T(\mathbf{v}) = \mathbf{v}\} \qquad \text{and} \qquad U_2 = \{\mathbf{v} \mid T(\mathbf{v}) = -\mathbf{v}\}$$

(a) Show that $V = U_1 \oplus U_2$.

(b) If $\dim V = n$, find a basis B of V such that $M_B(T) = \begin{bmatrix} I_k & 0 \\ 0 & -I_{n-k} \end{bmatrix}$ for some k.

(c) Conclude that, if A is an $n \times n$ matrix such that $A^2 = I$, then A is similar to $\begin{bmatrix} I_k & 0 \\ 0 & -I_{n-k} \end{bmatrix}$ for some k.

Solution

(a) The verification that U_1 and U_2 are subspaces of V is left to the reader. If $\mathbf{v}$ lies in $U_1 \cap U_2$, then $\mathbf{v} = T(\mathbf{v}) = -\mathbf{v}$, and it follows that $\mathbf{v} = \mathbf{0}$. Hence $U_1 \cap U_2 = 0$. Given $\mathbf{v}$ in V, write

$$\mathbf{v} = \tfrac{1}{2}\{[\mathbf{v} + T(\mathbf{v})] + [\mathbf{v} - T(\mathbf{v})]\}$$

Then $\mathbf{v} + T(\mathbf{v})$ lies in U_1, because $T[\mathbf{v} + T(\mathbf{v})] = T(\mathbf{v}) + T^2(\mathbf{v}) = T(\mathbf{v}) + \mathbf{v} = \mathbf{v} + T(\mathbf{v})$. Similarly $\mathbf{v} - T(\mathbf{v})$ lies in U_2, and it follows that $V = U_1 + U_2$. This proves part (a).

(b) U_1 and U_2 are clearly T-invariant, so the result follows from Theorem 8 if bases $B_1 = \{\mathbf{e}_1, \ldots, \mathbf{e}_k\}$ and $B_2 = \{\mathbf{e}_{k+1}, \ldots, \mathbf{e}_n\}$ of U_1 and U_2 can be found such that $M_{B_1}(T) = I_k$ and $M_{B_2}(T) = -I_{n-k}$. But this is true for *any* choice of B_1 and B_2:

$$\begin{aligned} M_{B_1}(T) &= [C_{B_1}[T(\mathbf{e}_1)] \quad C_{B_1}[T(\mathbf{e}_2)] \quad \cdots \quad C_{B_1}[T(\mathbf{e}_k)]] \\ &= [C_{B_1}(\mathbf{e}_1) \quad C_{B_1}(\mathbf{e}_2) \quad \cdots \quad C_{B_1}(\mathbf{e}_k)] \\ &= I_k \end{aligned}$$

A similar argument shows that $M_{B_2}(T) = -I_{n-k}$, so part (b) follows with $B = \{\mathbf{e}_1, \mathbf{e}_2, \ldots, \mathbf{e}_n\}$.

(c) Given A such that $A^2 = I$, consider $T_A : \mathbb{R}^n \to \mathbb{R}^n$. Then $(T_A)^2 = T_{A^2} = T_I = 1$ (by Theorem 6§7.3), so by part (a) there exists a basis B of $\mathbb{R}^n$ such that

$$M_B(T_A) = \begin{bmatrix} I_r & 0 \\ 0 & -I_{n-r} \end{bmatrix}$$

But Theorem 4§7.5 shows that $M_B(T_A) = P^{-1}AP$ for some invertible matrix P, and this proves part (c).

◆◆◆

Note that the passage from the result for operators to the analogous result for matrices is routine and can be carried out in any situation, as in the verification of part (c). The key is the analysis of the operators. In this case, the involutions are just the operators satisfying $T^2 = 1$, and the simplicity of this condition means that the invariant subspaces U_1 and U_2 are easy to find.

Unfortunately, not every linear operator $T : V \to V$ is reducible. In fact, the linear operator in Example 4 has *no* invariant subspaces except 0 and V. On the other hand, one might expect that this is the only type of nonreducible operator; that is, if the operator *has* an invariant subspace that is not 0 or V, then *some* invariant complement must exist. The next example shows that even this is not valid.

EXAMPLE 11

Consider the operator $T : \mathbb{R}^2 \to \mathbb{R}^2$ given by $T\begin{bmatrix} a \\ b \end{bmatrix} = \begin{bmatrix} a + b \\ b \end{bmatrix}$. Show that $U_1 = \mathbb{R}\begin{bmatrix} 1 \\ 0 \end{bmatrix}$ is T-invariant but that U_1 has no T-invariant complement in $\mathbb{R}^2$.

Solution   Because $U_1 = \text{span}\left\{ \begin{bmatrix} 1 \\ 0 \end{bmatrix} \right\}$, and $T\begin{bmatrix} 1 \\ 0 \end{bmatrix} = \begin{bmatrix} 1 \\ 0 \end{bmatrix}$, it follows (by Example 3) that U_1 is T-invariant. Now assume, if possible, that U_1 has a T-invariant complement U_2 in $\mathbb{R}^2$. Then $U_1 \oplus U_2 = \mathbb{R}^2$ and $T(U_2) \subseteq U_2$. Now

$$2 = \dim \mathbb{R}^2 = \dim U_1 + \dim U_2 = 1 + \dim U_2$$

so $\dim U_2 = 1$. Let $U_2 = \mathbb{R}\begin{bmatrix} p \\ q \end{bmatrix}$. Then $\begin{bmatrix} p \\ q \end{bmatrix}$ is not in U_1 (because $U_1 \cap U_2 = 0$) and hence $q \neq 0$. On the other hand, $T\begin{bmatrix} p \\ q \end{bmatrix}$ lies in $U_2 = \mathbb{R}\begin{bmatrix} p \\ q \end{bmatrix}$, say

$$\begin{bmatrix} p + q \\ q \end{bmatrix} = T\begin{bmatrix} p \\ q \end{bmatrix} = \lambda\begin{bmatrix} p \\ q \end{bmatrix}, \quad \lambda \text{ in } \mathbb{R}$$

Hence $p + q = \lambda p$ and $q = \lambda q$. Because $q \neq 0$, the second of these equations implies that $\lambda = 1$, whence the first implies that $q = 0$, a contradiction. So a T-invariant complement of U_1 does not exist.

◆◆◆

EXERCISES 7.6

1. Let T be a linear operator on V. If U and U_1 are T-invariant, show that $U \cap U_1$ and $U + U_1$ are also T-invariant.

2. If $T : V \to V$ is any linear operator, show that ker T and im T are T-invariant subspaces.

3. Let S and T be linear operators on V and assume that $ST = TS$.
   (a) Show that im S and ker S are T-invariant.
   ◆(b) If U is T-invariant, show that $S(U)$ is T-invariant.

4. Let $T : V \to V$ be a linear operator. Given **e** in V, let U denote the set of vectors in V that lie in every T-invariant subspace that contains **e**.
   (a) Show that U is a T-invariant subspace of V containing **e**.
   (b) Show that U is contained in every T-invariant subspace of V that contains **e**.

5. (a) Show that every subspace is T-invariant if T is a scalar operator (Example 3§7.1).

(b) Conversely, if every subspace is T-invariant, show that T is scalar.

◆ **6.** Show that the only subspaces of V that are T-invariant for every operator $T : V \to V$ are 0 and V. Assume that V is finite dimensional. [*Hint:* Theorem 4§7.1.]

7. Suppose that $T : V \to V$ is a linear operator and that U is a T-invariant subspace of V. If S is an invertible operator, put $T' = STS^{-1}$. Show that $S(U)$ is a T'-invariant subspace.

8. In each case, show that U is T-invariant, use it to find a block upper triangular matrix for T, and use that to compute $c_T(x)$.
  (a) $T : \mathbf{P}_2 \to \mathbf{P}_2$, $T(a + bx + cx^2) = (-a + 2b + c) + (a + 3b + c)x + (a + 4b)x^2$, $U = \text{span}\{1, x + x^2\}$
  ◆**(b)** $T : \mathbf{P}_2 \to \mathbf{P}_2$, $T(a + bx + cx^2) = (5a - 2b + c) + (5a - b + c)x + (a + 2c)x^2$, $U = \text{span}\{1 - 2x^2, x + x^2\}$

9. In each case, show that $T_A : \mathbb{R}^2 \to \mathbb{R}^2$ has no invariant subspaces except 0 and $\mathbb{R}^2$.
  (a) $A = \begin{bmatrix} 1 & 2 \\ -1 & -1 \end{bmatrix}$
  ◆**(b)** $A = \begin{bmatrix} \cos\theta & \sin\theta \\ -\sin\theta & \cos\theta \end{bmatrix}$, $0 < \theta < \pi$.

10. In each case, show that $V = U \oplus W$.
  (a) $V = \mathbb{R}^4$, $U = \text{span}\{(1,1,0,0), (0,1,1,0)\}$, $W = \text{span}\{(0,1,0,1), (0,0,1,1)\}$
  ◆**(b)** $V = \mathbb{R}^4$, $U = \{(a,a,b,b) \mid a, b \text{ in } \mathbb{R}\}$, $W = \{(c,d,c,-d) \mid c, d \text{ in } \mathbb{R}\}$
  (c) $V = \mathbf{P}_3$, $U = \{a + bx \mid a, b \text{ in } \mathbb{R}\}$, $W = \{ax^2 + bx^3 \mid a, b \text{ in } \mathbb{R}\}$
  ◆**(d)** $V = \mathbf{M}_{22}$, $U = \left\{ \begin{bmatrix} a & a \\ b & b \end{bmatrix} \,\middle|\, a, b \text{ in } \mathbb{R} \right\}$, $W = \left\{ \begin{bmatrix} a & b \\ -a & b \end{bmatrix} \,\middle|\, a, b \text{ in } \mathbb{R} \right\}$

11. Let $U = \text{span}\{(1,0,0,0), (0,1,0,0)\}$ in $\mathbb{R}^4$. Show that $\mathbb{R}^4 = U \oplus W_1$ and $\mathbb{R}^4 = U \oplus W_2$, where $W_1 = \text{span}\{(0, 0, 1, 0), (0, 0, 0, 1)\}$ and $W_2 = \text{span}\{(1, 1, 1, 1), (1, 1, 1, -1)\}$.

12. Let U be a subspace of V, and suppose that $V = U \oplus W_1$ and $V = U \oplus W_2$ hold for subspaces W_1 and W_2. Show that $\dim W_1 = \dim W_2$.

13. If U and W denote the subspaces of even and odd polynomials in $\mathbf{P}_n$, respectively, show that $\mathbf{P}_n = U \oplus W$. (See Exercise 7§5.3.) [*Hint:* $f(x) + f(-x)$ is even.]

◆**14.** Let E be a 2×2 matrix such that $E^2 = E$. Show that $\mathbf{M}_{22} = U \oplus W$, where $U = \{A \mid AE = A\}$ and $W = \{B \mid BE = 0\}$. [*Hint:* XE lies in U for every matrix X.]

15. If U and W are the subspaces of symmetric and skew-symmetric $n \times n$ matrices, show that $\mathbf{M}_{nn} = U \oplus W$. [*Hint:* $X + X^T$ is symmetric.]

16. Let $V \xrightarrow{T} W \xrightarrow{S} V$ be linear transformations, and assume that $\dim V$ and $\dim W$ are finite.
  (a) If $ST = 1_V$, show that $W = \text{im } T \oplus \text{ker } S$ (see Section 7.2). [*Hint:* Given $\mathbf{w}$ in W, show that $\mathbf{w} - TS(\mathbf{w})$ lies in ker S.]
  (b) Illustrate with $\mathbb{R}^2 \xrightarrow{T} \mathbb{R}^3 \xrightarrow{S} \mathbb{R}^2$ where $T(x, y) = (x, y, 0)$ and $S(x, y, z) = (x, y)$.

17. Let V be a finite dimensional vector space, and let U and W be subspaces such that $U \cap W = 0$. If $\dim U + \dim W = \dim V$, show that $V = U \oplus W$.

18. Let $A = \begin{bmatrix} 0 & 1 \\ 0 & 0 \end{bmatrix}$ and consider $T_A : \mathbb{R}^2 \to \mathbb{R}^2$.
  (a) Show that the only eigenvalue of T_A is $\lambda = 0$.
  ◆**(b)** Show that ker $(T_A) = \mathbb{R}\begin{bmatrix} 1 \\ 0 \end{bmatrix}$ is the unique T_A-invariant subspace of $\mathbb{R}^2$ (except for 0 and $\mathbb{R}^2$).

19. If $A = \begin{bmatrix} 2 & -5 & 0 & 0 \\ 1 & -2 & 0 & 0 \\ 0 & 0 & -1 & -2 \\ 0 & 0 & 1 & 1 \end{bmatrix}$, show that $T_A : \mathbb{R}^4 \to \mathbb{R}^4$ has two-dimensional T-invariant subspaces U and W such that $\mathbb{R}^4 = U \oplus W$, but A has no real eigenvalue.

◆**20.** Let $T : V \to V$ be a linear operator where $\dim V = n$. If U is a T-invariant subspace of V, let $T_1 : U \to U$ denote the restriction of T to U (so $T_1(\mathbf{u}) = T(\mathbf{u})$ for all $\mathbf{u}$ in U). Show that $c_T(x) = c_{T_1}(x) \cdot q(x)$ for some polynomial $q(x)$. [*Hint:* Theorem 1.]

21. Let $T : V \to V$ be a linear operator where $\dim V = n$. Show that V has a basis of eigenvectors if and only if V has a basis B such that $M_B(T)$ is diagonal.

22. In each case, show that $T^2 = 1$ and find (as in Example 10) an ordered basis B such that $M_B(T)$ has the form given.
  (a) $T : \mathbf{M}_{22} \to \mathbf{M}_{22}$ where $T(A) = A^T$, $M_B(T) = \begin{bmatrix} I_3 & 0 \\ 0 & -1 \end{bmatrix}$
  ◆**(b)** $T : \mathbf{P}_3 \to \mathbf{P}_3$ where $T[p(x)] = p(-x)$, $M_B(T) = \begin{bmatrix} I_2 & 0 \\ 0 & -I_2 \end{bmatrix}$
  (c) $T : \mathbb{C} \to \mathbb{C}$ where $T(a + bi) = a - bi$, $M_B(T) = \begin{bmatrix} 1 & 0 \\ 0 & -1 \end{bmatrix}$
  ◆**(d)** $T : \mathbb{R}^3 \to \mathbb{R}^3$ where $T(a, b, c) = (-a + 2b + c, b + c, -c)$, $M_B(T) = \begin{bmatrix} 1 & 0 \\ 0 & -I_2 \end{bmatrix}$

(e) $T : V \to V$ where $T(\mathbf{v}) = -\mathbf{v}$, dim $V = n$, $M_B(T) = -I_n$

23. Let U and W denote subspaces of a vector space V.
  (a) If $V = U \oplus W$, define $T : V \to V$ by $T(\mathbf{v}) = \mathbf{w}$ where $\mathbf{v}$ is written (uniquely) as $\mathbf{v} = \mathbf{u} + \mathbf{w}$ with $\mathbf{u}$ in U and $\mathbf{w}$ in W. Show that T is a linear transformation, $U = \ker T$, $W = \operatorname{im} T$, and $T^2 = T$.
  (b) Conversely, if $T : V \to V$ is a linear transformation such that $T^2 = T$, show that $V = \ker T \oplus \operatorname{im} T$. [*Hint:* $\mathbf{v} - T(\mathbf{v})$ lies in $\ker T$ for all $\mathbf{v}$ in V.]

24. Let $T : V \to V$ be a linear operator satisfying $T^2 = T$ (such operators are called **idempotents**). Define $U_1 = \{\mathbf{v} \mid T(\mathbf{v}) = \mathbf{v}\}$ and $U_2 = \ker T = \{\mathbf{v} \mid T(\mathbf{v}) = \mathbf{0}\}$.
  (a) Show that $V = U_1 \oplus U_2$.
  (b) If dim $V = n$, find a basis B of V such that $M_B(T) = \begin{bmatrix} I_r & 0 \\ 0 & 0 \end{bmatrix}$, where $r = \operatorname{rank} T$.
  (c) If A is an $n \times n$ matrix such that $A^2 = A$, show that A is similar to $\begin{bmatrix} I_r & 0 \\ 0 & 0 \end{bmatrix}$, where $r = \operatorname{rank} A$. [*Hint:* Example 10.]

25. In each case, show that $T^2 = T$ and find (as in the preceding exercise) an ordered basis B such that $M_B(T)$ has the form given (0_k is the $k \times k$ zero matrix).
  (a) $T : \mathbf{P}_2 \to \mathbf{P}_2$ where $T(a + bx + cx^2) = (a - b + c)$ $(1 + x + x^2)$, $M_B(T) = \begin{bmatrix} 1 & 0 \\ 0 & 0_2 \end{bmatrix}$
  ◆(b) $T : \mathbb{R}^3 \to \mathbb{R}^3$ where $T(a, b, c) = (a + 2b, 0, 4b + c)$, $M_B(T) = \begin{bmatrix} I_2 & 0 \\ 0 & 0 \end{bmatrix}$
  (c) $T : \mathbf{M}_{22} \to \mathbf{M}_{22}$ where $T \begin{bmatrix} a & b \\ c & d \end{bmatrix} = \begin{bmatrix} -5 & -15 \\ 2 & 6 \end{bmatrix} \begin{bmatrix} a & b \\ c & d \end{bmatrix}$, $M_B(T) = \begin{bmatrix} I_2 & 0 \\ 0 & 0_2 \end{bmatrix}$

26. Let $T : V \to V$ be an operator satisfying $T^2 = cT$, $c \neq 0$.
  (a) Show that $V = U \oplus \ker T$, where $U = \{\mathbf{u} \mid T(\mathbf{u}) = c\mathbf{u}\}$. [*Hint:* Compute $T(\mathbf{v} - \frac{1}{c} T(\mathbf{v}))$.]
  (b) If dim $V = n$, show that V has a basis B such that $M_B(T) = \begin{bmatrix} cI_r & 0 \\ 0 & 0 \end{bmatrix}$, where $r = \operatorname{rank} T$.
  (c) If A is any $n \times n$ matrix of rank r such that $A^2 = cA$, $c \neq 0$, show that A is similar to $\begin{bmatrix} cI_r & 0 \\ 0 & 0 \end{bmatrix}$.

27. Let $T : V \to V$ be an operator such that $T^2 = c^2$, $c \neq 0$.
  (a) Show that $V = U_1 \oplus U_2$, where $U_1 = \{\mathbf{v} \mid T(\mathbf{v}) = c\mathbf{v}\}$ and $U_2 = \{\mathbf{v} \mid T(\mathbf{v}) = -c\mathbf{v}\}$. [*Hint:* $\mathbf{v} = \frac{1}{2c} \{[T(\mathbf{v}) + c\mathbf{v}] - [T(\mathbf{v}) - c\mathbf{v}]\}$.]
  (b) If dim $V = n$, show that V has a basis B such that $M_B(T) = \begin{bmatrix} cI_k & 0 \\ 0 & -cI_{n-k} \end{bmatrix}$ for some k.
  (c) If A is an $n \times n$ matrix such that $A^2 = c^2 I$, $c \neq 0$, show that A is similar to $\begin{bmatrix} cI_k & 0 \\ 0 & -cI_{n-k} \end{bmatrix}$ for some k.

28. If P is a fixed $n \times n$ matrix, define $T : \mathbf{M}_{nn} \to \mathbf{M}_{nn}$ by $T(A) = PA$. Let U_j denote the subspace of $\mathbf{M}_{nn}$ consisting of all matrices with all columns zero except possibly column j.
  (a) Show that each U_j is T-invariant.
  (b) Show that $\mathbf{M}_{nn}$ has a basis B such that $M_B(T)$ is block diagonal with each block on the diagonal equal to P.

29. Let V be a vector space. If $f : V \to \mathbb{R}$ is a linear transformation and $\mathbf{z}$ is a vector in V, define $T_{f, \mathbf{z}} : V \to V$ by $T_{f, \mathbf{z}}(\mathbf{v}) = f(\mathbf{v})\mathbf{z}$ for all $\mathbf{v}$ in V. Assume that $f \neq 0$ and $\mathbf{z} \neq \mathbf{0}$.
  (a) Show that $T_{f, \mathbf{z}}$ is a linear operator of rank 1.
  ◆(b) Show that $T_{f, \mathbf{z}}$ is an idempotent if and only if $f(\mathbf{z}) = 1$. (Recall that $T : V \to V$ is an idempotent if $T^2 = T$.)
  (c) Show that every idempotent $T : V \to V$ of rank 1 has the form $T = T_{f, \mathbf{z}}$ for some $f : V \to \mathbb{R}$ and some $\mathbf{z}$ in V with $f(\mathbf{z}) = 1$. [*Hint:* Write $\operatorname{im} T = \mathbb{R}\mathbf{z}$ and show that $T(\mathbf{z}) = \mathbf{z}$. Then use Exercise 23.]

30. Let U be a fixed $n \times n$ matrix, and consider the operator $T : \mathbf{M}_{nn} \to \mathbf{M}_{nn}$ given by $T(A) = UA$.
  (a) Show that λ is an eigenvalue of T if and only if it is an eigenvalue of U.
  ◆(b) If λ is an eigenvalue of T, show that $E_\lambda(T)$ consists of all matrices whose columns lie in $E_\lambda(U)$:
  $$E_\lambda(T) = \{[\mathbf{p}_1 \ \mathbf{p}_2 \ \cdots \ \mathbf{p}_n] \mid \mathbf{p}_i \text{ in } E_\lambda(U) \text{ for each } i\}$$
  (c) Show that if dim$[E_\lambda(U)] = d$, then dim$[E_\lambda(T)] = nd$. [*Hint:* If $B = \{\mathbf{e}_1, \ldots, \mathbf{e}_d\}$ is a basis of $E_\lambda(U)$, consider the set of all matrices with one column from B and the other columns zero.]

31. Let $T : V \to V$ be a linear operator where V is finite dimensional. If $U \subseteq V$ is a subspace, let $\bar{U} = \{\mathbf{u}_0 + T(\mathbf{u}_1) + T^2(\mathbf{u}_2) + \cdots + T^k(\mathbf{u}_k) \mid \mathbf{u}_i \text{ in } U, k \geq 0\}$. Show that $\bar{U}$ is the smallest T-invariant subspace containing U (that is, it is T-invariant, contains U, and is contained in every such subspace).

32. Let $U_1, \ldots, U_m$ be subspaces of V and assume that $V = U_1 + \cdots + U_m$; that is, every $\mathbf{v}$ in V can be written (in at least one way) in the form $\mathbf{v} = \mathbf{u}_1 + \cdots + \mathbf{u}_m$, $\mathbf{u}_i$ in U_i. Show that the following conditions are equivalent.

   (i) If $\mathbf{u}_1 + \cdots + \mathbf{u}_m = \mathbf{0}$, $\mathbf{u}_1$ in U_i, then $\mathbf{u}_i = \mathbf{0}$ for each i.

   (ii) If $\mathbf{u}_1 + \cdots + \mathbf{u}_m = \mathbf{u}'_1 + \cdots + \mathbf{u}'_m$, $\mathbf{u}_i$ and $\mathbf{u}'_i$ in U_i, then $\mathbf{u}_i = \mathbf{u}'_i$ for each i.

   (iii) $U_i \cap (U_1 + \cdots + U_{i-1} + U_{i+1} + \cdots + U_m) = 0$ for each $i = 1, 2, \ldots, m$.

   (iv) $U_i \cap (U_{i+1} + \cdots + U_m) = 0$ for each $i = 1, 2, \ldots, m - 1$.

When these conditions are satisfied, we say that V is the **direct sum** of the subspaces U_i, and write $V = U_1 \oplus U_2 \oplus \cdots \oplus U_m$.

33. (a) Let B be a basis of V and let $B = B_1 \cup B_2 \cup \cdots \cup B_m$ where the B_i are pairwise disjoint, nonempty subsets of B. If $U_i = \text{span } B_i$ for each i, show that $V = U_1 \oplus U_2 \oplus \cdots \oplus U_m$ (preceding exercise).

(b) Conversely if $V = U_1 \oplus \cdots \oplus U_m$ and B_i is a basis of U_i for each i, show that $B = B_1 \cup \cdots \cup B_m$ is a basis of V as in **(a)**.

Section 7.7 Block Triangular Form

We have shown (Theorem 5§6.4) that any $n \times n$ matrix A with every eigenvalue real is similar to an upper triangular matrix U. The following theorem shows that U can be chosen in a special way.

THEOREM 1

Block Triangulation Theorem

Let A be an $n \times n$ matrix with real eigenvalues and let

$$c_A(x) = (x - \lambda_1)^{m_1}(x - \lambda_2)^{m_2} \cdots (x - \lambda_k)^{m_k}$$

where the λ_i are distinct. Then an invertible matrix P exists such that

$$P^{-1}AP = \begin{bmatrix} U_1 & 0 & 0 & \cdots & 0 \\ 0 & U_2 & 0 & \cdots & 0 \\ 0 & 0 & U_3 & \cdots & 0 \\ \vdots & \vdots & \vdots & & \vdots \\ 0 & 0 & 0 & \cdots & U_k \end{bmatrix}$$

where U_i is an $m_i \times m_i$ upper triangular matrix with every entry on the main diagonal equal to λ_i.

The proof is given at the end of this section. For now, we focus on a method for finding the matrix P. The key concept is the following.

DEFINITION

If A is as in Theorem 1, the **generalized eigenspace** $G_{\lambda_i}(A)$ is defined by

$$G_{\lambda_i}(A) = \text{null } [(\lambda_i I - A)^{m_i}].$$

Observe that $E_{\lambda_i}(A) = \text{null}(\lambda_i I - A)$ is a subspace of $G_{\lambda_i}(A)$. We need three technical results.

LEMMA 1

Using the notation of Theorem 1, we have dim $[G_{\lambda_i}(A)] = m_i$.

Proof   Write $A_i = (\lambda_i I - A)^{m_i}$ for convenience and let P be as in Theorem 1. The spaces $G_{\lambda_i}(A) = \text{null}(A_i)$ and $\text{null}(P^{-1}A_iP)$ are isomorphic via $X \leftrightarrow P^{-1}X$ so we show that dim $[\text{null}(P^{-1}A_iP)] = m_i$. Now $P^{-1}A_iP = (\lambda_i I - P^{-1}AP)^{m_i}$; in block form this is

$$
P^{-1}A_iP = \begin{bmatrix} \lambda_i I - U_1 & 0 & \cdots & 0 \\ 0 & \lambda_i I - U_2 & \cdots & 0 \\ \vdots & \vdots & & \vdots \\ 0 & 0 & \cdots & \lambda_i I - U_k \end{bmatrix}^{m_i}
$$

$$
= \begin{bmatrix} (\lambda_i I - U_1)^{m_i} & 0 & \cdots & 0 \\ 0 & (\lambda_i I - U_2)^{m_i} & \cdots & 0 \\ \vdots & \vdots & & \vdots \\ 0 & 0 & \cdots & (\lambda_i I - U_k)^{m_i} \end{bmatrix}
$$

The matrix $(\lambda_i I - U_j)^{m_i}$ is invertible if $j \neq i$ and zero if $j = i$ because U_i is an $m_i \times m_i$ upper triangular matrix with each entry on the main diagonal equal to λ_i. It follows that $m_i = \dim[\text{null}(P^{-1}A_iP)]$, as required.   ◆

LEMMA 2

If P is as in Theorem 1, denote the columns of P as follows:

$$
P_{11}, P_{12}, \ldots, P_{1m_1}; P_{21}, P_{22}, \ldots, P_{2m_2}; \ldots; P_{k1}, P_{k2}, \ldots, P_{km_k}
$$

Then $\{P_{i1}, P_{i2}, \ldots, P_{im_i}\}$ is a basis of $G_{\lambda_i}(A)$.

Proof   It suffices by Lemma 1 to show that each P_{ij} is in $G_{\lambda_i}(A)$. Write the matrix in Theorem 1 as $P^{-1}AP = \text{diag}(U_1, U_2, \ldots, U_k)$. Then

$$
AP = P \,\text{diag}(U_1, U_2, \ldots, U_k)
$$

Comparing columns gives, successively:

$$
\begin{aligned}
AP_{11} &= \lambda_1 P_{11}, & \text{so } (\lambda_1 I - A)P_{11} &= 0 \\
AP_{12} &= uP_{11} + \lambda_1 P_{12}, & \text{so } (\lambda_1 I - A)^2 P_{12} &= 0 \\
AP_{13} &= wP_{11} + vP_{12} + \lambda_1 P_{13}, & \text{so } (\lambda_1 I - A)^3 P_{13} &= 0 \\
&\;\;\vdots & &\;\;\vdots
\end{aligned}
$$

where u, v, w are in $\mathbb{R}$. In general, $(\lambda_1 I - A)^j P_{1j} = 0$ for $j = 1, 2, \ldots, m_1$, so P_{1j} is in $G_{\lambda_1}(A)$. Similarly, P_{ij} is in $G_{\lambda_i}(A)$ for each i and j. ◆

LEMMA 3

If B_i is any basis of $G_{\lambda_i}(A)$, then $B = B_1 \cup B_2 \cup \ldots \cup B_k$ is a basis of $\mathbb{R}^n$.

Proof   It suffices by Lemma 1 to show that B is independent. If a linear combination from B vanishes, let X_i be the sum of the terms from B_i. Then $x_1 + \cdots + x_k = 0$. But $x_i = \Sigma_j r_{ij} P_{ij}$ by Lemma 2, so $\Sigma_{i,j} r_{ij} P_{ij} = 0$. Hence each $X_i = 0$, so each coefficient in X_i is zero. ◆

Lemma 2 suggests an algorithm for finding the matrix P in Theorem 1. Observe that there is an ascending chain of subspaces leading from $E_{\lambda_i}(A)$ to $G_{\lambda_i}(A)$:

$$E_{\lambda_i}(A) = \text{null}[(\lambda_i I - A)] \subseteq \text{null}[(\lambda_i I - A)^2] \subseteq \cdots \subseteq \text{null}[(\lambda_i I - A)^{m_i}] = G_{\lambda_i}(A)$$

We construct a basis for $G_{\lambda i}(A)$ by climbing up this chain.

TRIANGULATION ALGORITHM

Suppose A has characteristic polynomial

$$c_A(x) = (x - \lambda_1)^{m_1}(x - \lambda_2)^{m_2} \cdots (x - \lambda_k)^{m_k}$$

1. Choose a basis of $\text{null}[(\lambda_1 I - A)]$; enlarge it by adding vectors (possibly none) to a basis of $\text{null}[(\lambda_1 I - A)^2]$; enlarge that to a basis of null $[(\lambda_1 I - A)^3]$, and so on. Continue to obtain an ordered basis $\{P_{11}, P_{12}, \ldots, P_{1m_1}\}$ of $G_{\lambda_1}(A)$.

2. As in (1) choose a basis of $\{P_{i1}, P_{i2}, \ldots, P_{im_i}\}$ of $G_{\lambda_i}(A)$ for each i.

3. Let $P = [P_{11}P_{12} \cdots P_{im_1}; P_{21}P_{22} \cdots P_{2m_2}; \cdots; P_{k1}P_{k2} \cdots P_{km_k}]$ be the matrix with these basis vectors (in order) as columns.

Then $P^{-1}AP = \text{diag}(U_1, U_2, \ldots, U_k)$ as in Theorem 1.

Proof   Lemma 3 guarantees that $B = \{P_{11}, \ldots, P_{km_k}\}$ is a basis of $\mathbb{R}^n$, and Theorem 4§7.5 shows that $P^{-1}AP = M_B(T_A)$. Now $G_{\lambda_i}(A)$ is T_A-invariant for each i because

$$(\lambda_i I - A)^{m_i} X = 0 \qquad \text{implies} \qquad (\lambda_i I - A)^{m_i}(AX) = A(\lambda_i I - A)^{m_i} X = 0$$

By Theorem 8§7.6 (and induction), we have $P^{-1}AP = M_B(T_A) = \text{diag}(U_1, U_2, \ldots, U_k)$ where U_i is the matrix of the restriction of T_A to $G_{\lambda_i}(A)$, and it remains to show that U_i has the desired upper triangular form. Given s, let P_{ij} be a basis vector in null$[(\lambda_i I - A)^{s+1}]$. Then $(\lambda_i I - A)P_{ij}$ is in null$[(\lambda_i I - A)^s]$, and therefore is a linear combination of the basis vectors P_{it} coming *before* P_{ij}. Hence

$$T_A(P_{ij}) = AP_{ij} = \lambda_i P_{ij} - (\lambda_i I - A)P_{ij}$$

shows that the column of U_i corresponding to P_{ij} has λ_i on the main diagonal and zeros below the main diagonal. This is what we wanted.   ◆

EXAMPLE 1

If $A = \begin{bmatrix} 2 & 0 & 0 & 1 \\ 0 & 2 & 0 & -1 \\ -1 & 1 & 2 & 0 \\ 0 & 0 & 0 & 2 \end{bmatrix}$, find P such that $P^{-1}AP$ is block triangular.

Solution $c_A(x) = \det[xI - A] = (x - 2)^4$, so $\lambda_1 = 2$ is the only eigenvalue and we are in the case $k = 1$ of Theorem 1. Compute:

$$(2I - A) = \begin{bmatrix} 0 & 0 & 0 & -1 \\ 0 & 0 & 0 & 1 \\ 1 & -1 & 0 & 0 \\ 0 & 0 & 0 & 0 \end{bmatrix} \qquad (2I - A)^2 = \begin{bmatrix} 0 & 0 & 0 & 0 \\ 0 & 0 & 0 & 0 \\ 0 & 0 & 0 & -2 \\ 0 & 0 & 0 & 0 \end{bmatrix} \qquad (2I - A)^3 = 0$$

By Gaussian elimination find a basis $\{P_{11}, P_{12}\}$ of null$(2I - A)$; then extend in any way to a basis $\{P_{11}, P_{12}, P_{13}\}$ of null$[(2I - A)^2]$; and finally get a basis $\{P_{11}, P_{12}, P_{13}, P_{14}\}$ of null$[(2I - A)^3] = \mathbb{R}^4$. One choice is

$$P_{11} = \begin{bmatrix} 1 \\ 1 \\ 0 \\ 0 \end{bmatrix} \qquad P_{12} = \begin{bmatrix} 0 \\ 0 \\ 1 \\ 0 \end{bmatrix} \qquad P_{13} = \begin{bmatrix} 0 \\ 1 \\ 0 \\ 0 \end{bmatrix} \qquad P_{14} = \begin{bmatrix} 0 \\ 0 \\ 0 \\ 1 \end{bmatrix}$$

Hence $P = [P_{11} \ \ P_{12} \ \ P_{13} \ \ P_{14}] = \begin{bmatrix} 1 & 0 & 0 & 0 \\ 1 & 0 & 1 & 0 \\ 0 & 1 & 0 & 0 \\ 0 & 0 & 0 & 1 \end{bmatrix}$ gives $P^{-1}AP = \begin{bmatrix} 2 & 0 & 0 & 1 \\ 0 & 2 & 1 & 0 \\ 0 & 0 & 2 & -2 \\ 0 & 0 & 0 & 2 \end{bmatrix}$ ◆◆◆

EXAMPLE 2

If $A = \begin{bmatrix} 2 & 0 & 1 & 1 \\ 3 & 5 & 4 & 1 \\ -4 & -3 & -3 & -1 \\ 1 & 0 & 1 & 2 \end{bmatrix}$, find P such that $P^{-1}AP$ is block triangular.

Solution   The eigenvalues are $\lambda_1 = 1$ and $\lambda_2 = 2$ because

$$c_A(x) = \begin{vmatrix} x-2 & 0 & -1 & -1 \\ -3 & x-5 & -4 & -1 \\ 4 & 3 & x+3 & 1 \\ -1 & 0 & -1 & x-2 \end{vmatrix} = \begin{vmatrix} x-1 & 0 & 0 & -x+1 \\ -3 & x-5 & -4 & -1 \\ 4 & 3 & x+3 & 1 \\ -1 & 0 & -1 & x-2 \end{vmatrix}$$

$$= \begin{vmatrix} x-1 & 0 & 0 & 0 \\ -3 & x-5 & -4 & -4 \\ 4 & 3 & x+3 & 5 \\ -1 & 0 & -1 & x-3 \end{vmatrix} = (x-1)\begin{vmatrix} x-5 & -4 & -4 \\ 3 & x+3 & 5 \\ 0 & -1 & x-3 \end{vmatrix}$$

$$= (x-1)\begin{vmatrix} x-5 & -4 & 0 \\ 3 & x+3 & -x+2 \\ 0 & -1 & x-2 \end{vmatrix} = (x-1)\begin{vmatrix} x-5 & -4 & 0 \\ 3 & x+2 & 0 \\ 0 & -1 & x-2 \end{vmatrix}$$

$$= (x-1)(x-2)\begin{vmatrix} x-5 & -4 \\ 3 & x+2 \end{vmatrix} = (x-1)^2(x-2)^2$$

By solving equations, we find null $(I - A) = \text{span}\{P_{11}\}$ and null $(I - A)^2 = \text{span}\{P_{11}, P_{12}\}$ where

$$P_{11} = \begin{bmatrix} 1 \\ 1 \\ -2 \\ 1 \end{bmatrix} \quad \text{and} \quad P_{12} = \begin{bmatrix} 0 \\ 3 \\ -4 \\ 1 \end{bmatrix}$$

As $\lambda_1 = 1$ has multiplicity 2 as a root of $c_A(x)$, dim $G_{\lambda_1}(A) = 2$ by Lemma 1. Since P_{11} and P_{12} both lie in $G_{\lambda_1}(A)$, we have $G_{\lambda_1}(A) = \text{span}\{P_{11}, P_{12}\}$. Turning to $\lambda_2 = 2$, we find that null $(2I - A) = \text{span}\{P_{21}\}$ and null $[(2I - A)^2] = \text{span}\{P_{21}, P_{22}\}$ where

$$P_{21} = \begin{bmatrix} 1 \\ 0 \\ -1 \\ 1 \end{bmatrix} \quad \text{and} \quad P_{22} = \begin{bmatrix} 0 \\ -4 \\ 3 \\ 0 \end{bmatrix}$$

Again, dim $G_{\lambda_2}(A) = 2$ as λ_2 has multiplicity 2, so $G_{\lambda_2}(A) = \text{span}\{P_{21}, P_{22}\}$.

Hence $P = \begin{bmatrix} 1 & 0 & 1 & 0 \\ 1 & 3 & 0 & -4 \\ -2 & -4 & -1 & 3 \\ 1 & 1 & 1 & 0 \end{bmatrix}$ gives $P^{-1}AP = \begin{bmatrix} 1 & -3 & 0 & 0 \\ 0 & 1 & 0 & 0 \\ 0 & 0 & 2 & 3 \\ 0 & 0 & 0 & 2 \end{bmatrix}.$

♦♦♦

If $p(x)$ is a polynomial and A is an $n \times n$ matrix, then $p(A)$ is also an $n \times n$ matrix if we interpret $A^0 = I_n$. For example, if $p(x) = x^2 - 2x + 3$, then $p(A) = A^2 - 2A + 3I$. With this we can state an important consequence of Theorem 1. As before, let $c_A(x)$ denote the characteristic polynomial of A.

THEOREM 2
Cayley–Hamilton[2] Theorem
If A is a square matrix with real eigenvalues, then $c_A(A) = 0$.

Proof   Write $c_A(x) = (x - \lambda_1)^{m_1} \cdots (x - \lambda_k)^{m_k}$, and let $D = P^{-1}AP = \text{diag}(U_1, U_2, \ldots, U_k)$ as in Theorem 1. Then $c_A(U_i) = 0$ for each i because $(U_i - \lambda_i I)^{m_i} = 0$. Hence

$$P^{-1}c_A(A)P = c_A(D) = c_A[\text{diag}(U_1, \ldots, U_k)]$$
$$= \text{diag}[c_A(U_1), \ldots, c_A(U_k)]$$
$$= 0$$

It follows that $c_A(A) = 0$.   ◆

EXAMPLE 3

If $A = \begin{bmatrix} 1 & 3 \\ -1 & 2 \end{bmatrix}$, then $c_A(x) = \det \begin{bmatrix} x-1 & -3 \\ 1 & x-2 \end{bmatrix} = x^2 - 3x + 5$. Then $c_A(A) =$

$A^2 - 3A + 5I_2 = \begin{bmatrix} -2 & 9 \\ -3 & 1 \end{bmatrix} - \begin{bmatrix} 3 & 9 \\ -3 & 6 \end{bmatrix} + \begin{bmatrix} 5 & 0 \\ 0 & 5 \end{bmatrix} = \begin{bmatrix} 0 & 0 \\ 0 & 0 \end{bmatrix}$

◆◆◆

Theorem 1 can be refined even further. The diagonal blocks are all upper triangular square matrices with each main diagonal entry equal to a fixed number λ. Given such a matrix U, it can be shown that U is similar to a block diagonal matrix

$$\begin{bmatrix} J_1 & 0 & \cdots & 0 \\ 0 & J_2 & \cdots & 0 \\ \vdots & \vdots & \ddots & \vdots \\ 0 & 0 & \cdots & J_t \end{bmatrix}$$

where each block J_i is one of the matrices

---

[2]Named after the English mathematician Arthur Cayley (1821–1895) (see page 33) and William Rowan Hamilton (1805–1865), an Irish mathematician famous for his work on physical dynamics.

$$[\lambda], \quad \begin{bmatrix} \lambda & 1 \\ 0 & \lambda \end{bmatrix}, \quad \begin{bmatrix} \lambda & 1 & 0 \\ 0 & \lambda & 1 \\ 0 & 0 & \lambda \end{bmatrix}, \quad \begin{bmatrix} \lambda & 1 & 0 & 0 \\ 0 & \lambda & 1 & 0 \\ 0 & 0 & \lambda & 1 \\ 0 & 0 & 0 & \lambda \end{bmatrix}, \quad \text{and so on}$$

These matrices are called **Jordan blocks** corresponding to λ. Combining this with Theorem 1, we see that every matrix A with real eigenvalues is similar to a block diagonal matrix with Jordan blocks (corresponding to various eigenvalues) on the main diagonal. This latter matrix is called the **Jordan canonical form** of the matrix A.[3] The interested reader can find a proof in I. N. Herstein and D. J. Winter, *Matrix Theory and Linear Algebra*, New York: Macmillan, 1988.

Proof of Theorem 1   The proof requires the following fact, the proof of which we leave to the reader.

LEMMA 4

If $\{\mathbf{v}_1, \mathbf{v}_2, \ldots, \mathbf{v}_n\}$ is a basis of a vector space V, so also is $\{\mathbf{v}_1 + s\mathbf{v}_2, \mathbf{v}_2, \ldots, \mathbf{v}_n\}$ for any scalar s.

Now let A be as in Theorem 1, and let $T = T_A : \mathbb{R}^n \to \mathbb{R}^n$ be the matrix transformation. For convenience, call a matrix a λ-m-ut matrix if it is an $m \times m$ upper triangular matrix and every diagonal entry equals λ. Then we must find a basis B of $\mathbb{R}^n$ such that $M_B(T) = \text{diag}(U_1, U_2, \ldots, U_k)$ where U_i is a λ_i-m_i-ut matrix for each i. We proceed by induction on n. If $n = 1$, take $B = \{\mathbf{v}\}$ where $\mathbf{v}$ is any eigenvector of T. If $n > 1$, let $\mathbf{v}_1$ be a λ_1-eigenvector of T, and let $B_0 = \{\mathbf{v}_1, \mathbf{w}_1, \ldots, \mathbf{w}_{n-1}\}$ be any basis of $\mathbb{R}^n$ containing $\mathbf{v}_1$. Then

$$M_{B_0}(T) = \begin{bmatrix} \lambda_1 & X \\ 0 & A_1 \end{bmatrix}$$

in block form where A_1 is $(n-1) \times (n-1)$. Moreover, A and $M_{B_0}(T)$ are similar, so

$$c_A(x) = c_{M_{B_0}(T)}(x) = (x - \lambda_1)c_{A_1}(x)$$

Hence $c_{A_1}(x) = (x - \lambda_1)^{m_1 - 1}(x - \lambda_2)^{m_2} \cdots (x - \lambda_k)^{m_k}$, so (by induction) let

$$Q^{-1}A_1Q = \text{diag}(Z_1, U_2, \ldots, U_k)$$

where Z_1 is a λ_1-$(m_1 - 1)$-ut matrix and U_i is a λ_i-m_i-ut matrix for each $i > 1$.

---

[3]Given in 1870 by the French mathematician Camille Jordan (1838–1922) in his monumental *Traité des substitutions et des équations algébriques*.

If $P = \begin{bmatrix} 1 & 0 \\ 0 & Q \end{bmatrix}$, then $P^{-1} M_{B_0}(T) P = \begin{bmatrix} \lambda_1 & XQ \\ 0 & Q^{-1} A_1 Q \end{bmatrix}$, so (by Theorem 3 §7.5) there is a basis B_1 of $\mathbb{R}^n$ such that $M_{B_1}(T) = P^{-1} M_{B_0}(T) P$ has the form

$$M_{B_1}(T) = \left[\begin{array}{cc|cccccc} \lambda_1 & X_1 & & & Y & & \\ \hline 0 & Z_1 & 0 & 0 & \cdots & 0 \\ \hline & & U_2 & 0 & \cdots & 0 \\ & & 0 & U_3 & \cdots & 0 \\ & 0 & \vdots & \vdots & & \vdots \\ & & 0 & 0 & \cdots & U_k \end{array}\right] \tag{$*$}$$

If we write $U_1 = \begin{bmatrix} \lambda_1 & X_1 \\ 0 & Z_1 \end{bmatrix}$, the basis B_1 fulfills our needs except that the row matrix Y may not be zero.

We remedy this defect as follows. Observe that the first vector in B_1 is a λ_1 eigenvector of T, which we continue to denote as $\mathbf{v}_1$. The idea is to add suitable scalar multiples of $\mathbf{v}_1$ to the other vectors in B_1. This results in a new basis by Lemma 4, and the multiples can be chosen so that the new matrix of T is the same as $(*)$ except that $Y = 0$. Let $\{\mathbf{w}_1, \ldots, \mathbf{w}_{m_2}\}$ be the vectors in B_1 corresponding to λ_2 (giving rise to U_2 in $(*)$). Write

$$U_2 = \begin{bmatrix} \lambda_2 & u_{12} & u_{13} & \cdots & u_{1m_2} \\ 0 & \lambda_2 & u_{23} & \cdots & u_{2m_2} \\ 0 & 0 & \lambda_2 & \cdots & u_{3m_2} \\ \vdots & \vdots & \vdots & & \vdots \\ 0 & 0 & 0 & \cdots & \lambda_2 \end{bmatrix} \quad \text{and} \quad Y = [y_1 \quad y_2 \ldots]$$

We first replace $\mathbf{w}_1$ by $\mathbf{w}_1' = \mathbf{w}_1 + s\mathbf{v}_1$ where s is to be determined. Then $(*)$ gives

$$\begin{aligned} T(\mathbf{w}_1') &= T(\mathbf{w}_1) + sT(\mathbf{v}_1) \\ &= (y_1\mathbf{v}_1 + \lambda_2\mathbf{w}_1) + s\lambda_1\mathbf{v}_1 \\ &= y_1\mathbf{v}_1 + \lambda_2(\mathbf{w}_1' - s\mathbf{v}_1) + s\lambda_1\mathbf{v}_1 \\ &= \lambda_2\mathbf{w}_1' + [y_1 - s(\lambda_2 - \lambda_1)]\mathbf{v}_1 \end{aligned}$$

Because $\lambda_2 \neq \lambda_1$ we can choose s such that $T(\mathbf{w}_1') = \lambda_2\mathbf{w}_1'$. Similarly, let $\mathbf{w}_2' = \mathbf{w}_2 + t\mathbf{v}_1$ where t is to be chosen. Then, as before,

$$\begin{aligned} T(\mathbf{w}_2') &= T(\mathbf{w}_2) + tT(\mathbf{v}_1) \\ &= (y_2\mathbf{v}_1 + u_{12}\mathbf{w}_1 + \lambda_2\mathbf{w}_2) + t\lambda_1\mathbf{v}_1 \\ &= u_{12}\mathbf{w}_1' + \lambda_2\mathbf{w}_2' + [(y_2 - u_{12}s) - t(\lambda_2 - \lambda_1)]\mathbf{v}_1 \end{aligned}$$

Again, t can be chosen so that $T(\mathbf{w}_2') = u_{12}\mathbf{w}_1' + \lambda_2\mathbf{w}_2'$. Continue in this way to eliminate $y_1, \ldots, y_{m_2}$. This procedure also works for $\lambda_3, \lambda_4, \ldots$ and so produces a new basis B such that $M_B(T)$ is as in $(*)$ but with $Y = 0$.   ◆

EXERCISES 7.7

1. In each case, find a matrix P such that $P^{-1}AP$ is in block triangular form as in Theorem 1.

(a) $\begin{bmatrix} 2 & 3 & 2 \\ -1 & -1 & -1 \\ 1 & 2 & 2 \end{bmatrix}$ ◆ **(b)** $\begin{bmatrix} -5 & 3 & 1 \\ -4 & 2 & 1 \\ -4 & 3 & 0 \end{bmatrix}$

(c) $\begin{bmatrix} 0 & 1 & 1 \\ 2 & 3 & 6 \\ -1 & -1 & -2 \end{bmatrix}$ ◆ **(d)** $\begin{bmatrix} -3 & -1 & 0 \\ 4 & -1 & 3 \\ 4 & -2 & 4 \end{bmatrix}$

(e) $\begin{bmatrix} -1 & -1 & -1 & 0 \\ 3 & 2 & 3 & -1 \\ 2 & 1 & 3 & -1 \\ 2 & 1 & 4 & -2 \end{bmatrix}$ ◆ **(f)** $\begin{bmatrix} -3 & 6 & 3 & 2 \\ -2 & 3 & 2 & 2 \\ -1 & 3 & 0 & 1 \\ -1 & 1 & 2 & 0 \end{bmatrix}$

2. Show that the following conditions are equivalent for a linear operator T on a finite dimensional space V.

(1) $M_B(T)$ is upper triangular for some ordered basis V of E.

(2) A basis $\{e_1, \ldots, e_n\}$ of V exists such that, for each i, $T(e_i)$ is a linear combination of $e_1, \ldots, e_i$.

(3) There exist T-invariant subspaces $V_1 \subseteq V_2 \subseteq \cdots \subseteq V_n = V$ such that $\dim V_i = i$ for each i.

3. If A is an $n \times n$ invertible matrix, show that $A^{-1} = r_0 I + r_1 A + \cdots + r_{n-1} A^{n-1}$ for some scalars $r_0, r_1, \ldots, r_{n-1}$. [*Hint:* Cayley–Hamilton theorem.]

◆ **4.** If $T : V \to V$ is a linear operator where V is finite dimensional, show that $C_T(T) = 0$. [*Hint:* Exercise 26§7.4.]

5. Define $T : \mathbf{P} \to \mathbf{P}$ by $T[p(x)] = xp(x)$. Show that:

(a) T is linear and $f(T)[p(x)] = f(x)p(x)$ for all polynomials $f(x)$.

(b) Conclude that $f(T) \neq 0$ for all nonzero polynomials $f(x)$.

6. Let $T : V \to V$ be a nilpotent operator.

(a) If $T^{m+1} = 0$ but $T^m \neq 0$, show that $B = \{T^m(\mathbf{v}), T^{m-1}(\mathbf{v}), \ldots, T(\mathbf{v}), \mathbf{v}\}$ is linearly independent for some $\mathbf{v}$ in V.

◆ **(b)** Show that $W = \text{span } B$ is T-invariant and $M_B(T)$ is a Jordan block.

7. Let $T : V \to V$ be an operator where $\dim V = n$ and assume that T has a unique real eigenvalue λ. Hence $(\lambda - T)^n = c_T(T) = 0$ by Exercise 4; assume that $(\lambda - T)^{n-1} \neq 0$. Show that a basis B of V exists such that $M_B(T)$ is a Jordan block. (*Remark:* This is the beginning of the proof of the Jordan canonical form.) [*Hint:* Exercise 6.]

Section 7.8   An Application to Linear Recurrence Relations (Optional)[4]

In many applications it is required to compute numbers $x_0, x_1, x_2, \ldots, x_n, \ldots$ having the property that each is determined by those that come before. The simplest example of such a situation occurs when each x_{n+1} is a fixed multiple of x_n — say, $x_{n+1} = 3x_n$ for all $n = 0, 1, 2, \ldots$. If x_0 is given, the remaining x_n can be computed successively.

$$x_1 = 3x_0$$
$$x_2 = 3x_1 = 3^2 x_0$$
$$x_3 = 3x_2 = 3^3 x_0$$

Clearly $x_n = 3^n x_0$ holds for all $n \geq 0$ and gives an explicit formula for x_n as a function of n, provided that x_0 is stipulated.

Other situations are possible. For example, each x_n could be determined by the two preceding numbers in the sequence:

$$x_{n+2} = 6x_n - x_{n+1} \qquad \text{for all } n \geq 0$$

Then the whole sequence $x_0, x_1, x_2, x_3, \ldots$ is determined once x_0 and x_1 are given. For example, if $x_0 = 1$ and $x_1 = 2$, then

---

[4]This section requires only Sections 7.1–7.3.

$$x_2 = 6x_0 - x_1 = 4$$
$$x_3 = 6x_1 - x_2 = 8$$
$$x_4 = 6x_2 - x_3 = 16$$

In this case it appears that $x_n = 2^n$ holds for all $n \geq 0$. This certainly works for $0 \leq n \leq 4$, and it satisfies the recurrence formula:

$$6x_n - x_{n+1} = 6 \cdot 2^n - 2^{n+1} = 2^{n+1}(3 - 1) = 2^{n+2} = x_{n+2}$$

However, the reader should not get the idea that it is always easy to guess the formula for x_n. For example, if $x_0 = 1$ and $x_1 = 1$, then $x_{n+2} = 6x_n - x_{n+1}$ generates the sequence

$$1, 1, 5, 1, 29, -23, 197, \ldots$$

No formula for the nth term of *this* sequence is apparent! Nonetheless, it is possible to use vector-space techniques to analyze such sequences.

Sequences will be considered entities in their own right, so it is useful to have a special notation for them. Let

$$[x_n) \quad \text{denote the sequence } x_0, x_1, x_2, \ldots, x_n, \ldots$$

EXAMPLE 1

$[n)$	is the sequence $0, \ 1, \ 2, \ 3, \ldots$
$[n + 1)$	is the sequence $1, \ 2, \ 3, \ 4, \ldots$
$[2^n)$	is the sequence $1, \ 2, \ 2^2, \ 2^3, \ldots$
$[(-1)^n)$	is the sequence $1, \ -1, \ 1, \ -1, \ldots$
$[5)$	is the sequence $5, \ 5, \ 5, \ 5, \ldots$

◆◆◆

Sequences of the form $[c)$ for a fixed number c will be referred to as **constant sequences**, and those of the form $[\lambda^n)$, λ some number, are **power sequences**.

Two sequences are regarded as **equal** when they are identical:

$$[x_n) = [y_n) \quad \text{means} \quad x_n = y_n \quad \text{for all } n = 0, 1, 2, \ldots$$

Addition and scalar multiplication of sequences are defined by

$$[x_n) + [y_n) = [x_n + y_n)$$
$$r[x_n) = [rx_n)$$

These operations are analogous to the addition and scalar multiplication in $\mathbb{R}^n$, and it is easy to check that the vector-space axioms are satisfied. The zero vector is the constant sequence $[0)$, and the negative of a sequence $[x_n)$ is given by $-[x_n) = [-x_n)$.

Now suppose k real numbers $r_0, r_1, \ldots, r_{k-1}$ are given, and consider the **linear recurrence relation** determined by these numbers.

$$x_{n+k} = r_0 x_n + r_1 x_{n+1} + \cdots + r_{k-1} x_{n+k-1} \tag{$*$}$$

When $r_0 \neq 0$, we say this recurrence has **length** k. (We shall usually assume that $r_0 \neq 0$; otherwise, we are essentially dealing with a recurrence of shorter length than k.) For example, the relation $x_{n+2} = 2x_n + x_{n+1}$ is of length 2.

A sequence $[x_n]$ is said to **satisfy** the relation $(*)$ if $(*)$ holds for all $n \geq 0$. Let V denote the set of all sequences that satisfy the relation. In symbols,

$$V = \{[x_n] \mid x_{n+k} = r_0 x_n + r_1 x_{n+1} + \cdots + r_{k-1} x_{n+k-1} \quad \text{holds for all } n \geq 0\}$$

It is easy to see that the constant sequence $[0]$ lies in V and that V is closed under addition and scalar multiplication of sequences. Hence V is vector space (being a subspace of the space of all sequences). The following important observation about V is needed (it was used implicitly earlier): If the first k terms of two sequences agree, then the sequences are identical. More formally,

LEMMA

Let $[x_n]$ and $[y_n]$ denote two sequences in V. Then

$$[x_n] = [y_n] \text{ if and only if } x_0 = y_0, x_1 = y_1, \ldots, x_{k-1} = y_{k-1}$$

Proof If $[x_n] = [y_n]$, then $x_n = y_n$ for *all* $n = 0, 1, 2, \ldots$. Conversely, if $x_i = y_i$ for all $i = 0, 1, \ldots, k - 1$, use the recurrence $(*)$ for $n = 0$.

$$x_k = r_0 x_0 + r_1 x_1 + \cdots + r_{k-1} x_{k-1} = r_0 y_0 + r_1 y_1 + \cdots + r_{k-1} y_{k-1} = y_k$$

Next the recurrence for $n = 1$ establishes $x_{k+1} = y_{k+1}$. The process continues to show that $x_{n+k} = y_{n+k}$ holds for *all* $n \geq 0$ by induction on n. Hence $[x_n] = [y_n]$. ◆

This shows that a sequence in V is completely determined by its first k terms. In particular, given a k-tuple

$$\mathbf{v} = (v_0, v_1, \ldots, v_{k-1})$$

In $\mathbb{R}^k$, define $T(\mathbf{v})$ to be the sequence in V whose first k terms are $v_0, v_1, \ldots, v_{k-1}$. The rest of the sequence $T(\mathbf{v})$ is determined by the recurrence, so $T : \mathbb{R}^k \to V$ is a function. In fact, it is an isomorphism.

THEOREM 1

Given real numbers $r_0, r_1, \ldots, r_{k-1}$, let

$$V = \{[x_n] \mid x_{n+k} = r_0 x_n + r_1 x_{n+1} + \cdots + r_{k-1} x_{n+k-1}, \text{ for all } n \geq 0\}$$

denote the vector space of all sequences satisfying the linear recurrence relation determined by $r_0, r_1, \ldots, r_{k-1}$. Then the function

$$T : \mathbb{R}^k \to V$$

defined above is an isomorphism. In particular:

1. dim $V = k$.

2. If $\{\mathbf{v}_1, \ldots, \mathbf{v}_k\}$ is any basis of $\mathbb{R}^k$, then $\{T(\mathbf{v}_1), \ldots, T(\mathbf{v}_k)\}$ is a basis of V.

Proof (1) and (2) will follow from Theorems 1 and 2, §7.3 as soon as we show that T is an isomorphism. Given $\mathbf{v}$ and $\mathbf{w}$ in $\mathbb{R}^k$, write $\mathbf{v} = (v_0, v_1, \ldots, v_{k-1})$ and $\mathbf{w} = (w_0, w_1, \ldots, w_{k-1})$. The first k terms of $T(\mathbf{v})$ and $T(\mathbf{w})$ are $v_0, v_1, \ldots, v_{k-1}$ and $w_0, w_1, \ldots, w_{k-1}$, respectively, so the first k terms of $T(\mathbf{v}) + T(\mathbf{w})$ are $v_0 + w_0, v_1 + w_1, \ldots, v_{k-1} + w_{k-1}$. Because these terms agree with the first k terms of $T(\mathbf{v} + \mathbf{w})$, the lemma implies that $T(\mathbf{v} + \mathbf{w}) = T(\mathbf{v}) + T(\mathbf{w})$. The proof that $T(r\mathbf{v}) = rT(\mathbf{v})$ is similar, so T is linear.

Now let $[x_n]$ be any sequence in V, and let $\mathbf{v} = (x_0, x_1, \ldots, x_{k-1})$. Then the first k terms of $[x_n]$ and $T(\mathbf{v})$ agree, so $T(\mathbf{v}) = [x_n]$. Hence T is onto. Finally, if $T(\mathbf{v}) = [0]$ is the zero sequence, then the first k terms of $T(\mathbf{v})$ are all zero (*all* terms of $T(\mathbf{v})$ are zero!) so $\mathbf{v} = \mathbf{0}$. This means that ker $T = 0$, so T is one-to-one. ◆

EXAMPLE 2

Show that the sequences $[1]$, $[n]$, and $[(-1)^n]$ are a basis of the space V of all solutions of the recurrence

$$x_{n+3} = -x_n + x_{n+1} + x_{n+2}$$

Then find the solution satisfying $x_0 = 1, x_1 = 2, x_2 = 5$.

Solution

The verifications that these sequences satisfy the recurrence (and hence lie in V) are left to the reader. They are a basis because $[1] = T(1, 1, 1)$, $[n] = T(0, 1, 2)$, and $[(-1)^n] = T(1, -1, 1)$; and $\{(1, 1, 1), (0, 1, 2), (1, -1, 1)\}$ is a basis of $\mathbb{R}^3$. Finally, the sequence $[x_n]$ in V satisfying $x_0 = 1, x_1 = 2, x_2 = 5$ is a linear combination of this basis:

$$[x_n) = t_1[1] + t_2[n] + t_3[(-1)^n)$$

The nth term is $x_n = t_1 + nt_2 + (-1)^n t_3$, so taking $n = 0, 1, 2$ gives

$$1 = x_0 = t_1 + 0 + t_3$$
$$2 = x_1 = t_1 + t_2 - t_3$$
$$5 = x_2 = t_1 + 2t_2 + t_3$$

This has the solution $t_1 = t_3 = \frac{1}{2}, t_2 = 2$, so $x_n = \frac{1}{2} + 2n + \frac{1}{2}(-1)^n$.

◆◆◆

This technique clearly works for any linear recurrence of length k: Simply take your favorite basis $\{\mathbf{v}_1, \ldots, \mathbf{v}_k\}$ of $\mathbb{R}^k$—perhaps the standard basis—and compute

$T(\mathbf{v}_1), \ldots, T(\mathbf{v}_k)$. This is a basis of V all right, but the nth term of $T(\mathbf{v}_i)$ is not usually given as an explicit function of n. (The basis in Example 2 was carefully chosen so that the nth terms of the three sequences were 1, n, and $(-1)^n$, respectively, each a simple function of n).

It turns out that an explicit basis of V can be given in the general situation. Given the recurrence

$$x_{n+k} = r_0 x_n + r_1 x_{n+1} + \cdots + r_{k-1} x_{n+k-1} \tag{$*$}$$

the idea is to look for numbers λ such that the power sequence $[\lambda^n)$ satisfies $(*)$. This happens if and only if

$$\lambda^{n+k} = r_0 \lambda^n + r_1 \lambda^{n+1} + \cdots + r_{k-1} \lambda^{n+k-1}$$

holds for all $n \geq 0$. This is true just when the case $n = 0$ holds; that is,

$$\lambda^k = r_0 + r_1 \lambda + \cdots + r_{k-1} \lambda^{k-1}$$

The polynomial

$$p(x) = x^k - r_{k-1} x^{k-1} - \cdots - r_1 x - r_0$$

is called the polynomial **associated** with the linear recurrence $(*)$. Thus every root λ of $p(x)$ provides a sequence $[\lambda^n)$ satisfying $(*)$. If there are k distinct roots, the power sequences provide a basis. Incidentally, if $\lambda = 0$, the sequence $[\lambda^n)$ is $1, 0, 0, \ldots$; that is, we accept the convention that $0^0 = 1$.

THEOREM 2

Let $r_0, r_1, \ldots, r_{k-1}$ be real numbers; let

$$V = \{[x_n) \mid x_{n+k} = r_0 x_n + \cdots + r_{k-1} x_{n+k-1} \text{ for all } n \geq 0\}$$

denote the vector space of all sequences satisfying the linear recurrence relation determined by $r_0, r_1, \ldots, r_{k-1}$; and let

$$p(x) = x^k - r_{k-1} x^{k-1} - \cdots - r_1 x - r_0$$

denote the polynomial associated with the recurrence relation. Then

1. $[\lambda^n)$ lies in V if and only if λ is a root of $p(x)$.
2. If $\lambda_1, \lambda_2, \ldots, \lambda_k$ are distinct real roots of $p(x)$, then $\{[\lambda_1^n), [\lambda_2^n), \ldots, [\lambda_k^n)\}$ is a basis of V.

Proof   It remains to prove (2). But $[\lambda_i^n) = T(\mathbf{v}_i)$ where $\mathbf{v}_i = (1, \lambda_i, \lambda_i^2, \ldots, \lambda_i^{k-1})$, so (2) follows by Theorem 1, provided that $\{\mathbf{v}_1, \mathbf{v}_2, \ldots, \mathbf{v}_n\}$ is a basis of $\mathbb{R}^k$. This is true provided that the matrix

$$\begin{bmatrix} 1 & 1 & \cdots & 1 \\ \lambda_1 & \lambda_2 & \cdots & \lambda_k \\ \lambda_1^2 & \lambda_2^2 & \cdots & \lambda_k^2 \\ \vdots & \vdots & & \vdots \\ \lambda_1^{k-1} & \lambda_2^{k-1} & \cdots & \lambda_k^{k-1} \end{bmatrix}$$

is invertible. But this is a Vandermonde matrix and so is invertible if the λ_i are distinct (Theorem 2§3.3). This proves (2). ◆

EXAMPLE 3

Find the solution of $x_{n+2} = 2x_n + x_{n+1}$ that satisfies $x_0 = a$, $x_1 = b$.

Solution   The associated polynomial is $p(x) = x^2 - x - 2 = (x - 2)(x + 1)$. The roots are $\lambda_1 = 2$ and $\lambda_2 = -1$, so the sequences $[2^n)$ and $[(-1)^n)$ are a basis for the space of solutions by Theorem 2. Hence every solution $[x_n)$ is a linear combination

$$[x_n) = t_1[2^n) + t_2[(-1)^n)$$

This means that $x_n = t_1 2^n + t_2(-1)^n$ holds for $n = 0, 1, 2 \ldots$, so (taking $n = 0, 1$) $x_0 = a$ and $x_1 = b$ give

$$t_1 + t_2 = a$$
$$2t_1 - t_2 = b$$

These are easily solved: $t_1 = \tfrac{1}{3}(a + b)$ and $t_2 = \tfrac{1}{3}(2a - b)$, so

$$x_n = \tfrac{1}{3}[(a + b)2^n + (2a - b)(-1)^n]$$

◆◆◆

The next example has historical interest.

EXAMPLE 4

How many pairs of rabbits will be produced in a year, beginning with a single pair, if in every month each pair brings forth a new pair that becomes productive from the second month on? Assume no pairs die.

Solution   Let x_n be the number of pairs after n months. Then $x_0 = 1$ and $x_1 = 1$ (the first pair does not reproduce in the first month). Moreover,

$$x_{n+2} = x_{n+1} + x_n$$

holds for all $n \geq 0$ because the x_{n+2} pairs at the end of month $n + 2$ are made up of x_{n+1} pairs alive at the end of the preceding month, together with x_n babies. The associated polynomial is $p(x) = x^2 - x - 1$, and $\lambda_1 = \tfrac{1}{2}(1 + \sqrt{5})$ and $\lambda_2 = \tfrac{1}{2}(1 - \sqrt{5})$ are the roots. These are distinct, so $\{[\lambda_1^n], [\lambda_2^n]\}$ is a basis of the space of solutions. Hence $x_n = t_1\lambda_1^n + t_2\lambda_2^n$ holds for some constants t_1 and t_2, and taking $n = 0, 1$ gives

$$\begin{aligned} t_1 \quad + t_2 &= x_0 = 1 \\ t_1\lambda_1 + t_2\lambda_2 &= x_1 = 1 \end{aligned}$$

The solution is $t_1 = \lambda_1/\sqrt{5}$, $t_2 = -\lambda_2/\sqrt{5}$, so

$$x_n = \frac{1}{\sqrt{5}}\left[\left(\frac{1 + \sqrt{5}}{2}\right)^{n+1} - \left(\frac{1 - \sqrt{5}}{2}\right)^{n+1}\right] \tag{$*$}$$

It is worth noting that $\frac{1}{2}(1 - \sqrt{5}) = -0.618$; large powers of this number will be very small. So for values of n larger than 10, x_n is approximated by

$$x_n = \frac{1}{\sqrt{5}}\left[\frac{1 + \sqrt{5}}{2}\right]^{n+1}.$$ This gives $x_{12} = 233$ pairs, and after 60 months the population

is $x_{60} = 2.5$ million million! Clearly the stipulation that none of these pairs die would be invalid by then!

◆◆◆

The numbers obtained in Example 4 are called the **Fibonacci sequence** and are usually denoted $f_0, f_1, f_2, \ldots$. They are defined by the conditions

$$\begin{aligned} f_0 &= f_1 = 1 \\ f_{n+2} &= f_n + f_{n+1} \qquad n = 0, 1, 2, \ldots \end{aligned}$$

The first few numbers in the sequence are 1, 1, 2, 3, 5, 8, 13, 21, 34, ... They are named after Leonardo of Pisa (c. 1180–1250), better known as Fibonacci or "son of Bonaccio." Fibonacci was one of the few outstanding European mathematicians of the middle ages, and Example 4 was stated in his main work *Liber Abaci* (1202). The formula $(*)$ in Example 4 is called the **Binet formula**.

If $p(x)$ is the polynomial associated with a linear recurrence relation, and if $p(x)$ has k distinct roots $\lambda_1, \lambda_2, \ldots, \lambda_k$ then $p(x)$ factors completely:

$$p(x) = (x - \lambda_1)(x - \lambda_1) \cdots (x - \lambda_k)$$

Each root λ_i provides a sequence $[\lambda_i^n]$ satisfying the recurrence, and they are a basis of V by Theorem 2. In this case, each λ_i has multiplicity 1 as a root of $p(x)$. In general a root λ has **multiplicity** m if $p(x) = (x - \lambda)^m q(x)$, where $q(\lambda) \neq 0$. In this case, there are fewer than k distinct roots and so fewer than k sequences $[\lambda^n]$ satisfying the recurrence. However, we can still obtain a basis because if λ has multiplicity m (and $\lambda \neq 0$), it provides m linearly independent sequences that satisfy the recurrence. To prove this, it is convenient to give another way to describe the space V of all sequences satisfying a given linear recurrence relation.

Let **S** denote the vector space of *all* sequences and define a function

$$S : \mathbf{S} \to \mathbf{S} \qquad \text{by} \qquad S[x_n) = [x_{n+1}) = [x_1, x_2, x_3, \ldots)$$

S is clearly a linear transformation and is called the **shift operator** on **S**. Note that powers of S shift the sequence further: $S^2[x_n] = S[x_{n+1}] = [x_{n+2}]$. In general,

$$S^k[x_n] = [x_{n+k}] = [x_k, x_{k+1}, \ldots)   \text{for each } k = 0, 1, 2, \ldots$$

But then a linear recurrence relation

$$x_{n+k} = r_0 x_n + r_1 x_{n+1} + \cdots + r_{k-1} x_{n+k-1}   \text{for } n = 0, 1, \ldots$$

can be written

$$S^k[x_n] = r_0[x_n] + r_1 S[x_n] + \cdots + r_{k-1} S^{k-1}[x_n] \tag{$*$}$$

Now let $p(x) = x^k - r_{k-1} x^{k-1} - \cdots - r_1 x - r_0$ denote the polynomial associated with the recurrence relation. The set $L[S, S]$ of all linear transformations from **S** to itself is a vector space (Exercise 19§7.4) that is closed under composition. In particular,

$$p(S) = S^k - r_{k-1} S^{k-1} - \cdots - r_1 S - r_0$$

is a linear transformation called the **evaluation** of p at S. The point is that condition ($*$) can be written as

$$p(S)[x_n] = 0$$

In other words, the space V of all sequences satisfying the recurrence relation is just $\ker[p(S)]$. This is the first assertion in the following theorem.

THEOREM 3

Let $r_0, r_1, \ldots, r_{k-1}$ be real numbers, and let

$$V = \{[x_n] \mid x_{n+k} = r_0 x_n + r_1 x_{n+1} + \cdots + r_{k-1} x_{n+k-1}   \text{for all } n \geq 0\}$$

denote the space of all sequences satisfying the linear recurrence relation determined by $r_0, r_1, \ldots, r_{k-1}$. Let

$$p(x) = x^k - r_{k-1} x^{k-1} - \cdots - r_1 x - r_0$$

denote the corresponding polynomial. Then:

1. $V = \ker[p(S)]$, where S is the shift operator.
2. If $p(x) = (x - \lambda)^m q(x)$, where $\lambda \neq 0$ and $m > 1$, then the sequences

$$\{[\lambda^n], [n\lambda^n], [n^2\lambda^n], \ldots, [n^{m-1}\lambda^n]\}$$

all lie in V and are linearly independent.

Proof (Sketch)   It remains to prove (2). If $\binom{n}{k} = \dfrac{n(n-1) \cdots (n-k+1)}{k!}$

denotes the binomial coefficient, the idea is to use (1) to show that the sequence

$s_k = \left[\binom{n}{k}\lambda^n\right]$ is a solution for each $k = 0, 1, \ldots, m - 1$. Then (2) of Theorem 1 can be applied to show that $\{s_0, s_1, \ldots, s_{m-1}\}$ is linearly independent. Finally, the sequences $t_k = [n^k\lambda^n]$, $k = 0, 1, \ldots, m - 1$, in the present theorem can be given by $t_k = \sum_{j=0}^{m-1} a_{kj}s_j$, where $A = [a_{ij}]$ is an invertible matrix. Then (2) follows.   ◆

This theorem combines with Theorem 2 to give a basis for V when $p(x)$ has k real roots (not necessarily distinct) none of which is zero. This last requirement means $r_0 \neq 0$, a condition that is unimportant in practice (see Remark 1 later).

THEOREM 4

Let $r_0, r_1, \ldots, r_{k-1}$ be real numbers with $r_0 \neq 0$; let

$$V = \{[x_n] \mid x_{n+k} = r_0 x_n + r_1 x_{n+1} + \cdots + r_{k-1}x_{n+k-1} \quad \text{for all } n \geq 0\}$$

denote the space of all sequences satisfying the linear recurrence relation of length k determined by $r_0, \ldots, r_{n-1}$; and assume that the polynomial

$$p(x) = x^k - r_{k-1}x^{k-1} - \cdots - r_1 x - r_0$$

factors completely as

$$p(x) = (x - \lambda_1)^{m_1}(x - \lambda_2)^{m_2} \ldots (x - \lambda_p)^{m_p}$$

where $\lambda_1, \lambda_2, \ldots, \lambda_p$ are distinct real numbers and each $m_i \geq 1$. Then $\lambda_i \neq 0$ for each i, and

$$\{[\lambda_1^n], [n\lambda_1^n], \ldots, [n^{m_1-1}\lambda_1^n]; \ [\lambda_2^n], [n\lambda_2^n], \ldots, [n^{m_2-1}\lambda_2^n]; \ldots;$$
$$[\lambda_p^n], [n\lambda_p^n], \ldots, [n^{m_p-1}\lambda_p^n]\}$$

is a basis of V.

Proof (Partial) There are $m_1 + m_2 + \cdots + m_p = k$ sequences in all, so because $\dim V = k$, it suffices to show that they are linearly independent. The assumption that $r_0 \neq 0$, implies that 0 is not a root of $p(x)$. Hence each $\lambda_i \neq 0$, so $\{[\lambda_i^n], [n\lambda_i^n], \ldots, [n^{m_i-1}\lambda_i^n]\}$ is linearly independent by Theorem 3. The proof that the whole set of sequences is linearly independent is omitted.   ◆

EXAMPLE 5

Find a basis for the space V of all sequences $[x_n]$ satisfying

$$x_{n+3} = -9x_n - 3x_{n+1} + 5x_{n+2}$$

Solution   The associated polynomial is $p(x) = x^3 - 5x^2 + 3x + 9 = (x - 3)^2 (x + 1)$. Hence $\lambda = 3$ is a double root, so $[3^n]$ and $[n3^n]$ both lie in V by Theorem 3 (the reader should verify this). Similarly, $\lambda = -1$ is a root of multiplicity 1, so $[(-1)^n]$ lies in V. Hence $\{[3^n], [n3^n], [(-1)^n]\}$ is a basis by Theorem 4.

$\blacklozenge\blacklozenge\blacklozenge$

Remark 1:   If $r_0 = 0$ [so $p(x)$ has 0 as a root], the recurrence reduces to one of shorter length. For example, consider

$$x_{n+4} = 0x_n + 0x_{n+1} + 3x_{n+2} + 2x_{n+3} \qquad (*)$$

If we set $y_n = x_{n+2}$, this recurrence becomes $y_{n+2} = 3y_n + 2y_{n+1}$, which has solutions $[3^n]$ and $[(-1)^n]$. These give the following solutions to $(*)$:

$$[0, \ 0, \ 1, \ 3, \ 3^2, \ldots)$$
$$[0, \ 0, \ 1, \ -1, \ (-1)^2, \ldots)$$

In addition, it is easy to verify that

$$[1, \ 0, \ 0, \ 0, \ 0, \ldots)$$
$$[0, \ 1, \ 0, \ 0, \ 0, \ldots)$$

are also solutions to $(*)$. The space of all solutions of $(*)$ has dimension 4 (Theorem 1), so these sequences are a basis. This technique works whenever $r_0 = 0$.

Remark 2:   Theorem 4 completely describes the space V of sequences that satisfy a linear recurrence relation for which the associated polynomial $p(x)$ has all real roots. However, in many cases of interest, $p(x)$ has complex roots that are not real. If $p(\mu) = 0$, μ complex, then $p(\bar{\mu}) = 0$ too ($\bar{\mu}$ the conjugate), and the main observation is that $[\mu^n + \bar{\mu}^n]$ and $[i(\mu^n - \bar{\mu}^n))$ are *real* solutions. Analogues of the preceding theorems can then be proved.

EXERCISES 7.8

1. Find a basis for the space V of sequences $[x_n]$ satisfying the following recurrences, and use it to find the sequence satisfying $x_0 = 1, x_1 = 2, x_2 = 1$.
   (a) $x_{n+3} = -2x_n + x_{n+1} + 2x_{n+2}$
 ◆ **(b)** $x_{n+3} = -6x_n + 7x_{n+1}$
   (c) $x_{n+3} = -36x_n + 7x_{n+2}$

2. In each case, find a basis for the space V of all sequences $[x_n]$ satisfying the recurrence, and use it to find x_n if $x_0 = 1, x_1 = -1$, and $x_2 = 1$.
   (a) $x_{n+3} = x_n + x_{n+1} - x_{n+2}$
 ◆ **(b)** $x_{n+3} = -2x_n + 3x_{n+1}$
   (c) $x_{n+3} = -4x_n + 3x_{n+2}$
 ◆ **(d)** $x_{n+3} = x_n - 3x_{n+1} + 3x_{n+2}$
   (e) $x_{n+3} = 8x_n - 12x_{n+1} + 6x_{n+2}$

3. Find a basis for the space V of sequences $[x_n]$ satisfying each of the following recurrences.
   (a) $x_{n+2} = -a^2 x_n + 2ax_{n+1}, a \neq 0$
 ◆ **(b)** $x_{n+2} = -abx_n + (a + b)x_{n+1}, (a \neq b)$

4. Define the **Lucas sequence** $x_0, x_1, x_2, \ldots$ by $x_0 = 1$, $x_1 = 3$, $x_{n+2} = x_n + x_{n+1}$, $n \geq 0$. Show that $x_n = \left[\dfrac{1 + \sqrt{5}}{2}\right]^{n+1} + \left[\dfrac{1 - \sqrt{5}}{2}\right]^{n+1}$ for $n = 0, 1, 2, \ldots$.

5. The **generalized Fibonacci sequence** $g_0, g_1, \ldots$ is defined by $g_0 = a$, $g_1 = b$, $g_{n+2} = g_n + g_{n+1}$. Show that $g_n = \frac{1}{\sqrt{5}}[(b - a\lambda_2)\lambda_1^n - (b - a\lambda_1)\lambda_2^n]$, where $\lambda_1 = \frac{1}{2}(1 + \sqrt{5})$ and $\lambda_2 = \frac{1}{2}(1 - \sqrt{5})$.

6. In each case, find a basis of V.

  (a) $V = \{[x_n] \mid x_{n+4} = 2x_{n+2} - x_{n+3}$ for $n \geq 0\}$

  ◆**(b)** $V = \{[x_n] \mid x_{n+4} = -x_{n+2} - 2x_{n+3}$ for $n \geq 0\}$

7. Suppose that $[x_n]$ satisfies a linear recurrence relation of length k. If

$$\{\mathbf{e}_0 = (1, 0, \ldots, 0), \mathbf{e}_1 = (0, 1, \ldots, 0), \ldots,$$
$$\mathbf{e}_{k-1} = (0, 0, \ldots, 1)\}$$

is the standard basis of $\mathbb{R}^k$, show that $x_n = x_0 T(\mathbf{e}_0) + x_1 T(\mathbf{e}_1) + \cdots + x_{k-1} T(\mathbf{e}_{k-1})$ holds for all $n \geq k$. (Here T is as in Theorem 1.)

8. Show that the shift operator S is onto but not one-to-one. Find ker S.

◆**9.** Find a basis for the space V of all sequences $[x_n]$ satisfying $x_{n+2} = -x_n$.

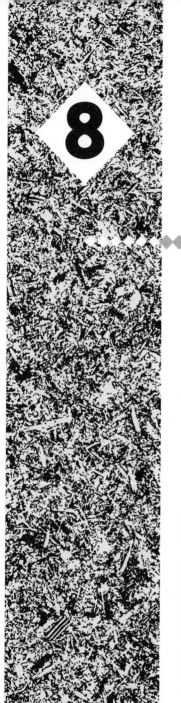

Inner Product Spaces

Section 8.1
Inner Products and Norms

The dot product was introduced in $\mathbb{R}^n$ to provide a natural generalization of the geometrical notions of length and orthogonality that were so important in Chapter 4. The plan in this chapter is to define an *inner product* on an arbitrary vector space V (of which the dot product is an example in $\mathbb{R}^n$) and use it to introduce these concepts in V.

DEFINITION

An **inner product** on a vector space V is a function that assigns a number $\langle \mathbf{v}, \mathbf{w} \rangle$ to every pair $\mathbf{v}, \mathbf{w}$ of vectors in V in such a way that the following axioms are satisfied.

P1. $\langle \mathbf{v}, \mathbf{w} \rangle$ is a real number for all $\mathbf{v}$ and $\mathbf{w}$ in V.

P2. $\langle \mathbf{v}, \mathbf{w} \rangle = \langle \mathbf{w}, \mathbf{v} \rangle$ for all $\mathbf{v}$ and $\mathbf{w}$ in V.

P3. $\langle \mathbf{v} + \mathbf{w}, \mathbf{u} \rangle = \langle \mathbf{v}, \mathbf{u} \rangle + \langle \mathbf{w}, \mathbf{u} \rangle$ for all $\mathbf{u}$, $\mathbf{v}$, and $\mathbf{w}$ in V.

P4. $\langle r\mathbf{v}, \mathbf{w} \rangle = r\langle \mathbf{v}, \mathbf{w} \rangle$ for all $\mathbf{v}$ and $\mathbf{w}$ in V and all r in $\mathbb{R}$.

P5. $\langle \mathbf{v}, \mathbf{v} \rangle > 0$ for all $\mathbf{v} \neq \mathbf{0}$ in V.

A vector space V with an inner product $\langle \; , \; \rangle$ will be called an **inner product space.**

EXAMPLE 1

$\mathbb{R}^n$ is an inner product space with the dot product as inner product: $\langle \mathbf{v}, \mathbf{w} \rangle = \mathbf{v} \cdot \mathbf{w}$. This is also called the **Euclidean** inner product.

EXAMPLE 2

If A and B are $m \times n$ matrices, define $\langle A, B \rangle = \text{tr}(AB^T)$ where $\text{tr}(X)$ is the trace of the square matrix X. Show that $\langle \, , \, \rangle$ is an inner product in $\mathbf{M}_{mn}$.

Solution   Checking axioms P1–P4 is left as Exercise 19. If $R_1, R_2, \ldots, R_n$ denote the rows of A, the (i, j)-entry of AA^T is $R_i \cdot R_j$, so

$$\langle A, A \rangle = \text{tr}(AA^T) = R_1 \cdot R_1 + R_2 \cdot R_2 + \cdots + R_n \cdot R_n$$

But $R_j \cdot R_j$ is the sum of the squares of the entries of R_j, so this shows that $\langle A, A \rangle$ is the sum of the squares of all nm entries of A. Axiom P5 follows.

◆◆◆

The next example is important in analysis.

EXAMPLE 3[1]

Let $\mathbf{C}[a, b]$ denote the vector space of **continuous functions** from $[a, b]$ to $\mathbb{R}$, a subspace of $\mathbf{F}[a, b]$. Show that

$$\langle f, g \rangle = \int_a^b f(x)g(x) \, dx$$

defines an inner product on $\mathbf{C}[a, b]$.

Solution   Axioms P1 and P2 are clear. As to axiom P4,

$$\langle rf, g \rangle = \int_a^b rf(x)g(x) \, dx = r \int_a^b f(x)g(x) \, dx = r\langle f, g \rangle$$

Axiom P3 is similar. Finally, $\langle f, f \rangle = \int_a^b f(x)^2 \, dx$, and it is a theorem of calculus that this is zero if and only if $f(x)$ is the zero function. This gives axiom P5.

◆◆◆

If $\mathbf{v}$ is any vector, then, using axiom P3, we get

$$\langle \mathbf{0}, \mathbf{v} \rangle = \langle \mathbf{0} + \mathbf{0}, \mathbf{v} \rangle = \langle \mathbf{0}, \mathbf{v} \rangle + \langle \mathbf{0}, \mathbf{v} \rangle$$

and it follows that the number $\langle \mathbf{0}, \mathbf{v} \rangle$ must be zero. This observation is recorded for reference in the following theorem, along with several other properties of inner products. The other proofs are left as Exercise 20.

THEOREM 1

Let $\langle \, , \, \rangle$ be an inner product on a space V, let $\mathbf{v}$, $\mathbf{u}$, and $\mathbf{w}$ denote vectors in V, and let r denote a real number.

1. $\langle \mathbf{u}, \mathbf{v} + \mathbf{w} \rangle = \langle \mathbf{u}, \mathbf{v} \rangle + \langle \mathbf{u}, \mathbf{w} \rangle$

---

[1]This example (and others later that refer to it) can be omitted with no loss of continuity by students with no calculus background.

2. $\langle \mathbf{v}, r\mathbf{w} \rangle = r\langle \mathbf{v}, \mathbf{w} \rangle = \langle r\mathbf{v}, \mathbf{w} \rangle$

3. $\langle \mathbf{v}, \mathbf{0} \rangle = 0 = \langle \mathbf{0}, \mathbf{v} \rangle$

4. $\langle \mathbf{v}, \mathbf{v} \rangle = 0$ if and only if $\mathbf{v} = \mathbf{0}$

If $\langle \ , \ \rangle$ is an inner product on a space V, then, given $\mathbf{u}$, $\mathbf{v}$, and $\mathbf{w}$ in V,

$$\langle r\mathbf{u} + s\mathbf{v}, \mathbf{w} \rangle = \langle r\mathbf{u}, \mathbf{w} \rangle + \langle s\mathbf{v}, \mathbf{w} \rangle = r\langle \mathbf{u}, \mathbf{w} \rangle + s\langle \mathbf{v}, \mathbf{w} \rangle$$

for all r and s in $\mathbb{R}$ by axioms P3 and P4. Moreover, there is nothing special about the fact that there are two terms in the linear combination or that it is in the first component:

$$\langle r_1\mathbf{v}_1 + r_2\mathbf{v}_2 + \cdots + r_n\mathbf{v}_n, \mathbf{w} \rangle = r_1\langle \mathbf{v}_1, \mathbf{w} \rangle + r_2\langle \mathbf{v}_2, \mathbf{w} \rangle + \cdots + r_n\langle \mathbf{v}_n, \mathbf{w} \rangle$$

and

$$\langle \mathbf{v}, s_1\mathbf{w}_1 + s_2\mathbf{w}_2 + \cdots + s_m\mathbf{w}_m \rangle = s_1\langle \mathbf{v}, \mathbf{w}_1 \rangle + s_2\langle \mathbf{v}, \mathbf{w}_2 \rangle + \cdots + s_m\langle \mathbf{v}, \mathbf{w}_m \rangle$$

hold for all r_i and s_i in $\mathbb{R}$ and all $\mathbf{v}$, $\mathbf{w}$, $\mathbf{v}_i$, and $\mathbf{w}_j$ in V. These results are described by saying that inner products "preserve" linear combinations. Moreover, two applications of these formulas give

$$\langle \mathbf{u} + \mathbf{v}, \mathbf{w} + \mathbf{p} \rangle = \langle \mathbf{u}, \mathbf{w} + \mathbf{p} \rangle + \langle \mathbf{v}, \mathbf{w} + \mathbf{p} \rangle$$
$$= \langle \mathbf{u}, \mathbf{w} \rangle + \langle \mathbf{u}, \mathbf{p} \rangle + \langle \mathbf{v}, \mathbf{w} \rangle + \langle \mathbf{v}, \mathbf{p} \rangle$$

In other words: To compute $\langle \mathbf{u} + \mathbf{v}, \mathbf{w} + \mathbf{p} \rangle$ take the inner product of each term in the first component with each term in the second component, and add the results. (The analogue in ordinary algebra is $(x + y) \cdot (z + t) = xz + xt + yz + yt$.) Thus $\langle \mathbf{u} + \mathbf{v}, \mathbf{w} + \mathbf{p} + \mathbf{y} \rangle$ would expand to a sum of six inner products.

EXAMPLE 4

If $\mathbf{u}$ and $\mathbf{v}$ are vectors in an inner product space, expand $\langle 2\mathbf{u} - \mathbf{v}, 3\mathbf{u} + 2\mathbf{v} \rangle$.

Solution

$$\langle 2\mathbf{u} - \mathbf{v}, 3\mathbf{u} + 2\mathbf{v} \rangle = \langle 2\mathbf{u}, 3\mathbf{u} \rangle + \langle 2\mathbf{u}, 2\mathbf{v} \rangle + \langle -\mathbf{v}, 3\mathbf{u} \rangle + \langle -\mathbf{v}, 2\mathbf{v} \rangle$$
$$= 6\langle \mathbf{u}, \mathbf{u} \rangle + 4\langle \mathbf{u}, \mathbf{v} \rangle - 3\langle \mathbf{v}, \mathbf{u} \rangle - 2\langle \mathbf{v}, \mathbf{v} \rangle$$
$$= 6\langle \mathbf{u}, \mathbf{u} \rangle + \langle \mathbf{u}, \mathbf{v} \rangle - 2\langle \mathbf{v}, \mathbf{v} \rangle$$

◆◆◆

If A is a symmetric $n \times n$ matrix and X and Y are columns in $\mathbb{R}^n$, we regard the 1×1 matrix $X^T A Y$ as a number. If we write

$$\langle X, Y \rangle = X^T A Y \qquad \text{for all columns } X, Y \text{ in } \mathbb{R}^n$$

then axioms P1–P4 follow from matrix arithmetic (only P2 requires that A is symmetric). Axiom P5 reads

$$X^T A X > 0 \qquad \text{for all columns } X \neq 0 \text{ in } \mathbb{R}^n$$

and this condition characterizes the positive definite matrices (Theorem 2§6.5). This proves the first assertion in the next theorem.

THEOREM 2

If A is any $n \times n$ positive definite matrix, then

$$\langle X, Y \rangle = X^T A Y \qquad \text{for all columns } X, Y \text{ in } \mathbb{R}^n$$

defines an inner product on $\mathbb{R}^n$, and every inner product arises in this way.

Proof Given an inner product $\langle \, , \, \rangle$ on $\mathbb{R}^n$, let $\{E_1, E_2, \ldots, E_n\}$ be the standard basis of $\mathbb{R}^n$. Given two vectors $X = \sum_{i=1}^{n} x_i E_i$ and $Y = \sum_{j=1}^{n} y_j E_j$, compute $\langle X, Y \rangle$ by taking the inner product of each term $x_i E_i$ with each term $y_j E_j$. The result is a double sum.

$$\langle X, Y \rangle = \sum_{i=1}^{n} \sum_{j=1}^{n} \langle x_i E_i, y_j E_j \rangle = \sum_{i=1}^{n} \sum_{j=1}^{n} x_i \langle E_i, E_j \rangle y_j$$

This is a matrix product:

$$\langle X, Y \rangle = [x_1 \quad x_2 \quad \cdots \quad x_n] \begin{bmatrix} \langle E_1, E_1 \rangle & \langle E_1, E_2 \rangle & \cdots & \langle E_1, E_n \rangle \\ \langle E_2, E_1 \rangle & \langle E_2, E_2 \rangle & \cdots & \langle E_2, E_n \rangle \\ \vdots & \vdots & & \vdots \\ \langle E_n, E_1 \rangle & \langle E_n, E_2 \rangle & \cdots & \langle E_n, E_n \rangle \end{bmatrix} \begin{bmatrix} y_1 \\ y_2 \\ \vdots \\ y_n \end{bmatrix}$$

Hence $\langle X, Y \rangle = X^T A Y$, where A is the $n \times n$ matrix whose (i, j)-entry is $\langle E_i, E_j \rangle$. The fact that $\langle E_i, E_j \rangle = \langle E_j, E_i \rangle$ shows that A is symmetric. ◆

Thus, just as every linear operator $\mathbb{R}^n \to \mathbb{R}^n$ corresponds to an $n \times n$ matrix, every inner product on $\mathbb{R}^n$ corresponds to a positive definite $n \times n$ matrix.

◆◆◆

Remark If we refer to the inner product space $\mathbb{R}^n$ without specifying the inner product, we mean that the dot product is to be used.

◆◆◆

EXAMPLE 5

Let the inner product $\langle \, , \, \rangle$ be defined on $\mathbb{R}^2$ by

$$\left\langle \begin{bmatrix} v_1 \\ v_2 \end{bmatrix}, \begin{bmatrix} w_1 \\ w_2 \end{bmatrix} \right\rangle = 2v_1 w_1 - v_1 w_2 - v_2 w_1 + v_2 w_2$$

Find a symmetric 2×2 matrix A such that $\langle X, Y \rangle = X^T A Y$ for all X, Y in $\mathbb{R}^2$.

Solution The (i, j)-entry of the matrix A is the coefficient of $v_i w_j$ in the expression, so

$$A = \begin{bmatrix} 2 & -1 \\ -1 & 1 \end{bmatrix}. \text{ Incidentally, if } X = \begin{bmatrix} x \\ y \end{bmatrix}, \text{ then}$$

$$\langle X, X \rangle = 2x^2 - 2xy + y^2 = x^2 + (x - y)^2 \geq 0$$

for all X, and $\langle X, X \rangle = 0$ implies $X = 0$. Hence $\langle\ ,\ \rangle$ is indeed an inner product, so A is positive definite.

◆◆◆

Let $\langle\ ,\ \rangle$ be an inner product on $\mathbb{R}^n$ given as in Theorem 2 by a positive definite matrix A. If $X = [x_1 \quad x_2 \quad \cdots \quad x_n]^T$, then $\langle X, X \rangle = X^T A X$ is an expression in the variables $x_1\ x_2, \ldots, x_n$ called a **quadratic form**. These are studied in detail in Section 6.9.

Norms and Distance

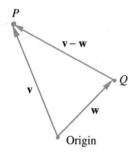

P

$\mathbf{v} - \mathbf{w}$

Q

$\mathbf{v}$

$\mathbf{w}$

Origin

FIGURE 8.1

If $P(x, y, z)$ is a point in space, the vector $\mathbf{v} = (x, y, z)$ in $\mathbb{R}^3$ is called the position vector of P and is thought of geometrically as the "arrow" from the origin to P (see Chapter 4). By Pythagoras's theorem, $\mathbf{v}$ has length

$$\|\mathbf{v}\| = \sqrt{x^2 + y^2 + z^2} = \sqrt{\mathbf{v} \cdot \mathbf{v}}$$

Moreover, if $Q(x_1, y_1, z_1)$ is another point with position vector $\mathbf{w} = (x_1, y_1, z_1)$, then $\mathbf{v} - \mathbf{w} = (x - x_1, y - y_1, z - z_1)$ is the vector from Q to P (see Figure 8.1), so the distance between P and Q is given by

$$d(\mathbf{v}, \mathbf{w}) = \|\mathbf{v} - \mathbf{w}\|$$

These formulas suggest the following definitions in any inner product space.

DEFINITION

If $\langle\ ,\ \rangle$ is an inner product on a space V, the **norm** or **length** $\|\mathbf{v}\|$ of a vector $\mathbf{v}$ in V is defined by

$$\|\mathbf{v}\| = \sqrt{\langle \mathbf{v}, \mathbf{v} \rangle}$$

Define the **distance** between vectors $\mathbf{v}$ and $\mathbf{w}$ in an inner product space V to be

$$d(\mathbf{v}, \mathbf{w}) = \|\mathbf{v} - \mathbf{w}\|$$

Note that axiom P5 guarantees that $\langle \mathbf{v}, \mathbf{v} \rangle \geq 0$, so $\|\mathbf{v}\|$ is a real number.

EXAMPLE 6

The length of $\mathbf{v} = (v_1, v_2, \ldots, v_n)$ in $\mathbb{R}^n$ (using the dot product) is

$$\|\mathbf{v}\| = \sqrt{v_1^2 + v_2^2 + \cdots + v_n^2}$$

The distance between $\mathbf{v}$ and $\mathbf{w} = (w_1, w_2, \ldots, w_n)$ is

$$d(\mathbf{v}, \mathbf{w}) = \sqrt{(v_1 - w_1)^2 + (v_2 - w_2)^2 + \cdots + (v_n - w_n)^2}$$

◆◆◆

EXAMPLE 7

The norm of a continuous function $f = f(x)$ in $\mathbf{C}[a, b]$ (with the inner product from Example 3) is given by

$$\|f\| = \sqrt{\int_a^b f(x)^2 \, dx}$$

◆◆◆

EXAMPLE 8

Show that $\langle \mathbf{u} + \mathbf{v}, \mathbf{u} - \mathbf{v} \rangle = \|\mathbf{u}\|^2 - \|\mathbf{v}\|^2$ in any inner product space.

Solution

$$\langle \mathbf{u} + \mathbf{v}, \mathbf{u} - \mathbf{v} \rangle = \langle \mathbf{u}, \mathbf{u} \rangle - \langle \mathbf{u}, \mathbf{v} \rangle + \langle \mathbf{v}, \mathbf{u} \rangle - \langle \mathbf{v}, \mathbf{v} \rangle$$
$$= \|\mathbf{u}\|^2 - \langle \mathbf{u}, \mathbf{v} \rangle + \langle \mathbf{u}, \mathbf{v} \rangle - \|\mathbf{v}\|^2$$
$$= \|\mathbf{u}\|^2 - \|\mathbf{v}\|^2$$

◆◆◆

A vector $\mathbf{v}$ in an inner product space V is called a **unit vector** if $\|\mathbf{v}\| = 1$. The set of all unit vectors in V is called the **unit ball** in V. For example, if $V = \mathbb{R}^2$ (with the dot product) and $\mathbf{v} = (x, y)$, then

$$\|\mathbf{v}\| = 1 \quad \text{if and only if} \quad x^2 + y^2 = 1$$

Hence the unit ball in $\mathbb{R}^2$ is the **unit circle** $x^2 + y^2 = 1$ with center at the origin and radius 1. However, the shape of the unit ball varies with the choice of inner product.

EXAMPLE 9

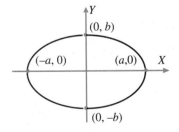

FIGURE 8.2

If $a > 0$ and $b > 0$, define an inner product on $\mathbb{R}^2$ by

$$\langle \mathbf{v}, \mathbf{w} \rangle = \frac{x\, x_1}{a^2} + \frac{y\, y_1}{b^2}$$

where $\mathbf{v} = (x, y)$ and $\mathbf{w} = (x_1, y_1)$. The reader can verify (Exercise 5) that this is indeed an inner product. In this case

$$\|\mathbf{v}\| = 1 \quad \text{if and only if} \quad \frac{x^2}{a^2} + \frac{y^2}{b^2} = 1$$

so the unit ball is the ellipse shown in Figure 8.2.

◆◆◆

Example 9 graphically illustrates the fact that norms and distances in an inner product space V vary with the choice of inner product in V.

The proof (Example 2§6.3) of the next result for the dot product works in general.

THEOREM 3

If $\mathbf{v} \neq \mathbf{0}$ is any vector in an inner product space V, then

$$\hat{\mathbf{v}} = \frac{1}{\|\mathbf{v}\|}\,\mathbf{v}$$

is the unique unit vector that is a positive multiple of $\mathbf{v}$.

The next theorem reveals an important and useful fact about the relationship between norms and inner products.

THEOREM 4

Schwarz Inequality[2]

If $\mathbf{v}$ and $\mathbf{w}$ are two vectors in an inner product space V, then

$$\langle \mathbf{v}, \mathbf{w} \rangle^2 \leq \|\mathbf{v}\|^2\,\|\mathbf{w}\|^2$$

Moreover, equality occurs if and only if one of $\mathbf{v}$ and $\mathbf{w}$ is a scalar multiple of the other.

Proof   If either $\mathbf{v} = \mathbf{0}$ or $\mathbf{w} = \mathbf{0}$, the inequality holds (in fact, it is an equality). Otherwise, write $\|\mathbf{v}\| = a > 0$ and $\|\mathbf{w}\| = b > 0$. Then

$$\|b\mathbf{v} - a\mathbf{w}\|^2 = 2ab(ab - \langle \mathbf{v}, \mathbf{w} \rangle) \quad \text{and} \quad \|b\mathbf{v} + a\mathbf{w}\|^2 = 2ab(ab + \langle \mathbf{v}, \mathbf{w} \rangle) \quad (*)$$

as the reader can verify. As $\|\mathbf{u}\|^2 \geq 0$ for all $\mathbf{u}$ in V, these give $-ab \leq \langle \mathbf{v}, \mathbf{w} \rangle \leq ab$, and the Schwarz inequality follows.

If equality holds—$\langle \mathbf{v}, \mathbf{w} \rangle^2 = a^2 b^2$—then $\langle \mathbf{v}, \mathbf{w} \rangle = ab$ or $\langle \mathbf{v}, \mathbf{w} \rangle = -ab$, so equations $(*)$ give $b\mathbf{v} = a\mathbf{w}$ or $b\mathbf{v} = -a\mathbf{w}$. Hence one of $\mathbf{v}$ and $\mathbf{w}$ is a scalar multiple of the other (even if $a = 0$ or $b = 0$). ◆

---

[2] Herman Amandus Schwarz (1843–1921) was a German mathematician at the University of Berlin. He had strong geometric intuition, which he applied with great ingenuity to particular problems. A version of the inequality appeared in 1885.

EXAMPLE 10

Cauchy Inequality[3]

If $v_1, v_2, \ldots, v_n$ and $w_1, w_2, \ldots, w_n$ are any two sequences of real numbers, then

$$(v_1w_1 + v_2w_2 + \cdots + v_nw_n)^2 \le (v_1^2 + v_2^2 + \cdots + v_n^2)(w_1^2 + w_2^2 + \cdots + w_n^2)$$

Solution   If $\mathbf{v} = (v_1, v_2, \ldots, v_n)$ and $\mathbf{w} = (w_1, w_2, \ldots, w_n)$ in $\mathbb{R}^n$, this is just the Schwarz inequality (using the dot product).   ◆◆◆

EXAMPLE 11

If f and g are continuous functions on the interval $[a, b]$, then (see Example 7)

$$\left\{ \int_a^b f(x)g(x)\, dx \right\}^2 \le \int_a^b f(x)^2\, dx \int_a^b g(x)^2\, dx$$

◆◆◆

Another famous inequality, the so-called *triangle inequality*, also comes from the Schwarz inequality. It is included in the following list of basic properties of the norm of a vector.

THEOREM 5

If V is an inner product space, the norm $\|\cdot\|$ has the following properties.

1. $\|\mathbf{v}\| \ge 0$ for every vector $\mathbf{v}$ in V
2. $\|\mathbf{v}\| = 0$ if and only if $\mathbf{v} = \mathbf{0}$
3. $\|r\mathbf{v}\| = |r|\, \|\mathbf{v}\|$ for every $\mathbf{v}$ in V and every r in $\mathbb{R}$
4. $\|\mathbf{v} + \mathbf{w}\| \le \|\mathbf{v}\| + \|\mathbf{w}\|$ for all $\mathbf{v}$ and $\mathbf{w}$ in V (**triangle inequality**)

Proof   Because $\|\mathbf{v}\| = \sqrt{\langle \mathbf{v}, \mathbf{v} \rangle}$, properties (1) and (2) follow immediately from (3) and (4) of Theorem 1. As to (3), compute

$$\|r\mathbf{v}\|^2 = \langle r\mathbf{v}, r\mathbf{v} \rangle = r^2 \langle \mathbf{v}, \mathbf{v} \rangle = r^2 \|\mathbf{v}\|^2$$

Hence (3) follows by taking positive square roots. Finally, the fact that $\langle \mathbf{v}, \mathbf{w} \rangle \le \|\mathbf{v}\|\, \|\mathbf{w}\|$ by the Schwarz inequality gives

$$\begin{aligned}
\|\mathbf{v} + \mathbf{w}\|^2 &= \langle \mathbf{v} + \mathbf{w}, \mathbf{v} + \mathbf{w} \rangle = \|\mathbf{v}\|^2 + 2\langle \mathbf{v}, \mathbf{w} \rangle + \|\mathbf{w}\|^2 \\
&\le \|\mathbf{v}\|^2 + 2\|\mathbf{v}\|\, \|\mathbf{w}\| + \|\mathbf{w}\|^2 \\
&= (\|\mathbf{v}\| + \|\mathbf{w}\|)^2
\end{aligned}$$

Hence (4) follows by taking positive square roots.   ◆

---

[3] Augustin Louis Cauchy (1789–1857) was born in Paris and became a professor at the École Polytechnique at the age of 26. He was one of the great mathematicians and produced 789 papers. He is best remembered for his work in analysis in which he established modern standards of rigor that carry down to today's calculus texts.

It is worth noting that the usual triangle inequality for absolute values, $|r + s| \leq |r| + |s|$ for all real numbers r and s, is a special case of (4) where $V = \mathbb{R} = \mathbb{R}^1$ and the dot product $\langle r, s \rangle = rs$ is used.

In many calculations in an inner product space, it is required to show that some vector $\mathbf{v}$ is zero. This is often accomplished most easily by showing that its norm $\|\mathbf{v}\|$ is zero. Here is an example.

EXAMPLE 12

Let $\{\mathbf{v}_1, \ldots, \mathbf{v}_n\}$ be a spanning set for an inner product space V. If $\mathbf{v}$ in V satisfies $\langle \mathbf{v}, \mathbf{v}_i \rangle = 0$ for each $i = 1, 2, \ldots, n$, show that $\mathbf{v} = \mathbf{0}$.

Solution   Write $\mathbf{v} = r_1\mathbf{v}_1 + \cdots + r_n\mathbf{v}_n$, r_i in $\mathbb{R}$. To show that $\mathbf{v} = \mathbf{0}$, we show that $\|\mathbf{v}\|^2 = \langle \mathbf{v}, \mathbf{v} \rangle = 0$. Compute:

$$\langle \mathbf{v}, \mathbf{v} \rangle = \langle \mathbf{v}, r_1\mathbf{v}_1 + \cdots + r_n\mathbf{v}_n \rangle = r_1\langle \mathbf{v}, \mathbf{v}_1 \rangle + \cdots + r_n\langle \mathbf{v}, \mathbf{v}_n \rangle = 0$$

by hypothesis, and the result follows.

◆◆◆

The norm properties in Theorem 5 translate to the following properties of distance familiar from geometry. The proof is Exercise 21.

THEOREM 6

Let V be an inner product space.

1. $d(\mathbf{v}, \mathbf{w}) \geq 0$ for all $\mathbf{v}, \mathbf{w}$ in V
2. $d(\mathbf{v}, \mathbf{w}) = 0$ if and only if $\mathbf{v} = \mathbf{w}$
3. $d(\mathbf{v}, \mathbf{w}) = d(\mathbf{w}, \mathbf{v})$ for all $\mathbf{v}$ and $\mathbf{w}$ in V
4. $d(\mathbf{v}, \mathbf{w}) \leq d(\mathbf{v}, \mathbf{u}) + d(\mathbf{u}, \mathbf{w})$ for all $\mathbf{v}, \mathbf{u}$, and $\mathbf{w}$ in V

EXERCISES 8.1

1. In each case, determine which of axioms P1–P5 fail to hold.
   (a) $V = \mathbb{R}^2$, $\langle (x_1, y_1), (x_2, y_2) \rangle = x_1 y_1 x_2 y_2$
   ◆(b) $V = \mathbb{R}^3$, $\langle (x_1, x_2, x_3), (y_1, y_2, y_3) \rangle = x_1 y_1 - x_2 y_2 + x_3 y_3$
   (c) $V = \mathbb{C}$, $\langle z, w \rangle = z\overline{w}$, where $\overline{w}$ is complex conjugation
   ◆(d) $V = \mathbf{P}_3$, $\langle p(x), q(x) \rangle = p(1)q(1)$
   (e) $V = \mathbf{M}_{22}$, $\langle A, B \rangle = \det(AB)$
   ◆(f) $V = \mathbf{F}[0, 1]$, $\langle f, g \rangle = f(1)g(0) + f(0)g(1)$

2. Verify that the dot product on $\mathbb{R}^n$ satisfies axioms P1–P5.

3. In each case, find a scalar multiple of $\mathbf{v}$ that is a unit vector.
   (a) $\mathbf{v} = (1, -1, 2, 0)$ in $\mathbb{R}^4$ (dot product)
   ◆(b) $\mathbf{v} = (2, -3, 1, 1)$ in $\mathbb{R}^4$ (dot product)
   (c) $\mathbf{v} = \begin{bmatrix} 1 \\ 3 \end{bmatrix}$ in $\mathbb{R}^2$ $\left(\langle \mathbf{v}, \mathbf{w} \rangle = \mathbf{v}^T \begin{bmatrix} 1 & 1 \\ 1 & 2 \end{bmatrix} \mathbf{w} \right)$
   ◆(d) $\mathbf{v} = \begin{bmatrix} 3 \\ -1 \end{bmatrix}$ in $\mathbb{R}^2$ $\left(\langle \mathbf{v}, \mathbf{w} \rangle = \mathbf{v}^T \begin{bmatrix} 1 & -1 \\ -1 & 2 \end{bmatrix} \mathbf{w} \right)$

4. In each case, find the distance between $\mathbf{u}$ and $\mathbf{v}$.
   (a) $\mathbf{u} = (1, 2, 2), \mathbf{v} = (2, -2, 1)$
   ◆(b) $\mathbf{u} = (1, 1, 0), \mathbf{v} = (0, 1, -1)$
   (c) $\mathbf{u} = (3, -1, 2, 0), \mathbf{v} = (1, 1, 1, 3)$
   ◆(d) $\mathbf{u} = (1, 2, -1, 2), \mathbf{v} = (2, 1, -1, 3)$

5. Let $a_1, a_2, \ldots, a_n$ be positive numbers. Given $\mathbf{v} = (v_1, v_2, \ldots, v_n)$ and $\mathbf{w} = (w_1, w_2, \ldots, w_n)$, define $\langle \mathbf{v}, \mathbf{w} \rangle = a_1 v_1 w_1 + \cdots + a_n v_n w_n$. Show that this is an inner product on $\mathbb{R}^n$.

6. If $\{\mathbf{b}_1, \ldots, \mathbf{b}_n\}$ is a basis of V and if $\mathbf{v} = v_1\mathbf{b}_1 + \cdots + v_n\mathbf{b}_n$ and $\mathbf{w} = w_1\mathbf{b}_1 + \cdots + w_n\mathbf{b}_n$ are vectors in V, define $\langle \mathbf{v}, \mathbf{w} \rangle = v_1 w_1 + \cdots + v_n w_n$. Show that this is an inner product on V.

7. If $p = p(x)$ and $q = q(x)$ are polynomials in $\mathbf{P}_n$, define

$$\langle p, q \rangle = p(0)q(0) + p(1)q(1) + \cdots + p(n)q(n)$$

Show that this is an inner product on $\mathbf{P}_n$. [*Hint for P5:* If $p(0) = p(1) = \cdots = p(n) = 0$, then $p = 0$—Corollary of Theorem 3§5.6.]

◆8. Let $\mathbf{D}_n$ denote the space of all functions from the set $\{1, 2, 3, \ldots, n\}$ to $\mathbb{R}$ with pointwise addition and scalar multiplication (see Exercise 36§5.3). Show that $\langle \ , \ \rangle$ is an inner product on $\mathbf{D}_n$ if $\langle f, g \rangle = f(1)g(1) + f(2)g(2) + \cdots + f(n)g(n)$.

9. Let re(z) denote the real part of the complex number z. Show that $\langle \ , \ \rangle$ is an inner product on $\mathbb{C}$ if $\langle z, w \rangle = \mathrm{re}(z\overline{w})$.

10. If $T : V \to V$ is an isomorphism of the inner product space V, show that $\langle \mathbf{v}, \mathbf{w} \rangle_1 = \langle T(\mathbf{v}), T(\mathbf{w}) \rangle$ defines an inner product $\langle \ , \ \rangle_1$ on V.

11. Show that every inner product $\langle \ , \ \rangle$ on $\mathbb{R}^n$ has the form $\langle X, Y \rangle = (UX) \cdot (UY)$ for some upper triangular matrix U with positive diagonal entries. [*Hint:* Theorem 4§6.5.]

12. In each case, show that $\langle \mathbf{v}, \mathbf{w} \rangle = \mathbf{v}^T A \mathbf{w}$ defines an inner product on $\mathbb{R}^2$ and hence show that A is positive definite.

   (a) $A = \begin{bmatrix} 2 & 1 \\ 1 & 1 \end{bmatrix}$ **◆(b)** $A = \begin{bmatrix} 5 & -3 \\ -3 & 2 \end{bmatrix}$

   (c) $A = \begin{bmatrix} 3 & 2 \\ 2 & 3 \end{bmatrix}$ **◆(d)** $A = \begin{bmatrix} 3 & 4 \\ 4 & 6 \end{bmatrix}$

13. In each case, find a symmetric matrix A such that $\langle \mathbf{v}, \mathbf{w} \rangle = \mathbf{v}^T A \mathbf{w}$.

   (a) $\left\langle \begin{bmatrix} v_1 \\ v_2 \end{bmatrix}, \begin{bmatrix} w_1 \\ w_2 \end{bmatrix} \right\rangle = v_1 w_1 + 2v_1 w_2 + 2v_2 w_1 + 5v_2 w_2$

   ◆(b) $\left\langle \begin{bmatrix} v_1 \\ v_2 \end{bmatrix}, \begin{bmatrix} w_1 \\ w_2 \end{bmatrix} \right\rangle = v_1 w_1 - v_1 w_2 - v_2 w_1 + 2v_2 w_2$

   (c) $\left\langle \begin{bmatrix} v_1 \\ v_2 \\ v_3 \end{bmatrix}, \begin{bmatrix} w_1 \\ w_2 \\ w_3 \end{bmatrix} \right\rangle = 2v_1 w_1 + v_2 w_2 + v_3 w_3 - v_1 w_2 - v_2 w_1 + v_2 w_3 + v_3 w_2$

   ◆(d) $\left\langle \begin{bmatrix} v_1 \\ v_2 \\ v_3 \end{bmatrix}, \begin{bmatrix} w_1 \\ w_2 \\ w_3 \end{bmatrix} \right\rangle = v_1 w_1 + 2v_2 w_2 + 5v_3 w_3 - 2v_1 w_3 - 2v_3 w_1$

14. If A is symmetric and $X^T A X = 0$ for all columns X in $\mathbb{R}^n$, show that $A = 0$. [*Hint:* Consider $\langle X + Y, X + Y \rangle$ where $\langle X, Y \rangle = X^T A X$.]

15. Show that the sum of two inner products on V is again an inner product.

16. Let $\|\mathbf{u}\| = 1$, $\|\mathbf{v}\| = 2$, $\|\mathbf{w}\| = \sqrt{3}$, $\langle \mathbf{u}, \mathbf{v} \rangle = -1$, $\langle \mathbf{u}, \mathbf{w} \rangle = 0$ and $\langle \mathbf{v}, \mathbf{w} \rangle = 3$. Compute:

   (a) $\langle \mathbf{v} + \mathbf{w}, 2\mathbf{u} - \mathbf{v} \rangle$

   ◆(b) $\langle \mathbf{u} - 2\mathbf{v} - \mathbf{w}, 3\mathbf{w} - \mathbf{v} \rangle$

17. Given the data in Exercise 16, show that $\mathbf{u} + \mathbf{v} = \mathbf{w}$.

18. Show that no vectors exist such that $\|\mathbf{u}\| = 1$, $\|\mathbf{v}\| = 2$, and $\langle \mathbf{u}, \mathbf{v} \rangle = -3$.

19. Complete Example 2.

◆20. Prove Theorem 1.

21. Prove Theorem 6.

22. Let $\mathbf{u}$ and $\mathbf{v}$ be vectors in an inner product space V.

   (a) Expand $\langle 2\mathbf{u} - 7\mathbf{v}, 3\mathbf{u} + 5\mathbf{v} \rangle$.

   ◆(b) Expand $\langle 3\mathbf{u} - 4\mathbf{v}, 5\mathbf{u} + \mathbf{v} \rangle$.

   (c) Show that $\|\mathbf{u} + \mathbf{v}\|^2 = \|\mathbf{u}\|^2 + 2\langle \mathbf{u}, \mathbf{v} \rangle + \|\mathbf{v}\|^2$.

   (d) Show that $\|\mathbf{u} - \mathbf{v}\|^2 = \|\mathbf{u}\|^2 - 2\langle \mathbf{u}, \mathbf{v} \rangle + \|\mathbf{v}\|^2$.

23. Show that $\|\mathbf{v}\|^2 + \|\mathbf{w}\|^2 = \frac{1}{2}\{\|\mathbf{v} + \mathbf{w}\|^2 + \|\mathbf{v} - \mathbf{w}\|^2\}$ for any $\mathbf{v}$ and $\mathbf{w}$ in an inner product space.

24. Let $\langle \ , \ \rangle$ be an inner product on a vector space V. Show that the corresponding distance function is translation invariant. That is, show that $d(\mathbf{v}, \mathbf{w}) = d(\mathbf{v} + \mathbf{u}, \mathbf{w} + \mathbf{u})$ for all $\mathbf{v}$, $\mathbf{w}$, and $\mathbf{u}$ in V.

25. **(a)** Show that $\langle \mathbf{u}, \mathbf{v} \rangle = \frac{1}{4}[\|\mathbf{u} + \mathbf{v}\|^2 - \|\mathbf{u} - \mathbf{v}\|^2]$ for all $\mathbf{u}, \mathbf{v}$ in an inner product space V.

   (b) If $\langle \ , \ \rangle$ and $\langle \ , \ \rangle'$ are two inner products on V that have equal associated norm functions, show that $\langle \mathbf{u}, \mathbf{v} \rangle = \langle \mathbf{u}, \mathbf{v} \rangle'$ holds for all $\mathbf{u}$ and $\mathbf{v}$.

26. Let $\mathbf{v}$ denote a vector in an inner product space V.

   (a) Show that $W = \{\mathbf{w} \mid \mathbf{w}$ in V, $\langle \mathbf{v}, \mathbf{w} \rangle = 0\}$ is a subspace of V.

   ◆(b) If $V = \mathbb{R}^3$ with the dot product, and if $\mathbf{v} = (1, -1, 2)$, find a basis for W.

27. Given vectors $\mathbf{w}_1, \mathbf{w}_2, \ldots, \mathbf{w}_n$ and $\mathbf{v}$, assume that $\langle \mathbf{v}, \mathbf{w}_i \rangle = 0$ for each i. Show that $\langle \mathbf{v}, \mathbf{w} \rangle = 0$ for all $\mathbf{w}$ in span $\{\mathbf{w}_1, \mathbf{w}_2, \ldots, \mathbf{w}_n\}$.

◆28. If $V = \mathrm{span}\ \{\mathbf{v}_1, \mathbf{v}_2, \ldots, \mathbf{v}_n\}$ and $\langle \mathbf{v}, \mathbf{v}_i \rangle = \langle \mathbf{w}, \mathbf{v}_i \rangle$ holds for each i, show that $\mathbf{v} = \mathbf{w}$.

29. Use the Cauchy inequality to prove that:

   (a) $(r_1 + r_2 + \cdots + r_n)^2 \leq n(r_1^2 + r_2^2 + \cdots + r_n^2)$ for all r_i in $\mathbb{R}$.

   (b) $r_1 r_2 + r_1 r_3 + r_2 r_3 \leq r_1^2 + r_2^2 + r_3^2$ for all r_1, r_2, r_3 in $\mathbb{R}$. [*Hint:* See part (a).]

30. Use the Schwarz inequality in an inner product space to show that:
  (a) If $\|\mathbf{u}\| \leq 1$, then $\langle \mathbf{u}, \mathbf{v} \rangle^2 \leq \|\mathbf{v}\|^2$ for all $\mathbf{v}$ in V.
  (b) $(x \cos \theta + y \sin \theta)^2 \leq x^2 + y^2$ for all real x, y, and θ.
  (c) $\|r_1\mathbf{v}_1 + \cdots + r_n\mathbf{v}_n\|^2 \leq \{r_1\|\mathbf{v}_1\| + \cdots + r_n\|\mathbf{v}_n\|\}^2$ for all vectors $\mathbf{v}_i$, and all $r_i > 0$ in $\mathbb{R}$.

31. If A is a $2 \times n$ matrix, let $\mathbf{u}$ and $\mathbf{v}$ denote the rows of A.
  (a) Show that $AA^T = \begin{bmatrix} \|\mathbf{u}\|^2 & \mathbf{u} \cdot \mathbf{v} \\ \mathbf{u} \cdot \mathbf{v} & \|\mathbf{v}\|^2 \end{bmatrix}$.
  (b) Show that $\det(AA^T) \geq 0$.

32. (a) If $\mathbf{v}$ and $\mathbf{w}$ are nonzero vectors in an inner product space V, show that $-1 \leq \dfrac{\langle \mathbf{v}, \mathbf{w} \rangle}{\|\mathbf{v}\| \, \|\mathbf{w}\|} \leq 1$, and hence that a unique angle θ exists such that $\dfrac{\langle \mathbf{v}, \mathbf{w} \rangle}{\|\mathbf{v}\| \, \|\mathbf{w}\|} = \cos \theta$ and $0 \leq \theta \leq \pi$. This angle θ is called the **angle between** $\mathbf{v}$ and $\mathbf{w}$.

  ◆**(b)** Find the angle between $\mathbf{v} = (1, 2, -1, 1, 3)$ and $\mathbf{w} = (2, 1, 0, 2, 0)$ in $\mathbb{R}^5$ with the dot product.
  (c) If θ is the angle between $\mathbf{v}$ and $\mathbf{w}$, show that the **law of cosines** is valid: $\|\mathbf{v} - \mathbf{w}\|^2 = \|\mathbf{v}\|^2 + \|\mathbf{w}\|^2 - 2\|\mathbf{v}\| \, \|\mathbf{w}\| \cos \theta$.

33. Let $\{\mathbf{e}_1, \mathbf{e}_2, \ldots, \mathbf{e}_n\}$ be an orthogonal basis of V, and let θ_i be the angle between $\mathbf{v} \neq \mathbf{0}$ and $\mathbf{e}_i$ for each i (preceding exercise). Show that $\cos^2\theta_1 + \cos^2\theta_2 + \cdots + \cos^2\theta_n = 1$.

34. If $V = \mathbb{R}^2$, define $\|(x, y)\| = |x| + |y|$.
  (a) Show that $\|\cdot\|$ satisfies the conditions in Theorem 5.
  (b) Show that $\|\cdot\|$ does not arise from an inner product on $\mathbb{R}^2$. [*Hint:* If it did, use Theorem 2 to find numbers a, b, and c such that $\|(x, y)\|^2 = ax^2 + bxy + cy^2$ for all x and y.]

Section 8.2   Orthogonal Sets of Vectors

The idea that two lines can be perpendicular is fundamental in geometry, and this section is devoted to introducing this notion into a general inner product space V. To motivate the definition, recall that two nonzero geometric vectors $\mathbf{v}$ and $\mathbf{w}$ in $\mathbb{R}^3$ are perpendicular (or orthogonal) if and only if $\mathbf{v} \cdot \mathbf{w} = 0$.

DEFINITION

Two vectors $\mathbf{v}$ and $\mathbf{w}$ in an inner product space V are said to be **orthogonal** if

$$\langle \mathbf{v}, \mathbf{w} \rangle = 0$$

A set $\{\mathbf{e}_1, \mathbf{e}_2, \ldots, \mathbf{e}_n\}$ of vectors is called an **orthogonal set of vectors** if

1. Each $\mathbf{e}_i \neq \mathbf{0}$.

2. $\langle \mathbf{e}_i, \mathbf{e}_j \rangle = 0$ for all $i \neq j$.

If, in addition, $\|\mathbf{e}_i\| = 1$ for each i, the set $\{\mathbf{e}_1, \mathbf{e}_2, \ldots, \mathbf{e}_n\}$ is called an **orthonormal** set.

These definitions agree with those in Section 6.3 for $\mathbb{R}^n$ where several examples are given.

EXAMPLE 1

$\{\sin x, \cos x\}$ is orthogonal in $C[-\pi, \pi]$ because

$$\int_{-\pi}^{\pi} \sin x \, \cos x \, dx = \left[-\tfrac{1}{4} \cos 2x\right]_{-\pi}^{\pi} = 0$$

◆◆◆

The first result about orthogonal sets extends the Pythogorean theorem in $\mathbb{R}^n$ (Theorem 2§6.3).

THEOREM 1

The Pythagorean Theorem

If $\{\mathbf{e}_1, \mathbf{e}_2, \ldots, \mathbf{e}_n\}$ is an orthogonal set of vectors, then

$$\|\mathbf{e}_1 + \mathbf{e}_2 + \cdots + \mathbf{e}_n\|^2 = \|\mathbf{e}_1\|^2 + \|\mathbf{e}_2\|^2 + \cdots + \|\mathbf{e}_n\|^2$$

Proof

$$\|\mathbf{e}_1 + \mathbf{e}_2 + \cdots + \mathbf{e}_n\|^2 = \langle \mathbf{e}_1 + \mathbf{e}_2 + \cdots + \mathbf{e}_n, \mathbf{e}_1 + \mathbf{e}_2 + \cdots + \mathbf{e}_n \rangle$$

$$= \langle \mathbf{e}_1, \mathbf{e}_1 \rangle + \langle \mathbf{e}_2, \mathbf{e}_2 \rangle + \cdots + \langle \mathbf{e}_n, \mathbf{e}_n \rangle + \sum_{i \neq j} \langle \mathbf{e}_i, \mathbf{e}_j \rangle$$

The result follows because $\langle \mathbf{e}_i, \mathbf{e}_i \rangle = \|\mathbf{e}_i\|^2$ for each i and $\langle \mathbf{e}_i, \mathbf{e}_j \rangle = 0$ if $i \neq j$. ◆

The proof of the next result is left to the reader.

THEOREM 2

Let $\{\mathbf{e}_1, \mathbf{e}_2, \ldots, \mathbf{e}_n\}$ be an orthogonal set of vectors.

1. $\{r_1\mathbf{e}_1, r_2\mathbf{e}_2, \ldots, r_n\mathbf{e}_n\}$ is also orthogonal for any $r_i \neq 0$ in $\mathbb{R}$.

2. $\left\{ \dfrac{1}{\|\mathbf{e}_1\|} \mathbf{e}_1, \dfrac{1}{\|\mathbf{e}_2\|} \mathbf{e}_2, \ldots, \dfrac{1}{\|\mathbf{e}_n\|} \mathbf{e}_n \right\}$ is an orthonormal set.

As before, the process of passing from an orthogonal set to an orthonormal one is called **normalizing** the orthogonal set.

THEOREM 3

Every orthogonal set of vectors is linearly independent.

Proof   The proof of Theorem 3§6.3 goes through.   ◆

EXAMPLE 2

Show that $\left\{ \begin{bmatrix} 2 \\ -1 \\ 0 \end{bmatrix}, \begin{bmatrix} 0 \\ 1 \\ 1 \end{bmatrix}, \begin{bmatrix} 0 \\ -1 \\ 2 \end{bmatrix} \right\}$ is an orthogonal basis of $\mathbb{R}^3$ with inner product $\langle \mathbf{v}, \mathbf{w} \rangle$

$= \mathbf{v}^T A \mathbf{w}$, where $A = \begin{bmatrix} 1 & 1 & 0 \\ 1 & 2 & 0 \\ 0 & 0 & 1 \end{bmatrix}$.

Solution   We have

$$\left\langle \begin{bmatrix} 2 \\ -1 \\ 0 \end{bmatrix}, \begin{bmatrix} 0 \\ 1 \\ 1 \end{bmatrix} \right\rangle = \begin{bmatrix} 2 & -1 & 0 \end{bmatrix} \begin{bmatrix} 1 & 1 & 0 \\ 1 & 2 & 0 \\ 0 & 0 & 1 \end{bmatrix} \begin{bmatrix} 0 \\ 1 \\ 1 \end{bmatrix} = \begin{bmatrix} 1 & 0 & 0 \end{bmatrix} \begin{bmatrix} 0 \\ 1 \\ 1 \end{bmatrix} = 0$$

and the reader can verify that the other pairs are orthogonal too. Hence the set is orthogonal, so it is linearly independent by Theorem 3. Because dim $\mathbb{R}^3 = 3$, it is a basis.
◆◆◆

The proof of Theorem 4§6.3 generalizes to give the following:

THEOREM 4

Expansion Theorem

Let $\{\mathbf{e}_1, \mathbf{e}_2, \ldots, \mathbf{e}_n\}$ be an orthogonal basis of an inner product space V. If $\mathbf{v}$ is any vector in V, then

$$\mathbf{v} = \frac{\langle \mathbf{v}, \mathbf{e}_1 \rangle}{\|\mathbf{e}_1\|^2} \mathbf{e}_1 + \frac{\langle \mathbf{v}, \mathbf{e}_2 \rangle}{\|\mathbf{e}_2\|^2} \mathbf{e}_2 + \cdots + \frac{\langle \mathbf{v}, \mathbf{e}_n \rangle}{\|\mathbf{e}_n\|^2} \mathbf{e}_n$$

is the expansion of $\mathbf{v}$ as a linear combination of the basis vectors.

The coefficients $\dfrac{\langle \mathbf{v}, \mathbf{e}_1 \rangle}{\|\mathbf{e}_1\|^2}, \dfrac{\langle \mathbf{v}, \mathbf{e}_2 \rangle}{\|\mathbf{e}_2\|^2}, \ldots, \dfrac{\langle \mathbf{v}, \mathbf{e}_n \rangle}{\|\mathbf{e}_n\|^2}$ in the expansion theorem are sometimes called the **Fourier coefficients** of $\mathbf{v}$ with respect to the orthogonal basis $\{\mathbf{e}_1, \mathbf{e}_2, \ldots, \mathbf{e}_n\}$. This is in honor of the French mathematician J. B. J. Fourier (1768–1830). His original work was with a particular orthogonal set in the space $C[a, b]$, and we will have more to say about that in Section 8.5.

EXAMPLE 3

If $a_0, a_1, \ldots, a_n$ are distinct numbers and $p(x)$ and $q(x)$ are in $\mathbf{P}_n$, define

$$\langle p(x), q(x) \rangle = p(a_0)q(a_0) + p(a_1)q(a_1) + \cdots + p(a_n)q(a_n)$$

This is an inner product on $\mathbf{P}_n$. (Axioms P1–P4 are routinely verified, and P5 holds because 0 is the only polynomial of degree n with $n + 1$ distinct roots.) Define the **Lagrange polynomials** $\delta_0(x), \delta_1(x), \ldots, \delta_n(x)$ relative to these numbers as follows:

$$\delta_k(x) = \frac{\prod_{i \neq k}(x - a_i)}{\prod_{i \neq k}(a_k - a_i)} \qquad k = 0, 1, 2, \ldots, n$$

where $\prod_{i \neq k}(x - a_i)$ means the product of all the terms $(x - a_0), (x - a_1), (x - a_2)$, $\ldots, (x - a_n)$ except that the kth term is omitted. Then $\{\delta_0(x), \delta_1(x), \ldots, \delta_n(x)\}$ is orthonormal with respect to $\langle \, , \, \rangle$ because $\delta_k(a_i) = 0$ if $i \neq k$ and $\delta_k(a_k) = 1$. These facts also show that $\langle p(x), \delta_k(x) \rangle = p(a_k)$, so the expansion theorem gives

$$p(x) = p(a_0)\delta_0(x) + p(a_1)\delta_1(x) + \cdots + p(a_n)\delta_n(x)$$

for each $p(x)$ in $\mathbf{P}_n$. This is the **Lagrange interpolation expansion** of $p(x)$, which is important in numerical integration.

◆◆◆

ORTHOGONAL LEMMA

Let $\{\mathbf{e}_1, \mathbf{e}_2, \ldots, \mathbf{e}_m\}$ be an orthogonal set of vectors in an inner product space V, and let $\mathbf{v}$ be any vector *not* in span$\{\mathbf{e}_1, \mathbf{e}_2, \ldots, \mathbf{e}_m\}$. Define

$$\mathbf{e}_{m+1} = \mathbf{v} - \frac{\langle \mathbf{v}, \mathbf{e}_1 \rangle}{\|\mathbf{e}_1\|^2}\mathbf{e}_1 - \frac{\langle \mathbf{v}, \mathbf{e}_2 \rangle}{\|\mathbf{e}_2\|^2}\mathbf{e}_2 - \cdots - \frac{\langle \mathbf{v}, \mathbf{e}_m \rangle}{\|\mathbf{e}_m\|^2}\mathbf{e}_m$$

Then $\{\mathbf{e}_1, \mathbf{e}_2, \ldots, \mathbf{e}_m, \mathbf{e}_{m+1}\}$ is an orthogonal set of vectors.

The proof of this result (and the next) is the same as for the dot product in $\mathbb{R}^n$.

THEOREM 5

Gram–Schmidt Orthogonalization Algorithm

Let V be an inner product space and let $\{\mathbf{v}_1, \mathbf{v}_2, \ldots, \mathbf{v}_n\}$ be any basis of V. Define vectors $\mathbf{e}_1, \mathbf{e}_2, \ldots, \mathbf{e}_n$ in V successively as follows:

$$\mathbf{e}_1 = \mathbf{v}_1$$

$$\mathbf{e}_2 = \mathbf{v}_2 - \frac{\langle \mathbf{v}_2, \mathbf{e}_1 \rangle}{\|\mathbf{e}_1\|^2}\mathbf{e}_1$$

$$e_3 = v_3 - \frac{\langle v_3, e_1 \rangle}{\|e_1\|^2} e_1 - \frac{\langle v_3, e_2 \rangle}{\|e_2\|^2} e_2$$

$$\vdots \qquad \vdots$$

$$e_k = v_k - \frac{\langle v_k, e_1 \rangle}{\|e_1\|^2} e_1 - \frac{\langle v_k, e_2 \rangle}{\|e_2\|^2} e_2 - \cdots - \frac{\langle v_k, e_{k-1} \rangle}{\|e_{k-1}\|^2} e_{k-1}$$

for each $k = 2, 3, \ldots, n$. Then

1. $\{e_1, e_2, \ldots, e_n\}$ is an orthogonal basis of V.

2. span $\{e_1, e_2, \ldots e_k\}$ = span $\{v_1, v_2, \ldots, v_k\}$ holds for each $k = 1, 2,$ $\ldots, n$.

In particular, Theorem 5 shows that every finite dimensional inner product space *has* an orthogonal basis.

EXAMPLE 4

Consider $V = \mathbf{P}_3$ with the inner product $\langle p, q \rangle = \int_{-1}^{1} p(x)q(x)\ dx$. If the Gram–Schmidt algorithm is applied to the basis $\{1, x, x^2, x^3\}$, show that the result is the orthogonal basis.

$$\left\{1,\ x,\ \tfrac{1}{3}(3x^2 - 1),\ \tfrac{1}{5}(5x^3 - 3x)\right\}$$

Solution

Take $e_1 = 1$. Then the algorithm gives

$$e_2 = x - \frac{\langle x, e_1 \rangle}{\|e_1\|^2} e_1 = x - \frac{0}{2} e_1 = x$$

$$e_3 = x^2 - \frac{\langle x^2, e_1 \rangle}{\|e_1\|^2} e_1 - \frac{\langle x^2, e_2 \rangle}{\|e_2\|^2} e_2$$

$$= x^2 - \frac{\frac{2}{3}}{2} 1 - \frac{0}{\frac{2}{3}} x$$

$$= \tfrac{1}{3}(3x^2 - 1)$$

The verification that $e_4 = \tfrac{1}{5}(5x^3 - 3x)$ is omitted.

◆◆◆

The polynomials in Example 4 are such that the leading coefficient is 1 in each case. In other contexts (the study of differential equations, for example), it is customary to take multiples of these polynomials $p(x)$ such that $p(1) = 1$. The resulting orthogonal basis of $\mathbf{P}_3$ is

$$\left\{ 1, \; x, \; \tfrac{1}{2}(3x^2 - 1), \; \tfrac{1}{2}(5x^3 - 3x^2) \right\}$$

and these are the first four **Legendre polynomials**, so called to honor the French mathematician A. M. Legendre (1752–1833). They are important in the study of differential equations.

The orthogonal complement of a subspace U of $\mathbb{R}^n$ was defined (in Chapter 6) to be the set of all vectors in $\mathbb{R}^n$ that are orthogonal to every vector in U. This notion has a natural extension in an arbitrary inner product space.

DEFINITION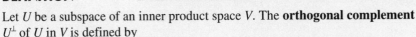

Let U be a subspace of an inner product space V. The **orthogonal complement** $U^\perp$ of U in V is defined by

$$U^\perp = \{ \mathbf{v} \mid \mathbf{v} \text{ in } V, \langle \mathbf{v}, \mathbf{u} \rangle = 0 \text{ for all } \mathbf{u} \text{ in } U \}$$

THEOREM 6

Let U be a finite dimensional subspace of an inner product space V.

1. $U^\perp$ is a subspace of V and $V = U \oplus U^\perp$.
2. If $\dim V = n$, then $\dim U + \dim U^\perp = n$.
3. If $\dim V = n$, then $U^{\perp\perp} = U$.

Proof

1. $U^\perp$ is a subspace by Theorem 1§8.1. If $\mathbf{v}$ is in $U \cap U^\perp$ then $\langle \mathbf{v}, \mathbf{v} \rangle = 0$. Hence $U \cap U^\perp = 0$, and $U + U^\perp = V$ follows from the orthogonal lemma.
2. This follows from Theorem 7§7.6.
3. $U \subseteq U^{\perp\perp}$ always holds (verify) and $\dim U^{\perp\perp} = n - \dim U^\perp = \dim U$, where (2) was used twice.   ◆

We digress briefly and consider a subspace U of an arbitrary vector space. If W is any complement of U in V—that is, $V = U \oplus W$—each vector $\mathbf{v}$ in V has a *unique* representation as a sum $\mathbf{v} = \mathbf{u} + \mathbf{w}$ where $\mathbf{u}$ is in U and $\mathbf{w}$ is in W. Hence we may define a function $T : V \to V$ as follows:

$$T(\mathbf{v}) = \mathbf{u} \quad \text{where } \mathbf{v} = \mathbf{u} + \mathbf{w}, \mathbf{u} \text{ in } U, \mathbf{w} \text{ in } W$$

Thus, to compute $T(\mathbf{v})$, express $\mathbf{v}$ in any way at all as the sum of a vector $\mathbf{u}$ in U and a vector in W; then $T(\mathbf{v}) = \mathbf{u}$.

This function T is a linear operator on V. Indeed, if $\mathbf{v}_1 = \mathbf{u}_1 + \mathbf{w}_1$ where $\mathbf{u}_1$ is in U and $\mathbf{w}_1$ is in W, then $\mathbf{v} + \mathbf{v}_1 = (\mathbf{u} + \mathbf{u}_1) + (\mathbf{w} + \mathbf{w}_1)$, so

$$T(\mathbf{v} + \mathbf{v}_1) = \mathbf{u} + \mathbf{u}_1 = T(\mathbf{v}) + T(\mathbf{v}_1)$$

because $\mathbf{u} + \mathbf{u}_1$ is in U and $\mathbf{w} + \mathbf{w}_1$ is in W. Similarly, $T(a\mathbf{v}) = aT(\mathbf{v})$ for all a in $\mathbb{R}$, so T is a linear operator. Furthermore, im $T = U$ and ker $T = W$ as the reader can verify, and T is called the **projection on U with kernel W**.

If U is a subspace of V, there are many projections on U, one for each complementary subspace W with $V = U \oplus W$. If V is an inner product space, we single out one for special attention.

DEFINITION

Let U be a finite subspace of an inner product space V. The projection on U with kernel $U^\perp$ is called the **orthogonal projection** on U (or simply the **projection** on U) and is denoted $\mathrm{proj}_U : V \to V$.

THEOREM 7

Projection Theorem

Let U be a finite dimensional subspace of an inner product space V and let $\mathbf{v}$ be a vector in V.

1. $\mathrm{proj}_U : V \to V$ is a linear operator with image U and kernel $U^\perp$.
2. $\mathrm{proj}_U(\mathbf{v})$ is in U and $\mathbf{v} - \mathrm{proj}_U(\mathbf{v})$ is in $U^\perp$.
3. If $\{\mathbf{e}_1, \mathbf{e}_2, \ldots, \mathbf{e}_m\}$ is any orthogonal basis of U, then

$$\mathrm{proj}_U(\mathbf{v}) = \frac{\langle \mathbf{v}, \mathbf{e}_1 \rangle}{\|\mathbf{e}_1\|^2}\mathbf{e}_1 + \frac{\langle \mathbf{v}, \mathbf{e}_2 \rangle}{\|\mathbf{e}_2\|^2}\mathbf{e}_2 + \cdots + \frac{\langle \mathbf{v}, \mathbf{e}_m \rangle}{\|\mathbf{e}_m\|^2}\mathbf{e}_m$$

Proof   Only (3) remains to be proved. If $\{\mathbf{e}_{m+1}, \ldots, \mathbf{e}_n\}$ is an orthogonal basis of $U^\perp$, then $\{\mathbf{e}_1, \ldots, \mathbf{e}_n\}$ is an orthogonal basis of V (Theorem 6§7.6) and the expansion theorem gives

$$\mathrm{proj}_U(\mathbf{v}) = \frac{\langle \mathbf{v}, \mathbf{e}_1 \rangle}{\|\mathbf{e}_1\|^2}\mathbf{e}_1 + \cdots + \frac{\langle \mathbf{v}, \mathbf{e}_m \rangle}{\|\mathbf{e}_m\|^2}\mathbf{e}_m + \frac{\langle \mathbf{v}, \mathbf{e}_{m+1} \rangle}{\|\mathbf{e}_{m+1}\|^2}\mathbf{e}_{m+1} + \cdots + \frac{\langle \mathbf{v}, \mathbf{e}_n \rangle}{\|\mathbf{e}_n\|^2}$$

Then (3) follows because the first m terms are in U and the last $n - m$ terms are in $U^\perp$.   ◆

EXAMPLE 5

Let U be a subspace of the finite dimensional inner product space V. Show that $\mathrm{proj}_{U^\perp}(\mathbf{v}) = \mathbf{v} - \mathrm{proj}_U(\mathbf{v})$ for all $\mathbf{v}$ in V.

Solution 1   The proof of Theorem 7 shows that $\mathbf{v} = \mathrm{proj}_U(\mathbf{v}) + \mathrm{proj}_{U^\perp}(\mathbf{v})$.

Solution 2   We have $V = U^{\perp} \oplus U^{\perp\perp}$ by Theorem 6. If we write $\mathbf{p} = \text{proj}_U(\mathbf{v})$, then $\mathbf{v} = (\mathbf{v} - \mathbf{p}) + \mathbf{p}$ where $\mathbf{v} - \mathbf{p}$ is in $U^{\perp}$ and $\mathbf{p}$ is in $U = U^{\perp\perp}$ by Theorem 7. Hence $\text{proj}_{U^{\perp}}(\mathbf{v}) = \mathbf{v} - \mathbf{p}$.

◆◆◆

The vectors $\mathbf{v}$, $\text{proj}_U(\mathbf{v})$, and $\mathbf{v} - \text{proj}_U(\mathbf{v})$ in Theorem 7 can be visualized geometrically as in Figure 8.3. This suggests that $\text{proj}_U(\mathbf{v})$ is the vector in U closest to $\mathbf{v}$. This is in fact the case.

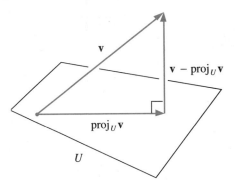

FIGURE 8.3

THEOREM 8

Approximation Theorem

Let U be a finite dimensional subspace of an inner product space V. If $\mathbf{v}$ is any vector in V, then $\text{proj}_U(\mathbf{v})$ is the vector in U that is closest to $\mathbf{v}$. Here *closest* means that

$$\|\mathbf{v} - \text{proj}_U(\mathbf{v})\| < \|\mathbf{v} - \mathbf{u}\|$$

for all $\mathbf{u}$ in U, $\mathbf{u} \neq \text{proj}_U(\mathbf{v})$.

Proof   Write $\mathbf{p} = \text{proj}_U(\mathbf{v})$, and consider $\mathbf{v} - \mathbf{u} = (\mathbf{v} - \mathbf{p}) + (\mathbf{p} - \mathbf{u})$. Because $\mathbf{v} - \mathbf{p}$ is in $U^{\perp}$ and $\mathbf{p} - \mathbf{u}$ is in U, the Pythagorean theorem gives

$$\|\mathbf{v} - \mathbf{u}\|^2 = \|\mathbf{v} - \mathbf{p}\|^2 + \|\mathbf{p} - \mathbf{u}\|^2 > \|\mathbf{v} - \mathbf{p}\|^2$$

because $\mathbf{p} - \mathbf{u} \neq \mathbf{0}$. The result follows.   ◆

EXAMPLE 6

Consider the space $\mathbf{C}[-1, 1]$ of real-valued continuous functions on the interval $[-1, 1]$ with inner product $\langle f, g \rangle = \int_{-1}^{1} f(x)g(x) \, dx$. Find the polynomial $p = p(x)$ of degree at most 2 that best approximates the absolute-value function f given by $f(x) = |x|$.

Solution

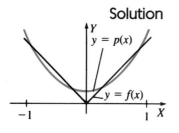

Here we want the vector p in the subspace $U = \mathbf{P}_2$ that is closest to f. In Example 4 the Gram–Schmidt algorithm was applied to give an orthogonal basis $\{e_1 = 1, e_2 = x, e_3 = 3x^2 - 1\}$ of $\mathbf{P}_2$ (where, for convenience, we have changed e_3 by a numerical factor). Hence the required polynomial is

$$p = \text{proj}_{\mathbf{P}_2}(f)$$

$$= \frac{\langle f,\ e_1 \rangle}{\|e_1\|^2}\, e_1 + \frac{\langle f,\ e_2 \rangle}{\|e_2\|^2}\, e_2 + \frac{\langle f,\ e_3 \rangle}{\|e_3\|^2}\, e_3$$

$$= \tfrac{1}{2}e_1 + 0e_2 + \frac{1/2}{8/5}\, e_3$$

$$= \tfrac{3}{16}(5x^2 + 1)$$

The graphs of $p(x)$ and $f(x)$ are given in the diagram.

◆◆◆

If polynomials of degree at most n are allowed in Example 6, the polynomial in $\mathbf{P}_n$ is $\text{proj}_{\mathbf{P}_n}(f)$, and it is calculated in the same way. Because the subspaces $\mathbf{P}_n$ get larger as n increases, it turns out that the approximating polynomials $\text{proj}_{\mathbf{P}_n}(f)$ get closer and closer to f. In fact, solving many practical problems comes down to approximating some interesting vector $\mathbf{v}$ (often a function) in an infinite dimensional inner product space V by vectors in finite dimensional subspaces (which can be computed). If $U_1 \subseteq U_2$ are finite dimensional subspaces of V, then

$$\|\mathbf{v} - \text{proj}_{U_2}(\mathbf{v})\| \leq \|\mathbf{v} - \text{proj}_{U_1}(\mathbf{v})\|$$

by Theorem 8 (because $\text{proj}_{U_1}(\mathbf{v})$ lies in U_1 and hence in U_2). Thus $\text{proj}_{U_2}(\mathbf{v})$ is a better approximation to $\mathbf{v}$ than $\text{proj}_{U_1}(\mathbf{v})$. Hence a general method in approximation theory might be described as follows: Given $\mathbf{v}$, use it to construct a sequence of finite dimensional subspaces

$$U_1 \subseteq U_2 \subseteq U_3 \subseteq \cdots$$

of V in such a way that $\|\mathbf{v} - \text{proj}_{U_k}(\mathbf{v})\|$ approaches zero as k increases. Then $\text{proj}_{U_k}(\mathbf{v})$ is a suitable approximation to $\mathbf{v}$ if k is large enough. For more information, the interested reader may wish to consult *Interpolation and Approximation* by Philip J. Davis (New York: Blaisdell, 1963).

EXERCISES 8.2

Use the dot product in $\mathbb{R}^n$ unless otherwise instructed.

1. In each case, verify that B is an orthogonal basis of V with the given inner product and use the expansion theorem to express $\mathbf{v}$ as a linear combination of the basis vectors.

(a) $\mathbf{v} = \begin{bmatrix} a \\ b \end{bmatrix}$, $B = \left\{ \begin{bmatrix} 1 \\ -1 \end{bmatrix}, \begin{bmatrix} 1 \\ 0 \end{bmatrix} \right\}$, $V = \mathbb{R}^2$, $\langle \mathbf{v}, \mathbf{w} \rangle = \mathbf{v}^T A \mathbf{w}$

where $A = \begin{bmatrix} 2 & 2 \\ 2 & 5 \end{bmatrix}$

◆**(b)** $\mathbf{v} = \begin{bmatrix} a \\ b \\ c \end{bmatrix}$, $B = \left\{ \begin{bmatrix} 1 \\ 1 \\ 1 \end{bmatrix}, \begin{bmatrix} -1 \\ 0 \\ 1 \end{bmatrix}, \begin{bmatrix} 1 \\ -6 \\ 1 \end{bmatrix} \right\}$, $V = \mathbb{R}^3$, $\langle \mathbf{v}, \mathbf{w} \rangle =$

$\mathbf{v}^T A \mathbf{w}$ where $A = \begin{bmatrix} 2 & 0 & 1 \\ 0 & 1 & 0 \\ 1 & 0 & 2 \end{bmatrix}$

(c) $\mathbf{v} = a + bx + cx^2$, $B = \{1, x, 2 - 3x^2\}$, $V = \mathbf{P}_2$,
$\langle p, q \rangle = p(0)q(0) + p(1)q(1) + p(-1)q(-1)$

(d) $\mathbf{v} = \begin{bmatrix} a & b \\ c & d \end{bmatrix}$,

$B = \left\{ \begin{bmatrix} 1 & 0 \\ 0 & 1 \end{bmatrix}, \begin{bmatrix} 1 & 0 \\ 0 & -1 \end{bmatrix}, \begin{bmatrix} 0 & 1 \\ 1 & 0 \end{bmatrix}, \begin{bmatrix} 0 & 1 \\ -1 & 0 \end{bmatrix} \right\}$,

$V = \mathbf{M}_{22}$, $\langle X, Y \rangle = \text{tr}(XY^T)$

2. Let $\mathbb{R}^3$ have the inner product $\langle (x, y, z), (x', y', z') \rangle = 2xx' + yy' + 3zz'$. In each case, use the Gram–Schmidt algorithm to transform B into an orthogonal basis.
(a) $B = \{(1, 1, 0), (1, 0, 1), (0, 1, 1)\}$
(b) $B = \{(1, 1, 1), (1, -1, 1), (1, 1, 0)\}$

3. Let $\mathbf{M}_{22}$ have the inner product $\langle X, Y \rangle = \text{tr}(XY^T)$. In each case, use the Gram–Schmidt algorithm to transform B into an orthogonal basis.
(a) $B = \left\{ \begin{pmatrix} 1 & 1 \\ 0 & 0 \end{pmatrix}, \begin{pmatrix} 1 & 0 \\ 1 & 0 \end{pmatrix}, \begin{pmatrix} 0 & 1 \\ 0 & 1 \end{pmatrix}, \begin{pmatrix} 1 & 0 \\ 0 & 1 \end{pmatrix} \right\}$
(b) $B = \left\{ \begin{pmatrix} 1 & 1 \\ 0 & 1 \end{pmatrix}, \begin{pmatrix} 1 & 0 \\ 1 & 1 \end{pmatrix}, \begin{pmatrix} 1 & 0 \\ 0 & 1 \end{pmatrix}, \begin{pmatrix} 1 & 0 \\ 0 & 0 \end{pmatrix} \right\}$

4. In each case, use the Gram–Schmidt process to convert the basis $B = \{1, x, x^2\}$ into an orthogonal basis of $\mathbf{P}_2$.
(a) $\langle p, q \rangle = p(0)q(0) + p(1)q(1) + p(2)q(2)$
(b) $\langle p, q \rangle = \int_0^2 p(x)q(x)dx$

5. Show that $\left\{ 1, x - \frac{1}{2}, x^2 - x + \frac{1}{6} \right\}$ is an orthogonal basis of $\mathbf{P}_2$ with the inner product $\langle p, q \rangle = \int_0^1 p(x)q(x)\,dx$, and find the corresponding orthonormal basis.

6. In each case find $U^\perp$ and compute $\dim U$ and $\dim U^\perp$.
(a) $U = \text{span}\{[1\ 1\ 2\ 0], [3\ -1\ 2\ 1], [1\ -3\ -2\ 1]\}$ in $\mathbb{R}^4$
(b) $U = \text{span}\{[1\ 1\ 0\ 0]\}$ in $\mathbb{R}^4$
(c) $U = \text{span}\{1, x\}$ in $\mathbf{P}_2$ with $\langle p, q \rangle = p(0)q(0) + p(1)q(1) + p(2)q(2)$
(d) $U = \text{span}\{x\}$ in $\mathbf{P}_2$ with $\langle p, q \rangle = \int_0^1 p(x)q(x)\,dx$
(e) $U = \text{span}\left\{ \begin{pmatrix} 1 & 0 \\ 0 & 1 \end{pmatrix}, \begin{pmatrix} 1 & 1 \\ 0 & 0 \end{pmatrix} \right\}$ in $\mathbf{M}_{22}$ with $\langle X, Y \rangle = \text{tr}(XY^T)$
(f) $U = \text{span}\left\{ \begin{pmatrix} 1 & 1 \\ 0 & 0 \end{pmatrix}, \begin{pmatrix} 1 & 0 \\ 1 & 0 \end{pmatrix}, \begin{pmatrix} 1 & 0 \\ 1 & 1 \end{pmatrix} \right\}$ in $\mathbf{M}_{22}$ with $\langle X, Y \rangle = \text{tr}(XY^T)$

7. Let $\langle X, Y \rangle = \text{tr}(XY^T)$ in $\mathbf{M}_{22}$. In each case find the matrix in U closest to A.
(a) $U = \text{span}\left\{ \begin{pmatrix} 1 & 0 \\ 0 & 1 \end{pmatrix}, \begin{pmatrix} 1 & 1 \\ 1 & 1 \end{pmatrix} \right\}$; $A = \begin{pmatrix} 1 & -1 \\ 2 & 3 \end{pmatrix}$
(b) $U = \text{span}\left\{ \begin{pmatrix} 1 & 0 \\ 0 & 1 \end{pmatrix}, \begin{pmatrix} 1 & 1 \\ 1 & -1 \end{pmatrix}, \begin{pmatrix} 1 & 1 \\ 0 & 0 \end{pmatrix} \right\}$;

$A = \begin{pmatrix} 2 & 1 \\ 3 & 2 \end{pmatrix}$

8. Let $\langle p(x), q(x) \rangle = p(0)q(0) + p(1)q(1) + p(2)q(2)$ in $\mathbf{P}_2$. In each case find the polynomial in U closest to $f(x)$.
(a) $U = \text{span}\{1 + x, x^2\}, f(x) = 1 + x^2$
(b) $U = \text{span}\{1, 1 + x^2\}; f(x) = x$

9. Using the inner product $\langle p, q \rangle = \int_0^1 p(x)q(x)\ dx$ on $\mathbf{P}_2$, write $\mathbf{v}$ as the sum of a vector in U and a vector in $U^\perp$.
(a) $\mathbf{v} = x^2, U = \text{span}\{x + 1, 9x - 5\}$
(b) $\mathbf{v} = x^2 + 1, U = \text{span}\{1, 2x - 1\}$

10. (a) Show that $\{\mathbf{u}, \mathbf{v}\}$ is orthogonal if and only if $\|\mathbf{u} + \mathbf{v}\|^2 = \|\mathbf{u}\|^2 + \|\mathbf{v}\|^2$.
(b) If $\mathbf{u} = \mathbf{v} = (1, 1)$ and $\mathbf{w} = (-1, 0)$, show that $\|\mathbf{u} + \mathbf{v} + \mathbf{w}\|^2 = \|\mathbf{u}\|^2 + \|\mathbf{v}\|^2 + \|\mathbf{w}\|^2$ but $\{\mathbf{u}, \mathbf{v}, \mathbf{w}\}$ is *not* orthogonal. Hence the converse to the Pythagorean theorem need not hold for more than two vectors.

11. Let $\mathbf{v}$ and $\mathbf{w}$ be vectors in an inner product space V. Show that:
(a) $\mathbf{v}$ is orthogonal to $\mathbf{w}$ if and only if $\|\mathbf{v} + \mathbf{w}\| = \|\mathbf{v} - \mathbf{w}\|$.
(b) $\mathbf{v} + \mathbf{w}$ and $\mathbf{v} - \mathbf{w}$ are orthogonal if and only if $\|\mathbf{v}\| = \|\mathbf{w}\|$.

12. Let U and W be subspaces of an n-dimensional inner product space V. If $\dim U + \dim W = n$ and $\langle \mathbf{u}, \mathbf{w} \rangle = 0$ for all $\mathbf{u}$ in U and $\mathbf{w}$ in W, show that $U^\perp = W$.

13. If U and W are subspaces of an inner product space, show that $(U + W)^\perp = U^\perp \cap W^\perp$.

14. If X is any set of vectors in an inner product space V, define $X^\perp = \{\mathbf{v} \mid \mathbf{v} \text{ in } V, \langle \mathbf{v}, \mathbf{x} \rangle = 0 \text{ for all } \mathbf{x} \text{ in } X\}$.
(a) Show that $X^\perp$ is a subspace of V.
(b) If $U = \text{span}\{\mathbf{u}_1, \mathbf{u}_2, \ldots, \mathbf{u}_m\}$, show that $U^\perp = \{\mathbf{u}_1, \ldots, \mathbf{u}_m\}^\perp$.
(c) If $X \subseteq Y$, show that $Y^\perp \subseteq X^\perp$.
(d) Show that $X^\perp \cap Y^\perp = (X \cup Y)^\perp$.

15. If $\dim V = n$, show that $\dim\{\mathbf{v} \mid \mathbf{v} \text{ in } V, \langle \mathbf{v}, \mathbf{w} \rangle = 0\} = n - 1$ for any $\mathbf{w} \neq \mathbf{0}$ in V.

16. If the Gram–Schmidt process is used on an orthogonal basis $\{\mathbf{v}_1, \ldots, \mathbf{v}_n\}$ of V, show that $\mathbf{e}_k = \mathbf{v}_k$ holds for each $k = 1, 2, \ldots, n$. That is, show that the algorithm reproduces the same basis.

17. If $\{\mathbf{e}_1, \mathbf{e}_2, \ldots, \mathbf{e}_{n-1}\}$ is orthonormal in an inner product space of dimension n, prove that there are exactly two vectors $\mathbf{e}_n$ such that $\{\mathbf{e}_1, \mathbf{e}_2, \ldots, \mathbf{e}_{n-1}, \mathbf{e}_n\}$ is an orthonormal basis.

18. Let U be a finite dimensional subspace of an inner product space V, and let $\mathbf{v}$ be a vector in V.
(a) Show that $\mathbf{v}$ lies in U if and only if $\mathbf{v} = \text{proj}_U(\mathbf{v})$.
(b) If $V = \mathbb{R}^3$, show that $(-5, 4, -3)$ lies in $\text{span}\{(3, -2, 5), (-1, 1, 1)\}$ but that $(-1, 0, 2)$ does not.

19. Let $\mathbf{n} \neq \mathbf{0}$ and $\mathbf{w} \neq \mathbf{0}$ be nonparallel vectors in $\mathbb{R}^3$ (as in Chapter 4).

(a) Show that $\left\{ \mathbf{n}, \ \mathbf{n} \times \mathbf{w}, \ \mathbf{w} - \dfrac{\mathbf{n} \cdot \mathbf{w}}{\|\mathbf{n}\|^2} \, \mathbf{n} \right\}$ is an orthogonal basis of $\mathbb{R}^3$.

(b) Show that $\text{span} \left\{ \mathbf{n} \times \mathbf{w}, \ \mathbf{w} - \dfrac{\mathbf{n} \cdot \mathbf{w}}{\|\mathbf{n}\|^2} \, \mathbf{n} \right\}$ is the plane through the origin with normal $\mathbf{n}$.

20. Let $E = \{\mathbf{e}_1, \mathbf{e}_2, \ldots, \mathbf{e}_n\}$ be an orthonormal basis of V.

(a) Show that $\langle \mathbf{v}, \mathbf{w} \rangle = C_E(\mathbf{v}) \cdot C_E(\mathbf{w})$ for all $\mathbf{v}, \mathbf{w}$ in V.

(b) If $P = [p_{ij}]$ is an $n \times n$ matrix, define $\mathbf{b}_i = p_{i1}\mathbf{e}_1 + \cdots + p_{in}\mathbf{e}_n$ for each i. Show that $B = \{\mathbf{b}_1, \mathbf{b}_2, \ldots, \mathbf{b}_n\}$ is an orthonormal basis if and only if P is an orthogonal matrix.

21. Let $\{\mathbf{e}_1, \ldots, \mathbf{e}_n\}$ be an orthogonal basis of V. If $\mathbf{v}$ and $\mathbf{w}$ are in V, show that

$$\langle \mathbf{v}, \mathbf{w} \rangle = \frac{\langle \mathbf{v}, \mathbf{e}_1 \rangle \langle \mathbf{w}, \mathbf{e}_1 \rangle}{\|\mathbf{e}_1\|^2} + \cdots + \frac{\langle \mathbf{v}, \mathbf{e}_n \rangle \langle \mathbf{w}, \mathbf{e}_n \rangle}{\|\mathbf{e}_n\|^2}.$$

22. Let $\{\mathbf{e}_1, \ldots, \mathbf{e}_n\}$ be an orthonormal basis of V, and let $\mathbf{v} = v_1\mathbf{e}_1 + \cdots + v_n\mathbf{e}_n$ and $\mathbf{w} = w_1\mathbf{e}_1 + \cdots + w_n\mathbf{e}_n$. Show that $\langle \mathbf{v}, \mathbf{w} \rangle = v_1w_1 + \cdots + v_nw_n$ and that $\|\mathbf{v}\|^2 = v_1^2 + \cdots + v_n^2$ (**Parseval's formula**).

23. Let $\mathbf{v}$ be a vector in an inner product space V.

(a) Show that $\|\mathbf{v}\| \geq \|\text{proj}_U(\mathbf{v})\|$ holds for all finite dimensional subspaces U. [*Hint:* Pythagorean theorem.]

(b) If $\{\mathbf{e}_1, \mathbf{e}_2, \ldots, \mathbf{e}_m\}$ is any orthogonal set in V, prove **Bessel's inequality**:

$$\frac{\langle \mathbf{v}, \mathbf{e}_1 \rangle^2}{\|\mathbf{e}_1\|^2} + \cdots + \frac{\langle \mathbf{v}, \mathbf{e}_m \rangle^2}{\|\mathbf{e}_m\|^2} \leq \|\mathbf{v}\|^2$$

24. (a) Let S denote a set of vectors in a finite dimensional inner product space V, and suppose that $\langle \mathbf{u}, \mathbf{v} \rangle = 0$ for all $\mathbf{u}$ in S implies $\mathbf{v} = \mathbf{0}$. Show that $V = \text{span } S$. [*Hint:* Write $U = \text{span } S$ and use Theorem 6.]

(b) Let $A_1, A_2, \ldots, A_k$ be $n \times n$ matrices. Show that the following are equivalent.

(i) If $A_i\mathbf{b} = \mathbf{0}$ for all i (where $\mathbf{b}$ is a column in $\mathbb{R}^n$), then $\mathbf{b} = \mathbf{0}$.

(ii) The set of all rows of the matrices A_i spans $\mathbb{R}^n$.

Section 8.3   Orthogonal Diagonalization

There is a natural way to define a symmetric linear operator T on a finite dimensional inner product space V. If T is such an operator, it is shown in this section that V has an orthogonal basis consisting of eigenvectors of T. This yields another proof of the principal axes theorem.

THEOREM 1

Let $T : V \to V$ be a linear operator on a finite dimensional space V. Then the following conditions are equivalent.

1. V has a basis consisting of eigenvectors of T.
2. There exists a basis B of V such that $M_B(T)$ is diagonal.

Proof   We have $M_B(T) = [C_B[T(\mathbf{e}_1)] \quad C_B[T(\mathbf{e}_2)] \quad \cdots \quad C_B[T(\mathbf{e}_n)]]$ where $B = \{\mathbf{e}_1, \mathbf{e}_2, \ldots, \mathbf{e}_n\}$ is any basis of V. Hence

$$M_B(T) = \begin{bmatrix} \lambda_1 & 0 & \cdots & 0 \\ 0 & \lambda_2 & \cdots & 0 \\ \vdots & \vdots & & \vdots \\ 0 & 0 & \cdots & \lambda_n \end{bmatrix} \quad \text{if and only if} \quad T(\mathbf{e}_i) = \lambda_i\mathbf{e}_i \text{ for each } i$$

Theorem 1 follows by comparing columns. $\blacklozenge$

> **DEFINITION**
>
> A linear operator T on a finite dimensional space V is called **diagonalizable** if V has a basis of eigenvectors of T.

EXAMPLE 1

Let $T : \mathbf{P}_2 \to \mathbf{P}_2$ be given by

$$T(a + bx + cx^2) = (a + 4c) - 2bx + (3a + 2c)x^2$$

Find the eigenspaces of T and hence find a basis of eigenvectors.

Solution   If $B_0 = \{1, x, x^2\}$, then

$$M_{B_0}(T) = \begin{bmatrix} 1 & 0 & 4 \\ 0 & -2 & 0 \\ 3 & 0 & 2 \end{bmatrix}$$

so $c_T(x) = (x + 2)^2(x - 5)$, and the eigenvalues of T are $\lambda = -2$ and $\lambda = 5$. Thus

$$\left\{ \begin{bmatrix} 0 \\ 1 \\ 0 \end{bmatrix}, \begin{bmatrix} 4 \\ 0 \\ -3 \end{bmatrix}, \begin{bmatrix} 1 \\ 0 \\ 1 \end{bmatrix} \right\}$$ is a basis of eigenvectors of $M_{B_0}(T)$, so $B = \{x, 4 - 3x^2, 1 + x^2\}$

is a basis of $\mathbf{P}_2$ consisting of eigenvectors of T.

◆◆◆

If V is an inner product space, the expansion theorem gives a simple formula for the matrix of a linear operator with respect to an orthogonal basis.

> **THEOREM 2**
>
> Let $T : V \to V$ be a linear operator on an inner product space V. If $B = \{\mathbf{e}_1, \mathbf{e}_2, \ldots, \mathbf{e}_n\}$ is an orthogonal basis of V, then
>
> $$M_B(T) = \left[\frac{\langle \mathbf{e}_i, T(\mathbf{e}_j) \rangle}{\|\mathbf{e}_i\|^2} \right]$$

Proof   Write $M_B(T) = [a_{ij}]$. The jth column of $M_B(T)$ is $C_B[T(\mathbf{e}_j)]$, so

$$T(\mathbf{e}_j) = a_{1j}\mathbf{e}_1 + \cdots + a_{ij}\mathbf{e}_i + \cdots + a_{nj}\mathbf{e}_n$$

On the other hand, the expansion theorem gives

$$\mathbf{v} = \frac{\langle \mathbf{e}_1, \mathbf{v} \rangle}{\|\mathbf{e}_1\|^2} \mathbf{e}_1 + \cdots + \frac{\langle \mathbf{e}_i, \mathbf{v} \rangle}{\|\mathbf{e}_i\|^2} \mathbf{e}_i + \cdots + \frac{\langle \mathbf{e}_n, \mathbf{v} \rangle}{\|\mathbf{e}_n\|^2} \mathbf{e}_n$$

for any $\mathbf{v}$ in V. The result follows by taking $\mathbf{v} = T(\mathbf{e}_j)$.

◆

EXAMPLE 2

Let $T: \mathbb{R}^3 \to \mathbb{R}^3$ be given by

$$T(a, b, c) = (a + 2b - c, 2a + 3c, -a + 3b + 2c)$$

If the dot product in $\mathbb{R}^3$ is used, find the matrix of T with respect to the standard basis $B = \{\mathbf{e}_1, \mathbf{e}_2, \mathbf{e}_3\}$ where $\mathbf{e}_1 = (1, 0, 0)$, $\mathbf{e}_2 = (0, 1, 0)$, $\mathbf{e}_3 = (0, 0, 1)$.

Solution   The basis B is orthonormal, so Theorem 2 gives

$$M_B(T) = \begin{bmatrix} \mathbf{e}_1 \cdot T(\mathbf{e}_1) & \mathbf{e}_1 \cdot T(\mathbf{e}_2) & \mathbf{e}_1 \cdot T(\mathbf{e}_3) \\ \mathbf{e}_2 \cdot T(\mathbf{e}_1) & \mathbf{e}_2 \cdot T(\mathbf{e}_2) & \mathbf{e}_2 \cdot T(\mathbf{e}_3) \\ \mathbf{e}_3 \cdot T(\mathbf{e}_1) & \mathbf{e}_3 \cdot T(\mathbf{e}_2) & \mathbf{e}_3 \cdot T(\mathbf{e}_3) \end{bmatrix} = \begin{bmatrix} 1 & 2 & -1 \\ 2 & 0 & 3 \\ -1 & 3 & 2 \end{bmatrix}$$

Of course, this can also be found in the usual way.

$\blacklozenge\blacklozenge\blacklozenge$

It is not difficult to verify that an $n \times n$ matrix A is symmetric if and only if $\mathbf{v} \cdot (A\mathbf{w}) = (A\mathbf{v}) \cdot \mathbf{w}$ holds for all columns $\mathbf{v}$ and $\mathbf{w}$ in $\mathbb{R}^n$. The analogue for operators is as follows:

THEOREM 3

Let V be a finite dimensional inner product space. The following conditions are equivalent for a linear operator $T: V \to V$.

1. $\langle \mathbf{v}, T(\mathbf{w}) \rangle = \langle T(\mathbf{v}), \mathbf{w} \rangle$ for all $\mathbf{v}$ and $\mathbf{w}$ in V.
2. The matrix of T is symmetric with respect to every orthonormal basis of V.
3. The matrix of T is symmetric with respect to some orthonormal basis of V.
4. There is an orthonormal basis $B = \{\mathbf{e}_1, \mathbf{e}_2, \ldots, \mathbf{e}_n\}$ of V such that $\langle \mathbf{e}_i, T(\mathbf{e}_j) \rangle = \langle T(\mathbf{e}_i), \mathbf{e}_j \rangle$ holds for all i and j.

Proof

(1) *implies* (2). Given (1), let $B = \{\mathbf{e}_1, \ldots, \mathbf{e}_n\}$ be an orthonormal basis of V and write $M_B(T) = [a_{ij}]$. The aim is to prove that $a_{ij} = a_{ji}$ for all i and j. The fact that $\langle \mathbf{v}, \mathbf{w} \rangle = \langle \mathbf{w}, \mathbf{v} \rangle$ for all $\mathbf{v}$ and $\mathbf{w}$, together with Theorem 2, gives

$$a_{ij} = \langle \mathbf{e}_i, T(\mathbf{e}_j) \rangle = \langle T(\mathbf{e}_i), \mathbf{e}_j \rangle = \langle \mathbf{e}_j, T(\mathbf{e}_i) \rangle = a_{ji}$$

(2) *implies* (3). This is clear.

(3) *implies* (4). Assume that $B = \{\mathbf{e}_1, \ldots, \mathbf{e}_n\}$ is an orthonormal basis of V such that $M_B(T)$ is symmetric. We have $M_B(T) = [\langle \mathbf{e}_i, T(\mathbf{e}_j) \rangle]$ by Theorem 2, so $\langle \mathbf{e}_i, T(\mathbf{e}_j) \rangle = \langle \mathbf{e}_j, T(\mathbf{e}_i) \rangle$ holds for all i and j. Then (4) follows because $\langle \, , \rangle$ is symmetric.

(4) *implies* (1). Let $\mathbf{v}$ and $\mathbf{w}$ be vectors in V and write them as $\mathbf{v} = \sum_{i=1}^{n} v_i \mathbf{e}_i$, $\mathbf{w} = \sum_{j=1}^{n} w_j \mathbf{e}_j$. Then

$$\langle \mathbf{v}, \ T(\mathbf{w}) \rangle = \left\langle \sum_i v_i \mathbf{e}_i, \ \sum_j w_j T(\mathbf{e}_j) \right\rangle = \sum_i \sum_j v_i w_j \langle \mathbf{e}_i, \ T(\mathbf{e}_j) \rangle$$

$$= \sum_i \sum_j v_i w_j \langle T(\mathbf{e}_i), \ \mathbf{e}_j \rangle$$

$$= \left\langle \sum_i v_i T(\mathbf{e}_i), \ \sum_j w_j \mathbf{e}_j \right\rangle$$

$$= \langle T(\mathbf{v}), \ \mathbf{w} \rangle$$

This proves (1). ◆

DEFINITION

A linear operator T on an inner product space V is called **symmetric** if $\langle \mathbf{v}, T(\mathbf{w}) \rangle = \langle T(\mathbf{v}), \mathbf{w} \rangle$ holds for all $\mathbf{v}$ and $\mathbf{w}$ in V.

EXAMPLE 3

If A is an $n \times n$ matrix, let $T_A : \mathbb{R}^n \to \mathbb{R}^n$ be the matrix operator given by $T_A(\mathbf{v}) = A\mathbf{v}$ for all columns $\mathbf{v}$. If the dot product is used in $\mathbb{R}^n$, then T_A is a symmetric operator if and only if A is a symmetric matrix.

Solution   If B is the standard basis of $\mathbb{R}^n$, then B is orthonormal if the dot product is used. We have $M_B(T_A) = A$ (by Example 4§7.4), so the result follows immediately from Theorem 3.

◆◆◆

It is important to note that whether an operator is symmetric depends on which inner product is being used (see Exercise 2.)

If V is a finite dimensional inner product space, the eigenvalues of an operator $T : V \to V$ are the same as those of $M_B(T)$ for any orthonormal basis B. If T is symmetric, $M_B(T)$ is a symmetric matrix and so has real eigenvalues by Theorem 2§6.1. Hence we have the following:

THEOREM 4

A symmetric linear operator on a finite dimensional inner product space has a real eigenvalue.

If U is a subspace of an inner product space V, recall that its orthogonal complement is the subspace $U^\perp$ of V defined by

$$U^\perp = \{ \mathbf{v} \text{ in } V \mid \langle \mathbf{v}, \mathbf{u} \rangle = 0 \text{ for all } \mathbf{u} \text{ in } U \}$$

THEOREM 5

Let $T : V \rightarrow V$ be a symmetric linear operator on an inner product space V, and let U be a T-invariant subspace of V. Then:

1. The restriction of T to U is a symmetric linear operator on U.
2. $U^{\perp}$ is also T-invariant.

Proof

1. U is itself an inner product space using the same inner product, and the condition that T is symmetric is clearly preserved.
2. If $\mathbf{v}$ is in $U^{\perp}$, our task is to show that $T(\mathbf{v})$ is also in $U^{\perp}$; that is, $\langle T(\mathbf{v}), \mathbf{u} \rangle = 0$ for all $\mathbf{u}$ in U. But if $\mathbf{u}$ is in U, then $T(\mathbf{u})$ also lies in U because U is T-invariant, so

$$\langle T(\mathbf{v}), \mathbf{u} \rangle = \langle \mathbf{v}, T(\mathbf{u}) \rangle = 0$$

using the symmetry of T and the definition of $U^{\perp}$. ◆

We can now give another proof of the following theorem, which is probably the most important result in this book.

THEOREM 6

Principal Axes Theorem

The following conditions are equivalent for a linear operator T on a finite dimensional inner product space V.

1. T is symmetric.
2. V has an orthogonal basis consisting of eigenvectors of T.

Proof

(1) *implies* (2). Assume that T is symmetric and proceed by induction on $n = \dim V$. If $n = 1$, *every* nonzero vector in V is an eigenvector of T, so there is nothing to prove. If $n \geq 2$, assume inductively that the theorem holds for spaces of dimension less than n. Let λ_1 be a real eigenvalue of T (by Theorem 4) and choose an eigenvector $\mathbf{e}_1$ corresponding to λ_1. Then $U = \mathbb{R}\mathbf{e}_1$ is T-invariant, so $U^{\perp}$ is also T-invariant by Theorem 5 (T is symmetric). Because $\dim U^{\perp} = n - 1$ (Theorem 6§8.2) and because the restriction of T to $U^{\perp}$ is a symmetric operator (Theorem 5), it follows by induction that $U^{\perp}$ has an orthogonal basis $\{\mathbf{e}_2, \ldots, \mathbf{e}_n\}$ of eigenvectors of T. Hence $B = \{\mathbf{e}_1, \mathbf{e}_2, \ldots, \mathbf{e}_n\}$ is an orthogonal basis of V, which proves (2).

(2) *implies* (1). If $B = \{\mathbf{e}_1, \ldots, \mathbf{e}_n\}$ is a basis as in (2), then $M_B(T)$ is symmetric (indeed diagonal), so T is symmetric by Theorem 3. ◆

The matrix version of the principal axes theorem is an immediate consequence of Theorem 6. If A is an $n \times n$ symmetric matrix, then $T_A : \mathbb{R}^n \to \mathbb{R}^n$ is a symmetric operator, so let B be an orthonormal basis of $\mathbb{R}^n$ consisting of eigenvectors of T_A (and hence of A). Then $P^T A P$ is diagonal where P is the orthogonal matrix whose columns are the vectors in B.

Similarly, let $T : V \to V$ be a symmetric linear operator on the n-dimensional inner product space V and let B_0 be any convenient orthonormal basis of V. Then an orthonormal basis of eigenvectors of T can be computed from $M_{B_0}(T)$. In fact, if $P^T M_{B_0}(T) P$ is diagonal where P is orthogonal, let $B = \{\mathbf{e}_1, \ldots, \mathbf{e}_n\}$ be the vectors in V such that $C_{B_0}(\mathbf{e}_j)$ is column j of P for each j. Then B consists of eigenvectors by Theorem 3§7.6, and they are orthonormal because B_0 is orthonormal. Indeed

$$\langle \mathbf{e}_i, \mathbf{e}_j \rangle = C_{B_0}(\mathbf{e}_i) \cdot C_{B_0}(\mathbf{e}_j)$$

holds for all i and j, as the reader can verify. Here is an example.

EXAMPLE 4

Let $T : \mathbf{P}_2 \to \mathbf{P}_2$ be given by

$$T(a + bx + cx^2) = (8a - 2b + 2c) + (-2a + 5b + 4c)x$$
$$+ (2a + 4b + 5c)x^2$$

Using the inner product $\langle a + bx + cx^2, a' + b'x + c'x^2 \rangle = aa' + bb' + cc'$, show that T is symmetric and find an orthonormal basis of $\mathbf{P}_2$ consisting of eigenvectors.

Solution   If $B_0 = \{1, x, x^2\}$, then $M_{B_0}(T) = \begin{bmatrix} 8 & -2 & 2 \\ -2 & 5 & 4 \\ 2 & 4 & 5 \end{bmatrix}$ is symmetric, so T is symmetric. This

matrix was analyzed in Example 5§6.4, where it was found that an *orthonormal* basis of eigenvectors is

$$\left\{ \tfrac{1}{3}\begin{bmatrix} 1 \\ 2 \\ -2 \end{bmatrix}, \ \tfrac{1}{3}\begin{bmatrix} 2 \\ 1 \\ 2 \end{bmatrix}, \ \tfrac{1}{3}\begin{bmatrix} -2 \\ 2 \\ 1 \end{bmatrix} \right\}$$

Because B_0 is orthonormal, the corresponding orthonormal basis of $\mathbf{P}_2$ is

$$B = \left\{ \tfrac{1}{3}(1 + 2x - 2x^2), \ \tfrac{1}{3}(2 + x + 2x^2), \ \tfrac{1}{3}(-2 + 2x + x^2) \right\}$$

◆◆◆

EXERCISES 8.3

1. In each case, show that T is symmetric by calculating $M_B(T)$ for some orthonormal basis B.

   (a) $T : \mathbb{R}^3 \to \mathbb{R}^3$; $T(a, b, c) = (a - 2b, -2a + 2b + 2c, 2b - c)$; dot product

   ◆(b) $T : \mathbf{M}_{22} \to \mathbf{M}_{22}$; $T\begin{bmatrix} a & b \\ c & d \end{bmatrix} = \begin{bmatrix} c - a & d - b \\ a + 2c & b + 2d \end{bmatrix}$;

   inner product $\left\langle \begin{bmatrix} x & y \\ z & w \end{bmatrix}, \begin{bmatrix} x' & y' \\ z' & w' \end{bmatrix} \right\rangle = xx' + yy' + zz' + ww'$

   (c) $T : \mathbf{P}_2 \to \mathbf{P}_2$; $T(a + bx + cx^2) = (b + c) + (a + c)x + (a + b)x^2$; inner product $\langle a + bx + cx^2, a' + b'x + c'x^2 \rangle = aa' + bb' + cc'$

2. Let $T : \mathbb{R}^2 \to \mathbb{R}^2$ be given by $T(a,b) = (2a + b, a - b)$.

   (a) Show that T is symmetric if the dot product is used.

   (b) Show that T is *not* symmetric if $\langle X, Y \rangle = XAY^T$, where $A = \begin{bmatrix} 1 & 1 \\ 1 & 2 \end{bmatrix}$.

   [*Hint:* Check that $B = \{(1, 0), (1, -1)\}$ is an orthonormal basis.]

3. Let $T : \mathbb{R}^2 \to \mathbb{R}^2$ be given by $T(a, b) = (a - b, b - a)$. Use the dot product in $\mathbb{R}^2$.

   (a) Show that T is symmetric.

   (b) Show that $M_B(T)$ is *not* symmetric if the orthogonal basis $B = \{(1, 0), (0, 2)\}$ is used. Why does this not contradict Theorem 3?

4. Let V be an n-dimensional inner product space, and let T and S denote symmetric linear operators on V. Show that:

   (a) The identity operator is symmetric.

   ◆(b) rT is symmetric for all r in $\mathbb{R}$.

   (c) $S + T$ is symmetric.

   (d) If $ST = TS$, then ST is symmetric.

5. In each case, show that T is symmetric and find an orthonormal basis of eigenvectors of T.

   (a) $T : \mathbb{R}^3 \to \mathbb{R}^3$; $T(a, b, c) = (2a + 2c, 3b, 2a + 5c)$; use the dot product

   ◆(b) $T : \mathbb{R}^3 \to \mathbb{R}^3$; $T(a, b, c) = (7a - b, -a + 7b, 2c)$; use the dot product

   (c) $T : \mathbf{P}_2 \to \mathbf{P}_2$; $T(a + bx + cx^2) = 3b + (3a + 4c)x + 4bx^2$; inner product $\langle a + bx + cx^2, a' + b'x + c'x^2 \rangle = aa' + bb' + cc'$

   ◆(d) $T : \mathbf{P}_2 \to \mathbf{P}_2$; $T(a + bx + cx^2) = (c - a) + 3bx + (a - c)x^2$; inner product as in part (c)

6. If A is any $n \times n$ matrix, let $T_A : \mathbb{R}^n \to \mathbb{R}^n$ be given by $T_A(X) = AX$. Suppose an inner product on $\mathbb{R}^n$ is given by $\langle X, Y \rangle = X^T PY$, where P is a positive definite matrix.

   (a) Show that T_A is symmetric if and only if $PA = A^T P$.

   (b) Use part (a) to deduce Example 3.

7. Let $T : \mathbf{M}_{22} \to \mathbf{M}_{22}$ be given by $T(X) = AX$, where A is a fixed 2×2 matrix.

   (a) Compute $M_B(T)$, where $B = \left\{ \begin{bmatrix} 1 & 0 \\ 0 & 0 \end{bmatrix}, \begin{bmatrix} 0 & 0 \\ 1 & 0 \end{bmatrix}, \begin{bmatrix} 0 & 1 \\ 0 & 0 \end{bmatrix}, \begin{bmatrix} 0 & 0 \\ 0 & 1 \end{bmatrix} \right\}$. Note the order!

   ◆(b) Show that $c_T(x) = [c_A(x)]^2$.

   (c) If the inner product on $\mathbf{M}_{22}$ is $\langle X, Y \rangle = \text{tr}(XY^T)$, show that T is symmetric if and only if A is a symmetric matrix.

8. Let $T : \mathbb{R}^2 \to \mathbb{R}^2$ be given by $T(a, b) = (b - a, a + 2b)$. Show that T is symmetric if the dot product is used in $\mathbb{R}^2$ but that it is not symmetric if the following inner product is used: $\langle X, Y \rangle = XAY^T$, $A = \begin{bmatrix} 1 & -1 \\ -1 & 2 \end{bmatrix}$.

9. If $T : V \to V$ is symmetric, show that $T(U)^\perp = T^{-1}(U^\perp)$ holds for every subspace U of V. Here $T^{-1}(W) = \{\mathbf{v} \mid T(\mathbf{v}) \text{ is in } W\}$.

10. Let $T : \mathbf{M}_{22} \to \mathbf{M}_{22}$ be defined by $T(X) = PXQ$, where P and Q are nonzero 2×2 matrices. Use the inner product $\langle X, Y \rangle = \text{tr}(XY^T)$. Show that T is symmetric if and only if either P and Q are both symmetric or both are scalar multiples of $\begin{bmatrix} 0 & 1 \\ -1 & 0 \end{bmatrix}$. [*Hint:* If B is as in part (a) of Exercise 7, then $M_B(T) = \begin{bmatrix} aP & cP \\ bP & dP \end{bmatrix}$ in block form, where $Q = \begin{bmatrix} a & b \\ c & d \end{bmatrix}$. If $B_0 = \left\{ \begin{bmatrix} 1 & 0 \\ 0 & 0 \end{bmatrix}, \begin{bmatrix} 0 & 1 \\ 0 & 0 \end{bmatrix}, \begin{bmatrix} 0 & 0 \\ 1 & 0 \end{bmatrix}, \begin{bmatrix} 0 & 0 \\ 0 & 1 \end{bmatrix} \right\}$, then $M_{B_0}(T) = \begin{bmatrix} pQ^T & qQ^T \\ rQ^T & sQ^T \end{bmatrix}$, where $P = \begin{bmatrix} p & q \\ r & s \end{bmatrix}$. Use the fact that $cP = bP^T$ implies that $(c^2 - b^2)P = 0$.]

11. Let $T : V \to W$ be any linear transformation and let $B = \{\mathbf{b}_1, \ldots, \mathbf{b}_n\}$ and $D = \{\mathbf{d}_1, \ldots, \mathbf{d}_m\}$ be bases of V and W, respectively. If W is an inner product space and D is orthogonal, show that

$$M_{DB}(T) = \left[\frac{\langle \mathbf{d}_i,\ T(\mathbf{b}_j) \rangle}{\|\mathbf{d}_i\|^2} \right]$$

This is a generalization of Theorem 2.

12. Let $T : V \to V$ be a linear operator on an inner product space V of finite dimension. Show that the following are equivalent.

(1) $\langle \mathbf{v}, T(\mathbf{w}) \rangle = -\langle T(\mathbf{v}),\mathbf{w} \rangle$ for all $\mathbf{v}$ and $\mathbf{w}$ in V.

(2) $M_B(T)$ is skew-symmetric for every orthonormal basis B.

(3) $M_B(T)$ is skew-symmetric for some orthonormal basis B.

Such operators T are called **skew-symmetric** operators.

13. Let $T : V \to V$ be a linear operator on an n-dimensional inner product space V.

(a) Show that T is symmetric if and only if it satisfies the following two conditions.

   (i) $c_T(x)$ factors completely over $\mathbb{R}$.

   (ii) If U is a T-invariant subspace of V, then $U^{\perp}$ is also T-invariant.

(b) Using the standard inner product in $\mathbb{R}^2$, show that $T : \mathbb{R}^2 \to \mathbb{R}^2$ with $T(a, b) = (a, a + b)$ satisfies condition **(i)** and that $S : \mathbb{R}^2 \to \mathbb{R}^2$ with $S(a, b) = (b, -a)$ satisfies condition **(ii)** but that neither is symmetric. (Example 4§7.6 is useful for S.)

[*Hint for part* **(a)**: If conditions **(i)** and **(ii)** hold, proceed by induction on n. By condition **(i)**, let $\mathbf{e}_1$ be an eigenvector of T. If $U = \mathbb{R}\mathbf{e}_1$, then $U^{\perp}$ is T-invariant by condition **(ii)**, so show that the restriction of T to $U^{\perp}$ satisfies conditions **(i)** and **(ii)**. (Theorem 1§7.6 is helpful for part **(i)**). Then apply induction to show that V has an orthogonal basis of eigenvectors (as in Theorem 6).]

14. Let $B = \{\mathbf{e}_1, \mathbf{e}_2, \ldots, \mathbf{e}_n\}$ be an orthonormal basis of an inner product space V. Given $T : V \to V$, define $T' : V \to V$ by

$$T'(\mathbf{v}) = \langle \mathbf{v},\ T(\mathbf{e}_1) \rangle \mathbf{e}_1 + \langle \mathbf{v},\ T(\mathbf{e}_2) \rangle \mathbf{e}_2 + \cdots +$$

$$\langle \mathbf{v},\ T(\mathbf{e}_n) \rangle \mathbf{e}_n = \sum_{i=1}^{n}\ \langle \mathbf{v},\ T(\mathbf{e}_i) \rangle \mathbf{e}_i$$

(a) Show that $(aT)' = aT'$.

(b) Show that $(S + T)' = S' + T'$.

◆**(c)** Show that $M_B(T')$ is the transpose of $M_B(T)$.

(d) Show that $(T')' = T$, using part **(c)**. [*Hint:* $M_B(S) = M_B(T)$ implies that $S = T$.]

(e) Show that $(ST)' = T'S'$, using part **(c)**.

(f) Show that T is symmetric if and only if $T = T'$. [*Hint:* Use the expansion theorem and Theorem 3.]

(g) Show that $T + T'$ and TT' are symmetric, using parts **(b)** through **(e)**.

(h) Show that $T'(\mathbf{v})$ is independent of the choice of orthonormal basis B. [*Hint:* If $D = \{\mathbf{f}_1, \ldots, \mathbf{f}_n\}$ is also orthonormal, use the fact that $\mathbf{e}_i = \sum_{i=1}^{n}\langle \mathbf{e}_i, \mathbf{f}_j \rangle \mathbf{f}_j$ for each i.]

15. Let V be a finite dimensional inner product space. Show that the following conditions are equivalent for a linear operator $T : V \to V$.

(a) T is symmetric and $T^2 = T$.

(b) $M_B(T) = \begin{bmatrix} I_r & 0 \\ 0 & 0 \end{bmatrix}$ for some orthonormal basis B of V.

An operator is called a **projection** if it satisfies these conditions. [*Hint:* If $T^2 = T$ and $T(\mathbf{v}) = \lambda\mathbf{v}$, apply T to get $\lambda\mathbf{v} = \lambda^2\mathbf{v}$. Hence show that $0, 1$ are the only eigenvalues of T.]

16. Let V denote a finite dimensional inner product space. Given a subspace U, define $\text{proj}_U : V \to V$ as in Theorem 7§8.2.

(a) Show that proj_U is a projection in the sense of Exercise 15.

(b) If T is any projection, show that $T = \text{proj}_U$, where $U = \text{im}\ T$. [*Hint:* Use $T^2 = T$ to show that $V = \text{im}\ T \oplus \ker T$ and $T(\mathbf{u}) = \mathbf{u}$ for all $\mathbf{u}$ in $\text{im}\ T$. Use the fact that T is symmetric to show that $\ker T \subseteq (\text{im}\ T)^{\perp}$ and hence that these are equal because they have the same dimension.]

Section 8.4   **Isometries**

Once the study of linear algebra has progressed to the point where both inner products and linear transformations have been discussed, the question of which transformations preserve certain geometrical properties is almost inevitable. In this section we analyze operators that preserve distance. It turns out that they are characterized as those operators that preserve other properties.

THEOREM 1

Let $T : V \to V$ be a linear operator on a finite dimensional inner product space V. The following conditions are equivalent.

1. $\|T(\mathbf{v})\| = \|\mathbf{v}\|$ for all $\mathbf{v}$ in V. (T preserves norms.)
2. $\|T(\mathbf{v}) - T(\mathbf{v}_1)\| = \|\mathbf{v} - \mathbf{v}_1\|$ for all $\mathbf{v}$ and $\mathbf{v}_1$ in V. (T preserves distance.)
3. $\langle T(\mathbf{v}), T(\mathbf{v}_1) \rangle = \langle \mathbf{v}, \mathbf{v}_1 \rangle$ for all $\mathbf{v}$ and $\mathbf{v}_1$ in V. (T preserves inner products.)
4. If $\{\mathbf{e}_1, \dots, \mathbf{e}_n\}$ is any orthonormal basis in V, then $\{T(\mathbf{e}_1), \dots, T(\mathbf{e}_n)\}$ is an orthonormal basis. (T preserves orthonormal bases.)
5. T carries some orthonormal basis to an orthonormal basis.

Proof

(1) *implies* (2). This is because $T(\mathbf{v}) - T(\mathbf{v}_1) = T(\mathbf{v} - \mathbf{v}_1)$.

(2) *implies* (3). (2) implies that

$$\|\mathbf{v} + \mathbf{v}_1\| = \|\mathbf{v} - (-\mathbf{v}_1)\| = \|T(\mathbf{v}) - T(-\mathbf{v}_1)\| = \|T(\mathbf{v}) + T(\mathbf{v}_1)\|$$

Hence (3) follows from the identity $\langle \mathbf{v}, \mathbf{v}_1 \rangle = \frac{1}{4}\{\|\mathbf{v} + \mathbf{v}_1\|^2 - \|\mathbf{v} - \mathbf{v}_1\|^2\}$.

(3) *implies* (4). (3) implies that $\{T(\mathbf{e}_1), \dots, T(\mathbf{e}_n)\}$ is orthonormal provided that each $T(\mathbf{e}_i) \neq \mathbf{0}$. In fact, $T(\mathbf{v}) = \mathbf{0}$ implies that $\|\mathbf{v}\|^2 = \langle \mathbf{v}, \mathbf{v} \rangle = \langle T(\mathbf{v}), T(\mathbf{v}) \rangle = 0$, so $\mathbf{v} = \mathbf{0}$.

(4) *implies* (5). This needs no proof.

(5) *implies* (1). Let $\{\mathbf{e}_1, \dots, \mathbf{e}_n\}$ be an orthonormal basis such that $\{T(\mathbf{e}_1), \dots, T(\mathbf{e}_n)\}$ is orthonormal. Given $\mathbf{v} = v_1 \mathbf{e}_1 + \cdots + v_n \mathbf{e}_n$ in V, we have $T(\mathbf{v}) = v_1 T(\mathbf{e}_1) + \cdots + v_n T(\mathbf{e}_n)$, so the Pythagorean theorem gives $\|T(\mathbf{v})\|^2 = v_1^2 + \cdots + v_n^2 = \|\mathbf{v}\|^2$. Hence $\|T(\mathbf{v})\| = \|\mathbf{v}\|$. ◆

DEFINITION

If V is an inner product space, a linear operator $T : V \to V$ is called an **isometry** if it satisfies one (and hence all) of the conditions in Theorem 1.

Before giving examples, we note two consequences of Theorem 1. The proofs are left to the reader.

COROLLARY

1. Every isometry of a finite dimensional space is an isomorphism.
2. The composite of two isometries is an isometry.

EXAMPLE 1

If V is any finite dimensional inner product space, the identity operator $1_V : V \to V$ is an isometry as is the negation map $T : V \to V$ with $T(\mathbf{v}) = -\mathbf{v}$.   ◆◆◆

EXAMPLE 2

Rotations of $\mathbb{R}^2$ about the origin are isometries, as are reflections about lines through the origin: They clearly preserve distance and are linear by Example 2§7.1. Similarly, rotations about lines through the origin and reflections about planes through the origin are isometries of $\mathbb{R}^3$.   ◆◆◆

EXAMPLE 3

Let $T : \mathbf{M}_{nn} \to \mathbf{M}_{nn}$ be the transposition operator: $T(A) = A^T$. Then T is an isometry if the inner product is $\langle A, B \rangle = \text{tr}(AB^T) = \sum_{i,j} a_{ij} b_{ij}$. In fact T permutes the basis consisting of all matrices with one entry 1 and the other entries 0.   ◆◆◆

The proof of the next result requires the fact that, if B is an orthonormal basis, then $\langle \mathbf{v}, \mathbf{w} \rangle = C_B(\mathbf{v}) \cdot C_B(\mathbf{w})$ for all vectors $\mathbf{v}$ and $\mathbf{w}$.

THEOREM 2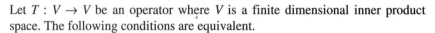

Let $T : V \to V$ be an operator where V is a finite dimensional inner product space. The following conditions are equivalent.

1. T is an isometry.
2. $M_B(T)$ is an orthogonal matrix for every orthonormal basis B.
3. $M_B(T)$ is an orthogonal matrix for some orthonormal basis B.

Proof

(1) *implies* (2). Let $B = \{\mathbf{e}_1, \ldots, \mathbf{e}_n\}$ be an orthonormal basis. Then the jth column of $M_B(T)$ is $C_B[T(\mathbf{e}_j)]$, and we have

$$C_B[T(\mathbf{e}_j)] \cdot C_B[T(\mathbf{e}_k)] = \langle T(\mathbf{e}_j), T(\mathbf{e}_k) \rangle = \langle \mathbf{e}_j, \mathbf{e}_k \rangle$$

using (1). Hence the columns of $M_B(T)$ are orthonormal in $\mathbb{R}^n$, which proves (2).

(2) *implies* (3). This is clear.

(3) *implies* (1). Let $B = \{\mathbf{e}_1, \ldots, \mathbf{e}_n\}$ be as in (3). Then, as before,

$$\langle T(\mathbf{e}_j), T(\mathbf{e}_k) \rangle = C_B[T(\mathbf{e}_j)] \cdot C_B[T(\mathbf{e}_k)]$$

so $\{T(\mathbf{e}_1), \ldots, T(\mathbf{e}_n)\}$ is orthonormal by (3) and Theorem 1 applies.   ◆

It is important that B is *orthonormal* in Theorem 2. For example, $T : V \rightarrow V$ given by $T(\mathbf{v}) = 2\mathbf{v}$ preserves *orthogonal* sets but is not an isometry, as is easily checked.

If P is an orthogonal square matrix, then $P^{-1} = P^{T}$. Taking determinants yields $(\det P)^{2} = 1$, so $\det P = \pm 1$. Hence:

COROLLARY

If $T : V \rightarrow V$ is an isometry where V is a finite dimensional inner product space, then $\det T = \pm 1$.

EXAMPLE 4

If A is any $n \times n$ matrix, the matrix operator $T_{A} : \mathbb{R}^{n} \rightarrow \mathbb{R}^{n}$ is an isometry if and only if A is orthogonal (use the dot product in $\mathbb{R}^{n}$). Indeed, if B is the standard basis of $\mathbb{R}^{n}$, then $M_{B}(T_{A}) = A$.

◆◆◆

Rotations and reflections that fix the origin are isometries in $\mathbb{R}^{2}$ and $\mathbb{R}^{3}$ (Example 2); we are going to show that these isometries (and compositions of them in $\mathbb{R}^{3}$) are the only possibilities. In fact this will follow from a general structure theorem for isometries. Surprisingly enough, much of the work involves the two dimensional case.

THEOREM 3

Let $T : V \rightarrow V$ be an isometry on the two dimensional inner product space V. Then there are two possibilities.

Either (1) There is an orthonormal basis B of V such that

$$M_{B}(T) = \begin{bmatrix} \cos \theta & -\sin \theta \\ \sin \theta & \cos \theta \end{bmatrix}, \quad 0 \leq \theta < 2\pi$$

or   (2) There is an orthonormal basis B of V such that

$$M_{B}(T) = \begin{bmatrix} 1 & 0 \\ 0 & -1 \end{bmatrix}$$

Furthermore, type (1) occurs if and only if $\det T = 1$, and type (2) occurs if and only if $\det T = -1$.

Proof   The final statement follows from the rest because $\det T = \det[M_{B}(T)]$ for any basis B. Let $B_{0} = \{\mathbf{e}_{1}, \mathbf{e}_{2}\}$ be any ordered orthonormal basis of V and write

$$A = M_{B_0}(T) = \begin{bmatrix} a & b \\ c & d \end{bmatrix}; \qquad \text{that is,} \qquad \begin{matrix} T(\mathbf{e}_1) = a\mathbf{e}_1 + c\mathbf{e}_2 \\ T(\mathbf{e}_2) = b\mathbf{e}_1 + d\mathbf{e}_2 \end{matrix}$$

Then A is orthogonal by Theorem 2, so its columns (and rows) are orthonormal. Hence $a^2 + c^2 = 1 = b^2 + d^2$, so (a, c) and (d, b) lie on the unit circle. Thus angles θ and φ exist such that

$$a = \cos\theta, \qquad c = \sin\theta \qquad 0 \le \theta < 2\pi$$
$$d = \cos\varphi, \qquad b = \sin\varphi \qquad 0 \le \varphi < 2\pi$$

Then $\sin(\theta + \varphi) = cd + ab = 0$, so $\theta + \varphi = k\pi$ for some integer k. This gives $d = \cos(k\pi - \theta) = (-1)^k \cos\theta$ and $b = \sin(k\pi - \theta) = (-1)^{k+1}\sin\theta$. Finally

$$A = \begin{bmatrix} \cos\theta & (-1)^{k+1}\sin\theta \\ \sin\theta & (-1)^k\cos\theta \end{bmatrix}$$

If k is even, we are in type (1) with $B = B_0$, so assume k is odd. Then $A = \begin{bmatrix} a & c \\ c & -a \end{bmatrix}$. If $a = -1$ and $c = 0$, we are in type (2) with $B = \{\mathbf{e}_2, \mathbf{e}_1\}$. Otherwise A has eigenvalues 1 and -1 with (nonzero) eigenvectors $\begin{bmatrix} 1 + a \\ c \end{bmatrix}$ and $\begin{bmatrix} -c \\ 1 + a \end{bmatrix}$ respectively. Hence

$$\mathbf{f}_1 = (1 + a)\mathbf{e}_1 + c\mathbf{e}_2 \qquad \text{and} \qquad \mathbf{f}_2 = -c\mathbf{e}_1 + (1 + a)\mathbf{e}_2$$

are orthogonal eigenvectors of T and we are in type (2) with $B = \left\{ \dfrac{1}{\|\mathbf{f}_1\|}\mathbf{f}_1, \dfrac{1}{\|\mathbf{f}_2\|}\mathbf{f}_2 \right\}$. ◆

If $V = \mathbb{R}^2$, this theorem has a satisfying geometrical interpretation: The type 1 and type 2 isometries are just rotations and reflections, respectively. In fact, if B_0 is the standard basis, then the clockwise rotation R_θ about the origin through an angle θ has matrix

$$M_{B_0}(R_\theta) = \begin{bmatrix} \cos\theta & -\sin\theta \\ \sin\theta & \cos\theta \end{bmatrix}$$

(see Example 12§7.1). On the other hand, if $S : \mathbb{R}^2 \to \mathbb{R}^2$ is the reflection in a line through the origin (called the **fixed line** of the reflection), let $\mathbf{f}_1$ be a unit vector pointing along the fixed line and let $\mathbf{f}_2$ be a unit vector perpendicular to the fixed line. Then $B = \{\mathbf{f}_1, \mathbf{f}_2\}$ is an orthonormal basis, $S(\mathbf{f}_1) = \mathbf{f}_1$ and $S(\mathbf{f}_2) = -\mathbf{f}_2$, so

$$M_B(S) = \begin{bmatrix} 1 & 0 \\ 0 & -1 \end{bmatrix}$$

Thus S is of type 2. Note that, in this case, 1 is an eigenvalue of S, and any eigenvector corresponding to 1 is a direction vector for the fixed line.

EXAMPLE 5

In each case, determine whether $T_A : \mathbb{R}^2 \to \mathbb{R}^2$ is a rotation or a reflection, and then find the angle or fixed line:

(a) $A = \frac{1}{2} \begin{bmatrix} 1 & \sqrt{3} \\ -\sqrt{3} & 1 \end{bmatrix}$

(b) $A = \frac{1}{5} \begin{bmatrix} -3 & 4 \\ 4 & 3 \end{bmatrix}$

Solution   Both matrices are orthogonal, so (because $M_{B_0}(T_A) = A$, where B_0 is the standard basis) T_A is an isometry in both cases. In the first case, $\det A = 1$, so T_A is a counterclockwise rotation through θ, where $\cos \theta = \frac{1}{2}$ and $\sin \theta = -\frac{\sqrt{3}}{2}$. Thus $\theta = \frac{5\pi}{3}$ (or $300°$). In (b), $\det A = -1$, so T_A is a reflection in this case. We verify that $\mathbf{d} = \begin{bmatrix} 1 \\ 2 \end{bmatrix}$ is an eigenvector corresponding to the eigenvalue 1. Hence the fixed line $\mathbb{R}\mathbf{d}$ has equation $y = 2x$.

◆◆◆

We now give a structure theorem for isometries. The proof requires three preliminary results, each of interest in its own right.

LEMMA 1

Let $T : V \to V$ be an isometry of a finite dimensional inner product space V. If U is a T-invariant subspace of V, then $U^\perp$ is also T-invariant.

Proof   Let $\mathbf{w}$ lie in $U^\perp$. We are to prove that $T(\mathbf{w})$ is also in $U^\perp$; that is, $\langle T(\mathbf{w}), \mathbf{u} \rangle = 0$ for all $\mathbf{u}$ in U. At this point, observe that the restriction of T to U is an isometry $U \to U$ and so is an isomorphism by the corollary to Theorem 1. Hence each $\mathbf{u}$ in U can be written in the form $\mathbf{u} = T(\mathbf{u}_1)$ for some $\mathbf{u}_1$ in U, so

$$\langle T(\mathbf{w}), \mathbf{u} \rangle = \langle T(\mathbf{w}), T(\mathbf{u}_1) \rangle = \langle \mathbf{w}, \mathbf{u}_1 \rangle = 0$$

because $\mathbf{w}$ is in $U^\perp$. This is what we wanted.   ◆

To employ Lemma 1 to analyze an isometry $T : V \to V$ when $\dim V = n$, it is necessary to show that a T-invariant subspace U exists such that $U \neq 0$ and $U \neq V$. We will show, in fact, that such a subspace U can always be found of dimension 1 or 2. If T has a real eigenvalue λ, this is no problem because, if $\mathbf{u}$ is any eigenvector, $U = \mathbb{R}\mathbf{u}$ is T-invariant. But in case (1) of Theorem 3 the eigenvalues of T are $e^{i\theta}$ and $e^{-i\theta}$ (the reader should check this), and these are nonreal if $\theta \neq 0$ and $\theta \neq \pi$. It turns out (Lemma 2 below) that every complex eigenvalue λ of T has absolute value 1; and (Lemma 3), if λ is not real, that U has a T-invariant subspace of dimension 2.

LEMMA 2

Let $T : V \to V$ be an isometry of the finite dimensional inner product space V. If λ is a complex eigenvalue of T, then $|\lambda| = 1$.

Proof   Choose an orthonormal basis B of V, and let $A = M_B(T)$. Then A is a real orthogonal matrix so, using the standard inner product $\langle X, Y \rangle = X^T \overline{Y}$ in $\mathbb{C}^n$, we get

$$\|AX\|^2 = (AX)^T \, \overline{(A X)} = X^T A^T A \overline{X} = \|X\|^2$$

for all X in $\mathbb{C}^n$. But $AX = \lambda X$ for some $X \neq 0$, whence $\|X\|^2 = \|\lambda X\|^2 = |\lambda|^2 \|X\|^2$. This gives $|\lambda| = 1$, as required.   ◆

LEMMA 3

Let $T : V \to V$ be an isometry of the n-dimensional inner product space V. If T has a nonreal eigenvalue, then V has a two-dimensional T-invariant subspace.

Proof   Let B be an orthonormal basis of V, let $A = M_B(T)$, and let $\lambda = e^{i\alpha}$ be a nonreal eigenvalue of A, say $AX = \lambda X$ where $X \neq 0$ in $\mathbb{C}^n$. Because A is real, complex conjugation gives $A\overline{X} = \overline{\lambda}\overline{X}$, so $\overline{\lambda}$ is also an eigenvalue. Moreover $\lambda \neq \overline{\lambda}$ (λ is nonreal), so $\{X, \overline{X}\}$ is linearly independent in $\mathbb{C}^n$ (the argument in the proof of Theorem 2§6.2 works). Now define

$$E_1 = X + \overline{X} \qquad \text{and} \qquad E_2 = i(X - \overline{X})$$

Then E_1 and E_2 lie in $\mathbb{R}^n$, and $\{E_1, E_2\}$ is linearly independent over $\mathbb{R}$ because $\{X, \overline{X}\}$ is linearly independent over $\mathbb{C}$. Moreover

$$X = \tfrac{1}{2}(E_1 - iE_2) \qquad \text{and} \qquad \overline{X} = \tfrac{1}{2}(E_1 + iE_2)$$

Now $\lambda + \overline{\lambda} = 2 \cos \alpha$ and $\lambda - \overline{\lambda} = 2i \sin \alpha$, and a routine computation gives

$$AE_1 = \cos \alpha \, E_1 + \sin \alpha \, E_2$$
$$AE_2 = -\sin \alpha \, E_1 + \cos \alpha \, E_2$$

Finally, let $\mathbf{e}_1$ and $\mathbf{e}_2$ in V be such that $E_1 = C_B(\mathbf{e}_1)$ and $E_2 = C_B(\mathbf{e}_2)$. Then

$$C_B[T(\mathbf{e}_1)] = AC_B(\mathbf{e}_1) = AE_1 = C_B(\cos \alpha \, \mathbf{e}_1 + \sin \alpha \, \mathbf{e}_2)$$

using Theorem 2§7.4. Because C_B is one-to-one, this gives the first of the following equations (the other is similar):

$$T(\mathbf{e}_1) = \cos \alpha \, \mathbf{e}_1 + \sin \alpha \, \mathbf{e}_2$$
$$T(\mathbf{e}_2) = -\sin \alpha \, \mathbf{e}_1 + \cos \alpha \, \mathbf{e}_2$$

Thus $U = \text{span} \{\mathbf{e}_1, \mathbf{e}_2\}$ is T-invariant and two-dimensional.   ◆

We can now prove the structure theorem for isometries.

THEOREM 4

Let $T : V \to V$ be an isometry of the n-dimensional inner product space V. Given an angle θ, write $R(\theta) = \begin{bmatrix} \cos\theta & -\sin\theta \\ \sin\theta & \cos\theta \end{bmatrix}$. Then there exists an orthonormal basis B of V such that $M_B(T)$ has one of the following block diagonal forms, classified for convenience by whether n is even or odd:

$$n = 2k+1 \quad \begin{bmatrix} 1 & 0 & \cdots & 0 \\ 0 & R(\theta_1) & \cdots & 0 \\ \vdots & \vdots & \ddots & \vdots \\ 0 & 0 & \cdots & R(\theta_k) \end{bmatrix} \quad \text{or} \quad \begin{bmatrix} -1 & 0 & \cdots & 0 \\ 0 & R(\theta_1) & \cdots & 0 \\ \vdots & \vdots & \ddots & \vdots \\ 0 & 0 & \cdots & R(\theta_k) \end{bmatrix}$$

$$n = 2k \quad \begin{bmatrix} R(\theta_1) & 0 & \cdots & 0 \\ 0 & R(\theta_2) & \cdots & 0 \\ \vdots & \vdots & \ddots & \vdots \\ 0 & 0 & \cdots & R(\theta_k) \end{bmatrix}$$

$$\text{or} \quad \begin{bmatrix} -1 & 0 & 0 & \cdots & 0 \\ 0 & 1 & 0 & \cdots & 0 \\ 0 & 0 & R(\theta_1) & \cdots & 0 \\ \vdots & \vdots & \vdots & \ddots & \vdots \\ 0 & 0 & 0 & \cdots & R(\theta_{k-1}) \end{bmatrix}$$

Proof   We show first, by induction on n, that an orthonormal basis B of V can be found such that $M_B(T)$ is a block diagonal matrix of the following form:

$$M_B(T) = \begin{bmatrix} I_r & 0 & 0 & \cdots & 0 \\ 0 & -I_s & 0 & \cdots & 0 \\ 0 & 0 & R(\theta_1) & \cdots & 0 \\ \vdots & \vdots & \vdots & \ddots & \vdots \\ 0 & 0 & 0 & \cdots & R(\theta_t) \end{bmatrix}$$

where the identity matrix I_r, the matrix $-I_s$, or the matrices $R(\theta_i)$ may be missing. If $n = 1$ and $V = \mathbb{R}\mathbf{v}$, this holds because $T(X) = \lambda\mathbf{v}$ and $\lambda = \pm1$ by Lemma 2. If $n = 2$, this follows from Theorem 3. If $n \geq 3$, either T has a real eigenvalue and therefore has a one-dimensional T-invariant subspace $U = \mathbb{R}\mathbf{u}$ for any eigenvector $\mathbf{u}$, or T has no real eigenvalue and therefore has a two-dimensional T-invariant subspace U by Lemma 3. In either case $U^{\perp}$ is T-invariant and $\dim U^{\perp} = n - \dim U < n$. Hence, by induction, let B_1 and B_2 be orthonormal bases of U and $U^\perp$ such that $M_{B_1}(T)$ and $M_{B_2}(T)$ have the form given. Then $B = B_1 \cup B_2$ is an orthonormal basis of V, and $M_B(T)$ has the desired form with a suitable ordering of the vectors in B.

Now observe that $R(0) = \begin{bmatrix} 1 & 0 \\ 0 & 1 \end{bmatrix}$ and $R(\pi) = \begin{bmatrix} -1 & 0 \\ 0 & -1 \end{bmatrix}$. It follows that an even number of 1's or -1's can be written as $R(\theta_1)$–blocks. Hence, with a suitable reordering of B, the theorem follows. ◆

As in the dimension-2 situation, these possibilities can be given a geometric interpretation when $V = \mathbb{R}^3$ is taken as Euclidean space. As before, this entails looking carefully at reflections and rotations in $\mathbb{R}^3$. If $R : \mathbb{R}^3 \to \mathbb{R}^3$ is any reflection in a plane through the origin (called the **fixed plane** of the reflection), take $\{E_2, E_3\}$ to be any orthonormal basis of the fixed plane and take E_1 to be a unit vector perpendicular to the fixed plane. Then $R(E_1) = -E_1$, whereas $R(E_2) = E_2$ and $R(E_3) = E_3$. Hence $B = \{E_1, E_2, E_3\}$ is an orthonormal basis such that

$$M_B(R) = \begin{bmatrix} -1 & 0 & 0 \\ 0 & 1 & 0 \\ 0 & 0 & 1 \end{bmatrix}$$

Similarly, suppose that $T : \mathbb{R}^3 \to \mathbb{R}^3$ is any rotation about a line through the origin (called the **axis** of the rotation), and let E_1 be a unit vector pointing along the axis, so $T(E_1) = E_1$. Now the plane through the origin perpendicular to the axis is a T-invariant subspace of $\mathbb{R}^2$ of dimension 2, and the restriction of T to this plane is a rotation. Hence, by Theorem 3, there is an orthonormal basis $B_1 = \{E_2, E_3\}$ of this plane such that $M_{B_1}(T) = \begin{bmatrix} \cos\theta & -\sin\theta \\ \sin\theta & \cos\theta \end{bmatrix}$. But then $B = \{E_1, E_2, E_3\}$ is an orthonormal basis such that the matrix of T is

$$M_B(T) = \begin{bmatrix} 1 & 0 & 0 \\ 0 & \cos\theta & -\sin\theta \\ 0 & \sin\theta & \cos\theta \end{bmatrix}$$

However, Theorem 4 shows that there are isometries T_1 in $\mathbb{R}^3$ of a third type: those with a matrix of the form

$$M_B(T_1) = \begin{bmatrix} -1 & 0 & 0 \\ 0 & \cos\theta & -\sin\theta \\ 0 & \sin\theta & \cos\theta \end{bmatrix}$$

If $B = \{E_1, E_2, E_3\}$, let R be the reflection in the plane spanned by E_2 and E_3 and let T be the rotation corresponding to θ about the line spanned by E_1. Then $M_B(R)$ and $M_B(T)$ are as before, and their product is $M_B(T_1)$. Hence

$$M_B(T_1) = M_B(R)M_B(T) = M_B(RT)$$

so (because M_B is one-to-one) it follows that $T_1 = RT$. A similar argument shows that also $T_1 = TR$, and we have Theorem 5.

THEOREM 5

If $T: \mathbb{R}^3 \to \mathbb{R}^3$ is an isometry, there are three possibilities.

(a) T is a rotation, and $M_B(T) = \begin{bmatrix} 1 & 0 & 0 \\ 0 & \cos\theta & -\sin\theta \\ 0 & \sin\theta & \cos\theta \end{bmatrix}$ for some orthonormal basis B.

(b) T is a reflection, and $M_B(T) = \begin{bmatrix} -1 & 0 & 0 \\ 0 & 1 & 0 \\ 0 & 0 & 1 \end{bmatrix}$ for some orthonormal basis B.

(c) $T = QR = RQ$ where Q is a reflection, R is a rotation about an axis perpendicular to the fixed plane of Q and $M_B(T) = \begin{bmatrix} -1 & 0 & 0 \\ 0 & \cos\theta & -\sin\theta \\ 0 & \sin\theta & \cos\theta \end{bmatrix}$ for some orthonormal basis B.

Hence T is a rotation if and only if $\det T = 1$.

Proof   It remains only to verify the final observation that T is a rotation if and only if $\det T = 1$. But clearly $\det T = -1$ in parts (b) and (c). ◆

A useful way of analyzing a given isometry $T: \mathbb{R}^3 \to \mathbb{R}^3$ comes from computing the eigenvalues of T. Because the characteristic polynomial of T has degree 3, it must have a real root. Hence, there must be at least one real eigenvalue, and the only possible real eigenvalues are ± 1. Thus Table 8.1 includes all possibilities.

TABLE 8.1

Eigenvalues of T	Action of T
(1) 1, no other real eigenvalues	Rotation about the line $\mathbb{R}E$ where E is an eigenvector corresponding to 1. [Case (a) of Theorem 5.]
(2) -1, no other real eigenvalues	Rotation about the line $\mathbb{R}E$ followed by reflection in the plane $(\mathbb{R}E)^{\perp}$ where E is an eigenvector corresponding to -1. [Case (c) of Theorem 5.]
(3) $-1, 1, 1$	Reflection in the plane $(\mathbb{R}E)^{\perp}$ where E is an eigenvector corresponding to -1. [Case (b) of Theorem 5.]
(4) $1, -1, -1$	This is as in (1) with a rotation of π.
(5) $-1, -1, -1$	Here $T(X) = -X$ for all X. This is (2) with a rotation of π.
(6) $1, 1, 1$	Here T is the identity isometry.

EXAMPLE 6

Analyze the isometry $T : \mathbb{R}^3 \to \mathbb{R}^3$ given by $T\begin{bmatrix} x \\ y \\ z \end{bmatrix} = \begin{bmatrix} y \\ z \\ -x \end{bmatrix}$.

Solution   If B_0 is the standard basis of $\mathbb{R}^3$, then $M_{B_0}(T) = \begin{bmatrix} 0 & 1 & 0 \\ 0 & 0 & 1 \\ -1 & 0 & 0 \end{bmatrix}$, so $c_T(x) = x^3 + 1 =$

$(x + 1)(x^2 - x + 1)$. This is (2) in Table 8.1.

Write:

$$E_1 = \frac{1}{\sqrt{3}}\begin{bmatrix} 1 \\ -1 \\ 1 \end{bmatrix} \qquad E_2 = \frac{1}{\sqrt{6}}\begin{bmatrix} 1 \\ 2 \\ 1 \end{bmatrix} \qquad E_3 = \frac{1}{\sqrt{2}}\begin{bmatrix} 1 \\ 0 \\ -1 \end{bmatrix}$$

Here E_1 is a unit eigenvector corresponding to -1, so T is a rotation (through an angle θ) about the line $L = \mathbb{R}E_1$, followed by reflection in the plane U through the origin perpendicular to E_1 (with equation $x - y + z = 0$). Moreover, $\{E_2, E_3\}$ is an orthonormal basis of U, so $B = \{E_1, E_2, E_3\}$ is an orthonormal basis of $\mathbb{R}^3$ and

$$M_B(T) = \begin{bmatrix} -1 & 0 & 0 \\ 0 & \frac{1}{2} & -\frac{\sqrt{3}}{2} \\ 0 & \frac{\sqrt{3}}{2} & \frac{1}{2} \end{bmatrix}$$

Hence θ is given by $\cos\theta = \frac{1}{2}$, $\sin\theta = \frac{\sqrt{3}}{2}$, so $\theta = \frac{\pi}{3}$.

◆◆◆

Let V be an n-dimensional inner product space. A subspace of V of dimension $n-1$ is called a **hyperplane** in V. Thus the hyperplanes in $\mathbb{R}^3$ and $\mathbb{R}^2$ are, respectively, the planes and lines through the origin. An isometry $Q : V \to V$ is called a **reflection** if there exists an orthonormal basis $B = \{e_1, e_2, \ldots, e_n\}$ such that

$$M_B(T) = \begin{bmatrix} -1 & 0 \\ 0 & I_{n-1} \end{bmatrix}$$

Thus $Q(e_1) = -e_1$ whereas $Q(u) = u$ for each u in $U = \text{span}\{e_2, \ldots, e_n\}$. Hence U is called the **fixed hyperplane** of Q. Note that each hyperplane in V is the fixed hyperplane of a (unique) reflection of V. Clearly, reflections in $\mathbb{R}^2$ and $\mathbb{R}^3$ are reflections in this more general sense.

Continuing the analogy with $\mathbb{R}^2$ and $\mathbb{R}^3$, an isometry $T : V \to V$ is called a **rotation** if there exists an orthonormal basis $\{e_1, \ldots, e_n\}$ such that

$$M_B(T) = \begin{bmatrix} I_r & 0 & 0 \\ 0 & R(\theta) & 0 \\ 0 & 0 & I_s \end{bmatrix}$$

in block form, where $R(\theta) = \begin{bmatrix} \cos\theta & -\sin\theta \\ \sin\theta & \cos\theta \end{bmatrix}$, and where either I_r or I_s (or both) may be missing. If $R(\theta)$ occupies columns i and $i+1$ of $M_B(T)$, and if $W = \text{span}\{e_i, e_{i+1}\}$, then W is T-invariant and the matrix of $T : W \to W$ with respect to $\{e_i, e_{i+1}\}$ is $R(\theta)$. Clearly, if W is viewed as a copy of $\mathbb{R}^2$, then T is a rotation in W. Moreover, $T(\mathbf{u}) = \mathbf{u}$ holds for all vectors $\mathbf{u}$ in the $(n-2)$-dimensional subspace $U = \text{span}\{e_1, \ldots, e_{i-1}, e_{i+2}, \ldots, e_n\}$, and U is called the **fixed axis** of the rotation T. In $\mathbb{R}^3$, the axis of any rotation is a line (one-dimensional), whereas in $\mathbb{R}^2$ the axis is $U = \{0\}$.

With these definitions, the following theorem is an immediate consequence of Theorem 5 (the details are left to the reader).

THEOREM 6

Let $T : V \to V$ be an isometry of a finite dimensional inner product space V. Then there exist isometries $T_1, \ldots, T_k$ such that

$$T = T_k T_{k-1} \cdots T_2 T_1$$

where each T_i is either a rotation or a reflection, at most one is a reflection, and $T_i T_j = T_j T_i$ holds for all i and j. Furthermore, T is a composite of rotations if and only if $\det T = 1$.

EXERCISES 8.4

Throughout these exercises, V denotes a finite dimensional inner product space.

1. Show that the following linear operators are isometries.
   (a) $T : \mathbb{C} \to \mathbb{C}$; $T(z) = \overline{z}$; $\langle z, w \rangle = \text{re}(z\overline{w})$
   (b) $T : \mathbb{R}^n \to \mathbb{R}^n$; $T(a_1, a_2, \ldots, a_n) = (a_n, a_{n-1}, \ldots, a_2, a_1)$; dot product
   (c) $T : M_{22} \to M_{22}$, $T\begin{bmatrix} a & b \\ c & d \end{bmatrix} = \begin{bmatrix} c & d \\ b & a \end{bmatrix}$; $\langle A, B \rangle = \text{tr}(AB^T)$
   (d) $T : \mathbb{R}^3 \to \mathbb{R}^3$, $T[a, b, c] = \frac{1}{9}(2a + 2b - c, 2a + 2c - b, 2b + 2c - a)$; dot product

2. In each case, show that T is an isometry of $\mathbb{R}^2$, determine whether it is a rotation or a reflection, and find the angle or the fixed line. Use the dot product.
   (a) $T\begin{bmatrix} a \\ b \end{bmatrix} = \begin{bmatrix} -a \\ b \end{bmatrix}$ ◆(b) $T\begin{bmatrix} a \\ b \end{bmatrix} = \begin{bmatrix} -a \\ -b \end{bmatrix}$
   (c) $T\begin{bmatrix} a \\ b \end{bmatrix} = \begin{bmatrix} b \\ -a \end{bmatrix}$ ◆(d) $T\begin{bmatrix} a \\ b \end{bmatrix} = \begin{bmatrix} -b \\ -a \end{bmatrix}$
   (e) $T\begin{bmatrix} a \\ b \end{bmatrix} = \frac{1}{\sqrt{2}}\begin{bmatrix} a + b \\ b - a \end{bmatrix}$ ◆(f) $T\begin{bmatrix} a \\ b \end{bmatrix} = \frac{1}{\sqrt{2}}\begin{bmatrix} a - b \\ a + b \end{bmatrix}$

3. In each case, show that T is an isometry of $\mathbb{R}^3$, determine the type (Theorem 5), and find the axis of any rotations and the fixed plane of any reflections involved.
   (a) $T\begin{bmatrix} a \\ b \\ c \end{bmatrix} = \begin{bmatrix} a \\ -b \\ c \end{bmatrix}$ ◆(b) $T\begin{bmatrix} a \\ b \\ c \end{bmatrix} = \frac{1}{2}\begin{bmatrix} \sqrt{3}c - a \\ \sqrt{3}a + c \\ 2b \end{bmatrix}$
   (c) $T\begin{bmatrix} a \\ b \\ c \end{bmatrix} = \begin{bmatrix} b \\ c \\ a \end{bmatrix}$ ◆(d) $T\begin{bmatrix} a \\ b \\ c \end{bmatrix} = \begin{bmatrix} a \\ -b \\ -c \end{bmatrix}$
   (e) $T\begin{bmatrix} a \\ b \\ c \end{bmatrix} = \frac{1}{2}\begin{bmatrix} a + \sqrt{3}b \\ b - \sqrt{3}a \\ 2c \end{bmatrix}$ ◆(f) $T\begin{bmatrix} a \\ b \\ c \end{bmatrix} = \frac{1}{\sqrt{2}}\begin{bmatrix} a + c \\ -\sqrt{2}b \\ c - a \end{bmatrix}$

4. Let $T : \mathbb{R}^2 \to \mathbb{R}^2$ be an isometry. A vector X in $\mathbb{R}^2$ is said to be **fixed** by T if $T(X) = X$. Let E_1 denote the set of all vectors in $\mathbb{R}^2$ fixed by T. Show that:
   (a) E_1 is a subspace of $\mathbb{R}^2$.
   (b) $E_1 = \mathbb{R}^2$ if and only if $T = 1$ is the identity map.
   (c) $\dim E_1 = 1$ if and only if T is a reflection (about the line E_1).
   (d) $E_1 = 0$ if and only if T is a rotation ($T \neq 1$).

5. Let $T : \mathbb{R}^3 \to \mathbb{R}^3$ be an isometry, and let E_1 be the subspace of all fixed vectors in $\mathbb{R}^3$ (see Exercise 4). Show that:

  (a) $E_1 = \mathbb{R}^3$ if and only if $T = 1$.

  (b) $\dim E_1 = 2$ if and only if T is a reflection (about the plane E_1).

  (c) $\dim E_1 = 1$ if and only if T is a rotation ($T \neq 1$) (about the line E_1).

  (d) $\dim E_1 = 0$ if and only if T is a reflection followed by a (nonidentity) rotation.

6. Let **I** denote the set of all isometries $V \to V$. Show that:

  (a) 1_V is in **I**.

  ◆**(b)** If S and T are in **I**, then ST is also in **I**.

  (c) If S is in **I**, then S is invertible and S^{-1} is in **I**.

  (The set **I** is called a **group** of operators by virtue of these properties, and such sets of operators are important in geometry. In fact, geometry itself can be fruitfully viewed as the study of those properties of a vector space that are preserved by some group of invertible linear operators.)

7. If T is an isometry, show that aT is an isometry if and only if $a = \pm 1$.

◆**8.** Show that every isometry preserves the angle between any pair of nonzero vectors (see Exercise 32§8.1). Must an angle-preserving isomorphism be an isometry? Support your answer.

9. If $T = V \to V$ is an isometry, show that $T^2 = 1_V$ if and only if the only complex eigenvalues of T are 1 and -1.

10. Let $T : V \to V$ be a linear operator. Show that any two of the following conditions implies the third:

  (1) T is symmetric.

  (2) T is an involution.

  (3) T is an isometry.

  [*Hint:* In all cases, use the definition $\langle \mathbf{v}, T(\mathbf{w}) \rangle = \langle T(\mathbf{v}), \mathbf{w} \rangle$ of a symmetric operator. For part **(1)** + part **(3)** implies part **(2)**, use the fact that, if $\langle T^2(\mathbf{v}) - \mathbf{v}, \mathbf{w} \rangle = 0$ for all $\mathbf{w}$, then $T^2(\mathbf{v}) = \mathbf{v}$.]

11. If B and D are any orthonormal bases of V, show that there is an isometry $T : V \to V$ that carries B to D.

12. Show that the following are equivalent for a linear transformation $S : V \to V$ where V is finite dimensional:

  (1) $\langle S(\mathbf{v}), S(\mathbf{w}) \rangle = 0$ whenever $\langle \mathbf{v}, \mathbf{w} \rangle = 0$;

  (2) $S = aT$ for some isometry $T : V \to V$.

  [*Hint:* Given (1), show that $\|S(\mathbf{e})\| = \|S(\mathbf{f})\|$ for all unit vectors $\mathbf{e}$ and $\mathbf{f}$ in V.]

13. Let V be an inner product space and let $T : V \to V$ be a function (not assumed to be linear) such that $T(\mathbf{0}) = \mathbf{0}$ and $\|T(\mathbf{v}) - T(\mathbf{w})\| = \|\mathbf{v} - \mathbf{w}\|$ holds for all $\mathbf{v}$ and $\mathbf{w}$ in V. Assume that $\dim V = n$.

  (a) Show that $\|T(\mathbf{v})\| = \|\mathbf{v}\|$ for all $\mathbf{v}$.

  (b) Show that $\langle T(\mathbf{v}), T(\mathbf{w}) \rangle = \langle \mathbf{v}, \mathbf{w} \rangle$ for all $\mathbf{v}$ and $\mathbf{w}$. [*Hint:* Compute $\|T(\mathbf{v}) - T(\mathbf{w})\|^2 = \|\mathbf{v} - \mathbf{w}\|^2$.]

  (c) If $\{\mathbf{e}_1, \ldots, \mathbf{e}_n\}$ is an orthonormal basis of V, show that $\{T(\mathbf{e}_1), \ldots, T(\mathbf{e}_n)\}$ is orthonormal too.

  ◆**(d)** Show that T is linear and hence is an isometry. [*Hint:* Use part **(b)** to show that $\langle T(r\mathbf{v}) - rT(\mathbf{v}), T(\mathbf{w}) \rangle = 0$ and $\langle T(\mathbf{u} + \mathbf{v}) - T(\mathbf{u}) - T(\mathbf{v}), T(\mathbf{w}) \rangle = 0$ hold for all $\mathbf{u}$, $\mathbf{v}$, and $\mathbf{w}$ in V and all r in $\mathbb{R}$. Then use part **(c)**.]

14. Let V be an inner product space. A function $F : V \to V$ is said to **preserve distance** if $\|F(\mathbf{v}) - F(\mathbf{w})\| = \|\mathbf{v} - \mathbf{w}\|$ for all $\mathbf{v}$ and $\mathbf{w}$ in V. Show that:

  (a) If $T : V \to V$ is an isometry and $\mathbf{v}_0$ in V is fixed, the map $F : V \to V$ defined by $F(\mathbf{v}) = T(\mathbf{v}) + \mathbf{v}_0$ preserves distance.

  (b) Every map $F : V \to V$ that preserves distance is given as in part **(a)** for some isometry T and some vector $\mathbf{v}_0$. [*Hint:* Define $\mathbf{v}_0 = F(\mathbf{0})$ and $T : V \to V$ by $T(\mathbf{v}) = F(\mathbf{v}) - \mathbf{v}_0$. Show that T preserves distance and that $T(\mathbf{0}) = \mathbf{0}$, and apply the preceding exercise.]

15. If V is an inner product space and $\mathbf{v}_0$ is a fixed vector in V, the map $G : V \to V$ defined by $G(\mathbf{v}) = \mathbf{v} + \mathbf{v}_0$ is called the **translation** by $\mathbf{v}_0$. Show that every distance-preserving map of V is the composite of an isometry followed by a translation. [*Hint:* Exercise 14.]

███████ Section 8.5   **An Application to Fourier Approximation**

In this section we shall investigate an important orthogonal set in the space $\mathbf{C}[-\pi, \pi]$ of continuous functions on the interval $[-\pi, \pi]$, using the inner product.

$$\langle f, g \rangle = \int_{-\pi}^{\pi} f(x)g(x)\, dx$$

Of course calculus will be needed. The orthogonal set in question is

$$\{1, \sin x, \cos x, \sin 2x, \cos 2x, \sin 3x, \cos 3x, \ldots\}$$

and the first such investigation was carried out by Jean Baptiste Joseph Fourier (1768–1830), who used these functions in 1822 to study the conduction of heat in solids.

Standard techniques of integration give

$$\|1\|^2 = \int_{-\pi}^{\pi} 1^2 \, dx = 2\pi$$

$$\|\sin kx\|^2 = \int_{-\pi}^{\pi} \sin^2(kx) \, dx = \pi \qquad \text{for any } k = 1, 2, 3, \dots$$

$$\|\cos kx\|^2 = \int_{-\pi}^{\pi} \cos^2(kx) \, dx = \pi \qquad \text{for any } k = 1, 2, 3, \dots$$

We leave the verifications to the reader, together with the task of showing that these functions are orthogonal:

$$\langle \sin(kx), \sin(mx) \rangle = 0 = \langle \cos(kx), \cos(mx) \rangle \qquad \text{if } k \neq m$$

and

$$\langle \sin(kx), \cos(mx) \rangle = 0 \qquad \text{for all } k \geq 0 \text{ and } m \geq 0$$

(Note that $1 = \cos(0x)$, so the function 1 is included.)

Now define the following subspace of $\mathbf{C}[-\pi, \pi]$:

$$T_n = \text{span}\{1, \sin x, \cos x, \sin 2x, \cos 2x, \dots, \sin nx, \cos nx\}$$

The aim is to use the approximation theorem (Theorem 8§8.2); so, given a function f in $\mathbf{C}[-\pi, \pi]$, define the **Fourier coefficients** of f by

$$a_0 = \frac{\langle f(x), 1 \rangle}{\|1\|^2} = \frac{1}{2\pi} \int_{-\pi}^{\pi} f(x) \, dx$$

$$a_k = \frac{\langle f(x), \cos(kx) \rangle}{\|\cos(kx)\|^2} = \frac{1}{\pi} \int_{-\pi}^{\pi} f(x) \cos(kx) \, dx \qquad k = 1, 2, \dots$$

$$b_k = \frac{\langle f(x), \sin(kx) \rangle}{\|\sin(kx)\|^2} = \frac{1}{\pi} \int_{-\pi}^{\pi} f(x) \sin(kx) \, dx \qquad k = 1, 2, \dots$$

Then the approximation theorem gives Theorem 1.

THEOREM 1

Let f be any continuous real-valued function defined on the interval $[-\pi, \pi]$. If $a_0, a_1, \dots,$ and $b_0, b_1, \dots$ are the Fourier coefficients of f, then

$$t_{nf}(x) = a_0 + a_1 \cos x + b_1 \sin x + a_2 \cos 2x + b_2 \sin 2x + \cdots$$
$$+ a_n \cos nx + b_n \sin nx$$

is a function in T_n that is closest to f in the sense that

$$\|f - t_{nf}\| \leq \|f - t\|$$

holds for all functions t in T_n.

The function t_{nf} is called the nth **Fourier approximation** to the function f.

EXAMPLE 1

Find the fifth Fourier approximation to the function $f(x)$ defined on $[-\pi, \pi]$ as follows:

$$f(x) = \begin{cases} \pi + x & -\pi \le x < 0 \\ \pi - x & 0 \le x \le \pi \end{cases}$$

Solution   The graph of $y = f(x)$ appears in the diagram. The Fourier coefficients are computed as follows. The details of the integrations (usually by parts) are omitted.

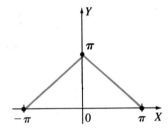

$$a_0 = \frac{1}{2\pi} \int_{-\pi}^{\pi} f(x) \, dx = \frac{\pi}{2}$$

$$a_k = \frac{1}{\pi} \int_{-\pi}^{\pi} f(x) \, \cos(kx) \, dx = \frac{2}{\pi k^2} [1 - \cos(k\pi)] = \begin{cases} 0 & \text{if } k \text{ is even} \\ \dfrac{4}{\pi k^2} & \text{if } k \text{ is odd} \end{cases}$$

$$b_k = \frac{1}{\pi} \int_{-\pi}^{\pi} f(x) \, \sin(kx) \, dx = 0 \qquad \text{for all } k = 1, 2, \ldots$$

Hence the fifth Fourier approximation is

$$t_{5f}(x) = \frac{\pi}{2} + \frac{4}{\pi} \left\{ \cos x + \frac{1}{3^2} \cos(3x) + \frac{1}{5^2} \cos(5x) \right\}$$

◆◆◆

EXAMPLE 2

Find the fourth Fourier approximation for the function $f(x) = x$ on the interval $[-\pi, \pi]$.

Solution   We have $a_k = 0$ for all $k \ge 0$ in this case, whereas

$$b_k = \frac{1}{\pi} \int_{-\pi}^{\pi} x \, \sin(kx) \, dx = \frac{2}{k\pi} [-\pi \cos (k\pi)]$$

$$= \frac{2}{k} (-1)^{k+1} \quad \text{for all } k = 1, 2, 3, \ldots$$

Again, we omit the details of the integration by parts. Hence the Fourier approximation is

$$t_{4f}(x) = 2 \left\{ \sin x - \frac{\sin(2x)}{2} + \frac{\sin(3x)}{3} - \frac{\sin(4x)}{4} \right\}$$

◆◆◆

We say that a function f is an **even function** if $f(x) = f(-x)$ holds for all x; f is called an **odd function** if $f(-x) = -f(x)$ holds for all x. Examples of even functions are constant functions, the even powers x^2, x^4, ..., and $\cos(kx)$; and these functions are characterized by the fact that the graph of $y = f(x)$ is symmetric about the Y axis. Examples of odd functions are the odd powers x, x^3, ... and $\sin(kx)$ where $k > 0$, and the graph of $y = f(x)$ is symmetric about the origin if f is odd. The usefulness of these functions stems from the fact that

$$\int_{-\pi}^{\pi} f(x)\, dx = 0 \qquad\qquad \text{if } f \text{ is odd}$$

$$\int_{-\pi}^{\pi} f(x)\, dx = 2\int_{0}^{\pi} f(x)\, dx \qquad \text{if } f \text{ is even}$$

These facts often simplify the computations of the Fourier coefficients. For example:

1. The Fourier sine coefficients b_k all vanish if f is even.
2. The Fourier cosine coefficients a_k all vanish if f is odd.

This is because $f(x)\sin(kx)$ is odd in the first case and $f(x)\cos(kx)$ is odd in the second case. These observations are illustrated in the two foregoing examples.

    The functions 1, $\cos(kx)$, and $\sin(kx)$ that occur in the Fourier approximation for $f(x)$ are all easy to generate as an electrical voltage (when x is time). By summing these signals (with the amplitudes given by the Fourier coefficients), it is possible to produce an electrical signal with (the approximation to) $f(x)$ as the voltage. Hence these Fourier approximations play a fundamental role in electronics.

    Finally, the Fourier approximations t_{1f}, t_{2f}, ... of a function f get better and better as n increases. The reason is that the subspaces T_n increase:

$$T_1 \subseteq T_2 \subseteq T_3 \subseteq \cdots \subseteq T_n \subseteq \cdots$$

So, because $t_{nf} = \text{proj}_{T_n}(f)$, we get (see the discussion following Example 6§8.2)

$$\|f - t_{1f}\| \geq \|f - t_{2f}\| \geq \cdots \geq \|f - t_{nf}\| \geq \cdots$$

This draws our attention to the infinite series

$$a_0 + a_1 \cos x + b_1 \sin x + a_2 \cos 2x + b_2 \sin 2x + \cdots \qquad (*)$$

where the a_k and b_k are the Fourier coefficients of f. This is called the **Fourier series** for $f(x)$, and the question arises immediately whether such an infinite sum makes any sense at all. This leads to the notion of *convergence*, which will not be dealt with here. However, whether the series $(*)$ makes sense or not, two observations about it can be made: First, the sum of the first $2n + 1$ terms is just t_{nf}, so these partial sums get closer and closer to f as n increases. Second, if f happens to lie in T_n for some n, then $a_k = b_k = 0$ for all $k > n$, so the sum is actually finite and it adds to f.

    It turns out that $(*)$ converges to f for every function f in $\mathbf{C}[-\pi, \pi]$ such that $f(-\pi) = f(\pi)$. The proof of this is given in books on Fourier analysis. [For example, R. V. Churchill and J. W. Brown, *Fourier Series and Boundary Value Problems;* 4th Ed. (New York: McGraw-Hill, 1987).] This subject not only had great historical impact on the development of mathematics but has become one of the standard tools in science and engineering.

EXERCISES 8.5

1. In each case, find the Fourier approximation t_{5f} of the given function in $C[-\pi, \pi]$.

   (a) $f(x) = \pi - x$

   ◆**(b)** $f(x) = |x| = \begin{cases} x & \text{if } 0 \le x \le \pi \\ -x & \text{if } -\pi \le x < 0 \end{cases}$

   (c) $f(x) = x^2$

   ◆**(d)** $f(x) \begin{cases} 0 & \text{if } -\pi \le x < 0 \\ x & \text{if } 0 \le x \le \pi \end{cases}$

2. **(a)** Find t_{5f} for the even function on $[-\pi, \pi]$ satisfying $f(x) = x$ for $0 \le x \le \pi$.

   ◆**(b)** Find t_{6f} for the even function on $[-\pi, \pi]$ satisfying $f(x) = \sin x$ for $0 \le x \le \pi$.

[*Hint:* If $k > 1$, $\int \sin x \cos(kx)\, dx = \frac{1}{2}\left[\dfrac{\cos[(k-1)x]}{k-1} - \dfrac{\cos[(k+1)x]}{k+1}\right].$]

3. **(a)** Prove that $\int_{-\pi}^{\pi} f(x)\, dx = 0$ if f is odd and that $\int_{-\pi}^{\pi} f(x)\, dx = 2\int_{0}^{\pi} f(x)\, dx$ if f is even.

   (b) Prove that $\frac{1}{2}[f(x) + f(-x)]$ is even and that $\frac{1}{2}[f(x) - f(-x)]$ is odd for any function f. Note that they sum to $f(x)$.

◆**4.** Show that $\{1, \cos x, \cos(2x), \cos(3x), \ldots\}$ is an orthogonal set in $C[0, \pi]$ with respect to the inner product $\langle f, g \rangle = \int_{0}^{\pi} f(x)g(x)\, dx$.

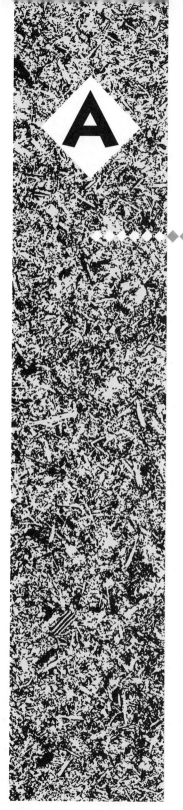

Complex Numbers

The fact that the square of every real number is positive shows that the equation $x^2 + 1 = 0$ has no real root; in other words, there is no real number u such that $u^2 = -1$. So the set of real numbers is inadequate for finding all roots of all polynomials. This kind of problem arises with other number systems as well. The set of integers contains no solution of the equation $3x + 2 = 0$, and the rational numbers had to be invented to solve such equations. But the set of rational numbers is also incomplete because, for example, it contains no root of the polynomial $x^2 - 2$. Hence the real numbers were invented. In the same way, the set of complex numbers was invented, which contains all real numbers together with a root of the equation $x^2 + 1 = 0$. However, the process ends here: the complex numbers have the property that every polynomial with complex coefficients has a (complex) root. This fact is known as the fundamental theorem of algebra.

One pleasant aspect of the complex numbers is that, whereas describing the real numbers in terms of the rationals is a rather complicated business, the complex numbers are quite easy to describe in terms of real numbers. Every **complex number** has the form

$$a + bi$$

where a and b are real numbers, and i is a root of the polynomial $x^2 + 1$. Here a and b are called the **real part** and the **imaginary part** of the complex number, respectively. The real numbers are now regarded as special complex numbers of the form $a + 0i = a$, with zero imaginary part. The complex numbers of the form $0 + bi = bi$ with zero real part are called **pure imaginary** numbers. The complex number i itself is called the **imaginary unit** and is distinguished by the fact that

$$i^2 = -1$$

As the terms *complex* and *imaginary* suggest, these numbers met with some resistance when they were first used. This has changed; now they are essential in science and engineering as well as mathematics, and they are used extensively. The names persist, however, and continue to be a bit misleading: These numbers are no more complex than the real numbers, and the number i is no more imaginary than -1.

Much as for polynomials, two complex numbers are declared to be **equal** if and only if they have the same real parts and the same imaginary parts. In symbols,

$$a + bi = a' + b'i \qquad \text{if and only if} \qquad a = a' \text{ and } b = b'$$

The addition and subtraction of complex numbers is accomplished by adding and subtracting real and imaginary parts:

$$(a + bi) + (a' + b'i) = (a + a') + (b + b')i$$
$$(a + bi) - (a' + b'i) = (a - a') + (b - b')i$$

This is analogous to these operations for linear polynomials $a + bx$ and $a' + b'x$, and the multiplication of complex numbers is also analogous with one difference: $i^2 = -1$. The definition is

$$(a + bi)(a' + b'i) = (aa' - bb') + (ab' + ba')i$$

With these definitions of equality, addition, and multiplication, the complex numbers satisfy all the basic arithmetical axioms adhered to by the real numbers (the verifications are omitted). One consequence of this is that they can be manipulated in the obvious fashion, except that i^2 is replaced by -1 wherever it occurs, and the rule for equality must be observed.

EXAMPLE 1

If $z = 2 - 3i$ and $w = -1 + i$, write each of the following in the form $a + bi$: $z + w$, $z - w$, zw, $\frac{1}{3}z$, and z^2.

Solution

$$z + w = (2 - 3i) + (-1 + i) = (2 - 1) + (-3 + 1)i = 1 - 2i$$
$$z - w = (2 - 3i) - (-1 + i) = (2 + 1) + (-3 - 1)i = 3 - 4i$$
$$zw = (2 - 3i)(-1 + i) = (-2 - 3i^2) + (2 + 3)i = 1 + 5i$$
$$\tfrac{1}{3}z = \tfrac{1}{3}(2 - 3i) = \tfrac{2}{3} - i$$
$$z^2 = (2 - 3i)(2 - 3i) = (4 + 9i^2) + (-6 - 6)i = -5 - 12i \qquad \blacklozenge\blacklozenge\blacklozenge$$

EXAMPLE 2

Find all complex numbers z such as that $z^2 = i$.

Solution

Write $z = a + bi$; we must determine a and b. Now $z^2 = (a^2 - b^2) + (2ab)i$, so the condition $z^2 = i$ becomes

$$(a^2 - b^2) + (2ab)i = 0 + i$$

Equating real and imaginary parts, we find that $a^2 = b^2$ and $2ab = 1$. The solution is $a = b = \pm\frac{1}{\sqrt{2}}$, so the complex numbers required are $z = \frac{1}{\sqrt{2}} + \frac{1}{\sqrt{2}}i$ and $z = -\frac{1}{\sqrt{2}} - \frac{1}{\sqrt{2}}i$.

$\blacklozenge\blacklozenge\blacklozenge$

As for real numbers, it is possible to divide by every nonzero complex number z. That is, there exists a complex number w such that $wz = 1$. As in the real case, this number w is called the **inverse** of z and is denoted by z^{-1} or $\frac{1}{z}$. Moreover, if $z = a + bi$, the fact that $z \neq 0$ means that $a \neq 0$ or $b \neq 0$. Hence $a^2 + b^2 \neq 0$, and an explicit formula for the inverse is

$$\frac{1}{z} = \frac{a}{a^2 + b^2} - \frac{b}{a^2 + b^2} \cdot i$$

In actual calculation, the work is facilitated by two useful notions: the conjugate and the absolute value of a complex number. The next example illustrates the technique.

EXAMPLE 3

Write $\dfrac{3 + 2i}{2 + 5i}$ in the form $a + bi$.

Solution

Multiply top and bottom by the complex number $2 - 5i$ (obtained from the denominator by negating the imaginary part). The result is

$$\frac{3 + 2i}{2 + 5i} = \frac{(2 - 5i)(3 + 2i)}{(2 - 5i)(2 + 5i)} = \frac{(6 + 10) + (4 - 15)i}{(4 + 25) + 0i} = \frac{16}{29} - \frac{11}{29}i$$

Hence the simplified form is $\frac{16}{29} - \frac{11}{29}i$, as required.

◆◆◆

The key to this technique is that the product $(2 - 5i)(2 + 5i) = 29$ in the denominator turned out to be a *real* number. The situation in general leads to the following notation: If $z = a + bi$ is a complex number, the **conjugate** of z is the complex number, denoted $\bar{z}$, given by

$$\bar{z} = a - bi \qquad \text{where } z = a + bi$$

Hence $\bar{z}$ is obtained from z by negating the imaginary part. For example, $\overline{(2 + 3i)} = 2 - 3i$ and $\overline{(1 - i)} = 1 + i$. If we multiply z by $\bar{z}$, we obtain

$$z\bar{z} = a^2 + b^2 \qquad \text{where } z = a + bi$$

The real number $a^2 + b^2$ is always nonnegative, so we can state the following definition: The **absolute value** or **modulus** of a complex number $z = a + bi$, denoted by $|z|$, is the positive square root $\sqrt{a^2 + b^2}$; that is,

$$|z| = \sqrt{a^2 + b^2} \qquad \text{where } z = a + bi$$

For example,

$$|2 - 3i| = \sqrt{2^2 + (-3)^2} = \sqrt{13} \qquad \text{and} \qquad |1 + i| = \sqrt{1^2 + 1^2} = \sqrt{2}$$

Note that if a real number a is viewed as the complex number $a + 0i$, its absolute value (as a complex number) is $|a| = \sqrt{a^2}$, which agrees with its absolute value as a *real* number.

With these notions in hand, we can describe the technique applied in Example 3 as follows: When converting a quotient z/w of complex numbers to the form $a + bi$, multiply top and bottom by the conjugate $\bar{w}$ of the denominator.

The following list contains the most important properties of conjugates and absolute values. Throughout, z and w denote complex numbers.

C1. $\overline{z \pm w} = \bar{z} \pm \bar{w}$

C2. $\overline{zw} = \bar{z}\,\bar{w}$

C3. $\overline{(z/w)} = \bar{z}/\bar{w}$

C4. $\overline{(\bar{z})} = z$

C5. z is real if and only if $\bar{z} = z$

C6. $z\bar{z} = |z|^2$

C7. $\dfrac{1}{z} = \dfrac{\bar{z}}{|z|^2}$

C8. $|z| \geq 0$ for all complex numbers z

C9. $|z| = 0$ if and only if $z = 0$

C10. $|zw| = |z|\,|w|$

C11. $\left|\dfrac{z}{w}\right| = \dfrac{|z|}{|w|}$

C12. $|z + w| \leq |z| + |w|$ **(triangle inequality)**

All these properties (except property C12) can (and should) be verified by the reader for arbitrary complex numbers $z = a + bi$ and $w = c + di$. They are not independent; for example, property C10 follows from properties C2 and C6.

The triangle inequality, as its name suggests, comes from a geometric representation of the complex numbers analogous to identification of the real numbers with the points of a line. The representation is achieved as follows: Introduce a rectangular coordinate system in the plane (Figure A.1), and identify the complex number $a + bi$ with the point (a, b). When this is done, the plane is called the **complex plane.** Note that the point $(a, 0)$ on the X axis now represents the *real* number $a = a + 0i$, and for this reason, the X axis is called the **real axis.** Similarly, the Y axis is called the **imaginary axis.** The identification $(a, b) = a + bi$ of the geometric point (a, b) and the complex number $a + bi$ will be used later without comment. For example, the origin will be referred to as 0.

This representation of the complex numbers in the complex plane gives a useful way of describing the absolute value and conjugate of a complex number $z = a + bi$. The absolute value $|z| = \sqrt{a^2 + b^2}$ is just the distance from z to the origin. This makes properties C8 and C9 quite obvious. The conjugate $\bar{z} = a - bi$ of z is just the reflection of z in the real axis (X axis), a fact that makes properties C4 and C5 clear.

Given two complex numbers $z_1 = a_1 + b_1 i = (a_1, b_1)$ and $z_2 = a_2 + b_2 i = (a_2, b_2)$, the absolute value of their difference

$$|z_1 - z_2| = \sqrt{(a_1 - a_2)^2 + (b_1 - b_2)^2}$$

is just the distance between them. This gives the **complex distance formula:**

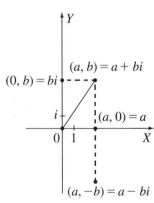

FIGURE A.1

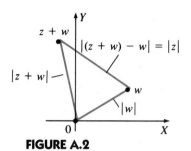

FIGURE A.2

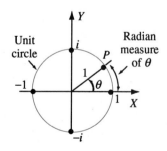

FIGURE A.3

FIGURE A.4

$|z_1 - z_2|$ is the distance between z_1 and z_2

This useful fact yields a simple verification of the triangle inequality, property C12. Suppose z and w are given complex numbers. Consider the triangle in Figure A.2 whose vertices are 0, w, and $z + w$. The three sides have lengths $|z|$, $|w|$, and $|z + w|$ by the complex distance formula, so the inequality

$$|z + w| \le |z| + |w|$$

expresses the obvious geometric fact that the sum of the lengths of two sides of a triangle is at least as great as the length of the third side.

The representation of complex numbers as points in the complex plane has another very useful property: It enables us to give a geometric description of the sum and product of two complex numbers. To obtain the description for the sum, let

$$z = a + bi = (a,\ b)$$
$$w = c + di = (c,\ d)$$

denote two complex numbers. We claim that the four points 0, z, w, and $z + w$ form the vertices of a parallelogram. In fact, in Figure A.3 the lines from 0 to z and from w to $z + w$ have slopes

$$\frac{b - 0}{a - 0} = \frac{b}{a} \quad \text{and} \quad \frac{(b + d) - d}{(a + c) - c} = \frac{b}{a}$$

respectively, so these lines are parallel. (If it happens that $a = 0$, then both these lines are vertical.) Similarly, the lines from z to $z + w$ and from 0 to w are also parallel, so the figure with vertices 0, z, w, and $z + w$ is indeed a parallelogram. Hence, the complex number $z + w$ can be obtained geometrically from z and w by *completing* the parallelogram. This is sometimes called the **parallelogram law** of complex addition. Readers who have studied mechanics will recall that velocities and accelerations add in the same way; in fact, these are all special cases of *vector* addition.

The geometric description of what happens when two complex numbers are multiplied is at least as elegant as the parallelogram law of addition, but it requires that the complex numbers be represented in polar form. Before discussing this, we pause to recall the general definition of the trigonometric functions sine and cosine. An angle θ in the complex plane is in **standard position** if it is measured counterclockwise from the real axis as indicated in Figure A.4. Rather than using degrees to measure angles, it is more natural to use radian measure. This is defined as follows: The circle with its center at the origin and radius 1 (called the **unit circle**) is drawn in Figure A.4. It has circumference 2π, and the **radian measure** of θ is the length of the arc on the unit circle from 1 to the point P on the unit circle determined by θ. Hence $90° = \frac{\pi}{2}$, $45° = \frac{\pi}{4}$, $180° = \pi$, and a full circle has the angle $360° = 2\pi$.

If an acute angle θ (that is, $0 \le \theta \le \frac{\pi}{2}$) is plotted in standard position as in Figure A.4, it determines a unique point P on the unit circle, and P has coordinates $(\cos \theta, \sin \theta)$ by elementary trigonometry. However, *any* angle θ (acute or not) determines a unique point on the unit circle, so we *define* the **cosine** and **sine** of θ (written $\cos \theta$ and $\sin \theta$) to be the X and Y coordinates of this point. For example, the points

$$1 = (1, 0) \qquad i = (0, 1) \qquad -1 = (-1, 0) \qquad -i = (0, -1)$$

plotted in Figure A.4 are determined by the angles $0, \frac{\pi}{2}, \pi, \frac{3\pi}{2}$, respectively. Hence

$$\cos 0 = 1 \qquad \cos \tfrac{\pi}{2} = 0 \qquad \cos \pi = -1 \qquad \cos \tfrac{3\pi}{2} = 0$$
$$\sin 0 = 0 \qquad \sin \tfrac{\pi}{2} = 1 \qquad \sin \pi = 0 \qquad \sin \tfrac{3\pi}{2} = -1$$

Now we can describe the polar form of a complex number. Let $z = a + bi$ be a complex number, and write the absolute value of z as

$$r = |z| = \sqrt{a^2 + b^2}$$

If $z \neq 0$, the angle θ shown in Figure A.5 is called an **argument** of z and is denoted

$$\theta = \arg z$$

This angle is not unique ($\theta + 2\pi k$ would do as well for any $k = 0, \pm 1, \pm 2, \ldots$). However, there is only one argument θ in the range $-\pi < \theta \leq \pi$, and this is sometimes called the **principal argument** of z.

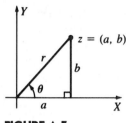

FIGURE A.5

Returning to Figure A.5, we find that the real and imaginary parts a and b of z are related to r and θ by

$$a = r \cos \theta$$
$$b = r \sin \theta$$

Hence the complex number $z = a + bi$ has the form

$$z = r(\cos \theta + i \sin \theta) \qquad r = |z|, \, \theta = \arg(z)$$

The combination $\cos \theta + i \sin \theta$ is so important that a special notation is used. This is

$$e^{i\theta} = \cos \theta + i \sin \theta$$

With this notation, z is written

$$z = r e^{i\theta} \qquad r = |z|, \, \theta = \arg(z)$$

This is a **polar form** of the complex number z. Of course it is not unique, because the argument can be changed by adding a multiple of 2π.

EXAMPLE 4

Write $z_1 = -2 + 2i$ and $z_2 = -i$ in polar form.

Solution The two numbers are plotted in the complex plane in Figure A.6. The absolute values are

$$r_1 = |-2 + 2i| = \sqrt{(-2)^2 + 2^2} = 2\sqrt{2}$$
$$r_2 = |-i| = \sqrt{0^2 + (-1)^2} = 1$$

By inspection of Figure A.6, arguments of z_1 and z_2 are

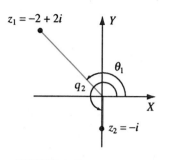

$z_1 = -2 + 2i$

$z_2 = -i$

FIGURE A.6

$$\theta_1 = \arg(-2 + 2i) = \tfrac{3\pi}{4}$$

$$\theta_2 = \arg(-i) = \tfrac{3\pi}{2}$$

The corresponding polar forms are $z_1 = -2 + 2i = 2\sqrt{2}e^{3\pi i/4}$ and $z_2 = -i = e^{3\pi i/2}$. Of course, we could have taken the argument $-\tfrac{\pi}{2}$ for z_2 and obtained the polar form $z_2 = e^{-\pi i/2}$.

◆◆◆

In the notation $e^{i\theta} = \cos\theta + i\sin\theta$, the number e is the familiar constant $e = 2.71828\ldots$ from calculus. The reason for using e will not be given here; the reason why $\cos\theta + i\sin\theta$ is written as an *exponential* function of θ is that the **law of exponents** holds:

$$e^{i\theta} \cdot e^{i\phi} = e^{i(\theta+\phi)}$$

where θ and ϕ are any two angles. In fact, this is an immediate consequence of the addition identities for $\sin(\theta + \phi)$ and $\cos(\theta + \phi)$:

$$
\begin{aligned}
e^{i\theta}e^{i\phi} &= (\cos\theta + i\sin\theta)(\cos\phi + i\sin\phi) \\
&= (\cos\theta\cos\phi - \sin\theta\sin\phi) + i(\cos\theta\sin\phi + \sin\theta\cos\phi) \\
&= \cos(\theta + \phi) + i\sin(\theta + \phi) \\
&= e^{i(\theta+\phi)}
\end{aligned}
$$

This is analogous to the rule $e^a e^b = e^{a+b}$, which holds for real numbers a and b, so it is not unnatural to use the exponential notation $e^{i\theta}$ for the expression $\cos\theta + i\sin\theta$. In fact, a whole theory exists wherein functions such as e^z, $\sin z$, and $\cos z$ are studied, where z is a *complex* variable. Many deep and beautiful theorems can be proved in this theory, one of which is the so-called fundamental theorem of algebra mentioned later (Theorem 5). We shall not pursue this here.

The geometric description of the multiplication of two complex numbers follows from the law of exponents.

THEOREM 1

Multiplication Rule

If $z_1 = r_1 e^{i\theta_1}$ and $z_2 = r_2 e^{i\theta_2}$ are complex numbers in polar form, then

$$z_1 z_2 = r_1 r_2 e^{i(\theta_1+\theta_2)}$$

In other words, to multiply two complex numbers, simply multiply the absolute values and add the arguments. This simplifies calculations considerably, particularly when we observe that it is valid for *any* arguments θ_1 and θ_2.

EXAMPLE 5

Multiply $(1 - i)(1 + \sqrt{3}i)$.

Solution   We have $|1 - i| = \sqrt{2}$ and $|1 + \sqrt{3}i| = 2$ and, from Figure A.7,

$$1 - i = \sqrt{2}e^{-i\pi/4}$$
$$1 + \sqrt{3}i = 2e^{i\pi/3}$$

Hence, by the multiplication rule,

$$(1 - i)(1 + \sqrt{3}i) = (\sqrt{2}e^{-i\pi/4})(2e^{i\pi/3})$$
$$= 2\sqrt{2}e^{i(-\pi/4+\pi/3)}$$
$$= 2\sqrt{2}e^{i\pi/12}$$

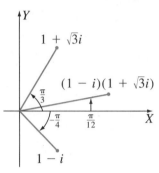

FIGURE A.7

This gives the required product in polar form. Of course, direct multiplication gives $(1 - i)(1 + \sqrt{3}i) = (\sqrt{3} + 1) + (\sqrt{3} - 1)i$. Hence, equating real and imaginary parts gives $\cos(\pi/12) = (\sqrt{3} + 1)/2\sqrt{2}$ and $\sin(\pi/12) = (\sqrt{3} - 1)/2\sqrt{2}$.

◆◆◆

If a complex number $z = re^{i\theta}$ is given in polar form, the powers assume a particularly simple form. In fact, $z^2 = (re^{i\theta})(re^{i\theta}) = r^2e^{2i\theta}$, $z^3 = z^2 \cdot z = (r^2e^{2i\theta})(re^{i\theta}) = r^3e^{3i\theta}$, and so on. Continuing in this way, it follows by induction that the following theorem holds for any positive integer n. The name honors Abraham De Moivre (1667–1754).

THEOREM 2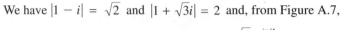

De Moivre's Theorem

If θ is any angle, then $(e^{i\theta})^n = e^{in\theta}$ holds for all integers n.

Proof   The case $n > 0$ has been discussed, and the reader can verify the result for $n = 0$. To derive it for $n < 0$, first observe that

$$\text{if} \qquad z = re^{i\theta} \neq 0 \qquad \text{then} \qquad z^{-1} = \tfrac{1}{r}\,e^{-i\theta}$$

In fact, $(re^{i\theta})(\tfrac{1}{r}e^{-i\theta}) = 1e^{i0} = 1$ by the multiplication rule. Now assume that n is negative and write it as $n = -m$, $m > 0$. Then

$$(re^{i\theta})^n = [(re^{i\theta})^{-1}]^m = \left(\tfrac{1}{r}\,e^{-i\theta}\right)^m = r^{-m}e^{i(-m\theta)} = r^ne^{in\theta}$$

If $r = 1$, this is De Moivre's theorem for negative n.   ◆

EXAMPLE 6

Verify that $(-1 + \sqrt{3}i)^3 = 8$.

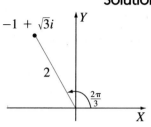

$-1 + \sqrt{3}i$

2

$\frac{2\pi}{3}$

FIGURE A.8

Solution   We have $|-1 + \sqrt{3}i| = 2$, so $-1 + \sqrt{3}i = 2e^{2\pi i/3}$ (see Figure A.8). Hence De Moivre's theorem gives

$$(-1 + \sqrt{3}i)^3 = (2e^{2\pi i/3})^3 = 8e^{3(2\pi/3)i} = 8e^{2\pi i} = 8.$$

◆◆◆

De Moivre's theorem can be used to find nth roots of complex numbers where n is positive. The next example illustrates this technique.

EXAMPLE 7

Find the cube roots of unity; that is, find all complex numbers z such that $z^3 = 1$.

Solution   First write $z = re^{i\theta}$ and $1 = 1e^{i \cdot 0}$ in polar form. We must use the condition $z^3 = 1$ to determine r and θ. Because $z^3 = r^3 e^{3i\theta}$ by De Moivre's theorem, this requirement becomes

$$r^3 \theta^{3\theta i} = 1e^{0 \cdot i}$$

These two complex numbers are equal, so their absolute values must be equal and the arguments must either be equal or differ by an integral multiple of 2π:

$$r^3 = 1$$
$$3\theta = 0 + 2k\pi, \quad k \text{ some integer}$$

Because r is real and positive, the condition $r^3 = 1$ implies that $r = 1$. However,

$$\theta = \tfrac{2k\pi}{3}, \quad k \text{ some integer}$$

seems at first glance to yield infinitely many different angles for z. However, choosing $k = 0, 1, 2$ gives three possible arguments θ (where $0 \le \theta < 2\pi$), and the corresponding roots are

$$1e^{0i} = 1$$
$$1e^{2\pi i/3} = -\tfrac{1}{2} + \tfrac{\sqrt{3}}{2} i$$
$$1e^{4\pi i/3} = -\tfrac{1}{2} - \tfrac{\sqrt{3}}{2} i$$

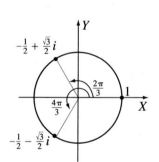

$-\tfrac{1}{2} + \tfrac{\sqrt{3}}{2}i$

$\frac{2\pi}{3}$

$\frac{4\pi}{3}$

1

$-\tfrac{1}{2} - \tfrac{\sqrt{3}}{2}i$

FIGURE A.9

These are displayed in Figure A.9. All other values of k yield values of θ that differ from one of these by a multiple of 2π—and so do not give new roots. Hence we have found all the roots.

◆◆◆

The same type of calculation gives all complex **nth roots of unity**; that is, all complex numbers z such that $z^n = 1$. As before, write $1 = 1e^{0 \cdot i}$ and

$$z = re^{i\theta}$$

in polar form. Then $z^n = 1$ takes the form

$$r^n e^{n\theta i} = 1e^{0i}$$

using De Moivre's theorem. Comparing absolute values and arguments yields

$$r^n = 1$$
$$n\theta = 0 + 2k\pi, \quad k \text{ some integer}$$

Hence $r = 1$, and the n values

$$\theta = \tfrac{2k\pi}{n}, \quad k = 0, 1, 2, \ldots, n-1$$

of θ all lie in the range $0 \le \theta < 2\pi$. As before, *every* choice of k yields a value of θ that differs from one of these by a multiple of 2π, so these give the arguments of *all* the possible roots.

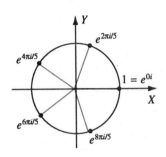

FIGURE A.10

THEOREM 3

nth Roots of Unity

If $n \ge 1$ is an integer, the nth roots of unity (that is, the solutions to $z^n = 1$) are given by

$$z = e^{2\pi ki/n}, \quad k = 0, 1, 2, \ldots, n-1.$$

The nth roots of unity can be found geometrically as the points on the unit circle that cut the circle into n equal sectors, starting at 1. The case $n = 5$ is shown in Figure A.10, where the five fifth roots of unity are plotted.

The method just used to find the nth roots of unity works equally well to find the nth roots of any complex number in polar form. We give one example.

EXAMPLE 8

Find the fourth roots of $\sqrt{2} + \sqrt{2}i$.

Solution   First write $\sqrt{2} + \sqrt{2}i = 2e^{\pi i/4}$ in polar form. If $z = re^{i\theta}$ satisfies $z^4 = \sqrt{2} + \sqrt{2}i$, then

$$r^4 e^{i(4\theta)} = 2e^{\pi i/4}$$

Hence $r^4 = 2$ and $4\theta = \tfrac{\pi}{4} + 2k\pi$, k an integer. We obtain four distinct roots (and hence all) by

$$r = \sqrt[4]{2}, \quad \theta = \tfrac{\pi}{16} + \tfrac{8k\pi}{16}, k = 0, 1, 2, 3$$

Thus the four roots are

$$\sqrt[4]{2}e^{\pi i/16} \qquad \sqrt[4]{2}e^{9\pi i/16} \qquad \sqrt[4]{2}e^{17\pi i/16} \qquad \sqrt[4]{2}e^{25\pi i/16}$$

Of course, reducing these roots to the form $a + bi$ would require the computation of $\sqrt[4]{2}$ and the sine and cosine of the various angles.

$$\blacklozenge\blacklozenge\blacklozenge$$

An expression of the form $ax^2 + bx + c$, where the coefficients $a \neq 0$, b, and c are real numbers, is called a **real quadratic.** A complex number u is called a **root** of the quadratic if $au^2 + bu + c = 0$. The roots are given by the famous **quadratic formula:**

$$u = \frac{-b \pm \sqrt{b^2 - 4ac}}{2a}$$

The quantity $d = b^2 - 4ac$ is called the **discriminant** of the quadratic $ax^2 + bx + c$, and there is no real root if and only if $d < 0$. In this case the quadratic is said to be **irreducible.** Moreover, the fact that $d < 0$ means that $\sqrt{d} = i\sqrt{|d|}$, so the two (complex) roots are conjugates of each other:

$$u = \tfrac{1}{2a}(-b + i\sqrt{|d|}) \qquad \text{and} \qquad \bar{u} = \tfrac{1}{2a}(-b - i\sqrt{|d|})$$

The converse of this is true too: Given any nonreal complex number u, then u and $\bar{u}$ are the roots of some real irreducible quadratic. Indeed, the quadratic

$$x^2 - (u + \bar{u})x + u\bar{u} = (x - u)(x - \bar{u})$$

has real coefficients ($u\bar{u} = |u|^2$ and $u + \bar{u}$ is twice the real part of u) and so is irreducible because its roots u and $\bar{u}$ are not real.

EXAMPLE 9

Find a real irreducible quadratic with $u = 3 - 4i$ as a root.

Solution

We have $u + \bar{u} = 6$ and $|u|^2 = 25$, so $x^2 - 6x + 25$ is irreducible with u and $\bar{u} = 3 + 4i$ as roots.

$$\blacklozenge\blacklozenge\blacklozenge$$

The quadratic formula works for quadratics with complex coefficients. If $p \neq 0$ and v and w are complex numbers, the equation

$$px^2 + vx + w = 0.$$

can be solved by an old technique called **completing the square** (it was known to the Moslem mathematician Al-Khowarizmi in the ninth century A.D.). The idea is to write the equation as $x^2 + \frac{v}{p}x = -\frac{w}{p}$ and then complete the square on the left by adding $(\frac{v}{2p})^2$ to each side:

$$\left(x + \frac{v}{2p}\right)^2 = x^2 + \frac{v}{p}x + \left(\frac{v}{2p}\right)^2 = -\frac{w}{p} + \left(\frac{v}{2p}\right)^2 = \frac{v^2 - 4pw}{4p^2}$$

Taking square roots gives the complex version of the quadratic formula

$$x = \frac{-v \pm \sqrt{v^2 - 4pw}}{2p}$$

Of course, the discriminant $v^2 - 4pw$ is now a complex number, and we need the foregoing methods to find its square roots. Here is an example.

EXAMPLE 10

Find all complex numbers z such that $z^2 - iz + (1 + 3i) = 0$.

Solution   The quadratic formula gives

$$z = \tfrac{1}{2}[i \pm \sqrt{i^2 - 4(1+3i)}] = \tfrac{1}{2}[i \pm \sqrt{-5 - 12i}] = \tfrac{1}{2}[i \pm w]$$

where $w = \sqrt{-5 - 12i}$. Hence $w^2 = -5 - 12i$; so if $w = a + bi$, equating real and imaginary parts gives $a^2 - b^2 = -5$ and $2ab = -12$. Hence $b = -6/a$, so $a^2 - 36/a^2 = -5$. This gives a quadratic in a^2: $a^4 + 5a^2 - 36 = 0$, which factors as $(a^2 - 4)(a^2 + 9) = 0$. Thus $a = \pm 2$ and $b = -6/a = \mp 3$, so $w = \pm(2 - 3i)$. Finally,

$$z = \tfrac{1}{2}[i \pm w] = \tfrac{1}{2}[i \pm (2 - 3i)]$$

Hence the roots are $z = 1 - i$ and $-1 + 2i$.

◆◆◆

If one root of a quadratic equation $px^2 + vx + w = 0$ is known, it is easy to find the other root. Because we can divide both sides of the equation by p, we state the result for quadratics with 1 as the coefficient of x^2.

THEOREM 4

If u_1 and u_2 are the roots of the quadratic equation

$$x^2 + vx + w = 0$$

then $u_1 + u_2 = -v$ and $u_1 u_2 = w$.

Proof   Because u_1 and u_2 are roots of $x^2 + vx + w = 0$ the factor theorem asserts that the quadratic factors as

$$x^2 + vx + w = (x - u_1)(x - u_2)$$

The right side is $x^2 - (u_1 + u_2)x + u_1 u_2$, so the result follows because corresponding coefficients must be equal.   ◆

EXAMPLE 11

Show that $u_1 = 1 + i$ is a root of $x^2 + (1 - 2i)x - (3 + i) = 0$ and then find the other root.

Solution

$u_1^2 + (1 - 2i)u_1 - (3 + i) = (2i) + (3 - i) - (3 + i) = 0$, so u_1 is a root. If u_2 is the other root, then $u_1 + u_2 = -(1 - 2i)$ by Theorem 4, so $u_2 = -(1 - 2i) - u_1 = -2 + i$. Of course, this also follows from $u_1u_2 = -(3 + i)$.

◆◆◆

As we mentioned earlier, the complex numbers are the culmination of a long search by mathematicians to find a set of numbers large enough to contain a root of every polynomial. The fact that the complex numbers have this property was first proved by Gauss in 1797 when he was 20 years old. The proof is omitted.

THEOREM 5

Fundamental Theorem of Algebra

Every polynomial of positive degree with complex coefficients has a complex root.

If $f(x)$ is a polynomial with complex coefficients, and if u_1 is a root, then the factor theorem asserts that

$$f(x) = (x - u_1)g(x)$$

where $g(x)$ is a polynomial with complex coefficients and with degree one less than the degree of $f(x)$. Suppose that u_2 is a root of $g(x)$, again by the fundamental theorem. Then $g(x) = (x - u_2)h(x)$, so

$$f(x) = (x - u_1)(x - u_2)h(x)$$

This process continues until the last polynomial to appear is linear. Thus $f(x)$ has been expressed as a product of linear factors. The last of these factors can be written in the form $u(x - u_n)$, where u and u_n are complex (verify this), so the fundamental theorem takes the following form.

THEOREM 6

Every complex polynomial $f(x)$ of degree $n \geq 1$ has the form

$$f(x) = u(x - u_1)(x - u_2) \cdots (x - u_n)$$

where $u, u_1, \ldots, u_n$ are complex numbers and $u \neq 0$. The numbers $u_1, u_2, \ldots, u_n$ are the roots of $f(x)$ (and need not all be distinct), and u is the coefficient of x^n.

This form of the fundamental theorem, when applied to a polynomial $f(x)$ with *real* coefficients, can be used to deduce the following result.

THEOREM 7

Every polynomial $f(x)$ of positive degree with real coefficients can be factored as a product of linear and irreducible quadratic factors.

In fact, suppose $f(x)$ has the form

$$f(x) = a_n x^n + a_{n-1} x^{n-1} + \cdots + a_1 x + a_0$$

where the coefficients a_i are real. If u is a complex root of $f(x)$, then we claim first that $\bar{u}$ is also a root. In fact, we have $f(u) = 0$, so

$$\begin{aligned}
0 = \bar{0} = \overline{f(u)} &= \overline{a_n u^n + a_{n-1} u^{n-1} + \cdots + a_1 u + a_0} \\
&= \overline{a_n u^n} + \overline{a_{n-1} u^{n-1}} + \cdots + \overline{a_1 u} + \overline{a_0} \\
&= \bar{a}_n \bar{u}^n + \bar{a}_{n-1} \bar{u}^{n-1} + \cdots + \bar{a}_1 \bar{u} + \bar{a}_0 \\
&= a_n \bar{u}^n + a_{n-1} \bar{u}^{n-1} + \cdots + a_1 \bar{u} + a_0 \\
&= f(\bar{u})
\end{aligned}$$

where $\bar{a}_i = a_i$ for each i because the coefficients a_i are real. Thus if u is a root of $f(x)$, so is its conjugate $\bar{u}$. Of course some of the roots of $f(x)$ may be real (and so equal their conjugates), but the nonreal roots come in pairs, u and $\bar{u}$. We can thus write $f(x)$ as a product:

$$f(x) = a_n(x - r_1) \cdots (x - r_k)(x - u_1)(x - \bar{u}_1) \cdots (x - u_m)(x - \bar{u}_m) \quad (*)$$

where a_n is the coefficient of x^n in $f(x)$; $r_1, r_2, \ldots, r_k$ are the real roots; and $u_1, \bar{u}_1, u_2, \bar{u}_2, \ldots, u_m, \bar{u}_m$ are the nonreal roots. But the product

$$(x - u_j)(x - \bar{u}_j) = x^2 - (u + \bar{u}_j)x + u_j \bar{u}_j$$

is a real irreducible quadratic for each j (see the discussion preceding Example 9). Hence $(*)$ shows that $f(x)$ is a product of linear and irreducible quadratic factors, each with real coefficients. This is the conclusion in Theorem 7.

EXERCISES A

1. Solve each of the following for the real number x.
   (a) $x - 4i = (2 - i)^2$ ◆(b) $(2 + xi)(3 - 2i) = 12 + 5i$
   (c) $(2 + xi)^2 = 4$ ◆(d) $(2 + xi)(2 - xi) = 5$

   ◆(d) $\dfrac{3 - 2i}{1 - i} - \dfrac{3 - 7i}{2 - 3i}$

   (e) i^{131} ◆(f) $(2 - i)^3$
   (g) $(1 + i)^4$ ◆(h) $(1 - i)^2(2 + i)^2$

2. Convert each of the following to the form $a + bi$.
   (a) $(2 - 3i) - 2(2 - 3i) + 9$
   ◆(b) $(3 - 2i)(1 + i) + |3 + 4i|$
   (c) $\dfrac{1 + i}{2 - 3i} + \dfrac{1 - i}{-2 + 3i}$

3. In each case, find the complex number z.
   (a) $iz - (1 + i)^2 = 3 - i$
   ◆(b) $(i + z) - 3i(2 - z) = iz + 1$
   (c) $z^2 = -i$
   ◆(d) $z^2 = 3 - 4i$

(e) $z(1 + i) = \overline{z} + (3 + 2i)$

◆**(f)** $z(2 - i) = (\overline{z} + 1)(1 + i)$

4. In each case, find the roots of the real quadratic equation.
   (a) $x^2 - 2x + 3 = 0$ ◆**(b)** $x^2 - x + 1 = 0$
   (c) $3x^2 - 4x + 2 = 0$ ◆**(d)** $2x^2 - 5x + 2 = 0$

5. Find all numbers x in each case.
   (a) $x^3 = 8$ ◆**(b)** $x^3 = -8$
   (c) $x^4 = 16$ ◆**(d)** $x^4 = 64$

6. In each case, find a real quadratic with u as a root, and find the other root.
   (a) $u = 1 + i$ ◆**(b)** $u = 2 - 3i$
   (c) $u = -i$ ◆**(d)** $u = 3 - 4i$

7. Find the roots of $x^2 - 2 \cos \theta\, x + 1 = 0$, θ any angle.

◆**8.** Find a real polynomial of degree 4 with $2 - i$ and $3 - 2i$ as roots.

9. Let re z and im z denote, respectively, the real and imaginary parts of z. Show that:
   (a) $\mathrm{im}(iz) = \mathrm{re}\ z$ **(b)** $\mathrm{re}(iz) = -\mathrm{im}\ z$
   (c) $z + \overline{z} = 2\,\mathrm{re}\ z$ **(d)** $z - \overline{z} = 2\,\mathrm{im}\ z$
   (e) $\mathrm{re}(z + w) = \mathrm{re}\ z + \mathrm{re}\ w$, and $\mathrm{re}(tz) = t \cdot \mathrm{re}\ z$ if t is real
   (f) $\mathrm{im}(z + w) = \mathrm{im}\ z + \mathrm{im}\ w$, and $\mathrm{im}(tz) = t \cdot \mathrm{im}\ z$ if t is real

10. In each case, show that u is a root of the quadratic equation, and find the other root.
   (a) $x^2 - 3ix + (-3 + i) = 0$; $u = 1 + i$
   ◆**(b)** $x^2 + ix - (4 - 2i) = 0$; $u = -2$
   (c) $x^2 - (3 - 2i)x + (5 - i) = 0$; $u = 2 - 3i$
   ◆**(d)** $x^2 + 3(1 - i)x - 5i = 0$; $u = -2 + i$

11. Find the roots of each of the following complex quadratic equations.
   (a) $x^2 + 2x + (1 + i) = 0$
   ◆**(b)** $x^2 - x + (1 - i) = 0$
   (c) $x^2 - (2 - i)x + (3 - i) = 0$
   ◆**(d)** $x^2 - 3(1 - i)x - 5i = 0$

12. In each case, describe the graph of the equation (where z denotes a complex number).
   (a) $|z| = 1$ ◆**(b)** $|z - 1| = 2$
   (c) $z = i\overline{z}$ ◆**(d)** $z = -\overline{z}$
   (e) $z = |z|$ ◆**(f)** $\mathrm{im}\ z = m \cdot \mathrm{re}\ z$, m a real number

13. **(a)** Verify $|zw| = |z| \cdot |w|$ directly for $z = a + bi$ and $w = c + di$.
   (b) Deduce **(a)** from properties C2 and C6.

14. Prove that $|w + z|^2 = |w|^2 + |z|^2 + w\overline{z} + \overline{w}z$ for all complex numbers w and z.

15. If zw is a real and $z \neq 0$, show that $w = a\overline{z}$ for some real number a.

16. If $zw = \overline{z}v$ and $z \neq 0$, show that $w = uv$ for some u in $\mathbb{C}$ with $|u| = 1$.

17. Use property C5 to show that $(1 + i)^n + (1 - i)^n$ is real for all n.

18. Express each of the following in polar form (use the principal argument).
   (a) $3 - 3i$ ◆**(b)** $-4i$
   (c) $-\sqrt{3} + i$ ◆**(d)** $-4 + 4\sqrt{3}i$
   (e) $-7i$ ◆**(f)** $-6 + 6i$

19. Express each of the following in the form $a + bi$.
   (a) $3e^{\pi i}$ ◆**(b)** $e^{7\pi i/3}$
   (c) $2e^{3\pi i/4}$ ◆**(d)** $\sqrt{2}e^{-\pi i/4}$
   (e) $e^{5\pi i/4}$ ◆**(f)** $2\sqrt{3}e^{-2\pi i/6}$

20. Express each of the following in the form $a + bi$.
   (a) $(-1 + \sqrt{3}i)^2$ ◆**(b)** $(1 + \sqrt{3}i)^{-4}$
   (c) $(1 + i)^8$ ◆**(d)** $(1 - i)^{10}$
   (e) $(1 - i)^6(\sqrt{3} + i)^3$ ◆**(f)** $(\sqrt{3} - i)^9(2 - 2i)^5$

21. Use De Moivre's theorem to show that:
   (a) $\cos 2\theta = \cos^2\theta - \sin^2\theta$; $\sin 2\theta = 2 \cos \theta \sin \theta$
   (b) $\cos 3\theta = \cos^3\theta - 3 \cos \theta \sin^2\theta$;
         $\sin 3\theta = 3 \cos^2\theta \sin \theta - \sin^3\theta$

22. **(a)** Find the fourth roots of unity.
    (b) Find the sixth roots of unity.

23. Find all complex numbers z such that:
   (a) $z^4 = -1$ ◆**(b)** $z^4 = 2(\sqrt{3}i - 1)$
   (c) $z^3 = -27i$ ◆**(d)** $z^6 = -64$

24. If $z = re^{i\theta}$ in polar form, show that:
   (a) $\overline{z} = re^{-i\theta}$ **(b)** $z^{-1} = \frac{1}{r}e^{-i\theta}$

25. Show that the sum of the nth roots of unity is zero. [*Hint:* $1 - z^n = (1 - z)(1 + z + z^2 + \cdots + z^{n-1})$ for any complex number z.]

26. **(a)** Suppose z_1, z_2, z_3, z_4, and z_5 are equally spaced around the unit circle. Show that $z_1 + z_2 + z_3 + z_4 + z_5 = 0$. [*Hint:* $(1 - z)(1 + z + z^2 + z^3 + z^4) = 1 - z^5$ for any complex number z.]
   ◆**(b)** Repeat **(a)** for any $n \geq 2$ points equally spaced around the unit circle.
    (c) If $|w| = 1$, show that the sum of the roots of $z^n = w$ is zero.

27. If $z = a + bi$, show that $|a| + |b| \leq \sqrt{2} \cdot |z|$. [*Hint:* $(|a| - |b|)^2 \geq 0$]

28. Let $z \neq 0$ be a complex number. If t is real, describe tz geometrically in terms of z if: **(a)** $t > 0$; **(b)** $t < 0$

29. If z and w are nonzero complex numbers, show that $|z + w| = |z| + |w|$ if and only if one is a positive real multiple of the other. [*Hint:* Consider the parallelogram with vertices 0, w, z, and $z + w$. Use the preceding exercise and the fact that if t is real, $|1 + t| = 1 + |t|$ is impossible if $t < 0$.]

30. If a and b are *rational* numbers, let p and q denote numbers of the form $a + b\sqrt{2}$. If $p = a + b\sqrt{2}$, define $\tilde{p} = a - b\sqrt{2}$ and $[p] = a^2 - 2b^2$. Show that each of the following holds.

(a) $a + b\sqrt{2} = a_1 + b_1\sqrt{2}$ only if $a = a_1$ and $b = b_1$

(b) $\widetilde{p \pm q} = \tilde{p} \pm \tilde{q}$

(c) $\widetilde{pq} = \tilde{p}\tilde{q}$

(d) $[p] = p\tilde{p}$

(e) $[pq] = [p][q]$

(f) If $f(x)$ is a polynomial with rational coefficients and $p = a + b\sqrt{2}$ is a root of $f(x)$, then $\tilde{p}$ is also a root of $f(x)$.

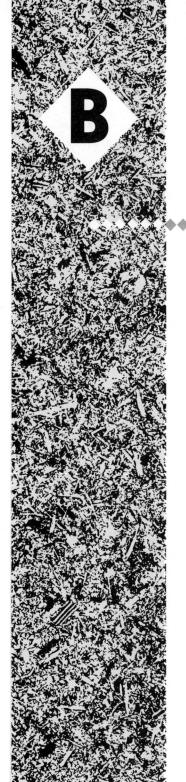

B ◆ Introduction to Linear Programming

Many important problems involve *linear inequalities* rather than *linear equations.* For example, a condition on the variables x and y might take the form of an inequality $2x - 3y \leq 4$ rather than that of an equality $2x - 3y = 4$. Linear programming is a method of finding a solution to a system of such inequalities that maximizes a function of the form $p = ax + by$, where a and b are constants. The general method of solving such problems (called the simplex method) involves Gaussian elimination techniques, so it is natural to include a discussion of it here. However, the proofs of the main theorems are omitted. The interested reader should consult a text on linear programming. [For example, S. I. Gass, *Linear Programming,* 4th ed. (New York: McGraw-Hill, 1975) gives a thorough treatment. J. G. Kemeny, J. L. Snell, and G. L. Thompson, *Introduction to Finite Mathematics* (Englewood Cliffs, N.J.: Prentice Hall, 1974) gives a more elementary treatment and relates linear programming to the theory of games.]

Section B.1
Graphical Methods

When only two variables are present, there is a geometric method of solution that, although it is not useful as a practical tool when more variables are involved, *is* useful in illustrating how solutions to these problems arise. Before giving an example, we must clarify what an inequality of the form

$$2x_1 + 3x_2 \leq 5$$

means in geometric terms. (We use x_1 and x_2 in place of x and y to conform to later notations.) Of course, the graph of the corresponding equation $2x_1 + 3x_2 = 5$ is well known; it is a line and consists of all points $P(x_1, x_2)$ in the plane whose coordinates x_1 and x_2 satisfy the equation. The lines parallel to this one all have equations $2x_1 + 3x_2 = c$ for some value of c, so the points $P(x_1, x_2)$ whose coordinates satisfy the *inequality* $2x_1 + 3x_2 < 5$ are just those points lying on one side of the line $2x_1 + 3x_2 = 5$

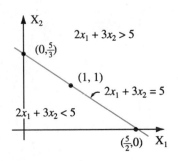

FIGURE B.1

(points on the other side are those satisfying $2x_1 + 3x_2 > 5$). The situation is illustrated in Figure B.1. The points $P(x_1, x_2)$ satisfying $2x_1 + 3x_2 \leq 5$ are just the points on or below the line. In general, the region on or to one side of a straight line is called a **half-plane.** We shall loosely speak of the half-plane $ax_1 + bx_2 \leq c$ (where a, b, and c are constants). The line $ax_1 + bx_2 = c$ will be called the **boundary** of the half-plane.

When two or more inequalities are given, the set of points that satisfies all of them is the region common to all the half-planes involved, and such regions are important in linear programming. Here is an example.

EXAMPLE 1

Determine the region in the plane given by

$$x_1 \geq 0$$
$$x_1 - x_2 \leq 0$$
$$x_1 + x_2 \leq 4$$

Solution

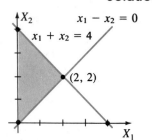

The half-plane $x_1 \geq 0$ consists of all points on or to the right of the X_2 axis, the half-plane $x_1 - x_2 \leq 0$ consists of all points on or above the line $x_2 = x_1$, and the half-plane $x_1 + x_2 \leq 4$ consists of all points on or below the line $x_2 = -x_1 + 4$. The lines in question are plotted in the diagram, and the region common to all these half-planes is just the shaded portion.

◆◆◆

The general linear programming problem (with two variables) can now be stated: Suppose a region in the plane (called the **feasible** region) is given as in Example 1 by a set of linear inequalities (called **constraints**) in two variables x_1 and x_2 (so the feasible region consists of all points common to all the corresponding half-planes). The problem is to find all points in this region at which a linear function of the form $p = ax_1 + bx_2$ is as large as possible. This function p is called the **objective function,** and these points are said to **maximize** p over the feasible region. In applications, p might be the profit in some commercial venture, or some other quantity that is desired to be large. The precise nature of p plays no part in the solution, except that it should be a linear function of the variables x_1 and x_2 (hence the name *linear* programming). The following example illustrates how the method works.

EXAMPLE 2

Find the point (or points) $P(x_1, x_2)$ in the region

$$4x_1 + x_2 \leq 16$$
$$x_1 + x_2 \leq 6$$

$$x_1 + 3x_2 \leq 15$$
$$x_1 \geq 0$$
$$x_2 \geq 0$$

for which the quantity

$$p = 2x_1 + 3x_2$$

is as large as possible.

Solution   The lines $4x_1 + x_2 = 16$, $x_1 + x_2 = 6$, and $x_1 + 3x_2 = 15$ are plotted in the first diagram (see figure), and the corresponding half-planes lie below these lines. Hence the feasible region in question is the shaded part. The second diagram exhibits this region again (but with a larger scale) and also shows the line $2x_1 + 3x_2 = p$, plotted for various values of p. Because p has the same value at any point on one of these lines, they are sometimes called **level lines** for p. The aim is to find a point in the shaded region at which p has as large a value as possible. These values increase as the level lines rise, so the vertex $(\frac{3}{2}, \frac{9}{2})$—the intersection of the lines $x_1 + 3x_2 = 15$ and $x_1 + x_2 = 6$—clearly gives the largest value of p. This value is

$$p = 2(\tfrac{3}{2}) + 3(\tfrac{9}{2}) = 16.5$$

and it is the desired maximal value.

◆◆◆

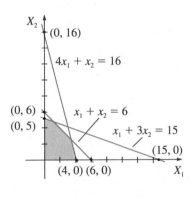
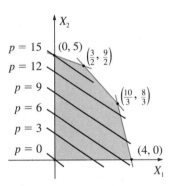

It is quite clear that the method used in Example 2 will work in a variety of similar situations. However, before we attempt to say anything in general, consider the following example.

EXAMPLE 3

Maximize $p = 2x_1 + x_2$ over the region

$$x_1 + x_2 \geq 3$$
$$-x_1 + x_2 \leq 3$$
$$-x_1 + 2x_2 \geq 0$$

Solution   The region in question is sketched in the diagram, where the main difference from the previous example emerges: The feasible region in this case is unbounded. Three of the level lines (corresponding to $p = 6$, 9, and 12) are also plotted, and it is clear that there are points in the feasible region at which the objective function p is as large as we please. Consequently p has *no maximum* over the feasible region in question.

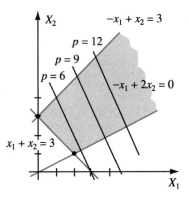

With this example in mind, the reader might guess that, in general, the objective function has no maximum only when the feasible region is unbounded. But this is not the case. Exercise 2 gives a situation wherein the feasible region is unbounded but the objective function does indeed have a maximum (the reader should try to construct such an example before working this exercise).

What is true is that if the feasible region is bounded (that is, can be contained in some circle), then the objective function p has a maximum value. In fact, the level lines for p form a set of parallel lines, each corresponding to a fixed value of p. If p were to increase continuously, the corresponding level line would move continuously in a direction perpendicular to itself. As this moving level line crosses the feasible region, it is clear that a largest value of p can be found so that the corresponding line intersects the feasible region. Because the edges of the feasible region are line segments, even more can be said: Either this level line corresponding to the largest value of p will intersect the feasible region at a vertex or, possibly, the intersection will consist of an entire edge of the feasible region. Either way, the maximum value of p will be achieved at a vertex of the feasible region. This is good news. There are at most a finite number of vertices (there are only finitely many constraints), so only a finite number of feasible points (the vertices) need be looked at. If p is evaluated at each of them, the vertex yielding the largest value of p will be the desired feasible point.

Finally, the same argument shows that p achieves a minimum value over a bounded feasible region, and that minimum is found at a vertex. The following theorem summarizes this discussion.

THEOREM 1

Let p be a linear function of two variables x_1 and x_2:

$$p = ax_1 + bx_2$$

If a finite set of linear inequalities in x_1 and x_2 determine a bounded feasible region, then there is a point in this feasible region (in fact, a vertex point) that maximizes p, and there is a vertex point that minimizes p.

EXAMPLE 4

A manufacturer wants to make two types of toys. The large toy requires 4 square feet of plywood and 50 milliliters (ml) of paint, whereas the smaller toy requires 3 square feet of plywood and only 20 ml of paint. There are 1800 square feet of plywood and 16 liters of paint available. If the large toys sell for a profit of $21 each, and each of the small toys yields an $18 profit, determine the number of toys of each size that the manufacturer must make to maximize the total profit.

Solution

Let x_1 and x_2 denote the number of large and small toys, respectively, to be made. These variables satisfy the following constraints:

$$4x_1 + 3x_2 \leq 1800 \quad \text{(plywood)}$$
$$5x_1 + 2x_2 \leq 1600 \quad \text{(paint)}$$
$$x_1 \geq 0$$
$$x_2 \geq 0$$

The feasible region corresponding to these inequalities is plotted in the diagram. The four vertices have coordinates $(0, 0)$, $(0, 600)$, $(\frac{1200}{7}, \frac{2600}{7})$, and $(320, 0)$. The total profit p of the enterprise is

$$p = 21x_1 + 18x_2$$

and the values of p at each of the vertices are

$$p = 0 \qquad \text{at } (0, 0)$$
$$p = 10{,}800 \qquad \text{at } (0, 600)$$
$$p = 10{,}285.71 \qquad \text{at } \left(\frac{1200}{7}, \frac{2600}{7}\right)$$
$$p = 6{,}720 \qquad \text{at } (320, 0)$$

Hence the manufacturer will maximize profits by producing 600 small toys and no large toys at all.

◆◆◆

Theorem 1 can be extended. First of all, the objective function p may very well have a maximum over the feasible region even if that region is unbounded. Moreover, the argument leading to Theorem 1 can be modified to show that, *if the objective function p has a maximum over the feasible region, that maximum will be attained at a vertex.*

Second, Theorem 1 can be extended to more than two variables $x_1, x_2, \ldots, x_n$. A function of the form

$$p = c_1 x_1 + c_2 x_2 + \cdots + c_n x_n$$

is called a **linear function** of these variables, and a condition of the form

$$a_1 x_1 + a_2 x_2 + \cdots + a_n x_n \leq b$$

is called a **linear constraint** on the variables. The set of n-tuples $(x_1, x_2, \ldots, x_n)$ satisfying a finite number of such linear constraints is called the **feasible region** determined by these constraints, and the n-tuples themselves are called **feasible points.** Consider the equation

$$a_1 x_1 + a_2 x_2 + \cdots + a_n x_n = b$$

which was obtained from the foregoing constraint by replacing the inequality by equality. The set of all n-tuples $(x_1, x_2, \ldots, x_n)$ satisfying this equation is called a **bounding hyperplane** for the feasible region. In the case of two variables these are lines; they are actual planes when $n = 3$. By analogy with the two-variable situation, an n-variable feasible point is called an **extreme point** (or **corner point**) of the feasible region if it lies on n (or more) bounding hyperplanes.

The general linear programming problem is to find a feasible point such that the **objective function** p is as large as possible, in which case the point is said to **maximize** p. (Similarly, we could seek a feasible point that **minimizes** p.) The extended theorem is stated below. The proof (though similar in spirit to that of Theorem 1) is omitted. The feasible region is said to be **bounded** if there exists a number M such that $|x_i| \leq M$ holds for every feasible point $(x_1, x_2, \ldots, x_n)$ and each $i = 1, 2, \ldots, n$.

THEOREM 2

Let p be a linear function of the variables $x_1, x_2, \ldots, x_n$:

$$p = c_1 x_1 + c_2 x_2 + \cdots + c_n x_n$$

and consider the feasible region determined by a finite number of linear constraints on these variables.

1. If p has a maximum value in the feasible region, then that maximum occurs at an extreme point.

2. If p has a minimum value in the feasible region, then that minimum occurs at an extreme point.

3. If the feasible region is bounded, then p has both a maximum and a minimum.

EXAMPLE 5

Find the maximum and minimum value of

$$p = 4x_1 - 3x_2 + 7x_3$$

subject to the following constraints;

$$5x_1 + 2x_2 + 4x_3 \leq 20$$
$$x_1 \geq 0$$
$$x_2 \geq 0$$
$$x_3 \geq 0$$

Solution These constraints are sufficiently simple that a picture of the feasible region can easily be drawn. The bounding hyperplanes in this case are ordinary planes. In the diagram, $x_1 \geq 0$ represents the region in front of the X_2X_3-plane, $x_2 \geq 0$ gives the region to the right of the X_1X_3-plane, and $x_3 \geq 0$ yields the region above the X_1X_2-plane. The fourth bounding hyperplane is

$$5x_1 + 2x_2 + 4x_3 = 20$$

and the intersections of this plane with the X_1, X_2, and X_3 axes are plotted as P_1, P_2, and P_3, respectively. The constraint

$$5x_1 + 2x_2 + 4x_3 \leq 20$$

determines the region below this plane, so the feasible region is the tetrahedron with vertices P_1, P_2, P_3, and the origin. Hence these are the extreme points. If the objective function P is evaluated at the extreme points, the results are

$$p = \quad 0 \quad \text{at the origin}$$
$$p = \quad 16 \quad \text{at } P_1(4, 0, 0)$$
$$p = -30 \quad \text{at } P_2(0, 10, 0)$$
$$p = \quad 35 \quad \text{at } P_3(0, 0, 5)$$

Because the feasible region is bounded, p has both a maximum and a minimum; they are $p = 35$ [at $P_3(0, 0, 5)$] and $p = -30$ [at $P_2(0, 10, 0)$].

◆◆◆

The procedure we just used has two drawbacks. First, it is not always easy to determine whether a maximum exists. But even when this is known (if the feasible set is bounded, for example), the number of extreme points can be very large and the amount of computation required to find them can be excessive, even for a computer.

This is why the method is not pursued here.

A much more efficient procedure exists that reduces the number of extreme points that must be examined. The idea is quite simple. To get some insight into how it works, consider the general linear programming problem with three variables. Then the bounding hyperplanes are real planes, and the edges of the feasible region are the lines of intersection of pairs of bounding planes. Now suppose the objective function p is evaluated at some extreme point. Choose an edge emanating from this point along which the function p increases (or decreases if a minimum is desired). If no such edge exists, it can be shown that the extreme point gives the maximum. Otherwise there are two possibilities: (1) We encounter another extreme point on that edge at which p is larger and then repeat the process. (2) There is no other vertex along this edge so p increases without bound, and no maximum exists. At each stage we either discover there is no maximum, or we find the maximum, or we are led to another extreme point at which p is larger. Clearly the same extreme point cannot be encountered twice in this fashion (p increases), and so, because there are only finitely many extreme points, the process is effective: It either shows that no maximum exists, or, if there *is* a maximum, it leads to an extreme point yielding that maximum. Furthermore, in the types of problems usually found in practice, the method converges quickly.

This description of the algorithm is geometric in nature. However, the whole thing can be cast in algebraic form. This will be described in the next section.

EXERCISES B.1

1. In each case, find the maximum and minimum values of p by finding the feasible region and examining p at the vertex points.

   (a) $p = 3x_1 + 2x_2$
   $3x_1 + x_2 \leq 9$
   $2x_1 + 5x_2 \leq 10$
   $x_1 \geq 0$
   $x_2 \geq 0$

   ◆(b) $p = 4x_1 + 3x_2$
   $x_2 \leq 5$
   $x_1 + x_2 \leq 8$
   $x_1 - x_2 \leq 2$
   $x_1 \geq 0$
   $x_2 \geq 0$

   (c) $p = 2x_1 - x_2$
   $x_1 + x_2 \geq 2$
   $x_1 + 2x_2 \leq 14$
   $-x_1 + x_2 \leq 4$
   $x_1 \leq 6$
   $x_1 \geq 0$
   $x_2 \geq 0$

   ◆(d) $p = x_2 - x_1$
   $3x_1 + 2x_2 \geq 12$
   $x_2 - x_1 \leq 6$
   $2x_1 + 3x_2 \leq 38$
   $x_1 \leq 10$
   $x_1 \geq 0$
   $x_2 \geq 0$

2. Consider the problem of maximizing $p = x_2 - 2x_1$ subject to:
   $$-x_1 + x_2 \leq 6$$
   $$x_1 - 2x_2 \leq 0$$
   $$x_1 \geq 0$$
   $$x_2 \geq 0$$

   Show that the feasible region is unbounded but that p still has a maximum.

3. Show that there are no points in the following feasible region.
   $$x_1 + x_2 \leq 4$$
   $$x_2 - x_1 \geq 4$$
   $$3x_1 + x_2 \geq 12$$

4. In Example 4, assume that small toys continue to earn a profit of $18 per toy but that profits for large toys increase. Find the number of toys that should be produced to maximize profits in each of the following cases.
   (a) Large-toy profit is $23 per toy.
   ◆(b) Large-toy profit is $24 per toy.
   (c) Large-toy profit is $25 per toy.

5. A man wishes to invest a portion of $100,000 in two stocks A and B. He feels that at most $70,000 should go into A and at most $60,000 should go into B. If A and B pay 10% and 8% dividends, respectively, how much should he invest in each stock to maximize his dividends?

◆6. Repeat Exercise 5 where A pays 8% and B pays 10%.

7. A vitamin pill manufacturer uses two ingredients P and Q. The amounts of vitamins A, C, and D per gram of ingredient are given in the table. The ingredients are mixed with at least 85 grams of filler to make batches of 100 grams that are then pressed into pills. The law requires that each batch contain at least 12 units of vitamin A, at least 12 units of vitamin C, and at least 10 units of vitamin D. If P costs $5 per gram and Q costs $2 per gram, how many grams of each should be used per batch to minimize the cost?

| | VITAMIN | | |
	A	C	D
P	2 units	1 unit	5 units
Q	2 units	4 units	1 unit

◆ 8. Repeat Exercise 7 where P costs $2 per gram and Q costs $3 per gram.

9. An oil company produces two grades of heating oil, grade 1 and grade 2, and makes a profit of $8 per barrel on grade 1 oil and $5 per barrel on grade 2 oil. The refinery operates 100 hours per week. Grade 1 oil takes $\frac{1}{4}$ hour per barrel to produce, whereas grade 2 oil takes only $\frac{1}{8}$ hour per barrel. The pipeline into the refinery can supply only enough crude to make 500 barrels of oil (either grade). Finally, warehouse constraints dictate that no more than 400 barrels of either type of oil can be produced per week. What production levels should be maintained to maximize profit?

◆ 10. Repeat Exercise 9 where the profits are $7 on grade 1 oil and $6 on grade 2 oil.

11. A small bakery makes white bread and brown bread in batches, and it has the capacity to make 8 batches per day. Each batch of white bread requires 1 unit of yeast, but the brown bread takes 2 units of yeast per batch. On the other hand, white bread costs $30 per day for marketing, whereas brown bread only costs $10 per day. If $180 per day are available for marketing, and if 13 units of yeast are available per day, find the number of batches of each type that the bakery should make per day to maximize profits if it makes $300 profit per batch of white bread and $200 profit per batch of brown bread.

Section B.2   The Simplex Algorithm

The simplex algorithm is a simple, straightforward method for solving linear programming problems that was discovered in the 1940s by George Dantzig. The idea is to identify certain "basic" feasible points and to prove that the maximum value (if it exists) of the objective function p occurs at one of these points. Then the algorithm proceeds roughly as follows: If a basic feasible point is at hand, a procedure is given for deciding whether it yields the maximum value of the objective function and, if not, for finding a basic feasible point that produces a larger value of the objective function. The process continues until a maximum is reached (or until it is established that there is no maximum).

We will develop the algorithm only for the **standard** linear programming problem:

Maximize the linear **objective function**

$$p = c_1 x_1 + c_2 x_2 + \cdots + c_n x_n$$

of the variables $x_1, x_2, \ldots, x_n$ subject to a finite collection of **constraints:**

$$a_{11}x_1 + a_{12}x_2 + \cdots + a_{1n}x_n \leq b_1$$
$$a_{21}x_1 + a_{22}x_2 + \cdots + a_{2n}x_n \leq b_2$$
$$\vdots \qquad \vdots \qquad \qquad \vdots \qquad \vdots$$
$$a_{m1}x_1 + a_{m2}x_2 + \cdots + a_{mn}x_n \leq b_m$$

Furthermore, the variables x_i and the constants b_j are all required to be nonnegative:

$$x_i \geq 0 \qquad \text{for } i = 1,\ 2, \ldots, n$$
$$b_j \geq 0 \qquad \text{for } j = 1,\ 2, \ldots, m$$

The requirement that p be a linear function of the variables is vital (nonlinear programming is much more difficult), but the condition that the x_i be nonnegative and the fact that we are maximizing p (not minimizing) are not severe restrictions. On the other hand, the requirement that the constants b_j be nonnegative is a serious restriction (although it is satisfied in many practical applications). We refer the reader to texts on linear programming for ways in which the algorithm can be used in the non-standard case.

The various steps in the algorithm are best explained by working a specific example in detail.

Prototype Example

Maximize $p = 2x_1 + 3x_2 - x_3$ subject to:

$$x_1 + 2x_2 + 2x_3 \leq 6$$
$$3x_1 - x_2 + x_3 \leq 9$$
$$2x_1 + 3x_2 + 5x_3 \leq 20$$
$$x_i \geq 0 \qquad \text{for } i = 1,\ 2,\ 3$$

The first step in the procedure is to convert the constraints from inequalities to equalities. This is achieved by introducing new variables x_4, x_5, and x_6 (called **slack variables**), one for each constraint. The new problem is to maximize $p = 2x_1 + 3x_2 - x_3 + 0x_4 + 0x_5 + 0x_6$ subject to:

$$x_1 + 2x_2 + 2x_3 + x_4 \qquad\qquad = 6$$
$$3x_1 - x_2 + x_3 \qquad + x_5 \qquad = 9$$
$$2x_1 + 3x_2 + 5x_3 \qquad\qquad + x_6 = 20$$
$$x_i \geq 0 \qquad \text{for } i = 1,\ 2,\ 3,\ 4,\ 5,\ 6$$

The claim is that if $(x_1, x_2, x_3, x_4, x_5, x_6)$ is a solution to this problem, then (x_1, x_2, x_3) is a solution to the original problem. In fact, the constraints are satisfied (because $x_4 \geq 0$, $x_5 \geq 0$, and $x_6 \geq 0$), so (x_1, x_2, x_3) is a feasible solution for the original problem. Moreover, if (x_1', x_2', x_3') were another feasible point for the original problem yielding a larger value of p, then taking

$$x_4' = 6 - x_1' - 2x_2' - 2x_3'$$
$$x_5' = 9 - 3x_1' + x_2' - x_3'$$
$$x_6' = 20 - 2x_1' - 3x_2' - 5x_3'$$

would give a feasible point $(x_1', x_2', x_3', x_4', x_5', x_6')$ for the new problem yielding a larger value of p, which is a contradiction.

So it suffices to solve the new problem. To do so, write the relationship $p = 2x_1 + 3x_2 - x_3$ as a fourth equation to get

$$x_1 + 2x_2 + 2x_3 + x_4 \qquad\qquad = 6$$
$$3x_1 - x_2 + x_3 \qquad + x_5 \qquad\qquad = 9$$
$$2x_1 + 3x_2 + 5x_3 \qquad\qquad + x_6 \qquad = 20$$
$$-2x_1 - 3x_2 + x_3 \qquad\qquad\qquad + p = 0$$

This amounts to considering p as yet another variable. The augmented matrix (Section 1.1) for this system of equations is

$$
\begin{array}{ccccccc}
x_1 & x_2 & x_3 & \boxed{x_4} & \boxed{x_5} & \boxed{x_6} & p
\end{array}
$$

$$
\begin{bmatrix}
1 & 2 & 2 & 1 & 0 & 0 & 0 & 6 \\
3 & -1 & 1 & 0 & 1 & 0 & 0 & 9 \\
2 & 3 & 5 & 0 & 0 & 1 & 0 & 20 \\
-2 & -3 & 1 & 0 & 0 & 0 & 1 & 0
\end{bmatrix}
$$

This is called the initial **simplex tableau** for the problem. The idea is to use elementary row operations to create a sequence of such tableaux (keeping p in the bottom row) that will lead to a solution. This is analogous to the modification of the augmented matrix in Gaussian elimination, except that here we allow only feasible solutions.

Note that the columns corresponding to the slack variables x_4, x_5, and x_6 all consist of zeros and a single 1 and that the 1's are in different rows (the way these variables were introduced guarantees this). These will be called **basic columns,** and the slack variables are called the **basic variables** in the initial tableau. (They are indicated by a box.) Because of this, there is one obvious solution to the equations: Set all the nonbasic variables equal to zero and solve for the basic variables: $x_4 = 6$, $x_5 = 9$, $x_6 = 20$ (and $p = 0$). In other words,

$$(0, 0, 0, 6, 9, 20) \text{ is a feasible solution yielding } p = 0$$

Such a solution (with all nonbasic variables zero) is called a **basic feasible solution.** Note the role of the last column in all this: The numbers 6, 9, and 20 are the constants in the original constraints. It is the fact that these are *positive* that makes $(0, 0, 0, 6, 9, 20)$ a *feasible* solution. Also, the bottom entry in the last column is the value of p at the basic feasible solution (0 in this case).

The key to the whole algorithm is the following theorem. The proof is not difficult but would require some preliminary discussion of convex sets. Hence we omit it and refer the reader to texts on linear programming, such as S. I. Gass, *Linear Programming,* 4th ed. (New York: McGraw-Hill, 1975).

THEOREM 1

If a standard linear programming problem has a solution, then there is a basic feasible solution that yields the maximum value of the objective function. (Such basic feasible solutions are called **optimal.**)

Hence our goal is to find an optimal basic feasible point. Our construction using slack variables guarantees an initial basic feasible solution; the next step is to see whether it is optimal.

The bottom row of the initial tableau gives p in terms of the nonbasic variables (the original expression for p in this case):

$$p = 2x_1 + 3x_2 - x_3$$

The fact that some of the coefficients here are positive suggests that this value of p is *not* optimal because increasing x_1 or x_2 at all will increase p. In fact, it would seem better to try to increase x_2; it has the larger of the two positive coefficients (equivalently the most *negative* entry in the last row of the tableau). This in turn suggests that we try to modify the tableau so that x_2 becomes a new basic variable. For this reason, x_2 is called the **entering** variable. Its column is called the **pivot column.**

This is accomplished by doing elementary row operations to convert the pivot column into a basic column. The question is where to locate the 1. We do not put the 1 in the last row because we do not want to disturb p, but it can be placed at any other location in the pivot column where the present entry is nonzero (they all qualify in this example). The entry chosen is called the **pivot,** and it is chosen as follows:

1. The pivot entry must be positive.

2. Among the positive entries available, the pivot is the one that produces the smallest ratio when divided into the right-most entry in its row.

These are chosen so that the basic feasible solution in the tableau we are creating will indeed be feasible. (The situation where no unique pivot entry is determined by conditions 1 and 2 will be discussed later.)

Returning to the prototype example, we rewrite the initial tableau and circle the pivot entry. The ratios corresponding to the two positive entries in the pivot column (column 2 here) are shown at the right. No ratio is computed for row 2, because the corresponding entry in the pivot column is negative. Hence the pivot is 2 (circled).

$$
\begin{array}{ccccccc|c}
\boldsymbol{x_1} & \boldsymbol{x_2} & \boldsymbol{x_3} & \boxed{\boldsymbol{x_4}} & \boxed{\boldsymbol{x_5}} & \boxed{\boldsymbol{x_6}} & \boldsymbol{p} & \\
\end{array}
$$

$$
\begin{bmatrix}
1 & ② & 2 & 1 & 0 & 0 & 0 & 6 \\
3 & -1 & 1 & 0 & 1 & 0 & 0 & 9 \\
2 & 3 & 5 & 0 & 0 & 1 & 0 & 20 \\
-2 & -3 & 1 & 0 & 0 & 0 & 1 & 0
\end{bmatrix}
\qquad
\begin{array}{l}
\text{ratio} : 6/2 = 3 \\[1.2em]
\text{ratio} : 20/3 = 6.7
\end{array}
$$

Now do elementary row operations to convert the pivot to 1 and all other entries in its column to 0. The result is

$$
\begin{array}{ccccccc|c}
\boldsymbol{x_1} & \boxed{\boldsymbol{x_2}} & \boldsymbol{x_3} & \boldsymbol{x_4} & \boxed{\boldsymbol{x_5}} & \boxed{\boldsymbol{x_6}} & \boldsymbol{p} & \\
\end{array}
$$

$$
\begin{bmatrix}
\frac{1}{2} & 1 & 1 & \frac{1}{2} & 0 & 0 & 0 & 3 \\
\frac{7}{2} & 0 & 2 & \frac{1}{2} & 1 & 0 & 0 & 12 \\
\frac{1}{2} & 0 & 2 & \frac{-3}{2} & 0 & 1 & 0 & 11 \\
\frac{-1}{2} & 0 & 4 & \frac{3}{2} & 0 & 0 & 1 & 9
\end{bmatrix}
$$

Note that the former basic variable x_4 is no longer basic (this is because it had a 1 in the same row as the pivot), and it is sometimes called the **departing** variable. The new basic variables are x_2, x_5, and x_6, and the new basic feasible solution (taking the new nonbasic variables equal to zero) is $x_2 = 3$, $x_5 = 12$, $x_6 = 11$, and $p = 9$. In other words,

$$(0, 3, 0, 0, 12, 11) \text{ is the feasible solution yielding } p = 9$$

This is better than before; p has increased from 0 to 9.

Now repeat the process. The last row here yields

$$p = 9 + \tfrac{1}{2}x_1 - 4x_3 - \tfrac{3}{2}x_4$$

so there is still hope of increasing p by making x_1 basic (it has a positive coefficient). Hence the first column is the pivot column and all three entries (above the bottom row) are positive. The tableau is displayed once more, with the ratios given and the next pivot (with the smallest ratio) circled.

$$
\begin{array}{ccccccc|c}
\boldsymbol{x_1} & \boxed{\boldsymbol{x_2}} & \boldsymbol{x_3} & \boldsymbol{x_4} & \boxed{\boldsymbol{x_5}} & \boxed{\boldsymbol{x_6}} & \boldsymbol{p} & \\
\end{array}
$$

$$
\begin{bmatrix}
\frac{1}{2} & 1 & 1 & \frac{1}{2} & 0 & 0 & 0 & 3 \\
⑦\!\!\!\frac{7}{2} & 0 & 2 & \frac{1}{2} & 1 & 0 & 0 & 12 \\
\frac{1}{2} & 0 & 2 & \frac{-3}{2} & 0 & 1 & 0 & 11 \\
\frac{-1}{2} & 0 & 4 & \frac{3}{2} & 0 & 0 & 1 & 9
\end{bmatrix}
\qquad
\begin{array}{l}
\text{ratio} : \frac{3}{1/2} = 6 \\[0.8em]
\text{ratio} : \frac{12}{7/2} = \frac{24}{7} \\[0.8em]
\text{ratio} : \frac{11}{1/2} = 22
\end{array}
$$

Row operations give the third tableau with x_1, x_2, and x_6 as basic variables.

X_1	X_2	x_3	x_4	x_5	X_6	p

$$\begin{bmatrix} 0 & 1 & \frac{5}{7} & \frac{3}{7} & \frac{-1}{7} & 0 & 0 & \frac{9}{7} \\ 1 & 0 & \frac{4}{7} & \frac{1}{7} & \frac{2}{7} & 0 & 0 & \frac{24}{7} \\ 0 & 0 & \frac{12}{7} & \frac{-11}{7} & \frac{-1}{7} & 1 & 0 & \frac{65}{7} \\ 0 & 0 & \frac{30}{7} & \frac{11}{7} & \frac{1}{7} & 0 & 1 & \frac{75}{7} \end{bmatrix}$$

The corresponding basic feasible solution (setting $x_3 = x_4 = x_5 = 0$) is $x_1 = \frac{24}{7}$, $x_2 = \frac{9}{7}$, $x_6 = \frac{65}{7}$, and $p = \frac{75}{7}$. In other words,

$$\left(\tfrac{24}{7}, \tfrac{9}{7}, 0, 0, 0, \tfrac{65}{7} \right) \text{ is a feasible solution yielding } p = \tfrac{75}{7}$$

However, we claim that this is optimal. The last row of the third tableau gives

$$p = \tfrac{75}{7} - \tfrac{30}{7}x_3 - \tfrac{11}{7}x_4 - \tfrac{1}{7}x_5$$

so, because x_3, x_4, and x_5 are nonnegative, p can be no greater than $\frac{75}{7}$. The preceding solution achieves $p = \frac{75}{7}$, so it must be optimal. This completes the solution of the prototype example.

Note that the test for optimality is clear: If the last row of a tableau has only *positive* entries, then the corresponding basic feasible solution is optimal. If not, the column corresponding to the *most negative* entry in the last row is the pivot column, the corresponding variable is the entering variable, and a new tableau is constructed with that variable as a new basic variable.

Of course not every standard linear programming problem has a solution (see Exercise 1). It can happen that feasible points can be found that make the objective function p as large as we like. In this case p has no maximum and we say p is **unbounded.** The simplex algorithm provides a way to determine whether this is the case (step 3 in the flow chart that follows).

The algorithm works in exactly the same way for any standard linear programming problem. Suppose that

$$p = c_1 x_1 + c_2 x_2 + \cdots + c_n x_n$$

is to be maximized subject to the following m constraints:

$$\begin{aligned} a_{11}x_1 + a_{12}x_2 + \cdots + a_{1n}x_n &\leq b_1 \\ a_{21}x_1 + a_{22}x_2 + \cdots + a_{2n}x_n &\leq b_2 \qquad x_i \geq 0 \text{ for } i = 1, 2, \ldots, n \\ \vdots \qquad \vdots \qquad\qquad \vdots \qquad \vdots &\qquad\quad b_j \geq 0 \text{ for } j = 1, \ldots, m \\ a_{m1}x_1 + a_{m2}x_2 + \cdots + a_{mn}x_n &\leq b_m \end{aligned}$$

Introduce m slack variables $x_{n+1}, \ldots, x_{n+m}$ to make the inequalities into equalities. The new problem is to maximize

$$p = c_1 x_1 + c_2 x_2 + \cdots + c_n x_n + 0 x_{n+1} + \cdots + 0 x_{n+m}$$

subject to:

$$a_{11}x_1 + a_{12}x_2 + \cdots + a_{1n}x_n + x_{n+1} \qquad\qquad\qquad = b_1$$
$$a_{21}x_1 + a_{22}x_2 + \cdots + a_{2n}x_n + \cdots + x_{n+2} \qquad\qquad = b_2$$
$$\vdots \qquad \vdots \qquad\qquad \vdots \qquad\qquad\qquad\qquad\qquad \vdots$$
$$a_{m1}x_1 + a_{m2}x_2 + \cdots + a_{mn}x_n + \cdots \qquad\qquad + x_{n+m} = b_m$$
$$x_i \geq 0 \qquad \text{for } i = 1, 2, \ldots, n + m$$
$$b_j \geq 0 \qquad \text{for } j = 1, 2, \ldots, m$$

Any optimal solution $(x_1, \ldots, x_n, x_{n+1}, \ldots, x_{n+m})$ to this new problem yields an optimal solution $(x_1, \ldots, x_n)$ to the original problem (the argument in the prototype example works), so we solve the new problem. Because the constraints are now equations, some of the methods of Gaussian elimination apply. For convenience, write the expression for p as another equation:

$$-c_1x_1 - c_2x_2 - \cdots - c_nx_n + \cdots + p = 0$$

The augmented matrix for this larger system of equations is

x_1	x_2	$\cdots$	x_n	x_{n+1}	x_{n+2}	$\cdots$	x_{n+m}	p	
a_{11}	a_{12}	$\cdots$	a_{1n}	1	0	$\cdots$	0	0	b_1
a_{21}	a_{22}	$\cdots$	a_{2n}	0	1	$\cdots$	0	0	b_2
$\vdots$	$\vdots$		$\vdots$	$\vdots$	$\vdots$		$\vdots$	$\vdots$	$\vdots$
a_{m1}	a_{m2}	$\cdots$	a_{mn}	0	0	$\cdots$	1	0	b_m
$-c_1$	$-c_2$	$\cdots$	$-c_n$	0	0	$\cdots$	0	1	0

and this is called the initial **simplex tableau** for the problem. The slack variables are called **basic variables** because their columns are **basic columns** (all entries are zero except for a single one). The fact that $b_j \geq 0$ holds for each j means that we can obtain a feasible solution by setting all the nonbasic variables equal to zero. The result:

$$(0, 0, \ldots, 0, b_1, b_2, \ldots, b_m) \text{ is a feasible solution yielding } p = 0$$

This is the initial basic feasible solution. Of course, it may not be optimal.

Now the algorithm starts. Suppose a tableau has been constructed with m basic variables (that is, m variables corresponding to basic columns in the tableau) such that the last column contains no negative entries (for example, the initial tableau). Then, if the nonbasic variables are set equal to zero, the values of the basic variables are determined (they are just the entries of the last column, the value of p being the last entry). Hence this gives the **basic feasible solution** corresponding to the tableau.

The actual execution of the algorithm is best described in the following steps. For convenience, we display them first as a flow chart.

The Simplex Algorithm

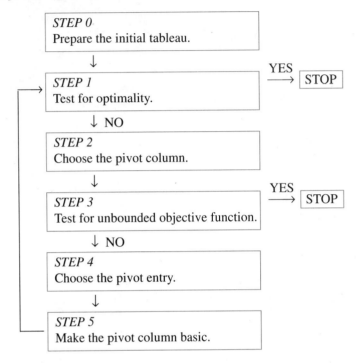

The details of the steps are as follows. As before, we assume that there are n variables and m constraints.

STEP 0.   Prepare the initial tableau.
Introduce slack variables, one for each constraint, and convert each inequality into an equation. Then write the expression for p as another equation (as before). The augmented matrix for the resulting system of $m + 1$ equations (the equation for p last) is the initial tableau.

STEP 1.   Test for optimality.
Given a tableau, the corresponding basic feasible solution is optimal if no entry in the last row (except the last) is negative. (The argument in the prototype example works.) In this case, stop — the maximum value of p is the lower right entry. Otherwise go on to step 2.

STEP 2.   Choose the pivot column.
This is the column (not the last) whose bottom entry is the most negative (the worst offender, as it were). If there is a tie, choose either possibility.

STEP 3.   Test for unbounded objective function.
This occurs if no entry in the pivot column is positive. (We omit the

proof.) In this case, stop —the objective function has no maximum. Otherwise go on to step 4.

STEP 4.   Choose the pivot entry.
Among the positive entries in the pivot column, choose the one that has the smallest ratio when divided into the last entry in its row. If two ratios are equal, choose either. (This may lead to cycling; see Remark 2, which follows Example 1.)

STEP 5.   Make the pivot column basic.
Use elementary row operations to make the pivot entry 1 and every other entry in the pivot column (including the last) zero.

EXAMPLE 1

Maximize $p = 3x_1 + x_2 + 2x_3$ subject to:

$$2x_1 - x_2 + 3x_3 \leq 2$$
$$3x_1 + x_2 + x_3 \leq 5 \qquad x_i \geq 0 \text{ for } i = 1, 2, 3$$

Solution   Introduce slack variables x_4 and x_5, and rewrite the equation for p.

$$2x_1 - x_2 + 3x_3 + x_4 \qquad\qquad = 2$$
$$3x_1 + x_2 + x_3 \qquad + x_5 \qquad = 5 \qquad x_i \geq 0 \text{ for } i = 1, 2, 3, 4, 5$$
$$-3x_1 - x_2 - 2x_3 \qquad\qquad + p = 0$$

Hence the initial tableau (with the basic variables boxed) is

x_1	x_2	x_3	$\boxed{x_4}$	$\boxed{x_5}$	p	
②	−1	3	1	0	0	2
3	1	1	0	1	0	5
−3	−1	−2	0	0	1	0

ratio: $2/2 = 1$
ratio: $5/3$

The basic feasible solution here is not optimal (the last row has negative entries), and the pivot column is the first (-3 is the most negative). The ratios are computed as before and the pivot is circled. Hence row operations give the next tableau (the new basic variables are boxed).

$\boxed{x_1}$	x_2	x_3	x_4	$\boxed{x_5}$	p	
1	$\frac{-1}{2}$	$\frac{3}{2}$	$\frac{1}{2}$	0	0	1
0	$\frac{5}{2}$	$\frac{-7}{2}$	$\frac{-3}{2}$	1	0	2
0	$\frac{-5}{2}$	$\frac{5}{2}$	$\frac{3}{2}$	0	1	3

This is still not optimal, and the pivot column is the second. Here the pivot is the only positive entry, so no ratios need to be computed. Row operations give

$$
\begin{array}{cccccc}
\boxed{x_1} & \boxed{x_2} & x_3 & x_4 & x_5 & p \\
\end{array}
$$

$$
\begin{bmatrix}
1 & 0 & \boxed{\tfrac{4}{5}} & \tfrac{1}{5} & \tfrac{1}{5} & 0 & \tfrac{7}{5} \\
0 & 1 & \tfrac{-7}{5} & \tfrac{-3}{5} & \tfrac{2}{5} & 0 & \tfrac{4}{5} \\
0 & 0 & -1 & 0 & 1 & 1 & 5
\end{bmatrix}
$$

Again no optimal solution exists, but p has increased to 5 (lower right entry). The next tableau is

$$
\begin{array}{cccccc}
x_1 & \boxed{x_2} & \boxed{x_3} & x_4 & x_5 & p \\
\end{array}
$$

$$
\begin{bmatrix}
\tfrac{5}{4} & 0 & 1 & \tfrac{1}{4} & \tfrac{1}{4} & 0 & \tfrac{7}{4} \\
\tfrac{7}{4} & 1 & 0 & \tfrac{-1}{4} & \tfrac{3}{4} & 0 & \tfrac{13}{4} \\
\tfrac{5}{4} & 0 & 0 & \tfrac{1}{4} & \tfrac{5}{4} & 1 & \tfrac{27}{4}
\end{bmatrix}
$$

Hence p has a maximum of $\tfrac{27}{4}$ when $x_1 = 0$, $x_2 = \tfrac{13}{4}$, and $x_3 = \tfrac{7}{4}$.   ◆◆◆

We conclude with two remarks on the algorithm.

◆◆

Remark 1:   Rationale for selecting the pivot entry (step 4).

◆◆

Suppose the rth column is the pivot column. Write it as

$$
\begin{bmatrix}
t_{1r} \\
\vdots \\
t_{qr} \\
\vdots \\
t_{mr} \\
s_r
\end{bmatrix}
$$

where $s_r < 0$ (from step 2) and at least one other entry is positive. Suppose we decide to take t_{qr} as the pivot. Write its row as

$$
(t_{q1}, t_{q2}, \ \ldots \ , t_{qr}, \ \ldots \ , 0, d_q)
$$

where the last entry $d_q \geq 0$ (because the tableau produced a feasible solution when all nonbasic variables were set equal to zero, so d_q is the value of one of the basic variables). We want to divide this row by t_{qr} (so $t_{qr} \neq 0$), and the last entry of the new row, d_q/t_{qr} is required to be nonnegative too (we want a new *tableau*). Hence the pivot t_{qr} must be *positive*.

Then we convert each other entry t_{ir} in the pivot column to zero by subtracting t_{ir} times the pivot row. If the right entry of row i is d_i, the new right entry is

$$d_i - t_{ir}\left[\frac{d_q}{t_{qr}}\right] = t_{ir}\left[\frac{d_i}{t_{ir}} - \frac{d_q}{t_{qr}}\right]$$

This is clearly positive if t_{ir} is negative or zero (use the left side). If $t_{ir} > 0$, it is positive if the ratio d_q/t_{qr} for the pivot is less than the ratio for t_{ir}. This shows why we choose the pivot with minimal ratio (in step 4).

◆◆

Remark 2:   Degeneracy and cycling.

◆◆

If two ratios in step 4 are equal, the argument in Remark 1 shows that in the next tableau, some entry in the last column will be zero (so some basic variable will take the value 0). In this case the algorithm is said to **degenerate,** and it may lead to cycling (that is, the sequence of basic feasible solutions we are creating may contain the same solution twice and so continue to loop indefinitely). This is rare in practical problems (computer round-off error tends to eliminate it), and algorithms exist to deal with it.

EXERCISES B.2

1. Consider the following standard linear programming problem: Maximize $p = x_1 + x_2$ subject to:

$$-x_1 + x_2 \leq 1$$
$$x_1 - 2x_2 \leq 2$$
$$x_1 \geq 0, x_2 \geq 0$$

(a) Using the methods of Section B.1, show that p is unbounded over the feasible region.

(b) Use the simplex algorithm to arrive at the same conclusion.

2. In each case, maximize p subject to the given constraints, and find values of the x_i that yield the maximum. Assume $x_i \geq 0$ for all i.

(a) $p = x_1 + 2x_2 + 3x_3$
$$3x_1 - x_2 - x_3 \leq 3$$
$$x_1 + x_2 + 2x_3 \leq 2$$

◆(b) $p = 2x_1 + x_2 + x_3$
$$3x_1 + x_2 + 2x_3 \leq 2$$
$$x_1 + x_2 + 3x_3 \leq 5$$

(c) $p = x_1 + 2x_2$
$$3x_1 + x_2 \leq 4$$
$$x_1 + 2x_2 \leq 3$$
$$2x_1 + 3x_2 \leq 5$$

◆(d) $p = 3x_1 + 2x_2$
$$4x_1 + 3x_2 \leq 5$$
$$x_1 + x_2 \leq 2$$
$$3x_1 + 4x_2 \leq 4$$

(e) $p = 3x_1 + x_2 + 2x_3$
$$2x_1 + x_2 - x_3 \leq 3$$
$$x_1 + x_2 + x_3 \leq 4$$
$$x_1 + 2x_2 + x_3 \leq 5$$

◆(f) $p = 2x_1 + 3x_2 + 2x_3$
$$2x_1 + 3x_2 - x_3 \leq 4$$
$$x_1 - x_2 + x_3 \leq 2$$
$$3x_1 + 4x_2 + x_3 \leq 5$$

(g) $p = x_1 + x_2 + 2x_3 + x_4$
$$3x_1 + x_2 - x_3 + 2x_4 \leq 5$$
$$x_1 + 2x_2 + x_3 - x_4 \leq 4$$

◆(h) $p = 2x_1 + x_2 + 3x_3 + 2x_4$
$$3x_1 + 4x_2 + x_3 - 2x_4 \leq 6$$
$$2x_1 + 3x_2 - x_3 + 3x_4 \leq 5$$

3. Can the maximum of the objective function ever be negative in a standard linear programming problem? Explain.

4. Suppose a standard linear programming problem has more variables than constraints. If the objective function has a maximum, show that this must have at least one optimal solution with one original variable zero.

5. An automobile company makes three types of cars: compact, sports, and full-size; the profits per unit are $500, $700, and $600, respectively. Transportation costs per vehicle are $300, $400, and $500, respectively. And labor costs are $500, $500, and $400, respectively. If the total transportation cost is not to exceed $40,000 and the total labor cost is not to exceed $30,000, find the maximum profit.

◆ **6.** A short-order restaurant sells three dinners—regular, diet, and super—on which it makes profits of $1.00, $2.00, and $1.50, respectively. The restaurant cannot serve more than 300 dinners daily. The three dinners require 2, 4, and 2 minutes to prepare, and at most 1000 minutes of preparation time are available per day. The dinners require 2, 0, and 3 minutes to cook, and 1000 minutes of cooking time are available daily. Finally, the dinners require 50, 300, and 100 grams of fresh produce, and this commodity is limited to 45 kilograms daily. Find the numbers of dinners of each type that will maximize profits.

7. A trucking company has 100 trucks, which are dispatched from three locations: A, B, and C. Each truck at A, B, and C uses 40, 30, and 30 units of fuel daily, and 2500 units per day are available. The costs of labor to operate and maintain each truck are $70, $80, and $70 per truck per day at the three locations, and $8000 per day is the maximum that the company can pay for labor. How many trucks should be allocated to each location if the daily profits per truck are $300, $250, and $200 at locations A, B, and C?

◆ **8.** A lawn mower company makes three models: standard, deluxe, and super. The construction of each mower involves three stages: motor construction, frame construction, and final assembly. The following table gives the number of hours of labor required per mower for each stage and the total number of hours of labor available per week for each stage. It also gives the profit per week. Find the weekly production schedule that maximizes profit.

	Standard	Deluxe	Super	Hours Available
motor	1	1	2	2500
frame	1	2	2	2000
assembly	1	1	1	1800
PROFIT	$30	$40	$55	

Mathematical Induction

Suppose one is presented with the following sequence of equations:

$$1 = 1$$
$$1 + 3 = 4$$
$$1 + 3 + 5 = 9$$
$$1 + 3 + 5 + 7 = 16$$
$$1 + 3 + 5 + 7 + 9 = 25$$

It is clear that there is a pattern. The numbers on the right side of the equations are the squares 1^2, 2^2, 3^2, 4^2, and 5^2 and, in the equation with n^2 on the right side, the left side is the sum of the first n odd numbers. The odd numbers are

$$1 = 2 \cdot 1 - 1$$
$$3 = 2 \cdot 2 - 1$$
$$5 = 2 \cdot 3 - 1$$
$$7 = 2 \cdot 4 - 1$$
$$9 = 2 \cdot 5 - 1$$

and from this it is clear that the nth odd number is $2n - 1$. Hence, at least for $n = 1$, 2, 3, 4, or 5, the following is true:

$$1 + 3 + \cdots + (2n - 1) = n^2 \qquad (S_n)$$

The question arises whether the statement (S_n) is true for *every* n. There is no hope of separately verifying all these statements because there are infinitely many of them. A more subtle approach is required.

The idea is as follows: Suppose it is verified that the statement S_{n+1} will be true whenever S_n is true. That is, suppose we prove that, *if* S_n is true, then it necessarily follows that S_{n+1} is also true. Then, if we can show that S_1 is true, it follows that S_2 is true, and from this that S_3 is true, hence that S_4 is true, and so on and on. This is the principle of induction. To express it more compactly, it is useful to have a short way to express the assertion "If S_n is true, then S_{n+1} is true." We write this assertion as

$$S_n \implies S_{n+1}$$

and read it as "S_n implies S_{n+1}." We can now state the principle of mathematical induction.

THE PRINCIPLE OF MATHEMATICAL INDUCTION

Suppose S_n is a statement about the natural number n for each $n = 1, 2, 3, \ldots$. Suppose further that:

1. S_1 is true.

2. $S_n \implies S_{n+1}$ for every $n \geq 1$.

Then S_n is true for every $n \geq 1$.

This is one of the most useful techniques in all of mathematics. It applies in a wide number of situations. The following examples illustrate this.

EXAMPLE 1

Show that $1 + 2 + \cdots + n = \frac{1}{2}n(n + 1)$ for $n \geq 1$.

Solution   Let S_n be the statement: $1 + 2 + \cdots + n = \frac{1}{2}n(n + 1)$. We apply induction.

1. S_1 is true. The statement S_1 is $1 = \frac{1}{2}1(1 + 1)$, which is true.

2. $S_n \implies S_{n+1}$. We *assume* that S_n is true for some $n \geq 1$—that is, that $1 + 2 + \cdots + n = \frac{1}{2}n(n + 1)$.

We must prove that the statement

$$S_{n+1}: \quad 1 + 2 + \cdots + (n + 1) = \tfrac{1}{2}(n + 1)(n + 2)$$

is also true, and we are entitled to use S_n to do so. Now the left side of S_{n+1} is the sum of the first $n + 1$ positive integers. Hence the second-to-last term is n, so we can write

$$1 + 2 + \cdots + (n + 1) = (1 + 2 + \cdots + n) + (n + 1)$$
$$= \tfrac{1}{2}n(n + 1) + (n + 1) \qquad \text{using } S_n$$
$$= \tfrac{1}{2}(n + 1)[n + 2]$$

This shows that S_{n+1} is true and so completes the induction.

◆◆◆

In the verification that $S_n \Rightarrow S_{n+1}$, we *assume* that S_n is true and use it to deduce that S_{n+1} is true. The assumption that S_n is true is sometimes called the **induction hypothesis.**

EXAMPLE 2

If x is any number such that $x \neq 1$, show that $1 + x + x^2 + \cdots + x^n = \dfrac{x^{n+1} - 1}{x - 1}$ for $n \geq 1$.

Solution   Let S_n be the statement: $1 + x + x^2 + \cdots + x^n = \dfrac{x^{n+1} - 1}{x - 1}$.

1. S_1 is true. S_1 reads $1 + x = \dfrac{x^2 - 1}{x - 1}$, which is true because $x^2 - 1 = (x - 1)(x + 1)$.

2. $S_n \Rightarrow S_{n+1}$. Assume the truth of S_n: $1 + x + x^2 + \cdots + x^n = \dfrac{x^{n+1} - 1}{x - 1}$.

We must *deduce* from this the truth of S_{n+1}: $1 + x + x^2 + \cdots + x^{n+1} = \dfrac{x^{n+2} - 1}{x - 1}$. Starting with the left side of S_{n+1}, we find

$$1 + x + x^2 + \cdots + x^{n+1} = (1 + x + x^2 + \cdots + x^n) + x^{n+1}$$
$$= \dfrac{x^{n+1} - 1}{x - 1} + x^{n+1}$$
$$= \dfrac{x^{n+1} - 1 + x^{n+1}(x - 1)}{x - 1}$$
$$= \dfrac{x^{n+2} - 1}{x + 1}$$

This shows that S_{n+1} is true and so completes the induction.

$\blacklozenge\blacklozenge\blacklozenge$

Both of these examples involve formulas for a certain sum, and it is often convenient to use summation notation. For example, $\sum_{k=1}^{n}(2k - 1)$ means that in the expression $(2k - 1)$, k is to be given the values $k = 1$, $k = 2$, $k = 3$, ..., $k = n$, and then the resulting n numbers are to be added. The same thing applies to other expressions involving k. For example,

$$\sum_{k=1}^{n} k^3 = 1^3 + 2^3 + \cdots + n^3$$

$$\sum_{k=1}^{5}(3k - 1) = (3 \cdot 1 - 1) + (3 \cdot 2 - 1) + (3 \cdot 3 - 1) + (3 \cdot 4 - 1) + (3 \cdot 5 - 1)$$

The next example involves this notation.

EXAMPLE 3

Show that $\sum_{k=1}^{n}(3k^2 - k) = n^2(n + 1)$ for each $n \geq 1$.

Solution   Let S_n be the statement: $\sum_{k=1}^{n}(3k^2 - k) = n^2(n + 1)$.

1. S_1 is true. S_1 reads $(3 \cdot 1^2 - 1) = 1^2(1 + 1)$, which is true.
2. $S_n \Rightarrow S_{n+1}$. Assume that S_n is true. We must prove S_{n+1}:

$$\sum_{k=1}^{n+1}(3k^2 - k) = \sum_{k=1}^{n}(3k^2 - k) + [3(n + 1)^2 - (n + 1)]$$
$$= n^2(n + 1) + (n + 1)[3(n + 1) - 1] \quad (\text{using } S_n)$$
$$= (n + 1)[n^2 + 3n + 2]$$
$$= (n + 1)[(n + 1)(n + 2)]$$
$$= (n + 1)^2(n + 2)$$

This proves that S_{n+1} is true.

◆◆◆

We now turn to examples wherein induction is used to prove propositions that do not involve sums.

EXAMPLE 4

Show that $7^n + 2$ is a multiple of 3 for all $n \geq 1$.

Solution
1. S_1 is true. $7^1 + 2 = 9$ is a multiple of 3.
2. $S_n \Rightarrow S_{n+1}$. Assume that $7^n + 2$ is a multiple of 3 for some $n \geq 1$; say, $7^n + 2 = 3m$ for some integer m. Then

$$7^{n+1} + 2 = 7(7^n) + 2 = 7(3m - 2) + 2 = 21m - 12 = 3(7m - 4)$$

so $7^{n+1} + 2$ is also a multiple of 3. This proves that S_{n+1} is true.

◆◆◆

In all the foregoing examples, we have used the principle of induction starting at 1; that is, we have verified that S_1 is true and that $S_n \Rightarrow S_{n+1}$ for each $n \geq 1$, and then we have concluded that S_n is true for every $n \geq 1$. But there is nothing special about 1 here. If m is some fixed integer and we verify that

1. S_m is true
2. $S_n \Rightarrow S_{n+1}$ for every $n \geq m$

then it follows that S_n is true for every $n \geq m$. This "extended" induction principle is just as plausible as the induction principle and can, in fact, be proved by induction. The next example will illustrate it. Recall that if n is a positive integer, the number $n!$ (which is read "n-factorial") is the product

$$n! = n(n - 1)(n - 2) \cdots 3 \cdot 2 \cdot 1$$

of all the numbers from n to 1. Thus $2! = 2$, $3! = 6$, and so on.

EXAMPLE 5

Show that $2^n < n!$ for all $n \geq 4$.

Solution   Observe that $2^n < n!$ is actually false if $n = 1, 2, 3$.

1. S_4 is true. $2^4 = 16 < 24 = 4!$.

2. $S_n \Rightarrow S_{n+1}$ if $n \geq 4$. Assume that S_n is true; that, is $2^n < n!$. Then

$$2^{n+1} = 2 \cdot 2^n$$
$$< 2 \cdot n! \qquad \text{because } 2^n < n!$$
$$< (n + 1)n! \qquad \text{because } 2 < n + 1$$
$$= (n + 1)!$$

Hence S_{n+1} is true.

◆◆◆

EXERCISES C

In Exercises 1–19, prove the given statement by induction for all $n \geq 1$.

1. $1 + 3 + 5 + 7 + \cdots + (2n - 1) = n^2$

2. $1^2 + 2^2 + \cdots + n^2 = \frac{1}{6}n(n + 1)(2n + 1)$

3. $1^3 + 2^3 + \cdots + n^3 = (1 + 2 + \cdots + n)^2$

4. $1 \cdot 2 + 2 \cdot 3 + \cdots + n(n + 1) = \frac{1}{3}n(n + 1)(n + 2)$

5. $1 \cdot 2^2 + 2 \cdot 3^2 + \cdots + n(n + 1)^2 = \frac{1}{12}n(n + 1)(n + 2)$
   $(3n + 5)$

◆ 6. $\dfrac{1}{1 \cdot 2} + \dfrac{1}{2 \cdot 3} + \cdots + \dfrac{1}{n(n + 1)} = \dfrac{n}{n + 1}$

7. $1^2 + 3^2 + \cdots + (2n - 1)^2 = \frac{n}{3}(4n^2 - 1)$

8. $\dfrac{1}{1 \cdot 2 \cdot 3} + \dfrac{1}{2 \cdot 3 \cdot 4} + \cdots + \dfrac{1}{n(n + 1)(n + 2)}$
   $= \dfrac{n(n + 3)}{4(n + 1)(n + 2)}$

9. $1 + 2 + 2^2 + \cdots + 2^{n-1} = 2^n - 1$

10. $3 + 3^3 + 3^5 + \cdots + 3^{2n-1} = \frac{3}{8}(9^n - 1)$

11. $\dfrac{1}{1^2} + \dfrac{1}{2^2} + \cdots + \dfrac{1}{n^2} \leq 2 - \dfrac{1}{n}$

12. $n < 2^n$

13. For any integer $m > 0$, $m! \, n! < (m + n)!$

◆ 14. $\dfrac{1}{\sqrt{1}} + \dfrac{1}{\sqrt{2}} + \cdots + \dfrac{1}{\sqrt{n}} \leq 2\sqrt{n} - 1$

15. $\dfrac{1}{\sqrt{1}} + \dfrac{1}{\sqrt{2}} + \cdots + \dfrac{1}{\sqrt{n}} \geq \sqrt{n}$

16. $n^3 + (n + 1)^3 + (n + 2)^3$ is a multiple of 9.

17. $5^n + 3$ is a multiple of 4.

18. $n^3 - n$ is a multiple of 3.

19. $3^{2n+1} + 2^{n+2}$ is a multiple of 7.

◆ 20. Let $B_n = 1 \cdot 1! + 2 \cdot 2! + 3 \cdot 3! + \cdots + n \cdot n!$ Find a formula for B_n and prove it.

21. Let $A_n = \left(1 - \frac{1}{2}\right)\left(1 - \frac{1}{3}\right)\left(1 - \frac{1}{4}\right) \cdots \left(1 - \frac{1}{n}\right)$. Find a formula for A_n and prove it.

22. Suppose S_n is a statement about n for each $n \geq 1$. Explain what must be done to prove that S_n is true for all $n \geq 1$ if:
    (a) It is known that $S_n \Rightarrow S_{n+2}$ for each $n \geq 1$.
    ◆ (b) It is known that $S_n \Rightarrow S_{n+8}$ for each $n \geq 1$.
    (c) It is known that $S_n \Rightarrow S_{n+1}$ for each $n \geq 10$.
    (d) It is known that both S_n and $S_{n+1} \Rightarrow S_{n+2}$ for each $n \geq 1$.

23. If S_n is a statement for each $n \geq 1$, argue that S_n is true for all $n \geq 1$ if it is known that the following two conditions hold:
    (a) $S_n \Rightarrow S_{n-1}$ for each $n \geq 2$.
    (b) S_n is true for infinitely many values of n.

24. Suppose a sequence $a_1, a_2, \ldots$ of numbers is given that satisfies:

(a) $a_1 = 2$

(b) $a_{n+1} = 2a_n$ for each $n \geq 1$

Formulate a theorem giving a_n in terms of n, and prove your result by induction.

25. Suppose a sequence $a_1, a_2, \ldots$ of numbers is given that satisfies:

(a) $a_1 = b$

(b) $a_{n+1} = ca_n + b$ for $n = 1, 2, 3, \ldots$

Formulate a theorem giving a_n in terms of n, and prove your result by induction.

26. (a) Show that $n^2 \leq 2^n$ for all $n \geq 4$.

(b) Show that $n^3 \leq 2^n$ for all $n \geq 10$.

Selected Answers

◆◆◆

Exercises 1.1
(Page 9)

2. (b) $x = t, y = \frac{1}{3}(1 - 2t)$ or $x = \frac{1}{2}(1 - 3s), y = s$ **(d)** $x = 1 + 2s - 5t, y = s, z = t$ or $x = s, y = t, z = \frac{1}{5}(1 - s + 2t)$ **4.** $x = \frac{1}{4}(3 + 2s), y = s, z = t$ **5. (a)** No solution if $b \neq 0$. If $b = 0$, *any x* is a solution.
(b) $x = \frac{b}{a}$

7. (b) $\begin{bmatrix} 1 & 2 & | & 0 \\ 0 & 1 & | & 1 \end{bmatrix}$ **(d)** $\begin{bmatrix} 1 & 1 & 0 & | & 1 \\ 0 & 1 & 1 & | & 0 \\ -1 & 0 & 1 & | & 2 \end{bmatrix}$ **8. (b)** $\begin{array}{r} 2x - y = -1 \\ -3x + 2y + z = 0 \\ y + z = 3 \end{array}$ or $\begin{array}{r} 2x_1 - x_2 = -1 \\ -3x_1 + 2x_2 + x_3 = 0 \\ x_2 + x_3 = 3 \end{array}$

9. (b) $x = -3, y = 2$ **(d)** $x = -17, y = 13$ **10. (b)** $x = \frac{1}{9}, y = \frac{10}{9}, z = \frac{-7}{3}$ **11. (b)** No solution
14. $x' = 5, y' = 1$, so $x = 23, y = -32$ **15.** $a = -\frac{1}{9}, b = -\frac{5}{9}, c = \frac{11}{9}$

Exercises 1.2
(Page 22)

1. (b) No, no   **(d)** No, yes   **(f)** No, no   **2. (b)** $\begin{bmatrix} 0 & 1 & -3 & 0 & 0 & 0 & 0 \\ 0 & 0 & 0 & 1 & 0 & 0 & -1 \\ 0 & 0 & 0 & 0 & 1 & 0 & 0 \\ 0 & 0 & 0 & 0 & 0 & 1 & 1 \end{bmatrix}$

3. (b) $x_1 = 2r - 2s - t + 1, x_2 = r, x_3 = -5s + 3t - 1, x_4 = s, x_5 = -6t + 1, x_6 = t$ **(d)** $x_1 = -4s - 5t - 4, x_2 = -2s + t - 2, x_3 = s, x_4 = 1, x_5 = t$ **4. (b)** $x = -\frac{1}{7}, y = -\frac{3}{7}$ **(d)** $x = \frac{1}{3}(t + 2), y = t$ **(f)** No solution
5. (b) $x = -15t - 21, y = -11t - 17, z = t$ **(d)** No solution   **(f)** $x = -7, y = -9, z = 1$
(h) $x = 4, y = 3 + 2t, z = t$ **6. (b)** $x_1 = 2t + 8, x_2 = -t - 19, x_3 = t. R_3 = 5R_1 - 4R_2$
7. (b) $x_1 = 0, x_2 = -t, x_3 = 0, x_4 = t$ **(d)** $x_1 = 1, x_2 = 1 - t, x_3 = 1 + t, x_4 = t$

8. (b) If $ab \neq 2$, unique solution $x = \dfrac{-2 - 5b}{2 - ab}, y = \dfrac{a + 5}{2 - ab}$. If $ab = 2$: no solution if $a \neq -5$; if $a = -5$, the solutions are

$x = -1 + \frac{2}{5}t, y = t$.   **(d)** If $a \neq 2$, unique solution $x = \dfrac{1 - b}{a - 2}, y = \dfrac{ab - 2}{a - 2}$. If $a = 2$, no solution if $b \neq 1$;

if $b = 1$, the solutions are $x = \frac{1}{2}(1 - t), y = t$.   **9. (b)** Unique solution $x = -2a + b + 5c, y = 3a - b - 6c$,
$z = -2a + b + 4c$, for any a, b, c.   **(d)** If $abc \neq -1$, unique solution $x = y = z = 0$; if $abc = -1$ the solutions are
$x = abt, y = -bt, z = t$.   **(f)** If $a = 1$, solutions $x = -t, y = t, z = -1$. If $a = 0$, there is no solution. If $a \neq 1$ and

$a \neq 0$, unique solution $x = \dfrac{a - 1}{a}, y = 0, z = \dfrac{-1}{a}$.   **10. (b)** 1   **(d)** 3   **(f)** 1

11. (b) 2   **(d)** 3   **(f)** 2 if $a = 0$ or $a = 2$; 3, otherwise.   **12. (b)** False. $A = \begin{bmatrix} 1 & 0 & | & 1 \\ 0 & 1 & | & 1 \\ 0 & 0 & | & 0 \end{bmatrix}$

(d) False. $A = \begin{bmatrix} 1 & 0 & | & 1 \\ 0 & 1 & | & 0 \\ 0 & 0 & | & 0 \end{bmatrix}$ **(f)** True. A has 3 rows, so there are at most 3 leading 1's.

16. (b) $x^2 + y^2 - 2x + 6y - 6 = 0$ **18.** $\frac{5}{20}$ in A, $\frac{7}{20}$ in B, $\frac{8}{20}$ in C

Exercises 1.3
(Page 26)

1. (b) False. $A = \begin{bmatrix} 1 & 0 & 1 & | & 0 \\ 0 & 1 & 1 & | & 0 \end{bmatrix}$ **(d)** False. $A = \begin{bmatrix} 1 & 0 & 1 & | & 1 \\ 0 & 1 & 1 & | & 0 \end{bmatrix}$ **(f)** False. $A = \begin{bmatrix} 1 & 0 & | & 0 \\ 0 & 1 & | & 0 \end{bmatrix}$ **(h)** False. $A = \begin{bmatrix} 1 & 0 & | & 0 \\ 0 & 1 & | & 0 \\ 0 & 0 & | & 0 \end{bmatrix}$

2. (b) $a = -3, x = 9t, y = -5t, z = t$ **(d)** $a = 1, x = -t, y = t, z = 0$; or $a = -1, x = t, y = 0, z = t$

4. (b) By Theorem 2§1.2, there are $n - r = 6 - 1 = 5$ parameters.   **(d)** If R is the row-echelon form of A, then R has a row of zeros and 4 rows in all. Hence R has $r = \text{rank } A = 1, 2,$ or 3 leading 1's. Thus there are $n - r = 6 - r = 5, 4,$ or 3 parameters.

Exercises 1.4
(Page 28)

1. (b) $\begin{aligned} f_1 &= 85 - f_4 - f_7 \\ f_2 &= 60 - f_4 - f_7 \\ f_3 &= -75 + f_4 + f_6 \\ f_5 &= 40 - f_6 + f_7 \end{aligned}$ **2. (b)** $f_5 = 15$ **3. (b)** CD

$25 \le f_4 \le 30$

Exercises 1.5
(Page 30)

2. $I_1 = -\frac{1}{5}, I_2 = \frac{3}{5}, I_3 = \frac{4}{5}$ **4.** $I_1 = 2, I_2 = 1, I_3 = \frac{1}{2}, I_4 = \frac{3}{2}, I_5 = \frac{3}{2}, I_6 = \frac{1}{2}$

Supplementary Exercises for Chapter 1
(Page 30)

1. (b) No. If the corresponding planes are parallel and distinct, there is no solution. Otherwise they either coincide or have a whole common line of solutions.   **2. (b)** $x_1 = \frac{1}{10}(-6s - 6t + 16), x_2 = \frac{1}{10}(4s - t + 1), x_3 = s, x_4 = t$

3. (b) If $a = 1$, no solution. If $a = 2, x = 2 - 2t, y = -t, z = t$. If $a \ne 1$ and $a \ne 2$, the unique solution is $x = \dfrac{8 - 5a}{3(a - 1)}$,

$y = \dfrac{-2 - a}{3(a - 1)}, z = \dfrac{a + 2}{3}$ **4.** $\begin{bmatrix} R_1 \\ R_2 \end{bmatrix} \to \begin{bmatrix} R_1 + R_2 \\ R_2 \end{bmatrix} \to \begin{bmatrix} R_1 + R_2 \\ -R_1 \end{bmatrix} \to \begin{bmatrix} R_2 \\ -R_1 \end{bmatrix} \to \begin{bmatrix} R_2 \\ R_1 \end{bmatrix}$

6. $a = 1, b = 2, c = -1$ **9. (b)** 5 of brand 1, 0 of brand 2, 3 of brand 3

Exercises 2.1
(Page 42)

1. (b) $(a, b, c, d) = (-2, -4, -6, 0) + t(1, 1, 1, 1)$, t arbitrary   **(d)** $a = b = c = d = t$, t arbitrary

2. (b) $\begin{bmatrix} -14 \\ -20 \end{bmatrix}$ **(d)** $(-12, 4, -12)$ **(f)** $\begin{bmatrix} 0 & 1 & -2 \\ -1 & 0 & 4 \\ 2 & -4 & 0 \end{bmatrix}$ **(h)** $\begin{bmatrix} 4 & -1 \\ -1 & -6 \end{bmatrix}$

3. (b) $\begin{bmatrix} 15 & -5 \\ 10 & 0 \end{bmatrix}$ **(d)** Impossible **(f)** $\begin{bmatrix} 5 & 2 \\ 0 & -1 \end{bmatrix}$ **(h)** Impossible **4. (b)** $\begin{bmatrix} 4 \\ \frac{1}{2} \end{bmatrix}$ **5. (b)** $A = -\frac{11}{3}B$

6. (b) $X = 4A - 3B$, $Y = 4B - 5A$ **7. (b)** $Y = (s, t)$, $X = \frac{1}{2}(1 + 5s, 2 + 5t)$; s and t arbitrary

8. (b) $20A - 7B + 2C$ **9. (b)** If $A = \begin{bmatrix} a & b \\ c & d \end{bmatrix}$, then $(p, q, r, s) = \frac{1}{2}(2d, a + b - c - d, a - b + c - d, -a + b + c + d)$. **11. (b)** If $A + A' = 0$ then $-A = -A + 0 = -A + (A + A') = (-A + A) + A' = 0 + A' = A'$

14. (b) $s = 1$ or $t = 0$ **(d)** $s = 0, t = 3$ **15. (b)** $\begin{bmatrix} 2 & 0 \\ 1 & -1 \end{bmatrix}$ **(d)** $\begin{bmatrix} 2 & 7 \\ -\frac{9}{2} & -5 \end{bmatrix}$ **16. (b)** $A = A^T$, so using Theorem 2§2.1, $(kA)^T = kA^T = kA$. **18. (c)** Suppose $A = S + W$, where $S = S^T$ and $W = -W^T$. Then $A^T = S^T + W^T = S - W$, so $A + A^T = 2S$ and $A - A^T = 2W$. Hence $S = \frac{1}{2}(A + A^T)$ and $W = \frac{1}{2}(A - A^T)$ are uniquely determined by A.

Exercises 2.2
(Page 54)

1. (b) $\begin{bmatrix} -1 & -6 & -2 \\ 0 & 6 & 10 \end{bmatrix}$ **(d)** $[-3 \ -15]$ **(f)** $[-23]$ **(h)** $\begin{bmatrix} 1 & 0 \\ 0 & 1 \end{bmatrix}$ **(j)** $\begin{bmatrix} aa' & 0 & 0 \\ 0 & bb' & 0 \\ 0 & 0 & cc' \end{bmatrix}$

2. (b) $BA = \begin{bmatrix} -1 & 4 & -10 \\ 1 & 2 & 4 \end{bmatrix}$ $B^2 = \begin{bmatrix} 7 & -6 \\ -1 & 6 \end{bmatrix}$ $CB = \begin{bmatrix} -2 & 12 \\ 2 & -6 \\ 1 & 6 \end{bmatrix}$ $AC = \begin{bmatrix} 4 & 10 \\ -2 & -1 \end{bmatrix}$ $CA = \begin{bmatrix} 2 & 4 & 8 \\ -1 & -1 & -5 \\ 1 & 4 & 2 \end{bmatrix}$

3. (b) $(a, b, a_1, b_1) = (3, 0, 1, 2)$ **4. (b)** $A^2 - A - 6I = \begin{bmatrix} 8 & 2 \\ 2 & 5 \end{bmatrix} - \begin{bmatrix} 2 & 2 \\ 2 & -1 \end{bmatrix} - \begin{bmatrix} 6 & 0 \\ 0 & 6 \end{bmatrix} = \begin{bmatrix} 0 & 0 \\ 0 & 0 \end{bmatrix}$

5. (b) $A(BC) = \begin{bmatrix} 1 & -1 \\ 0 & 1 \end{bmatrix} \begin{bmatrix} -9 & -16 \\ 5 & 1 \end{bmatrix} = \begin{bmatrix} -14 & -17 \\ 5 & 1 \end{bmatrix} = \begin{bmatrix} -2 & -1 & -2 \\ 3 & 1 & 0 \end{bmatrix} \begin{bmatrix} 1 & 0 \\ 2 & 1 \\ 5 & 8 \end{bmatrix} = (AB)C$

7. (b) $AX = B$, where $A = \begin{bmatrix} -1 & 2 & -1 & 1 \\ 2 & 1 & -1 & 2 \\ 3 & -2 & 0 & 1 \end{bmatrix}$, $X = \begin{bmatrix} x_1 \\ x_2 \\ x_3 \\ x_4 \end{bmatrix}$, and $B = \begin{bmatrix} 6 \\ 1 \\ 0 \end{bmatrix}$ **8. (b)** $\begin{bmatrix} -2 \\ 2 \\ 0 \end{bmatrix} + t \begin{bmatrix} 1 \\ -3 \\ 1 \end{bmatrix}$ **(d)** $\begin{bmatrix} 3 \\ -9 \\ -2 \\ 0 \end{bmatrix} + t \begin{bmatrix} -1 \\ 4 \\ 1 \\ 1 \end{bmatrix}$

10. (b) $s \begin{bmatrix} -7 \\ 15 \\ 7 \\ 2 \\ 0 \end{bmatrix} + t \begin{bmatrix} -7 \\ 17 \\ 9 \\ 0 \\ 4 \end{bmatrix}$ **12. (b)** $m \times n$ and $n \times m$ for some m and n **13. (b) (i)** $\begin{bmatrix} 1 & 0 \\ 0 & 1 \end{bmatrix}, \begin{bmatrix} 1 & 0 \\ 0 & -1 \end{bmatrix}, \begin{bmatrix} 1 & 1 \\ 0 & -1 \end{bmatrix}$

(ii) $\begin{bmatrix} 1 & 0 \\ 0 & 0 \end{bmatrix}, \begin{bmatrix} 1 & 0 \\ 0 & 1 \end{bmatrix}, \begin{bmatrix} 1 & 1 \\ 0 & 0 \end{bmatrix}$

16. (b) $AB = \begin{bmatrix} 2 & -1 & 3 & 1 \\ 1 & 0 & 1 & 2 \\ 0 & 0 & 1 & 0 \\ 0 & 0 & 0 & 1 \end{bmatrix} \begin{bmatrix} 1 & 2 & 0 \\ -1 & 0 & 0 \\ 0 & 5 & 1 \\ 1 & -1 & 0 \end{bmatrix} = \begin{bmatrix} \begin{bmatrix} 3 & 19 \\ 1 & 7 \\ 0 & 5 \end{bmatrix} + \begin{bmatrix} 1 & -1 \\ 2 & -2 \\ 0 & 0 \end{bmatrix} & \begin{bmatrix} 3 \\ 1 \\ 1 \end{bmatrix} + \begin{bmatrix} 0 \\ 0 \\ 0 \end{bmatrix} \\ (1-1) & 0 \end{bmatrix} = \begin{bmatrix} 4 & 18 & 3 \\ 3 & 5 & 1 \\ 0 & 5 & 1 \\ 1 & -1 & 0 \end{bmatrix}$

17. (b) $A^{2k} = \begin{bmatrix} 1 & -2k & 0 & 0 \\ 0 & 1 & 0 & 0 \\ 0 & 0 & 1 & 0 \\ 0 & 0 & 0 & 1 \end{bmatrix}$ for $k = 0, 1, 2, \ldots$, $A^{2k+1} = A^k A = \begin{bmatrix} 1 & -(2k+1) & 2 & -1 \\ 0 & 1 & 0 & 0 \\ 0 & 0 & -1 & 1 \\ 0 & 0 & 0 & 1 \end{bmatrix}$ for $k = 0, 1, 2, \ldots$

18. (b) $\begin{bmatrix} I & 0 \\ 0 & I \end{bmatrix} = I_{2k}$ **(d)** 0_k **(f)** $\begin{bmatrix} X^m & 0 \\ 0 & X^m \end{bmatrix}$ if $n = 2m$; $\begin{bmatrix} 0 & X^{m+1} \\ X^m & 0 \end{bmatrix}$ if $n = 2m + 1$.

20. (b) If Y is row i of the identity matrix I, then YA is row i of $IA = A$.   **22. (b)** $AB - BA$ **(d)** 0

24. (b) $(kA)C = k(AC) = k(CA) = C(kA)$

26. We have $A^T = A$ and $B^T = B$, so $(AB)^T = B^TA^T = BA$. Hence AB is symmetric if and only if $AB = BA$.

30. If $BC = I$, then $AB = 0$ gives $0 = 0C = (AB)C = A(BC) = AI = A$, contrary to assumption.

33. (b) If $A = [a_{ij}]$, then $\operatorname{tr}(kA) = \operatorname{tr}[ka_{ij}] = \sum_{i=1}^{n} ka_{ii} = k \sum_{i=1}^{n} a_{ii} = k \operatorname{tr}(A)$.

   (e) Write $A^T = [a'_{ij}]$, where $a'_{ij} = a_{ji}$. Then $AA^T = \left(\sum_{k=1}^{n} a_{ik}a'_{kj} \right)$, so $\operatorname{tr}(AA^T) = \sum_{i=1}^{n} \left[\sum_{k=1}^{n} a_{ik}a'_{ki} \right] = \sum_{i=1}^{n} \sum_{k=1}^{n} a_{ik}^2$.

35. (e) Observe that $PQ = P^2 + PAP - P^2AP = P$, so $Q^2 = PQ + APQ - PAPQ = P + AP - PAP = Q$.

37. (b) $(A + B)(A - B) = A^2 - AB + BA - B^2$, and $(A - B)(A + B) = A^2 + AB - BA - B^2$. These are equal if and only if $- AB + BA = AB - BA$; that is, $2BA = 2AB$; that is, $BA = AB$.

Exercises 2.3

(Page 65)

2. (b) $\frac{1}{5}\begin{bmatrix} 2 & -1 \\ -3 & 4 \end{bmatrix}$ **(d)** $\begin{bmatrix} 2 & -1 & 3 \\ 3 & 1 & -1 \\ 1 & 1 & -2 \end{bmatrix}$ **(f)** $\frac{1}{10}\begin{bmatrix} 1 & 4 & -1 \\ -2 & 2 & 2 \\ -9 & 14 & -1 \end{bmatrix}$ **(h)** $\frac{1}{4}\begin{bmatrix} 2 & 0 & -2 \\ -5 & 2 & 5 \\ -3 & 2 & -1 \end{bmatrix}$

(j) $\begin{bmatrix} 0 & 0 & 1 & -2 \\ -1 & -2 & -1 & -3 \\ 1 & 2 & 1 & 2 \\ 0 & -1 & 0 & 0 \end{bmatrix}$ **(l)** $\begin{bmatrix} 1 & -2 & 6 & -30 & 210 \\ 0 & 1 & -3 & 15 & -105 \\ 0 & 0 & 1 & -5 & 35 \\ 0 & 0 & 0 & 1 & -7 \\ 0 & 0 & 0 & 0 & 1 \end{bmatrix}$ **3. (b)** $\begin{bmatrix} x \\ y \end{bmatrix} = \frac{1}{5}\begin{bmatrix} 4 & -3 \\ 1 & -2 \end{bmatrix}\begin{bmatrix} 0 \\ 1 \end{bmatrix} = \frac{1}{5}\begin{bmatrix} -3 \\ -2 \end{bmatrix}$

(d) $\begin{bmatrix} x \\ y \\ z \end{bmatrix} = \frac{1}{5}\begin{bmatrix} 9 & -14 & 6 \\ 4 & -4 & 1 \\ -10 & 15 & -5 \end{bmatrix}\begin{bmatrix} 1 \\ -1 \\ 0 \end{bmatrix} = \frac{1}{5}\begin{bmatrix} 23 \\ 8 \\ -25 \end{bmatrix}$ **(f)** $\begin{bmatrix} x \\ y \\ z \\ w \end{bmatrix} = \frac{1}{2}\begin{bmatrix} 0 & 1 & -1 & 1 \\ 0 & 1 & 1 & -1 \\ 2 & -1 & -1 & -1 \\ 0 & -1 & 1 & 1 \end{bmatrix}\begin{bmatrix} 1 \\ 0 \\ -1 \\ 2 \end{bmatrix} = \frac{1}{2}\begin{bmatrix} 3 \\ -3 \\ 1 \\ 1 \end{bmatrix}$

4. (b) $B = A^{-1}(AB) = \begin{bmatrix} 4 & -2 & 1 \\ 7 & -2 & 4 \\ -1 & 2 & -1 \end{bmatrix}$ **5. (b)** $\frac{1}{10}\begin{bmatrix} 3 & -2 \\ 1 & 1 \end{bmatrix}$ **(d)** $\frac{1}{2}\begin{bmatrix} 0 & 1 \\ 1 & -1 \end{bmatrix}$ **(f)** $\frac{1}{2}\begin{bmatrix} 2 & 0 \\ -6 & 1 \end{bmatrix}$ **(h)** $-\frac{1}{2}\begin{bmatrix} 11 \\ 10 \end{bmatrix}$

6. (b) $A = \frac{1}{2}\begin{bmatrix} 2 & -1 & 3 \\ 0 & 1 & -1 \\ -2 & 1 & -1 \end{bmatrix}$ **8. (b)** $AB = I$ **9. (b)** False. $\begin{bmatrix} 1 & 0 \\ 0 & 1 \end{bmatrix} + \begin{bmatrix} 1 & 0 \\ 0 & -1 \end{bmatrix}$ **(d)** True. $A^{-1} = \frac{1}{3}A^3$

   (f) False. $A = B = \begin{bmatrix} 1 & 0 \\ 0 & 0 \end{bmatrix}$ **10.** $B = IB = (CA)B = C(AB) = C$ **11. (b) (ii)** $\begin{bmatrix} x_1 \\ x_2 \end{bmatrix} = \begin{bmatrix} 2 \\ -1 \end{bmatrix}$

15. (b) $B^4 = I$, so $B^{-1} = B^3 = \begin{bmatrix} 0 & 1 \\ -1 & 0 \end{bmatrix}$ **16. (b)** $\frac{1}{6 + c^2}\begin{bmatrix} 3 & c \\ -c & 2 \end{bmatrix}$ **(d)** $\begin{bmatrix} c^2 - 2 & -c & 1 \\ -c & 1 & 0 \\ 3 - c^2 & c & -1 \end{bmatrix}$

19. (b) If column j of A is zero, $AY = 0$ where Y is column j of the identity matrix. Proceed as in Example 10.
　　(d) If each column of A sums to 0, $XA = 0$ where X is the row of 1's. Proceed as in Example 10.

20. (b) (ii) $(-1, 1, 1)A = 0$　　　**23. (d) (ii)** $\begin{bmatrix} 2 & -1 & 0 \\ -5 & 3 & 0 \\ \hline 0 & 0 & -1 \end{bmatrix}$　**(iv)** $\begin{bmatrix} 3 & -4 & 0 & 0 \\ -2 & 3 & 0 & 0 \\ \hline 0 & 0 & 1 & 3 \\ 0 & 0 & 0 & -1 \end{bmatrix}$

24. (b) Verify that $= \begin{bmatrix} A & X \\ 0 & B \end{bmatrix}\begin{bmatrix} A^{-1} & -A^{-1}XB^{-1} \\ 0 & B^{-1} \end{bmatrix} = \begin{bmatrix} AA^{-1} & -XB^{-1} + XB^{-1} \\ 0 & BB^{-1} \end{bmatrix} = \begin{bmatrix} I & 0 \\ 0 & I \end{bmatrix}$.

　　Similarly: $\begin{bmatrix} A^{-1} & -A^{-1}XB^{-1} \\ 0 & B^{-1} \end{bmatrix}\begin{bmatrix} A & X \\ 0 & B \end{bmatrix} = \begin{bmatrix} I & 0 \\ 0 & I \end{bmatrix}$.　　**(c) (ii)** $\begin{bmatrix} 1 & -1 & 7 & -13 \\ -2 & 3 & -18 & 33 \\ \hline 0 & 0 & -1 & 2 \\ 0 & 0 & 3 & -5 \end{bmatrix}$

26. (b) $P = \begin{bmatrix} 1 & 0 \\ 0 & 0 \end{bmatrix}$, $Q = \begin{bmatrix} 1 & 0 \\ 0 & 0 \end{bmatrix}$, $R = \begin{bmatrix} 1 & 0 \\ 1 & 0 \end{bmatrix}$　　**27. (d)** If $A^n = 0$, $(I - A)^{-1} = I + A + \cdots + A^{n-1}$.

29. (b) $A[B(AB)^{-1}] = I = [(BA)^{-1}B]A$, so A is invertible by Exercise 10.

31. (a) Have $AC = CA$. Left-multiply by A^{-1} to get $C = A^{-1}CA$. Then right-multiply by A^{-1} to get $CA^{-1} = A^{-1}C$.

35. (b) $(I - 2P)^2 = I - 4P + 4P^2$, and this equals I if and only if $P^2 = P$.

37. (b) $(A^{-1} + B^{-1})^{-1} = B(A + B)^{-1}A$

Exercises 2.4
(Page 76)

1. (b) Interchange rows 1 and 3 of I. $E^{-1} = E$　　**(d)** Add (-2) times row 1 of I to row 2. $E^{-1} = \begin{bmatrix} 1 & 0 & 0 \\ 2 & 1 & 0 \\ 0 & 0 & 1 \end{bmatrix}$

　(f) Multiply row 3 of I by 5. $E^{-1} = \begin{bmatrix} 1 & 0 & 0 \\ 0 & 1 & 0 \\ 0 & 0 & \frac{1}{5} \end{bmatrix}$　　**2. (b)** $\begin{bmatrix} -1 & 0 \\ 0 & 1 \end{bmatrix}$　**(d)** $\begin{bmatrix} 1 & -1 \\ 0 & 1 \end{bmatrix}$　**(f)** $\begin{bmatrix} 0 & 1 \\ 1 & 0 \end{bmatrix}$

3. (b) The only possibilities for E are $\begin{bmatrix} 0 & 1 \\ 1 & 0 \end{bmatrix}$, $\begin{bmatrix} k & 0 \\ 0 & 1 \end{bmatrix}$, $\begin{bmatrix} 1 & 0 \\ 0 & k \end{bmatrix}$, $\begin{bmatrix} 1 & k \\ 0 & 1 \end{bmatrix}$, and $\begin{bmatrix} 1 & 0 \\ k & 1 \end{bmatrix}$. In each case, EA has a row different

　　from C.　　**5. (b)** 0 is not invertible.　　**6. (b)** $\begin{bmatrix} 1 & -2 \\ 0 & 1 \end{bmatrix}\begin{bmatrix} 1 & 0 \\ 0 & \frac{1}{2} \end{bmatrix}\begin{bmatrix} 1 & 0 \\ -5 & 1 \end{bmatrix} A = \begin{bmatrix} 1 & 0 & 7 \\ 0 & 1 & -3 \end{bmatrix}$. Alternatively,

$\begin{bmatrix} 1 & 0 \\ 0 & \frac{1}{2} \end{bmatrix}\begin{bmatrix} 1 & -1 \\ 0 & 1 \end{bmatrix}\begin{bmatrix} 1 & 0 \\ -5 & 1 \end{bmatrix} A = \begin{bmatrix} 1 & 0 & 7 \\ 0 & 1 & -3 \end{bmatrix}$.

　(d) $\begin{bmatrix} 1 & 2 & 0 \\ 0 & 1 & 0 \\ 0 & 0 & 1 \end{bmatrix}\begin{bmatrix} 1 & 0 & 0 \\ 0 & \frac{1}{5} & 0 \\ 0 & 0 & 1 \end{bmatrix}\begin{bmatrix} 1 & 0 & 0 \\ 0 & 1 & 0 \\ 0 & -1 & 1 \end{bmatrix}\begin{bmatrix} 1 & 0 & 0 \\ 0 & 1 & 0 \\ -2 & 0 & 1 \end{bmatrix}\begin{bmatrix} 1 & 0 & 0 \\ -3 & 1 & 0 \\ 0 & 0 & 1 \end{bmatrix}\begin{bmatrix} 0 & 0 & 1 \\ 0 & 1 & 0 \\ 1 & 0 & 0 \end{bmatrix} A = \begin{bmatrix} 1 & 0 & \frac{1}{5} & \frac{1}{5} \\ 0 & 1 & -\frac{7}{5} & -\frac{2}{5} \\ 0 & 0 & 0 & 0 \end{bmatrix}$

7. (b) $U = \begin{bmatrix} 1 & 1 \\ 1 & 0 \end{bmatrix} = \begin{bmatrix} 1 & 1 \\ 0 & 1 \end{bmatrix}\begin{bmatrix} 0 & 1 \\ 1 & 0 \end{bmatrix}$　　**8. (b)** $A = \begin{bmatrix} 0 & 1 \\ 1 & 0 \end{bmatrix}\begin{bmatrix} 1 & 0 \\ 2 & 1 \end{bmatrix}\begin{bmatrix} 1 & 0 \\ 0 & -1 \end{bmatrix}\begin{bmatrix} 1 & 2 \\ 0 & 1 \end{bmatrix}$

　(d) $A = \begin{bmatrix} 1 & 0 & 0 \\ 0 & 1 & 0 \\ -2 & 0 & 1 \end{bmatrix}\begin{bmatrix} 1 & 0 & 0 \\ 0 & 1 & 0 \\ 0 & 2 & 1 \end{bmatrix}\begin{bmatrix} 1 & 0 & -3 \\ 0 & 1 & 0 \\ 0 & 0 & 1 \end{bmatrix}\begin{bmatrix} 1 & 0 & 0 \\ 0 & 1 & 4 \\ 0 & 0 & 1 \end{bmatrix}$　　**14. (ii)** implies **(i)**. If $YA = 0$ and Y is $1 \times n$, then

　　$YAB = 0$, so $Y = 0$ (AB is invertible). Hence A is invertible (Theorem 8), so $B = A^{-1}(AB)$ is invertible (Theorem 2).

15. (b) $B\begin{bmatrix} -1 \\ 3 \\ -1 \end{bmatrix} = 0$, so B is not invertible by Theorem 10.　　**21. (b) (i)** $A \underset{r}{\sim} A$ because $A = IA$

(ii) If $A \; \underline{r} \; B$ then $A = UB$, U invertible, so $B = U^{-1}A$. Thus $B \; \underline{r} \; A$.   **(iii)** If $A \; \underline{r} \; B$ and $B \; \underline{r} \; C$, then $A = UB$ and $B = VC$,
    U and V invertible. Hence $A = U(VC) = (UV)C$, so $A \; \underline{r} \; C$.

23. (b) If $B \; \underline{r} \; A$, let $B = UA$, U invertible. If $U = \begin{bmatrix} a & b \\ c & d \end{bmatrix}$, $B = UA = \begin{bmatrix} 0 & 0 & b \\ 0 & 0 & d \end{bmatrix}$ where b and d are not both zero (as U is

    invertible). Every such matrix B arises in this way: Use $U = \begin{bmatrix} a & b \\ -b & a \end{bmatrix}$ — it is invertible by Example 4§2.3.

26. (b) Multiply column i by $1/k$.

Exercises 2.5
(Page 89)

1. (b) $\begin{bmatrix} 2 & 0 & 0 \\ 1 & -3 & 0 \\ -1 & 9 & 1 \end{bmatrix} \begin{bmatrix} 1 & 2 & 1 \\ 0 & 1 & -\frac{2}{3} \\ 0 & 0 & 0 \end{bmatrix}$ **(d)** $\begin{bmatrix} -1 & 0 & 0 & 0 \\ 1 & 1 & 0 & 0 \\ 1 & -1 & 1 & 0 \\ 0 & -2 & 0 & 1 \end{bmatrix} \begin{bmatrix} 1 & 3 & -1 & 0 & 1 \\ 0 & 1 & 2 & 1 & 0 \\ 0 & 0 & 0 & 0 & 0 \\ 0 & 0 & 0 & 0 & 0 \end{bmatrix}$

    (f) $\begin{bmatrix} 2 & 0 & 0 & 0 \\ 1 & -2 & 0 & 0 \\ 3 & -2 & 1 & 0 \\ 1 & 2 & 0 & 1 \end{bmatrix} \begin{bmatrix} 1 & 1 & -1 & 2 & 1 \\ 0 & 1 & -\frac{1}{2} & 0 & 0 \\ 0 & 0 & 0 & 0 & 0 \\ 0 & 0 & 0 & 0 & 0 \end{bmatrix}$

2. (b) $P = \begin{bmatrix} 0 & 0 & 1 \\ 1 & 0 & 0 \\ 0 & 1 & 0 \end{bmatrix}$ $PA = \begin{bmatrix} -1 & 2 & 1 \\ 0 & -1 & 2 \\ 0 & 0 & 4 \end{bmatrix} = \begin{bmatrix} -1 & 0 & 0 \\ 0 & -1 & 0 \\ 0 & 0 & 4 \end{bmatrix} \begin{bmatrix} 1 & -2 & -1 \\ 0 & 1 & -2 \\ 0 & 0 & 1 \end{bmatrix}$

    (d) $P = \begin{bmatrix} 1 & 0 & 0 & 0 \\ 0 & 0 & 1 & 0 \\ 0 & 0 & 0 & 1 \\ 0 & 1 & 0 & 0 \end{bmatrix}$ $PA = \begin{bmatrix} -1 & -2 & 3 & 0 \\ 1 & 1 & -1 & 3 \\ 2 & 5 & -10 & 1 \\ 2 & 4 & -6 & 5 \end{bmatrix} = \begin{bmatrix} -1 & 0 & 0 & 0 \\ 1 & -1 & 0 & 0 \\ 2 & 1 & -2 & 0 \\ 2 & 0 & 0 & 5 \end{bmatrix} \begin{bmatrix} 1 & 2 & -3 & 0 \\ 0 & 1 & -2 & -3 \\ 0 & 0 & 1 & -2 \\ 0 & 0 & 0 & 1 \end{bmatrix}$

3. (b) $Y = \begin{bmatrix} -1 \\ 0 \\ 0 \end{bmatrix}$ $X = \begin{bmatrix} -1 + 2t \\ -t \\ s \\ t \end{bmatrix}$ s and t arbitrary   **(d)** $Y = \begin{bmatrix} 2 \\ 8 \\ -1 \\ 0 \end{bmatrix}$ $X = \begin{bmatrix} 8 - 2t \\ 6 - t \\ -1 - t \\ t \end{bmatrix}$ t arbitrary

6. $\begin{bmatrix} R_1 \\ R_2 \end{bmatrix} \rightarrow \begin{bmatrix} R_1 + R_2 \\ R_2 \end{bmatrix} \rightarrow \begin{bmatrix} R_1 + R_2 \\ -R_1 \end{bmatrix} \rightarrow \begin{bmatrix} R_2 \\ -R_1 \end{bmatrix} \rightarrow \begin{bmatrix} R_2 \\ R_1 \end{bmatrix}$

7. (b) Let $A = LU = L_1U_1$ be LU-factorizations of the invertible matrix A. Then U and U_1 have no row of zeros and so (being row-echelon) are upper triangular with 1's on the main diagonal. Thus, using **(a)**, the diagonal matrix $D = UU_1^{-1}$ has 1's on the main diagonal. Thus $D = I$, $U = U_1$, and $L = L_1$.

8. If $A = \begin{bmatrix} a & 0 \\ X & A_1 \end{bmatrix}$ and $B = \begin{bmatrix} b & 0 \\ Y & B_1 \end{bmatrix}$ in block form, then $AB = \begin{bmatrix} ab & 0 \\ Xb + A_1Y & A_1B_1 \end{bmatrix}$, and A_1B_1 is lower triangular by induction.

Exercises 2.6
(Page 95)

1. (b) $\begin{bmatrix} t \\ 3t \\ t \end{bmatrix}$ **(d)** $\begin{bmatrix} 14t \\ 17t \\ 47t \\ 23t \end{bmatrix}$ **2.** $\begin{bmatrix} t \\ t \\ t \end{bmatrix}$ **4.** $P = \begin{bmatrix} bt \\ (1-a)t \end{bmatrix}$ is nonzero (for some t) unless $b = 0$ and $a = 1$. In that case,

    $\begin{bmatrix} 1 \\ 1 \end{bmatrix}$ is a solution. If the entries of A are positive, then $P = \begin{bmatrix} b \\ 1 - a \end{bmatrix}$ has positive entries.

7. (b) $\begin{bmatrix} 0.4 & 0.8 \\ 0.7 & 0.2 \end{bmatrix}$ **9. (b)** Use $P = \begin{bmatrix} 3 \\ 2 \\ 1 \end{bmatrix}$ in Theorem 2.

Exercises 2.7
(Page 104)

1. (b) Not regular   **2. (b)** $\frac{1}{3}\begin{bmatrix} 2 \\ 1 \end{bmatrix}, \frac{3}{8}$ **(d)** $\frac{1}{3}\begin{bmatrix} 1 \\ 1 \\ 1 \end{bmatrix}, .312$ **(f)** $\frac{1}{20}\begin{bmatrix} 5 \\ 7 \\ 8 \end{bmatrix}, .306$

4. (b) 50% middle, 25% upper, 25% lower   **6.** $\frac{7}{16}, \frac{9}{16}$

8. (a) $\frac{94}{450}$ **(b)** He spends equal time in compartments 3 and 4. The steady-state vector is $\frac{1}{18}\begin{bmatrix} 3 \\ 2 \\ 5 \\ 5 \\ 3 \end{bmatrix}$.

Supplementary Exercises for Chapter 2
(Page 105)

2. (b) $U^{-1} = \frac{1}{4}(U^2 - 5U + 11\,I)$.

4. (b) If $X_k = X_m$, then $Y + k(Y - Z) = Y + m(Y - Z)$. So $(k - m)(Y - Z) = 0$. But $Y - Z$ is not zero (because Y and Z are distinct), so $k - m = 0$ by Example 7§2.1.

6. (d) Using parts **(c)** and **(b)** gives $I_{pq}AI_{rs} = \sum_{i=1}^{n}\sum_{j=1}^{n} a_{ij}I_{pq}I_{ij}I_{rs}$. The only nonzero term occurs when $i = q$ and $j = r$, so $I_{pq}AI_{rs} = a_{qr}I_{ps}$.

Exercises 3.1
(Page 118)

1. (b) 0   **(d)** -1 **(f)** -39 **(h)** 0   **(j)** $2\,abc$ **(l)** 0   **(n)** -56 **(p)** $abcd$
5. (b) -17 **(d)** 106   **6. (b)** 0   **7. (b)** 12   **9. (b)** 35   **10. (b)** -6 **(d)** -6

12. (b) $-(x - 2)(x^2 + 2x - 12)$ **13. (b)** -7 **14. (b)** $\pm\dfrac{\sqrt{6}}{2}$ **(d)** $x = \pm y$

22. If A is $n \times n$, then $\det B = (-1)^{n(n-1)/2} \det A$.

Exercises 3.2
(Page 130)

1. (b) $\begin{bmatrix} 1 & -1 & -2 \\ -3 & 1 & 6 \\ -3 & 1 & 4 \end{bmatrix}$ **(d)** $\frac{1}{3}\begin{bmatrix} -1 & 2 & 2 \\ 2 & -1 & 2 \\ 2 & 2 & -1 \end{bmatrix}$ **2. (b)** $c \neq 0$ **(d)** any c **(f)** $c \neq -1$

3. (b) -2 **4. (b)** 1   **6. (b)** $\frac{4}{9}$ **7. (b)** 16   **8. (b)** $\frac{1}{11}\begin{bmatrix} 5 \\ 21 \end{bmatrix}$ **(d)** $\frac{1}{79}\begin{bmatrix} 12 \\ -37 \\ -2 \end{bmatrix}$ **9. (b)** $\frac{4}{51}$

10. (b) $\det A = 1, -1$ **(d)** $\det A = 1$ **(f)** $\det A = 0$ if n is odd; nothing can be said if n is even

20. **(b)** $\dfrac{1}{c^2}\begin{bmatrix} c & 0 & c \\ 0 & c^2 & c \\ -c & c^2 & c \end{bmatrix}$, $c \neq 0$ **(d)** $\dfrac{1}{2}\begin{bmatrix} 8 - c^2 & -c & c^2 - 6 \\ c & 1 & -c \\ c^2 - 10 & c & 8 - c^2 \end{bmatrix}$ **(f)** $\dfrac{1}{c^3 + 1}\begin{bmatrix} 1 - c & c^2 + 1 & -c - 1 \\ c^2 & -c & c + 1 \\ -c & 1 & c^2 - 1 \end{bmatrix}$, $c \neq -1$

24. $-\dfrac{1}{21}\begin{bmatrix} 3 & 0 & 1 \\ 0 & 2 & 3 \\ 3 & 1 & -1 \end{bmatrix}$ **30.** **(b)** Have $(\text{adj } A)A = (\det A)I$; so taking inverses, $A^{-1} \cdot (\text{adj } A)^{-1} = \frac{1}{\det A}I$.

On the other hand, $A^{-1} \text{adj}(A^{-1}) = \det(A^{-1})I = \frac{1}{\det A}I$. Comparison yields $A^{-1}(\text{adj } A)^{-1} = A^{-1}\text{adj}(A^{-1})$, and part **(b)** follows.

Exercises 3.3
(Page 136)

1. **(b)** $5 - 4x + 2x^2$ **2.** **(b)** $1 - \frac{5}{3}x + \frac{1}{2}x^2 + \frac{7}{6}x^3$ **3.** **(b)** $p(x) = 1 - 0.51x + 2.1x^2 - 1.1x^3$; 1.25
4. **(b)** $p(0.7) = 0.6432$ [$\sin(0.7) = 0.6442$]

Exercises 4.1
(Page 153)

3. **(b)** $\vec{FE} = \vec{FC} + \vec{CE} = \frac{1}{2}\vec{AC} + \frac{1}{2}\vec{CB} = \frac{1}{2}(\vec{AC} + \vec{CB}) = \frac{1}{2}\vec{AB}$ **4.** **(b)** v
5. **(b)** $(-3, 5, 13)$ **(d)** $(10, -12, -27)$ **(f)** $(0, 0, 0)$ **6.** **(b)** Yes   **(d)** Yes
7. **(b)** u   **(d)** $-(u + v)$ **8.** **(b)** $(-1, -1, 5)$ **(d)** $(0, 0, 0)$ **(f)** $(-2, 2, -2)$
9. **(b)** **(i)** $Q(5, -1, 2)$ **(ii)** $Q(1, 1, -4)$ **10.** **(b)** $x = u - 6v + 5w = (-16, 4, 9)$
11. **(b)** $(a, b, c) = (-5, 8, 6)$ **13.** **(b)** $\frac{1}{4}(5, -5, -2)$
16. **(b)** $(3, -1, 4) + t(2, -1, 5)$; $x = 3 + 2t$, $y = -1 - t$, $z = 4 + 5t$ **(d)** $(1, 1, 1) + t(1, 1, 1)$; $x = y = z = 1 + t$
    (f) $(2, -1, 1) + t(-1, 0, 1)$; $x = 2 - t$, $y = -1$, $z = 1 + t$
17. **(b)** P corresponds to $t = 2$; Q corresponds to $t = 5$.   **18.** **(b)** No intersection   **(d)** $P(2, -1, 3)$; $t = -2$, $s = -3$
24. $P(3, 1, 0)$ or $P(\frac{5}{3}, \frac{-1}{3}, \frac{4}{3})$ **28.** **(b)** $\vec{CP_k} = -\vec{CP}_{n+k}$ if $1 \leq k \leq n$, where there are $2n$ points.
30. $\vec{DA} = 2\vec{EA}$ and $2\vec{AF} = \vec{FC}$, so $2\vec{EF} = 2(\vec{EA} + \vec{AF}) = \vec{DA} + \vec{FC} = \vec{CB} + \vec{FC} = \vec{FC} + \vec{CB} = \vec{FB}$. Hence $\vec{EF} = \frac{1}{2}\vec{FB}$.

Exercises 4.2
(Page 163)

1. **(b)** $\sqrt{6}$ **(d)** $\sqrt{5}$ **(f)** $3\sqrt{6}$ **2.** **(b)** $\frac{1}{3}(-2, -1, 2)$ **4.** **(b)** $\sqrt{2}$ **(d)** 3
5. **(b)** 6 **(d)** 0 **(f)** 0 **6.** **(b)** π or $180°$ **(d)** $\frac{\pi}{3}$ or $60°$ **(f)** $\frac{2\pi}{3}$ or $120°$ **7.** **(b)** 1 or -17
8. **(b)** $t(-1, 1, 2)$ **(d)** $s(1, 2, 0) + t(0, 3, 1)$ **10.** **(b)** $29 + 57 = 86$ **12.** **(b)** $A = B = C = \frac{\pi}{3}$ or $60°$
14. **(b)** $\frac{11}{18}v$ **(d)** $-\frac{1}{2}v$ **15.** **(b)** $\frac{5}{21}(2, -1, -4) + \frac{1}{21}(53, 26, 20)$ **(d)** $\frac{27}{53}(6, -4, 1) + \frac{1}{53}(-3, 2, 26)$
16. **(b)** $\frac{1}{26}\sqrt{5642}$, $Q(\frac{71}{26}, \frac{15}{26}, \frac{34}{26})$
20. The four diagonals are (a, b, c), $(-a, b, c)$, $(a, -b, c)$, and $(a, b, -c)$ or their negatives. The dot products are $\pm(-a^2 + b^2 + c^2)$, $\pm(a^2 - b^2 + c^2)$, and $\pm(a^2 + b^2 - c^2)$.
25. **(b)** The sum of the squares of the lengths of the diagonals equals the sum of the squares of the lengths of the four sides.
29. **(b)** The angle θ between u and $(u + v + w)$ is given by $\cos\theta = \dfrac{u \cdot (u + v + w)}{\|u\| \, \|u + v + w\|} = \dfrac{\|u\|}{\sqrt{\|u\|^2 + \|v\|^2 + \|w\|^2}} = \frac{1}{\sqrt{3}}$,

       because $\|u\| = \|v\| = \|w\|$. Similar remarks apply to the other angles.
32. **(b)** This follows from **(a)** because $\|v\|^2 = a^2 + b^2 + c^2$.
39. Using the hint gives $a^2 + b^2 = hp + hq = h(p + q) = h^2$.

Exercises 4.3
(Page 175)

1. **(b)** $(0, 0, 0)$ **(d)** $(4, -15, 8)$ 4. **(b)** $\pm\frac{\sqrt{3}}{3}(1, -1, -1)$
5. **(b)** $-23x + 32y + 11z = 11$ **(d)** $2x - y + z = 5$ **(f)** $2x + 3y + 2z = 7$ **(h)** $2x - 7y - 3z = -1$
   (j) $x - y - z = 3$
6. **(b)** $(x, y, z) = (2, -1, 3) + t(2, 1, 0)$ **(d)** $(x, y, z) = (1, 1, -1) + t(1, 1, 1)$ **(f)** $(x, y, z) = (1, 1, 2) + t(4, 1, -5)$
7. **(b)** $\frac{\sqrt{6}}{3}, Q(\frac{7}{3}, \frac{2}{3}, -\frac{2}{3})$ 8. **(b)** Yes. The equation is $5x - 3y - 4z = 0$. 10. **(b)** $(-2, 7, 0) + t(3, -5, 2)$
11. **(b)** None **(d)** $P(\frac{9}{5}, 0, \frac{13}{5})$
12. **(b)** $3x + 2z = d, d$ arbitrary **(d)** $a(x - 3) + b(y - 2) + c(z + 4) = 0$ $a, b,$ and c not all zero
   (f) $ax + by + (b - a)z = a$ a and b not both zero **(h)** $ax + by + (a + 2b)z = 5a + 4b$ a and b not both zero
14. **(b)** $\sqrt{10}$ 15. **(b)** $\frac{\sqrt{14}}{2}, A(3, 1, 2), B(\frac{7}{2}, -\frac{1}{2}, 3)$ **(d)** $\frac{\sqrt{6}}{6}, A(\frac{19}{3}, 2, \frac{1}{3}), B(\frac{37}{6}, \frac{13}{6}, 0)$ 16. **(b)** 0 **(d)** $\sqrt{5}$
17. **(b)** 7 18. **(b)** distance is $\|\mathbf{p} - \mathbf{p}_0\|$
27. **(b)** If $\mathbf{u} = (u_1, u_2, u_3)$, $\mathbf{v} = (v_1, v_2, v_3)$, and $\mathbf{w} = (w_1, w_2, w_3)$, then

$$\mathbf{u} \times (\mathbf{v} + \mathbf{w}) = \det\begin{bmatrix} \mathbf{i} & \mathbf{j} & \mathbf{k} \\ u_1 & u_2 & u_3 \\ v_1 + w_1 & v_2 + w_2 & v_3 + w_3 \end{bmatrix} = \det\begin{bmatrix} \mathbf{i} & \mathbf{j} & \mathbf{k} \\ u_1 & u_2 & u_3 \\ v_1 & v_2 & v_3 \end{bmatrix} + \det\begin{bmatrix} \mathbf{i} & \mathbf{j} & \mathbf{k} \\ u_1 & u_2 & u_3 \\ w_1 & w_2 & w_3 \end{bmatrix} = (\mathbf{u} \times \mathbf{v}) + (\mathbf{u} \times \mathbf{w})$$

where we used Exercise 19§3.1.

34. By block multiplication, $\begin{bmatrix} \mathbf{u}A \\ \mathbf{v}A \\ \mathbf{w}A \end{bmatrix} = \begin{bmatrix} \mathbf{u} \\ \mathbf{v} \\ \mathbf{w} \end{bmatrix} A$. Take determinants and use Theorems 1§3.2 and 5§4.3.

Exercises 4.4
(Page 185)

1. **(b)** $y = \frac{64}{13} - \frac{6}{13}x$ **(d)** $y = -\frac{4}{10} - \frac{17}{10}x$ 2. **(b)** $(M^TM)^{-1} = \frac{1}{4248}\begin{bmatrix} 3348 & 642 & -426 \\ 642 & 571 & -187 \\ -426 & -187 & 91 \end{bmatrix}, y = 0.127 - 0.024x + 0.194x^2$

4. $s = 99.71 - 4.87x, g = 9.74$ (the true value of g is 9.81). If a quadratic in s is fit, $(M^TM)^{-1} = \frac{1}{4}\begin{bmatrix} 76 & -84 & 20 \\ -84 & 98 & -24 \\ 20 & -24 & 6 \end{bmatrix}$,

$s = 101 - \frac{3}{2}t - \frac{9}{2}t^2, g = 9$

Supplementary Exercises for Chapter 4
(Page 186)

4. 125 knots in a direction θ degrees east of north, where $\cos\theta = 0.6$ ($\theta = 53°$ or 0.93 radians).
6. $(12, 5)$. Actual speed 12 knots.

Exercises 5.1
(Page 196)

1. **(b)** $(-1, -8, -13, -6, 16)$ **(d)** $(12, 51, 51, 27, -42)$
2. **(b)** Not linearly independent (for example, $2\mathbf{u} - 3\mathbf{v} + \mathbf{w} = \mathbf{0}$) **(d)** Linearly independent
3. **(b)** There are no such numbers $a, b,$ and c.
4. **(b)** **A1.** $(x, -x) + (y, -y) = [x + y, -(x + y)]$ and so lies in V.
   A2. $(x, -x) + (y, -y) = [x + y, -(x + y)] = [y + x, -(y + x)] = (y, -y) + (x, -x)$

A3. $[(x,-x) + (y, -y)] + (z, -z) = (x + y + z, -x - y - z) = (x, -x) + [(y, -y) + (z, -z)]$

A4. $(x, -x) + (0, -0) = (x, -x)$ for all $(x, -x)$, so $(0, 0)$ is the zero vector.

A5. $(x, -x) + (-x, x) = (0, 0)$ so $(-x, x) = (-x, -(-x))$ in V is the negative of $(x, -x)$.

S1. $a(x, -x) = [ax, -(ax)]$ and so lies in V.

S2. $a[(x, -x) + (y, -y)] = (ax + ay, -ax - ay) = (ax, -ax) + (ay, -ay) = a(x, -x) + a(y, -y)$

S3. $(a + b)(x, -x) = [(a + b)x, -(a + b)x] = (ax, -ax) + (bx, -bx) = a(x, -x) + b(x, -x)$

S4. $(ab)(x, -x) = (abx, -abx) = a(bx, -bx) = a[b(x, -x)]$

S5. $1(x, -x) = (1x, -1x) = (x, -x)$

5. **(b)** No; S5 fails. **(d)** No; S4 and S5 fail.

6. **(b)** No; only A1 fails. **(d)** Yes **(f)** Yes **(h)** Yes **(j)** No; only S3 fails. **(l)** No; only S4 and S5 fail.

8. The zero vector is $(0, -1)$; the negative of (x, y) is $(-x, -2 - y)$. **11.** **(b)** $x = \frac{1}{7}(5u - 2v), y = \frac{1}{7}(4u - 3v)$

12. **(b)** $x = 11u - 3v - 7t, y = 8u - 2v - 5t, z = t$; t an arbitrary vector **(d)** $x = t, y = 19t, z = 4t$; t an arbitrary vector

13. **(b)** Equating entries gives

$$\begin{aligned} a & & + c &= 0 \\ & b + c &= 0 \\ & b + c &= 0 \\ a & & - c &= 0 \end{aligned}$$

The solution is $a = b = c = 0$

(d) If $a \sin x + b \cos y + c = 0$ in $F[0, \pi]$, then this must hold for *every* x in $[0, \pi]$. Taking $x = 0, \frac{\pi}{2}$, and π, respectively, gives

$$\begin{aligned} & b + c &= 0 \\ a & + c &= 0 \\ & -b + c &= 0 \end{aligned}$$

whence $a = b = c = 0$.

14. **(b)** $4w$

19. **(b)** $(-a)v + av = (-a + a)v = 0v = 0$ by Theorem 3. Because also $-(av) + av = 0$ (by the definition of $-(av)$ in axiom A.5), this means that $(-a)v = -(av)$ by cancellation. Alternatively, use Theorem 3(4) to give $(-a)v = [(-1)a]v = (-1)(av) = -(av)$.

Exercises 5.2
(Page 207)

1. **(b)** Yes **(d)** No; not closed under addition. **(f)** No; not closed under addition.

2. **(b)** Yes **(d)** Yes **(f)** No; not closed under addition or scalar multiplication, and 0 is not in the set.

3. **(b)** Yes **(d)** Yes **(f)** No; not closed under addition.

4. **(b)** No; not closed under addition. **(d)** No; not closed under scalar multiplication. **(f)** Yes

6. **(b)** If entry k of X is $x_k \neq 0$, and if Y is in $\mathbb{R}^n$, then $Y = AX$ where the only nonzero column of A is column $k = x_k^{-1}Y$.

7. **(b)** $-3(x + 1) + 0(x^2 + x) + 2(x^2 + 2)$ **(d)** $\frac{2}{3}(x + 1) + \frac{1}{3}(x^2 + x) - \frac{1}{3}(x^2 + 2)$

8. **(b)** $3(1, -1, 1) + 2(1, 0, 1) - 4(1, 1, 0)$ **(d)** $0(1, -1, 1) + 0(1, 0, 1) + 0(1, 1, 0)$

9. **(b)** Yes; $v = 3u - 2w$ **(d)** No **(f)** Yes; $v = 3u - w$

10. **(b)** Yes; $1 = \cos^2 x + \sin^2 x$ **(d)** No. If $1 + x^2 = a \cos^2 x + b \sin^2 x$, then taking $x = 0$ and $x = \pi$ gives $a = 1$ and $a = 1 + \pi^2$.

11. **(b)** Because $P_2 = \text{span}\{1, x, x^2\}$, it suffices to show that $\{1, x, x^2\} \subseteq \text{span}\{1 + 2x^2, 3x, 1 + x\}$. But $x = \frac{1}{3}(3x); 1 = (1 + x) - x$ and $x^2 = \frac{1}{2}[(1 + 2x^2) - 1]$.

13. **(b)** $u = (u + w) - w, v = -(u - v) + (u + w) - w$, and $w = w$ **16.** No

20. If P_0 has position vector p_0, the line through P_0 with direction vector d consists of all points with position vector $p = p_0 + td$ for some t in $\mathbb{R}$. If $p_0 = 0$, the line is the set $\{td \mid t \text{ in } \mathbb{R}\} = \text{span}\{d\}$.

Exercises 5.3
(Page 217)

2. (b) Independent   **(d)** Dependent; $(1, 1, 0, 0) - (1, 0, 1, 0) + (0, 0, 1, 1) - (0, 1, 0, 1) = (0, 0, 0, 0)$

(f) $3(x^2 - x + 3) - 2(2x^2 + x + 5) + (x^2 + 5x + 1) = 0$

(h) $2\begin{bmatrix} -1 & 0 \\ 0 & -1 \end{bmatrix} + \begin{bmatrix} 1 & -1 \\ -1 & 1 \end{bmatrix} + \begin{bmatrix} 1 & 1 \\ 1 & 1 \end{bmatrix} = \begin{bmatrix} 0 & 0 \\ 0 & 0 \end{bmatrix}$ **(j)** $\dfrac{5}{x^2 + x - 6} + \dfrac{1}{x^2 - 5x + 6} - \dfrac{6}{x^2 - 9} = 0$

3. (b) Dependent; $1 - \sin^2 x - \cos^2 x = 0$ **4. (b)** $x \ne -\frac{1}{3}$

5. (b) If $r(-1, 1, 1) + s(1, -1, 1) + t(1, 1, -1) = (0, 0, 0)$, then

$$-r + s + t = 0$$
$$r - s + t = 0$$
$$r + s - t = 0$$

and this implies that $r = s = t = 0$. This proves independence. To prove that they span $\mathbb{R}^3$, observe that

$$(0, 0, 1) = \tfrac{1}{2}[(-1, 1, 1) + (1, -1, 1)]$$

so $(0, 0, 1)$ lies in $\text{span}\{(-1, 1, 1), (1, -1, 1), (1, 1, -1)\}$. The proof is similar for $(0, 1, 0)$ and $(1, 0, 0)$.

(d) If $r(1 + x) + s(x + x^2) + t(x^2 + x^3) + ux^3 = 0$, then $r = 0, r + s = 0, s + t = 0$, and $t + u = 0$, so $r = s = t = u = 0$. This proves independence. To show that they span $\mathbf{P}_3$, observe that $x^2 = (x^2 + x^3) - x^3, x = (x + x^2) - x^2$, and $1 = (1 + x) - x$, so $\{1, x, x^2, x^3\} \subseteq \text{span}\{1 + x, x + x^2, x^2 + x^3, x^3\}$.

6. (b) $\{(1, 1, 1, 0), (1, -1, 0, 1)\}$; dimension $= 2$ **(d)** $\{(1, 0, 1, 0), (-1, 1, 0, 1), (0, 1, 0, 1)\}$; dimension $= 3$
(f) $\{(1, 0, 1, 0), (0, 1, 0, 1), (1, -1, 0, 0)\}$; dimension $= 3$ **(h)** $\{(1, -1, 3, 0), (4, 5, 0, -3)\}$; dimension $= 2$

7. (b) $\{1, x + x^2\}$; dimension $= 2$ **(d)** $\{1, x^2\}$; dimension $= 2$

8. (b) $\left\{ \begin{bmatrix} 1 & 1 \\ -1 & 0 \end{bmatrix}, \begin{bmatrix} 1 & 0 \\ 0 & 1 \end{bmatrix} \right\}$; dimension $= 2$ **(d)** $\left\{ \begin{bmatrix} 1 & 0 \\ 1 & 1 \end{bmatrix}, \begin{bmatrix} 0 & 1 \\ -1 & 0 \end{bmatrix} \right\}$; dimension $= 2$ **9. (b)** $\left\{ \begin{bmatrix} 1 & 0 \\ 0 & 0 \end{bmatrix}, \begin{bmatrix} 0 & 1 \\ 0 & 0 \end{bmatrix} \right\}$

10. (b) $\left\{ \begin{bmatrix} -1 \\ 7 \\ 10 \end{bmatrix} \right\}$ **(d)** $\left\{ \begin{bmatrix} -1 \\ 5 \\ 4 \\ 0 \end{bmatrix}, \begin{bmatrix} 1 \\ 7 \\ 0 \\ -4 \end{bmatrix} \right\}$ **11. (b)** $\dim V = 7$ **(c)** $\dim V = n^2 - n + 1$

12. (b) $\{x^2 - x, x(x^2 - x), x^2(x^2 - x), x^3(x^2 - x)\}$; $\dim V = 4$

13. (b) No. Any linear combination of such polynomials has $f(0) = 0$.

(d) No. $\left\{ \begin{bmatrix} 1 & 0 \\ 0 & 1 \end{bmatrix}, \begin{bmatrix} 1 & 1 \\ 0 & 1 \end{bmatrix}, \begin{bmatrix} 1 & 0 \\ 1 & 1 \end{bmatrix}, \begin{bmatrix} 0 & 1 \\ 1 & 1 \end{bmatrix} \right\}$ consists of invertible matrices.

(f) Yes. $0\mathbf{u} + 0\mathbf{v} + 0\mathbf{w} = \mathbf{0}$ for every set $\{\mathbf{u}, \mathbf{v}, \mathbf{w}\}$.

(h) Yes. $s\mathbf{u} + t(\mathbf{u} + \mathbf{v}) = \mathbf{0}$ gives $(s + t)\mathbf{u} + t\mathbf{v} = \mathbf{0}$, whence $s + t = 0 = t$.

(j) Yes. If $r\mathbf{u} + s\mathbf{v} = \mathbf{0}$, then $r\mathbf{u} + s\mathbf{v} + 0\mathbf{w} = \mathbf{0}$, so $r = 0 = s$.

16. If a linear combination of the subset vanishes, it is a linear combination of the vectors in the larger set (coefficients outside the subset are zero) so it is trivial.

22. Because $\{\mathbf{u}, \mathbf{v}\}$ is linearly independent, $s\mathbf{u}' + t\mathbf{v}' = \mathbf{0}$ is equivalent to $(s, t)\begin{bmatrix} a & b \\ c & d \end{bmatrix} = (0, 0)$. Now apply Theorem 4§2.4.

25. (b) Independent   **(d)** Dependent. For example, $(\mathbf{u} + \mathbf{v}) - (\mathbf{v} + \mathbf{w}) +$

$(\mathbf{w} + \mathbf{z}) - (\mathbf{z} + \mathbf{u}) = \mathbf{0}$.

37. (b) $\dim 0_n = \frac{n}{2}$ if n is even and $\dim 0_n = \frac{n+1}{2}$ if n is odd.

Exercises 5.4
(Page 225)

1. (b) $\{(0, 1, 1), (1, 0, 0), (0, 1, 0)\}$ **(d)** $\{x^2 - x + 1, 1, x\}$ **2. (b)** Any three except $\{x^2 + 3, x + 2, x^2 - 2x - 1\}$

3. (b) Add $(0, 1, 0, 0)$ and $(0, 0, 1, 0)$.   **(d)** Add 1 and x^3.

4. (b) If $z = a + bi$, then $a \neq 0$ and $b \neq 0$. If $rz + s\bar{z} = 0$, then $(r + s)a = 0$ and $(r - s)b = 0$. This means that $r + s = 0 = r - s$, so $r = s = 0$. Thus $\{z, \bar{z}\}$ is independent; it is a basis because $\dim \mathbb{C} = 2$.

7. (b) Not a basis   **(d)** Not a basis   **8. (b)** Yes; no

10. (b) $\{(1, 0, -1, 3), (2, 1, 0, -2), (-1, 1, 2, 1)\}$

21. (b) The set $\{(1, 0, 0, 0, \ldots), (0, 1, 0, 0, 0, \ldots), (0, 0, 1, 0, 0, \ldots), \ldots\}$ contains independent subsets of arbitrary size.

22. (b) $\mathbb{R}\mathbf{u} + \mathbb{R}\mathbf{w} = \{r\mathbf{u} + s\mathbf{w} \mid r, s \text{ in } \mathbb{R}\} = \text{span}\{\mathbf{u}, \mathbf{w}\}$

Exercises 5.5
(Page 236)

1. (b) Invertible   **2. (b)** Not independent

7. (b) $\left\{(2, -1, 1), (0, 0, 1)\right\}$; $\left\{\begin{bmatrix} 2 \\ -2 \\ 4 \\ -6 \end{bmatrix}, \begin{bmatrix} 1 \\ 1 \\ 3 \\ 0 \end{bmatrix}\right\}$; 2   **(d)** $\{(1, 2, -1, 3), (0, 0, 0, 1)\}$; $\left\{\begin{bmatrix} 1 \\ 0 \end{bmatrix}, \begin{bmatrix} 3 \\ -2 \end{bmatrix}\right\}$; 2

8. (b) $\{(1, 1, 0, 0, 0), (0, -2, 2, 5, 1), (0, 0, 2, -3, 6)\}$ **(d)** $\left\{\begin{bmatrix} 1 \\ 5 \\ -6 \end{bmatrix}, \begin{bmatrix} 0 \\ 1 \\ -1 \end{bmatrix}, \begin{bmatrix} 0 \\ 0 \\ 1 \end{bmatrix}\right\}$ **9. (b)** No; no   **(d)** No

12. $\text{col}(AV) = \text{row}[(AV)^T] = \text{row}[V^T A^T] \subseteq \text{row}(A^T) = \text{col } A$. If V is invertible, the reverse inclusion follows by applying this to $A = (AV)V^{-1}$.

15. (b) The basis is $\left\{\begin{bmatrix} 6 \\ 0 \\ -4 \\ 1 \\ 0 \end{bmatrix}, \begin{bmatrix} 5 \\ 0 \\ -3 \\ 0 \\ 1 \end{bmatrix}\right\}$, so the dimension is 2. Have rank $A = 3$ and $n - 3 = 2$.

16. (b) The r columns containing leading 1's are independent because the leading 1's are in different rows. If R is $m \times n$, col R is contained in the subspace of all columns in $\mathbb{R}^m$ with the last $m - r$ entries zero. This space has dimension r, so the (independent) columns containing leading 1's are a basis.

17. (b) $U = A^{-1}, V = I_2$; rank $A = 2$ **(d)** $U = \begin{bmatrix} -2 & 1 & 0 \\ 3 & -1 & 0 \\ 2 & -1 & 1 \end{bmatrix}$, $V = \begin{bmatrix} 1 & 0 & -1 & -3 \\ 0 & 1 & 1 & 4 \\ 0 & 0 & 1 & 0 \\ 0 & 0 & 0 & 1 \end{bmatrix}$; rank $A = 2$ **18. (b)** $n - 1$

25. (b) Let $\{U_1, \ldots, U_r\}$ be a basis of col(A). Then B is *not* in col(A), so $\{U_1, \ldots, U_r, B\}$ is linearly independent. Show that col$[A \quad B] = \text{span}\{U_1, \ldots, U_r, B\}$.

30. (b) Clearly $B = \{V^{-1}X_1, \ldots, V^{-1}X_k\} \subseteq N(AV)$; we show it is a basis. If $\sum r_i(V^{-1}X_i) = 0$, then $V^{-1}[\sum r_i X_i] = 0$, so $\sum r_i X_i = 0$, whence $r_i = 0$. Thus B is independent. Given Y in $N(AV)$, we have $AVY = 0$, so VY is in $N(A)$. Thus $VY = \sum r_i X_i$, so $Y = \sum r_i(V^{-1}X_i)$.

35. (b) $P = A, Q = I_2$ **(d)** $P = \begin{bmatrix} 1 & 1 \\ 3 & 2 \\ 1 & 0 \end{bmatrix}$, $Q = \begin{bmatrix} 1 & 0 & 1 & 3 \\ 0 & 1 & -1 & -4 \end{bmatrix}$

Exercises 5.6
(Page 243)

2. (b) $3 + 4(x - 1) + 3(x - 1)^2 + (x - 1)^3$ **(d)** $1 + (x - 1)^3$

6. (b) The polynomials are $(x - 1)(x - 2), (x - 1)(x - 3), (x - 2)(x - 3)$. Use $a_0 = 3, a_1 = 2,$ and $a_2 = 1$.

10. (b) If $r(x - a)^2 + s(x - a)(x - b) + t(x - b)^2 = 0$, then evaluation at $x = a$ ($x = b$) gives $t = 0$ ($r = 0$). Thus $s(x - a)(x - b) = 0$, so $s = 0$. Use Theorem 8§5.3.

11. (b) Suppose $\{p_0(x), p_1(x), \ldots, p_{n-2}(x)\}$ is a basis of $\mathbf{P}_{n-2}$. We show that

$$\{(x - a)(x - b)p_0(x), (x - a)(x - b)p_1(x), \ldots, (x - a)(x - b)p_{n-2}(x)\}$$

is a basis of U_n. It is a spanning set by part (a), so assume that a linear combination vanishes with coefficients $r_0, r_1, \ldots, r_{n-2}$. Then $(x - a)(x - b)[r_0p_0(x) + \cdots + r_{n-2}p_{n-2}(x)] = 0$, so $r_0p_0(x) + \cdots + r_{n-2}p_{n-2}(x) = 0$. This implies that $r_0 = \cdots = r_{n-2} = 0$.

Exercises 5.7
(Page 249)

1. (b) e^{1-x} **(d)** $\dfrac{e^{2x} - e^{-3x}}{e^2 - e^{-3}}$ **(f)** $2e^{2x}(1 + x)$ **(h)** $\dfrac{e^{ax} - e^{a(2-x)}}{1 - e^{2a}}$ **(j)** $e^{\pi - 2x}\sin x$

5. (b) $ce^{-x} + 2$, c a constant   **6. (b)** $ce^{-3x} + de^{2x} - \dfrac{x^3}{3}$ **7. (b)** $t = \dfrac{3\ln\left(\frac{1}{2}\right)}{\ln\left(\frac{4}{5}\right)} = 9.32$ hours

9. $k = \left(\dfrac{\pi}{15}\right)^2 = 0.044$

Exercises 6.1
(Page 259)

1. (b) $(x - 3)(x + 2)$; $E_3 = \text{span}\{(4, -1)^T\}$, $E_{-2} = \text{span}\{(1, 1)^T\}$ **(d)** $(x - 2)^3$; $E_2 = \text{span}\{(1, 1, 0)^T, (-3, 0, 1)^T\}$
   (f) $(x + 1)^2(x - 2)$; $E_{-1} = \text{span}\{(-1, 1, 0)^T, (-1, 0, 1)^T\}$ $E_2 = \text{span}\{(1, 1, 1)^T\}$
   (h) $(x - 1)^2(x - 3)$; $E_1 = \text{span}\{(-1, 0, 1)^T\}$, $E_3 = \text{span}\{(1, 0, 1)^T\}$

4. $A\mathbf{p} = \lambda\mathbf{p}$ if and only if $(A - \alpha I)\mathbf{p} = (\lambda - \alpha)\mathbf{p}$. Same eigenvectors.

8. (b) traces $= 2$, ranks $= 2$, but $\det A = -5$, $\det B = -1$ **(d)** ranks $= 2$, determinants $= 7$, but tr $A = 5$, tr $B = 4$
   (f) traces $= -5$, determinants $= 0$, but rank $A = 2$, rank $B = 1$

10. (b) If $B = P^{-1}AP$, then $B^{-1} = P^{-1}A^{-1}(P^{-1})^{-1} = P^{-1}A^{-1}P$.

11. (b) $P^{-1}AP = \begin{bmatrix} 1 & 0 \\ 0 & 2 \end{bmatrix}$, so $A^n = P\begin{bmatrix} 1 & 0 \\ 0 & 2^n \end{bmatrix}P^{-1} = \begin{bmatrix} 9 - 8 \cdot 2^n & 12(1 - 2^n) \\ 6(2^n - 1) & 9 \cdot 2^n - 8 \end{bmatrix}$.

12. (b) $p(A) = A^3 + 3A^2 + A - I = \begin{bmatrix} 0 & 0 \\ 0 & 21 \end{bmatrix}$. Hence $p(B) = P^{-1}p(A)P = \begin{bmatrix} 84 & 42 \\ -126 & -63 \end{bmatrix}$.

16. (b) $c_{rA}(x) = \det[xI - rA] = r^n \det\left[\dfrac{x}{r}I - A\right] = r^n c_A\left[\dfrac{x}{r}\right]$

18. (b) If $\lambda \neq 0$, $AX = \lambda X$ if and only if $A^{-1}X = \dfrac{1}{\lambda}X$. The result follows.

21. (b) If $A^m = 0$ and $AX = \lambda X$, $X \neq 0$, then $A^2X = A(\lambda X) = \lambda AX = \lambda^2 X$. In general, $A^kX = \lambda^kX$ for all $k \geq 1$. Hence, $\lambda^mX = A^mX = 0X = 0$, so $\lambda = 0$ (because $X \neq 0$.)

Exercises 6.2
(Page 269)

1. (b) Yes. $P = \begin{bmatrix} 2 & -2 \\ 1 & 1 \end{bmatrix}$, $P^{-1}AP = \begin{bmatrix} 0 & 0 \\ 0 & 4 \end{bmatrix}$ **(d)** Yes. $P = \begin{bmatrix} -1 & 0 & 6 \\ 0 & 1 & 0 \\ 1 & 0 & 5 \end{bmatrix}$, $P^{-1}AP = \begin{bmatrix} -3 & 0 & 0 \\ 0 & -3 & 0 \\ 0 & 0 & 8 \end{bmatrix}$

   (f) No. $c_A(x) = (x - 1)(x^2 + 1)$ has roots 1, i, and $-i$.

2. AB has only one eigenvalue, 14, and dim $(E_{14}) = 1$, so Theorem 4 applies.

5. (b) If $P^{-1}AP = D$ is diagonal, then $P^{-1}(kA)P = k(P^{-1}AP) = kD$ is also diagonal.

6. $\begin{bmatrix} 1 & 1 \\ 0 & 1 \end{bmatrix} = \begin{bmatrix} 2 & 1 \\ 0 & -1 \end{bmatrix} + \begin{bmatrix} -1 & 0 \\ 0 & 2 \end{bmatrix}$. Both $\begin{bmatrix} 2 & 1 \\ 0 & -1 \end{bmatrix}$ and $\begin{bmatrix} -1 & 0 \\ 0 & 2 \end{bmatrix}$ are diagonalizable by Theorem 3, but for $\begin{bmatrix} 1 & 1 \\ 0 & 1 \end{bmatrix}$, dim$(E_1) = 1$.
So Theorem 4 applies.

12. We have $P^{-1}AP = \lambda I$ by the diagonalization algorithm, so $A = P(\lambda I) P^{-1} = \lambda I$.

22. (b) $x_n = \frac{1}{3}[4 - (-2)^n]$

Exercises 6.3
(Page 279)

1. (b) $\left\{ \dfrac{1}{\sqrt{3}} [1 \ \ 1 \ \ 1], \ \ \dfrac{1}{\sqrt{42}} [4 \ \ 1 \ \ -5], \ \ \dfrac{1}{\sqrt{14}} [2 \ \ -3 \ \ 1] \right\}$

3. (b) $(a, b, c) = \frac{1}{2}(a - c)[1 \ \ 0 \ \ -1] + \frac{1}{18}(a + 4b + c)[1 \ \ 4 \ \ 1] + \frac{1}{9}(2a - b + 2c)[2 \ \ -1 \ \ 2]$
   (d) $(a, b, c) = \frac{1}{3}(a + b + c)[1 \ \ 1 \ \ 1] + \frac{1}{2}(a - b)[1 \ \ -1 \ \ 0] + \frac{1}{6}(a + b - 2c)[1 \ \ 1 \ \ -2]$

4. (b) $[14 \ \ 1 \ \ -8 \ \ 5] = 3[2 \ \ -1 \ \ 0 \ \ 3] + 4[2 \ \ 1 \ \ -2 \ \ -1]$ **5. (b)** $t[-1 \ \ 3 \ \ 10 \ \ 11]$, t in $\mathbb{R}$

7. (b) $\sqrt{29}$ **(d)** 19   **8. (b)** $\{[2 \ \ 1], \frac{3}{5}[-1 \ \ 2]\}$ **(d)** $\{[0 \ \ 1 \ \ 1], [1 \ \ 0 \ \ 0], \frac{1}{5}[0 \ \ -2 \ \ 2]\}$

9. (b) $X = \frac{1}{182}[271 \ \ -221 \ \ 1030] + \frac{1}{182}[93 \ \ 403 \ \ 62]$ **(d)** $X = \frac{1}{4}[1 \ \ 7 \ \ 11 \ \ 17] + \frac{1}{4}[7 \ \ -7 \ \ -7 \ \ 7]$
   (f) $X = \frac{1}{12}[5a - 5b + c - 3d, -5a + 5b - c + 3d, a - b + 11c + 3d, -3a + 3b + 3c + 3d] + \frac{1}{12}[7a + 5b - c + 3d,$
   $5a + 7b + c - 3d, -a + b + c - 3d, 3a - 3b - 3c + 9d]$

10. (a) $\frac{1}{10}[-9 \ \ 3 \ \ -21 \ \ 33] = \frac{3}{10}[-3 \ \ 1 \ \ -7 \ \ 11]$ **(c)** $\frac{1}{70}[-63 \ \ 21 \ \ -147 \ \ 231] = \frac{3}{10}[-3 \ \ 1 \ \ -7 \ \ 11]$

11. (b) $\{[1 \ \ -1 \ \ 0], \frac{1}{2}[-1 \ \ -1 \ \ 2]\}$; $\text{proj}_U(X) = [1 \ \ 0 \ \ -1]$
   (d) $\{[1 \ \ -1 \ \ 0 \ \ 1], [1 \ \ 1 \ \ 0 \ \ 0], \frac{1}{3}[-1 \ \ 1 \ \ 0 \ \ 2]\}$; $\text{proj}_U(X) = [2 \ \ 0 \ \ 0 \ \ 1]$

12. (b) $U^{\perp} = \text{span}\{[1 \ \ 3 \ \ 1 \ \ 0], [-1 \ \ 0 \ \ 0 \ \ 1]\}$

20. If X is orthogonal to U, it is orthogonal to each X_i because X_i is in U. Conversely, if $X \cdot X_i = 0$ for each i and $Y = r_1 X_1 + \cdots + r_m X_m$ is in U, then $X \cdot Y = r_1(X \cdot X_1) + \cdots + r_m (X \cdot X_m) = 0$.

31. (d) $E^T = A^T[(AA^T)^{-1}]^T (A^T)^T = A^T[(AA^T)^T]^{-1} A = A^T[AA^T]^{-1} A = E$
   $E^2 = A^T(AA^T)^{-1}AA^T(AA^T)^{-1}A = A^T(AA^T)^{-1}A = E$

Exercises 6.4
(Page 287)

1. (b) $\frac{1}{5}\begin{bmatrix} 3 & -4 \\ 4 & 3 \end{bmatrix}$ **(d)** $\dfrac{1}{\sqrt{a^2 + b^2}} \begin{bmatrix} a & b \\ -b & a \end{bmatrix}$ **(f)** $\begin{bmatrix} \frac{2}{\sqrt{6}} & \frac{1}{\sqrt{6}} & \frac{-1}{\sqrt{6}} \\ \frac{1}{\sqrt{3}} & \frac{-1}{\sqrt{3}} & \frac{1}{\sqrt{3}} \\ 0 & \frac{1}{\sqrt{2}} & \frac{1}{\sqrt{2}} \end{bmatrix}$ **(h)** $\frac{1}{7}\begin{bmatrix} 2 & 6 & -3 \\ 3 & 2 & 6 \\ -6 & 3 & 2 \end{bmatrix}$

5. (b) $\frac{1}{\sqrt{2}}\begin{bmatrix} 0 & 1 & 1 \\ \sqrt{2} & 0 & 0 \\ 0 & 1 & -1 \end{bmatrix}$ **(d)** $\frac{1}{3\sqrt{2}}\begin{bmatrix} 2\sqrt{2} & 3 & 1 \\ \sqrt{2} & 0 & -4 \\ 2\sqrt{2} & -3 & 1 \end{bmatrix}$ or $\frac{1}{3}\begin{bmatrix} 2 & -2 & 1 \\ 1 & 2 & 2 \\ 2 & 1 & -2 \end{bmatrix}$ **(f)** $\frac{1}{2}\begin{bmatrix} 1 & -1 & \sqrt{2} & 0 \\ -1 & 1 & \sqrt{2} & 0 \\ -1 & -1 & 0 & \sqrt{2} \\ 1 & 1 & 0 & \sqrt{2} \end{bmatrix}$

6. $P = \dfrac{1}{\sqrt{2k}}\begin{bmatrix} c\sqrt{2} & a & a \\ 0 & k & -k \\ -a\sqrt{2} & c & c \end{bmatrix}$ **12. (b)** If $B = P^T AP$, $P^T = P^{-1}$, then $B^2 = P^T APP^T AP = P^T A^2 P$.

14. (b) $\det\begin{bmatrix} \cos \theta & \sin \theta \\ -\sin \theta & \cos \theta \end{bmatrix} = 1$ and $\det\begin{bmatrix} \cos \theta & \sin \theta \\ \sin \theta & -\cos \theta \end{bmatrix} = -1$ (Remark: Exercise 22 shows that these give *all* possibilities.)

20. We have $AA^T = D$, where D is diagonal with main diagonal entries $\|R_1\|^2, \ldots, \|R_n\|^2$. Hence $A^{-1} = A^T D^{-1}$, and the result follows because D^{-1} has diagonal entries $1/\|R_1\|^2, \ldots, 1/\|R_n\|^2$.

22. (b) Because $I - A$ and $I + A$ commute, $PP^T = (I - A)(I + A)^{-1}[(I + A)^{-1}]^T(I - A)^T = (I - A)(I + A)^{-1}(I - A)^{-1}(I + A) = I$.

Exercises 6.5
(Page 294)

1. (b) $U^T = \frac{\sqrt{2}}{2}\begin{bmatrix} 2 & 0 \\ -1 & 1 \end{bmatrix}$ **(d)** $U^T = \frac{1}{30}\begin{bmatrix} 60\sqrt{5} & 0 & 0 \\ 12\sqrt{5} & 6\sqrt{30} & 0 \\ 15\sqrt{5} & 10\sqrt{30} & 5\sqrt{15} \end{bmatrix}$

Exercises 6.6
(Page 298)

1. (b) $L = \frac{1}{\sqrt{5}}\begin{bmatrix} 5 & 0 \\ 3 & 1 \end{bmatrix}$, $P = \frac{1}{\sqrt{5}}\begin{bmatrix} 2 & 1 \\ -1 & 2 \end{bmatrix}$, $(AA^T)^{-1} = \begin{bmatrix} 2 & -3 \\ -3 & 5 \end{bmatrix} = (L^{-1})^T L^{-1}$

(d) $L = \frac{1}{\sqrt{3}}\begin{bmatrix} 3 & 0 & 0 \\ 0 & 3 & 0 \\ -1 & 1 & 2 \end{bmatrix}$, $P = \frac{1}{\sqrt{3}}\begin{bmatrix} 1 & -1 & 0 & 1 \\ 1 & 0 & 1 & -1 \\ 0 & 1 & 1 & 1 \end{bmatrix}$, $(AA^T)^{-1} = \frac{1}{12}\begin{bmatrix} 5 & -1 & 3 \\ -1 & 5 & -3 \\ 3 & -3 & 9 \end{bmatrix} = (L^{-1})^T L^{-1}$

Exercises 6.7
(Page 301)

1. (b) Eigenvalues $4, -1$; eigenvectors $\begin{bmatrix} 2 \\ -1 \end{bmatrix}$, $\begin{bmatrix} 1 \\ -3 \end{bmatrix}$; $X_4 = \begin{bmatrix} 409 \\ -203 \end{bmatrix}$; $r_3 = 3.94$

(d) Eigenvalues $\lambda_1 = \frac{1}{2}(3 + \sqrt{13})$, $\lambda_2 = \frac{1}{2}(3 - \sqrt{13})$; eigenvectors $\begin{bmatrix} \lambda_1 \\ 1 \end{bmatrix}$, $\begin{bmatrix} \lambda_2 \\ 1 \end{bmatrix}$; $X_4 = \begin{bmatrix} 142 \\ 43 \end{bmatrix}$; $r_3 = 3.3027750$

(The true value is $\lambda_1 = 3.3027756$, to seven decimal places.)

2. (b) Eigenvalues $\lambda_1 = \frac{1}{2}(3 + \sqrt{13}) = 3.302776$, $\lambda_2 = \frac{1}{2}(3 - \sqrt{13}) = -0.302776$

$$A_1 = \begin{bmatrix} 3 & 1 \\ 1 & 0 \end{bmatrix}, \quad Q_1 = \frac{1}{\sqrt{10}}\begin{bmatrix} 3 & -1 \\ 1 & 3 \end{bmatrix}, \quad R_1 = \frac{1}{\sqrt{10}}\begin{bmatrix} 10 & 3 \\ 0 & -1 \end{bmatrix}$$

$$A_2 = \frac{1}{10}\begin{bmatrix} 33 & -1 \\ -1 & -3 \end{bmatrix}, \quad Q_2 = \frac{1}{\sqrt{1090}}\begin{bmatrix} 33 & 1 \\ -1 & 33 \end{bmatrix}, \quad R_2 = \frac{1}{\sqrt{1090}}\begin{bmatrix} 109 & -3 \\ 0 & -10 \end{bmatrix}$$

$$A_3 = \frac{1}{109}\begin{bmatrix} 360 & 1 \\ 1 & -33 \end{bmatrix} = \begin{bmatrix} 3.302775 & 0.009174 \\ 0.009174 & -0.302775 \end{bmatrix}$$

4. Use induction on k. If $k = 1$, $A_1 = A$. In general $A_{k+1} = Q_k^{-1}A_kQ_k = Q_k^TA_kQ_k$, so the fact that $A_k^T = A_k$ implies $A_{k+1}^T = A_{k+1}$. The eigenvalues of A are all real (Theorem 2§6.1), so the A_k converge to an upper triangular matrix T. But T must also be symmetric (it is the limit of symmetric matrices), so it is diagonal.

Exercises 6.8
(Page 311)

1. (b) $\sqrt{6}$ **(d)** $\sqrt{13}$ **2. (b)** Not orthogonal **(d)** Orthogonal
3. (b) Not a subspace. For example, $i(0, 0, 1) = (0, 0, i)$ is not in U. **(d)** This is a subspace.
4. (b) Basis $\{(i, 0, 2), (1, 0, -1)\}$; dimension 2 **(d)** Basis $\{(1, 0, -2i), (0, 1, 1 -i)\}$; dimension 2
5. (b) Normal only **(d)** Hermitian (and normal), not unitary **(f)** None **(h)** Unitary (and normal); Hermitian if z is real

6. (b) $U = \frac{1}{\sqrt{14}}\begin{bmatrix} -2 & 3 - i \\ 3 + i & 2 \end{bmatrix}$, $U^*ZU = \begin{bmatrix} -1 & 0 \\ 0 & 6 \end{bmatrix}$ **(d)** $U = \frac{1}{\sqrt{3}}\begin{bmatrix} 1 + i & 1 \\ -1 & 1 - i \end{bmatrix}$, $U^*ZU = \begin{bmatrix} 1 & 0 \\ 0 & 4 \end{bmatrix}$

(f) $U = \frac{1}{\sqrt{3}}\begin{bmatrix} \sqrt{3} & 0 & 0 \\ 0 & 1 + i & 1 \\ 0 & -1 & 1 - i \end{bmatrix}$, $U^*ZU = \begin{bmatrix} 1 & 0 & 0 \\ 0 & 0 & 0 \\ 0 & 0 & 3 \end{bmatrix}$

8. (b) $\|\lambda Z\|^2 = \langle \lambda Z, \lambda Z \rangle = \lambda \bar{\lambda} \langle Z, Z \rangle = |\lambda|^2 \|Z\|^2$

9. (b) If the (k, k)-entry of Z is z_{kk}, then the (k, k)-entry of $\bar{Z}$ is $\bar{z}_{kk}$, so the (k, k)-entry of $Z^* = (\bar{Z})^T$ is $\bar{z}_{kk}$. As $Z = Z^*$ we have $z_{kk} = \bar{z}_{kk}$, so z_{kk} is real.

12. (b) $(S^2)^* = S^*S^* = (-S)(-S) = S^2$; $(iS)^* = \bar{i}S^* = (-i)(-S) = iS$

(d) If $Z = H + S$, as given, then $Z^* = H^* + S^* = H - S$, so $H = \frac{1}{2}(Z + Z^*)$ and $S = \frac{1}{2}(Z - Z^*)$. Hence the representation is unique if it exists. But, always, $Z = \frac{1}{2}(Z + Z^*) + \frac{1}{2}(Z + Z^*)$, and these are Hermitian and skew-Hermitian, respectively.

14. (b) If $U^{-1} = U^*$, then $(U^{-1})^* = (U^*)^* = (U^{-1})^{-1}$.

19. (b) Let $U = \begin{bmatrix} a & b \\ c & d \end{bmatrix}$ be real and invertible, and assume that $U^{-1}AU = \begin{bmatrix} \lambda & \mu \\ 0 & v \end{bmatrix}$. Then $AU = U \begin{bmatrix} \lambda & \mu \\ 0 & v \end{bmatrix}$, and first column entries are $c = a\lambda$ and $-a = c\lambda$. Hence λ is real (c and a are both real and are not both 0), and $(1 + \lambda^2)a = 0$. Thus $a = 0, c = a\lambda = 0$, a contradiction.

Exercises 6.9
(Page 323)

1. (b) $A = \begin{bmatrix} 1 & 0 \\ 0 & 2 \end{bmatrix}$ **(d)** $A = \begin{bmatrix} 1 & 3 & 2 \\ 3 & 1 & -1 \\ 2 & -1 & 3 \end{bmatrix}$ **2. (b)** $P = \frac{1}{\sqrt{2}}\begin{bmatrix} 1 & 1 \\ 1 & -1 \end{bmatrix}$; $Y = \frac{1}{\sqrt{2}}\begin{bmatrix} x_1 + x_2 \\ x_1 - x_2 \end{bmatrix}$; $q = 3y_1^2 - y_2^2$; 1, 2

(d) $P = \frac{1}{3}\begin{bmatrix} 2 & 2 & -1 \\ 2 & -1 & 2 \\ -1 & 2 & 2 \end{bmatrix}$; $Y = \frac{1}{3}\begin{bmatrix} 2x_1 + 2x_2 - x_3 \\ 2x_1 - x_2 + 2x_3 \\ -x_1 + 2x_2 + 2x_3 \end{bmatrix}$; $q = 9y_1^2 + 9y_2^2 - 9y_3^2$; 2, 3

(f) $P = \frac{1}{3}\begin{bmatrix} -2 & 1 & 2 \\ 2 & 2 & 1 \\ 1 & -2 & 2 \end{bmatrix}$; $Y = \frac{1}{3}\begin{bmatrix} -2x_1 + 2x_2 + x_3 \\ x_1 + 2x_2 - 2x_3 \\ 2x_1 + x_2 + 2x_3 \end{bmatrix}$; $q = 9y_1^2 + 9y_2^2$; 2, 2

(h) $P = \frac{1}{\sqrt{6}}\begin{bmatrix} -\sqrt{2} & \sqrt{3} & 1 \\ \sqrt{2} & 0 & 2 \\ \sqrt{2} & \sqrt{3} & -1 \end{bmatrix}$; $Y = \frac{1}{\sqrt{6}}\begin{bmatrix} -\sqrt{2}x_1 + \sqrt{2}x_2 + \sqrt{2}x_3 \\ \sqrt{3}x_1 + \sqrt{3}x_3 \\ x_1 + 2x_2 - x_3 \end{bmatrix}$; $q = 2y_1^2 + y_2^2 - y_3^2$; 2, 3

3. (b) $x_1 = \frac{1}{\sqrt{5}}(2x - y)$, $y_1 = \frac{1}{\sqrt{5}}(x + 2y)$; $4x_1^2 - y_1^2 = 2$; hyperbola

(d) $x_1 = \frac{1}{\sqrt{5}}(x + 2y)$, $y_1 = \frac{1}{\sqrt{5}}(2x - y)$; $6x_1^2 + y_1^2 = 1$; ellipse

7. (b) $3y_1^2 + 5y_2^2 - y_3^2 - 3\sqrt{2}y_1 + \frac{11}{3}\sqrt{3}y_2 + \frac{2}{3}\sqrt{6}y_3 = 7$

$y_1 = \frac{1}{\sqrt{2}}(x_2 + x_3)$, $y_2 = \frac{1}{\sqrt{3}}(x_1 + x_2 - x_3)$, $y_3 = \frac{1}{\sqrt{6}}(2x_1 - x_2 + x_3)$

Exercises 6.10
(Page 330)

1. (b) $\frac{1}{36} = \begin{bmatrix} -60 \\ 138 \\ 285 \end{bmatrix}$, $(A^TA)^{-1} = \frac{1}{36}\begin{bmatrix} 24 & -30 & -54 \\ -30 & 42 & 72 \\ -54 & 72 & 129 \end{bmatrix}$

2. (b) $\frac{1}{68}[11x + 75x^2 + 33(-1)^x]$, $(M^TM)^{-1} = \frac{1}{68}\begin{bmatrix} 89 & -33 & -5 \\ -33 & 13 & 3 \\ -5 & 3 & 19 \end{bmatrix}$

3. (b) $3 + \frac{1}{4}x^2 - \frac{7}{2}\sin\left(\frac{\pi x}{2}\right)$, $(M^T M)^{-1} = \frac{1}{2}\begin{bmatrix} -24 & 7 & 35 \\ 7 & -2 & -10 \\ 35 & -10 & -49 \end{bmatrix}$

4. $y = -5.19 + 0.34x_1 + 0.51x_2 + 0.71x_3$ $(A^T A)^{-1} = \frac{1}{50160}\begin{bmatrix} 1035720 & -16032 & 10080 & -45300 \\ -16032 & 416 & -632 & 800 \\ 10080 & -632 & 2600 & -2180 \\ -45300 & 800 & -2180 & 3950 \end{bmatrix}$

6. (b) It suffices to show that the columns of $M = \begin{bmatrix} 1 & e^{x_1} \\ \vdots & \vdots \\ 1 & e^{x_n} \end{bmatrix}$ are independent. If $r_0\begin{bmatrix} 1 \\ \vdots \\ 1 \end{bmatrix} + r_1\begin{bmatrix} e^{x_1} \\ \vdots \\ e^{x_n} \end{bmatrix} = \begin{bmatrix} 0 \\ \vdots \\ 0 \end{bmatrix}$, then $r_0 + r_1 e^{x_i} = 0$ for each i. Thus $r_1(e^{x_i} - e^{x_j}) = 0$ for all i and j. Hence $r_1 = 0$ because two x_i are distinct. Then $r_0 = r_0 + r_1 e^{x_i} = 0$ too.

Exercises 6.11
(Page 337)

1. (b) $c_1\begin{bmatrix} 1 \\ 1 \end{bmatrix}e^{4x} + c_2\begin{bmatrix} 5 \\ -1 \end{bmatrix}e^{-2x}$; $c_1 = -\frac{2}{3}$, $c_2 = \frac{1}{3}$

(d) $c_1\begin{bmatrix} -8 \\ 10 \\ 7 \end{bmatrix}e^{-x} + c_2\begin{bmatrix} 1 \\ -2 \\ 1 \end{bmatrix}e^{2x} + c_3\begin{bmatrix} 1 \\ 0 \\ 1 \end{bmatrix}e^{4x}$; $c_1 = 0$, $c_2 = -\frac{1}{2}$, $c_3 = \frac{3}{2}$

Exercises 7.1
(Page 350)

1. (b) $T(\mathbf{v}) = A\mathbf{v}$ where $A = \begin{bmatrix} 2 & -3 & 5 \\ 0 & 0 & 0 \end{bmatrix}$ **(d)** $T(\mathbf{v}) = \mathbf{v}A$ where $A = \begin{bmatrix} 1 & 0 & 0 \\ 0 & 1 & 0 \\ 0 & 0 & -1 \end{bmatrix}$

(f) $T(A + B) = P(A + B)Q = PAQ + PBQ = T(A) + T(B)$; $T(rA) = P(rA)Q = rPAQ = rT(A)$

(h) $T[(p + q)(x)] = (p + q)(0) = p(0) + q(0) = T[p(x)] + T[q(x)]$
$T[(rp)(x)] = (rp)(0) = rp(0) = rT[p(x)]$

(j) $T(X + Y) = (X + Y) \cdot Z = X \cdot Z + Y \cdot Z = T(X) + T(Y)$, and $T(rX) = (rX) \cdot Z = r(X \cdot Z) = rT(X)$

(l) If $\mathbf{v} = r_1\mathbf{e}_1 + \cdots + r_n\mathbf{e}_n$ and $\mathbf{w} = s_1\mathbf{e}_1 + \cdots + s_n\mathbf{e}_n$, then $T(\mathbf{v} + \mathbf{w}) = r_1 + s_1 = T(\mathbf{v}) + T(\mathbf{w})$, and $T(r\mathbf{v}) = rr_1 = rT(\mathbf{v})$

2. (b) $\text{rank}(A + B) \neq \text{rank } A + \text{rank } B$ in general. For example, $A = \begin{bmatrix} 1 & 0 \\ 0 & 1 \end{bmatrix}$ and $B = \begin{bmatrix} 1 & 0 \\ 0 & -1 \end{bmatrix}$.

(d) $T(\mathbf{0}) = \mathbf{0} + \mathbf{u} = \mathbf{u} \neq \mathbf{0}$, so T is not linear by Theorem 2.

3. (b) $T(3\mathbf{v}_1 + 2\mathbf{v}_2) = 0$ **(d)** $T\begin{bmatrix} 1 \\ -7 \end{bmatrix} = \begin{bmatrix} -3 \\ 4 \end{bmatrix}$ **(f)** $T(2 - x + 3x^2) = 46$

4. (b) $T(x, y) = \frac{1}{3}(x - y, 3y, x - y)$; $T(-1, 2) = (-1, 2, -1)$ **(d)** $T\begin{bmatrix} a & b \\ c & d \end{bmatrix} = 3a - 3c + 2b$

5. (b) $T(\mathbf{v}) = \frac{1}{3}(7\mathbf{v} - 9\mathbf{w})$, $T(\mathbf{w}) = \frac{1}{3}(\mathbf{v} + 3\mathbf{w})$

18. (b) $A = \begin{bmatrix} -1 & 0 \\ 0 & 1 \end{bmatrix}$ **(d)** $A = \begin{bmatrix} -1 & 0 \\ 0 & -1 \end{bmatrix}$ **(f)** $A = \begin{bmatrix} 0 & 1 \\ -1 & 0 \end{bmatrix}$ **(h)** $A = \begin{bmatrix} 0 & 0 \\ 0 & 1 \end{bmatrix}$

19. (b) $T\begin{bmatrix} x \\ y \\ z \end{bmatrix} = \begin{bmatrix} x \\ y \\ -z \end{bmatrix}$; $A = \begin{bmatrix} 1 & 0 & 0 \\ 0 & 1 & 0 \\ 0 & 0 & -1 \end{bmatrix}$ **20. (b)** $\dfrac{1}{a^2 + b^2 + c^2}\begin{bmatrix} -a^2 + b^2 + c^2 & -2ab & -2ac \\ -2ab & a^2 - b^2 + c^2 & -2bc \\ -2ac & -2bc & a^2 + b^2 - c^2 \end{bmatrix}$

Exercises 7.2
(Page 360)

1. (b) $\left\{ \begin{bmatrix} -3 \\ 7 \\ 1 \\ 0 \end{bmatrix}, \begin{bmatrix} 1 \\ 1 \\ 0 \\ -1 \end{bmatrix} \right\}$; $\left\{ \begin{bmatrix} 1 \\ 0 \\ 1 \end{bmatrix}, \begin{bmatrix} 0 \\ 1 \\ -1 \end{bmatrix} \right\}$; 2, 2   **(d)** $\left\{ \begin{bmatrix} -1 \\ 2 \\ 1 \end{bmatrix} \right\}$; $\left\{ \begin{bmatrix} 1 \\ 0 \\ 1 \\ 1 \end{bmatrix}, \begin{bmatrix} 0 \\ 1 \\ -1 \\ -2 \end{bmatrix} \right\}$; 2, 1

2. (b) $\{x^2 - x\}$; $\{(1, 0), (0, 1)\}$ **(d)** $\{(0, 0, 1)\}$; $\{(1, 1, 0, 0), (0, 0, 1, 1)\}$ **(f)** $\left\{ \begin{bmatrix} 1 & 0 \\ 0 & -1 \end{bmatrix}, \begin{bmatrix} 0 & 1 \\ 0 & 0 \end{bmatrix}, \begin{bmatrix} 0 & 0 \\ 1 & 0 \end{bmatrix} \right\}$; $\{1\}$

(h) $\{(1, 0, 0, \ldots, 0, -1), (0, 1, 0, \ldots, 0, -1), \ldots, (0, 0, 0, \ldots, 1, -1)\}$; $\{1\}$

(j) $\left\{ \begin{bmatrix} 0 & 1 \\ 0 & 0 \end{bmatrix}, \begin{bmatrix} 0 & 0 \\ 0 & 1 \end{bmatrix} \right\}$; $\left\{ \begin{bmatrix} 1 & 1 \\ 0 & 0 \end{bmatrix}, \begin{bmatrix} 0 & 0 \\ 1 & 1 \end{bmatrix} \right\}$

3. (b) $T(\mathbf{v}) = \mathbf{0} = [0, 0]$ if and only if $P(\mathbf{v}) = 0$ and $Q(\mathbf{v}) = 0$; that is, if and only if $\mathbf{v}$ is in ker $P \cap$ ker Q.

4. (b) ker $T = \text{span}\{(-4, 1, 3)\}$; $B = \{(1, 0, 0), (0, 1, 0), (-4, 1, 3)\}$, im $T = \text{span}\{(1, 2, 0, 3), (1, -1, -3, 0)\}$

6. (b) Yes. $\dim(\text{im } T) = 5 - \dim(\text{ker } T) = 3$, so im $T = W$ as $\dim W = 3$.   **(d)** No. $T = 0 : \mathbb{R}^2 \to \mathbb{R}^2$

(f) No. $T : \mathbb{R}^2 \to \mathbb{R}^2$, $T(x, y) = (y, 0)$. Then ker $T = \text{im } T$

(h) Yes. $\dim V = \dim(\text{ker } T) + \dim(\text{im } T) \le \dim W + \dim W = 2 \dim W$

(j) No. $T : \mathbb{R}^2 \to \mathbb{R}^2$, $T(x, y) = (x, 0)$ **(l)** No. $T : \mathbb{R}^2 \to \mathbb{R}^2$, $T(x, y) = (x, 0)$

7. (b) Given $\mathbf{w}$ in W, let $\mathbf{w} = T(\mathbf{v})$, $\mathbf{v}$ in V, and write $\mathbf{v} = r_1\mathbf{v}_1 + \cdots + r_n\mathbf{v}_n$. Then $\mathbf{w} = T(\mathbf{v}) = r_1T(\mathbf{v}_1) + \cdots + r_nT(\mathbf{v}_n)$.

15. (b) $B = \{x - 1, \ldots, x^n - 1\}$ is independent (distinct degrees) and contained in ker T. Hence B is a basis of ker T by **(a)**.

28. (b) By Theorem 4§5.4, let $\{\mathbf{u}_1, \ldots, \mathbf{u}_m, \ldots, \mathbf{u}_n\}$ be a basis of V where $\{\mathbf{u}_1, \ldots, \mathbf{u}_m\}$ is a basis of U. By Theorem 4§7.1 there is a linear transformation $S : V \to V$ such that $S(\mathbf{u}_i) = \mathbf{u}_i$ for $1 \le i \le m$, and $S(\mathbf{u}_i) = \mathbf{0}$ if $i > m$. Because each $\mathbf{u}_i$ is in im S, $U \subseteq$ im S. But if $S(\mathbf{v})$ is in im S, write $\mathbf{v} = r_1\mathbf{u}_1 + \cdots + r_m\mathbf{u}_m + \cdots + r_n\mathbf{u}_n$. Then $S(\mathbf{v}) = r_1S(\mathbf{u}_1) + \cdots + r_mS(\mathbf{u}_m) = r_1\mathbf{u}_1 + \cdots + r_m\mathbf{u}_m$ is in U. So im $S \subseteq U$.

Exercises 7.3
(Page 369)

1. (b) T is onto because $T(1, -1, 0) = (1, 0, 0)$, $T(0, 1, -1) = (0, 1, 0)$, and $T(0, 0, 1) = (0, 0, 1)$. Use Theorem 3.

(d) T is one-to-one because $0 = T(X) = UXV$ implies that $X = 0$ (U and V are invertible). Use Theorem 3.

(f) T is one-to-one because $\mathbf{0} = T(\mathbf{v}) = k\mathbf{v}$ implies that $\mathbf{v} = \mathbf{0}$ (because $k \ne 0$). T is onto because $T(\frac{1}{k}\mathbf{v}) = \mathbf{v}$ for all $\mathbf{v}$. [Here Theorem 3 does not apply if dim V is not finite.]

(h) T is one-to-one because $T(A) = 0$ implies $A^T = 0$, whence $A = 0$. Use Theorem 3.

4. (b) $ST(x, y, z) = (x + y, 0, y + z)$, $TS(x, y, z) = (x, 0, z)$ **(d)** $ST\begin{bmatrix} a & b \\ c & d \end{bmatrix} = \begin{bmatrix} c & 0 \\ 0 & b \end{bmatrix}$, $TS\begin{bmatrix} a & b \\ c & d \end{bmatrix} = \begin{bmatrix} 0 & a \\ d & 0 \end{bmatrix}$

5. (b) $T^2(x, y) = T(x + y, 0) = (x + y, 0) = T(x, y)$. Hence $T^2 = T$.

(d) $T^2\begin{bmatrix} a & b \\ c & d \end{bmatrix} = \frac{1}{2} T\begin{bmatrix} a + c & b + d \\ a + c & b + d \end{bmatrix} = \frac{1}{2}\begin{bmatrix} a + c & b + d \\ a + c & b + d \end{bmatrix}$

6. (b) No inverse; $(1, -1, 1, -1)$ is in ker T.   **(d)** $T^{-1}\begin{bmatrix} a & b \\ c & d \end{bmatrix} = \frac{1}{5}\begin{bmatrix} 3a - 2c & 3b - 2d \\ a + c & b + d \end{bmatrix}$

(f) $T^{-1}(a, b, c) = \frac{1}{2}[2a + (b - c)x - (2a - b - c)x^2]$

7. (b) $T^2(x, y) = T(ky - x, y) = (ky - (ky - x), y) = (x, y)$ **(d)** $T^2(X) = A^2X = IX = X$

10. (b) $T^3(x, y, z, w) = (x, y, z, -w)$ so $T^6(x, y, z, w) = T^3[T^3(x, y, z, w)] = (x, y, z, w)$ Hence $T^{-1} = T^5$. So $T^{-1}(x, y, z, w) = (y - x, -x, z, -w)$.

12. (b) Given $\mathbf{u}$ in U, write $\mathbf{u} = S(\mathbf{w})$, $\mathbf{w}$ in W (because S is onto). Then write $\mathbf{w} = T(\mathbf{v})$, $\mathbf{v}$ in V (T is onto). Hence $\mathbf{u} = ST(\mathbf{v})$, so ST is onto.

15. (b) Given $\mathbf{w}$ in W, write $\mathbf{w} = ST(\mathbf{v})$, $\mathbf{v}$ in V (ST is onto). Then $\mathbf{w} = S[T(\mathbf{v})]$, $T(\mathbf{v})$ in U, so S is onto. But then im $S = W$, so $\dim U = \dim(\ker S) + \dim(\text{im } S) \geq \dim(\text{im } S) = \dim W$.

21. (b) $T(x, y) = (x, y + 1)$

29. (b) If $T(p) = 0$, then $p(x) = -xp'(x)$. We write $p(x) = a_0 + a_1x + a_2x^2 + \cdots + a_nx^n$, and this becomes

$$a_0 + a_1x + a_2x^2 + \cdots + a_nx^n = -a_1x - 2a_2x^2 - \cdots - na_nx^n$$

Equating coefficients yields $a_0 = 0$, $2a_1 = 0$, $3a_2 = 0$, $\ldots$, $(n + 1)a_n = 0$, whence $p(x) = 0$. This means that $\ker T = 0$, so T is one-to-one. But then T is an isomorphism by Theorem 3.

32. Let $B = \{\mathbf{e}_1, \ldots, \mathbf{e}_r, \mathbf{e}_{r+1}, \ldots, \mathbf{e}_n\}$ be a basis of V with $\{\mathbf{e}_{r+1}, \ldots, \mathbf{e}_n\}$ a basis of $\ker T$. If $\{T(\mathbf{e}_1), \ldots, T(\mathbf{e}_r) \mathbf{w}_{r+1}, \ldots, \mathbf{w}_n\}$ is a basis of V, define S by $S[T(\mathbf{e}_i)] = \mathbf{e}_i$ for $1 \leq i \leq r$, and $S(\mathbf{w}_j) = \mathbf{e}_j$ for $r + 1 \leq j \leq n$. Then S is an isomorphism by Theorem 1, and $TST(\mathbf{e}_i) = T(\mathbf{e}_i)$ clearly holds for $1 \leq i \leq r$. But if $i \geq r + 1$, then $T(\mathbf{e}_i) = 0 = TST(\mathbf{e}_i)$, so $T = TST$ by Theorem 3§7.1.

Exercises 7.4
(Page 377)

1. (b) $\begin{bmatrix} a \\ 2b - c \\ c - b \end{bmatrix}$ **(d)** $\frac{1}{2}\begin{bmatrix} a - b \\ a + b \\ -a + 3b + 2c \end{bmatrix}$

2. (b) Let $\mathbf{v} = a + bx + cx^2$. Then

$$C_D[T(\mathbf{v})] = M_{DB}(T)C_B(\mathbf{v}) = \begin{bmatrix} 2 & 1 & 3 \\ -1 & 0 & -2 \end{bmatrix}\begin{bmatrix} a \\ b \\ c \end{bmatrix} = \begin{bmatrix} 2a + b + 3c \\ -a - 2c \end{bmatrix}$$

Hence $T(\mathbf{v}) = (2a + b + 3c)(1, 1) + (-a - 2c)(0, 1) = (2a + b + 3c, a + b + c)$.

3. (b) $\begin{bmatrix} 1 & 0 & 0 & 0 \\ 0 & 0 & 1 & 0 \\ 0 & 1 & 0 & 0 \\ 0 & 0 & 0 & 1 \end{bmatrix}$ **(d)** $\begin{bmatrix} 1 & 1 & 1 \\ 0 & 1 & 2 \\ 0 & 0 & 1 \end{bmatrix}$

4. (b) $\begin{bmatrix} 1 & 2 \\ 5 & 3 \\ 4 & 0 \\ 1 & 1 \end{bmatrix}$; $C_D[T(a, b)] = \begin{bmatrix} 1 & 2 \\ 5 & 3 \\ 4 & 0 \\ 1 & 1 \end{bmatrix}\begin{bmatrix} b \\ a - b \end{bmatrix} = \begin{bmatrix} 2a - b \\ 3a + 2b \\ 4b \\ a \end{bmatrix}$

(d) $\frac{1}{2}\begin{bmatrix} 1 & 1 & -1 \\ 1 & 1 & 1 \end{bmatrix}$; $C_D[T(a + bx + cx^2)] = \frac{1}{2}\begin{bmatrix} 1 & 1 & -1 \\ 1 & 1 & 1 \end{bmatrix}\begin{bmatrix} a \\ b \\ c \end{bmatrix} = \frac{1}{2}\begin{bmatrix} a + b - c \\ a + b + c \end{bmatrix}$

(f) $\begin{bmatrix} 1 & 0 & 0 & 0 \\ 0 & 1 & 1 & 0 \\ 0 & 1 & 1 & 0 \\ 0 & 0 & 0 & 1 \end{bmatrix}$; $C_D\left(T\begin{bmatrix} a & b \\ c & d \end{bmatrix}\right) = \begin{bmatrix} 1 & 0 & 0 & 0 \\ 0 & 1 & 1 & 0 \\ 0 & 1 & 1 & 0 \\ 0 & 0 & 0 & 1 \end{bmatrix}\begin{bmatrix} a \\ b \\ c \\ d \end{bmatrix} = \begin{bmatrix} a \\ b + c \\ b + c \\ d \end{bmatrix}$

5. (b) $M_{ED}(S)M_{DB}(T) = \begin{bmatrix} 1 & 1 & 0 & 0 \\ 0 & 0 & 1 & -1 \end{bmatrix}\begin{bmatrix} 1 & 1 & 0 \\ 0 & 1 & 1 \\ 1 & 0 & 1 \\ -1 & 1 & 0 \end{bmatrix} = \begin{bmatrix} 1 & 2 & 1 \\ 2 & -1 & 1 \end{bmatrix} = M_{EB}(ST)$

(d) $M_{ED}(S)M_{DB}(T) = \begin{bmatrix} 1 & -1 & 0 \\ 0 & 0 & 1 \end{bmatrix}\begin{bmatrix} 1 & -1 & 0 \\ -1 & 0 & 1 \\ 0 & 1 & 0 \end{bmatrix} = \begin{bmatrix} 2 & -1 & -1 \\ 0 & 1 & 0 \end{bmatrix} = M_{EB}(ST)$

7. (b) $T^{-1}(a, b, c) = \frac{1}{2}(b + c - a, a + c - b, a + b - c)$ $M_{DB}(T) = \begin{bmatrix} 0 & 1 & 1 \\ 1 & 0 & 1 \\ 1 & 1 & 0 \end{bmatrix}$; $M_{BD}(T^{-1}) = \frac{1}{2}\begin{bmatrix} -1 & 1 & 1 \\ 1 & -1 & 1 \\ 1 & 1 & -1 \end{bmatrix}$

(d) $T^{-1}(a, b, c) = (a - b) + (b - c)x + cx^2$ $\quad M_{DB}(T) = \begin{bmatrix} 1 & 1 & 1 \\ 0 & 1 & 1 \\ 0 & 0 & 1 \end{bmatrix}$; $M_{BD}(T^{-1}) = \begin{bmatrix} 1 & -1 & 0 \\ 0 & 1 & -1 \\ 0 & 0 & 1 \end{bmatrix}$

8. (b) $M_{DB}(T^{-1}) = [M_{BD}(T)]^{-1} = \begin{bmatrix} 1 & 1 & 1 & 0 \\ 0 & 1 & 1 & 0 \\ 0 & 0 & 1 & 0 \\ 0 & 0 & 0 & 1 \end{bmatrix}^{-1} = \begin{bmatrix} 1 & -1 & 0 & 0 \\ 0 & 1 & -1 & 0 \\ 0 & 0 & 1 & 0 \\ 0 & 0 & 0 & 1 \end{bmatrix}$. Hence $C_B[T^{-1}(a, b, c, d)] = M_{BD}(T^{-1})C_D(a, b, c, d) =$

$\begin{bmatrix} 1 & -1 & 0 & 0 \\ 0 & 1 & -1 & 0 \\ 0 & 0 & 1 & 0 \\ 0 & 0 & 0 & 1 \end{bmatrix}\begin{bmatrix} a \\ b \\ c \\ d \end{bmatrix} = \begin{bmatrix} a - b \\ b - c \\ c \\ d \end{bmatrix}$, so $T^{-1}(a, b, c, d) = \begin{bmatrix} a - b & b - c \\ c & d \end{bmatrix}$.

12. Have $C_D[T(e_j)] = $ column j of I_n. Hence $M_{DB}(T) = [C_D[T(e_1)] \ C_D[T(e_2)] \ \cdots \ C_D[T(e_n)]] = I_n$.

21. (b) If w lies in im$(S + T)$, then $w = (S + T)(v)$ for some v in V. But then $w = S(v) + T(v)$, so w lies in im S + im T.

22. (b) If $X \subseteq X_1$, let T lie in X_1^0. Then $T(v) = 0$ for all v in X_1, whence $T(v) = 0$ for all v in X. Thus T is in X^0 and we have shown that $X_1^0 \subseteq X^0$.

Exercises 7.5
(Page 387)

1. (b) $\frac{1}{2}\begin{bmatrix} -3 & -2 & 1 \\ 2 & 2 & 0 \\ 0 & 0 & 2 \end{bmatrix}$ **4. (b)** $P_{B \leftarrow D} = \begin{bmatrix} 1 & 1 & -1 \\ 1 & -1 & 0 \\ 1 & 0 & 1 \end{bmatrix}$, $P_{D \leftarrow B} = \frac{1}{3}\begin{bmatrix} 1 & 1 & 1 \\ 1 & -2 & 1 \\ -1 & -1 & 2 \end{bmatrix}$, $P_{E \leftarrow D} = \begin{bmatrix} 1 & 0 & 1 \\ 1 & -1 & 0 \\ 1 & 0 & 0 \end{bmatrix}$, $P_{E \leftarrow B} = \begin{bmatrix} 0 & 0 & 1 \\ 0 & 1 & 0 \\ 1 & 0 & 0 \end{bmatrix}$

5. (b) $A = P_{D \leftarrow B}$, where $B = \{(1, 2, -1), (2, 3, 0), (1, 0, 2)\}$. Hence $A^{-1} = P_{B \leftarrow D} = \begin{bmatrix} 6 & -4 & -3 \\ -4 & 3 & 2 \\ 3 & -2 & -1 \end{bmatrix}$

7. (b) $P = \begin{bmatrix} 1 & 1 & 0 \\ 0 & 1 & 2 \\ -1 & 0 & 1 \end{bmatrix}$ **8. (b)** $B = \left\{\begin{bmatrix} 3 \\ 7 \end{bmatrix}, \begin{bmatrix} 2 \\ 5 \end{bmatrix}\right\}$

9. (b) $c_T(x) = x^2 - 6x - 1$ **(d)** $c_T(x) = x^3 + x^2 - 8x - 3$ **(f)** $c_T(x) = x^4$

16. If $b_j = \sum_{i=1}^n p_{ij}d_i$, then $C_D(b_j) = \begin{bmatrix} p_{1j} \\ p_{2j} \\ \vdots \\ p_{nj} \end{bmatrix} = $ column j of P. Hence $P_{D \leftarrow B} = [C_D(b_1) \ \cdots \ C_D(b_n)] = P$. Given v, write

$v = \sum_{i=1}^n v_j b_j$, so $C_B(v) = \begin{bmatrix} v_1 \\ \vdots \\ v_n \end{bmatrix}$. Then $v = \sum_{j=1}^n v_j\left[\sum_{i=1}^n p_{ij}d_i\right] = \sum_{i=1}^n\left[\sum_{j=1}^n p_{ij}v_i\right]d_i$. Hence $\sum_{j=1}^n p_{ij}v_j$ is the ith entry of $C_D(v)$.

But it is also the ith entry of $PC_B(v)$.

Exercises 7.6
(Page 399)

3. (b) If v is in $S(U)$, write $v = S(u)$, u in U. Then

$$T(v) = T[S(u)] = (TS)(u) = (ST)(u) = S[T(u)]$$

and this lies in $S(U)$ because $T(u)$ lies in U (U is T-invariant).

6. Suppose U is T-invariant for every T. If $U \neq 0$, choose $u \neq 0$ in U. Choose a basis $B = \{u, u_2, \ldots, u_n\}$ of V containing u.

Given any $\mathbf{v}$ in V, there is (by Theorem 4§7.1) a linear transformation $T : V \to V$ such that $T(\mathbf{u}) = \mathbf{v}$, $T(\mathbf{u}_2) = \cdots = T(\mathbf{u}_n) = \mathbf{0}$. Then $\mathbf{v} = T(\mathbf{u})$ lies in U because U is T-invariant. This shows that $V = U$.

8. (b) $T(1 - 2x^2) = 3 + 3x - 3x^2 = 3(1 - 2x^2) + 3(x + x^2)$ and $T(x + x^2) = -(1 - 2x^2)$, so both are in U. Hence U is

T-invariant by Example 3. If $B = \{1 - 2x^2, x + x^2, x^2\}$ then $M_B(T) = \begin{bmatrix} 3 & -1 & 1 \\ 3 & 0 & 1 \\ 0 & 0 & 3 \end{bmatrix}$, so

$$c_T(x) = \det \begin{bmatrix} x - 3 & 1 & -1 \\ -3 & x & -1 \\ 0 & 0 & x - 3 \end{bmatrix} = (x - 3)\det \begin{bmatrix} x - 3 & 1 \\ -3 & x \end{bmatrix} = (x - 3)(x^2 - 3x + 3)$$

9. (b) Suppose $\mathbb{R}\mathbf{u}$ is T_A-invariant where $\mathbf{u} \neq \mathbf{0}$. Then $T_A(\mathbf{u}) = r\mathbf{u}$ for some r in $\mathbb{R}$, so $(rI - A)\mathbf{u} = \mathbf{0}$. But $\det(rI - A) = (r - \cos \theta)^2 + \sin^2 \theta \neq 0$ because $0 < \theta < \pi$. Hence $\mathbf{u} = \mathbf{0}$, a contradiction.

14. The fact that U and W are subspaces is easily verified using the subspace test. If A lies in $U \cap V$, then $A = AE = 0$; that is, $U \cap V = 0$. To show that $\mathbf{M}_{22} = U + V$, choose any A in $\mathbf{M}_{22}$. Then $A = AE + (A - AE)$, and AE lies in U [because $(AE)E = AE^2 = AE$], and $A - AE$ lies in W [because $(A - AE)E = AE - AE^2 = 0$].

22. (b) $T^2[p(x)] = p[-(-x)] = p(x)$, so $T^2 = 1$; $B = \{1, x^2; x, x^3\}$
  (d) $T^2(a, b, c) = T(-a + 2b + c, b + c, -c) = (a, b, c)$, so $T^2 = 1$; $B = \{(1, 1, 0); (1, 0, 0), (0, -1, 2)\}$

25. (b) $T^2(a, b, c) = T(a + 2b, 0, 4b + c) = (a + 2b, 0, 4b + c) = T(a, b, c)$ so $T^2 = T$; $B = \{(1, 0, 0), (0, 0, 1); (2, -1, 4)\}$

Exercises 7.7
(Page 410)

1. (b) $c_A(x) = (x + 1)^3$; $P = \begin{bmatrix} 1 & 0 & 0 \\ 1 & 1 & 0 \\ 1 & -3 & 1 \end{bmatrix}$; $P^{-1}AP = \begin{bmatrix} -1 & 0 & 1 \\ 0 & -1 & 0 \\ 0 & 0 & -1 \end{bmatrix}$

  (d) $c_A(x) = (x - 1)^2(x + 2)$; $P = \begin{bmatrix} -1 & 0 & -1 \\ 4 & 1 & 1 \\ 4 & 2 & 1 \end{bmatrix}$; $P^{-1}AP = \begin{bmatrix} 1 & 1 & 0 \\ 0 & 1 & 0 \\ 0 & 0 & -2 \end{bmatrix}$

  (f) $c_A(x) = (x + 1)^2(x-1)^2$; $P = \begin{bmatrix} 1 & 1 & 5 & 1 \\ 0 & 0 & 2 & -1 \\ 0 & 1 & 2 & 0 \\ 1 & 0 & 1 & 0 \end{bmatrix}$; $P^{-1}AP = \begin{bmatrix} -1 & 1 & 0 & 0 \\ 0 & -1 & 0 & 0 \\ 0 & 0 & 1 & -2 \\ 0 & 0 & 0 & 1 \end{bmatrix}$

Exercises 7.8
(Page 419)

1. (b) $\{[1), [2^n), [(-3)^n)\}$; $x_n = \frac{1}{20}(15 + 2^{n+3} + (-3)^{n+1})$

2. (b) $\{[1), [n), [(-2)^n)\}$; $x_n = \frac{1}{9}(5 - 6n + (-2)^{n+2})$ **(d)** $\{[1), [n), [n^2)\}$; $x_n = 2(n - 1)^2 - 1$

3. (b) $\{[a^n), [b^n)\}$ **6. (b)** $[1, 0, 0, 0, 0, \ldots), [0, 1, 0, 0, 0, \ldots), [0, 0, 1, 1, 1, \ldots), [0, 0, 1, 2, 3, \ldots)$

9. By Remark 2,

$$[i^n + (-i)^n) = [2, 0, -2, 0, 2, 0, -2, 0, \ldots)$$

$$[i(i^n - (-i)^n)) = [0, -2, 0, 2, 0, -2, 0, 2, \ldots)$$

are solutions. They are linearly independent and so are a basis.

Exercises 8.1
(Page 429)

1. (b) P5 fails.   **(d)** P5 fails.   **(f)** P5 fails.

3. (b) $\frac{1}{\sqrt{15}}(2, -3, 1, 1)$ **(d)** $\frac{1}{\sqrt{17}}\begin{bmatrix} 3 \\ -1 \end{bmatrix}$ **4. (b)** $\sqrt{2}$ **(d)** $\sqrt{3}$

12. (b) $\langle \mathbf{v}, \mathbf{v} \rangle = 5v_1^2 - 6v_1v_2 + 2v_2^2 = \frac{1}{5}[(5v_1 - 3v_2)^2 + v_2^2]$ **(d)** $\langle \mathbf{v}, \mathbf{v} \rangle = 3v_1^2 + 8v_1v_2 + 6v_2^2 = \frac{1}{3}[(3v_1 + 4v_2)^2 + 2v_2^2]$

13. (b) $\begin{bmatrix} 1 & -1 \\ -1 & 2 \end{bmatrix}$ **(d)** $\begin{bmatrix} 1 & 0 & -2 \\ 0 & 2 & 0 \\ -2 & 0 & 5 \end{bmatrix}$ **16. (b)** -15 **22. (b)** $15\|\mathbf{u}\|^2 - 17\langle \mathbf{u}, \mathbf{v} \rangle - 4\|\mathbf{v}\|^2$

26. (b) $\{(1, 1, 0), (0, 2, 1)\}$ **28.** $\langle \mathbf{v} - \mathbf{w}, \mathbf{v}_i \rangle = \langle \mathbf{v}, \mathbf{v}_i \rangle - \langle \mathbf{w}, \mathbf{v}_i \rangle = 0$ for each i, so $\mathbf{v} = \mathbf{w}$ by Exercise 27.

Exercises 8.2
(Page 439)

1. (b) $\frac{1}{14}\left\{ (6a + 2b + 6c)\begin{bmatrix} 1 \\ 1 \\ 1 \end{bmatrix} + (7c - 7a)\begin{bmatrix} -1 \\ 0 \\ 1 \end{bmatrix} + (a - 2b + c)\begin{bmatrix} 1 \\ -6 \\ 1 \end{bmatrix} \right\}$

   (d) $\left(\frac{a + d}{2}\right)\begin{bmatrix} 1 & 0 \\ 0 & 1 \end{bmatrix} + \left(\frac{a - d}{2}\right)\begin{bmatrix} 1 & 0 \\ 0 & -1 \end{bmatrix} + \left(\frac{b + c}{2}\right)\begin{bmatrix} 0 & 1 \\ 1 & 0 \end{bmatrix} + \left(\frac{b - c}{2}\right)\begin{bmatrix} 0 & 1 \\ -1 & 0 \end{bmatrix}$

2. (b) $\{(1, 1, 1), (1, -5, 1), (3, 0, -2)\}$ **3. (b)** $\left\{ \begin{bmatrix} 1 & 1 \\ 0 & 1 \end{bmatrix}, \begin{bmatrix} 1 & -2 \\ 3 & 1 \end{bmatrix}, \begin{bmatrix} 1 & -2 \\ -2 & 1 \end{bmatrix}, \begin{bmatrix} 1 & 0 \\ 0 & -1 \end{bmatrix} \right\}$

4. (b) $\{1, x - 1, x^2 - 2x + \frac{2}{3}\}$

6. (b) $U^\perp = \text{span}\{[1 \quad -1 \quad 0 \quad 0], [0 \quad 0 \quad 1 \quad 0], [0 \quad 0 \quad 0 \quad 1]\}$, $\dim U^\perp = 3$, $\dim U = 1$

   (d) $U^\perp = \text{span}\{2 - 3x, 1 - 2x^2\}$, $\dim U^\perp = 2$, $\dim U = 1$ **(f)** $U^\perp = \text{span}\left\{ \begin{bmatrix} 1 & -1 \\ -1 & 0 \end{bmatrix} \right\}$, $\dim U^\perp = 1$, $\dim U = 3$

7. (b) $U = \text{span}\left\{ \begin{bmatrix} 1 & 0 \\ 0 & 1 \end{bmatrix}, \begin{bmatrix} 1 & 1 \\ 1 & -1 \end{bmatrix}, \begin{bmatrix} 0 & 1 \\ -1 & 0 \end{bmatrix} \right\}$; $\text{proj}_U(A) = \begin{bmatrix} 3 & 0 \\ 2 & 1 \end{bmatrix}$

8. (b) $U = \text{span}\{1, 5 - 3x^2\}$; $\text{proj}_U(x) = \frac{3}{13}(1 + 2x^2)$

9. (b) $B = \{1, 2x - 1\}$ is an orthogonal basis of U because $\int_0^1 (2x - 1)dx = 0$. Using it, we get $\text{proj}_U(x^2 + 1) = x + \frac{5}{6}$, so $x^2 + 1 = (x + \frac{5}{6}) + (x^2 - x + \frac{1}{6})$.

11. (b) This follows from $\langle \mathbf{v} + \mathbf{w}, \mathbf{v} - \mathbf{w} \rangle = \|\mathbf{v}\|^2 - \|\mathbf{w}\|^2$.

14. (b) $U^\perp \subseteq \{\mathbf{u}_1, \ldots, \mathbf{u}_m\}^\perp$ because each $\mathbf{u}_i$ is in U. Conversely, if $\langle \mathbf{v}, \mathbf{u}_i \rangle = 0$ for each i, and $\mathbf{u} = r_1\mathbf{u}_1 + \cdots + r_m\mathbf{u}_m$ is any vector in U, then $\langle \mathbf{v}, \mathbf{u} \rangle = r_1\langle \mathbf{v}, \mathbf{u}_1 \rangle + \cdots + r_m\langle \mathbf{v}, \mathbf{u}_m \rangle = 0$.

18. (b) $\text{proj}_U(-5, 4, -3) = (-5, 4, -3)$; $\text{proj}_U(-1, 0, 2) = \frac{1}{38}(-17, 24, 73)$

Exercises 8.3
(Page 447)

1. (b) $B = \left\{ \begin{bmatrix} 1 & 0 \\ 0 & 0 \end{bmatrix}, \begin{bmatrix} 0 & 1 \\ 0 & 0 \end{bmatrix}, \begin{bmatrix} 0 & 0 \\ 1 & 0 \end{bmatrix}, \begin{bmatrix} 0 & 0 \\ 0 & 1 \end{bmatrix} \right\}$; $M_B(T)\begin{bmatrix} -1 & 0 & 1 & 0 \\ 0 & -1 & 0 & 1 \\ 1 & 0 & 2 & 0 \\ 0 & 1 & 0 & 2 \end{bmatrix}$

4. (b) $\langle \mathbf{v}, (rT)(\mathbf{w}) \rangle = \langle \mathbf{v}, rT(\mathbf{w}) \rangle = r\langle \mathbf{v}, T(\mathbf{w}) \rangle = r\langle T(\mathbf{v}), \mathbf{w} \rangle = \langle rT(\mathbf{v}), \mathbf{w} \rangle = \langle (rT)(\mathbf{v}), \mathbf{w} \rangle$

5. (b) If $B_0 = \{(1, 0, 0), (0, 1, 0), (0, 0, 1)\}$, then $M_{B_0}(T) = \begin{bmatrix} 7 & -1 & 0 \\ -1 & 7 & 0 \\ 0 & 0 & 2 \end{bmatrix}$ has an orthonormal basis of eigenvectors

$\left\{ \frac{1}{\sqrt{2}}\begin{bmatrix} 1 \\ 1 \\ 0 \end{bmatrix}, \frac{1}{\sqrt{2}}\begin{bmatrix} 1 \\ -1 \\ 0 \end{bmatrix}, \begin{bmatrix} 0 \\ 0 \\ 1 \end{bmatrix} \right\}$. Hence an orthonormal basis of eigenvectors of T is $\left\{ \frac{1}{\sqrt{2}}(1, 1, 0), \frac{1}{\sqrt{2}}(1, -1, 0), (0, 0, 1) \right\}$.

(d) If $B_0 = \{1, x, x^2\}$, then $M_{B_0}(T) = \begin{bmatrix} -1 & 0 & 1 \\ 0 & 3 & 0 \\ 1 & 0 & -1 \end{bmatrix}$ has an orthonormal basis of eigenvectors $\left\{ \begin{bmatrix} 0 \\ 1 \\ 0 \end{bmatrix}, \frac{1}{\sqrt{2}}\begin{bmatrix} 1 \\ 0 \\ 1 \end{bmatrix}, \frac{1}{\sqrt{2}}\begin{bmatrix} 1 \\ 0 \\ -1 \end{bmatrix} \right\}$.

Hence an orthonormal basis of eigenvectors of T is $\left\{ x, \frac{1}{\sqrt{2}}(1 + x^2), \frac{1}{\sqrt{2}}(1 - x^2) \right\}$.

7. (b) $M_B(T) = \begin{bmatrix} A & 0 \\ 0 & A \end{bmatrix}$, so $c_T(x) = \det\begin{bmatrix} xI_2 - A & 0 \\ 0 & xI_2 - A \end{bmatrix} = [c_A(x)]^2$.

14. (c) The coefficients in the definition of $T'(e_j) = \sum_{i=1}^{n} \langle e_i, T(e_i) \rangle e_i$ are the entries in the jth column $C_B[T'(e_j)]$ of $M_B(T')$. Hence

$M_B(T') = [\langle e_j, T(e_i) \rangle]$, and this is the transpose of $M_B(T)$ by Theorem 2.

Exercises 8.4
(Page 459)

2. (b) Rotation through π **(d)** Reflection in the line $y = -x$ **(f)** Rotation through $\frac{\pi}{4}$

3. (b) $c_T(x) = (x - 1)(x^2 + \frac{3}{2}x + 1)$. If $E = [1 \quad \sqrt{3} \quad \sqrt{3}]^T$, then T is a rotation about $\mathbb{R}E$.

(d) $c_T(x) = (x - 1)(x + 1)^2$. Rotation (of π) about the X axis.

(f) $c_T(x) = (x + 1)(x^2 - \sqrt{2}x + 1)$. Rotation (of $\frac{3\pi}{4}$) about the Y axis followed by a reflection in the $X-Z$ plane.

6. (b) If S and T are isometries, $\|ST(v)\| = \|S(T(v))\| = \|T(v)\| = \|v\|$ for all v in V. So ST is an isometry.

Exercises 8.5
(Page 464)

1. (b) $\frac{\pi}{2} - \frac{4}{\pi}\left[\cos x + \frac{\cos 3x}{3^2} + \frac{\cos 5x}{5^2} \right]$

(d) $\frac{\pi}{4} + \left[\sin x - \frac{\sin 2x}{2} + \frac{\sin 3x}{3} - \frac{\sin 4x}{4} + \frac{\sin 5x}{5} \right] - \frac{2}{\pi}\left[\cos x + \frac{\cos 3x}{3^2} + \frac{\cos 5x}{5^2} \right]$

2. (b) $\frac{2}{\pi} - \frac{8}{\pi}\left[\frac{\cos 2x}{2^2 - 1} + \frac{\cos 4x}{4^2 - 1} + \frac{\cos 6x}{6^2 - 1} \right]$

4. $\int \cos kx \cos lx \, dx = \frac{1}{2}\left[\frac{\sin[(k + l)x]}{k + l} - \frac{\sin[(k - l)x]}{k - l} \right]_0^\pi = 0$ provided that $k \neq l$

Appendix A
(Page 478)

1. (b) $x = 3$ **(d)** $x = \pm 1$ **2. (b)** $10 + i$ **(d)** $\frac{11}{26} + \frac{23}{26}i$ **(f)** $2 - 11i$ **(h)** $8 - 6i$

3. (b) $\frac{11}{5} + \frac{3}{5}i$ **(d)** $\pm(2 - i)$ **(f)** $1 + i$ **4. (b)** $\frac{1}{2} \pm \frac{\sqrt{3}}{2}i$ **(d)** $2, \frac{1}{2}$

5. (b) $-2, 1 \pm \sqrt{3}i$ **(d)** $\pm 2\sqrt{2}, \pm 2\sqrt{2}i$ **6. (b)** $x^2 - 4x + 13; 2 + 3i$ **(d)** $x^2 - 6x + 25; 3 + 4i$

8. $x^4 - 10x^3 + 42x^2 - 82x + 65$

10. (b) $(-2)^2 + 2i - (4 - 2i) = 0; 2 - i$ **(d)** $(-2 + i)^2 + 3(1 - i)(-1 + 2i) - 5i = 0; -1 + 2i$
11. (b) $-i, 1 + i$ **(d)** $2 - i, 1 - 2i$
12. (b) Circle, center at 1, radius 2   **(d)** Imaginary axis   **(f)** Line $y = mx$
18. (b) $4e^{-\pi i/2}$ **(d)** $8e^{2\pi i/3}$ **(f)** $6\sqrt{2}e^{3\pi i/4}$ **19. (b)** $\frac{1}{2} + \frac{\sqrt{3}}{2}i$ **(d)** $1 - i$ **(f)** $\sqrt{3} - 3i$
20. (b) $-\frac{1}{32} + \frac{\sqrt{3}}{32}i$ **(d)** $-32i$ **(f)** $-2^{16}(1 + i)$
23. (b) $\pm\frac{\sqrt{2}}{2}(\sqrt{3} + i), \pm\frac{\sqrt{2}}{2}(-1 + \sqrt{3}i)$ **(d)** $\pm 2i, \pm(\sqrt{3} + i), \pm(\sqrt{3} - i)$

Appendix B.1
(Page 488)

1. (b) Minimum 0, maximum 29   **(d)** Minimum -10, maximum 6 (at two vertices)
4. (b) Either no large toys and 600 small toys, or 171 large toys and 372 small toys [*Note:* (172,372) fails to satisfy $5x_1 + 2x_2 \leq 1600$.]
6. $40,000 in A, 60,000 in B **8.** 4 grams of P, 2 grams of Q **10.** 300 barrels of grade 1, 200 barrels of grade 2

Appendix B.2
(Page 499)

2. (b) $p = 2; x_2 = 2, x_1 = x_3 = 0$ **(d)** $p = \frac{15}{4}; x_1 = \frac{5}{4}, x_2 = 0$ **(f)** $p = 7, x_1 = 0, x_2 = \frac{3}{5}, x_3 = \frac{13}{5}$
(h) $p = 106; x_1 = x_2 = 0, x_3 = 28, x_4 = 11$
6. No regular dinners, 75 diet dinners, and 225 super dinners. Profit: $487.50.
8. 1600 standard mowers, no deluxe mowers, and 200 super mowers, Profit: $59,000 per week.

Appendix C
(Page 505)

6. $\dfrac{n}{n + 1} + \dfrac{1}{(n + 1)(n + 2)} = \dfrac{n(n + 2) + 1}{(n + 1)(n + 2)} = \dfrac{(n + 1)^2}{(n + 1)(n + 2)} = \dfrac{1}{n + 2}$

14. $\left(2\sqrt{n} - 1\right) + \dfrac{1}{\sqrt{n + 1}} = \dfrac{2\sqrt{n^2 + n} + 1}{\sqrt{n + 1}} - 1 < \dfrac{2(n + 1)}{\sqrt{n + 1}} - 1 = 2\sqrt{n + 1} - 1$

20. $B_n = (n + 1)! - 1$ **22. (b)** Verify each of $S_1, S_2, \ldots, S_8$.

Index

IMPORTANT SYMBOLS